SAP PRESS e-books

Print or e-book, Kindle or iPad, workplace or airplane: Choose where and how to read your SAP PRESS books! You can now get all our titles as e-books, too:

- By download and online access
- For all popular devices
- And, of course, DRM-free

Convinced? Then go to www.sap-press.com and get your e-book today.

SAP® Build

SAP PRESS

SAP PRESS is a joint initiative of SAP and Rheinwerk Publishing. The know-how offered by SAP specialists combined with the expertise of Rheinwerk Publishing offers the reader expert books in the field. SAP PRESS features first-hand information and expert advice, and provides useful skills for professional decision-making.

SAP PRESS offers a variety of books on technical and business-related topics for the SAP user. For further information, please visit our website: *www.sap-press.com*.

Glavanovits, Koch, Krancz, Olzinger
Full Stack Development with SAP
2023, 635 pages, hardcover and e-book
www.sap-press.com/5733

Martin Koch, Siegfried Zeilinger
Cloud Connector for SAP
2023, 352 pages, hardcover and e-book
www.sap-press.com/5683

Dutta, Ghosh, Goon, Jana, Mukherjee, Rao, Rao, Sane, Veshala, Viswanathan
Workflow for SAP S/4HANA
2024, 686 pages, hardcover and e-book
www.sap-press.com/5697

Banda, Chandra, Gooi
SAP Business Technology Platform: An Introduction
2022, 570 pages, hardcover and e-book
www.sap-press.com/5440

Acharya, Bajaj, Dhar, Ghosh, Lahiri
Application Development with SAP Business Technology Platform
2023, 574 pages, hardcover and e-book
www.sap-press.com/5504

Rene Glavanovits, Gernot Haider, Martin Koch, Daniel Krancz

SAP® Build

No-Code Development, Centralized Access, and Process Automation

Editor Meagan White
Acquisitions Editor Hareem Shafi
Copyeditor Melinda Rankin
Cover Design Graham Geary
Photo Credit iStockphoto: 1423832061/© akinbostanci
Layout Design Vera Brauner
Production Hannah Lane
Typesetting III-satz, Germany
Printed and bound in the United States of America, on paper from sustainable sources

ISBN 978-1-4932-2481-4

© 2024 by Rheinwerk Publishing, Inc., Boston (MA)
1st edition 2024

Library of Congress Cataloging-in-Publication Control Number: 2024009911

All rights reserved. Neither this publication nor any part of it may be copied or reproduced in any form or by any means or translated into another language, without the prior consent of Rheinwerk Publishing, 2 Heritage Drive, Suite 305, Quincy, MA 02171.

Rheinwerk Publishing makes no warranties or representations with respect to the content hereof and specifically disclaims any implied warranties of merchantability or fitness for any particular purpose. Rheinwerk Publishing assumes no responsibility for any errors that may appear in this publication.

"Rheinwerk Publishing" and the Rheinwerk Publishing logo are registered trademarks of Rheinwerk Verlag GmbH, Bonn, Germany. SAP PRESS is an imprint of Rheinwerk Verlag GmbH and Rheinwerk Publishing, Inc.

All of the screenshots and graphics reproduced in this book are subject to copyright © SAP SE, Dietmar-Hopp-Allee 16, 69190 Walldorf, Germany.

SAP, ABAP, ASAP, Concur Hipmunk, Duet, Duet Enterprise, Expenselt, SAP ActiveAttention, SAP Adaptive Server Enterprise, SAP Advantage Database Server, SAP ArchiveLink, SAP Ariba, SAP Business ByDesign, SAP Business Explorer (SAP BEx), SAP BusinessObjects, SAP BusinessObjects Explorer, SAP BusinessObjects Web Intelligence, SAP Business One, SAP Business Workflow, SAP BW/4HANA, SAP C/4HANA, SAP Concur, SAP Crystal Reports, SAP EarlyWatch, SAP Fieldglass, SAP Fiori, SAP Global Trade Services (SAP GTS), SAP GoingLive, SAP HANA, SAP Jam, SAP Leonardo, SAP Lumira, SAP MaxDB, SAP NetWeaver, SAP PartnerEdge, SAPPHIRE NOW, SAP PowerBuilder, SAP PowerDesigner, SAP R/2, SAP R/3, SAP Replication Server, SAP Roambi, SAP S/4HANA, SAP S/4HANA Cloud, SAP SQL Anywhere, SAP Strategic Enterprise Management (SAP SEM), SAP SuccessFactors, SAP Vora, TripIt, and Qualtrics are registered or unregistered trademarks of SAP SE, Walldorf, Germany.

All other products mentioned in this book are registered or unregistered trademarks of their respective companies.

Contents at a Glance

PART I   Introduction

1 Introduction to Low-Code and No-Code Development 23
2 SAP Business Technology Platform 51

PART II   SAP Build Apps

3 Installing and Configuring SAP Build Apps 91
4 No-Code Development Environment 111
5 Developing Applications 135
6 Data Integration and Authentication 165
7 Visual Cloud Functions 217
8 Developing Extensions 243
9 Deploying Applications 281

PART III   SAP Build Work Zone

10 Introduction to SAP Build Work Zone 311
11 Installing and Configuring SAP Build Work Zone 337
12 UI Integration 373
13 External Integrations and Content Federation 403
14 Content Transport 425
15 Advanced Topics 455
16 Administration 481

PART IV   SAP Build Process Automation

17 Introduction to SAP Build Process Automation 507
18 Installing and Configuring SAP Build Process Automation 527
19 Processes 549
20 Rules and Decisions 627
21 Action Projects 655
22 Visibility Scenarios 687
23 Automation 715
24 My Inbox 755
25 Monitoring and Administration 765

Contents

Preface 17

PART I Introduction

1 Introduction to Low-Code and No-Code Development   23

1.1	Challenges in IT	26
1.2	Pro-Code versus Low-Code versus No-Code	30
	1.2.1 Pro-Code	30
	1.2.2 Low-Code	31
	1.2.3 No-Code	32
1.3	No-Code Use Cases	33
	1.3.1 Process Automation	33
	1.3.2 Software Modernization	34
	1.3.3 Mobile Apps	35
	1.3.4 Minimum Viable Products	37
	1.3.5 Websites/Intranet Sites	38
1.4	SAP Build	39
	1.4.1 SAP Build Apps	40
	1.4.2 SAP Build Work Zone	44
	1.4.3 SAP Build Process Automation	47
1.5	Summary	49

2 SAP Business Technology Platform   51

2.1	Overview	52
	2.1.1 SAP Business Technology Platform: Introduction	53
	2.1.2 Commercial Models	55
	2.1.3 Account Structure	57
	2.1.4 Service Overview	61

2.2	Security		65
	2.2.1	User Authentication	66
	2.2.2	User Authorization	68
	2.2.3	Single Sign-On	71
	2.2.4	Identity Authentication	72
2.3	Cloud Connector		78
	2.3.1	Initial Configuration	79
	2.3.2	Mapping	83
2.4	Summary		88

PART II SAP Build Apps

3 Installing and Configuring SAP Build Apps 91

3.1	Installation	91
3.2	Configuration	99
3.3	Security	105
3.4	Summary	110

4 No-Code Development Environment 111

4.1	Launching SAP Build Apps Composer	112
4.2	Managing Development Projects	116
4.3	SAP Build Apps Composer User Interface	117
4.4	SAP Build Apps Administration	128
4.5	Help and Documentation	131
4.6	Summary	134

5 Developing Applications 135

5.1	Developing a Basic App	135
5.2	Theming	145
5.3	Building and Testing	149

5.4	App Logic, Variables, and Data Binding	154
5.5	Lifecycle Management and Team Collaboration	161
5.6	Summary	163

6 Data Integration and Authentication — 165

6.1	User Authentication	166
6.2	Using Data from SAP Systems	168
	6.2.1 Cloud Systems	170
	6.2.2 On-Premise Systems	185
6.3	Non-SAP Systems	194
6.4	Local Data Storage	201
6.5	Summary	214

7 Visual Cloud Functions — 217

7.1	Introduction	217
7.2	Entity Modeling	220
	7.2.1 Native Entities	221
	7.2.2 Extended Data Entities	224
	7.2.3 Virtual Fields	226
	7.2.4 Data Browser	229
7.3	Functions	231
	7.3.1 Create a New Function	231
	7.3.2 Input Parameter	232
	7.3.3 Function Outcomes	234
	7.3.4 Defining the Logic Flow	237
7.4	Deployment	238
	7.4.1 Deployment Card	239
	7.4.2 Change Types	240
	7.4.3 Delete/Pause Deployment	241
7.5	Summary	242

8 Developing Extensions — 243

- 8.1 SAP SuccessFactors — 244
  - 8.1.1 Consume API and Activate Authentication — 244
  - 8.1.2 Build a List of Personal Data — 246
  - 8.1.3 Build a Detail Page for Employee Details — 254
- 8.2 SAP S/4HANA and SAP S/4HANA Cloud — 265
- 8.3 Summary — 279

9 Deploying Applications — 281

- 9.1 Build Configuration — 282
- 9.2 SAP Business Technology Platform — 283
- 9.3 Mobile Deployment — 292
  - 9.3.1 iOS — 293
  - 9.3.2 Android — 303
- 9.4 Summary — 306

PART III SAP Build Work Zone

10 Introduction to SAP Build Work Zone — 311

- 10.1 SAP Build Work Zone — 311
- 10.2 Standard Edition — 315
  - 10.2.1 Functionality — 316
  - 10.2.2 Working Environment — 318
  - 10.2.3 Installing SAP Build Work Zone, Standard Edition — 323
- 10.3 Advanced Edition — 327
- 10.4 Summary — 336

11 Installing and Configuring SAP Build Work Zone 337

11.1	Installation	337
11.2	Configuration	344
	11.2.1 Trust and Authorizations	344
	11.2.2 Identity Authentication	353
	11.2.3 Identity Provisioning	355
11.3	Summary	371

12 UI Integration 373

12.1	Content Packages	373
	12.1.1 Creating a Content Package	374
	12.1.2 Deploying a Content Package	377
12.2	UI Integration Cards	384
	12.2.1 Structure of Cards	385
	12.2.2 Card Types	386
	12.2.3 Developing a UI Card	389
	12.2.4 Uploading a UI Integration Card	393
	12.2.5 Adding a UI Integration Card to a Work Page	395
12.3	Widgets	398
12.4	Summary	402

13 External Integrations and Content Federation 403

13.1	Business Content	403
	13.1.1 Create Destinations	405
	13.1.2 Expose Content	407
	13.1.3 Add Content Channel	409
	13.1.4 Check Content and Add Roles	410
	13.1.5 Configure Site	411
	13.1.6 Assign Roles	413
13.2	Microsoft Teams	416
13.3	Summary	423

14 Content Transport — 425

14.1 Manual Transport — 427
- 14.1.1 Business Content — 427
- 14.1.2 Home Page — 430
- 14.1.3 Workspaces — 432
- 14.1.4 Workspace Templates — 436

14.2 SAP Cloud Transport Management — 438
- 14.2.1 Initial Setup — 438
- 14.2.2 Configuration — 448

14.3 Summary — 453

15 Advanced Topics — 455

15.1 Notifications — 455
- 15.1.1 Types of Notifications — 457
- 15.1.2 Push Notifications with Webhooks — 460

15.2 SAP HANA Enterprise Search — 462

15.3 SAP Task Center Integration — 468
- 15.3.1 Steps in SAP Build Work Zone — 468
- 15.3.2 Steps in SAP Task Center — 472

15.4 SAP Mobile Start — 474

15.5 Summary — 480

16 Administration — 481

16.1 Feature Management — 481

16.2 User Management — 486

16.3 Compliance — 488

16.4 Error Logs — 499

16.5 Summary — 504

PART IV SAP Build Process Automation

17 Introduction to SAP Build Process Automation — 507

17.1 Subscribing to SAP Build Process Automation — 507
17.2 Features and Components — 508
17.3 The Lobby — 514
17.4 Lifecycle Management — 519
17.5 Collaboration — 521
17.6 Roadmap — 524
17.7 Summary — 525

18 Installing and Configuring SAP Build Process Automation — 527

18.1 Installation — 527
18.2 Configuration — 536
18.3 Security — 543
18.4 Summary — 547

19 Processes — 549

19.1 Creating a Project with a Business Process — 549
19.2 Triggers — 555
    19.2.1 Form Triggers — 555
    19.2.2 API Triggers — 564
    19.2.3 Event Triggers — 583
19.3 Forms — 585
    19.3.1 Creating a New Form — 585
    19.3.2 Using Forms in the Example Process — 595
    19.3.3 Approval Forms — 604
19.4 Conditions and Branches — 613
    19.4.1 Conditions — 613
    19.4.2 Branches — 619

| 19.5 | Actions | 623 |
| 19.6 | Summary | 625 |

20 Rules and Decisions — 627

20.1	Decision Editor	627
20.2	Creating a Decision Artifact	629
20.3	Creating a Data Type	632
20.4	Creating a Text Rule	635
20.5	Using a Text Rule in a Process	642
20.6	Creating a Decision Table	646
20.7	Summary	653

21 Action Projects — 655

21.1	Create an Action Project	655
	21.1.1 Project Creation	655
	21.1.2 Input	659
	21.1.3 Output	662
	21.1.4 Test	664
21.2	Using the Action within a Process	674
21.3	Summary	685

22 Visibility Scenarios — 687

22.1	Creating a Visibility Scenario	687
22.2	Configuring a Visibility Scenario	690
	22.2.1 General	691
	22.2.2 Processes	691
	22.2.3 Correlation	695
	22.2.4 Phases	696
	22.2.5 State	698
	22.2.6 Status	699

		22.2.7	Attributes	700
		22.2.8	Actions	707
		22.2.9	Performance Indicators	710
	22.3	Summary		714

23 Automation 715

23.1	Desktop Agent	716
23.2	Capture Applications	722
23.3	Files	734
23.4	Environment Variables	745
23.5	Deployment	750
23.6	Summary	754

24 My Inbox 755

24.1	Using My Inbox		755
	24.1.1	Task Functionality	755
	24.1.2	Search Functionality	757
	24.1.3	Sorting	757
	24.1.4	Filter	758
	24.1.5	Grouping	760
24.2	Substitution		761
24.3	My Inbox and SAP Build Process Automation		764
24.4	Summary		764

25 Monitoring and Administration 765

25.1	Monitoring		765
	25.1.1	Process and Workflow Instances	766
	25.1.2	Automation Jobs	770
	25.1.3	Acquired Events	770
	25.1.4	Automation Overview	771
	25.1.5	Processes and Workflows	772

	25.1.6	Triggers	773
	25.1.7	Visibility Scenarios	774
25.2	**Administration via the Control Tower**		**774**
	25.2.1	Tenant Details	776
	25.2.2	Mail Server	777
	25.2.3	SAP Cloud ALM	778
	25.2.4	Destinations	779
	25.2.5	External Authentication	781
	25.2.6	Agent Configuration	783
	25.2.7	Cloud Studio Variables	785
	25.2.8	API Keys	786
	25.2.9	Alert Handlers	787
25.3	**Summary**		**789**

The Authors		791
Index		793

Preface

Embark on a journey to the forefront of the digital revolution with this comprehensive guide to SAP Build. As organizations seek agility in a fast-paced world, the demand for rapid and flexible application development has given rise to a new paradigm—low-code/no-code development. This guide is structured into four distinct parts, each offering an in-depth exploration of SAP Build's capabilities, woven together with practical insights.

Who This Book Is For

This book is a gateway to the world of SAP Build, specially crafted for a wide array of professionals—from IT decision makers who plot the course, developers and architects who bring it to life, to administrators who keep the gears turning, and project managers/agile coaches who guide the team to success. It's a rich reservoir of knowledge that caters to both the strategic planner and the hands-on expert.

How This Book Is Organized

While we suggest reading through the entire book to capture the full spectrum of insights, we understand that your time is precious. So, feel free to dive into specific parts or even individual chapters that are most relevant to you. Whether you're orchestrating an enterprise-wide implementation or focusing on a single aspect of SAP Build, this book is your flexible companion on the journey to digital mastery.

Part I lays the foundational bricks, starting with Chapter 1's walkthrough of low-code/no-code development—a primer that demystifies the concepts and prepares you for the creation of applications without deep coding knowledge. Chapter 2 serves as your introduction to SAP Business Technology Platform (SAP BTP), offering a comprehensive understanding of the platform's framework, essential for leveraging the power of SAP Build.

Part II on SAP Build Apps is where your creativity meets execution. Chapter 3 provides a step-by-step guide to installing and configuring SAP Build Apps, ensuring a solid ground for your building experience. Chapter 4 opens the doors to the no-code development environment, showcasing the user interface that makes app creation possible. In Chapter 5, you dive into the actual app development process, learning to harness the platform's features. Data integration and authentication are tackled in Chapter 6, ensuring your applications are both robust and secure. Chapter 7 explores the SAP Build Apps visual cloud functions, elevating your apps with visualizations and data manipulation. Chapter 8 and Chapter 9 guide you through developing extensions for SAP products and deploying your apps, turning them from concepts into real-world solutions.

Part III on SAP Build Work Zone focuses on creating collaborative and intelligent workplaces. Chapter 10 introduces you to SAP Build Work Zone and Chapter 11 teaches you about installation and configuration. Chapter 12 and Chapter 13 delve into UI and external integrations, opening up the world of content federation. Content transport is the focus of Chapter 14, providing insights into efficient content management, while Chapter 15 covers advanced topics. Chapter 16 on administration ensures that you can manage and maintain SAP Build Work Zone.

Part IV on SAP Build Process Automation is dedicated to enhancing operational efficiency through automation. With Chapter 17's introduction to SAP Build Process Automation, you're positioned to streamline business processes. Chapter 18 gives you a thorough understanding of installation and configuration, ensuring a smooth start. Chapter 19 through Chapter 23 offer a granular look into designing processes, setting rules and decisions, creating action projects, and designing visibility scenarios for comprehensive automations. Chapter 24 discusses My Inbox, a tool for task management, and Chapter 25 closes the loop with monitoring and administration techniques that keep your processes sharp and responsive.

Text Boxes

Throughout the book, we've also provided several elements that will help you access useful information:

Tips and Tricks

Boxes with this symbol provide you with recommendations as to how you can simplify your work.

Notes

Boxes marked with this symbol contain additional information or important contents that you should keep in mind.

Examples

Boxes marked with this symbol provide practical scenarios and explain, in detail, how particular functions can be applied.

> **Warnings** [!]
> Boxes with this symbol contain details worth considering. Moreover, it warns you of common errors or problems that might occur.

Conclusion

This guide is not merely a collection of chapters but a narrative that progresses from core concepts to advanced implementations in SAP Build. It's tailored to provide a structured and detailed pathway for you to harness the simplicity and power of low-code/no-code development, transforming the way your digital solutions are delivered. Whether you are building your first app or automating complex processes, these pages will accompany you to mastery. Welcome to the world of SAP Build, where innovation is at your fingertips!

PART I
Introduction

Chapter 1
Introduction to Low-Code and No-Code Development

The promise of no-code platforms is that they make software development as easy as using Word or PowerPoint. In this chapter, we show to what extent this promise is kept and introduce you to SAP's low- and no-code options.

Welcome to the world of modern IT development! The days when programming was the exclusive preserve of masters of code are long gone. Today, we can draw on a colorful palette of approaches that bring into the world of creation even those who were previously only confronted with mysterious signs and brackets. Imagine a world where you can develop complex software solutions without having to display a computer science degree certificate on your wall, a world where you can effortlessly build applications without having to run a marathon through the depths of source code: You heard right, we're talking about low-code and no-code. They are the heroes of the new IT world!

Hardly anyone has been unaware of the hype surrounding low-code and no-code platforms lately. The promise of no-code platforms is to simplify development to such an extent that average business users, in the form of so-called citizen developers, can implement projects. The beauty of this is that no engineering or development team is required, and thus development can be accomplished at no additional cost. Unlike no-code platforms, low-code platforms still require programming skills, but they promise to accelerate software development by enabling developers to work with prebuilt code components.

In a world where technology continues to advance, low-code and no-code can become your best allies. They help overcome skills shortages, master timelines, digitize processes, and bring business ideas to life. But let's take a closer look at the challenges companies face in IT. Because even in this magical world of low-code and no-code, there are still some obstacles to overcome. In Section 1.1, we'll explore what difficulties can be overcome when we unleash the powers of code.

Low code lets you put your development projects on the fast track by using visual tools and prebuilt building blocks. It's like using a LEGO kit to build your applications. Drag a button here, put a dropdown menu there, and voila—your application takes shape! It's like speed dating for developers and their ideas: fast, efficient, and sometimes surprisingly charming.

The term *low-code* first emerged in the late 2000s. Companies were looking for ways to speed up their development processes and address the skills shortage of the time. The idea was to use visual development tools and prebuilt components to speed up application development and reduce the amount of code required. By using *drag-and-drop* visual editors, developers could build applications faster by reusing prebuilt building blocks and business logic. This enabled faster time-to-market for applications and more efficient use of developer resources.

And then we have *no-code*, the ultimate superhero of software development. Here, even those who shy away from the magical world of code can conjure up their own applications. Imagine being Harry Potter and being able to create your own magical application with a single swish and flick. Drag and drop here, a bit of configuration there, and your digital dreams take shape. It's like preparing a delicate gourmet meal with just a few flicks of the wrist, but without the mess and worry of burnt fingers and dishes and the perfect mise en place.

The no-code approach emerged later as an extension of the low-code concept. No code aims to make application development accessible not only to developers, but also to nondevelopers. The idea was to create visual tools and user-friendly interfaces that would allow business users and departments to create their own applications without extensive programming knowledge. This opened up the opportunity for more innovation by allowing those closest to business processes and requirements to create and customize applications themselves without relying on developers. No-code platforms often offer a variety of prebuilt templates and features to facilitate application creation.

At this point, we need to introduce the term *citizen developer*. A citizen developer is a person who creates applications or software solutions despite having no formal education or professional experience as a professional developer. Citizen developers are usually business users or employees in specialist departments who have knowledge of specific business processes and have some technical understanding or willingness to learn low-code or no-code platforms. Citizen developers use visual development tools and user-friendly interfaces that enable them to build applications, automate workflows, and improve processes without extensive programming. These applications are often developed for specific business needs and are used to eliminate bottlenecks, automate manual tasks, or increase efficiency. However, it is important to note that citizen developers should perform their roles in accordance with corporate policies and guidelines. Proper training, support, and monitoring from IT or experienced developers are often required to ensure that the applications created are secure, efficient, and maintainable. Citizen developers can make a valuable contribution to agile development and addressing bottlenecks, provided adequate governance and support is provided.

But wait, valued readers, let's not forget the role of pro-code. *Pro-code* is the master architect of the digital world. This is true for those of you who are native speakers of the languages of bits and bytes and find a home in the depths of code. Yes, for those who want to express their visions with the finest nuances, there is still room for your creative expression with pro-code.

But as with any heroism, there are limits. As I'm sure you know from many comics, even the most powerful superpowers have their limitations. For particularly complex challenges, we sometimes require the use of the famous (and infamous) pro-code. However, we won't delve deeper into the world of handwritten code in this book. We'll first deal with an exciting duel: low-code versus no-code. In Section 1.2, we'll take a closer look at these two methods and highlight their strengths and differences. Together, we'll dive into the epic battle between efficiency and usability! There will also be a compact excursion into the world of pro-code, but don't worry, we won't overwhelm you with it.

Having already talked about how no-code enables developers and business users without extensive programming skills to create their own solutions, let's continue the fascinating journey through the world of no-code use cases. There are many concrete use cases where no-code can play to its strengths in the SAP environment. In Section 1.3, we'll explore the many possibilities of no-code in the SAP world, from user interface development to workflow and approval process automation to seamless data integration and business logic extension. We'll see how no-code helps companies use their SAP systems more efficiently and customize them to their specific needs. Join us on this exciting journey; let's find out together how no-code is revolutionizing the SAP world. In Section 1.3, we'll discover the almost limitless possibilities that no-code offers to shape the SAP landscape in a whole new way.

An important part of the SAP Build portfolio is SAP Build Apps. With this service, users can develop no-code applications using visual tools and prebuilt components. This allows applications to be created quickly and customized to meet specific business requirements. SAP Build Apps provides an intuitive user interface that enables users without programming skills to design their own applications. Another service in the SAP Build portfolio is SAP Build Work Zone. This service is available in various forms. In the form of the SAP Fiori launchpad, it serves as a central entry point to SAP Fiori apps. However, it can also serve as a collaboration platform on which teams can work together. In addition, SAP Build Process Automation provides a way to automate and optimize business processes. With this service, companies can analyze, model, and automate their existing processes. This leverages visual tools and workflow engines to increase efficiency and productivity. In Section 1.4, we'll look at the SAP Build portfolio in more detail, examining the individual services, their functions, and their use cases. There, we will see how SAP is responding to the growing demands in the area of no-code and low-code, helping companies to develop innovative solutions and drive their digital transformations.

1.1 Challenges in IT

In today's fast-paced and technology-driven business world, companies face a variety of IT challenges. From a shortage of skilled workers to tight deadlines and the digitization of business processes, it's often a demanding task to develop and implement innovative solutions. But fortunately, new ways are emerging for companies to successfully meet these challenges. The advent of low-code and no-code has revolutionized the way software is developed, opening new opportunities for companies to work quickly, efficiently, and in a user-centric way.

Low-code and no-code approaches offer a promising alternative to traditional software development by reducing the complexity of programming and providing a visual, intuitive environment. With *low-code*, developers can build applications faster by accessing prebuilt building blocks and components and easily combining them. *No-code*, on the other hand, allows people without programming skills to build their own applications by using visual drag-and-drop tools and configurations. These new approaches enable companies to shorten development times and accelerate time-to-market while responding effectively to increasing market demands.

Ahead, we'll take an in-depth look at the challenges organizations face in IT and how low-code and no-code solutions can serve as effective tools to address these challenges:

- **Skills shortage**
  The *shortage of skilled workers* is a widespread problem in the IT industry. The increasing demand for qualified developers and IT professionals often exceeds the supply of available talent. Companies face the challenge of filling open positions and finding highly skilled developers capable of developing and maintaining complex software solutions. This is where low-code and no-code come into play as promising solutions. With low-code, companies can free up their existing development teams by spending less time developing applications. Visual development tools and prebuilt components allow developers to build applications faster and to focus on complex problems instead of wasting time on repetitive, time-consuming coding tasks. This enables organizations to use existing resources more efficiently and tackle more projects with limited resources.

  No-code goes one step further and opens up the possibility for nondevelopers, such as business users, marketers, or project managers, to create their own applications. This reduces dependency on specialized developers and significantly expands the pool of potential application developers. By enabling people without extensive programming skills to create their own applications, companies can tap into more innovative ideas while reserving developer resources for more challenging tasks. The use of low-code and no-code mitigates the skills shortage to some extent, as companies are no longer solely dependent on a limited number of skilled developers. Instead, they can expand their existing teams by bringing in nondevelopers and expanding development opportunities for their employees. This also encourages collaboration between different departments and creates a culture of innovation and ownership. While the skills shortage remains a challenge, low-code and no-code offer companies a way to address this challenge and to use their development resources more effectively to meet their technology needs.

- **Digitalization of business processes**
The *digitization* of business processes is a key concern for companies today. Traditional, manual, and paper-based processes are often time-consuming, error-prone, and inefficient. Switching to digital processes is therefore crucial to increase efficiency, reduce costs, and improve the customer experience. This is where low-code and no-code come into play as valuable tools. With their visual development environment and easy-to-use tools, they enable companies to digitize business processes quickly and efficiently. Instead of laboriously writing complex lines of code, low-code and no-code allow users to use visual drag-and-drop editors to create workflows, data flows, and automations. These approaches enable companies to develop customized solutions tailored to their specific business processes. Users can create forms, approval workflows, and integrations to improve process efficiency. By automating routine tasks, employees can focus their time on value-added activities while reducing errors and delays.

    In addition, the digitization of business processes with low-code and no-code enables seamless integration with other systems and data sources. Companies can connect their existing IT infrastructure, such as ERP systems like SAP S/4HANA, to ensure a smooth flow of data and a consistent and shared information base. This helps improve collaboration. But it can also create more transparency, enabling companies to make informed decisions based on up-to-date and accurate data. Digitizing business processes with low-code and no-code enables companies to become more agile and efficient. They can eliminate manual steps, ensure data integrity, and increase productivity. It also provides the flexibility to adapt and evolve processes as needed to meet changing market and business requirements. By focusing on low-code and no-code, companies can drive digital transformation and reap the benefits of a modern, automated, and efficient way of working. The digitization of business processes is the key to competitiveness and enables companies to successfully meet the challenges of the digital age.

- **Legacy systems**
Legacy systems are a challenge for many organizations. These outdated systems are often difficult to maintain, extend, and integrate with modern applications. Dealing with them requires specialized knowledge and complex programming work, which can lead to high costs and time-consuming processes. At this point, low-code and no-code solutions offer a way to overcome the difficulties of dealing with legacy systems. Instead of having to make extensive code changes, companies can use low-code and no-code to integrate new functionality into their legacy systems. This is typically done by using interfaces or connectors that allow data to be read or written from the legacy systems.

With low-code, developers can use prebuilt building blocks and components to facilitate integration with legacy systems. They can use visual tools to create database queries, connect interfaces, or integrate data from different sources. In this way, development time is reduced and integration complexity is mitigated. No-code approaches go a step further and allow nondevelopers to extend their legacy systems or integrate with other applications. These users can use visual tools to create workflows, transform data, or even build small applications that interact with legacy systems. This creates opportunities for business departments to implement their specific requirements without relying on a large IT department.

By using low-code and no-code, companies can gradually modernize their legacy systems instead of having to tackle a complete redevelopment. They can add new functionality to improve efficiency and usability without compromising the stability of existing systems. This allows companies to modernize their IT landscapes while protecting the value of their existing investments. Thanks to the flexibility and adaptability of low-code and no-code, companies can successfully overcome the challenges of dealing with legacy systems. They can incrementally modernize their systems, add new functionality, and simplify integration with other applications. This leads to better system performance, more efficient operation, and seamless interaction between legacy systems and modern technologies.

- **Time restrictions**
  Time constraints are an ever-present challenge in today's busy world. Companies are under pressure to complete projects quickly, bring products to market, and respond to changing customer needs. Traditional software development methods that require a lot of time to write code often reach their limits. Low-code and no-code solutions offer a solution here to meet the time constraints. With low-code, developers can rely on prebuilt building blocks and components that already cover much of the functionality. These reusable modules make it possible to build applications faster because developers don't have to develop every detail from scratch. By using visual tools and drag-and-drop functionality, they can quickly assemble and customize applications. This significantly speeds up the development process by eliminating the need for lengthy code iterations.

  Accelerating development time with low-code and no-code has far-reaching benefits for companies. It enables them to respond more quickly to market demands and deliver their products and services in a timely manner. This time gain can also help gain competitive advantage by being first to market or responding quickly to changing customer needs. In addition, the reduced development time with low-code and no-code also enables faster iteration and customization of applications. Companies can gather feedback from users, implement improvements, and introduce new features faster. This promotes business agility and innovation. Time is a valuable commodity, and with low-code and no-code, companies can use this resource more effectively.

- **Agile development**
  Agile development has established itself as an effective way to meet the ever-changing demands of today's business world. Companies strive to be flexible, respond quickly to feedback, and continuously deliver high-quality solutions. In this context, low-code, and no-code play an important role in making agile development practical. By using low-code and no-code, companies can quickly prototype and create *minimal viable products* (MVPs). Developers and nondevelopers can use visual tools to quickly create basic functionality and user interfaces. This allows companies to validate their ideas, gather feedback from users, and make iterative improvements. Rapid prototyping and MVPs enable companies to gain early market insights and continuously adapt their products and solutions to meet requirements.

  In addition, low-code and no-code also support collaboration and communication within agile teams. Because these approaches are visual and user-friendly, different stakeholders, including developers, business users, and project managers, can collaborate effectively. They can quickly visualize their ideas, discuss requirements, and collaborate on development. This promotes smooth collaboration and ensures that all stakeholders are on the same level of knowledge. The flexibility of low-code and no-code also enables agile response to changes and evolving requirements. Because the applications are modular, changes and enhancements can be implemented quickly without requiring an extensive rewrite process. This enables companies to remain competitive in the market and continuously improve their solutions. By combining agile methods with low-code and no-code, companies can shorten their development cycles, improve collaboration, and respond quickly to change.

  Agile development helps companies drive innovation, leverage customer feedback, and continuously optimize their products and services. In an agile environment, it's important to act and react quickly to changes. Low-code and no-code offer the tools and possibilities to implement this agility in software development. By making their development processes agile and relying on low-code and no-code, companies can increase their competitiveness and develop innovative solutions efficiently.

Low-code and no-code solutions can address and provide a potential solution to various challenges in IT. They give companies ways to deal with skills shortages, meet timelines, digitize business processes, modernize legacy systems, and support agile development methodologies. By using low-code and no-code, companies can use their development resources more efficiently, foster innovation, and respond quickly to changing requirements. However, it's important to note the potential pitfalls, such as limited flexibility, security risks, and reliance on third-party vendors. Careful evaluation of requirements and capabilities, user training, and appropriate security measures are critical to reap the full benefits of low-code and no-code while overcoming potential challenges. With the right approach, companies can benefit from the many opportunities and potentials that low-code and no-code offer to successfully drive their digital transformations.

1.2 Pro-Code versus Low-Code versus No-Code

In a world where code seems to be a sacred mantra, we break into this chapter to challenge the status quo. Pro-code, low-code, and no-code: three approaches that seem like rival faiths. In the fast-paced digital world, companies face the challenge of developing innovative solutions that meet their unique needs. This raises questions such as: Should we rely on experienced programmers and develop custom code (pro-code)? Or is it better to rely on visual development tools and prebuilt components to work more efficiently (low-code)? Or can nondevelopers use no-code platforms to create their own applications without having to write a line of code?

In this section, we'll take an in-depth look at the advantages and disadvantages of these different approaches and provide a comprehensive comparison of pro-code, low-code, and no-code. We'll explore the strengths and areas of application of each approach to help you make the right choice for your business.

While pro-code, as presented in Section 1.2.1, is the traditional method of software development and requires deep programming knowledge, low-code, which is detailed in Section 1.2.2, offers faster development through the use of visual tools and prebuilt building blocks. No-code, on the other hand, allows nondevelopers to build their own applications and further reduce development time. We will discuss this in Section 1.2.3. We'll look at the different target groups, use cases, and challenges of each approach. We'll also look at scalability, flexibility, security, and the future of software development.

It's important for us to note, from our experience, that the boundaries between target groups are not always strict, and there is often overlap. For example, some developers may use low-code or no-code platforms to speed up development while also using pro-code development when needed. Ultimately, the choice of method depends on the project requirements, available skills and resources, and software development goals. Use cases are also not strictly delineated, and organizations often use a combination of these approaches, depending on project requirements and resource availability. The key is to select the right approach for the particular use case and to consider the strengths and limitations of the different methods.

1.2.1 Pro-Code

Pro-code refers to the traditional method of software development in which developers write source code manually. This involves understanding and using programming languages, frameworks, and tools to develop custom software solutions. Unlike low-code and no-code, pro-code offers the most freedom and flexibility, but also requires advanced technical knowledge and programming skills. Pro-code in the SAP context refers to the traditional method of application development, where programmers write custom code to develop complex applications or extensions based on SAP technologies.

This approach requires extensive knowledge of SAP programming languages and tools, such as ABAP or SAPUI5, to create customized solutions.

The pro-code method is aimed at professional developers and software engineers with extensive programming knowledge and experience. This target group has mastered various programming languages, frameworks and tools and needs full control over the source code to develop customized and complex software solutions. Pro-code offers maximum flexibility and customization options but is more suitable for experienced developers.

Pro-code is perfect for complex, customized applications. When it comes to sophisticated functionalities or specific requirements, the pro-code approach is the best choice. Developers with in-depth knowledge of programming languages can develop comprehensive and individualized solutions.

Major limitations of pro-code are its complexity and the required expertise. Development with pro-code requires extensive expertise and experience in programming languages. The complexity of development can lead to longer development cycles and higher costs. In addition, finding and retaining highly skilled developers is a challenge.

1.2.2 Low-Code

Low-code refers to a method of software development that uses visual development tools and a minimal amount of traditional handwritten code. Low-code platforms allow developers to build applications faster by accessing a visual user interface and dragging and dropping prebuilt building blocks and components to create functionality. These approaches are particularly well-suited for rapid prototyping, automating business processes, and deploying simple applications. In the SAP context, the low-code approach is often associated with SAP Fiori elements. SAP Fiori elements is a collection of design guidelines, templates, and tools that enable developers to build applications quickly and efficiently. It's a low-code development method based on the SAPUI5 framework.

The target audience for low-code platforms is primarily professional developers and corporate IT teams. This methodology appeals to people who have basic programming skills and want to accelerate the development of applications. Low-code enables developers to quickly prototype, automate simple applications, and make business processes more efficient.

Low-code is a good candidate for rapid application development. It's well-suited to organizations that need to develop applications in a short period of time. With prebuilt building blocks and visual development tools, developers can quickly assemble and customize applications without having to code from scratch. But low-code is also suitable for integrating systems. Companies that need to connect different systems or databases can use low-code to make integration more efficient. Visual interface design enables seamless connection between different applications and data sources.

Limited flexibility is one of the most important limitations of low-code. Although low-code speeds up development, a predefined structure and the use of prefabricated building blocks can limit flexibility. For complex requirements or specific customizations, low-code platforms may reach their limits. Using low-code platforms often means dependency on a vendor. Changes in a platform or the discontinuation of a vendor can lead to challenges and require migration or adaptation.

> **Gartner's Projections**
>
> Gartner's projections for the no-code/low-code market indicate a growth to $30 billion by 2025, with a significant adoption rate among medium to large companies. By 2025, it is expected that 70% of new enterprise applications will utilize low-code or no-code technologies, emphasizing their increasing significance.

1.2.3 No-Code

No-code goes one step further than low-code and enables people without programming skills to create software applications. No-code platforms allow users to create applications using visual drag-and-drop tools and configurations without having to write traditional code. These approaches are particularly user-friendly and democratize software development by allowing people without a technical background to create their own applications. In the SAP context, the no-code approach is covered by SAP Build Apps, formerly known as SAP AppGyver. SAP Build Apps provides a user-friendly, visual development environment where users can build applications using drag-and-drop tools and an intuitive user interface.

No-code platforms are aimed at a broader audience, including people without technical or programming skills. Business users, marketers, project managers, and other nontechnical people can use no-code tools to quickly create simple applications, websites, automations, and more. This methodology allows people with limited technical skills to implement their ideas without the need for traditional code.

No-code is perfect for developers from business users and specialist departments. It allows business users and nondevelopers to create their own applications without having to program. It's ideal for simple applications such as forms, workflows, and smaller automations that can be developed without technical expertise. However, no-code is also an ideal candidate for *prototyping* and MVPs, allowing them to be created quickly. Users can quickly visualize ideas, gather feedback, and make iterative improvements without having to go through long development cycles.

No-code allows nondevelopers to build applications, but there are natural limits to complex functionality and special requirements. The ability to customize and extend is limited, which can be problematic in complex scenarios. When users create applications without technical knowledge, there is a risk of security vulnerabilities and privacy issues. The lack of programming experience can lead to insecure implementations if appropriate security measures are not followed.

1.3 No-Code Use Cases

In today's technological age, the ability to respond quickly to changing business requirements is crucial to the success of companies. In this context, low-code and no-code solutions are becoming increasingly important. They offer an agile and efficient way to develop applications and create individual solutions without relying on extensive programming skills.

In this section, we'll look at the various use cases for low-code and no-code technologies in enterprises. From process automation and software modernization to the creation of mobile apps, websites, and self-service portals, these approaches offer a wide range of opportunities to optimize business processes, improve customer and employee experiences, and drive innovation.

In Section 1.3.1, you'll learn how companies can save time, minimize errors, and increase efficiency by automating workflows. In Section 1.3.2, we take a closer look at legacy modernization and how low-code and no-code solutions can help organizations achieve this. In Section 1.3.3, you'll discover how enterprises can use low-code and no-code technologies to develop custom mobile apps to deliver best-in-class mobile experiences to customers and employees. In Section 1.3.4, you'll learn how companies can use low-code and no-code tools to quickly prototype and create MVPs to validate new ideas and get early feedback. And in Section 1.3.5, we'll dive into the world of website creation with low-code and no-code platforms and learn how companies can improve their online presence and develop engaging websites.

1.3.1 Process Automation

Companies are often faced with the challenge of optimizing manual and time-consuming business processes to increase efficiency and reduce human error. This is exactly where low-code and no-code solutions come into play. Process automation enables companies to transform manual workflows into digital workflows. This is done by automating tasks, decisions, and communication between different actors. With low-code and no-code platforms, business users and nondevelopers can visually design workflows by linking elements such as forms, rules, notifications, and approval mechanisms.

The advantages of process automation are manifold. By using low-code and no-code solutions, companies can save time because manual activities are reduced or, ideally, even eliminated. Employees can focus on value-added tasks instead of wasting time on repetitive tasks. Automating processes minimizes human error and improves consistency, leading to high-quality results. This contributes to end user and customer satisfaction and increases confidence in business processes. Another benefit of process automation is the transparency and traceability of workflows. The digital capture of data and information creates a seamless audit trail. This enables companies to track the status of tasks, identify bottlenecks, and optimize throughput. In addition, dashboards and reports provide insight into process performance and provide continuous improvement information.

> **Multiview's 2021 Study**
>
> Multiview's 2021 study indicates a strong trend towards automation, with 73% of CFOs leveraging such technologies to enhance operational efficiency and financial processes. Automation is streamlining workflows, reducing manual errors, and enabling faster decision-making, showcasing the transformative potential of these technologies in business operations.

With process automation, different types of workflows can be automated, such as classic approval workflows, simple release procedures, and escalation processes. Companies can use low-code and no-code to adapt workflows to their specific business requirements and integrate rule-based decisions into the process.

In response to the growing demand for process automation, SAP has developed the SAP Build Process Automation product. SAP Build Process Automation is a low-code platform specifically designed to help companies automate business processes.

1.3.2 Software Modernization

Many companies face the challenge of updating outdated legacy systems and adapting them to the requirements of the digital age. Low-code and no-code solutions give companies the opportunity to modernize their software landscapes without having to perform extensive reprogramming. With the idea to "never stop a running system" in mind, it's often a challenge for companies to modernize outdated legacy systems. These systems may still be serving their purpose and are firmly entrenched in business processes. The thought of changing or replacing these systems can be perceived as risky and time-consuming. Software modernization encompasses various aspects, such as updating the user interface, integrating new functions, or migrating to a new technological platform. This is where low-code and no-code technologies come into play to help companies modernize their software more quickly, efficiently, and cost-effectively.

This enables companies to develop more agile applications that meet current business needs. These platforms provide a variety of prebuilt building blocks, templates, and components that enable developers to quickly assemble and customize applications. Users can visually define workflows, create data models, design user interfaces, and implement functionality without extensive programming skills.

In addition, software modernization helps to reduce maintenance costs. Outdated legacy systems often require expensive and time-consuming maintenance. In many cases, full knowledge of the legacy system is no longer available. By upgrading to modern technologies and reducing the amount of proprietary code, companies can significantly reduce their maintenance costs.

A major advantage of software modernization with low-code and no-code technologies is accelerated time-to-market. Because developers can rely on prebuilt components and templates, the effort required to develop code from scratch is eliminated. This enables companies to deploy their applications faster and implement new features and enhancements quickly. Another important aspect of software modernization is scalability. Low-code and no-code solutions can be used to develop applications that can be easily adapted to growing user numbers and increasing data volumes. This is especially relevant when companies need to be agile to respond to changing business requirements and market conditions.

Software modernization based on low-code and no-code solutions gives companies the chance to modernize their IT landscapes without having to make large investments in extensive development projects. By accelerating development, reducing maintenance costs, and improving scalability, companies can increase their competitiveness and offer their customers state-of-the-art applications and functions. Software modernization is thus a central use case for low-code and no-code technologies and platforms.

To support the software modernization use case, SAP offers suitable services in its portfolio. With SAP Build Process Automation and SAP Build Apps, companies can modernize their software landscapes and adapt them to the requirements of the digital age. The combination of SAP Build Process Automation and SAP Build Apps enables companies to approach their software modernization at different levels. They can automate their business processes while building custom apps that meet the specific needs of their organization. These services provide flexibility and efficiency, and they enable companies to modernize their software landscapes incrementally without disrupting operations.

1.3.3 Mobile Apps

In an increasingly digitized world, delivering user-friendly and feature-rich mobile apps is critical to delivering world-class user experiences to customers and employees. Low-code and no-code technologies play an essential role in this. Mobile app development has traditionally been a laborious process, requiring extensive programming skills for the relevant platforms, such as iOS or Android. However, with low-code and no-code platforms, companies can speed up the development process and enable non-developers to create mobile apps.

> **Businesswire's Research**
>
> Businesswire's research suggests an unprecedented proliferation of app development, with over 500 million cloud-based apps expected to be developed in the near future, surpassing the total number of apps developed in the previous 40 years. This reflects a rapid acceleration of digital transformation initiatives, with cloud platforms enabling the creation of customized applications that support and drive these efforts.

Low-code and no-code platforms offer a variety of prebuilt building blocks, templates, and UI components that allow users to design and customize mobile apps through simple drag-and-drop. These visual tools make it possible to design user interfaces, integrate functionality, and define app logic without extensive programming knowledge.

A major advantage of mobile app development with low-code and no-code technologies is accelerated time-to-market. Developers can make use of prefabricated components and templates. Therefore, the effort of developing code from scratch is eliminated. Companies can thus deploy their mobile apps faster and quickly provide their customers and employees with new features. In addition, low-code and no-code platforms enable integration with backend systems. This ensures a seamless connection between the mobile app and relevant data and processes in the enterprise. Citizen or professional developers can leverage existing APIs or create new integrations to retrieve data in real time, process user interactions, and trigger actions in other systems.

> **Citizen Developers**
>
> The landscape of software development is undergoing a significant shift, with Gartner predicting a quadrupling of citizen developers compared to professional developers at large enterprises by 2023. This surge is emblematic of a broader democratization of technology, where individuals with domain expertise but no traditional programming background are empowered to create applications, thereby accelerating digital transformation.

By using low-code and no-code technologies, companies can create custom mobile apps tailored to their specific user requirements. They can design engaging user interfaces; integrate features such as push notifications, location services, and offline access; and provide a seamless mobile experience for their customers and employees.

Mobile app development is therefore a key use case for low-code and no-code technologies, offering companies the opportunity to strengthen their mobile presence and provide cutting-edge mobile experiences to their customers and employees.

In response to the growing need for mobile app development tools, SAP has suitable services in its portfolio with both SAP Build Apps and the *Mobile Development Kit*

(MDK), part of SAP Mobile Services. SAP Build Apps provides a no-code solution that enables companies to develop custom mobile apps without extensive programming skills. With SAP Build Apps, users can intuitively navigate the design and development process to create engaging user interfaces, add functionality, and define app logic. For more advanced requirements and extended functionality, SAP offers the MDK as a low-code solution. The MDK enables developers to create custom mobile apps using a variety of prebuilt building blocks, templates, and UI components. It provides a visual development environment wherein developers can implement app logic, data integrations, and advanced features to create customized mobile solutions. With SAP Build Apps and the MDK, SAP offers companies the flexibility to use either a no-code or low-code solution, depending on the requirements and skill level of the developers. These services enable companies to develop their mobile apps quickly and efficiently while ensuring seamless integration with existing SAP systems and data.

1.3.4 Minimum Viable Products

An MVP is a basic first version of a product or even an application. It's developed to quickly gather feedback, test acceptance, and make improvements. Low-code and no-code technologies offer companies the opportunity to create MVPs quickly and cost-effectively and accelerate their innovation processes.

For the development of an MVP, low-code and no-code solutions offer companies several advantages. The main advantage is the fast time-to-market. By using visual development tools, prebuilt building blocks, and templates, developers and business users can create MVPs without extensive programming knowledge. This enables companies to quickly turn ideas and concepts into a working application and bring it to market within a very short time.

Another advantage of MVPs is that they allow companies to gather feedback from customers, future users, and other stakeholders early on. By providing a basic version of the application, companies can gain valuable insights into the user experience, functionality, and potential of the product. This feedback can be used to optimize the product to add new features or adjust before rolling it out at scale.

With low-code and no-code solutions, companies can develop MVPs in an agile manner and make iterative improvements. The visual development environment enables developers to quickly make changes and adjustments based on collected feedback. This shortens development cycles and makes it possible to adapt quickly to changing requirements and market conditions.

MVP development also fosters collaboration and a spirit of innovation within the organization. By involving expert users and nondevelopers, different teams and departments can contribute their ideas and expertise and work together to develop the MVP. This contributes to an agile and collaborative work culture and enables companies to respond quickly to new market opportunities.

In the SAP environment, SAP Build Apps offers a suitable solution for the development of MVPs. With SAP Build Apps, companies can create MVPs quickly and efficiently by implementing both the user interface and partially the backend for data storage. SAP Build Apps enables users to create custom apps using visual tools and prebuilt building blocks. With this service, companies can quickly and easily develop interactive prototypes and MVPs by using drag-and-drop capabilities to design the user interface and add the features they want. With SAP Build Apps, you can create data models, implement business logic, and access existing backend services to provide MVPs with the data and functionality they need. The combination of frontend and backend development in SAP Build Apps enables companies to implement MVPs fully or partially in SAP environments. This provides seamless integration with existing systems and enables companies to integrate their MVPs into the existing IT infrastructure at an early stage.

1.3.5 Websites/Intranet Sites

Appealing and user-friendly websites and intranet sites are essential for businesses today. They allow businesses to establish their presence in the digital space and present products or services. Low-code and no-code technologies offer companies the ability to create websites quickly and efficiently while maintaining high design quality and usability. Low-code and no-code solutions allow companies to develop websites without having extensive programming skills. They offer a variety of prebuilt templates, layouts, and design elements for users to choose from. Drag-and-drop functionality allows them to intuitively place and customize the content, images, and features they want to create their unique websites.

A major benefit of using low-code and no-code technologies for website development is accelerated time-to-market. Because developers and users can rely on prebuilt building blocks, the effort of developing code from scratch is eliminated. Organizations can quickly build, update, and customize their websites to keep up with current information and trends. In addition, websites can be optimized for different devices and screen sizes. The created websites are *responsive*, which means that they automatically adapt to the functional display size and provide an optimal user experience on desktops, tablets, and mobile devices. Low-code and no-code solutions can also simplify the integration of backend functionality into websites. They can access existing APIs to display dynamic content, process forms, or support user interactions. This enables companies to add relevant and personalized content to their websites. Using low-code and no-code technologies to develop websites also promotes collaboration and the agile development process. By involving expert users and nondevelopers, different teams and departments can participate in creating and updating a website without relying on extensive programming skills.

For website development needs, SAP has the right product in its portfolio: SAP Build Work Zone, advanced edition. SAP Build Work Zone, advanced edition is a powerful low-code and no-code solution that helps companies create engaging websites and content quickly and efficiently. With SAP Build Work Zone, advanced edition, users can access a variety of prebuilt templates, layouts, and design elements to create their custom websites. The platform offers an intuitive user interface and drag-and-drop functionality that allows users to easily place and customize the content, images, and features they want. Another major benefit of SAP Build Work Zone, advanced edition is its integration with other SAP solutions and backend systems. Companies can access existing functionality to include dynamic content, process forms, and enable personalized user interactions. This allows companies to add relevant information to their websites while ensuring seamless integration with their existing systems.

For the use cases we have looked at so far, SAP offers suitable solutions to implement them easily and efficiently. SAP has a wide range of low-code and no-code technologies and platforms to help companies address their individual challenges in the areas of process automation, software modernization, mobile apps, MVPs, and websites.

1.4 SAP Build

In today's rapidly evolving business landscape, you need to give your business experts the tools they need to drive innovation and deliver solutions quickly. With SAP Build's low-code tools, anyone, regardless of their skill set, can participate in the development and automation process. SAP Build offers a suite of tools, including SAP Build Apps, SAP Build Process Automation, and SAP Build Work Zone, that are designed to enable your business experts to easily create, enhance, and streamline enterprise apps, automate processes, and design business sites. The drag-and-drop functionality of these tools ensures that your business experts can leverage their unique domain knowledge and skills to build tailored solutions that meet your organization's specific needs.

Using SAP Build, you can transform your organization's development approach. Instead of relying solely on a small team of developers, you can tap into the collective skills and insights of your business experts. This approach not only accelerates the development process but also fosters a culture of innovation and collaboration. Your business experts become active contributors to the solution-building process, ensuring that the final product aligns with your business goals.

SAP Build offers a visual development environment that allows users to build applications and solutions by simply dragging and dropping prebuilt components. This low-code approach eliminates the need for manual coding, significantly reducing development time and effort. Section 1.4.1 will delve deeper into how SAP Build Apps enables users to build visually and create enterprise-grade apps tailored to their specific needs.

Collaboration between business and development teams is essential for successful application development. SAP Build recognizes this need and provides built-in governance and lifecycle management features. These features enable secure collaboration, ensuring that all stakeholders can contribute to the development process while adhering to established guidelines and standards. Section 1.4.2 will focus on the secure collaboration capabilities of SAP Build Work Zone, emphasizing how it fosters efficient teamwork and effective management of the application lifecycle.

In addition to its visual development capabilities, SAP Build provides seamless integration with a wide range of SAP and non-SAP systems. The platform offers prebuilt connectors and business content, enabling users to accelerate development by leveraging existing integrations. This integration-centric approach will be explored in detail in Section 1.4.3, showcasing how SAP Build Process Automation empowers businesses to leverage their existing IT landscapes while extending functionality through streamlined integrations.

1.4.1 SAP Build Apps

SAP Build Apps is a low-code platform designed to empower users at all skill levels to create custom applications without the need for extensive coding knowledge. With SAP Build Apps, you develop user interfaces with drag-and-drop functionality, without the need for complex coding and lengthy development cycles. Now you can compose enterprise-grade custom apps, utilizing a visual interface that allows you to easily arrange components and design engaging user experiences.

One of the most remarkable features of SAP Build Apps is its ability to create data models and business logic visually, without the need to write a single line of code. This enables users to build robust cloud services and backend functionalities that perfectly align with their business requirements. SAP Build Apps also provides seamless integration, enabling you to securely connect with both SAP and non-SAP solutions. With prebuilt components, connectors, and integrations, you can bring together different systems, leveraging existing assets and accelerating development. This means you can integrate your custom apps with other enterprise systems, ensuring smooth data flow and a cohesive user experience. SAP Build Apps removes the traditional barriers of application development, enabling professionals from various roles and skill levels to contribute to the app-creation process. SAP Build Apps is designed to scale: whether you're building a small application or developing an extensive enterprise solution, the platform offers the flexibility and power to support your growing needs.

In the world of application development, the ability to create robust solutions without writing a single line of code is a game-changer. With a powerful set of features designed to empower users at every skill level, full stack, no-code development has become a reality. Ahead, we'll explore the key features that make this possible, including a user-friendly drag-and-drop interface, visual app logic, visual cloud functions, seamless connection to SAP and non-SAP solutions, integration with APIs, and multiplatform native performance. Let's dive in and discover how these features revolutionize the way applications are built:

1.4 SAP Build

- **Drag-and-drop interface**
  SAP Build Apps contains an extensive array of reusable components to design engaging and uniform user interfaces with ease. The platform allows for rapid prototyping, with the capability to instantly preview changes, speeding up the design process. You can utilize the intuitive tools provided to improve your productivity and refine your UI development workflow. SAP Build Apps offers a practical solution for crafting visually appealing applications efficiently. The visual drag-and-drop editor is shown in Figure 1.1.

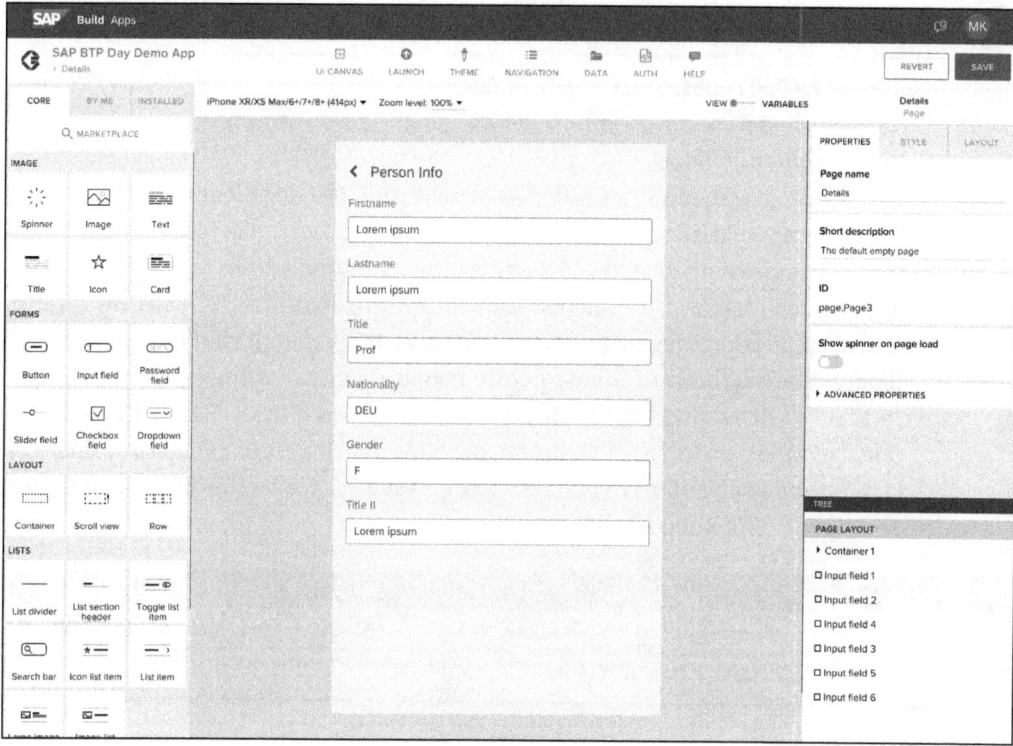

Figure 1.1 SAP Build Apps: Drag-and-Drop UI

- **Visual app logic**
  SAP Build Apps facilitates the creation of app logic through a visual interface, even for applications with complex functionalities. It provides a straightforward approach to development, allowing you to handle user interactions, app lifecycle events, native device capabilities, data operations, and error management effectively. The platform offers a library of over 400 formula functions, enabling you to conduct calculations, text formatting, list and object manipulation, dynamic styling, and more. These functions are applied to your app's data in real time, ensuring timely and accurate outcomes. SAP Build Apps simplifies the development process, allowing for the easy integration of advanced features. Figure 1.2 shows an example of the visual logic editor.

41

1   Introduction to Low-Code and No-Code Development

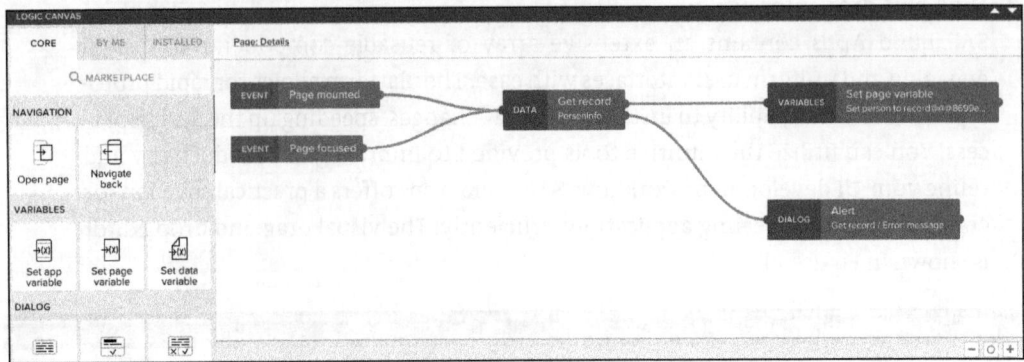

Figure 1.2 SAP Build Apps: Visual App Logic Editor

- **Visual cloud functions**
  SAP Build Apps streamlines serverless application development for developers by providing a visual interface for crafting business logic and data models. It simplifies the development process by abstracting away the complexities of server provisioning and maintenance. Developers can utilize this platform to focus on creating robust applications, leveraging the cloud's scalability and flexibility. This approach reduces the overhead of infrastructure management, allowing developers to concentrate on delivering high-quality, scalable solutions quickly. SAP Build Apps supports an agile development workflow, enabling developers to efficiently bring their cloud-based applications to market. Figure 1.3 shows an example of the visual cloud functions in SAP Build Apps.

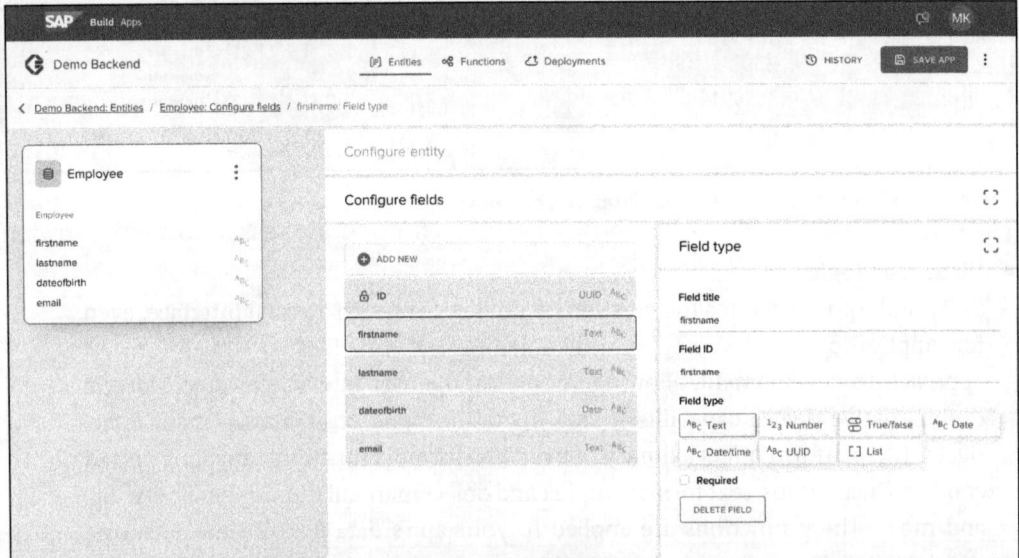

Figure 1.3 SAP Build Apps: Visual Cloud Functions

- **Connections to SAP and non-SAP solutions**
  You can integrate SAP Build Apps with SAP Business Technology Platform (SAP BTP) for secure access to SAP solutions. You can use the SAP Destination service to establish secure connections and seamlessly access SAP solutions such as SAP S/4HANA. With SAP Build Apps, you can expose one or more OData services, leveraging the content from these solutions in your applications. Extend the capabilities of your apps further by integrating with SAP Sales Cloud and SAP Service Cloud, enabling you to incorporate data and functionality from this cloud-based solution. In addition, SAP Integration Suite provides connectivity to non-SAP enterprise solutions. In short, with SAP Build Apps and SAP BTP, you can bridge the gap between SAP and non-SAP systems, ensuring a holistic and integrated approach to your application ecosystem.

- **Integration with APIs**
  Using the REST Integration Wizard of SAP Build Apps, you can incorporate modern APIs, including external APIs, into your applications. SAP Build Apps lets you effortlessly compose APIs using the same no-code paradigm employed for frontend development, revolutionizing the way you integrate data and functionality from various sources. With the REST Integration Wizard, you can rapidly establish connections, retrieve data, and orchestrate interactions with external systems, all without the need for complex coding. Figure 1.4 shows what an API integration looks like for an SAP SuccessFactors API.

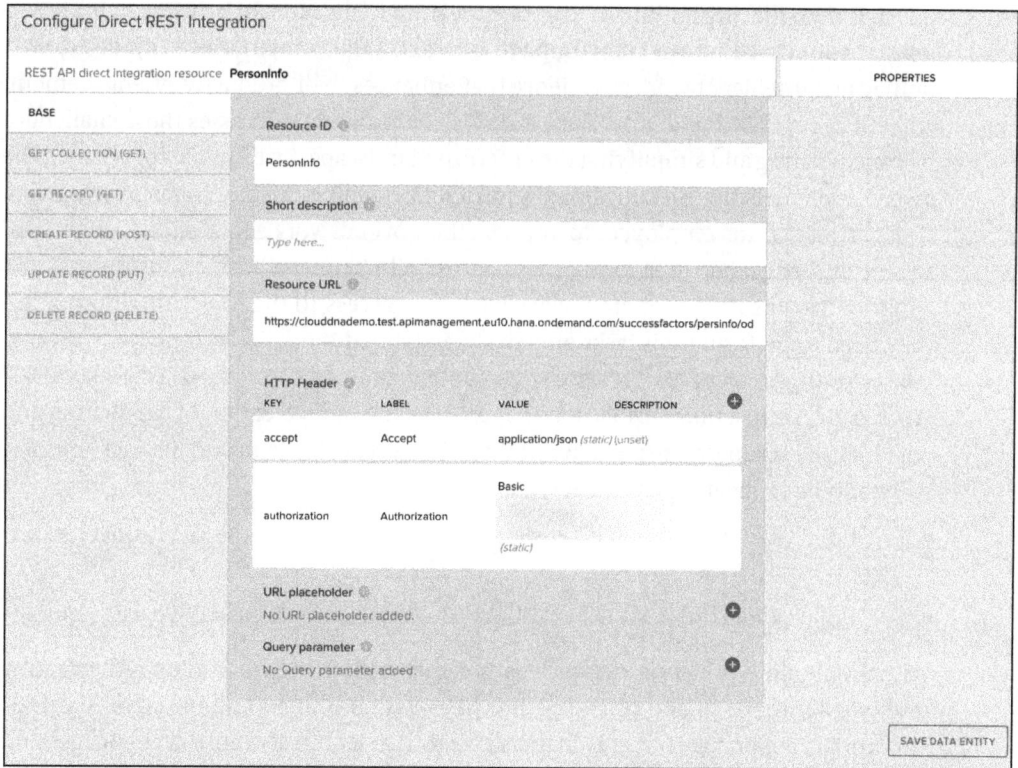

Figure 1.4 SAP Build Apps: API Integration

- **Multiplatform native performance**
  You can develop applications with SAP Build Apps and run them across multiple platforms. Whether it's on the web, iOS, or Android, your apps will work out of the box, delivering a native runtime optimized specifically for each platform. This optimization ensures a smooth and immersive user experience, allowing your applications to leverage the full capabilities of each platform. With SAP Build Apps, you can eliminate the hassle of building separate applications for different platforms and instead focus on creating a unified and consistent user experience across devices.

1.4.2 SAP Build Work Zone

In today's complex enterprise landscapes, organizations face the challenge of managing a heterogeneous and fragmented ecosystem comprising various content types, user interfaces, IT systems, content repositories, applications, and channels. This fragmented landscape often leads to difficulties for business users in finding relevant information and navigating between different applications, resulting in a disjointed employee experience.

Employees frequently express frustration when they struggle to locate the appropriate application to complete a given task, leading to a sense of being lost in the multitude of tools and systems. Decision-making and actions often take place without a comprehensive view, making it difficult to derive meaningful insights or provide accurate recommendations based on the complete context. What employees truly desire is a unified and intelligent work experience that enhances their productivity, engagement, and efficiency. SAP Build Work Zone is SAP's solution that addresses these challenges by consolidating and simplifying the enterprise landscape. By bringing together all the disparate elements into a cohesive platform, SAP Build Work Zone enables companies, IT departments, and employees to improve their overall work experience. It offers consistent and engaging user experiences across all channels, business processes, and applications, ensuring that employees have easy access to the tools and information they need to perform their tasks effectively. SAP Build Work Zone combines the capabilities of two earlier SAP BTP services—namely, the SAP Launchpad service and SAP Work Zone—into a unified solution that enables the rapid creation of captivating and personalized business sites. SAP Build Work Zone is available in two distinct editions, each tailored to meet specific needs:

- SAP Build Work Zone, standard edition (previously known as the SAP Launchpad service)
- SAP Build Work Zone, advanced edition (previously known as SAP Work Zone)

With SAP Build Work Zone, organizations can eliminate the frustration of navigating multiple applications and provide employees with a unified and intuitive interface. This unified experience not only increases productivity but also enhances engagement and efficiency, enabling employees to focus on their core responsibilities without the distraction of fragmented workflows. Organizations can create an environment where employees have a seamless and integrated experience across the entire enterprise landscape. With a centralized platform that harmonizes various systems and channels, employees can access relevant information, collaborate effectively, and make informed decisions based on a holistic view of the organization.

SAP Build Work Zone, Standard Edition

This edition provides a centralized access point, offering personalized and role-based access to both SAP and non-SAP applications. It simplifies the user experience by providing a central entry point for users to access the applications relevant to their roles and responsibilities. At a high level, the standard edition of SAP Build Work Zone is designed for individual users working independently within the system. In this edition, the primary emphasis is on enabling users to perform their tasks without being aware of or interacting with other users on the platform.

SAP Build Work Zone, standard edition offers a comprehensive set of features that empower organizations to enhance their digital workplace experience. Let's delve into the major capabilities that form the pillars of this edition:

- **Secure and centralized access point**
  With SAP Build Work Zone, standard edition, users gain a secure and centralized access point to SAP and non-SAP applications, tasks, and processes. This includes seamless access to a wide range of applications such as SAP S/4HANA apps, SAP BTP apps, and third-party apps, all from a single unified launchpad. By consolidating access in one place, users can simplify their workflows and navigate effortlessly across various systems, increasing productivity and efficiency.

- **Personalization and role-based setup**
  SAP Build Work Zone, standard edition offers a personalized and role-based setup that empowers users to tailor their experience according to their unique needs. Whether it's customizing their homepage, organizing their apps, or selecting preferred themes, users have the flexibility to personalize their work environment, enabling them to work more efficiently and effectively.

- **Customization and extension capabilities**
  Administrators can leverage the flexibility of the platform in SAP Build Work Zone, standard edition to customize and extend its functionality. This includes adapting the user interface to match corporate design guidelines, incorporating custom logos, and tailoring the platform to align with specific business requirements. With these customization options, organizations can create a branded and cohesive digital workplace experience that reflects their unique identity.

- **Seamless integration with SAP and third-party applications**
  SAP Build Work Zone, standard edition offers robust integration capabilities, enabling seamless connectivity to integration content from SAP as well as third-party business applications. This ensures smooth data exchange and process orchestration

across different systems, fostering collaboration and enabling users to access relevant information and functionality within their work environment.

SAP Build Work Zone, standard edition empowers organizations to create a unified and efficient digital workplace. With secure access, personalized experiences, customization options, and seamless integration, businesses can enhance productivity, streamline workflows, and drive user satisfaction.

SAP Build Workzone, Advanced Edition

This edition encompasses all the capabilities of the standard edition and goes beyond, incorporating additional functionalities. It empowers business users to manage unstructured content, create and publish content, and further personalize the user experience. SAP Build Work Zone, advanced edition takes user engagement to a new level by facilitating structured interactions and collaboration among users directly within the platform. This edition recognizes the importance of seamless collaboration and encourages users to engage, share insights, and work together to achieve common goals. By fostering interactive collaboration, the advanced edition promotes teamwork, knowledge sharing, and efficient decision-making within the platform.

SAP Build Work Zone, advanced edition introduces a range of powerful capabilities that empower organizations to create immersive digital experiences. Let's explore the key features that set this edition apart:

- **Powerful page building and content authoring**
  SAP Build Work Zone, advanced edition equips business users with an exceptional page-building experience and robust content-authoring tools. This empowers them to share information, create engaging content, and enable seamless knowledge sharing across the organization. With intuitive drag-and-drop interfaces and versatile content creation capabilities, users can effortlessly craft visually appealing pages and deliver impactful content.

- **Engaging workspaces and decentralized collaboration**
  Workspaces within SAP Build Work Zone, advanced edition provide decentralized topic ownership, enabling users to create interactive, one-stop-shop experiences. These engaging workspaces foster collaboration and knowledge sharing among team members, promoting decentralized ownership and allowing users to contribute their expertise in a structured and cohesive manner.

- **Blending of business data with structured and unstructured information**
  SAP Build Work Zone, advanced edition enables the blending of business data with structured and unstructured information across different formats and channels. Users can seamlessly integrate data from various sources, enriching their content with real-time insights and providing a holistic view of information. This fusion of data types enhances decision-making, improves user experiences, and drives organizational agility.

- **Extensibility framework for customized user experiences**
  SAP Build Work Zone, advanced edition offers a powerful extensibility framework that enables the creation of fully customized user experiences. Through content widgets, context-aware UI integration cards, and other flexible tools, organizations can integrate business data and tailor the user interface to align with specific requirements. This extensibility framework empowers businesses to deliver personalized experiences that cater to the unique needs of their users.

With SAP Build Work Zone, advanced edition, organizations can transform their digital experiences, foster collaboration, and unlock the full potential of their workforce. SAP Build Work Zone, advanced edition empowers users to create captivating and customized digital environments that enhance productivity, enable knowledge sharing, and drive meaningful outcomes.

1.4.3 SAP Build Process Automation

In the fast-paced world of business, automating workflow processes is crucial for efficiency, productivity, and staying ahead of the competition. With SAP Build Process Automation, you can unlock the power of automation without the need for coding expertise. This solution empowers business users to automate complex workflow processes and tasks, revolutionizing the way work gets done.

> **IDC**
> IDC reports that organizations adopting low-code and intelligent process automation solutions have seen an ROI of 509% over five years, with every enterprise reporting positive outcomes. This demonstrates the measurable benefits and efficiency gains from low-code platforms, which facilitate faster application development and process optimization.

By harnessing the power of workflow management, robotic process automation (RPA), and embedded AI capabilities, SAP Build Process Automation enables you to scale process automation to meet evolving business needs. With intuitive low-code and no-code tools, business users can automate workflows and processes using a simple drag-and-drop approach. Professional developers, on the other hand, can leverage visual tooling and prebuilt content to deliver results more efficiently and respond swiftly to new requirements with enhanced agility. They can create comprehensive workflows and automations that can be easily packaged for business users to adopt in their projects. SAP Build Process Automation seamlessly integrates with other SAP offerings, providing native integration and prebuilt process content that includes RPA bots specifically designed for SAP applications. In addition, AI-assisted intelligent document processing further enhances the capabilities of this solution. Built on a robust enterprise-grade platform, SAP Build Process Automation empowers organizations to innovate with

1 Introduction to Low-Code and No-Code Development

speed, simplicity, and confidence. With SAP Build Process Automation, businesses can unlock the potential for process optimization, efficiency gains, and improved customer experiences. Embrace the power of this solution to embark on a journey of transformative automation, where innovation becomes a reality with greater speed, simplicity, and confidence.

SAP Build Process Automation offers a range of key features that empower businesses to boost efficiency, streamline workflows, and leverage the value of AI in their automation initiatives. Let's explore these features in detail:

- **Boost the efficiency and agility of business processes**
  With SAP Build Process Automation, businesses can optimize and accelerate their processes, driving efficiency gains and improving overall agility. By automating manual and repetitive tasks, organizations can free up valuable time and resources, enabling employees to focus on higher-value activities. Figure 1.5 shows a process with a simple one-step approval.

- **Support automation and extensions with workflows**
  The solution provides robust workflow capabilities, allowing users to design, implement, and manage complex automation processes seamlessly. Whether it's simple task automation or comprehensive end-to-end workflows, SAP Build Process Automation offers the flexibility and scalability needed to support diverse business requirements.

- **Improve process efficiency by automating repetitive work**
  SAP Build Process Automation enables businesses to automate repetitive work, reducing errors, improving accuracy, and increasing productivity. By leveraging low-code and no-code tools, users can easily automate tasks and processes, eliminating time-consuming manual efforts and enhancing overall process efficiency.

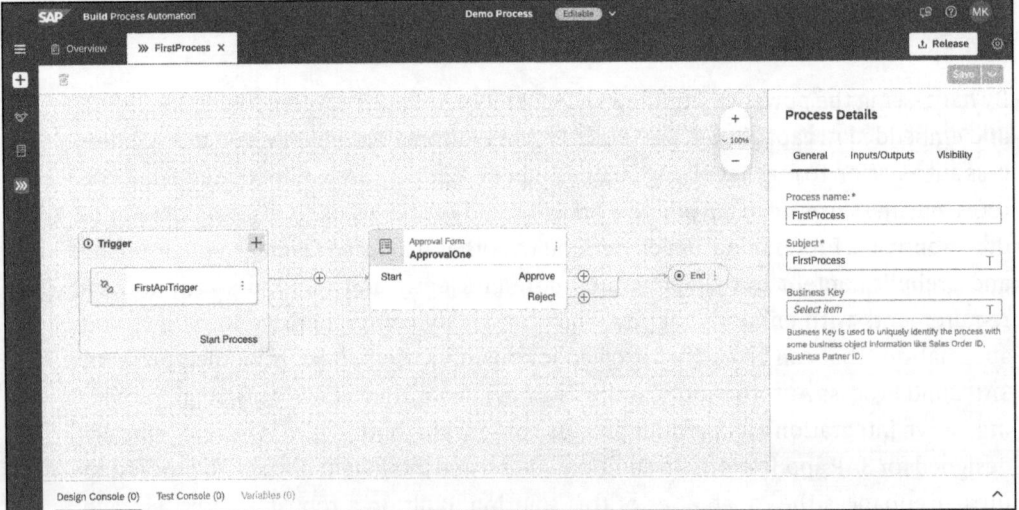

Figure 1.5 SAP Build Process Automation Process

- **Add AI to business process automation**
  SAP Build Process Automation integrates the power of AI, enabling businesses to leverage intelligent capabilities for enhanced automation. By incorporating AI-driven features such as intelligent document processing, organizations can extract valuable insights, streamline decision-making, and improve the overall effectiveness of their business processes.

1.5 Summary

In Section 1.1, we looked at the IT challenges that companies are facing today. These include a shortage of skilled workers, the digitization of business processes, the modernization of legacy systems, time constraints, and the need for agile development. These challenges require innovative solutions to be more efficient and competitive. This is where low-code and no-code come into play, as they enable companies to develop applications faster, overcome resource constraints, and increase flexibility. By using low-code and no-code, companies can effectively respond to the challenges and successfully drive their digital transformations.

In Section 1.2, we looked at the concepts of pro-code, low-code, and no-code. Pro-code refers to the traditional development of software by professional developers who write the code from scratch. Low-code is the use of visual tools and prebuilt components to speed up application development. No-code, on the other hand, allows users to create applications without *any* programming knowledge. The choice of pro-code, low-code, or no-code depends on an organization's unique requirements and resources. Pro-code offers the greatest flexibility and control but requires experienced developers and longer development times. Low-code allows for faster application development and requires less programming expertise but may have limitations when specific requirements are outside the scope. No-code is ideal for simple applications and allows nontechnical users to build applications on their own, but with limited options for customization and complexity.

Section 1.3 showed potential use cases for low-code and no-code. SAP provides companies with the necessary tools and platforms to implement these use cases easily and efficiently. This enables companies to automate their business processes, modernize their software landscape, develop innovative mobile apps and websites, and quickly receive feedback from users. With SAP as a partner, companies are well-equipped to successfully implement their digital transformations and remain competitive.

Finally, Section 1.4 provided a brief introduction into SAP Build, offering a high-level overview of the different SAP Build capabilities: SAP Build Apps, SAP Build Work Zone, and SAP Build Process Automation.

Chapter 2
SAP Business Technology Platform

SAP BTP is a cloud-based platform that helps companies drive their digital transformation by providing a variety of services and tools to optimize business processes and develop applications.

Welcome to the fascinating world of SAP Business Technology Platform (SAP BTP). In this chapter, we will dive into the possibilities and opportunities that the platform-as-a-service offering from SAP has in store for companies. From developing innovative applications to integrating data and systems to optimizing business processes, SAP BTP opens a wide range of solutions to drive your digital transformation and strengthen your competitiveness. In an increasingly connected and digitized world, this can be a decisive advantage. Dive into the world of SAP BTP with us and let's discover together how SAP BTP helps companies successfully move into the future.

In this context, we need to explain the terms *infrastructure-as-a-service* (IaaS), *platform-as-a-service*, (PaaS) and *software-as-a-service* (SaaS). IaaS, PaaS, and SaaS are three different models for cloud computing services, each offering different levels of responsibility and control. IaaS provides the basic IT infrastructure, including virtual machines, storage, and networks. The customer is responsible for configuring, managing, and maintaining this infrastructure. IaaS enables companies to run their own operating systems and applications in the cloud and provides flexibility, scalability, and control over the infrastructure. SAP does not have any IaaS offerings. PaaS goes beyond IaaS to provide a complete development and deployment environment for applications.

In these models, the cloud provider provides the platform with operating systems, development tools, databases, and middleware. The customer focuses on application development and deployment, while the provider is responsible for the infrastructure. PaaS enables faster application development and deployment by allowing developers to focus on application development and innovation without having to worry about the underlying infrastructure. SAP BTP is the PaaS offering from SAP. SaaS is the most comprehensive cloud computing model, where applications are delivered entirely over the internet. The cloud provider hosts, manages, and maintains the applications while customers access them over the internet. SaaS provides out-of-the-box deployment because no installation or configuration is required on the customer side. Customers use the applications via subscriptions and typically have limited customization and configuration options. SaaS offerings from SAP include SAP S/4HANA Cloud, SAP SuccessFactors, and SAP Ariba.

In this chapter, we'll provide a comprehensive overview of SAP BTP, including its many functionalities and benefits for enterprises. In Section 2.1, you'll learn about the basic structure and key components of SAP BTP to develop a solid understanding of this powerful platform.

Because security is critical in today's digital landscape, we devote Section 2.2 to looking at the security aspects of SAP BTP. We'll look at the robust security mechanisms and best practices provided by the platform to ensure protection of sensitive corporate data and compliance.

The *cloud connector* is the central component in hybrid landscapes. It's used as a gateway to access the OData service and SAP Fiori apps provided in on-premise systems out of SAP BTP. The cloud connector is covered in section Section 2.3.

2.1 Overview

SAP BTP unifies data and analytics, artificial intelligence, application development, automation, and integration within a single, cohesive environment.

SAP positions SAP BTP to empower businesses to personalize experiences for SAP applications, enabling them to innovate faster within a business context. By running on a trusted, enterprise-grade platform, SAP BTP provides the foundation for organizations to drive digital transformation, deliver exceptional user experiences, and accelerate innovation in a secure and scalable environment.

SAP BTP enables delivering innovations that seamlessly integrate with SAP applications. You can leverage the power of artificial intelligence and automation to enrich user interactions, providing tailored and intuitive experiences. With real-time access to comprehensive data, SAP BTP enables businesses to gain holistic insights, enabling informed decision-making and empowering organizations to unlock the full potential of their SAP applications.

With SAP BTP, businesses can innovate faster within a business context by leveraging a range of powerful tools. You can accelerate development processes with both no-code and code-first development tools, enabling rapid prototyping and deployment of applications. You can gain deeper insights by utilizing and analyzing data from SAP applications within the appropriate context and meaningful insights. Furthermore, you can jump-start projects with prebuilt industry content and use cases, leveraging proven best practices and reducing time-to-value. By embracing the innovative capabilities of SAP BTP, organizations can drive continuous innovation and stay ahead in today's dynamic business landscape.

SAP BTP runs on a trusted, enterprise-grade platform, providing organizations with confidence and reliability. You can deploy your applications in a mission-critical cloud environment that is managed by SAP, ensuring high availability, scalability, and robust security measures. Also, you can customize your business processes without the burden of maintenance, allowing you to focus on driving innovation and growth. With SAP BTP, you have the flexibility to choose your preferred cloud provider while seamlessly interoperating with your existing IT landscape, enabling smooth integration and leveraging your existing investments. Trust in the robust infrastructure of SAP BTP to run your critical business operations and unlock the full potential of your digital transformation journey.

Now that you have an overview of how SAP is positioning SAP BTP, it's time to dive deeper into this versatile platform. In this section, we provide a comprehensive insight into SAP BTP and its various facets. Let's start with an introduction to SAP BTP (Section 2.1.1) to gain a solid understanding of its importance and purpose. Learn how SAP BTP helps companies drive their digital transformation and develop innovative solutions.

Then, in Section 2.1.2, we'll look at the account structure of SAP BTP. Here you'll learn how accounts are structured in the platform and how they are managed. From the organizational level to individual services, we'll shed light on the components and their interrelationships.

Finally, in Section 2.1.3, we'll provide a comprehensive overview of the various services offered by SAP BTP. From development and integration tools to databases and analysis tools, SAP BTP offers a wide range of services that support companies with their specific requirements. We'll highlight the most important services and explain their basic functions.

2.1.1 SAP Business Technology Platform: Introduction

The development of SAP BTP began in October 2012 with the introduction of SAP NetWeaver Cloud. In May 2013, the platform was renamed *SAP HANA Cloud Platform*, with a focus on the integration of SAP HANA, a powerful in-memory database that enables fast data processing and analysis. Companies could develop innovative applications on the SAP HANA Cloud Platform, analyze data in real time, and optimize business processes. SAP HANA Cloud Platform offered a variety of services and tools for developers to build custom applications and deploy them in the cloud. The platform also supported integration with external systems and data sources to ensure seamless collaboration and data consistency. SAP HANA Cloud Platform played an important role in the development of today's SAP BTP. It paved the way for a cloud-based platform that helps companies address their business challenges and succeed in an increasingly digitized world. SAP HANA Cloud Platform was deployed exclusively on SAP's own infrastructure, known as the Neo environment. This means that SAP had full control over but also responsibility for the deployment, maintenance, and scaling of the platform. The Neo environment provided a stable and secure infrastructure to meet customers' requirements and offer them a reliable environment for their applications.

In February 2017, the platform was renamed again to *SAP Cloud Platform* to reflect its broader functionality and expanded ecosystem. One of the key enhancements was the integration of additional services and tools beyond just data integration. SAP Cloud Platform offers a wide range of services to help companies develop, integrate, and deploy applications. These include development tools, database services, integration services, analytics and predictive analytics capabilities, AI and IoT services, and mobile and portal solutions. SAP Cloud Platform continued to be deployed on SAP's own infrastructure.

The most recent renaming occurred in January 2021, when the platform was named *SAP Business Technology Platform*. This name change underscores the platform's strategic focus on business technology and clarifies its role as a comprehensive cloud-based solution for enterprises. With its ongoing evolution, SAP BTP is positioning itself as a key player in the business technology space, helping companies increase their competitiveness in an increasingly digitized world. As it evolved into SAP BTP, the platform became more flexible and enabled the use of hyperscalers and deployment in different environments such as SAP BTP, Neo environment and SAP BTP, Cloud Foundry environment.

The Neo and Cloud Foundry environments are two different deployment models within SAP BTP that offer different approaches to developing and running applications. The Neo environment was SAP's original deployment model and was based on the company's own infrastructure. It provides a stable and controlled environment that allows organizations to develop and run applications. With the Neo environment, customers can run their applications in a secure and reliable environment and have full control over their resources and data. On the other side is Cloud Foundry, an open and flexible platform developed by the Cloud Foundry Foundation.

The Cloud Foundry Foundation
The Cloud Foundry Foundation is a nonprofit organization dedicated to promoting and supporting the Cloud Foundry ecosystem. Cloud Foundry is an open-source PaaS that helps enterprises develop, deploy, and scale applications in the cloud. The foundation coordinates the evolution and maintenance of the Cloud Foundry platform, as well as collaboration among enterprises, developers, and cloud technology providers. Key members of the Cloud Foundry Foundation include companies such as IBM, VMware, SAP, Pivotal, SUSE, and Hewlett Packard Enterprise (HPE). Together, they drive the development and innovation of the Cloud Foundry platform and promote its use in the industry.

Cloud Foundry enables enterprises to deploy applications in a scalable and agile environment. It provides a modular and extensible framework for developing, deploying, and scaling applications across multiple cloud environments. Cloud Foundry provides flexibility in choosing a cloud provider and enables enterprises to run their applications in a hybrid or multicloud environment. By using Cloud Foundry, developers can develop their applications faster because they can focus on application development while the platform manages the underlying infrastructure. Cloud Foundry enables developers to deploy applications on SAP BTP in a variety of environments. This involves hosting the platform on so-called hyperscalers such as Amazon Web Services (AWS), Microsoft Azure, and Google Cloud Platform (GCP). This environment allows greater flexibility and choice for customers who want to host their applications in different cloud environments. The choice between Neo and Cloud Foundry depends on the organization's specific requirements, preferences, and integration goals.

What Is a Hyperscaler?

Hyperscalers are the big players among cloud service providers. These include AWS, Microsoft Azure, GCP, and others. They have immense infrastructure and resources to offer scalable and diverse cloud services. In terms of SAP BTP, integration with hyperscalers enables enterprises to deploy their applications and data in a broader range of cloud environments. This opens additional options for customers to host their SAP-based solutions on hyperscaler platforms and benefit from the specific features, scaling capabilities, and global presences of these hyperscalers. This enables companies to support their digital transformation initiatives and use their IT resources more efficiently by relying on the benefits of the hyperscalers in combination with the comprehensive services of SAP BTP.

2.1.2 Commercial Models

In this section, we delve into the different commercial models offered by SAP BTP. SAP BTP offers two distinct commercial models for enterprise accounts, providing flexibility to meet different business requirements. First, we explore the *consumption-based model*, in which organizations have the flexibility to access a comprehensive range of services, turning them on and off as needed. Then, we delve into the *subscription-based model*, which allows organizations to subscribe to specific services at a fixed cost, regardless of consumption. We will explore the benefits, considerations, and implementation details of each model, empowering you to make informed decisions that align with your organization's requirements and goals. Let's dive into the details of these innovative licensing models and discover the best fit for your business needs.

Depending on your business needs, you can choose to utilize either or both commercial models. They can be implemented within separate global accounts or within the same global account. To gather more detailed information about these commercial models and determine the best fit for your organization, it is recommended to reach out to your SAP account executive or sales representative.

2   SAP Business Technology Platform

Consumption-Based Commercial Model

The first model is the consumption-based commercial model, which grants your organization access to all current and future eligible services. With this model, you have the freedom to activate or deactivate services and switch between them as needed throughout the duration of your contract. There are two variations of this model: the Cloud Platform Enterprise Agreement (CPEA), and Pay-As-You-Go for SAP BTP. The CPEA offers a structured agreement with a defined commitment, while the Pay-As-You-Go option allows you to pay based on actual usage.

Note that CPEA and Pay-As-You-Go for SAP BTP cannot be mixed within the same global account. The following are some details of the CPEA licensing agreement:

- **Prepaid investment**
  Your organization makes a prepaid investment in cloud credits for the contract duration with an annual commitment to consume SAP BTP services.

- **Flexibility**
  This option is suitable for customers with well-established and planned use cases who want the flexibility to turn services on and off and switch between services without being tied to a single service throughout the contract.

- **Monthly balance statement**
  You receive a monthly statement detailing the usage consumption and costs of each service, with the total monthly cost deducted from your cloud credits balance.

- **Minimum investment and volume-based discounts**
  This model requires a minimum investment entry, and volume-based discounts are available.

- **Billing and top-ups**
  You are billed annually in advance, and any overages are billed in arrears at list price. You can top up your cloud credits at any time to prevent overages.

And now consider some Pay-As-You-Go for SAP BTP licensing model details:

- **Zero-commitment model**
  With no up-front payment, minimum usage requirement, or annual commitment, you have access to all services available in CPEA.

- **Usage-based payment**
  You pay only for SAP BTP services that you use, with monthly billing in arrears.

- **Nondiscountable service charges**
  Service charges under this model are nondiscountable.

- **Suitable for proof of concept**
  Ideal for customers with less defined use cases, interested in running a proof of concept in a productive environment. This model provides the flexibility to turn services on and off and switch between services throughout the contract.

- **Seamless transition to CPEA**
  If you have no other CPEA-based global accounts, a seamless transition to the CPEA model is available.

Subscription-Based Commercial Model

The second model is the subscription-based commercial model. Under this model, your organization subscribes only to the specific services you plan to use. With this model, your organization receives a fixed price and a predefined period, typically ranging from one to three years, ensuring consistent access to your subscribed SAP BTP services throughout the agreed-upon duration. The following are some subscription-based licensing model details:

- **Access to subscribed services**
  Enjoy the usage rights exclusively for the subscribed services, tailored to your specific needs.
- **Expand service portfolio**
  If the need arises for additional services beyond your initial subscription, you can modify your contract through your dedicated sales representative or account executive, allowing your organization to adapt and evolve.
- **Predictable fixed costs**
  Benefit from a fixed cost structure, regardless of the actual consumption of the subscribed services, providing financial predictability and simplifying budget planning.
- **Advance payment**
  Payment is made in advance at the start of the contract period, ensuring seamless service provisioning and reducing administrative efforts.
- **Renewal opportunities**
  At the end of the contract period, your organization has the option to renew the subscription, ensuring uninterrupted access to SAP BTP services and continued innovation.

2.1.3 Account Structure

Accounts in SAP BTP are organized into a hierarchical structure, consisting of global accounts, subaccounts, and directories, providing flexibility and efficient management.

A *global account* represents the contractual agreement between you or your company and SAP. It serves as the foundation for managing subaccounts, members, entitlements, and quotas. Entitlements and quotas for platform resources are assigned to the global account and then distributed to subaccounts for actual consumption. Two commercial models are available for global accounts: the consumption-based model and the subscription-based model. Global accounts are independent of regions and environments. Within a global account, you can manage multiple subaccounts, each specific to a particular region. Global accounts are not bound to specific regions or environments. They provide a region- and environment-independent framework for managing all subaccounts.

Each *subaccount*, in turn, is dedicated to a specific region, allowing for efficient management and organization within the global account structure. Subaccounts allow you to structure your global account based on your organization's and project's requirements, defining members, authorizations, and entitlements. Each subaccount operates independently, ensuring security and facilitating efficient management of data, member roles, integration, and overall landscape planning. Each subaccount is associated with a *region*, representing the physical location where applications, data, and services are hosted (see Figure 2.1). The region assignment for a subaccount doesn't have to align with your organization's physical location. For example, you can operate a subaccount in Europe while being in the United States.

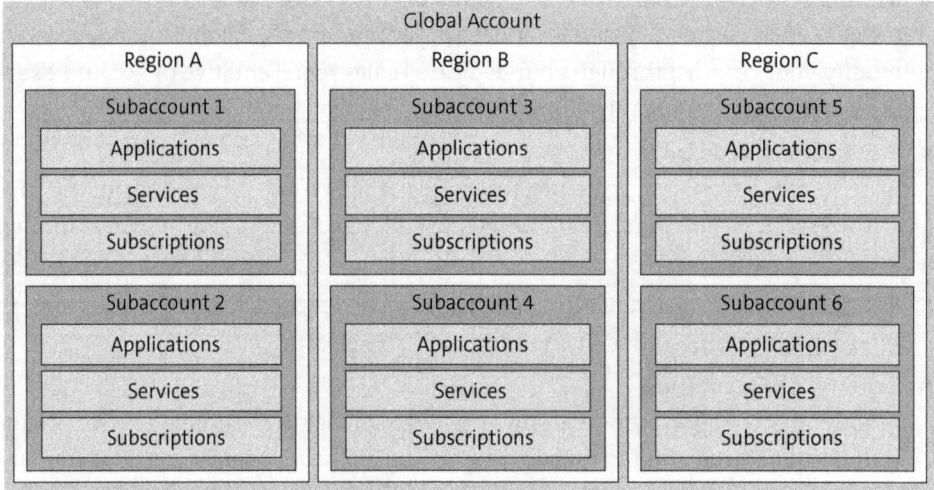

Figure 2.1 SAP BTP Account Structure

User accounts are separate entities from global accounts and subaccounts, maintaining independence and ensuring secure access control. When enabling the Cloud Foundry environment within a subaccount, a corresponding Cloud Foundry org is automatically created. The subaccount and org have a 1:1 relationship and share the same navigation level in the SAP BTP cockpit. Spaces can be created within the Cloud Foundry organization, further organizing your account model, and facilitating the use of services and functionalities within the Cloud Foundry environment (see Figure 2.2).

Directories provide the means to organize and manage subaccounts based on technical and business requirements. Directories can contain subaccounts and other directories, allowing for a hierarchical structure. However, creating directories to group subaccounts is optional, as subaccounts can be created directly under the global account (see

Figure 2.3). You have the flexibility to create a hierarchical structure up to seven levels deep, with the global account at the highest level and the subaccount at the lowest. This means that you can have up to five levels of directories between the global account and the lowest-level subaccount, providing a comprehensive organizational structure.

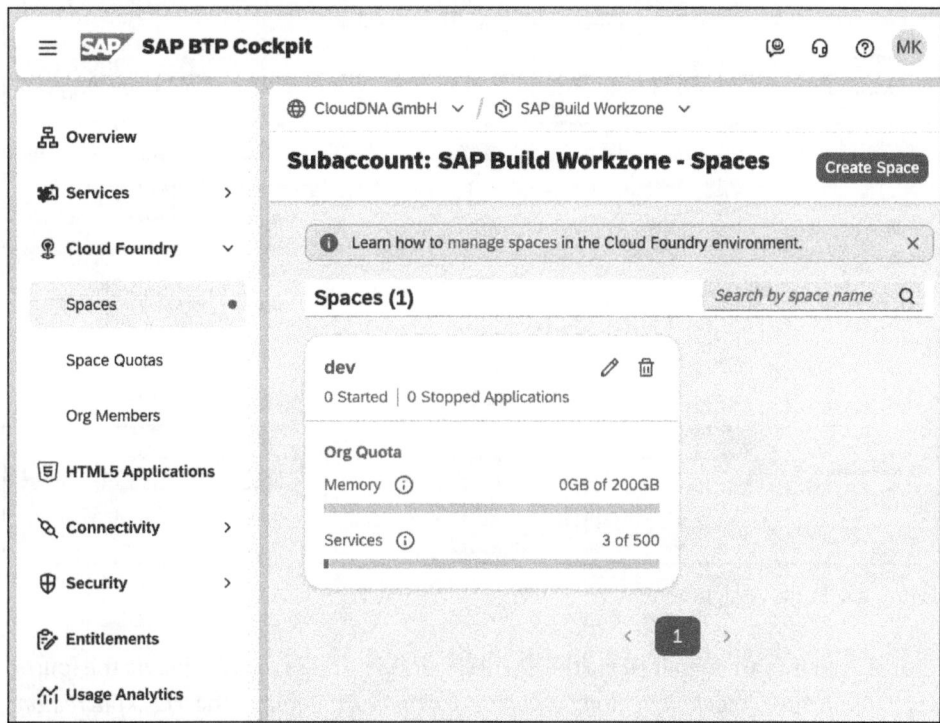

Figure 2.2 Cloud Foundry Spaces

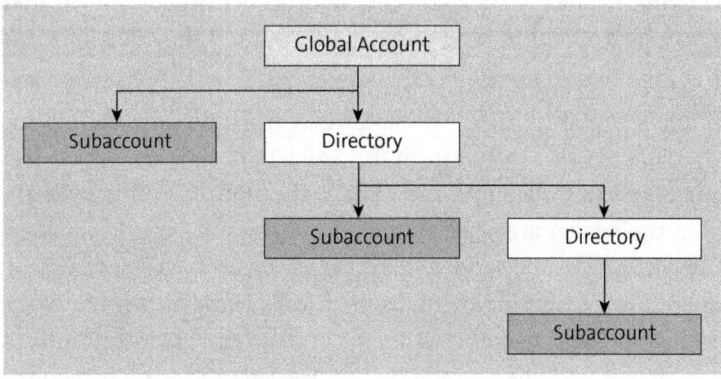

Figure 2.3 SAP BTP Directories

The entitlements and quotas assigned to a global account need to be allocated to individual subaccounts based on their specific needs. To do so, navigate to the **Entitlements • Entity Assignments** section in the SAP BTP cockpit (see Figure 2.4).

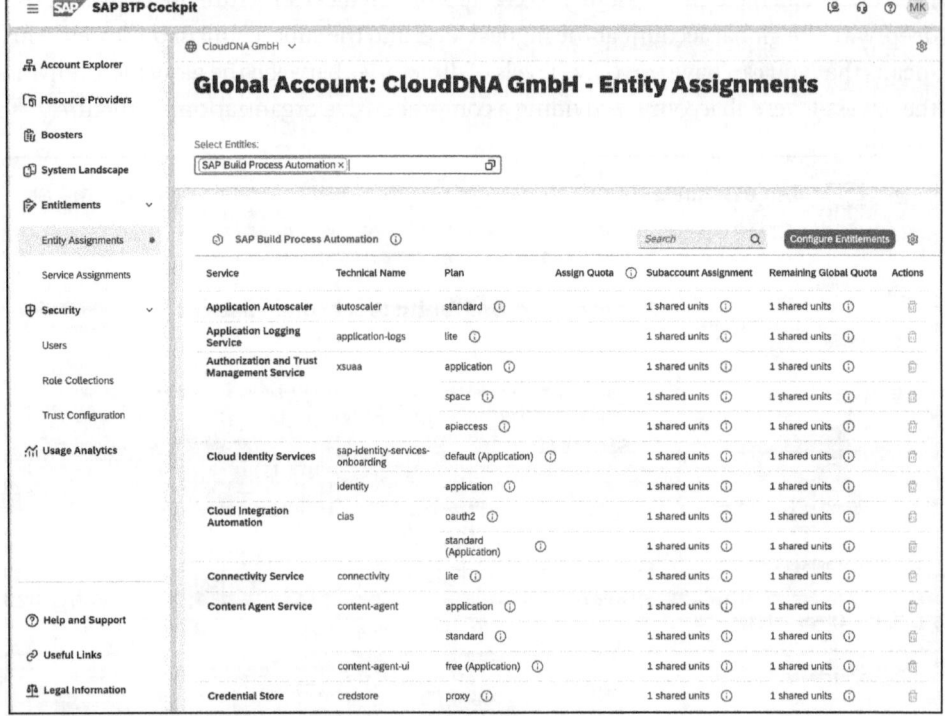

Figure 2.4 Entity Assignments

Global accounts are managed via the SAP BTP cockpit. Access the cockpit via the following URL: *https://<cockpit region>.cockpit.btp.cloud.sap*. Replace the *<cockpit region>* placeholder with *emea*, *amer*, or *apac*, depending on the geographical location of your global account. A complete URL therefore looks like this, for example: *https://emea.cockpit.btp.cloud.sap/*. In the SAP BTP cockpit, you can perform administrative tasks such as managing global accounts, directories, subaccounts, role collections, and users (see Figure 2.5).

From the global account view of the SAP BTP cockpit, you can create subaccounts and jump to them. Subaccounts are also managed in the SAP BTP cockpit. The most efficient way to navigate between subaccounts and back to the global account is via the breadcrumbs located in the header area (see Figure 2.6). In the subaccount, subscriptions are subsequently performed, and service instances are created. As you can see in the figure, SAP has tried to keep the side menu as identical as possible to that of the global account administration. This makes it easier for users to find their way around. In the subaccount, you can see the **Subaccount ID** in the **Overview** area. This is required in addition to valid user credentials so that the cloud connector can connect to the subaccount (for more about the cloud connector, see Section 2.3).

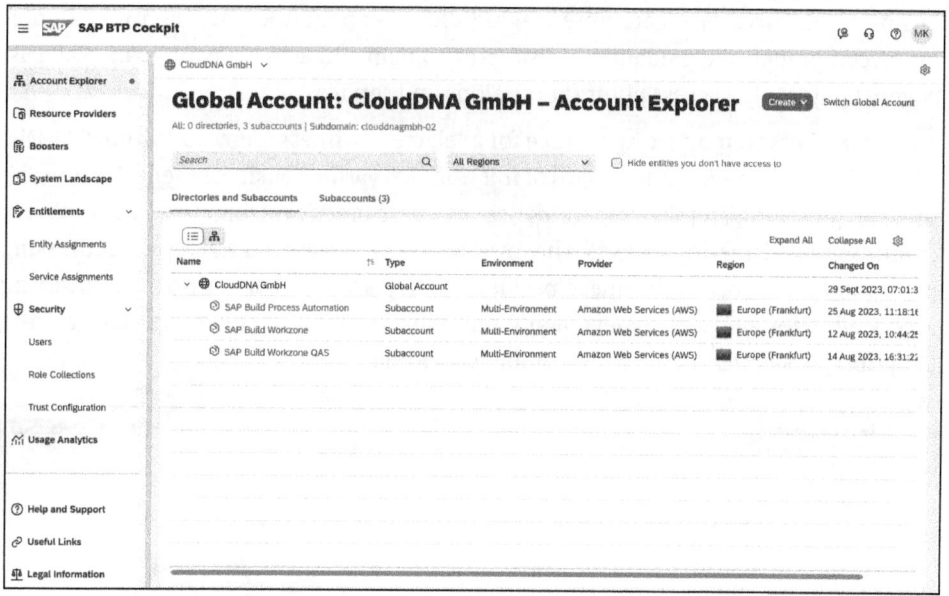

Figure 2.5 SAP BTP Cockpit

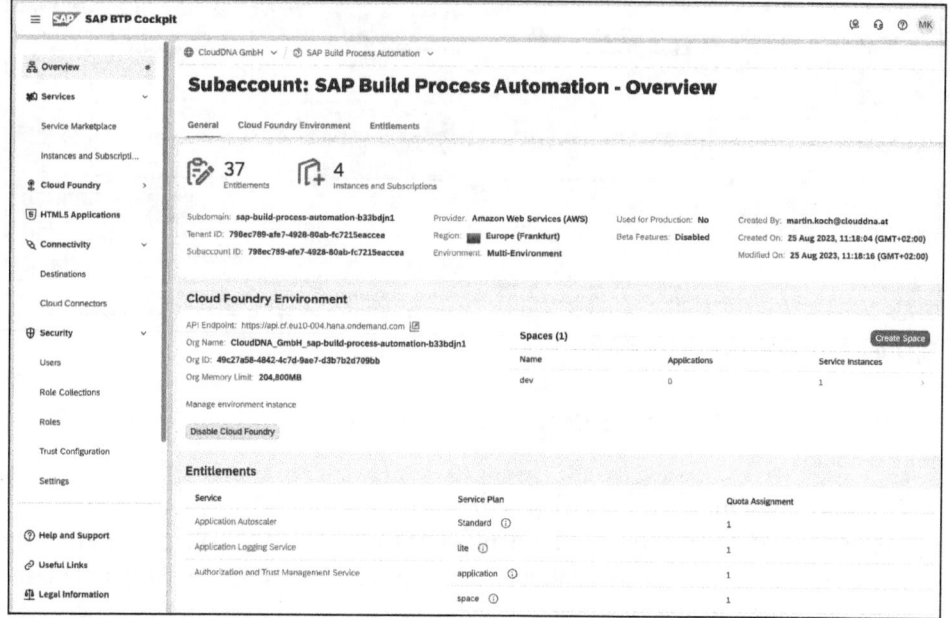

Figure 2.6 SAP BTP Cockpit: Subaccount

2.1.4 Service Overview

SAP Discovery Center (https://discovery-center.cloud.sap/) is designed to assist you in implementing your business scenarios. Whether you are seeking to optimize your business processes or create an engaging digital experience, SAP Discovery Center offers comprehensive guidance and support to help you achieve your goals. SAP Discovery Center is the access point for missions and services.

A *mission* offers step-by-step guidance for a selected business scenario. Within SAP Discovery Center, you will find a wealth of missions for various business scenarios (see Figure 2.7), providing step-by-step guidance and best practices, supported by a vibrant community and mission experts. The missions cover a wide range of topics, addressing pain points and solving business problems. They are available at no cost, ensuring accessibility for all users. By leveraging missions, your project team can navigate the adoption phases of SAP BTP with confidence and clarity.

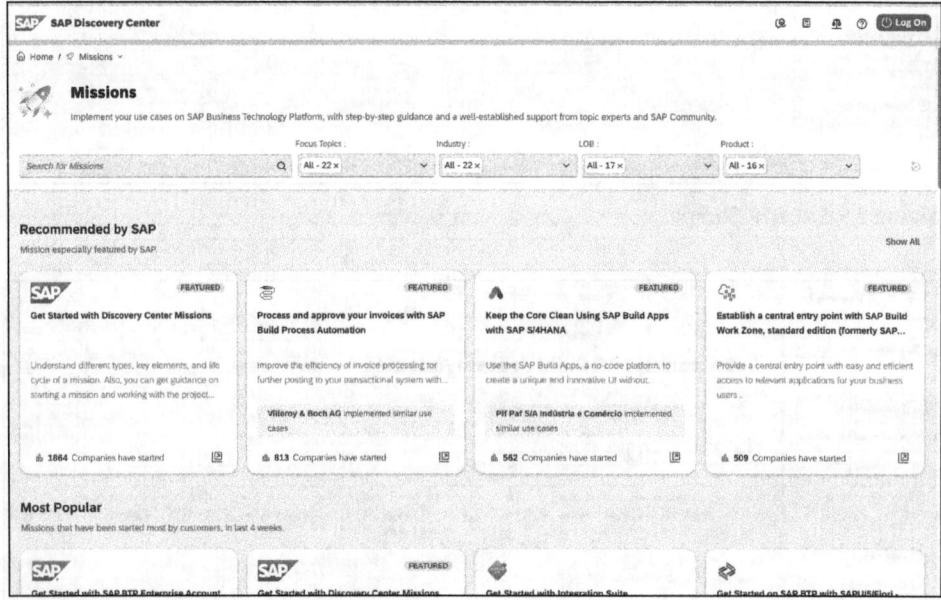

Figure 2.7 SAP Discovery Center: Missions

The missions in SAP Discovery Center encompass different types to cater to diverse requirements:

- **Platform missions**
  Created by SAP product teams, these missions address specific pain points and help solve business problems throughout the entire adoption lifecycle, ultimately leading to successful go-live outcomes.

- **Partner missions**
  These missions assist customers in implementing business use cases using the SAP Business Technology Platform, with guidance and support from SAP's trusted partners.

- **SAP enterprise support missions**
  Developed by the SAP Enterprise Support team and in collaboration with the SAP Enterprise Support Advisory Council (ESAC) program, these missions enable close

collaboration with SAP experts, allowing you to structure a tangible transition plan toward an intelligent enterprise.

- **SAP preferred success missions**
  Created by the SAP Preferred Success team, these missions are designed to jumpstart your success in the cloud and maximize your return on investment (ROI). Access to these missions is available to members of the SAP Preferred Success program.

Complementing the missions, SAP Discovery Center offers a range of services that facilitate and accelerate the development of business applications and other platform services on SAP BTP (see Figure 2.8). These services are designed to enable, facilitate, and optimize your business processes, creating an engaging digital experience for your customers. SAP BTP services and components are packaged as a bill of materials, providing you with the necessary resources to run missions successfully and drive your business forward.

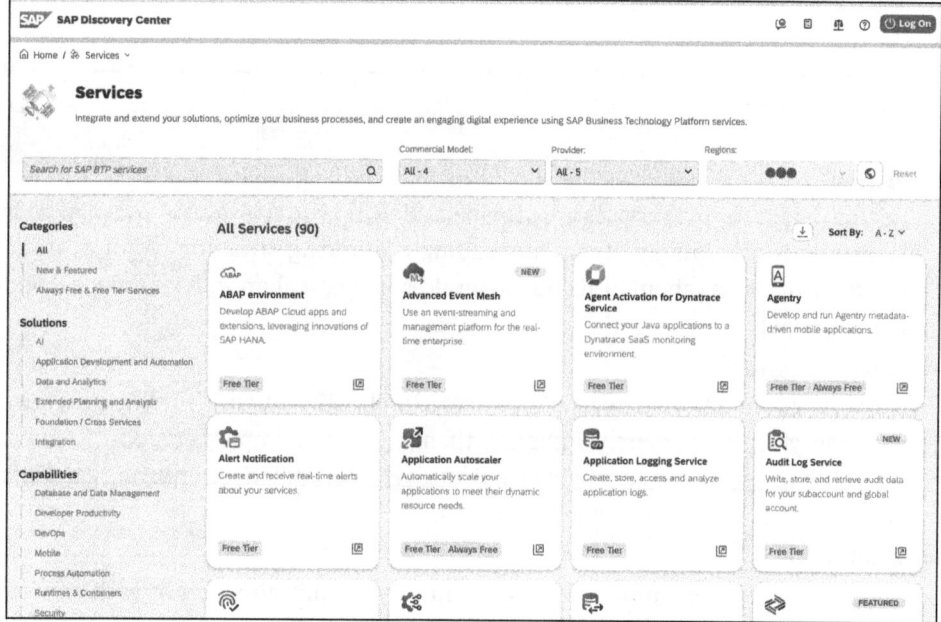

Figure 2.8 SAP Discovery Center Services

The services are designed to provide the necessary tools and resources to optimize your business processes, create engaging digital experiences, and drive your organization's success. As shown in Figure 2.9, the service details typically contain information in the following categories:

- **Overview**
  Each service in SAP Discovery Center offers a comprehensive overview, highlighting its key features, functionalities, and benefits. You can gain a deeper understanding of how each service can support your specific business needs and goals.

2   SAP Business Technology Platform

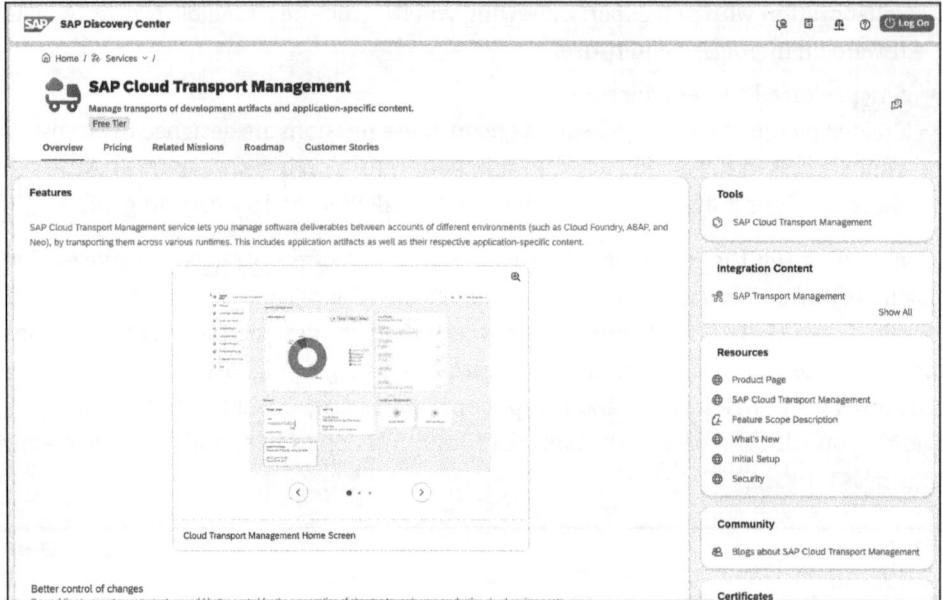

Figure 2.9 Service Details

- **Pricing**
  Pricing details for the services are provided, allowing you to evaluate and plan your investments accordingly. By understanding the pricing structure, you can make informed decisions about utilizing the services that best align with your budget and requirements.

- **Related missions**
  Services within SAP Discovery Center are closely aligned with specific missions, enabling you to seamlessly integrate them into your implementation journey. These missions provide step-by-step guidance and best practices, ensuring a holistic approach to achieving your business objectives.

- **Roadmap**
  The roadmap section provides insights into the future direction of each service, including planned enhancements, updates, and new features. This allows you to align your strategies and stay ahead of evolving business needs, leveraging upcoming capabilities.

- **Customer success stories**
  Discover real-world examples of how organizations have successfully leveraged the services within SAP BTP. These customer success stories highlight the benefits, challenges, and outcomes experienced by businesses that have embraced these services, offering valuable insights and inspiration for your own projects.

- **Links to tools, resources, community, certificates, and support**
  SAP Discovery Center provides direct links to a wide range of tools, resources, and

support channels that can further enhance your experience with the services. Access development tools, documentation, community forums, certification programs, and dedicated support channels to empower your journey with SAP BTP.

Back in the SAP BTP cockpit, within the subaccount, you can see all services assigned to the subaccount in the **Services** • **Service Marketplace** area (see Figure 2.10). You can instantiate or subscribe to them at this point.

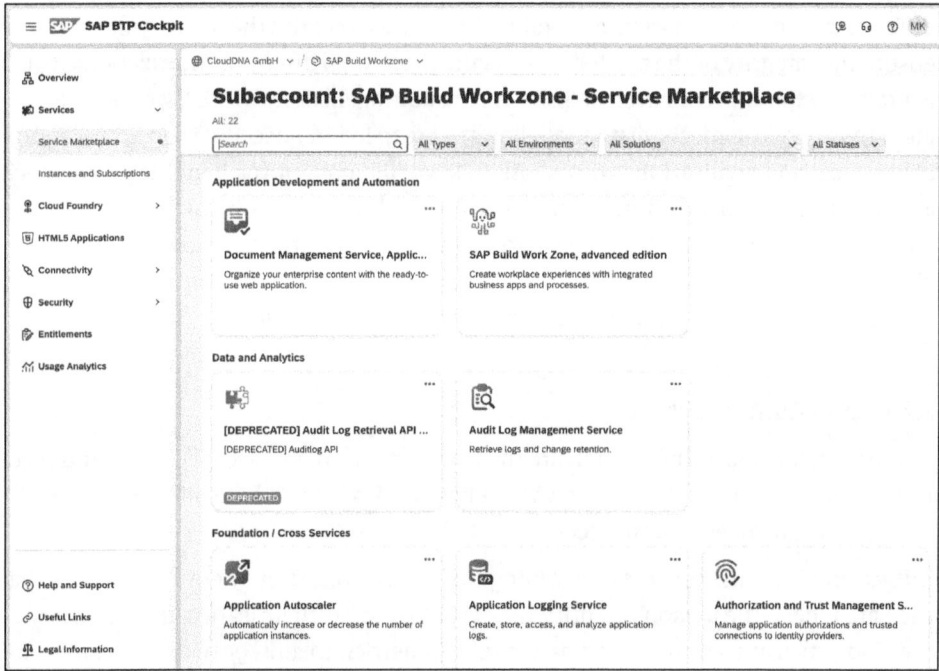

Figure 2.10 Cloud Foundry Service Marketplace

2.2 Security

In today's digital age, the security of corporate data is at the forefront. Security starts with the selection of secure data transmission technologies and continues with the identification of users through to the assignment of rights. The maintenance and configuration of user management should not be independent of the company's own infrastructure. Users should also be able to log into the cloud with the usual authorizations, and if a user is no longer valid, the cloud applications and services should also be aware of this. Integration at the technical level is therefore a prerequisite for seamless integration of the applications at the business level. To illustrate the importance of secure communication, we will use a trivial example that is very commonly used in the literature at this point. Imagine a simple scenario: you want to send a message by mail to a specific recipient. In this scenario, insecure communication would be equivalent to sending the message by postcard. A postcard can be read by any person involved in transit, such as the mail carrier. Sending the message to the recipient in an envelope corresponds to secure communication. Only the recipient can read the message, provided the envelope is not opened in transit.

In this section, we will take an in-depth look at the security aspects of SAP BTP. Let's start with user authentication (see Section 2.2.1) and examine how SAP BTP ensures that only authorized users have access to the platform and its services. We'll explore different authentication methods and best practices to verify the identity of users and ensure the integrity of the platform. In Section 2.2.2, we'll discuss authorizations. You'll get to know the concepts used in SAP BTP. In addition, in Section 2.2.3, we'll look at single sign-on (SSO), a mechanism that allows users to log in once and then access multiple services and applications within SAP BTP. We'll highlight the benefits of SSO, such as improved usability and more efficient authentication. Finally, in Section 2.2.4, we will look at Identity Authentication, part of SAP Cloud Identity Services, a solution specifically designed for managing user identities and protecting corporate resources. We'll examine its features and capabilities for strengthening security in SAP BTP.

2.2.1 User Authentication

Authentication and authorization are the two cornerstones of access control and security. In this section, you'll learn how these key concepts provide a solid foundation for secure and controlled system access.

Authentication is the process of verifying the identity of a user or entity to ensure that they are indeed the person or entity they claim to be. Authentication uses a variety of methods, such as user name and password, biometrics, tokens, or two-factor authentication (2FA). The goal of authentication is to ensure that the user or entity is legitimate and is allowed to access the desired resources.

Here's an example of authentication in an SAP system: Assume that a user wants to log onto an SAP system. He enters his user name and password. The system checks this information to determine whether the user has provided the correct credentials. If the authentication is successful, the user is recognized as a legitimate user in the system and can access his personalized data and functions.

Basically, there are several ways to authenticate. The most common methods are as follows:

- **User name and password**
  This is the classic method where a user enters a unique user name and password to log in. The password is matched against the stored user data to verify authentication.

- **Two-factor authentication/multifactor authentication (MFA)**
  This method combines two or more different authentication factors to provide a higher level of security. Typically, a combination of something the user knows (e.g.,

a password) and something the user has (e.g., a one-time password via SMS or a security token app) is used.

- **Biometric authentication**
  This uses a person's biometric characteristics, such as a fingerprint, facial recognition, an iris scan, or voice recognition. These characteristics are captured and compared with previously registered biometric data to confirm identity.

- **Certificates**
  This uses a digital certificate issued by a trusted certification authority. The certificate contains information about the user's identity and is used to ensure the authenticity and integrity of data.

- **Social login**
  This method allows users to confirm their identity through their existing social media accounts, such as Facebook, Google, or LinkedIn, instead of creating a separate account and password.

Authorization, on the other hand, refers to the process of setting access rights and permissions for an authenticated user or entity. Once authentication is successfully completed, authorization determines which actions, resources, or functions the user or entity can use. This is done by assigning roles, permissions, or access control lists (ACLs). Authorization determines what actions a user can perform, based on their specific rights and permissions.

Here's an example of authorization in an SAP system: After a user is successfully authenticated, authorization is performed to ensure that the user can only access the resources and functions for which he is authorized. Let's assume that the user has the role of a sales representative. Authorization specifies that the user has access to sales data, customer data, and sales reports. However, he does not have authorization to access financial data or personnel files. Authorization thus controls which specific areas of the SAP system the user can use, based on his assigned role and defined access rights.

In SAP BTP, a distinction must be made between the global account and subaccount in terms of authentication. Basically, the SAP ID service can be used for authentication at both levels via user identities, which you can either create or manage via SAP for Me (*https://me.sap.com/*; see Figure 2.11). However, this approach also has some limitations. One of its drawbacks is that you have limited or no control over certain configuration parameters, such as password policies.

Alternatively, you can use Identity Authentication instead of the SAP ID service. This gives you control over the administration and maintenance of users. Identity Authentication is included in the SAP BTP license and can therefore be used without additional license costs. Together, Identity Provisioning and Identity Authentication make up SAP Cloud Identity Services. The goal of SAP Cloud Identity Services is to enable seamless SSO across all systems while ensuring secure access to systems and data. Identity Authentication will be explained in more detail in Section 2.2.4.

2   SAP Business Technology Platform

Figure 2.11 SAP for Me: User Management

Alternatively, you can use a third-party SAML identity provider at the subaccount level to perform user authentication. A common example of this is Microsoft Entra ID (formerly Azure Active Directory), which is widely used in practice. By integrating Microsoft Entra ID as a SAML identity provider with your SAP subaccount, users can use their existing Microsoft Entra ID credentials to access resources and services within the SAP system. This seamless integration enables a great user experience. You also benefit from the extensive security features and controls that Microsoft Entra ID provides to enhance access security.

2.2.2   User Authorization

As already mentioned, authentication checks a user's credentials. In simplified terms, the system checks whether the user is who he or she claims to be. Because the user should not be able to access all functions without further checks after successful authentication, appropriate authorization checks are required. Authorizations are managed and assigned to users in SAP BTP based on role collections.

A *role* is an instance of a role template (see Figure 2.12). You can build a role based on a role template and assign the role to a role collection. The roles are defined in the respective services, such as SAP Integration Suite, and are provided in the subaccount when the services are activated. Roles are assigned to role collections which are assigned in turn to users or user groups.

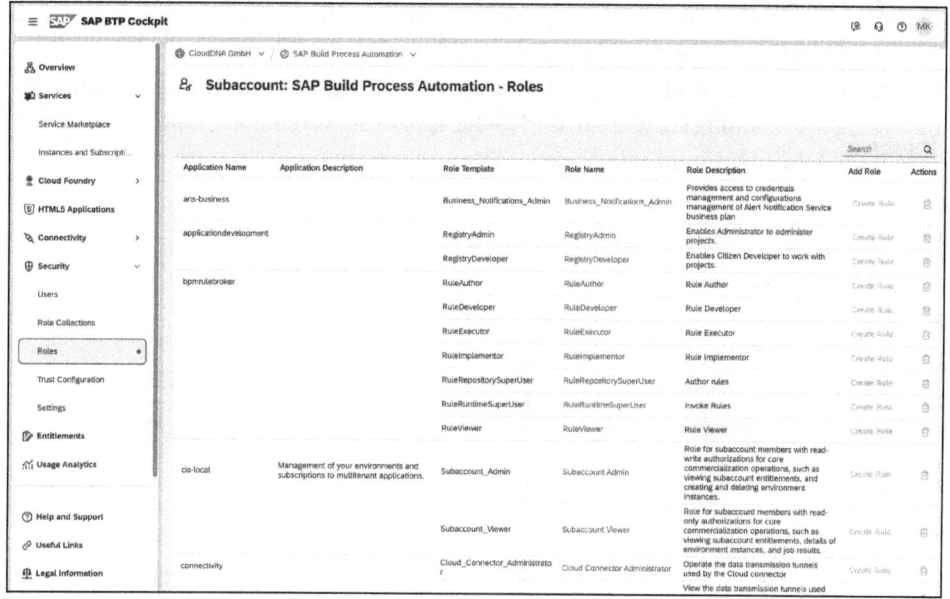

Figure 2.12 Subaccount Roles

In the SAP BTP cockpit, you can view and access information regarding the role collections that have been established, along with the roles contained within each role collection (see Figure 2.13).

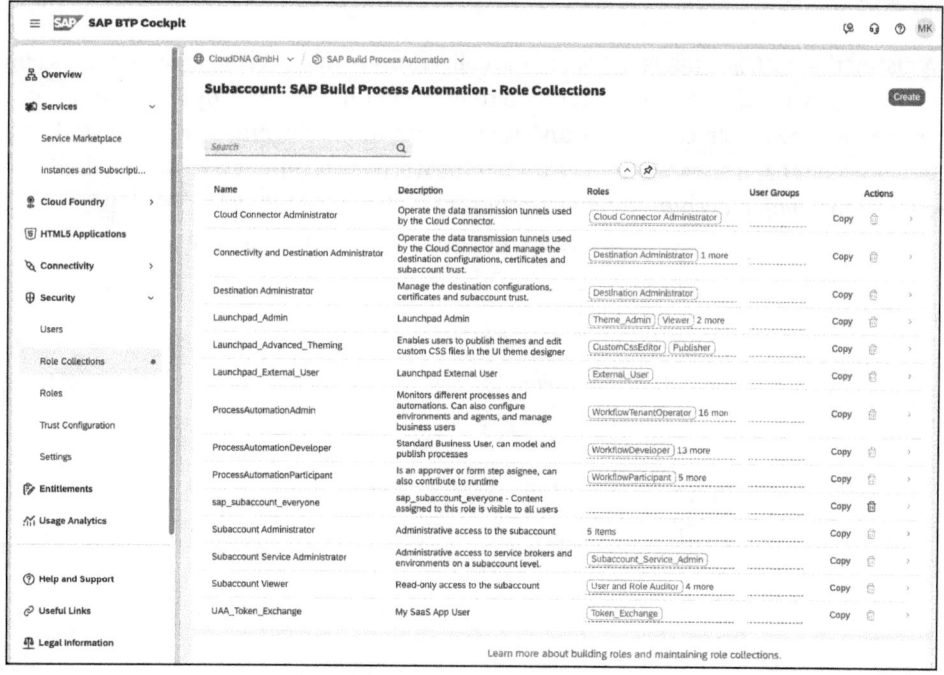

Figure 2.13 Subaccount Role Collections

2   SAP Business Technology Platform

In the details of the role collection, you can see which roles are assigned to this collection (see Figure 2.14). There you can also assign users to the role collection or map them to user groups of the underlying identity provider. In addition, you can perform *attribute mapping*. In doing so, you can derive the assignment to the role collection from an attribute of the identity provider, such as the department.

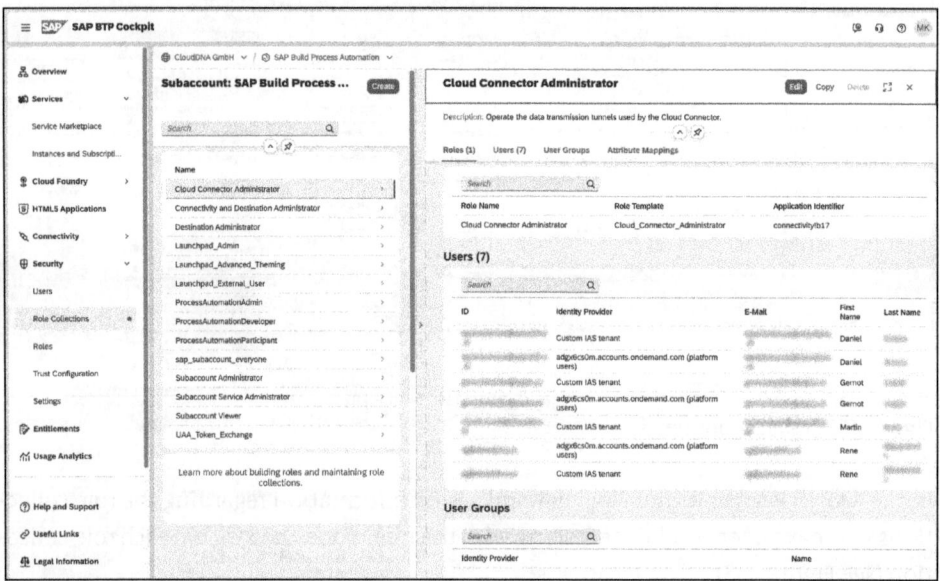

Figure 2.14 User Assignment

A distinction must be made between role collections at the level of global accounts and subaccounts. In addition to the standard role collections delivered by SAP, you can also create your own role collections and tailor them to your requirements on both the global account and subaccount levels. The global account comes with two standard role collections: *global account administrator* and *global account viewer* (see Figure 2.15). Basically, developers and end users do not need a user on the global account.

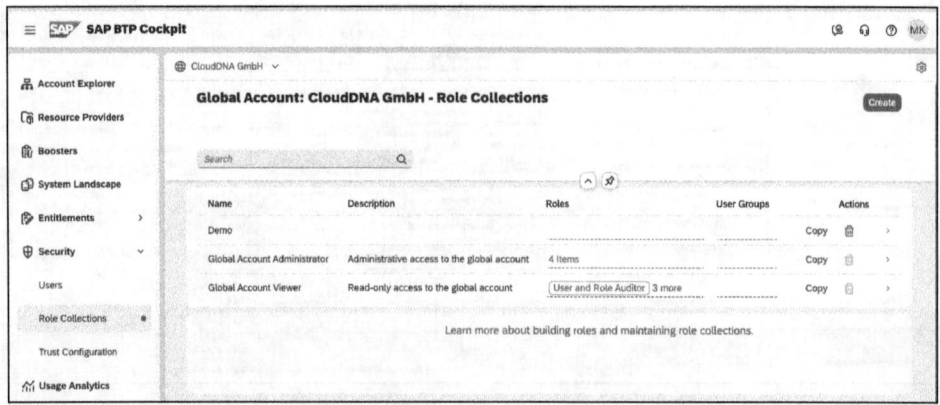

Figure 2.15 Global Account Role Collections

2.2.3 Single Sign-On

SSO is an authentication method where a user logs in once and then automatically gains access to multiple connected systems and applications without having to log in again. With SSO, users can access multiple services and resources with a single sign-on, increasing usability and productivity and reducing the number of passwords.

In the SSO process, logon is usually done via a central *identity provider* (IDP), which verifies the user's authentication information and creates a so-called SSO token. This token is then passed to the various connected systems and applications to enable access without the user having to provide additional credentials. Any application that relies on this identity provider is also called a service provider.

SSO enhances the user experience by allowing users to log in only once and then seamlessly access multiple applications and systems without having to log in repeatedly. SSO serves to increase efficiency by reducing the number of passwords users need to manage. But it also reduces the time needed to log in and switch between different applications. By centrally managing authentication information and the associated use of modern security standards, SSO can improve security and reduce the risk of password theft and unauthorized access.

In this context, the IDP and the *service provider* (SP) are two important components that play a crucial role in the context of authentication and authorization.

The identity provider is a component responsible for managing and verifying user identities. The IDP authenticates users and issues identity credentials in the form of digital tokens. The IDP assumes responsibility for verifying the user's identity and securely providing authentication information to the service provider. Examples of identity providers include Microsoft Entra ID and Okta.

The SP is an application, a service that users want to access. The SP is responsible for protecting the resources and ensuring that only authenticated and authorized users are granted access. The SP trusts the IDP to perform authentication and obtain information about the authenticated user. This allows the SP to verify access rights and release the appropriate resources. Examples of service providers include web applications, APIs, and cloud services that users want to access.

Various standards and protocols are used for SSO, such as Security Assertion Markup Language (SAML) and OpenID Connect (OIDC). These established standards ensure seamless collaboration between different systems and facilitate the smooth integration of SSO into existing IT infrastructures.

SAML is an XML-based authentication and authorization protocol commonly used for SSO. It enables the secure transfer of authentication information between an IDP and an SP. The IDP issues a SAML response containing the user's authentication status, while the SP verifies this response and grants access to the requested resources. SAML is based on the principle of digital signatures and enables secure and trusted communication between the parties involved.

OIDC is an OAuth 2.0–based identity protocol designed specifically for identity management on the web. It allows users to log into a web application using their existing identity providers (such as Google, Facebook, or Microsoft) instead of creating separate accounts and credentials. OIDC uses JSON Web Tokens (JWTs) to securely exchange information about the authenticated user between the identity provider and the relying party. It provides a simple and easy-to-use way to implement SSO for web applications and ensure that users are properly authenticated.

2.2.4 Identity Authentication

SAP offers SAP Cloud Identity Services in SAP BTP with different areas of responsibility. With this collection of services, you can map the entire security lifecycle of users or identities. The spectrum ranges from the onboarding of new users and the provision of these users in the various cloud applications to the detection of potential problems in the assignment of critical authorizations. The Identity Authentication service is also part of SAP Cloud Identity Services. This allows you to implement single sign-on scenarios and delegate authentication to third-party identity providers or identity providers in your on-premise systems.

In connection with SAP Cloud Identity Services, the Identity Directory plays an important role. The *Identity Directory* is the central component for the persistence of users and groups within SAP Cloud Identity Services. It's the central point of contact for users who have access to SAP cloud applications.

Identity Authentication is an SAML 2.0 and OpenID Connect identity provider for authenticating users in SAP BTP applications and in all SAP SaaS products. Optionally, the service can also be used for applications running in your on-premise landscape and for non-SAP applications. The Identity Authentication service enables you to authenticate and single sign-on users in the cloud. The service is provided for both the Neo and Cloud Foundry environments. It's a key element for cloud architecture security. SAP's strategy is to deliver its cloud solutions preconfigured with Identity Authentication.

Identity Authentication provides the following core functions (see Figure 2.16):

- **Authentication**
  The Identity Authentication service acts as an identity provider that validates credentials provided by users. Upon successful authentication, it issues a SAML assertion that is used by the target applications.

- **Identity federation and single sign-on**
  The service can delegate authentication to a third-party identity provider, enabling SSO across on-premise and cloud applications.

- **Risk-based authentication**
  The service enables multifactor authentication if stronger means of authentication are required to access specific business applications.

- **User management**
  The service enables management of its users.

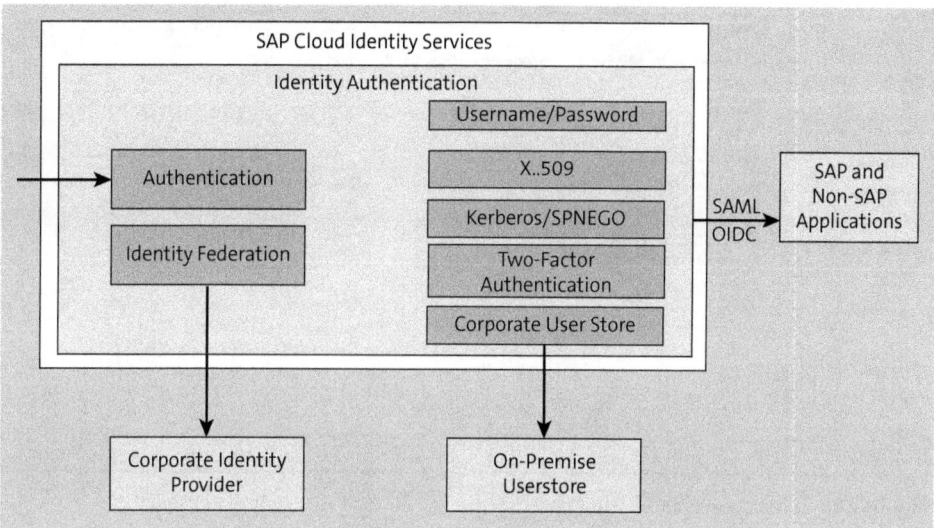

Figure 2.16 Identity Authentication Service

This service plan for the Identity Authentication service is included in the overall SAP BTP contract. You receive two instances of the Identity Authentication service, one for testing purposes and one for production use. These instances are called Identity Authentication *tenants*. In principle, you have the option to purchase additional tenants, regardless of whether they are used productively or for testing purposes. The license scope of the Identity Authentication service includes all SAP cloud logins. If you would also like to connect non-SAP products, then an additional license must be purchased. The metric for this is based on the number of logins.

After you purchase a license for the Identity Authentication service, you receive an email invitation to register the initial administrator. The email contains a link to the initial page of the Identity Authentication tenant administration console, where you can confirm the registration of the initial administration user.

You can log in to the Identity Authentication tenant using the following URL: *https:// <IdentityAuthenticationTenantID>.accounts.ondemand.com*. Replace the *<IdentityAuthenticationTenantID>* placeholder with the ID of your Identity Authentication tenant.

The tenant ID is automatically generated by SAP. The first administration user created for the client receives an activation email with a URL. This URL contains the tenant ID. However, you also have the option to jump directly from the SAP BTP cockpit to the Identity Authentication tenant.

2 SAP Business Technology Platform

In the Cloud Foundry subaccount, it's a prerequisite that a trust relationship has been established between the subaccount and the Identity Authentication tenant. To do this, navigate to the **Security • Trust Configuration** section and click the **Establish Trust** button (see Figure 2.17).

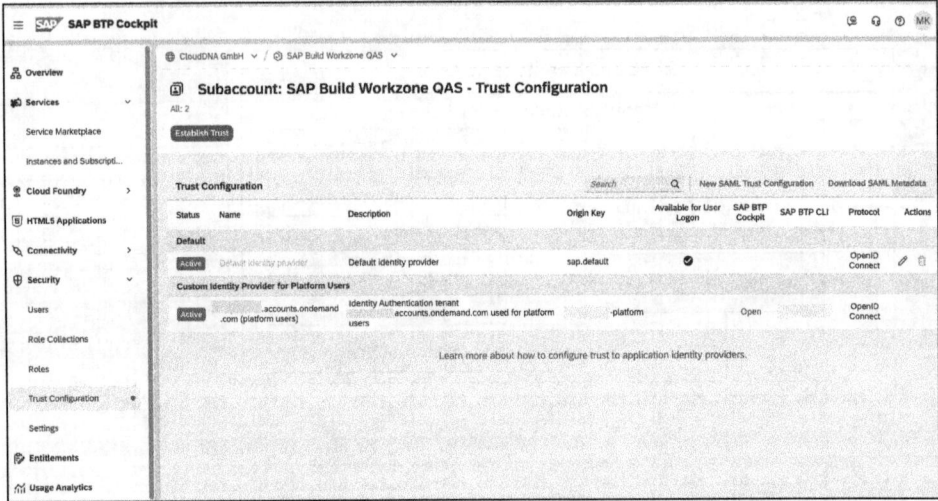

Figure 2.17 Subaccount: Establish Trust

You can then select an Identity Authentication tenant assigned to your global account (see Figure 2.18). You also can perform advanced configuration.

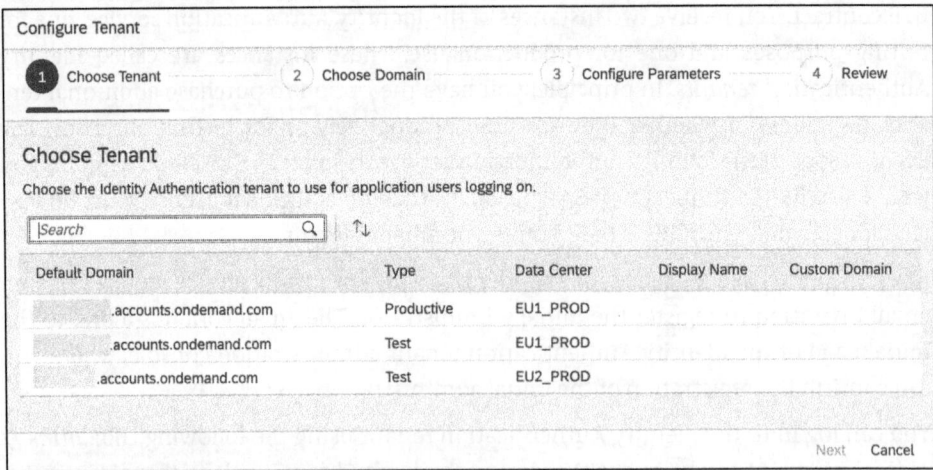

Figure 2.18 Select Identity Authentication Tenant

After the configuration has been performed, you can jump directly to the Identity Authentication tenant by clicking the link (see Figure 2.19). As you can also see in Figure 2.19, for both the connection to the Identity Authentication tenant and for the connection to the

2.2 Security

default identity provider (the SAP ID service), the previously mentioned OpenID Connect Protocol is used, not SAML 2.0.

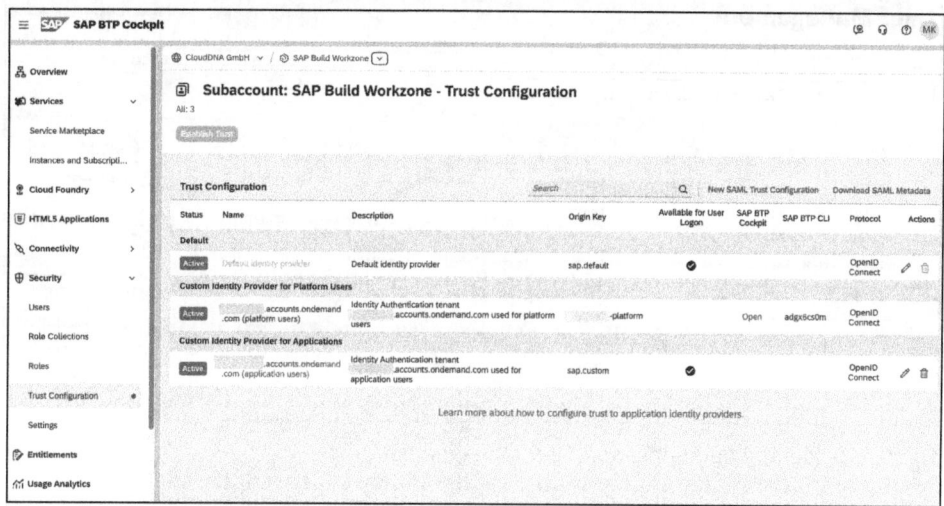

Figure 2.19 Custom Identity Provider

At this point, you can jump to the corresponding Identity Authentication tenant. You will first see the home area, where all functions can be called directly (see Figure 2.20).

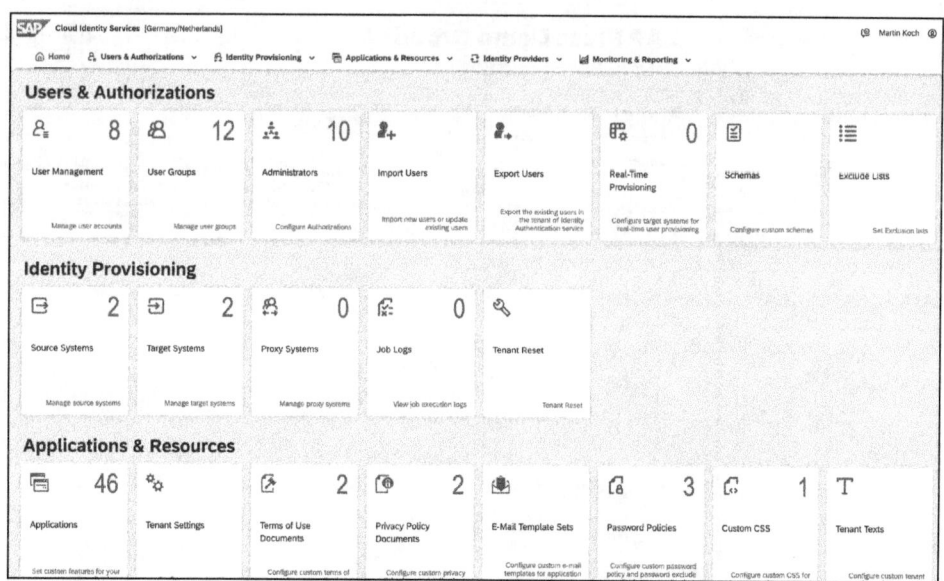

Figure 2.20 Identity Authentication: Overview

In the Identity Authentication tenant, you can manage users in the **User Management** area. For example, you can create, delete, and delimit users (see Figure 2.21).

2 SAP Business Technology Platform

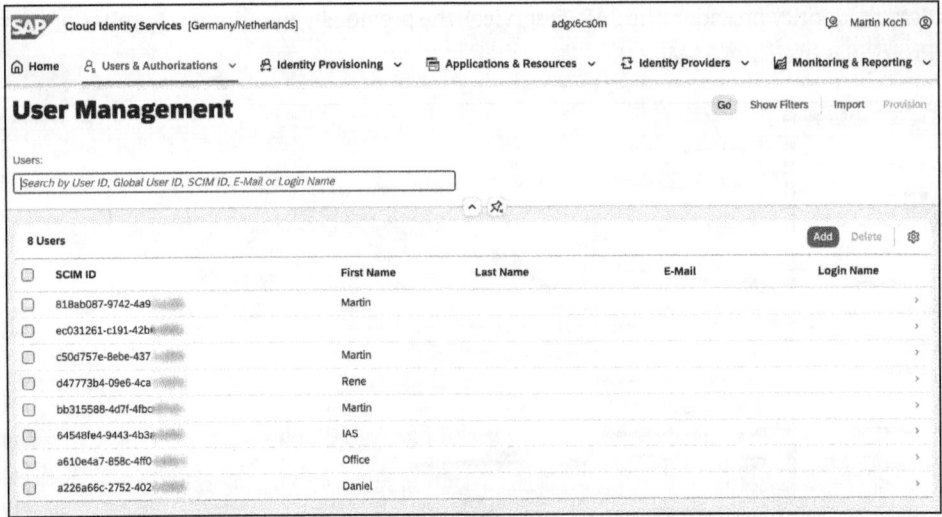

Figure 2.21 Identity Authentication: User Management

In the Identity Authentication tenant, as shown in Figure 2.22, you can also create groups and assign users to these groups. In SAP BTP, you can map the groups to role collections.

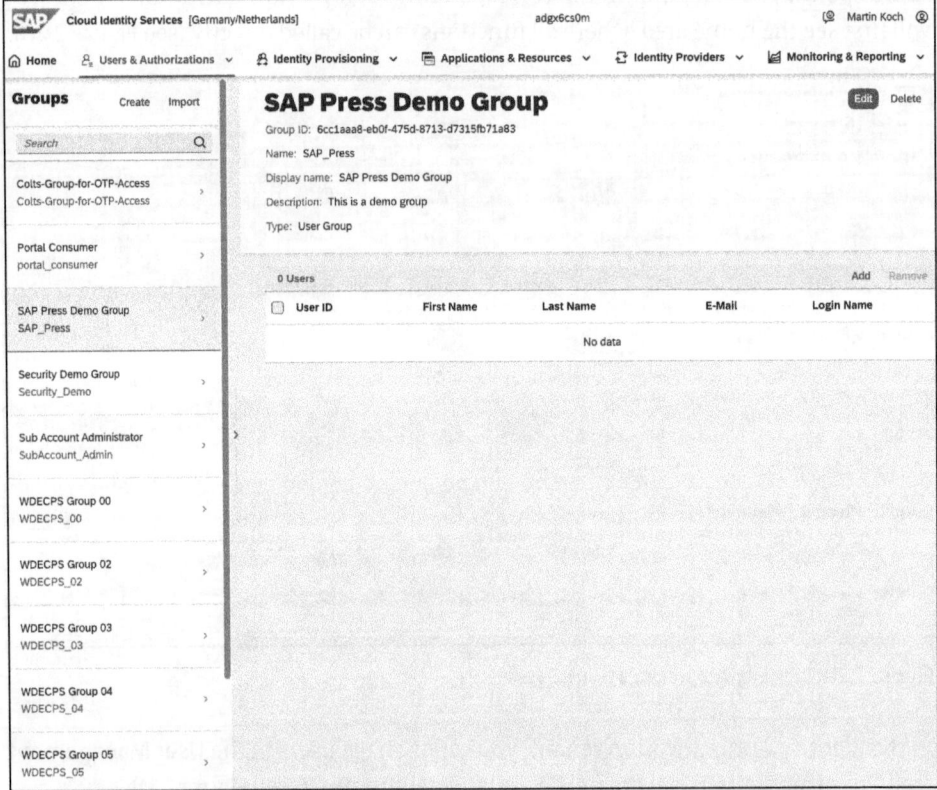

Figure 2.22 Groups in Identity Authentication

In addition to using the SAP ID service and the Identity Authentication service, it's possible for SAP BTP applications to delegate authentication and identity management to an existing identity provider within your company (a corporate identity provider). This can authenticate your company's employees against a corporate directory service, for example. This allows your employees and, if applicable, customers and partners to log into the cloud application using their usual user information. All information about a user required by SAP BTP can be securely passed on with the logon process based on a proven and standardized security protocol.

In this scenario, there is no need to manage additional systems that take care of synchronization or provisioning of user accounts between the corporate network and SAP BTP. All that needs to be done is to establish a trust relationship, simply called a *trust*, between an SAP BTP subaccount running the application and your corporate identity provider (see Figure 2.23).

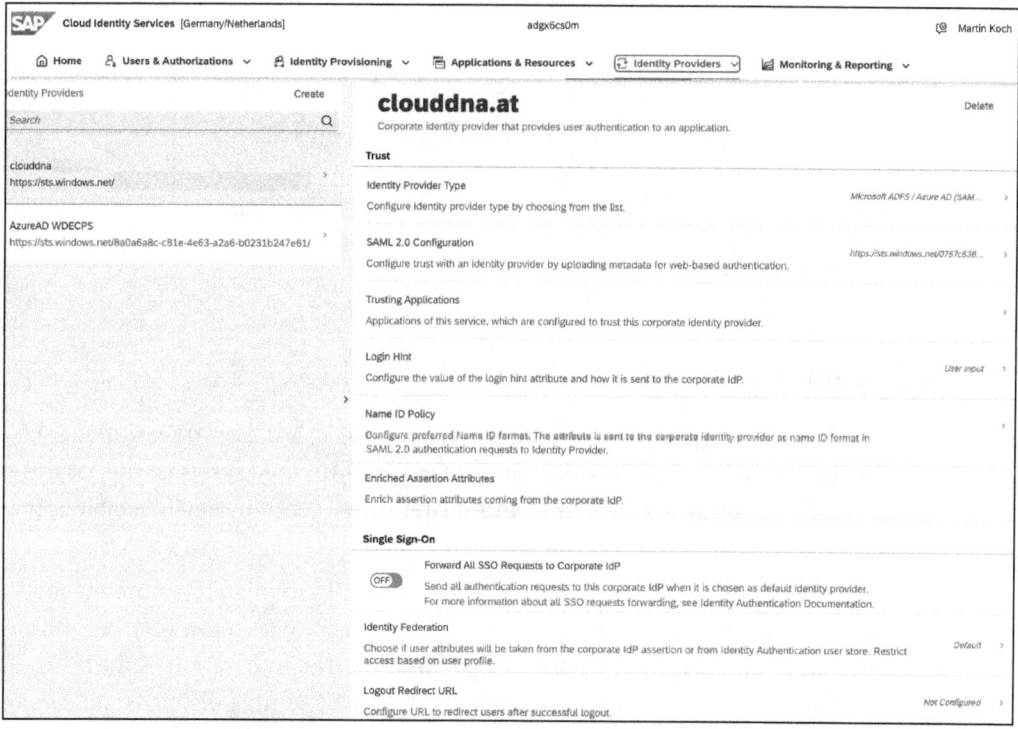

Figure 2.23 Corporate Identity Provider

Depending on your company's internal requirements, it may be necessary to authenticate certain user groups differently (see Figure 2.24). For example, you can set two-factor authentication for administrators. The Identity Authentication service supports you in such cases with risk-based authentication.

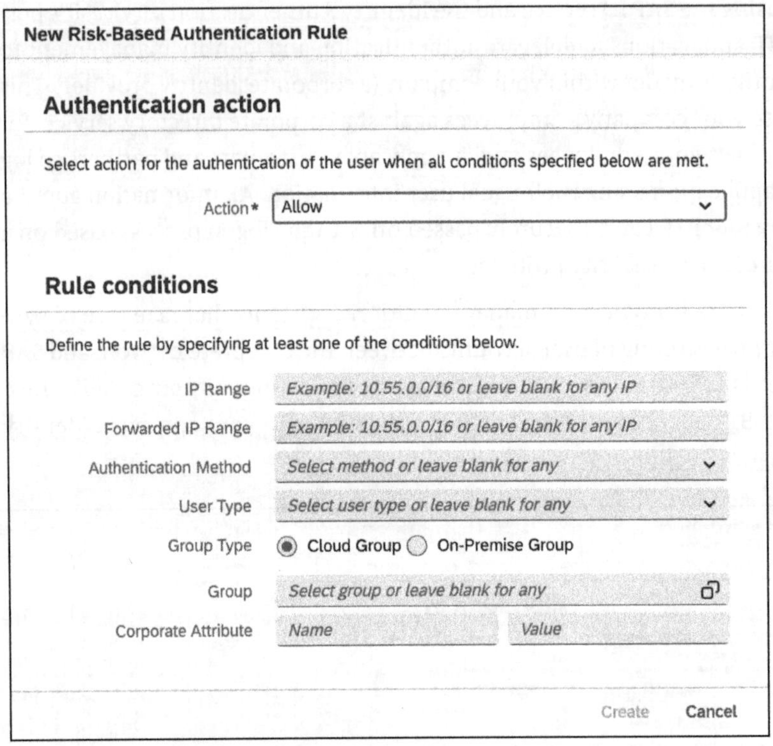

Figure 2.24 Risk-Based Authentication

2.3 Cloud Connector

The cloud connector (part of the SAP Connectivity service) is the central component for connecting your on-premise system landscape to the SAP cloud systems. The basic idea of the cloud connector is to be able to administer access centrally and to enable appropriate monitoring.

In a hybrid system landscape, the complexity of your system architecture inevitably increases. A few years ago, there was a trend to reintegrate systems, such as with the methods and tools of system landscape optimization. The integration of the business partner model into the world of SAP ERP was also an example of this trend. Today, cloud solutions add another, external layer to these integrated systems. This changes the requirements for system administration. For on-premise systems, issues such as the following play a frontline role:

- Where is your server located?
- How is this server virtualized?
- In which network or zone is this server located?

In the cloud world, this information is not necessarily transparent.

The cloud connector forms the central interface to the services of SAP BTP within your own system landscape. It acts as a link between cloud and on-premise systems and communicates with SAP BTP via secure connections.

At the architecture level, the goal is for system connections that send data to or receive data from SAP cloud solutions to be handled by the cloud connector. Multiple instances of the cloud connector can also be deployed. Distinct from these integration scenarios are connections from end devices that typically access cloud applications or services directly.

When a service in SAP BTP, such as SAP Build Apps, requires data from your local SAP systems, the cloud connector comes into play. Because users access SAP BTP services from the internet, you want the accesses into the local system to take place via a central tool rather than directly via the internet. To do this, you use the cloud connector. Figure 2.25 illustrates the architecture of this scenario.

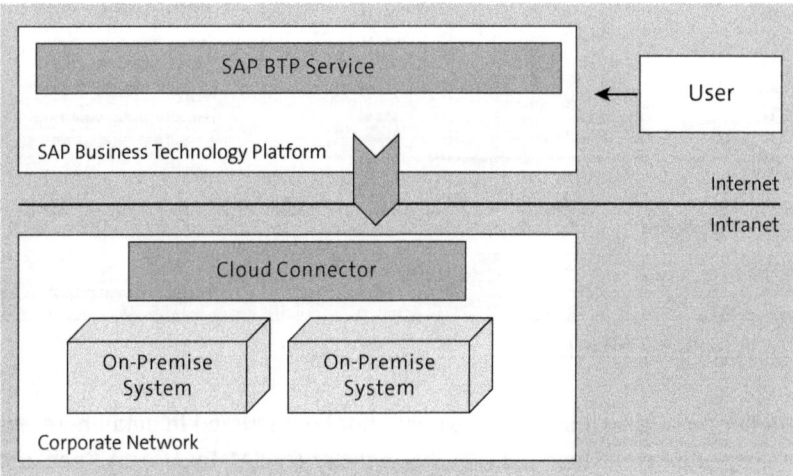

Figure 2.25 Cloud Connector

We'll show how to configure the cloud connector in Section 2.3.1. The most common scenario in combination with the SAP Build product portfolio is accessing data from an on-premise ABAP system (e.g., SAP Business Suite or SAP S/4HANA). This configuration is done via a mapping, which we'll describe in Section 2.3.2.

2.3.1 Initial Configuration

The cloud connector is offered for the Linux and Windows operating systems. There is also a version for MacOS, but this is not intended for productive use. As illustrated in Figure 2.26, all versions can be downloaded from *https://tools.hana.ondemand.com*. Preconfigured test versions (called a portable), which are aimed at developers, or versions suitable for production (refer to as the installer) can be downloaded there.

Developer versions enable testing of system connections to the cloud, for example, without having to contact the Basis department. However, it should be noted, for example, that this version cannot be run in the background. For productive installations, an installation program is provided via which you execute the installation. To install the cloud connector, a version of the Java Development Kit (JDK) must be installed on your local machine.

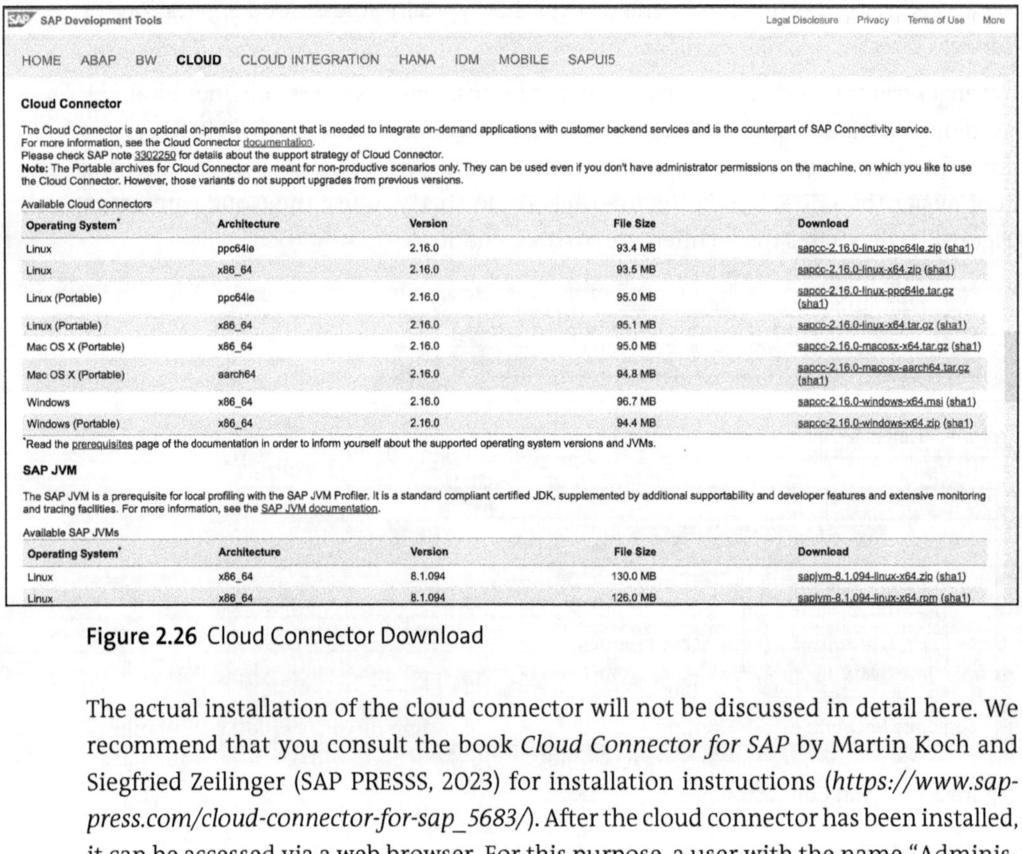

SAP Development Tools   Legal Disclosure | Privacy | Terms of Use | More

HOME ABAP BW CLOUD CLOUD INTEGRATION HANA IDM MOBILE SAPUI5

Cloud Connector

The Cloud Connector is an optional on-premise component that is needed to integrate on-demand applications with customer backend services and is the counterpart of SAP Connectivity service.
For more information, see the Cloud Connector documentation.
Please check SAP note 3302250 for details about the support strategy of Cloud Connector.
Note: The Portable archives for Cloud Connector are meant for non-productive scenarios only. They can be used even if you don't have administrator permissions on the machine, on which you like to use the Cloud Connector. However, those variants do not support upgrades from previous versions.

Available Cloud Connectors

Operating System*	Architecture	Version	File Size	Download
Linux	ppc64le	2.16.0	93.4 MB	sapcc-2.16.0-linux-ppc64le.zip (sha1)
Linux	x86_64	2.16.0	93.5 MB	sapcc-2.16.0-linux-x64.zip (sha1)
Linux (Portable)	ppc64le	2.16.0	95.0 MB	sapcc-2.16.0-linux-ppc64le.tar.gz (sha1)
Linux (Portable)	x86_64	2.16.0	95.1 MB	sapcc-2.16.0-linux-x64.tar.gz (sha1)
Mac OS X (Portable)	x86_64	2.16.0	95.0 MB	sapcc-2.16.0-macosx-x64.tar.gz (sha1)
Mac OS X (Portable)	aarch64	2.16.0	94.8 MB	sapcc-2.16.0-macosx-aarch64.tar.gz (sha1)
Windows	x86_64	2.16.0	96.7 MB	sapcc-2.16.0-windows-x64.msi (sha1)
Windows (Portable)	x86_64	2.16.0	94.4 MB	sapcc-2.16.0-windows-x64.zip (sha1)

*Read the prerequisites page of the documentation in order to inform yourself about the supported operating system versions and JVMs.

SAP JVM

The SAP JVM is a prerequisite for local profiling with the SAP JVM Profiler. It is a standard compliant certified JDK, supplemented by additional supportability and developer features and extensive monitoring and tracing facilities. For more information, see the SAP JVM documentation.

Available SAP JVMs

Operating System*	Architecture	Version	File Size	Download
Linux	x86_64	8.1.094	130.0 MB	sapjvm-8.1.094-linux-x64.zip (sha1)
Linux	x86_64	8.1.094	126.0 MB	sapjvm-8.1.094-linux-x64.rpm (sha1)

Figure 2.26 Cloud Connector Download

The actual installation of the cloud connector will not be discussed in detail here. We recommend that you consult the book *Cloud Connector for SAP* by Martin Koch and Siegfried Zeilinger (SAP PRESSS, 2023) for installation instructions (*https://www.sap-press.com/cloud-connector-for-sap_5683/*). After the cloud connector has been installed, it can be accessed via a web browser. For this purpose, a user with the name "Administrator" and the password "manage" is created in the standard.

It's mandatory to change the password when logging in for the first time (see Figure 2.27). You must also specify whether this is the master or shadow installation.

In the next step, you must connect to the desired subaccount in SAP BTP from the cloud connector. In addition to the subaccount ID, you also need the provider and the region in which the subaccount is provided. You can obtain this information by logging into the SAP BTP cockpit and navigating to the desired subaccount. There you must open the **Overview** area in the side menu as shown in Figure 2.28. There you will find the **Provider**, **Region**, and **Subaccount ID** attributes under the **General** tab.

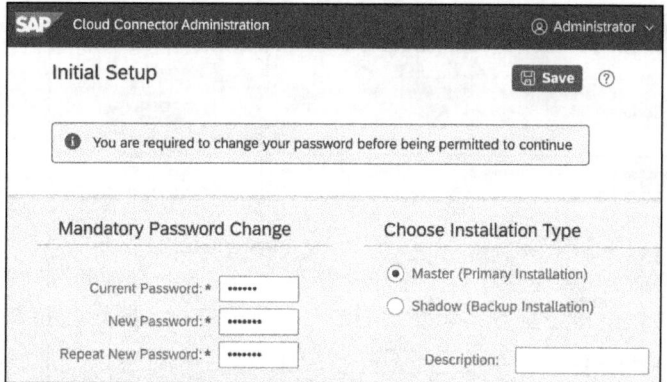

Figure 2.27 Initial Setup

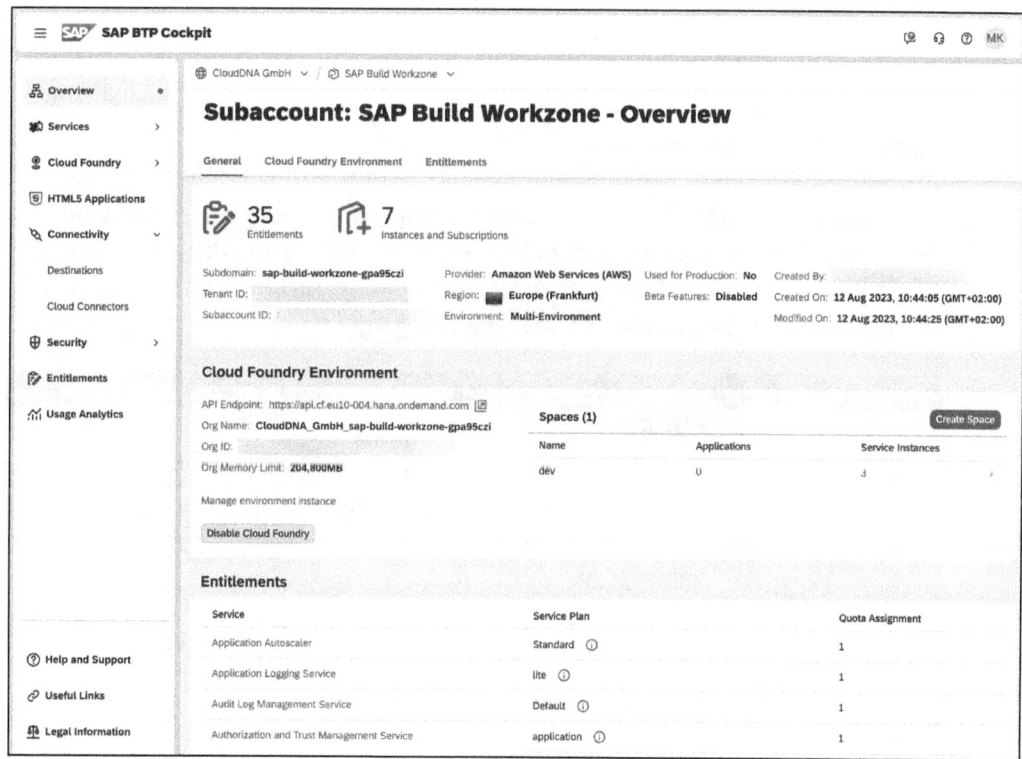

Figure 2.28 Determine Subaccount ID

In the cloud connector, you can now maintain the determined attributes as shown in Figure 2.29. You must also enter a subaccount user and the associated password. The subaccount user must be assigned to the subaccount as a user. After you have maintained the attributes, you need to save them by clicking the **Save** button. After that, the connection to the subaccount is established.

2   SAP Business Technology Platform

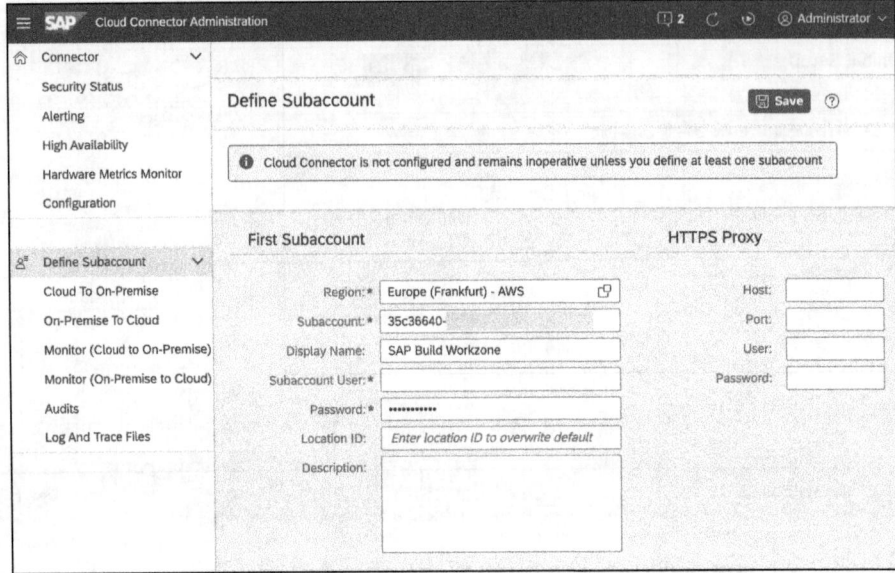

Figure 2.29 Subaccount Connection

In the cloud connector, the status of the connection is visible, as shown in Figure 2.30. Note that a client certificate is created for the user used to log into the subaccount. This certificate has a validity of one year and is subsequently used for logging in. Therefore, you must renew the certificate before it expires.

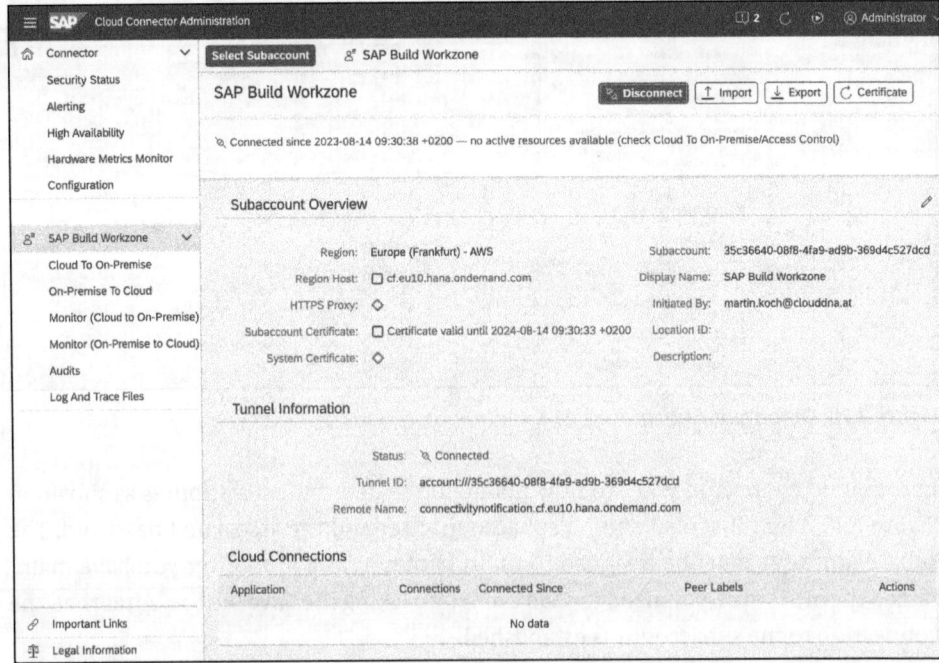

Figure 2.30 Connection Status in Cloud Connector

2.3 Cloud Connector

You can also check the connection status in the subaccount. To do this, navigate to the **Connectivity • Cloud Connectors** area in the SAP BTP cockpit, as shown in Figure 2.31.

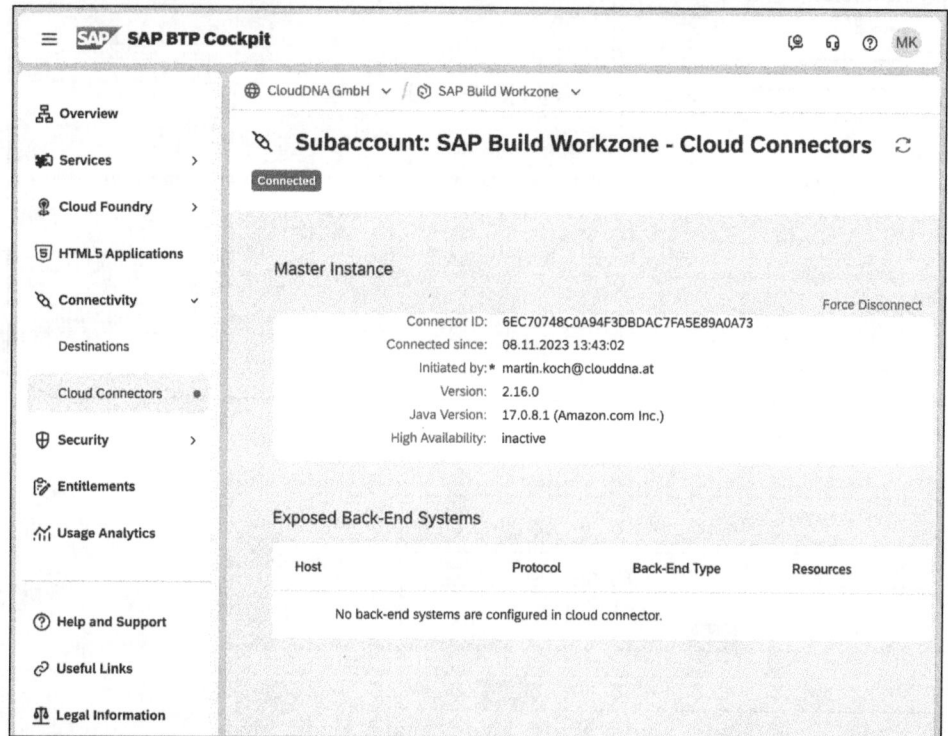

Figure 2.31 Connection Status in Subaccount

2.3.2 Mapping

All connections from SAP BTP to your on-premise systems must be explicitly enabled in the cloud connector. In addition to the type of connection (e.g., ABAP system), you specify the connection data of the system and also how authentication is to take place. Therefore, you define a so-called mapping. The mapping is used to hide the physical host name and port in SAP BTP and to use a virtual host name and port instead. As illustrated in Figure 2.32, the configuration is triggered in the **Cloud to On-Premise** section in the side menu. Click the **+** button to add a new mapping.

In the first step, you must select a **Backend Type**. In the case of an SAP Business Suite or an SAP S/4HANA system, you can select **ABAP System**, as shown in Figure 2.33.

In the next step, you need to select the protocol that will be used for communication between the cloud connector and the backend. If OData services or SAP Fiori apps are accessed, you will select **HTTP** or **HTTPS** in the **Protocol** field.

After that you must specify the **Internal Host** and the **Internal Port** (see Figure 2.34). These are the host name and port through which the cloud connector reaches the backend.

83

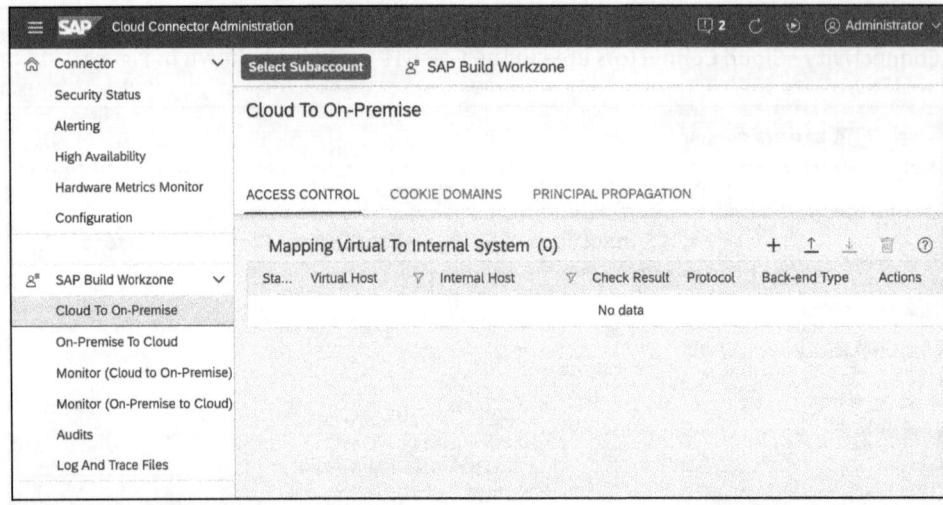

Figure 2.32 Mapping Overview

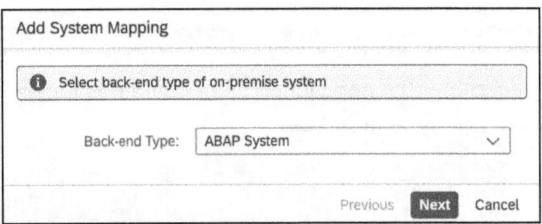

Figure 2.33 Backend Type Selection

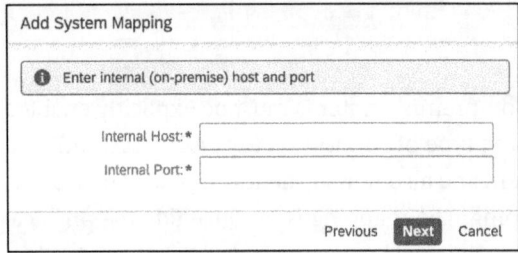

Figure 2.34 Internal Host Configuration

Then, as shown in Figure 2.35, the virtual host and the virtual port must be specified. This allows the backend to be addressed in the SAP BTP subaccount. SAP recommends choosing a **Virtual Host** name that is different from the physical hostname. The virtual hostname cannot be changed later.

On the next screen, you can then specify whether to **Allow Principal Propagation** using the checkbox. Principal propagation allows you to route the identity of the user from SAP BTP through to the backend. This is necessary, for example, when authorization checks take place in the backend.

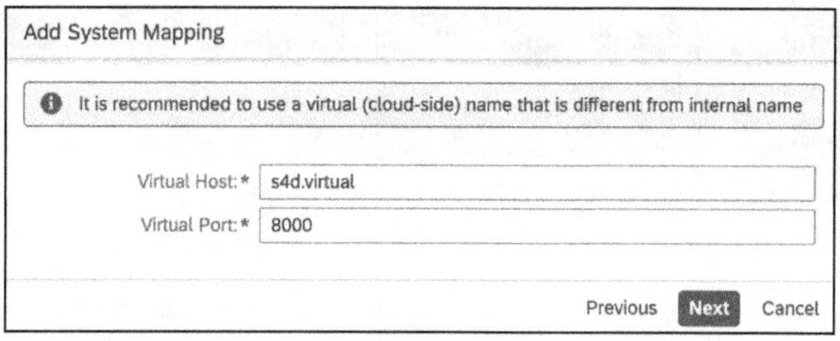

Figure 2.35 Virtual Host Configuration

After that, you still need to configure which host name should be used in the HOST HTTP request header field (see Figure 2.36). In most cases, it's sufficient that the virtual host is used.

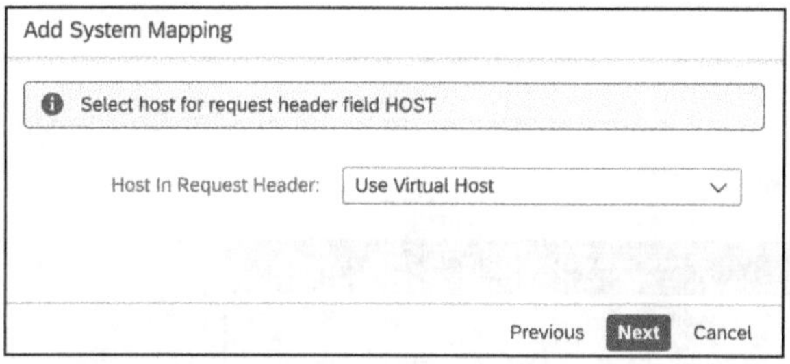

Figure 2.36 System Mapping: Request Header Configuration

In the next step, you can (optionally) assign a description. Finally, as shown in Figure 2.37, you will again see a summary of the entries made previously. In addition, you can specify whether a check should be made immediately after saving the mapping to determine whether the internal host can be reached from the cloud connector.

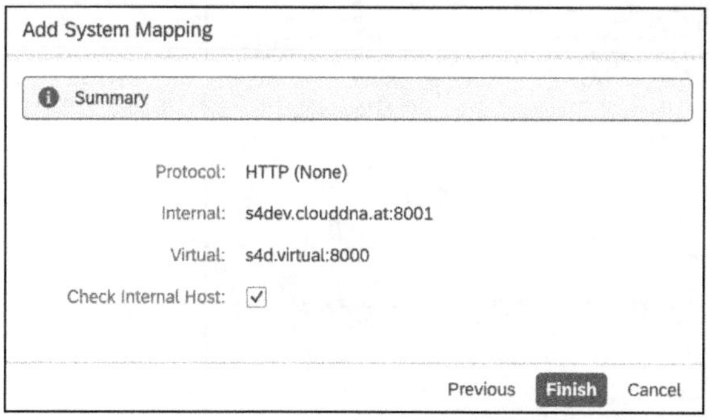

Figure 2.37 System Mapping Overview

If you use HTTPS as the communication protocol, an error will usually occur at this point (see Figure 2.38). The error occurs because the cloud connector checks if the SSL certificate of the backend system is in the allow list. This requires that either the certificate is imported into the cloud connector or that the check is turned off.

2 SAP Business Technology Platform

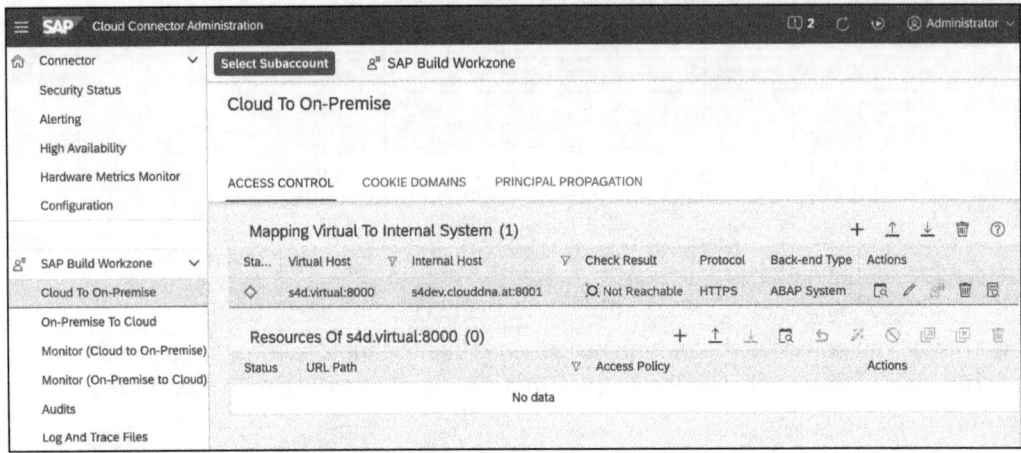

Figure 2.38 Connection Status

In the cloud connector, navigate to the **Connector • Configuration** section in the side menu to turn off the check (see Figure 2.39).

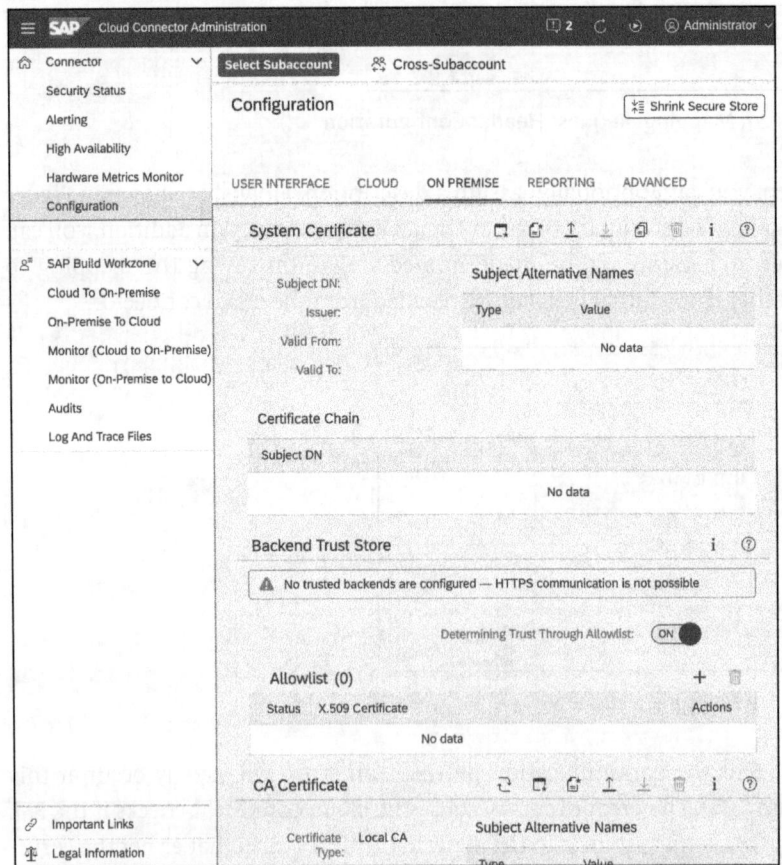

Figure 2.39 Trust Configuration

2.3 Cloud Connector

Disable **Determinining Trust through Allowlist**, as shown in Figure 2.40. Alternatively, you can import the corresponding certificate by clicking the **+** button in the **Allowlist** area.

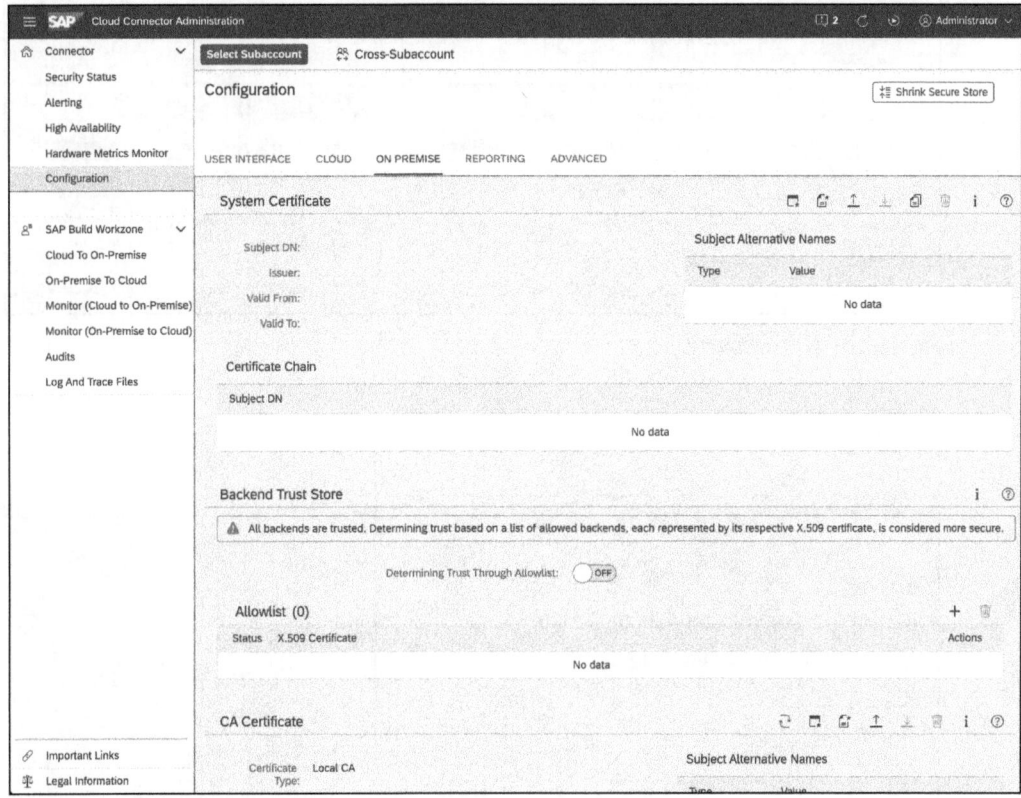

Figure 2.40 Trust Store Status

This means that you have created all the technical prerequisites for accessing the backend. When using OData services and accessing SAP Fiori apps, the Internet Communication Framework (ICF) is used in the backend. Therefore, as shown in Figure 2.41, in the cloud connector, you still need to configure which URL paths the cloud connector is allowed to access. To do this, click the **+** button in the **Resources Of** area.

You can now specify the desired path—for example, "/sap". In addition, as shown in Figure 2.42, you can configure whether only this path can be accessed explicitly or both this path and all paths below it (its subpaths).

2   SAP Business Technology Platform

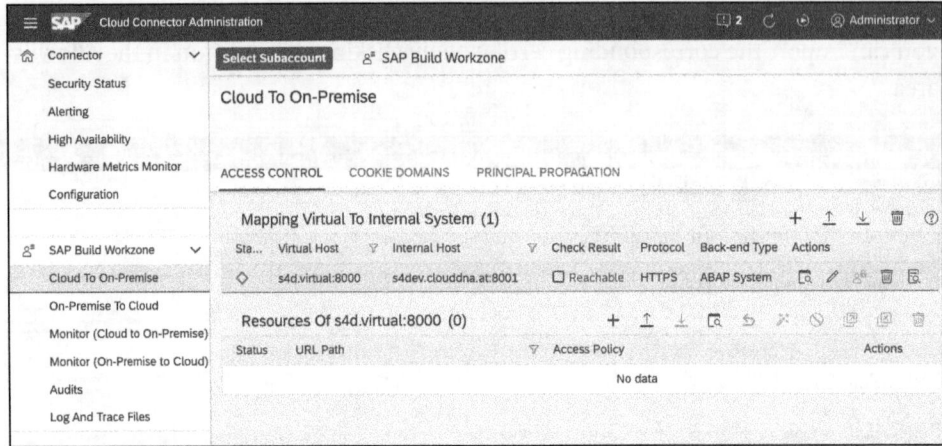

Figure 2.41 Add Resource

Figure 2.42 Resource Details

2.4 Summary

In this chapter, you received a general overview of SAP BTP in Section 2.1. This included the commercial model under which SAP BTP can be licensed. In addition, the account structure was presented, and an overview of the available services was given.

Section 2.2 then dealt with the topic of security. User authentication and user authorization were discussed. In addition, the topic of SSO was touched on superficially and the Identity Authentication service was explained in this context.

Finally, Section 2.3 examined the cloud connector, which represents the central link in hybrid system landscapes.

PART II
SAP Build Apps

Chapter 3
Installing and Configuring SAP Build Apps

Before you can get started with SAP Build Apps, it's necessary to perform a basic installation and configuration. With the use of a booster, this is a straightforward task.

SAP Build Apps is a no-code development tool. It was previously known under the awesome name SAP AppGyver. Before you can start using SAP Build App, you must install and configure it in your SAP BTP subaccount, and SAP provides a booster for this task.

Section 3.1 is dedicated to the installation phase for SAP Build Apps on SAP BTP. Here, you'll be introduced to the *booster*—a nifty tool designed to facilitate a smoother and faster installation journey. We'll provide step-by-step instructions on employing this tool effectively, aiming for an installation experience that's as close to seamless as possible.

Moving forward, Section 3.2 delves into configuration nuances. Although the booster undertakes this task automatically, it's crucial to comprehend the configurations it implements. Understanding these details equips you with the knowledge to troubleshoot any potential issues swiftly, ensuring the stability and reliability of your setup.

The pivotal topic of security is tackled in Section 3.3. We're committed to guiding you in establishing a formidable security framework for SAP Build Apps, with a particular focus on authentication protocols. The booster sets up a foundational security model, but we'll shed light on the intricacies of this setup, empowering you with a deep understanding of the protective measures in place.

3.1 Installation

SAP Discovery Center is an indispensable source for an initial orientation about SAP Build Apps. In addition to detailed service information and pricing information, it also provides insight into the geographical distribution of the various data centers used for service provisioning (see Figure 3.1). In addition, you will find the cornerstone features of the roadmap there.

3   Installing and Configuring SAP Build Apps

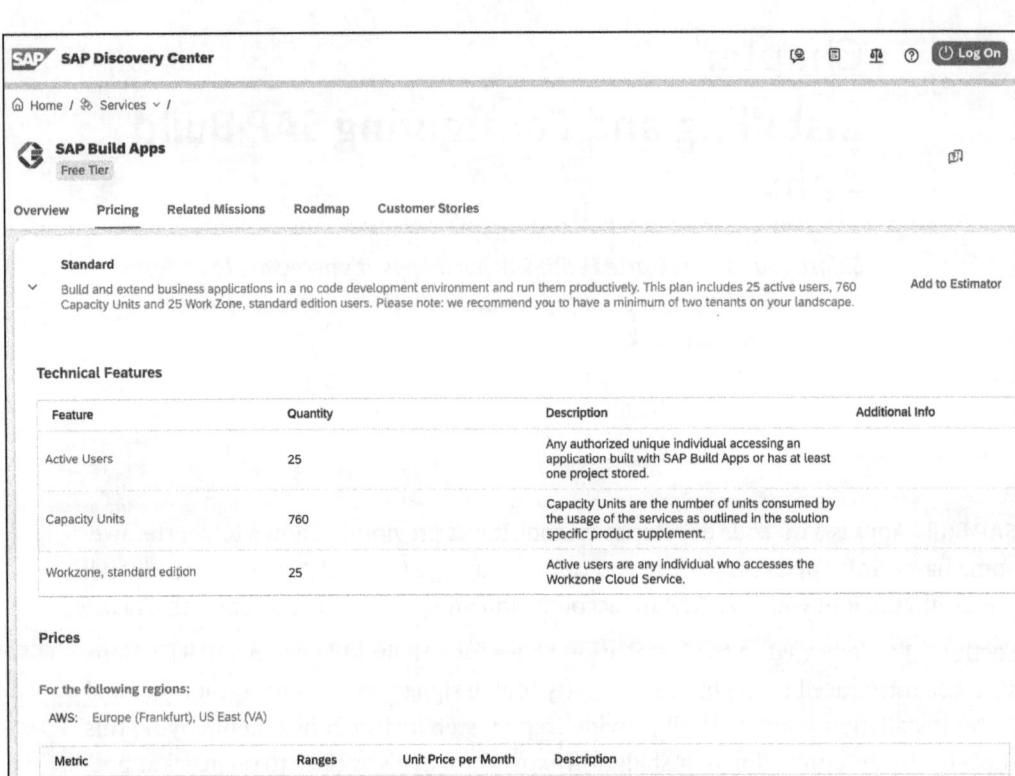

Figure 3.1 SAP Discovery Center for SAP Build Apps

SAP offers two plans for SAP Build Apps: a standard plan and a free plan. The standard plan is ready for productive use, encompassing the full range of SAP Build Apps capabilities, with the added assurance of enterprise-level support. The free plan comes with the following restrictions:

- **Number of builds**
  There is a limit of wo successful builds per app per platform.
- **Cloud runtime**
  Only development runtime is available; there is no production use.
- **Support**
  Only community support is available for free service plans, and these aren't subject to a service-level agreement (SLA).

In SAP BTP, boosters at the global account level facilitate the installation and configuration of services in a subaccount. These boosters act as automated assistants that perform standardized setups and configurations. They minimize manual effort and reduce sources of error during setup by automatically performing the steps required to activate a service. By using these boosters, developers and system administrators can save time and focus on more specific customizations and optimizations after the basic service setup is complete.

Within the SAP BTP cockpit for the global account, the boosters are accessible via the side menu. Select the **Booster** option to find the available automation tools. As shown in Figure 3.2, you can use the search function to filter specifically for "SAP Build". This will result in the display of tiles for two specific boosters for SAP Build Apps: one for **Detailed Account Setup**, which provides a comprehensive configuration of the account, and another for **Quick Account Setup**, which provides a quick and simplified setup. These boosters differ in their scope and level of detail, both aiming at efficient setup and configuration of SAP Build Apps in your subaccount. We will refer to the booster for the detailed account setup in this chapter, so click the corresponding tile to open the booster.

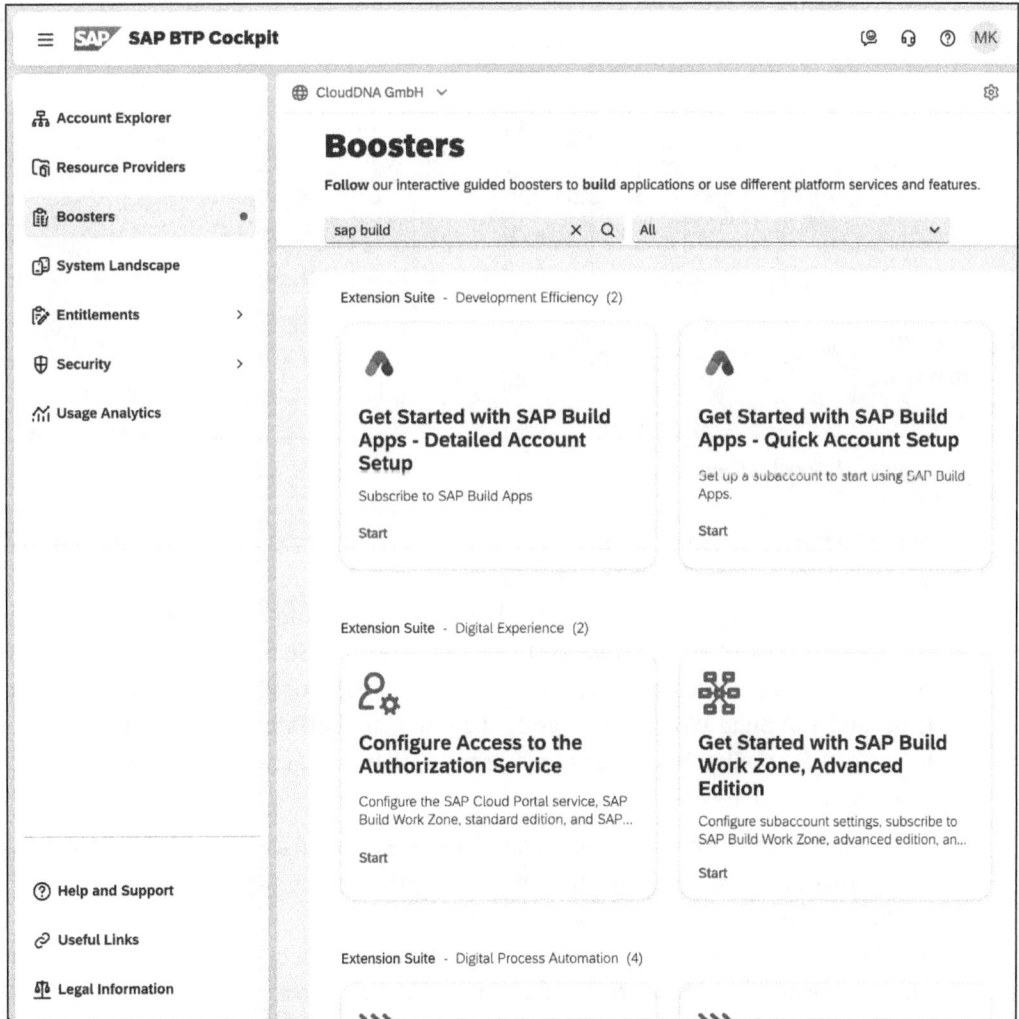

Figure 3.2 SAP Build Apps Booster

In the **Overview** tab, you will find an architecture overview, as shown in Figure 3.3. This contains all the components installed by the booster, including the optional components. The integration into other cloud and on-premise systems is also shown.

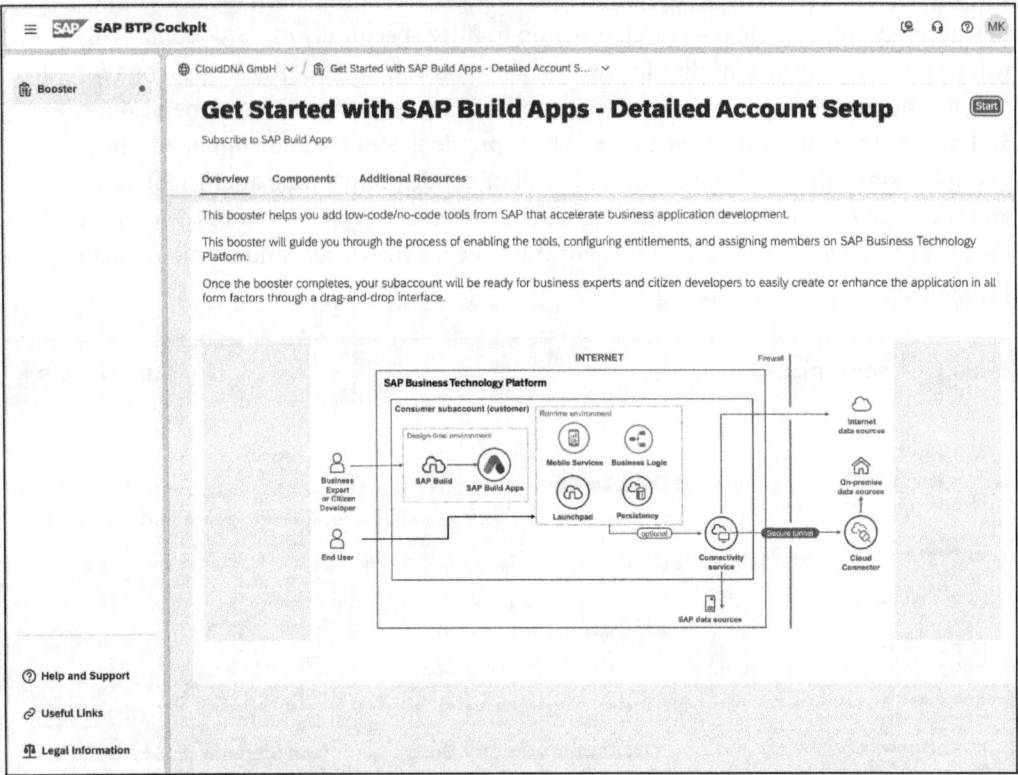

Figure 3.3 Booster Overview

In the **Components** tab, you will find an overview of all services and subscriptions that are installed by the booster. The mandatory label shows you whether this component is mandatory. Using the example of the booster for SAP Build Apps (see Figure 3.4), you can see that only SAP Build Apps and SAP Cloud Identity Services are mandatory. The remaining components, such as SAP Mobile Services; SAP BTP, Cloud Foundry runtime; and SAP Build Work Zone, standard edition, are optional. Note that the services incur costs as soon as they have been subscribed to in a subaccount. Click the **Start** button to start the booster.

In the first step of the booster, the prerequisites are checked as shown in Figure 3.5. This means that the booster checks whether you have the appropriate authorizations to run the booster, whether a custom identity provider is available, and whether the required entitlements are present. If all prerequisites have been checked successfully, you can jump to the next step of the configuration by clicking the **Next** button.

3.1 Installation

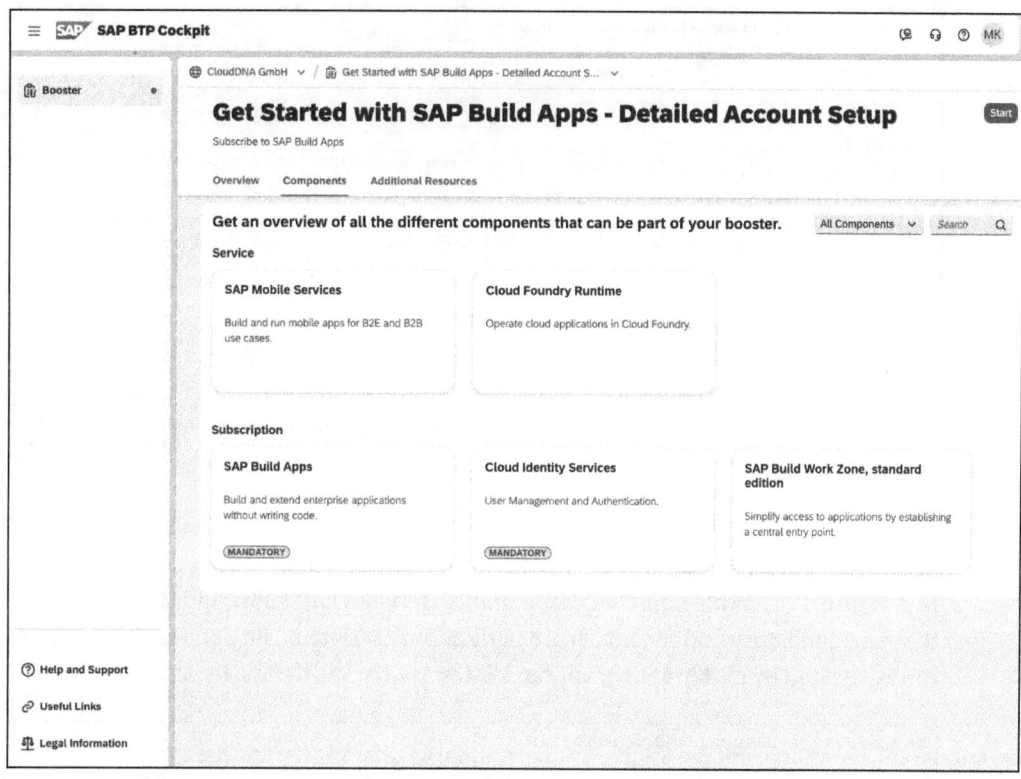

Figure 3.4 Booster Components

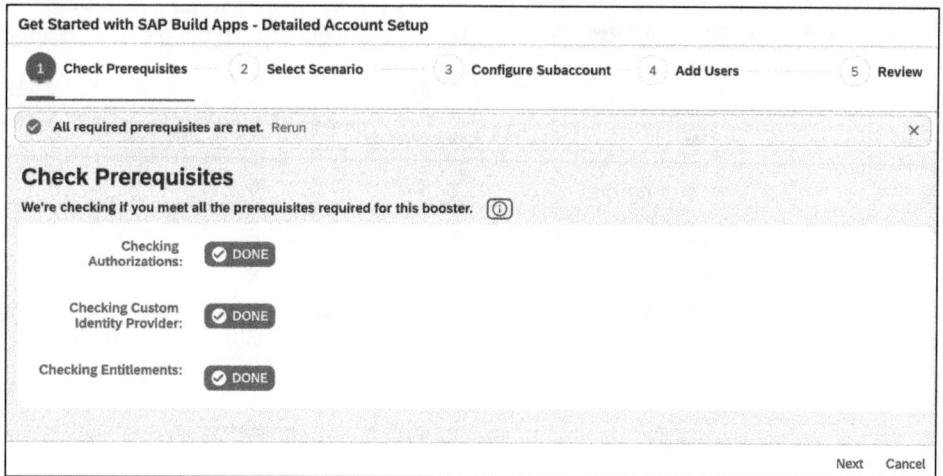

Figure 3.5 Prerequisite Check

In the **Select Scenario** step, as shown in Figure 3.6, you can choose whether an existing subaccount should be used or whether the booster should create a new subaccount for this use case. After selecting the appropriate action, click the **Next** button to continue.

3 Installing and Configuring SAP Build Apps

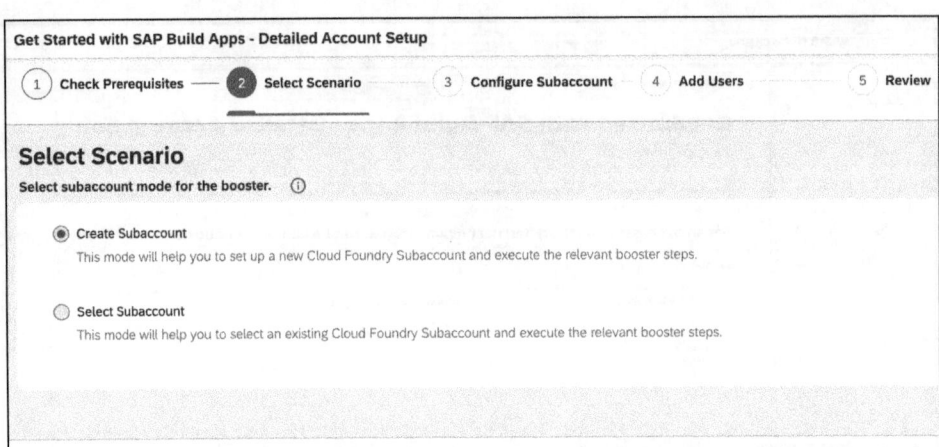

Figure 3.6 Subaccount Selection

In the **Configure Subaccount** step, you have the option of customizing the entitlements (see Figure 3.7). At this point, all components (services and subscriptions), both those that are mandatory and those that are optional, are selected. The optional components can be deleted by clicking the trash can button.

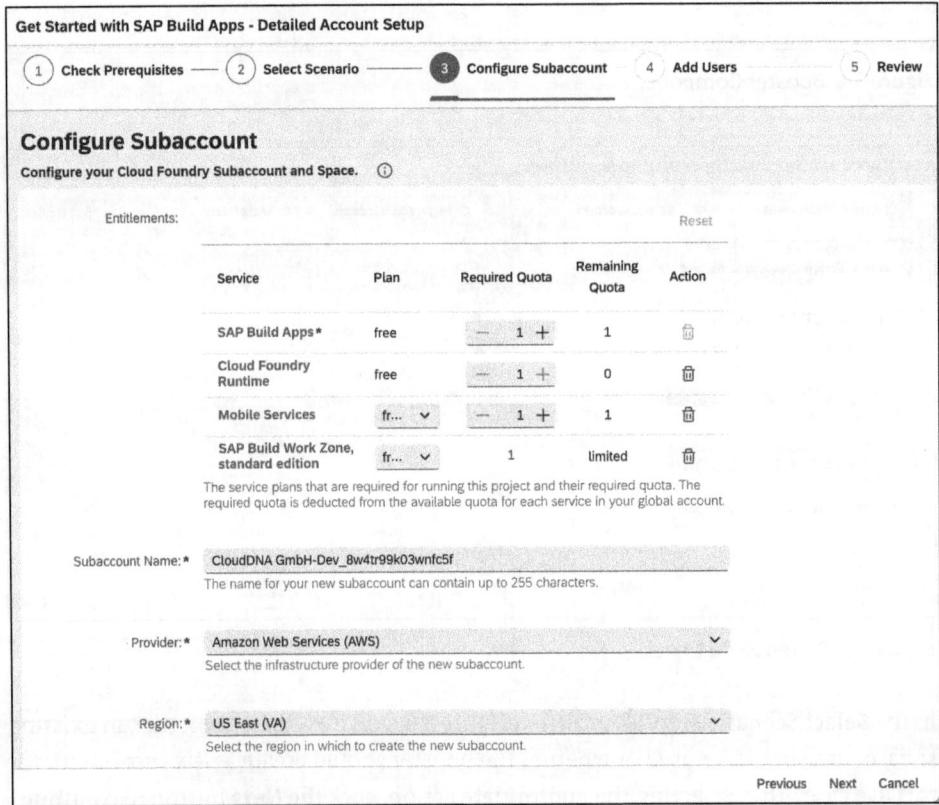

Figure 3.7 Configuration of Entitlements and Data Center

You can then change the **Subaccount Name**. By default, this is composed of the global account name in combination with a generated name. We recommend using a human-readable name at this point. We also advise against using spaces in the name. You then have the option of selecting the **Provider** (AWS, Azure, or Google) and an associated data center—that is, a **Region**. Note that SAP Build Apps is not available in all data centers of all providers. Click **Next** as usual to jump to the next step of the configuration.

In the **Add Users** step, you must select an **Application Identity Provider** and a **Platform Identity Provider**, as shown in Figure 3.8. The application identity provider is used to authenticate the developers who want to use SAP Build Apps. It is also responsible for authenticating end users if you deploy the created apps to this subaccount and make them available via SAP Mobile Services or SAP Build Work Zone, standard edition. The platform identity provider authenticates those users who administer the subaccount and, for example, create destinations there. You must then enter the email addresses of the users who are assigned the subaccount administrator and SAP Build Apps administrator roles in the **Administrators** field. These users are authenticated against the platform identity provider. In the **Developers** field, you must maintain the email addresses of those users who are used as developers in SAP Build Apps. You can also add new users for both user types in the subaccount later.

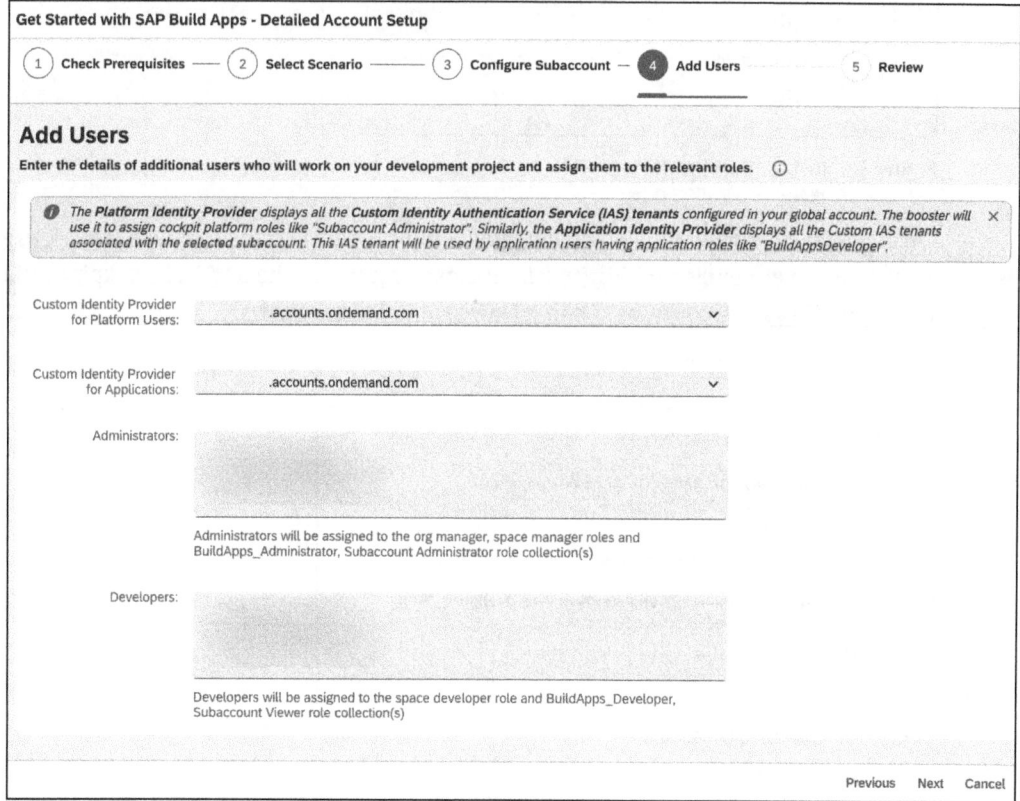

Figure 3.8 User Management

In the last step, **Review**, you will see a summary of the previously created configuration. Click **Finish** to start the installation and configuration of the services.

Once the installation and configuration have been started, you will see a list of the steps performed (see Figure 3.9). This step may take a few minutes. If an error occurs, you will receive corresponding error information that will help you to create a support ticket.

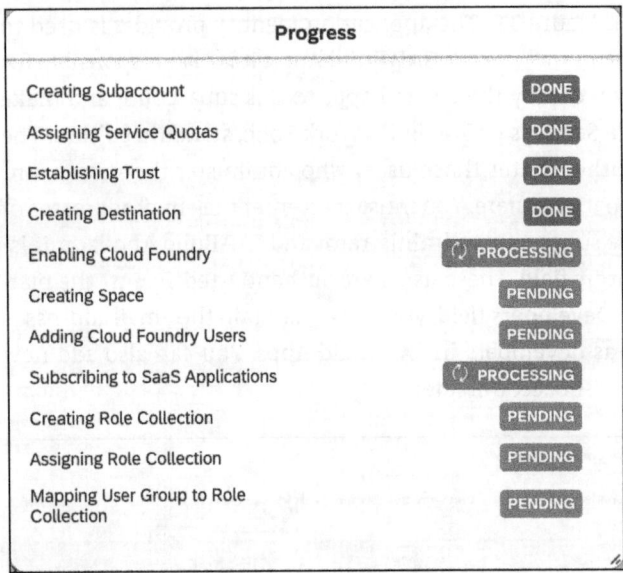

Figure 3.9 Installation Progress

Once the booster has run successfully, you will receive a success message (see Figure 3.10). You can either close this by clicking **Close** or jump to the SAP BTP cockpit of the corresponding subaccount by clicking **Navigate to Subaccount**.

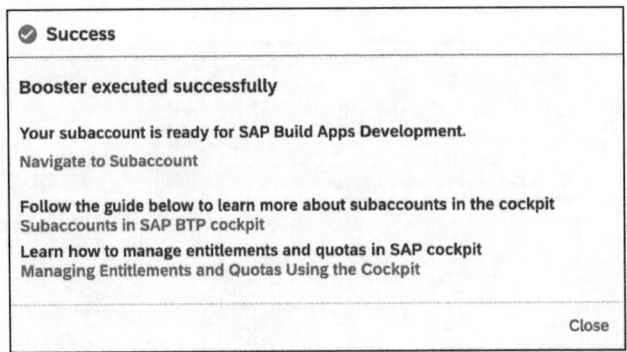

Figure 3.10 Success Message

To keep an eye on the license costs caused by SAP Build Apps and SAP Build Work Zone, you should occasionally take a look at usage analytics. To do this, navigate to the **Usage**

3.2 Configuration

Analytics area in the side menu (see Figure 3.11). There you can see the current and historical usage for all metrics.

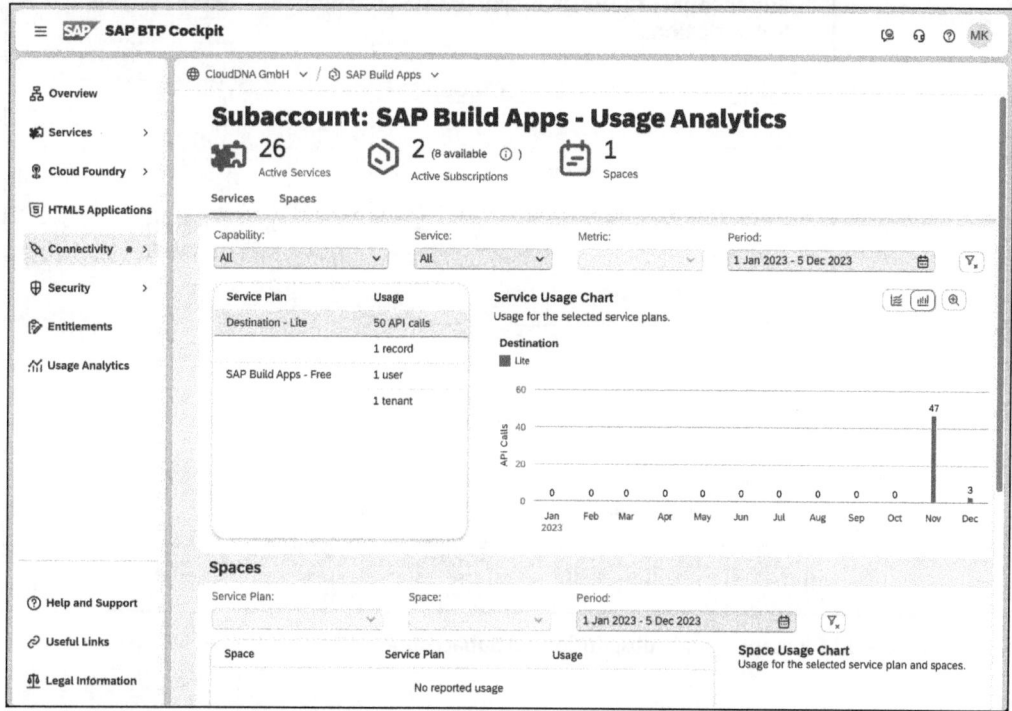

Figure 3.11 Usage Analytics

3.2 Configuration

The booster has carried out a whole range of activities in the background. These include subscribing and setting up the subscriptions and services. The SAP BTP cockpit provides you with an overview of the instances and subscriptions created. To do this, navigate to the **Services • Instances and Subscriptions** area in the side menu (see Figure 3.12). There you will find two subscriptions shown for the boost we performed previously, one for **SAP Build Apps** and one for **SAP Build Work Zone, standard edition**.

You should now test whether SAP Build Apps can be started as expected. To do this, click the link for the SAP Build Apps subscription (see Figure 3.12). You will then be redirected to Identity Authentication to log in (see Figure 3.13). You can see from the example that it's possible to store your own theme in Identity Authentication.

3   Installing and Configuring SAP Build Apps

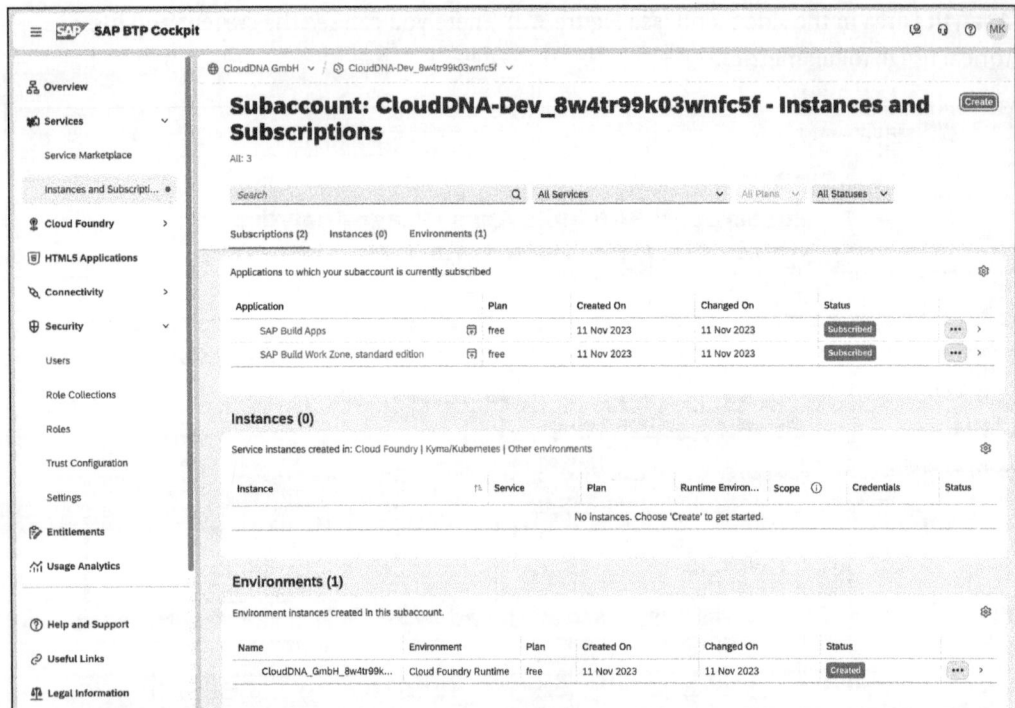

Figure 3.12 Instances and Subscriptions in Subaccount

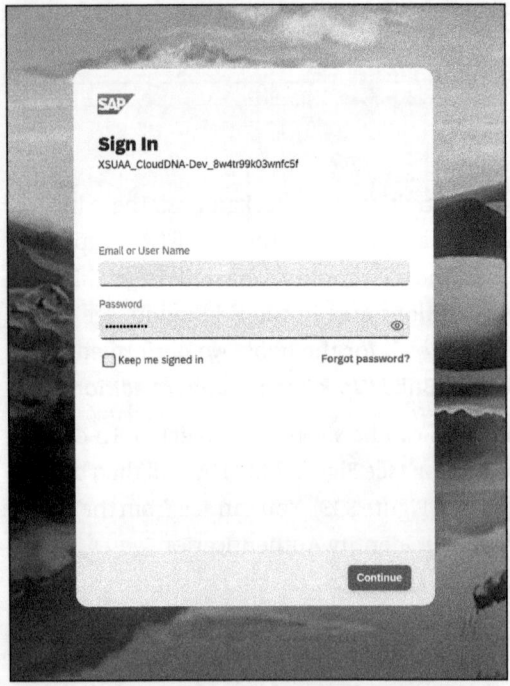

Figure 3.13 Identity Authentication Login

3.2 Configuration

After successful login, the SAP Build Apps lobby opens, as shown in Figure 3.14.

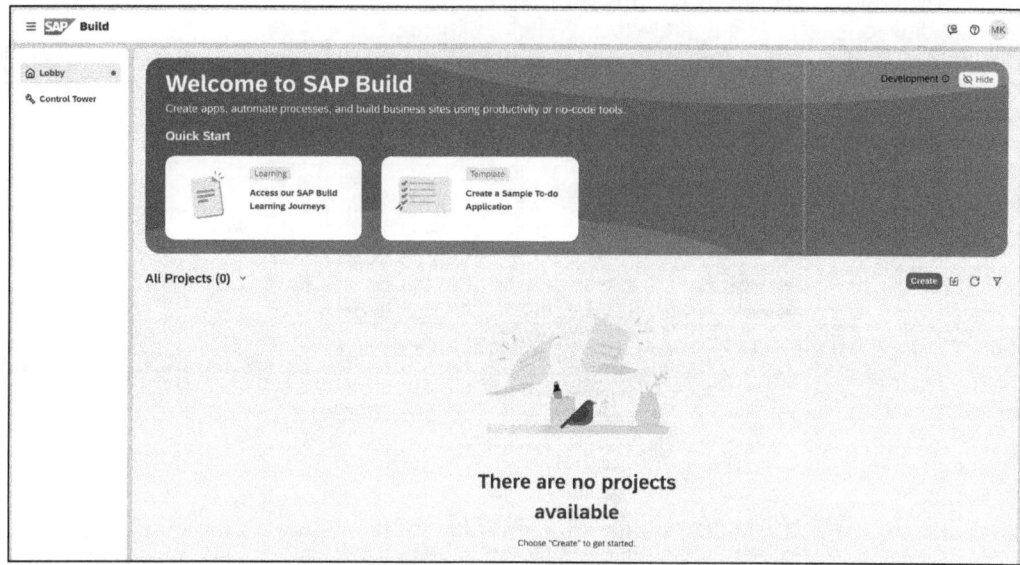

Figure 3.14 SAP Build Apps Lobby

Roles are created in SAP BTP with every subscription. However, roles cannot be assigned directly to users. Instead, the concept of role collections is used. A *role collection* is a collection of different, technically related roles. Navigate to the **Security Role Collections** area as shown in Figure 3.15 to see all the role collections available in the subaccount. The following role collections are available in each subaccount:

- Cloud connector administrator
- Connectivity and destination administrator
- Destination administrator
- Subaccount administrator
- Subaccount service administrator
- Subaccount viewer

The following role collections are delivered with the SAP Build Apps subscription:

- BuildApps_Administrator
- BuildApps_Developer ausgeliefert

You can also see the following role collections created for SAP Build Work Zone, standard edition in the illustration:

- Launchpad_Admin
- Launchpad_Advanced_Theming
- Launchpad_External_User

101

3 Installing and Configuring SAP Build Apps

If the available role collections do not meet your requirements, you can create your own role collections by clicking the **Create** button.

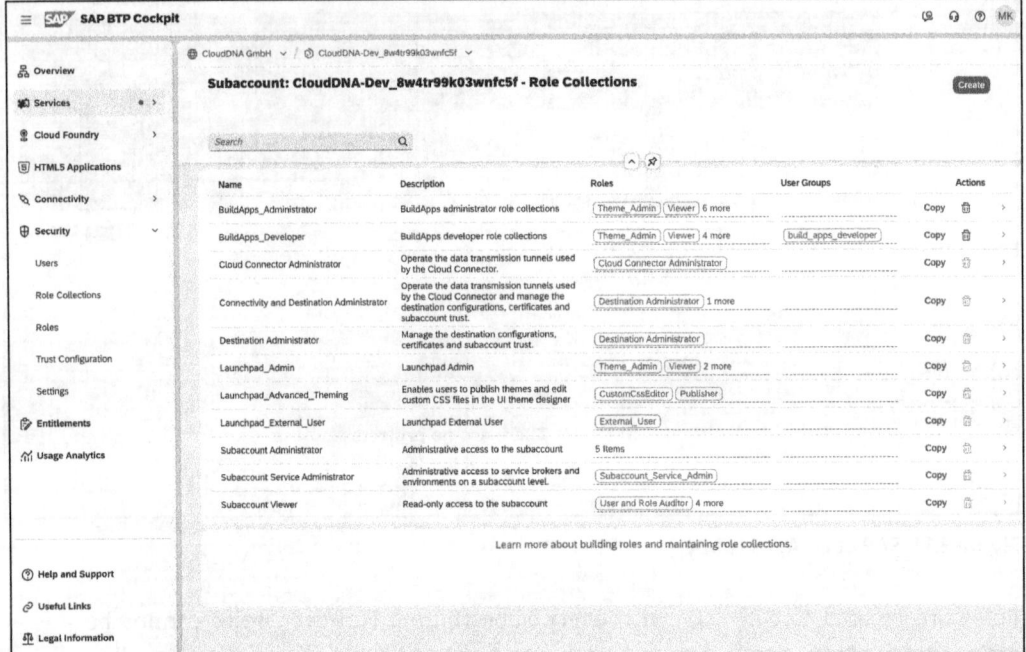

Figure 3.15 Role Collection Overview

Let's take a closer look at the two role collections delivered with SAP Build Apps. Click a role collection in the role collection overview to view the details. As already mentioned, a role collection is a collection of roles. These are assigned to users either directly or via user groups, which in turn are mapped to groups in the underlying identity provider. However, you also have the option of assigning the role collections via attribute mappings. The role collection is derived using certain attributes that are sent by the identity provider after successful registration. For example, this could be the department or country in which the user is located.

Figure 3.16 shows the `BuildApps_Administrator` role collection. This is made up of the following roles:

- `BuildAppsAdmin`
- `BuildAppsDeveloper`
- `Editor`
- `RegistryAdmin`
- `RegistryDeveloper`
- `Super_Admin`
- `Theme_Admin`
- `Viewer`

Right now, we're missing a clear breakdown of each individual role, and unfortunately, the official documentation isn't filling in the gaps. So, we're left to figure out what each role does based on their names, which will hopefully give us enough of a clue to piece things together.

3.2 Configuration

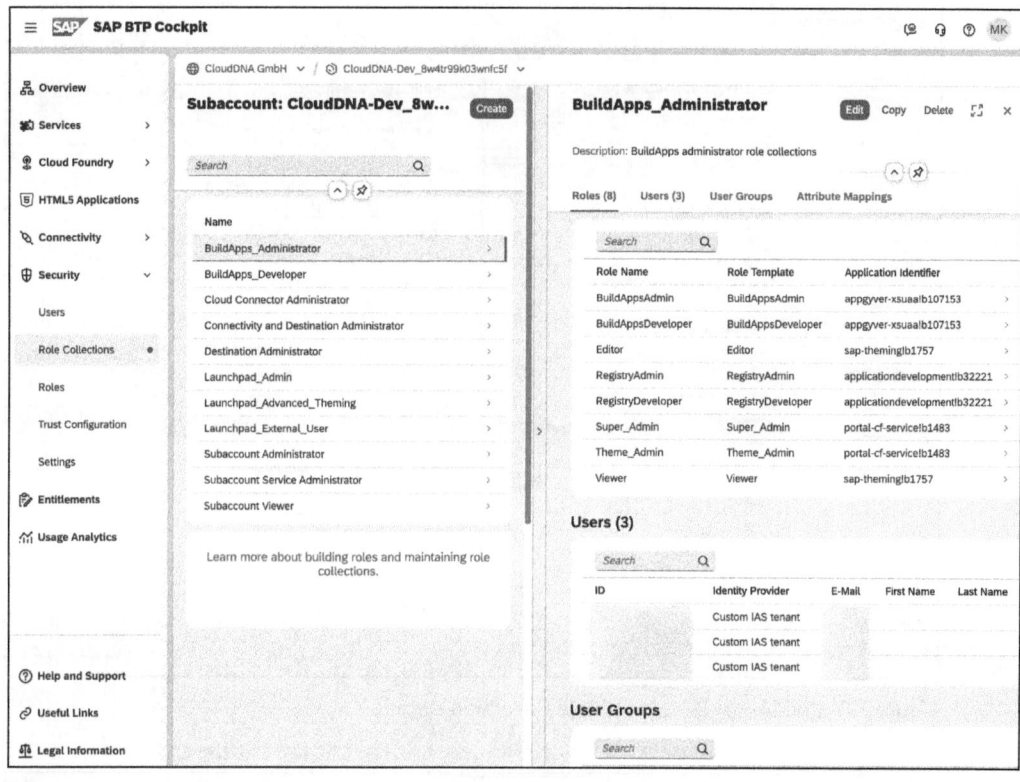

Figure 3.16 BuildApps_Administrator Role Collection: Details

Figure 3.17 shows the `BuildApps_Developer` role collection. This contains the following roles:

- `BuildApps_Developer`
- `Editor`
- `RegistryDeveloper`
- `Super_Admin`
- `Theme_Admin`
- `Viewer`

One might wonder why the `Super_Admin` role is included in the `BuildApps_Developer` role collection. Unfortunately, we're at a bit of a standstill with that question because there's no clear explanation available in the existing documentation. Another point to note is the role collection's connection to a user group called `build_apps_developer`. This link means that we can assign this user group in Identity Authentication, which is quite convenient. It spares us the hassle of having to individually assign role collections to each user in the SAP BTP cockpit at the subaccount level.

This discussion will not delve into the intricacies of the SAP Build Work Zone, standard edition role collections at this juncture, as a comprehensive exploration of this subject is reserved for Chapter 10, designed to give it the focused attention it warrants. The booster has also created a destination called *SAP-Build-Apps-Runtime* in the SAP BTP subaccount. This destination is required for the visual cloud functions. Navigate to the **Connectivity • Destinations** area in the side menu to view the details of the destination (see Figure 3.18).

103

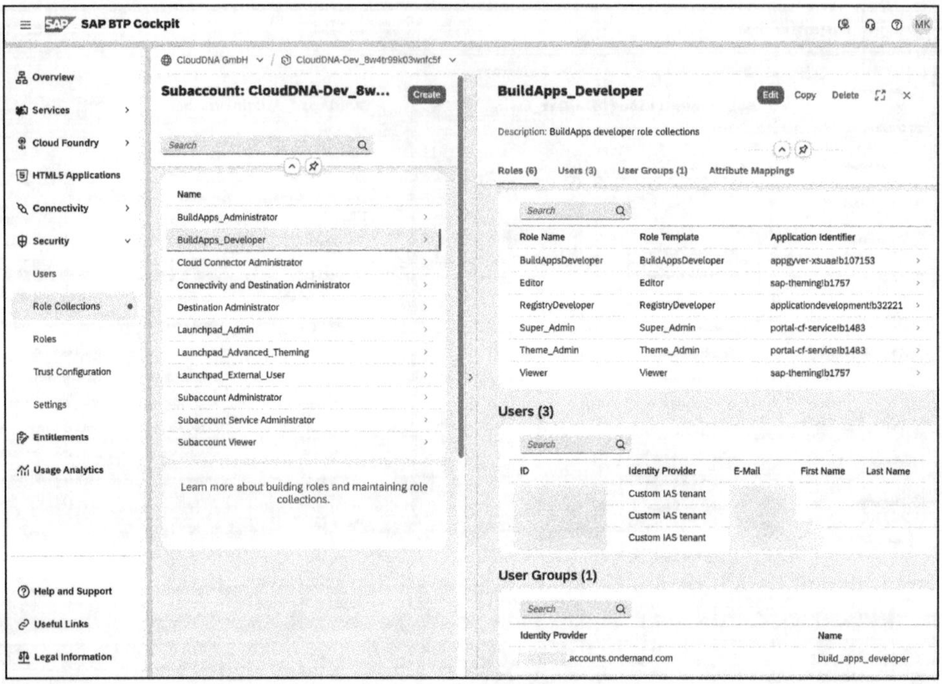

3   Installing and Configuring SAP Build Apps

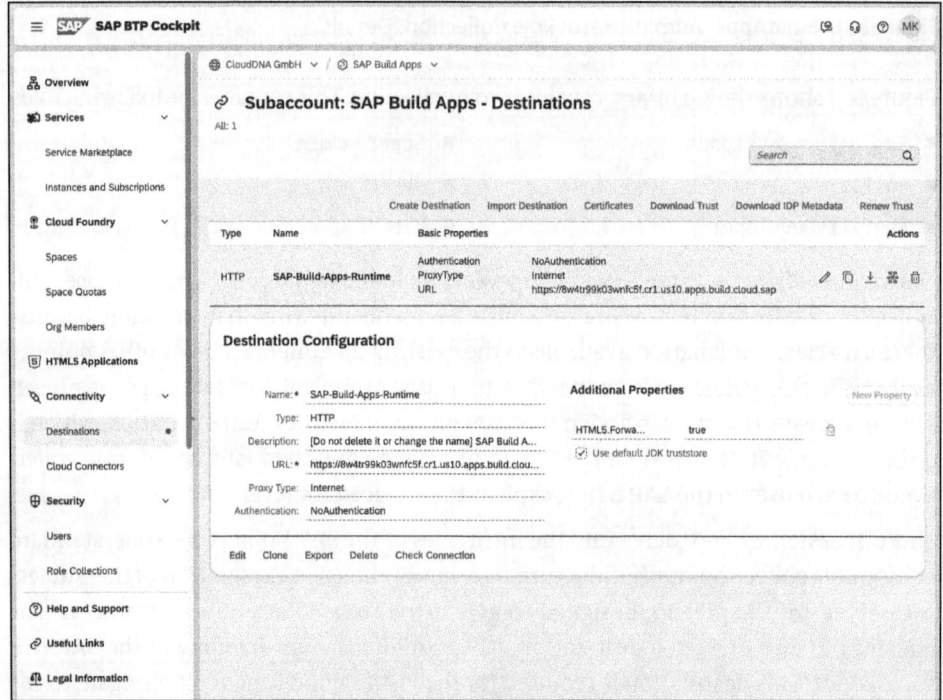

Figure 3.17 BuildApps_Developer Role Collection Details

Figure 3.18 Runtime Destination

3.3 Security

Security is a central element of SAP BTP. Although we do not provide a detailed introduction here, we would like to highlight some key aspects and pieces of information that are crucial for the effective use of SAP Build Apps.

In SAP BTP, users are authenticated by default via the SAP ID service, a central SAP user store in which, among other things, S-user accounts are managed. Although the SAP ID service serves as a kind of security network, it has some limitations in practice. For example, it is not possible to create your own password policies or integrate existing infrastructures such as Microsoft Entra ID (formerly Azure Active Directory).

SAP has recognized these limitations and offers a suitable alternative with Identity Authentication. Identity Authentication is part of SAP Cloud Identity Services, which also includes Identity Provisioning. These services are available to all SAP BTP customers as part of their license. Identity Authentication can be used as a platform identity provider for the authentication of users of a subaccount or global account and as an application identity provider for the authentication of users of services on SAP BTP. This includes end users who access SAP Build Work Zone, for example, as well as developers who want to create applications with SAP Build Apps.

In addition, Identity Authentication offers enhanced flexibility and security through customizable authentication mechanisms and integration options with other identity management systems, ensuring a seamless and secure user experience and meeting modern enterprise security requirements.

During implementation, the booster set up the selected Identity Authentication instance as a cloud identity provider for both platform users and applications. From a technical point of view, it should be emphasized that the modern OpenID Connect protocol is used here, which offers advantages over the older SAML standard. The corresponding configuration of the trust position is shown in Figure 3.19. To view this, navigate to the **Security** area in the SAP BTP cockpit of your subaccount and then to **Trust Configuration**. Here you can check the established trust relationship and the associated settings, which ensures a smooth identity exchange and secure authentication.

By clicking the name of the identity provider, you can jump to its details. Figure 3.20 shows this for the application identity provider. You will find further information there. Here you can also see and customize the text that is displayed to the end user if more than one identity provider is allowed to log in.

In the side menu of the SAP BTP cockpit, under the **Role Collection Mappings** section, you can view the mappings of role collections to Identity Authentication groups. Figure 3.21 shows an example of a mapping created by the booster. Here, the `build_apps_developer` Identity Authentication group is assigned to the `BuildApps_Developer` role collection within SAP BTP. Although this mapping was created correctly, it must be

3 Installing and Configuring SAP Build Apps

noted that the corresponding group has not yet been set up in Identity Authentication. This means that although the mapping is defined, the corresponding group in Identity Authentication that is used for authentication and authorization still needs to be created to make the mapping fully functional.

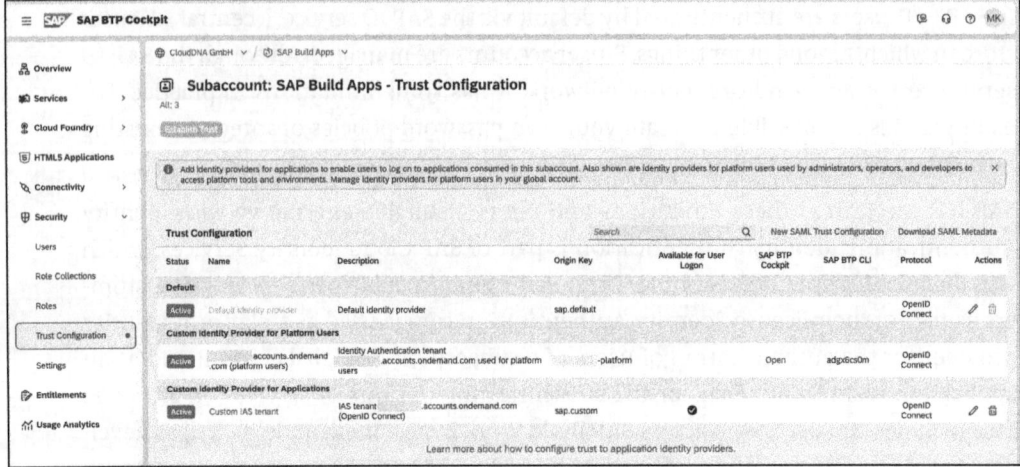

Figure 3.19 Trust Configuration Overview

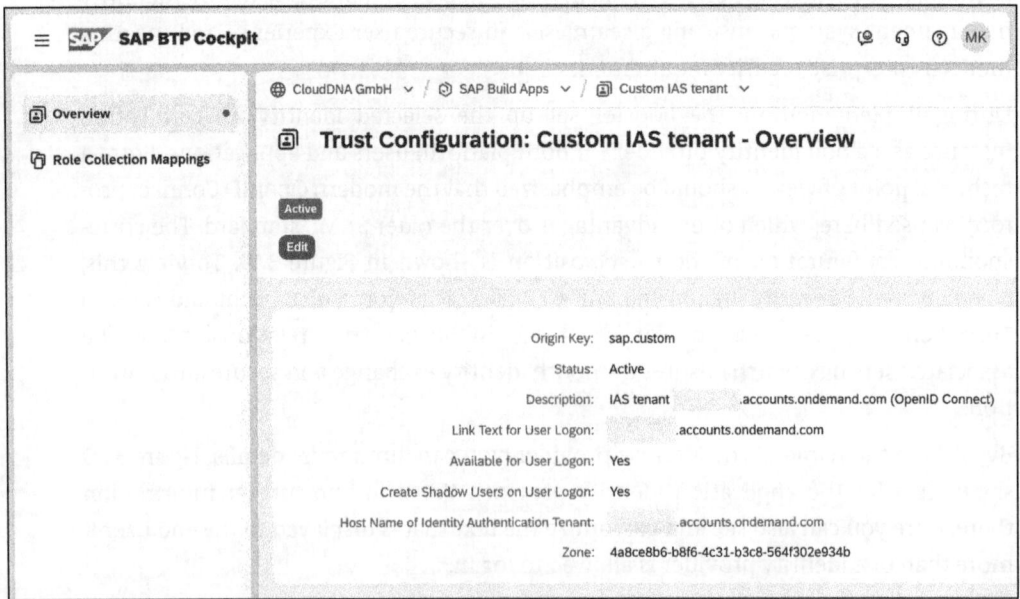

Figure 3.20 Identity Authentication Trust Configuration Details

3.3 Security

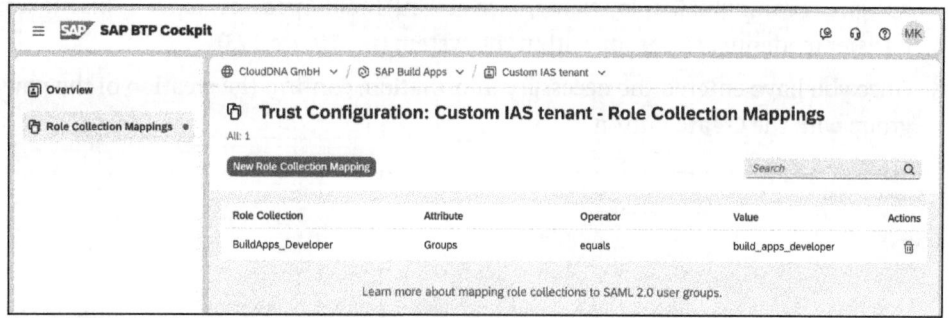

Figure 3.21 Role Collection Mapping to Identity Authentication Group

As the required group in Identity Authentication was not set up automatically by the booster, this step must be carried out manually. To do this, log in to Identity Authentication with an administrator account. You will then see various tiles on the start screen that reflect the scope of your authorizations. In the **Users & Authorizations** area, you will find the **Groups** tile (see Figure 3.22). Clicking this tile takes you to the area where you can create and manage new user groups. Here you can manually add the missing group to complete the required mapping and ensure correct role assignment.

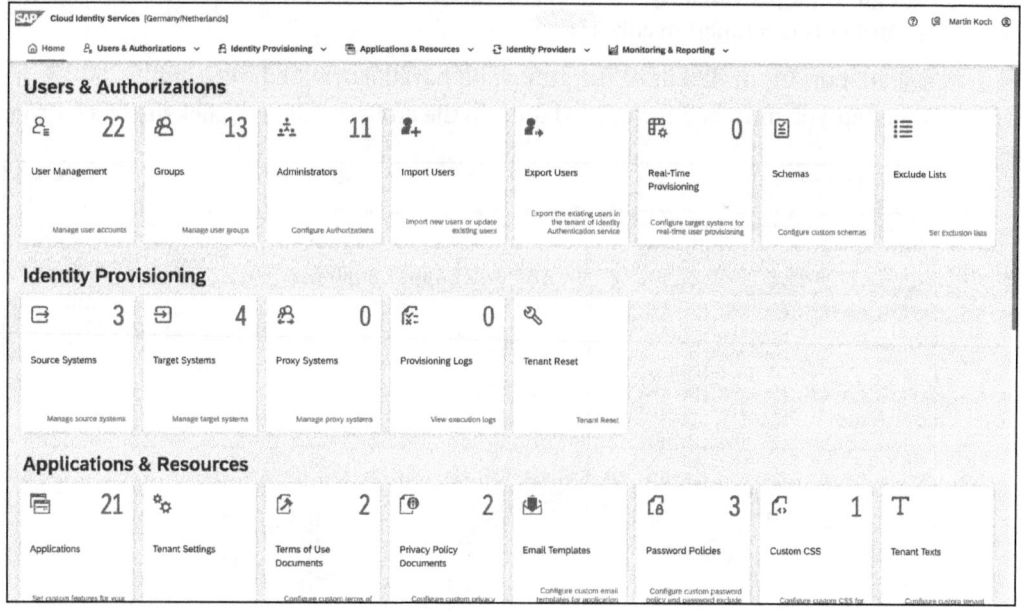

Figure 3.22 SAP Cloud Identity Services Overview

To create a new group in Identity Authentication, proceed as follows in the overview of all groups:

1. Select the option to add a new group.
2. Enter "build_apps_developer" as the **Name** for the group. This name is case sensitive and must be typed exactly as shown here.

3   Installing and Configuring SAP Build Apps

3. Assign a meaningful **Display Name**, such as "SAP Build Apps Developer". This makes it easier to identify the group within the system (see Figure 3.23).
4. Once you have entered the necessary information, confirm the creation of the new group with the **Create** button.

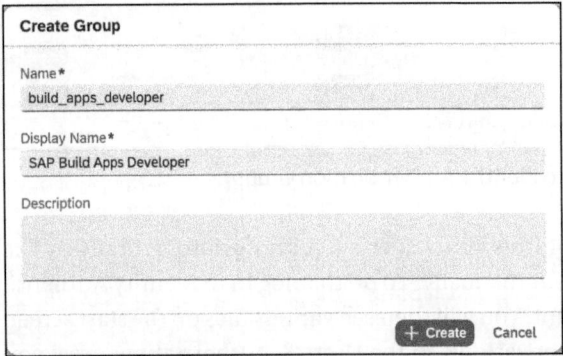

Figure 3.23 Create Group for SAP Build Apps Developers

With these steps, you have manually set up the missing group, which completes the configuration provided by the booster. This ensures that the role and authorization assignments can function correctly.

You are now in the details of the group you have just created (see Figure 3.24). In the next step, you must add the desired users to the group. To do this, click the **Add** button.

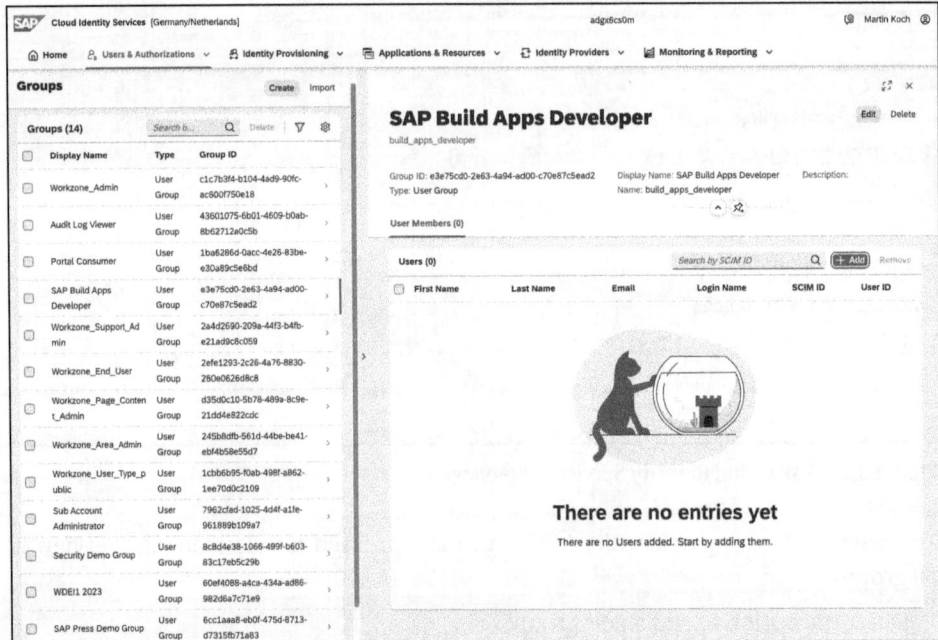

Figure 3.24 Identity Authentication to SAP Build Apps Developer Group Overview

You can now select from all users that exist in Identity Authentication (see Figure 3.25). You also have the option of using the search function, which is particularly useful if there are many users in Identity Authentication. Finally, you must click the **Add** button after you have selected the desired users.

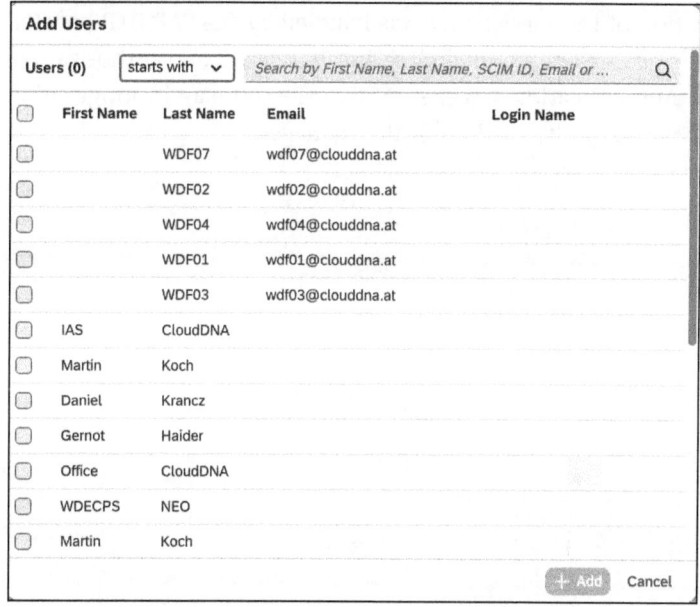

Figure 3.25 User Selection

After the users have been added to the group, you should see a result as shown in Figure 3.26. Note that the role assignment only works after the next user login.

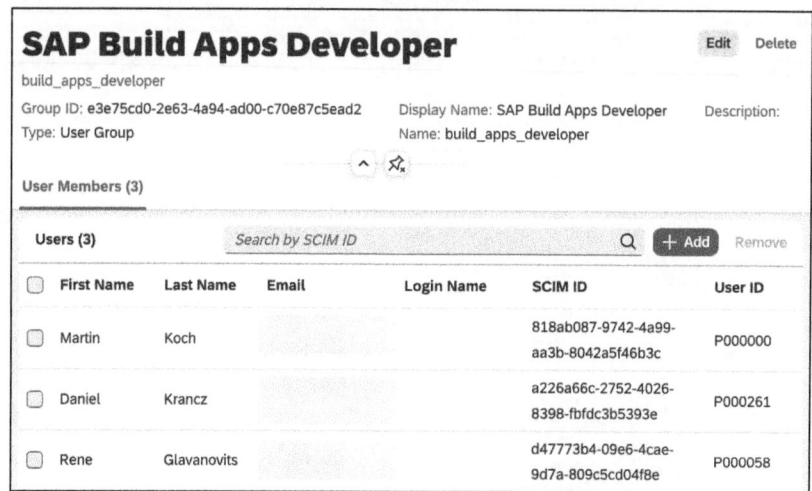

Figure 3.26 Identity Authentication Group User Assignment Result

3.4 Summary

This chapter covered the steps and considerations for setting up and managing SAP Build Apps within SAP BTP. In Section 3.1, we discussed the prerequisites, the installation process, and how to verify that the installation was successful. In Section 3.2, we discussed the configuration of the service that was handled by the SAP BTP booster. Finally, in Section 3.3, we focused on understanding authentication and authorization. This is done by Identity Authentication and ensures that the setup and configuration adheres to best practices in security, including the management of user roles and access rights.

Chapter 4
No-Code Development Environment

To be able to perform efficient development work, you should know and master the available tools as well as possible. In this chapter, we'll introduce the possibilities SAP Build Apps can offer you.

For SAP Build Apps developments, you can use the development environment called *SAP Build Apps Composer*. Originally, apps in SAP Build Apps were only developed in the Composer Pro development environment, which is called the *Community Edition* nowadays. At that time, SAP Build Apps was also called *SAP AppGyver*. Only after AppGyver was bought by SAP and thus became part of the ecosystem did the possibility to have no-code development in SAP Business Technology Platform (SAP BTP) with SAP Build Apps come into play. In this process, SAP also made some name changes, from SAP AppGyver to SAP Build Apps, although the same technological basis is still used.

SAP BTP is a platform-as-a-service (PaaS) offering from SAP that not only provides services for development environments, integrations, and APIs, but also provides portals or options for deployment. Meanwhile, a corresponding service on SAP BTP can also be used for development with SAP Build Apps. However, this requires more administration work than in the Community Edition, where you only need to register to gain access. If you want to learn more about licensing, access, and permissions in SAP BTP, we recommend you look back to Chapter 3, where we summarize the information needed for administration and security.

> **Free Community Edition of SAP Build Apps** [+]
>
> In this book, we focus on the SAP Build Apps service in SAP BTP, but there is also a Community Edition version of SAP Build Apps. This is available on the web for free and it can be accessed by simply registering at *https://appgyver.com/community*. Of course, this option doesn't come without its drawbacks. The website details which functions are not available for Community Edition, among other information.

In this chapter, we want to familiarize you with the development environment and introduce the tools this IDE provides. In Section 4.1, we will show you how to get to the SAP Build Apps Composer development environment. Section 4.2 goes into more detail about the creation of new projects, and we introduce the collaboration tools. In Section 4.3, you will see all the functionalities and possibilities that this development environment offers. Section 4.4 deals with the administration tasks that take place

4 No-Code Development Environment

either before or after development. These topics concern the import and export of SAP Build Apps projects, data integration, authentication, versioning of developments, and deployment. All of these topics are covered in detail in other SAP Build Apps chapters in Part II. Finally, from time to time, you simply need help from SAP or the community, which is why we offer further information, help pages, blogs, and sources for official documentation in Section 4.5.

> **Different Names for the Same Service**
>
> We would like to point out that although SAP Build Apps is the official name for the no-code environment development platform of SAP, this service has already undergone several name changes. So don't be surprised if in the course of not only this chapter but also the whole book, the name AppGyver is mentioned. Initially, the product's name was AppGyver until SAP bought the platform, when it became SAP AppGyver. In 2022 at TechEd, SAP Build was announced, and it was revealed that SAP AppGyver would be integrated into SAP Build as the service called SAP Build Apps.

4.1 Launching SAP Build Apps Composer

As already mentioned, SAP BTP is SAP's PaaS offering. A PaaS is characterized by the fact that, in addition to the infrastructure, a provider also provides a platform on which and with which development can take place, applications can be operated in runtime environments, or virtualization options can be used. Here you can license and use various services, just as we will do with SAP Build Apps.

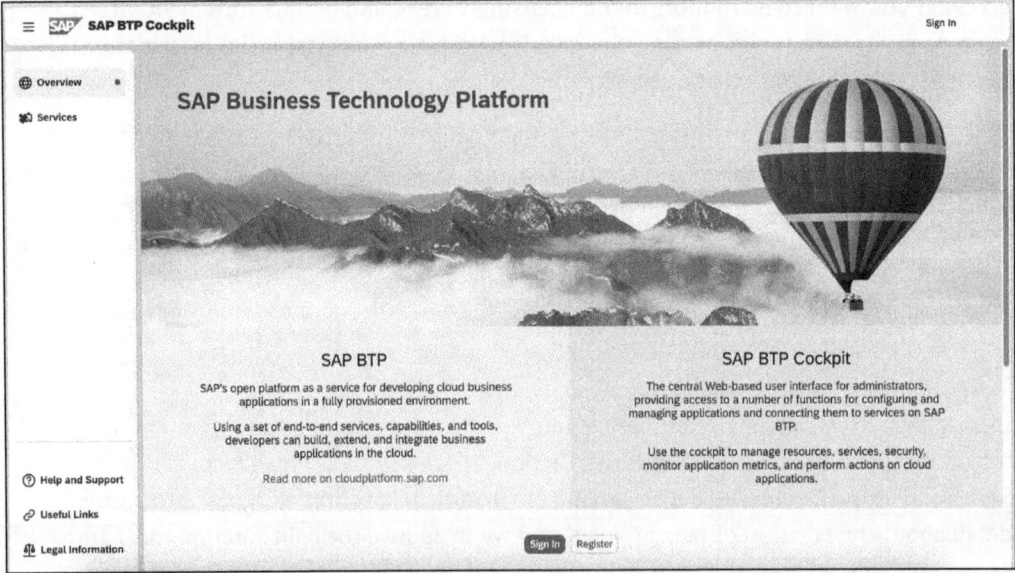

Figure 4.1 Launching SAP Business Technology Platform

4.1 Launching SAP Build Apps Composer

Figure 4.1 shows the start page of SAP BTP. As an SAP customer, you can simply log in here with your SAP credentials. You can access SAP BTP via the following URL: *https://www.hana.ondemand.com*.

Once the SAP Build Apps service is licensed and you have the appropriate permissions to access the service, you will find it in your subaccount at **Services • Service Marketplace** (see Figure 4.2). Normally, the process of licensing and authorization doesn't need to be taken care of by developers. Nevertheless, if you are interested in any information about the administrative tasks, we recommend that you refer to Chapter 3.

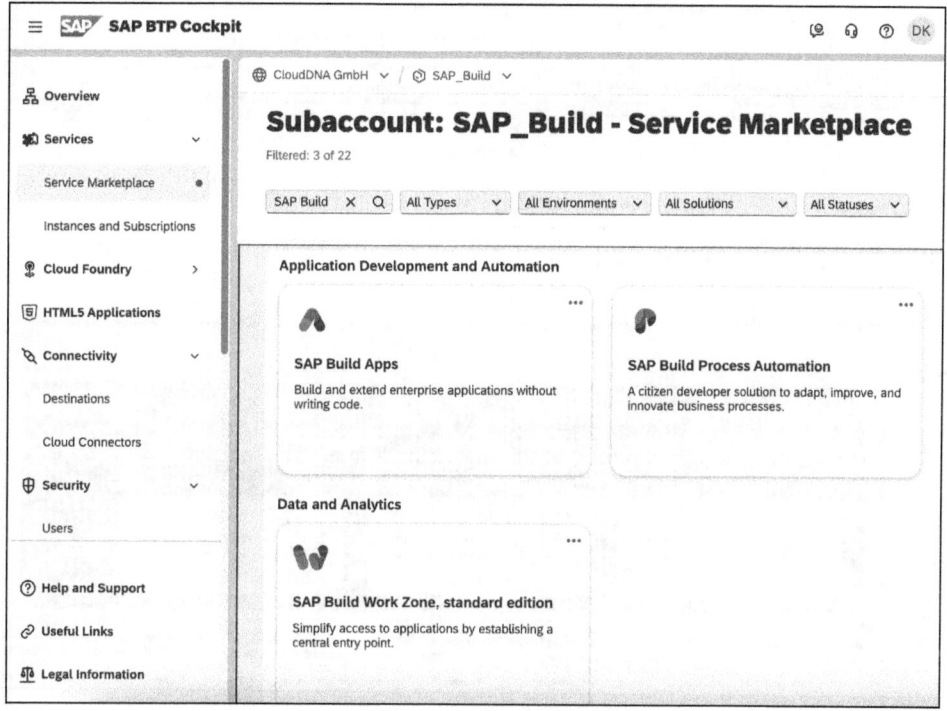

Figure 4.2 Searching for SAP Build Apps Service in SAP BTP

Clicking the service opens the associated details and also the option to start the app via the **Go to Application** button (see Figure 4.3). Launching the app in this case means that you will be redirected to a new page. You can save the URL of the new page as a bookmark in your browser. You will then always be able to use this bookmark to get back to the service without having to enter via SAP BTP again.

In SAP Build, you will be welcomed in the lobby. In Figure 4.4, you can see that on the one hand a list of already existing projects appears at the bottom, but on the other hand, by means of cards in the **Quick Start** area, new applications with sample implementations can be created. On the left-hand side, there are options for accessing public content like connectors or already existing and developed projects offered by the community.

4   No-Code Development Environment

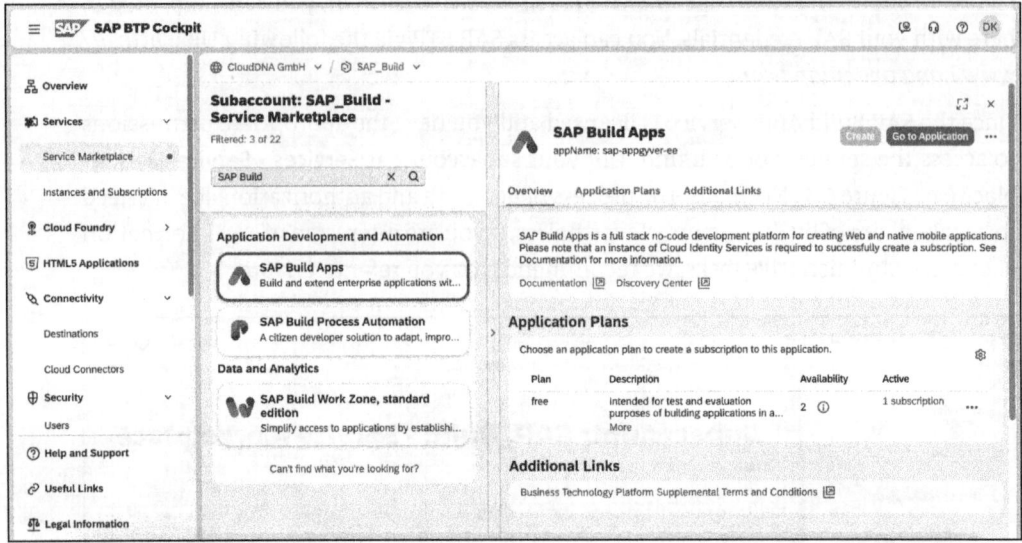

Figure 4.3 Details of SAP Build Apps Service

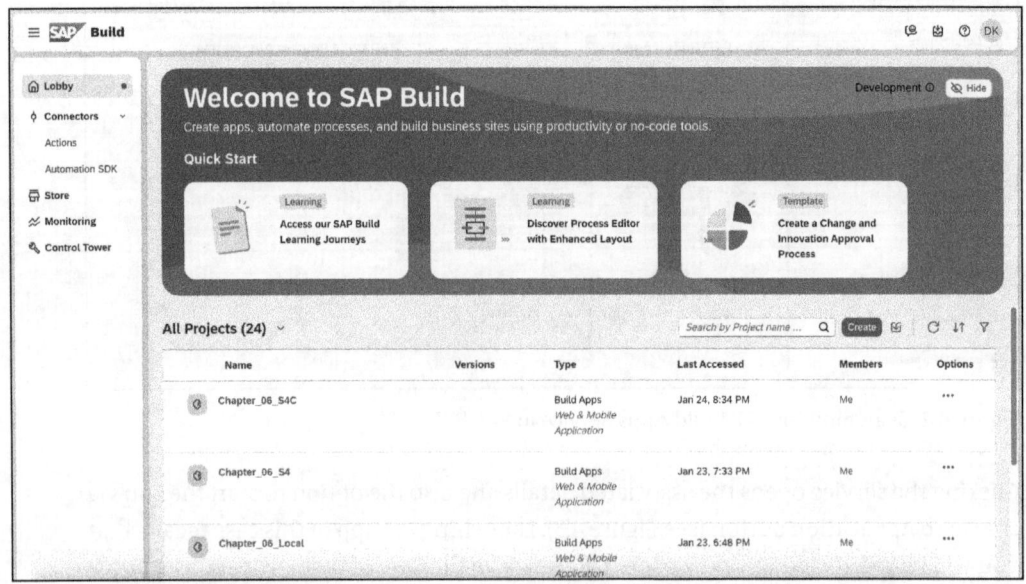

Figure 4.4 SAP Build Lobby

When you click the **Create** button, you have the option to create an SAP Build Apps project (**Build an Application**), to use the robotic process automation module (**Build an Automated Process**), or to build a new site for SAP Build Work Zone, advanced edition (**Build a Business Site**). The dialog with these three options is displayed in Figure 4.5. Part III and Part IV of this book will focus on SAP Build Work Zone and SAP Build Process Automation, but for now we will examine no-code application development.

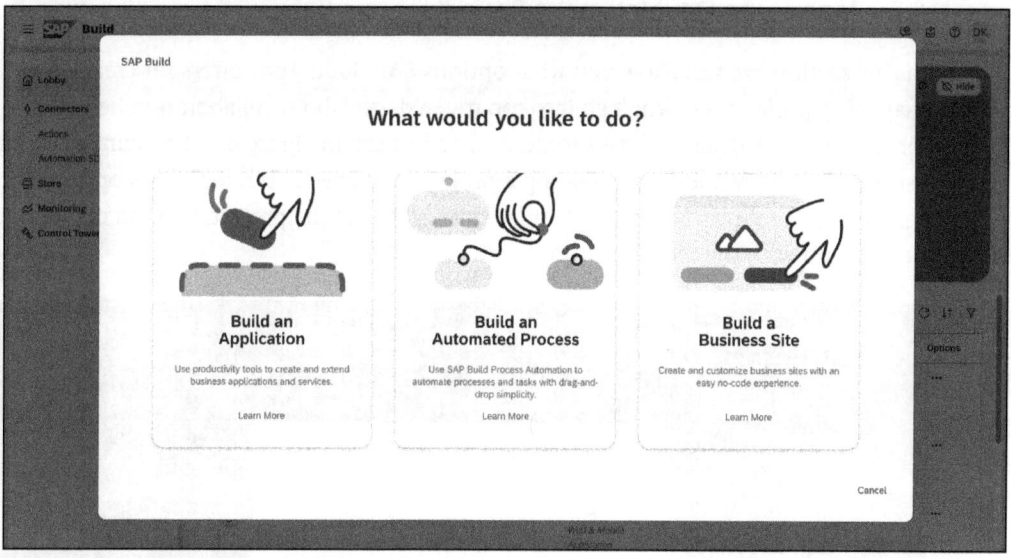

Figure 4.5 Creating New Project in SAP Build

Now that you have selected **Build an Application**, you must further distinguish whether you want to develop the user interface and application logic for the upcoming application (**Web & Mobile Application**) or the associated backend (**Application Backend**). The options for the development of different application types are displayed in Figure 4.6.

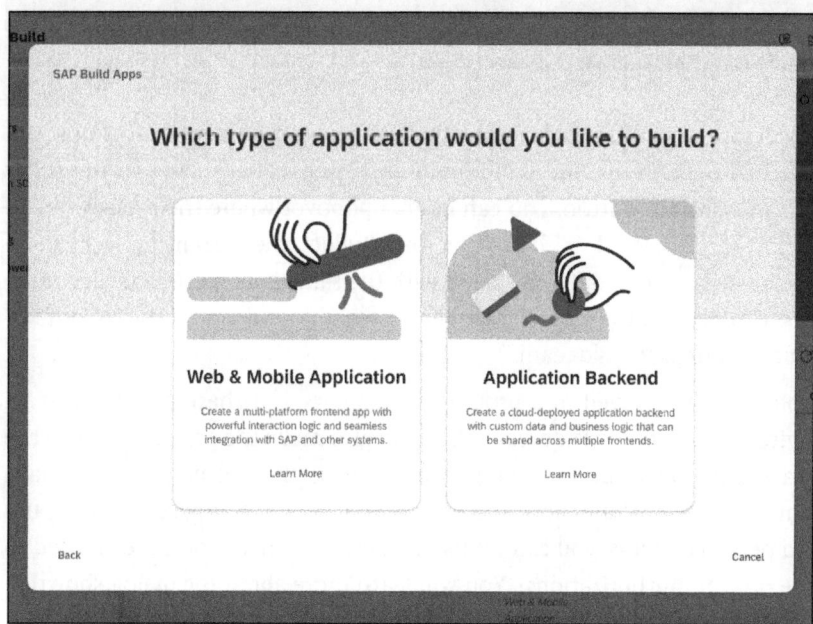

Figure 4.6 Creating New Project to Build Application

4.2 Managing Development Projects

In this section, we will show you what options SAP Build Apps offers for creating and managing projects. We won't go into too much depth about collaboration here; Chapter 5, Section 5.5 deals with this topic. As you can see in Figure 4.7, it is mandatory to assign a **Project Name** when creating a project. The project **Description** is optional; it can be added to include a more detailed description of the project. You can create the new project with the **Create** button.

Figure 4.7 Create New SAP Build Apps Project

After clicking **Create**, you will be taken directly to the project workspace in a new tab. We have switched back to the SAP Build lobby in Figure 4.8 to show you the other options. With the **Options** button, you can open a popover where the project can be renamed, exported, or deleted. With **Save as New Project**, the current project can be copied and recreated under a different name with the same content. If you click **Manage Members**, you can edit the project members or whether the project should be publicly accessible in your SAP Build team.

Several people can be involved in a project as members, but these people must be explicitly invited via their email addresses. You can also make the project public in the **General Access** dropdown options. Being public means that all persons who have access to your SAP Build instance and lobby can at least view the project. For both public access and personal access, you can set users' roles and what people are allowed to do within the project (**Authorizations**). You will learn more about the dialog shown in Figure 4.9 and what options you have for collaboration in Chapter 5, Section 5.5.

4.3 SAP Build Apps Composer User Interface

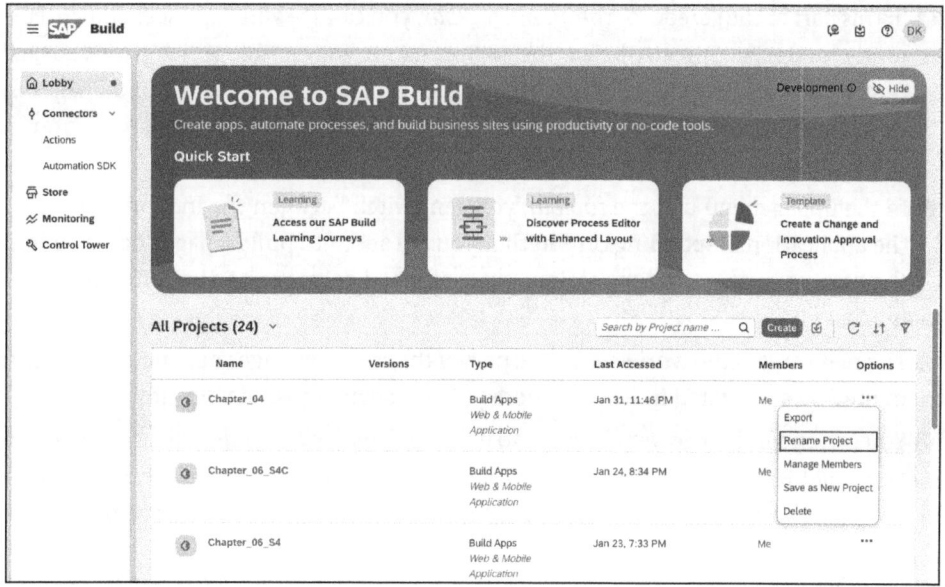

Figure 4.8 More Options for Development Project

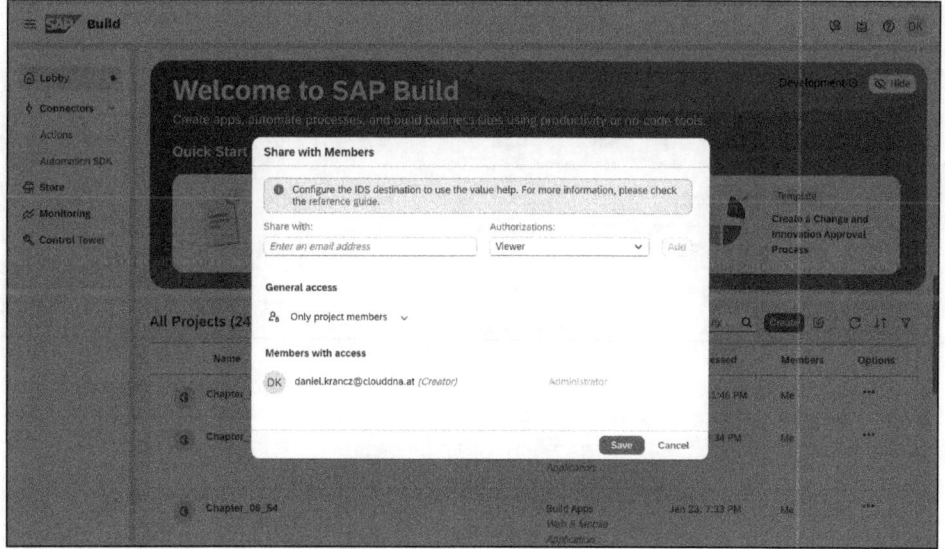

Figure 4.9 Manage Project Members

4.3 SAP Build Apps Composer User Interface

Once you are in a project, the first page of the SAP Build Apps Composer user interface opens. As you can see in Figure 4.10, the following actions and areas are available on this page:

4   No-Code Development Environment

❶ In this part of the screen, in the *global toolbar*, you can see which project you are currently located in. Directly below the project name, you can see which page you currently have opened. Clicking the page name will take you to the list of all pages in this project, where you can also create new pages. We'll show you this step later in this section.

❷ In the upper menu bar, the *toolbar*, you can switch between the individual parts of the application. Depending on whether you are about to build a page, load data into the app, or enable authentication, you will need to move to the different menu items.

❸ The main area is the *workbench*. This part of the screen changes depending on which menu you are located in. This is where you will spend most of your time.

❹ You can expand the lower black bar to get to the *application logic*. This area will help you build the appropriate business logic for your user interface. But don't worry: there will be a lot of no-code building blocks here as well. The application logic will be discussed in Chapter 5.

❺ For beginners, there are helpful tips and tutorials below the black bar, which are located behind the (in this case, already green) progress bar. By clicking the progress bar of the currently selected tutorial, you can expand and collapse the area. This progress bar can also be hidden in the settings.

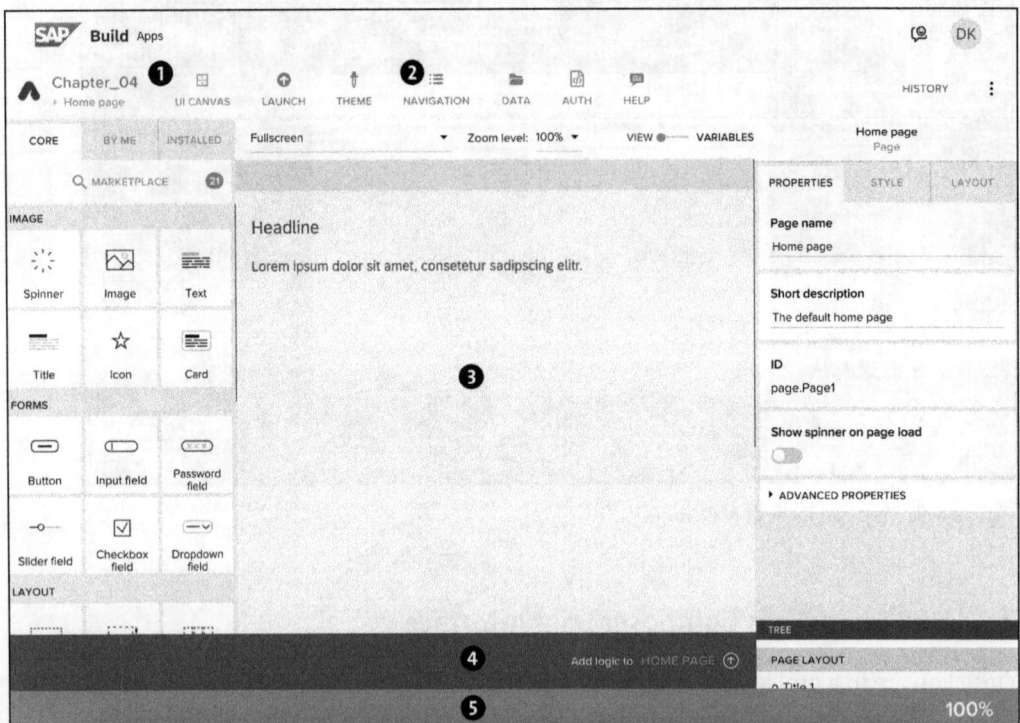

Figure 4.10 Main Parts of SAP Build Apps Composer User Interface

4.3 SAP Build Apps Composer User Interface

Instead of covering all menu items and areas of the application at once, we will briefly introduce them to you step by step in this section.

On the right-hand side, there is a slider between two options: **VIEW** and **VARIABLES**. If you have switched the slider to **VIEW**, you can edit a page (see Figure 4.11). Let's look at the main parts of the workbench in terms of the development of views:

❶ The left-hand scroll bar, the *view component library*, contains all the UI components that are either shipped by default (**CORE**) or that you have developed yourself (**BY ME**), or all the available UI components, including those developed by the community and added from the marketplace (**INSTALLED**). We'll discuss the different UI components in more detail in Chapter 5.

❷ The interactive workspace in the center is also called the *view canvas* and works on the principle of *what you see is what you get* (WYSISYG). It also serves as a real-time preview of what your app will look like later. Using drag and drop, you drag UI components from the left side onto the canvas and place them where you want them to appear on the page.

❸ Once you have selected a UI component on the page, information about the component is displayed in the right-hand scroll bar. This information includes the properties of a UI component (**PROPERTIES**), predefined designs (**STYLE**), and settings for formatting, alignment, and more (**LAYOUT**).

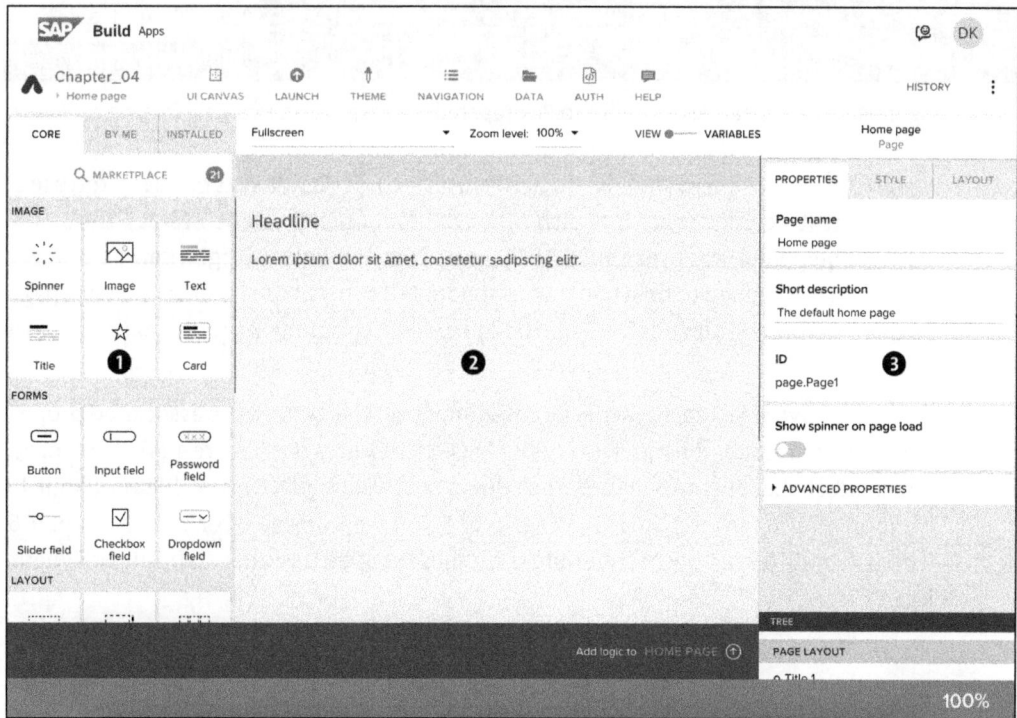

Figure 4.11 Main Parts of Workbench for View Development

Figure 4.12 shows that you can change the screen size of the workspace. This allows you to check how the application might look on various devices. In addition, you can zoom in and out of the workspace by selecting a percentage.

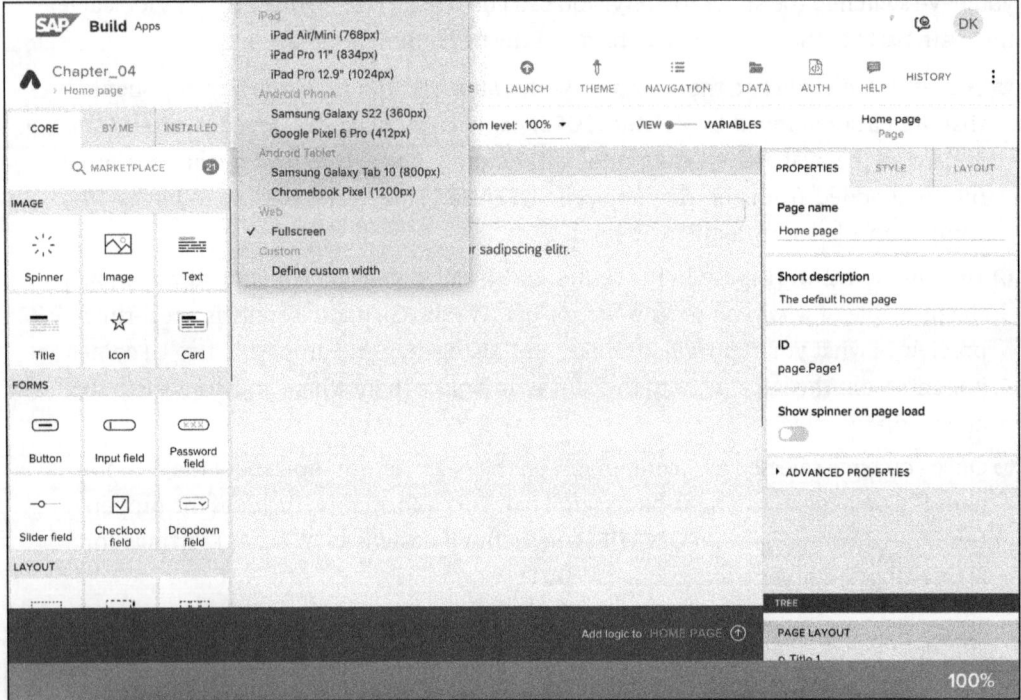

Figure 4.12 Changing Device Types to Be Prepared for Responsive Design

If you lose track of the layout due to many inserted UI components or due to the use of containers (UI components in which other UI components can be nested), the *layout tree* will help you. The area called **TREE**, located at the bottom right, can be dragged larger and is intended to help you find individual UI components that are on your page. To demonstrate this layout tree, we added some UI components to the page in Figure 4.13.

Once you have selected a specific UI component on the canvas, the associated properties are displayed on the right side in the **PROPERTIES** area (see Figure 4.14). These properties affect the UI component either directly or indirectly. You will learn about UI component properties primarily in Chapter 5, but we will also refer you to Chapter 6, where you will learn how to store values for these properties either statically or dynamically.

4.3 SAP Build Apps Composer User Interface

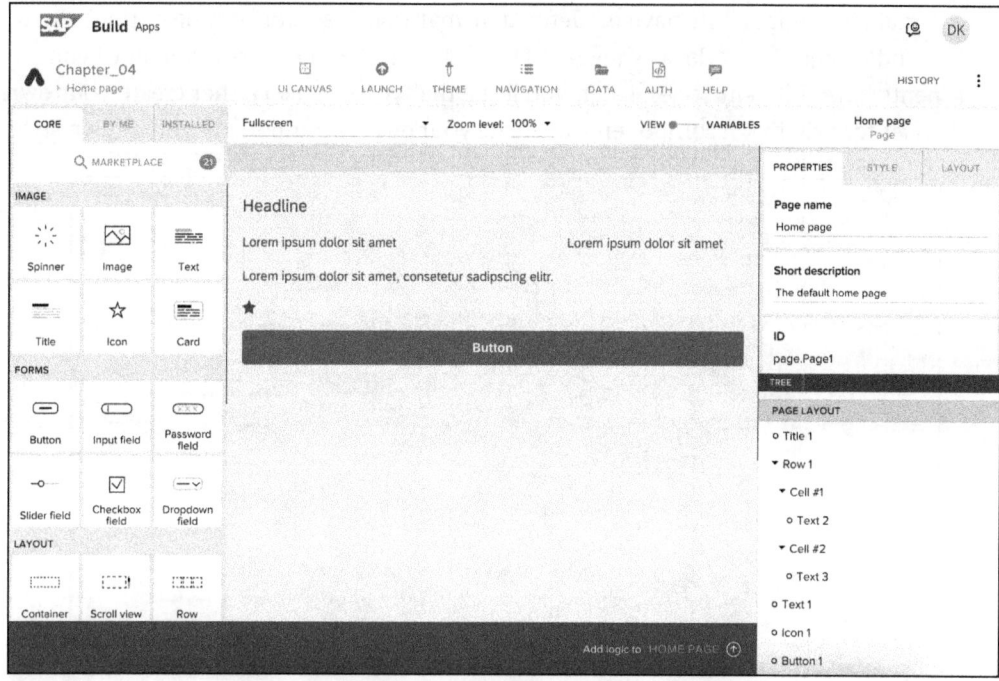

Figure 4.13 Layout Tree for Examining Page Layout

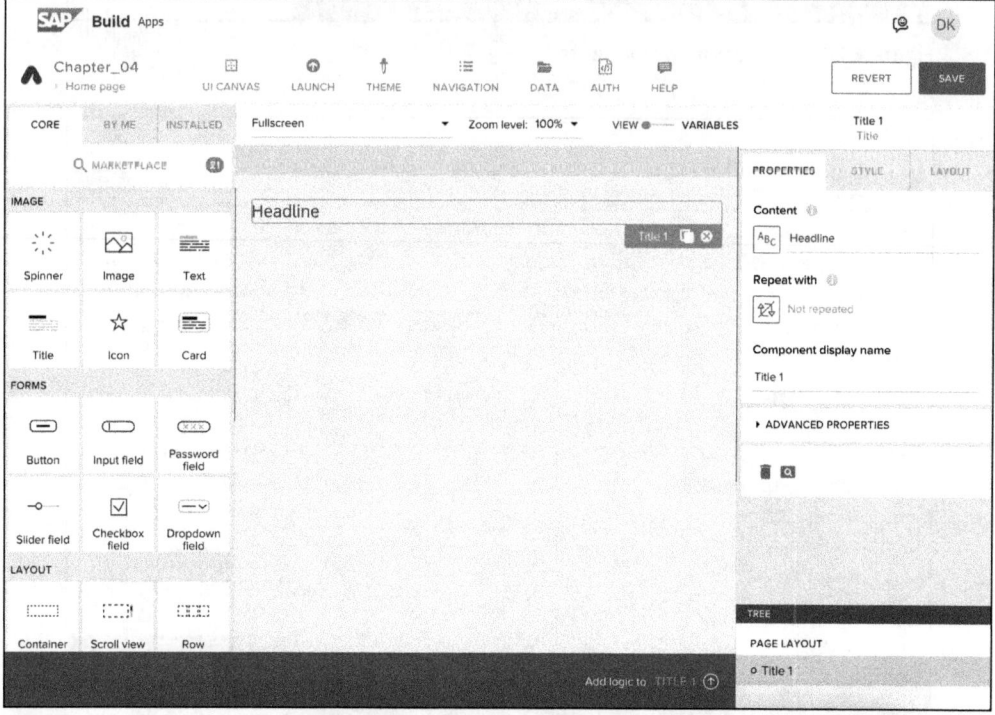

Figure 4.14 Properties of UI Components

4 No-Code Development Environment

Certain UI components have predefined formatting. These formatting options can be found on the right side, as shown in Figure 4.15, in the **STYLE** area. Not all UI components have different styles. Later, you'll see that you can also either create your own styles for existing UI components or create your own UI component with its own style.

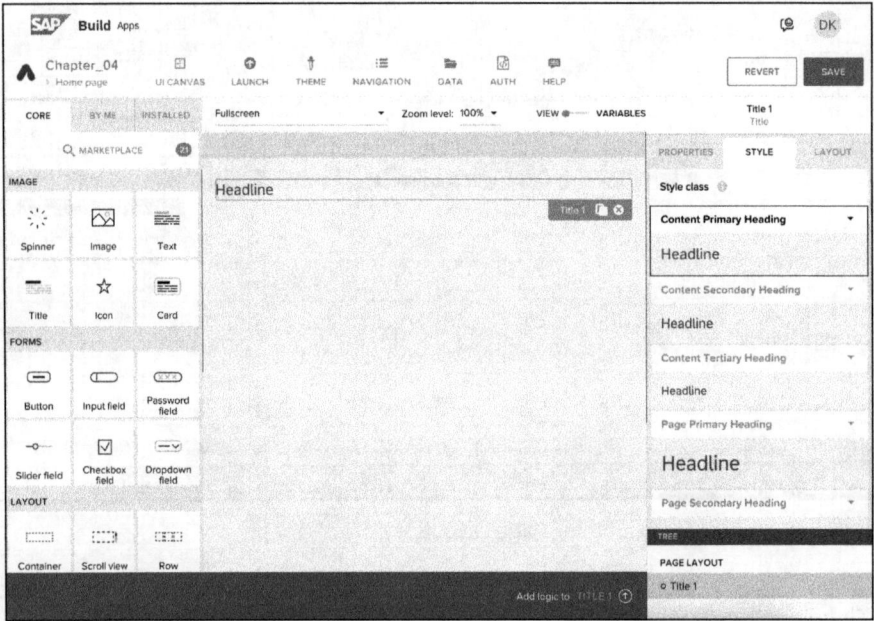

Figure 4.15 Styling of UI Components

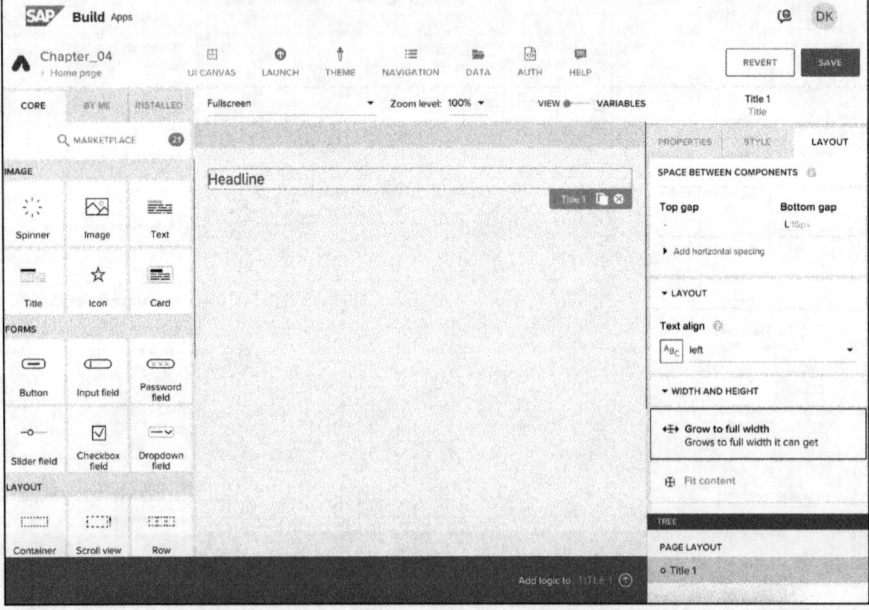

Figure 4.16 Layouts of UI Components

Alignment, width and height, spacing, and more can be set in the **LAYOUT** section (see Figure 4.16). In Chapter 5, formatting and certain layout changes will be covered.

As soon as a UI component is selected in the **UI CANVAS**, the **LOGIC CANVAS** can be opened via the lower black bar, which shows the text **Add Logic To ....** As you can see in Figure 4.17, in this area you have the ability to drag and drop the prepared logic blocks of the selected UI component from the left-hand side into the workspace of the business logic. This way, you can determine what should happen when an event is triggered either by the framework using lifecycle events or by user interaction. On the topic of the business logic, we'll go into more detail in Chapter 5, Chapter 6, and Chapter 7.

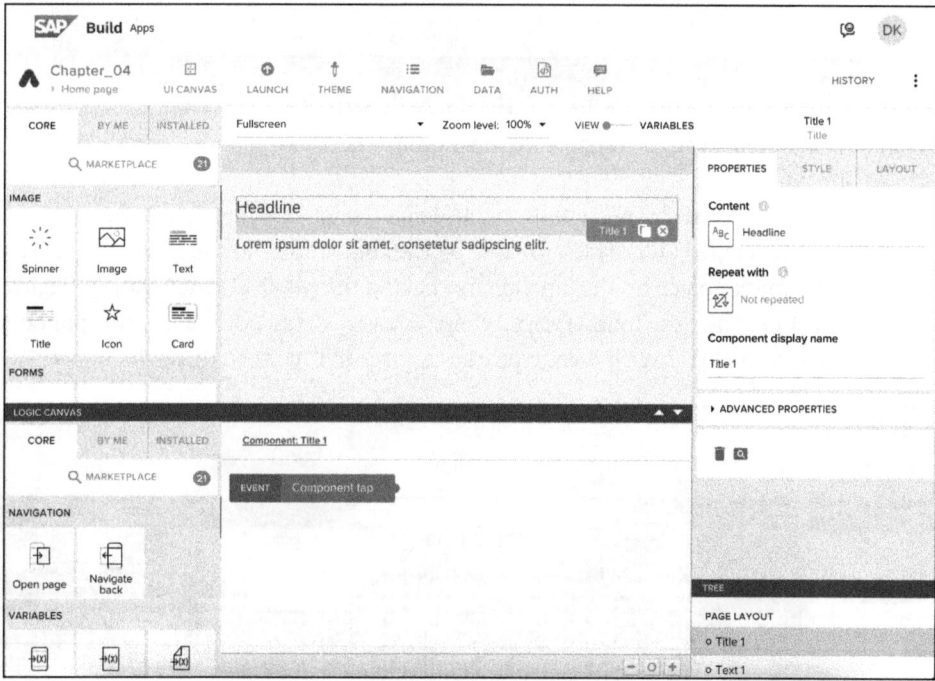

Figure 4.17 Logic Canvas of UI Components

If you switch the slider in the **UI CANVAS** from **VIEW** to **VARIABLES**, you have the ability to create variables of different types. The *type* provides information not only about the visibility, but also about the lifetime of these variables. On the left side, you can select these different types of variables. In the middle of the screen, the variables can be created visually and managed in a list. If you select a variable in the middle, it can be edited and customized on the right side, just like a UI component (see Figure 4.18). Variables help hold data for runtime, cache it, or prepare it so that it can be used by the view for visualization. Variables are not intended per se for long-term storage—that is, for storage beyond runtime. There are other options for this type of data storage, which we will discuss later in this chapter. We won't go into more detail about variables in this chapter, but we'll do so in Chapter 5 through Chapter 7.

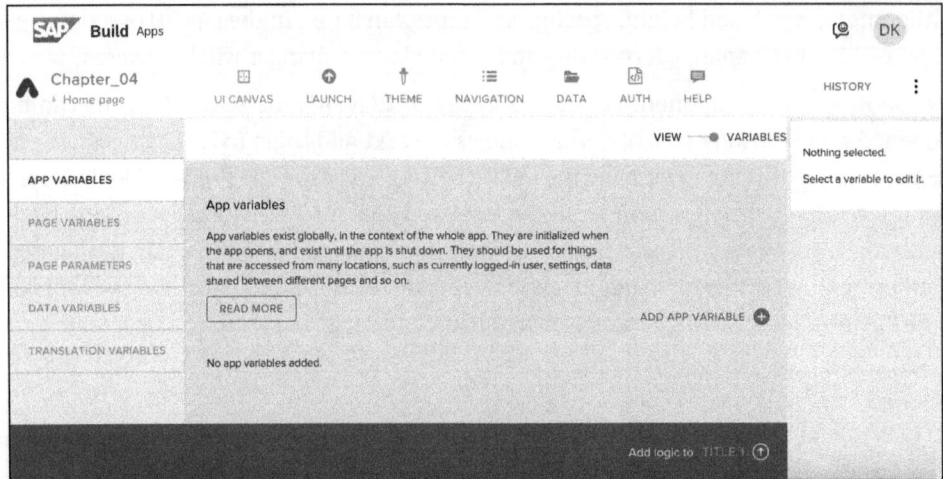

Figure 4.18 Different Types of Variables to Store Data in Runtime

If you aren't satisfied with the UI components delivered by default, you can search for UI components developed and made available by the community in the marketplace. You can open the marketplace on the left side by clicking the **MARKETPLACE** button (Figure 4.16). Figure 4.19 shows the marketplace, where you can install not only UI components but also application logic. We'll access the marketplace in Chapter 5 through Chapter 7.

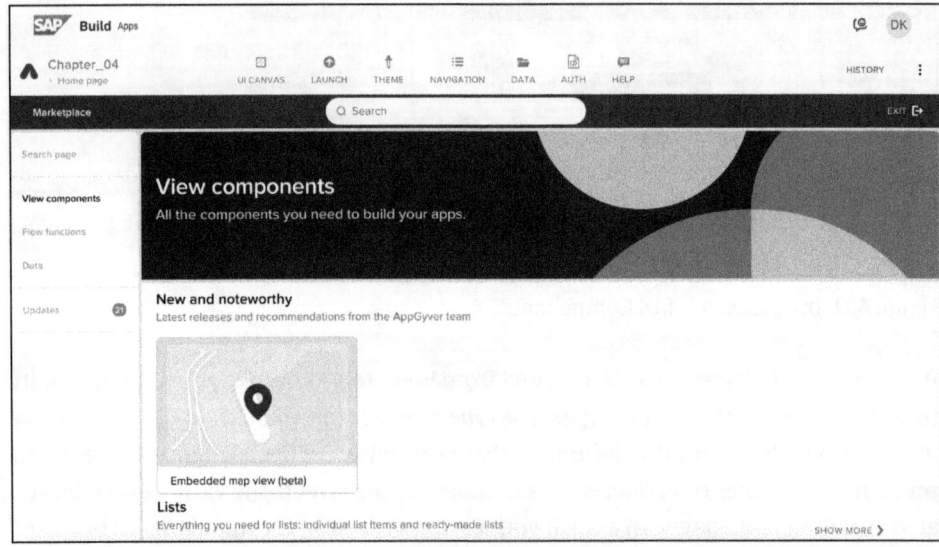

Figure 4.19 Marketplace: Additional Content Developed by SAP and Community

At the top left, clicking the page name (**Home Page** in this case) will take you to the list of all pages located in your application (see Figure 4.20). Here you can also create new pages (**ADD NEW PAGE**). We'll go into more detail about pages, the creation of pages, and the global canvas in Chapter 5.

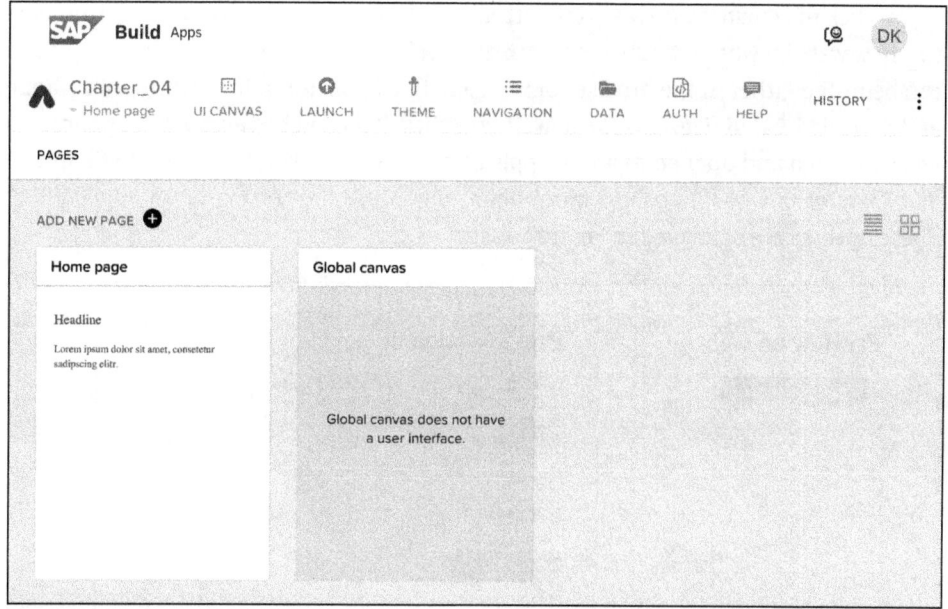

Figure 4.20 Global Canvas and List of All Pages in Application

Leaving the **UI CANVAS** tab and moving to **LAUNCH** will give the possibility to either test the application or to prepare the application for deployment by means of a build process (see Figure 4.21).

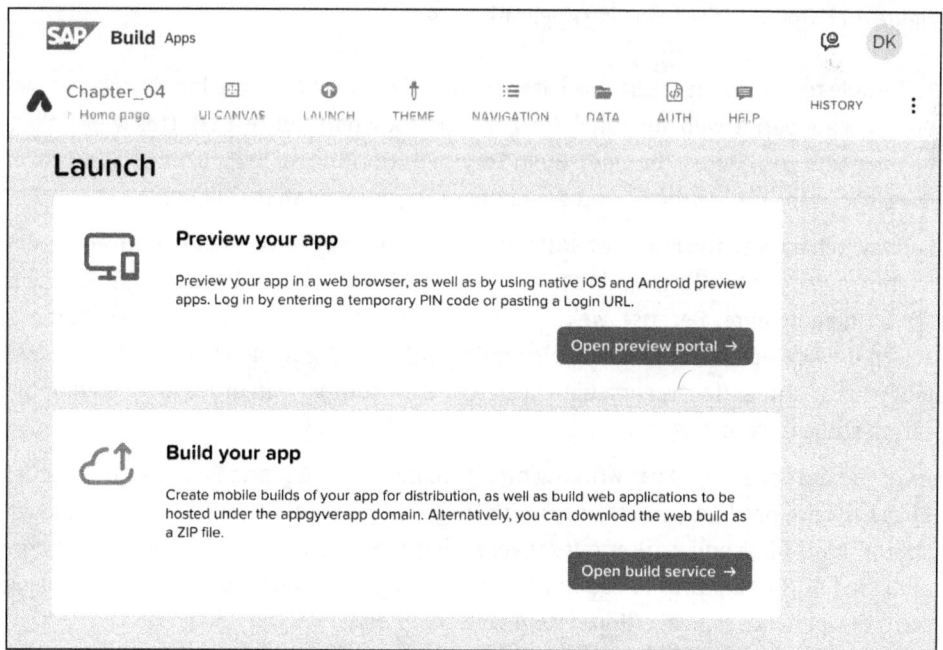

Figure 4.21 LAUNCH Tab for Testing and Building Applications

If you click the **Open Preview Portal** button shown in Figure 4.21, you will be redirected to a new website where the currently opened application can be tested. The application can be tested either in the browser or with a native application for your mobile phone and PC called *SAP Build Apps Preview*. If you click **Open Web Preview**, the application will be started and opened as a web application directly in your browser. (In Chapter 5 and Chapter 6, we will go into more detail regarding the testing of an application.) These options are displayed in Figure 4.22.

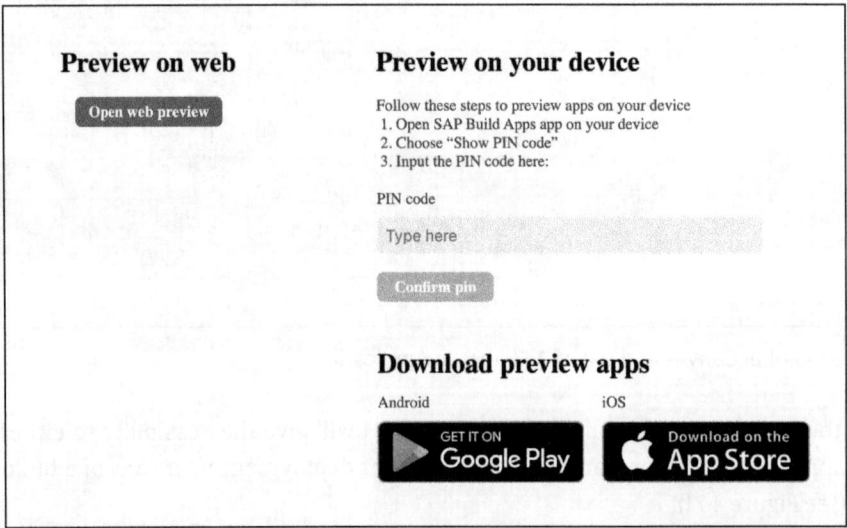

Figure 4.22 Options for Previewing Application

To be able to deploy an application, it must first be built. Again, the build process takes place via a separate website called *Build Service* (shown in Figure 4.23). This website will be opened if you choose the **Open Build Service** option shown in Figure 4.21. We'll cover the topic of the build process and deployment in more detail in Chapter 9.

Let's switch back to the editor of SAP Build Apps. You may have already noticed thanks to the font of the headings and the text elements in Figure 4.13 that a specific theme is set by default here. Because we're in the SAP environment, an SAP Fiori theme is selected by default and the colors also reflect this (see Figure 4.24). Within the **THEME** tab, you can integrate the corporate design of your company in order to let your application shine in your own colors.

As of the start of 2023, SAP is working on replacing the Quartz and Belize legacy themes in all systems, products, and services with the new theme called *Horizon*, which is available in both light and dark modes. Everything you need to know about this theme delivered by SAP is summarized in the official SAP Fiori design guidelines, available at *https://experience.sap.com/fiori-design-web/theming/*.

4.3 SAP Build Apps Composer User Interface

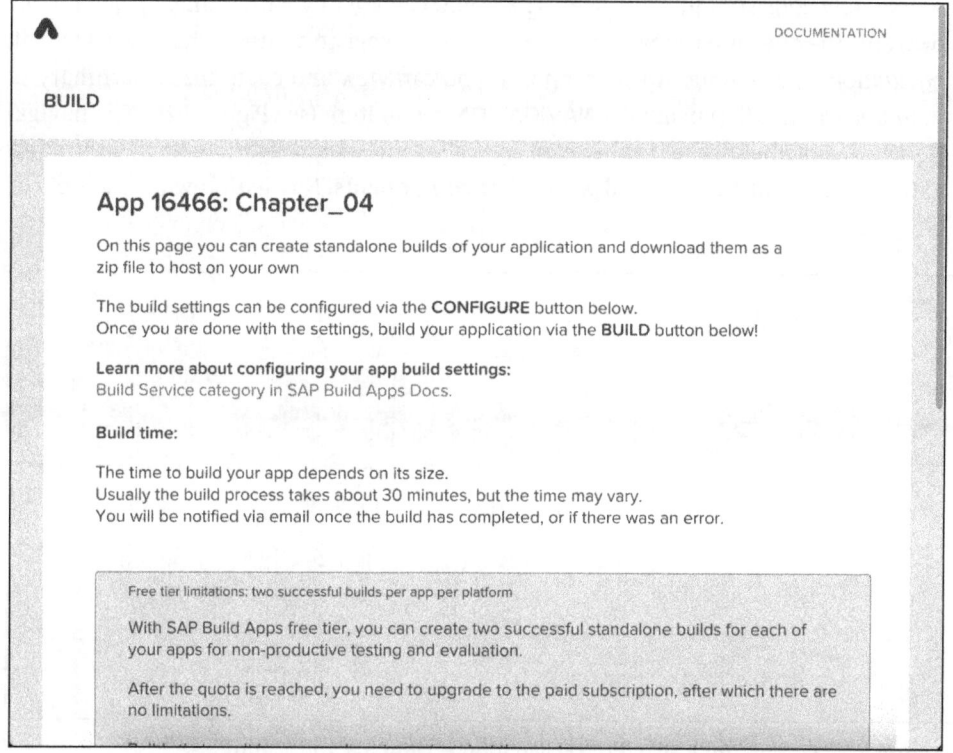

Figure 4.23 Build Service to Prepare Applications for Deployment

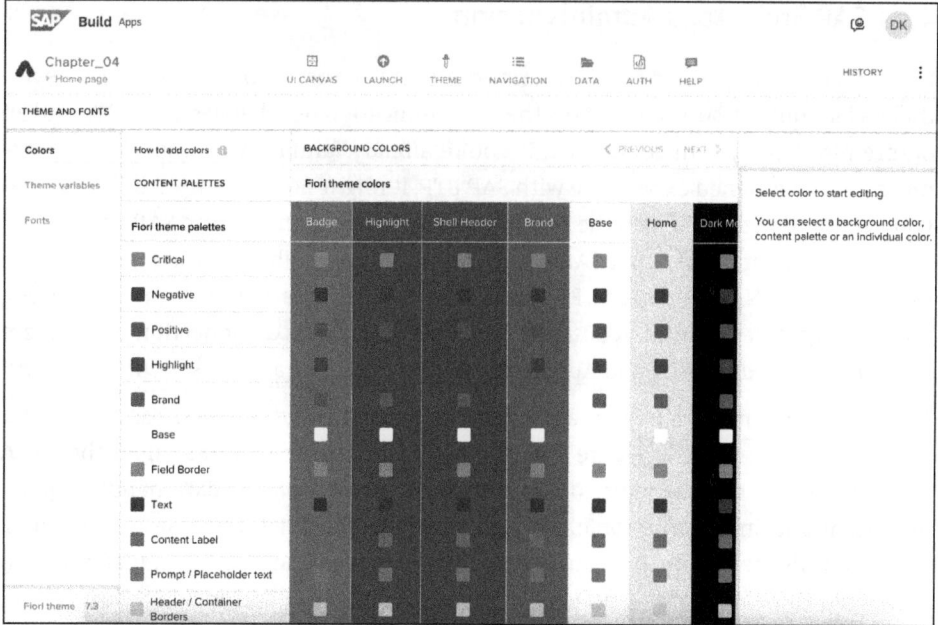

Figure 4.24 Theming: Themes Delivered by SAP or Your Own Corporate Design

4 No-Code Development Environment

Apps rarely consist of just a single page. If you can't get by with a single page in your own app, the question is how to get users from one page to another. This is what we call *navigation*. In SAP Build Apps Composer, you can view and customize a summary of your app's navigation using the **NAVIGATION** menu item (see Figure 4.25). The navigation you implement here is always only so-called internal navigation. Navigation to an external page must be realized with other components. Navigation will be a topic of Chapter 5.

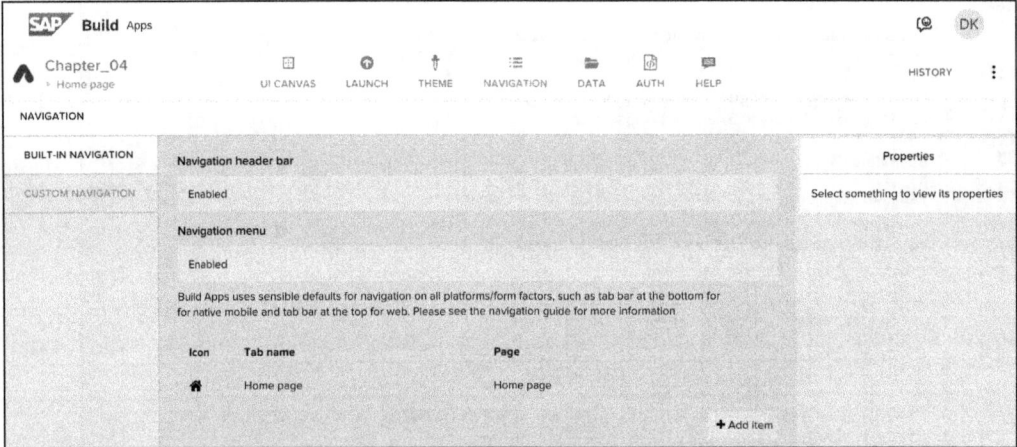

Figure 4.25 Settings for Navigation: Navigation Header Bar and Menu

4.4 SAP Build Apps Administration

In Chapter 2, we explained the most important information about SAP BTP, but nevertheless let's quickly summarize how the communication flow between SAP BTP and an on-premise SAP system could look. If you're already familiar with the SAP environment and have gained experience with SAP BTP, it might not be a surprise for you that you cannot directly establish the connection between an on-premise SAP system and your cloud system unless your SAP system is publicly accessible. This also applies if you have SAP S/4HANA Cloud in use. For this, you need additional components and, among other things, a destination. A *destination* in SAP BTP stores a connection to a system and can be reused in SAP Build Apps to access a system, like an SAP S/4HANA system.

To access and store data in your app, you need to create *data entities*, also called *data sources*. As you can see in Figure 4.26, these data sources are located under the **DATA** menu item and play a major role in SAP Build Apps. Because data handling is an important and interesting topic, we have devoted Chapter 6 to these data sources. There you will learn how to create and use these data entities, how to transfer data to individual UI components, and what options you have for caching data during runtime. We'll also cover there how to gain access to the SAP system and consume and manipulate data.

4.4 SAP Build Apps Administration

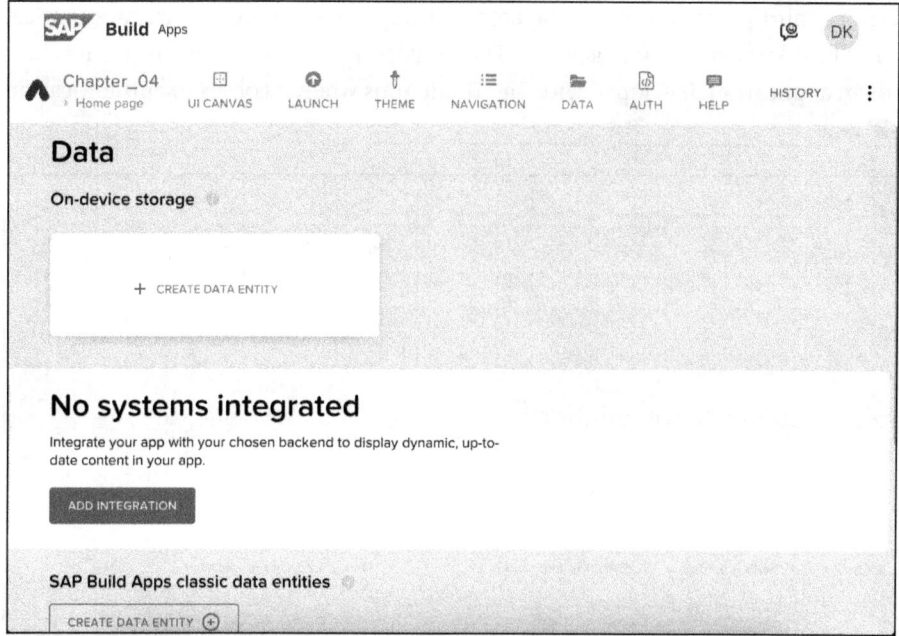

Figure 4.26 Data Handling and System Integration in SAP Build Apps

An application developed with SAP Build Apps is generally accessible without any restrictions after deployment. However, if you want to use authentication methods in the app itself and thus also check access within the web application again, you can activate these methods in the **AUTH** menu tab (see Figure 4.27).

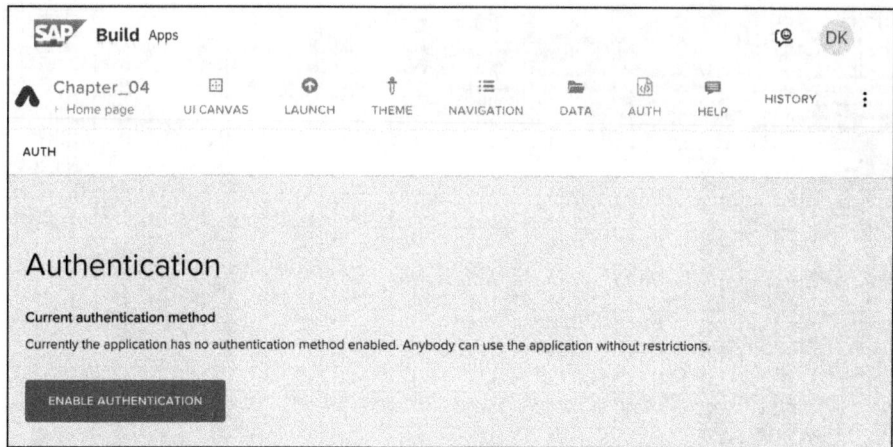

Figure 4.27 Enablement of Authentication Methods

The SAP Build Apps service in SAP BTP also has advantages when it comes to authentication compared to Community Edition. You can outsource authentication to SAP BTP and thus need no major effort to set up a third-party connection (see Figure 4.28). There

4   No-Code Development Environment

are various third-party identity management systems that can take care of authentication. The best-known example is Active Directory from Microsoft, which can also be a part of an application developed with SAP Build Apps when it comes to authentication restrictions.

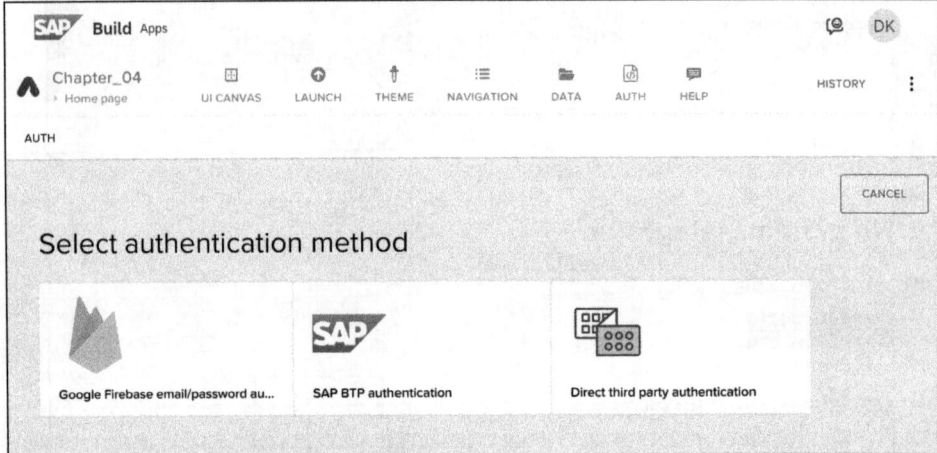

Figure 4.28 Different Authentication Methods Available in SAP Build Apps

After each save sequence, the current state of the application is saved in the history. You can open this history with the **HISTORY** button in the upper-right corner and also click **ROLLBACK** to roll back to an older version of the app if you notice that the changes were not successful (see Figure 4.29).

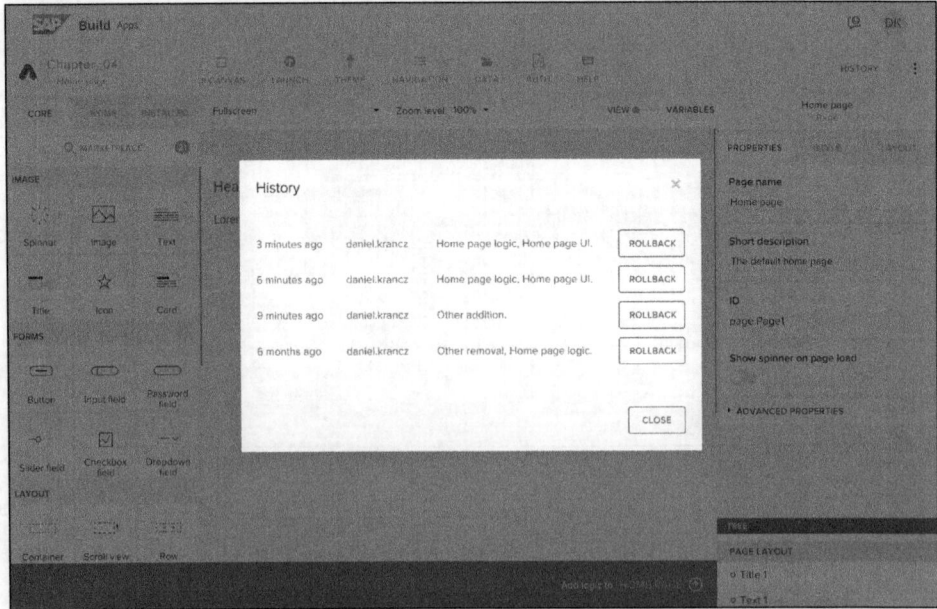

Figure 4.29 History Entries Created on Every Save

Next to the **HISTORY** button, as shown in Figure 4.30, you can open the settings and either export the project (**Export** button) or import an exported project that you have in your file system and replace the current application (**Replace** button).

> **[!] Replace Really Means Replace!**
>
> If you select the **Replace** option, then you will also be informed in a dialog that not only an import is taking place here: the current application is really being replaced. All deviations that you have in your current application compared to the imported application will be discarded and irrevocably deleted.

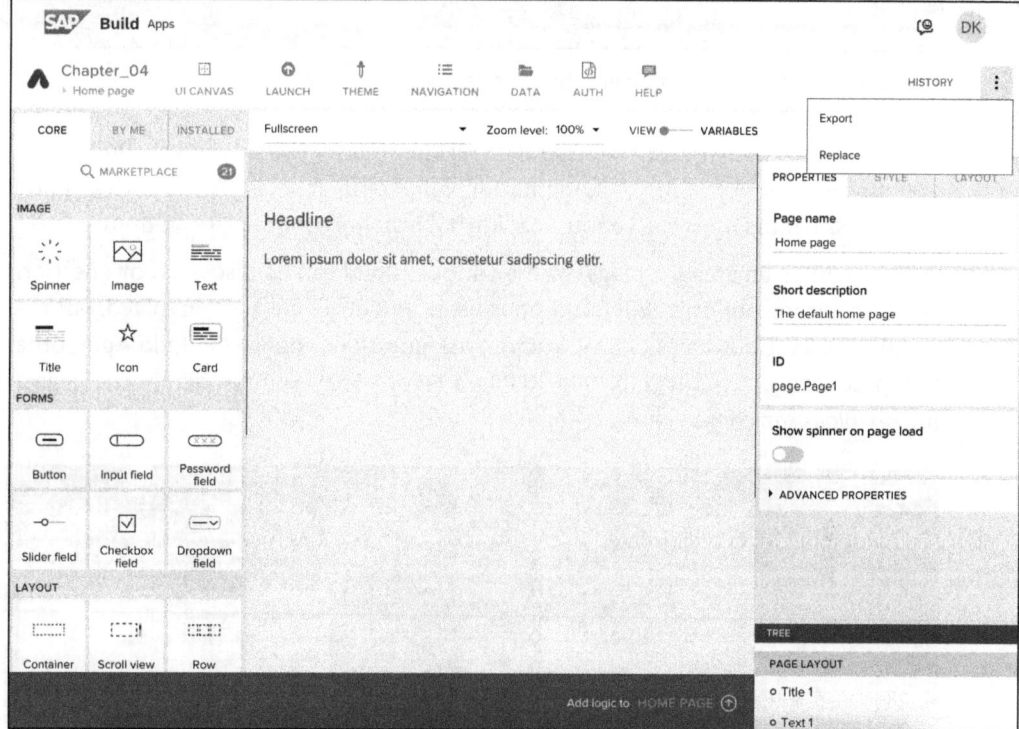

Figure 4.30 Export or Import (Replace) Current Project

4.5 Help and Documentation

Let's now examine both the resources available directly within SAP Build Apps and other websites where you can encounter help, get answers to your questions, and get in touch with the community. Under the **HELP** tab, you can find diverse help pages, status displays, and blogs (see Figure 4.31).

4   No-Code Development Environment

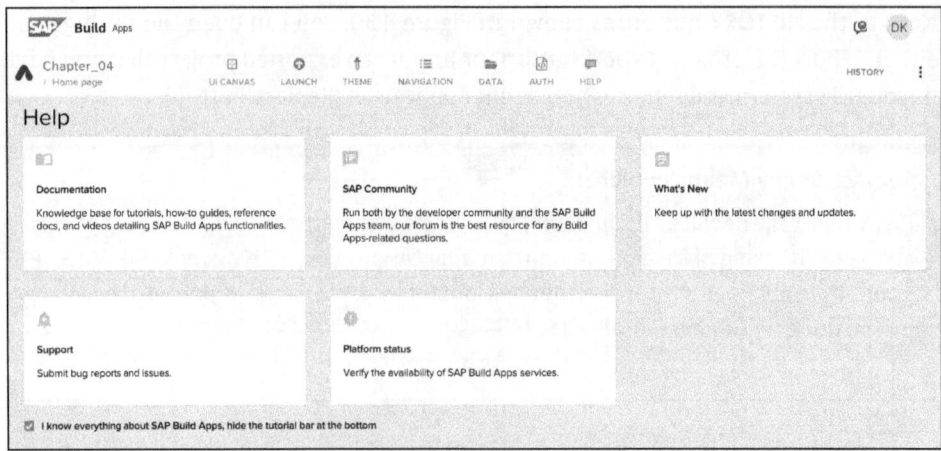

Figure 4.31 HELP Tab for Documentation, Blogs, and More

The official documentation is found on SAP Help Portal where you can find official announcements from SAP, documentation, and status reports. The SAP Help Portal page for SAP Build Apps can be found at *https://help.sap.com/docs/build-apps*.

SAP also offers an area for blogs, where various topics can be discussed or questions answered. Here, not only SAP Build Apps users and developers are involved, but also people from various areas of SAP. You can ask questions about SAP Build Apps, other SAP products, SAP services, or SAP BTP (see Figure 4.32). You can access the page at *https://blogs.sap.com/*.

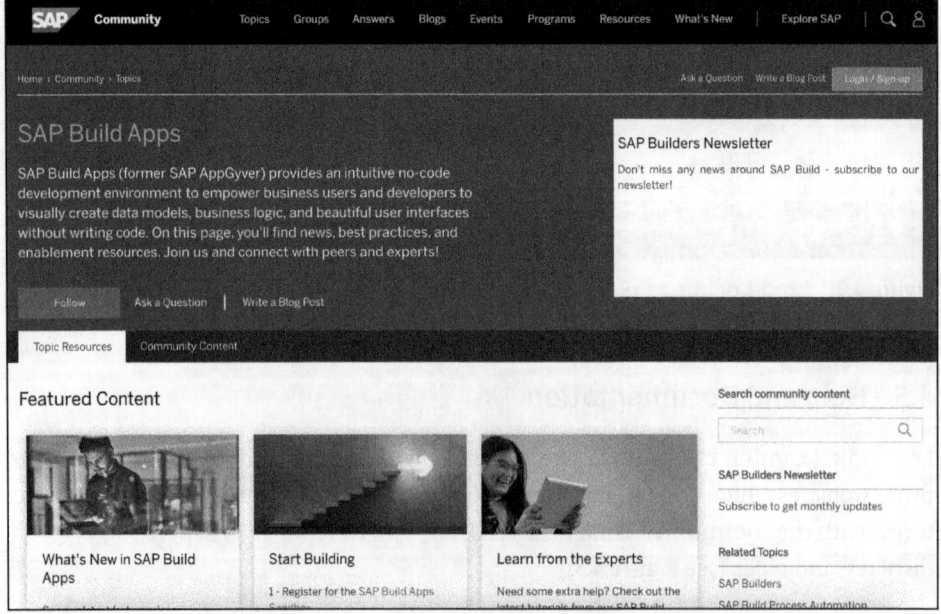

Figure 4.32 SAP Community for Various Topics: SAP Build Apps and More

In the SAP Learning platform, you will find many tutorials with step-by-step instructions and videos that can support you in your development with SAP Build Apps. You can access this page at *https://learning.sap.com/products/sap-build*.

If the development environment or the associated services are not available, you can check the status overview to see if there are currently any known issues with the platform. Here you can also see whether the individual functions and platforms should be accessible or whether there is so-called downtime. You can access the status overview, shown in Figure 4.33, at *https://www.sap.com/about/trust-center/cloud-service-status.html*.

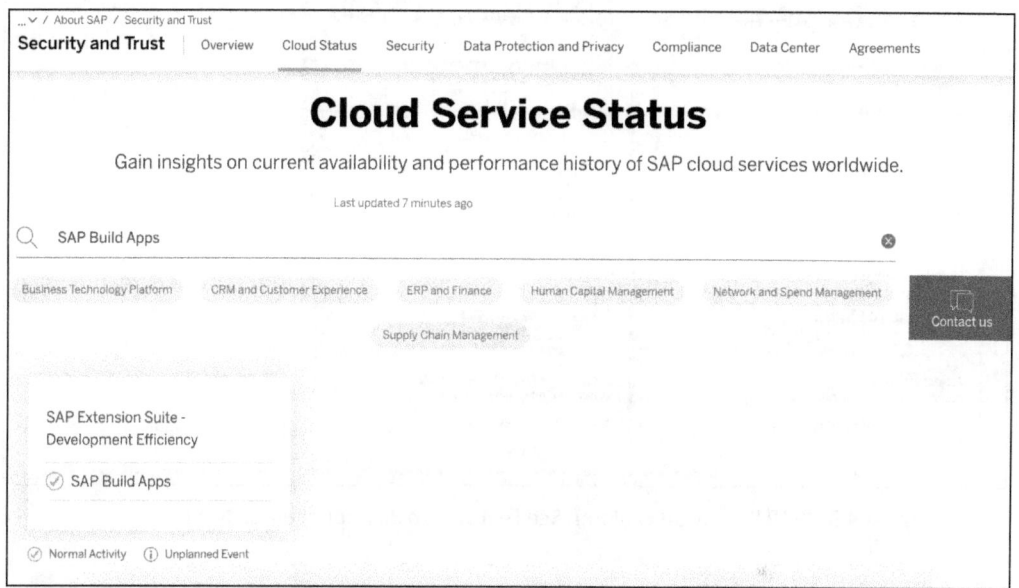

Figure 4.33 Cloud Service Status Overview

In the official SAP Road Map Explorer, you can see the planned features of all services. If you filter for SAP Build Apps, you can get an idea of which new features are planned and should be implemented on a quarterly basis (see Figure 4.34). This page can be accessed by logging in at *http://s-prs.co/v577200*.

If you are interested in the Community Edition version of SAP Build Apps, you can examine not only the past changes, but also the future ones. All planned changes and currently known errors and bugs in the framework can be found in the SAP AppGyver Tracker. On this page, you can also give feedback and come up with your ideas for Community Edition. If you click the **GIVE FEEDBACK** button, you can report bugs you have discovered or submit ideas for future features and change. You will be forwarded to this page by entering the following URL: *https://tracker.appgyver.com/*.

4   No-Code Development Environment

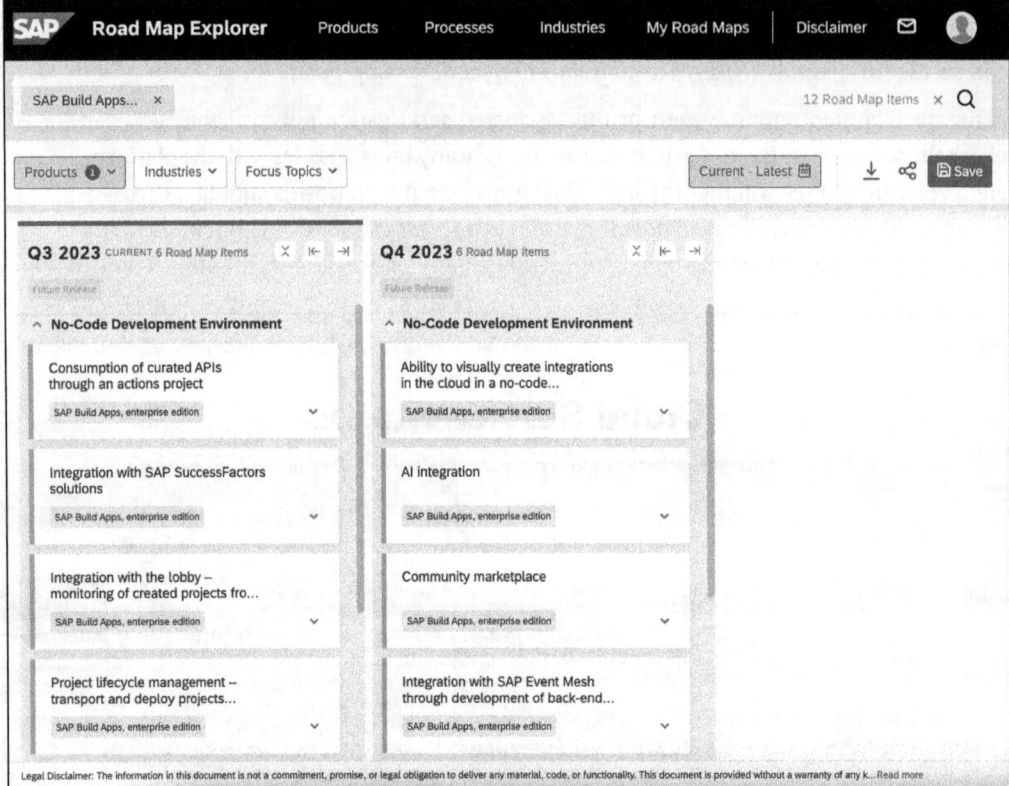

Figure 4.34 SAP Road Map Explorer: See Features to Be Implemented Next

4.6 Summary

In this chapter, we familiarized you with the user interface of SAP Build Apps and talked about the individual tabs and options in the context of development. Collaboration and project management was a small part of the overall repertoire of what is possible with SAP Build Apps. All these elements will be covered again in the coming chapters, and we will go into more depth, especially on the individual tabs and development options.

Chapter 5
Developing Applications

The development of applications can be very complex and sometimes complicated. You not only have to take care of the user interface, but also prepare the application logic, which is the magic behind the scenes. The interplay of user interface, application logic, and later also data and proper data management must be learned well. That sounds complicated, doesn't it? Don't worry: SAP Build Apps provides sophisticated no-code development objects that can usually be added by drag and drop.

App development and knowledge of the tools and techniques behind it play a major role in SAP Build Apps. Although SAP Build Apps delivers a number of UI elements that can of course be used directly via drag and drop, there are nonetheless various setting options, display options, and layout options that need to be applied at the right time.

In this chapter, we'll develop a simple app step by step to introduce the most important building blocks and UI elements. We'll also show you some tips for how to implement certain requirements with regard to user interface development. In addition to building forms and tables, application logic and the so-called logic flow will also play a major role. Section 5.1 deals with the first development steps toward a basic app. Corporate design and the integration of self-developed apps into the existing product range play a major role nowadays. For this reason, Section 5.2 also address the topic of theming. A developed app must of course be tested and, after a successful development test, also deployed and thus made available to end users. Building and testing will be discussed in Section 5.3. An app will rarely function without data and application logic, which is why we'll take a closer look at these details in Section 5.4. This section also discusses how the communication flow between the data and the UI takes place. Finally, Section 5.5 shows how you can work together with others on a project in SAP Build Apps and thus promote collaboration during development.

5.1 Developing a Basic App

Development starts in the SAP Build Apps lobby. Although before development can even begin, someone has to create a corresponding project. This can be either you or your colleague, who then invites you to collaborate. In our example, we'll create a project ourselves.

As you can see in Figure 5.1, we are greeted in the lobby with a table of existing projects. Above this, you will find templates and sample apps that you can try out as you wish. On the left, you can navigate between the lobby and the available extensions, such as connectors and the tenant configuration. The existing projects can be searched, filtered, and sorted. Next to the search box, you can click the **Create** button to create a new project.

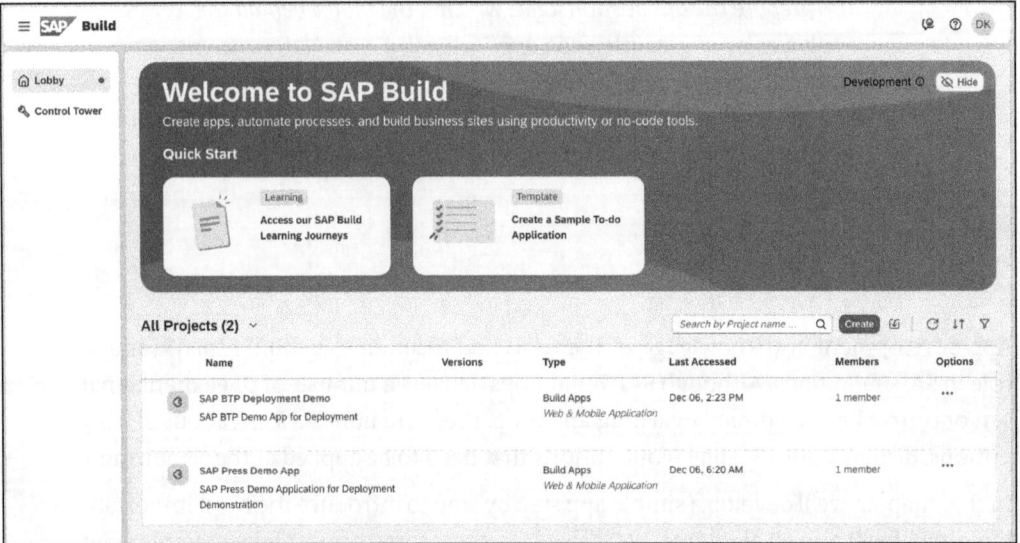

Figure 5.1 SAP Build Lobby as Starting Point for Developments

First, you're asked what type of project you want to create in SAP Build Apps. The following three options are available:

- **Build an Application**
  This involves the development of applications. These in turn are divided into two subcategories, which we'll discuss in a moment.

- **Build an Automated Process**
  You'll get to know SAP Build Process Automation in detail in Part IV of this book. The name is decisive here: the aim is to convert manual processes into automated processes in no time at all.

- **Build a Business Site**
  The third type of application allows you to bring content into SAP Build Work Zone. You can find out more about this in Part III of this book.

In this case, we want to develop a web application that falls into the **Build an Application** category, so select this type. As already mentioned, application development is divided into two parts:

- *Web and mobile applications* are applications that can be run in the browser (either be hosted in house on a web server) or made available to end users via SAP Build Work Zone. Mobile applications are installed directly on mobile devices and must normally also be developed in a native programming language such as Java or Kotlin for Android or Swift for iOS. In this case, SAP Build Apps offers a way to develop an app more generically and thus also develop a mobile app independently of these languages. This can either be installed directly or rolled out in a public app store or via an internal mobile device management system.
- *Application backend* refers to the corresponding no-code development of business logic. We'll show you how to work with these visual cloud functions in Chapter 7. This business logic can then be called up by the web and mobile application.

For now, to start developing a web application, select **Web & Mobile Application**.

Once you have assigned a name for the project in a dedicated dialog, you will be taken to the development environment. A first page has already been created, and you can start developing a web application. Changes to the user interface are made in the **UI CANVAS** tab menu, which is already preselected in the system. We'll gradually introduce the other menus.

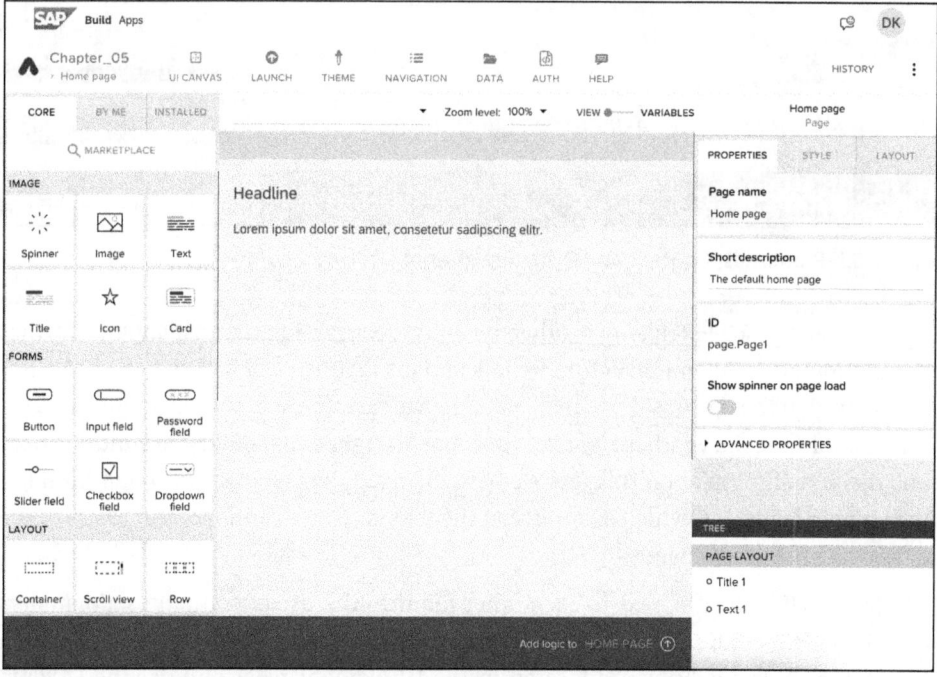

Figure 5.2 UI CANVAS Tab for Editing User Interface

The UI elements that are placed on a page can be selected and edited. As soon as you have selected an element, a blue border appears, which marks the selected element. Click the title UI element with the current content "Headline" and edit the **Content** property on the right-hand side under **PROPERTIES**. *Properties* describe a UI element in more detail and are decisive for what is to be displayed and how in the corresponding UI element. Change the content to read "This is my first Basic App" with a simple keyboard input (see Figure 5.3).

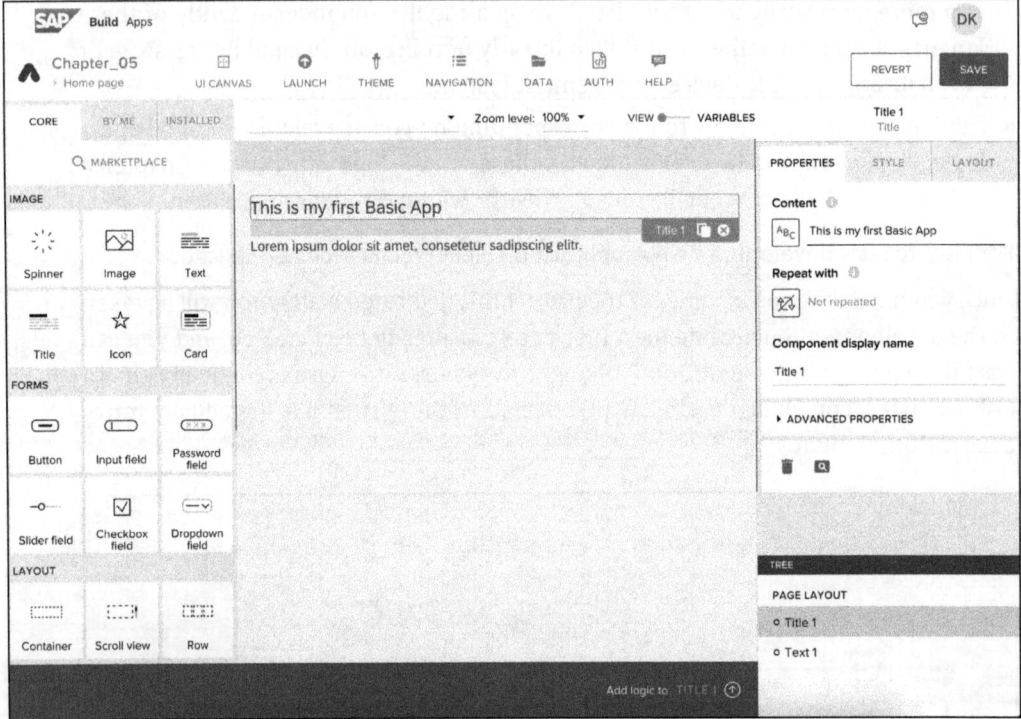

Figure 5.3 Properties: Content and Other Features of UI Element

In addition to **PROPERTIES**, two other tabs are decisive for the display and further description of a UI element. Before we talk about **STYLE**, we will first switch to **LAYOUT** and show you the options that can be used. In the case of the text UI element, you can, for example, set the width, height, and spacing (margin and padding) around the text, and the text alignment on this tab. We have changed the value of **Text Align** from left to center in Figure 5.4, which ensures that the text is always centered, regardless of the device size the user is using.

Displaying UI elements next to each other requires the use of containers. *Containers* are UI elements that in turn contain UI elements and can be displayed and formatted together as a group. They are located in the UI element palette under the **LAYOUT** group. One such container is a **Row**, which allows you to define several cells next to each other as a container. You will need this UI element, as the name suggests, if you want to display rows in a table. There is no UI element for a table per se; you have to get creative and create your own table with the **Row** UI element.

5.1 Developing a Basic App

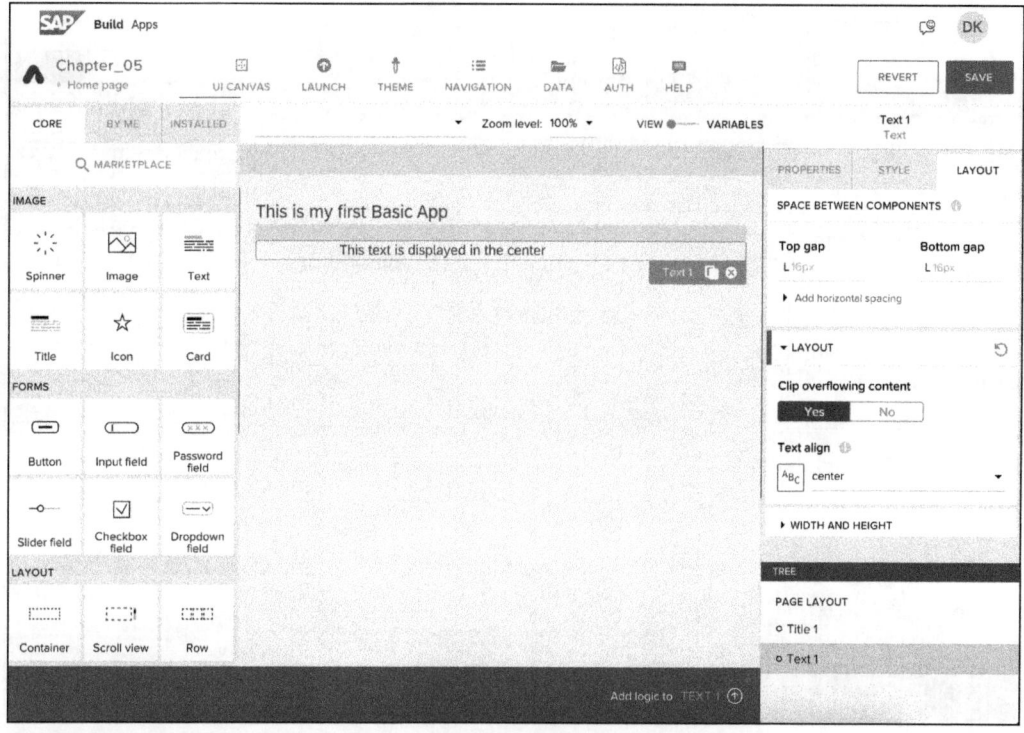

Figure 5.4 Layout Settings for Alignment, Width, and Height of UI Elements

As you can see in Figure 5.5, we have dragged a row from the left from the UI elements palette and dropped it into the workspace. Compare the changes to the **PROPERTIES**, **STYLE**, and **LAYOUT** options to a title or text we already used. Under **LAYOUT**, for example, the **ROW CELLS** option can be used to set how many cells you want to display in the currently selected row. You can also change the horizontal alignment in the cells and the individual cell widths. In this case, we have currently set two cells, each with 50% of the width.

> **Manually Assigned Width of UI Elements**
> We recommend that you work with relative widths instead of absolute widths. The advantage of relative specifications is that even if the user is using different device sizes, the ratios of distances and widths of and between the UI elements remain constant. For example, 50% width will always be relative to 50% of the corresponding screen. In contrast, 500 px could appear small on a desktop device and take up the entire width on a mobile device.

Additional UI elements can now be dragged and dropped into the corresponding cells. A *cell* is a generic container that can contain all possible UI elements. During the drag-and-drop process, you can see from the border where the UI element will end up and whether you have just hit the cell or whether the UI element will be placed outside the cell on the page (see Figure 5.6).

139

5   Developing Applications

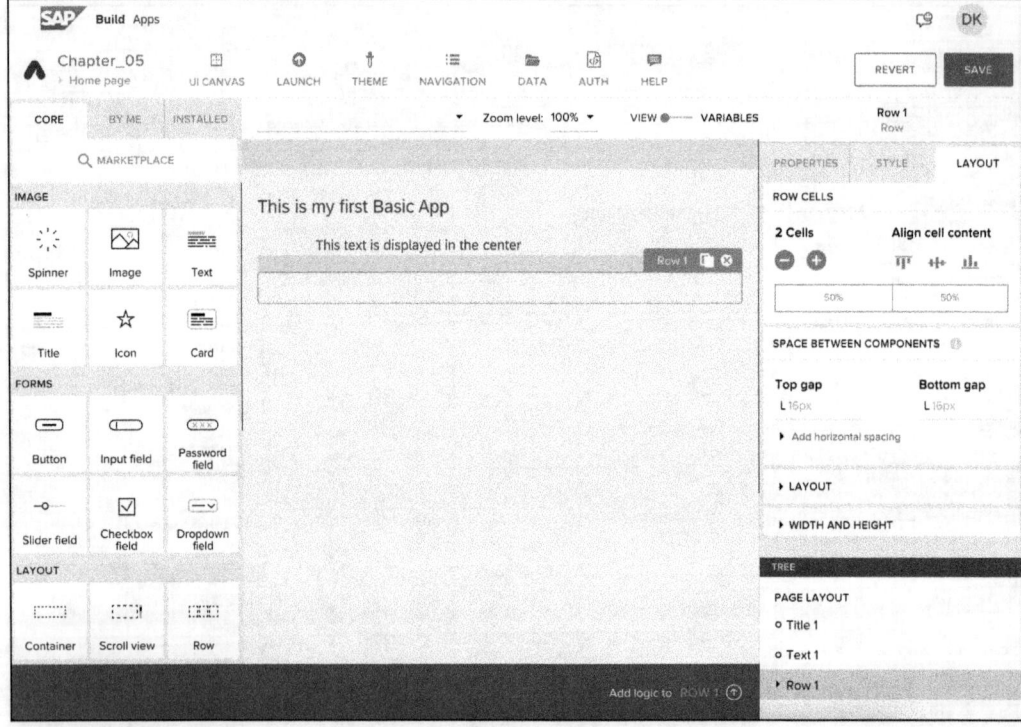

Figure 5.5 Rows: Display Multiple UI Elements Next to Each Other

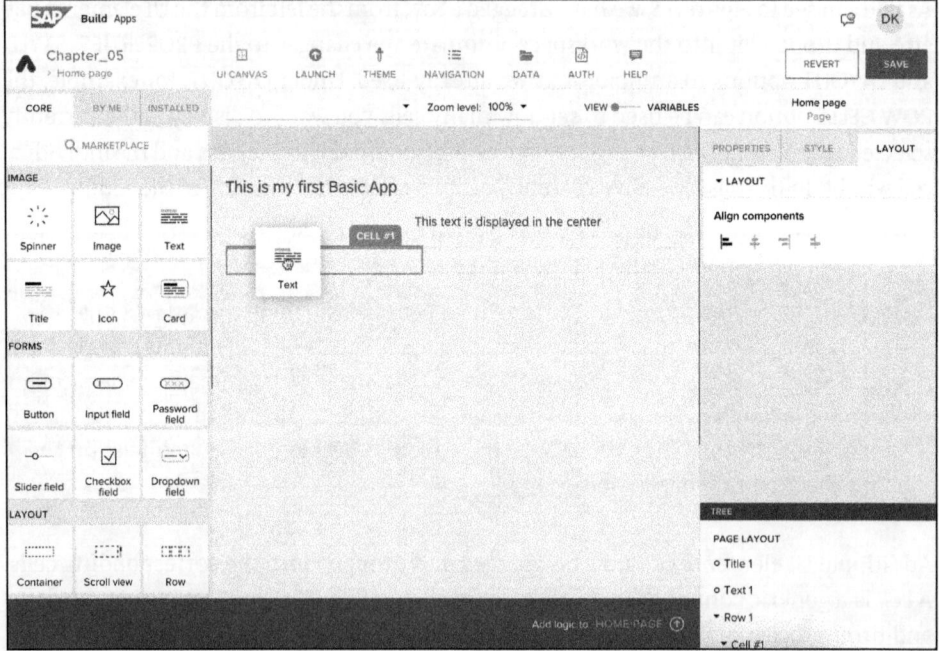

Figure 5.6 Adding Content to Cells

Displaying text is a good start, but you will often encounter the requirement to enable data entry. Again, there are various options for input, but let's start with the simplest and most obvious, an input field (added with the **Input Field** element from the **Forms** area).

An input field offers other options under **PROPERTIES**. **Value**, for example, determines which value should be in the input field. In other words, the value entered by a user input is also saved in this property. **Label** can be used to determine the heading of the input field, whereby the value of the **Placeholder Text** property is displayed in the input field and is intended to provide the user with an input aid or reminder. **Disabled** can be used to determine whether the input field can be filled by the user or whether it is only available in read-only mode.

Other, less frequently used property settings determine, for example, whether the input should always be capital letters (**Auto-capitalize**) or whether the input should be restricted to certain characters such as numbers, letters, or a telephone number (**Keyboard Type**).

You can see in Figure 5.7 that we have changed both the label and the placeholder text. We have not assigned anything to the **Value** property, so this input field will be empty by default.

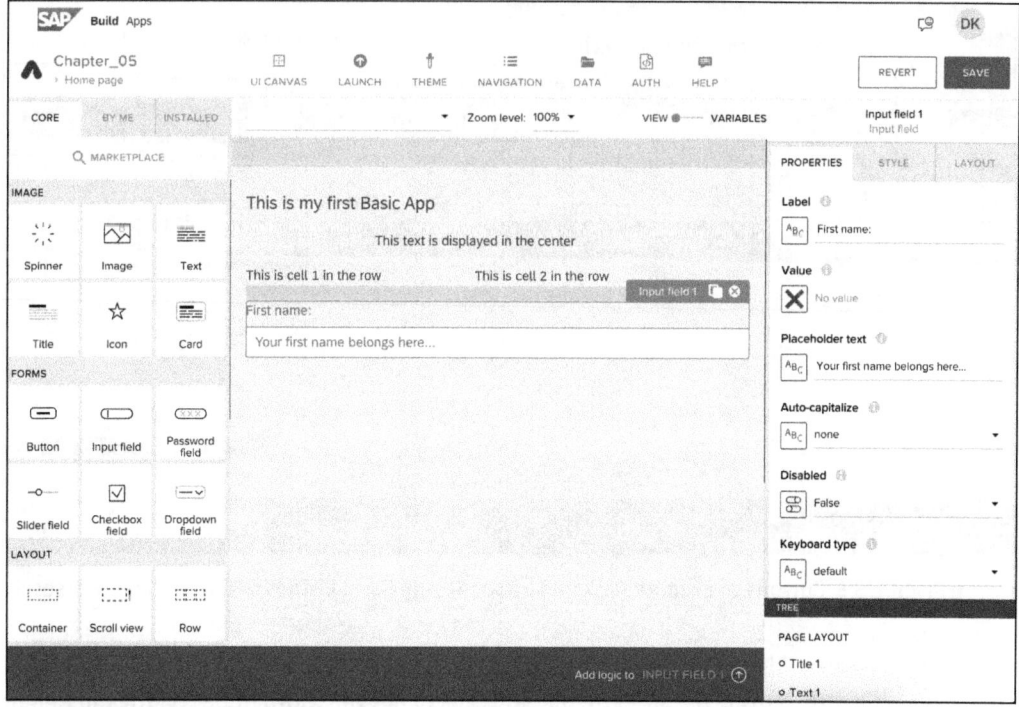

Figure 5.7 Add Input Field and Change Its Properties

Next, let's add a **Button** to the page (see Figure 5.8). Buttons allow users to perform actions and have a trigger to initiate the corresponding application logic behind them. You will find out what options you have for the application logic in the course of this chapter, and we will also cover it in more detail in Chapter 6.

A button can and should have an appropriate style (set in the **STYLE** menu tab) depending on the corresponding action. Section 5.2 will cover color options in more detail, but for now, let's use a practical example to explain how the color scheme should be set. Semantic colors are intended to give users a sense of the impact an action could have.

For example, the color green could stand for save, red is usually for cancel or delete, and more neutral colors such as blue are for information buttons or for editing. Some of these semantic colors are already provided and can also be selected under **STYLE**. But you will see in the next section how you can change these and possibly also adapt them to your corporate colors.

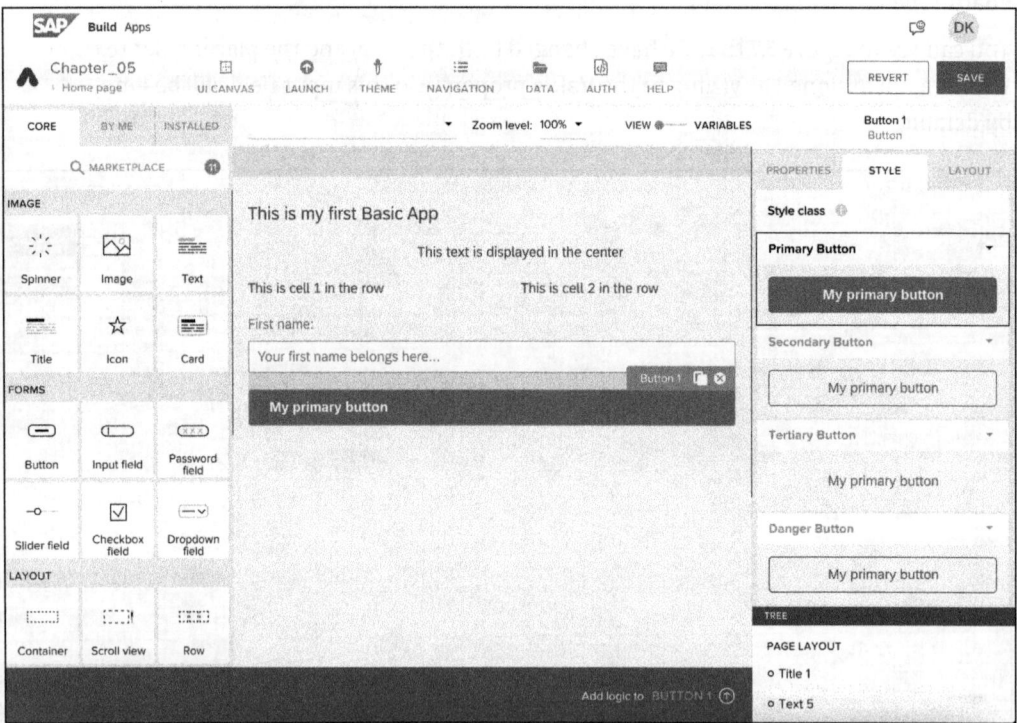

Figure 5.8 Buttons to Execute Actions Such as Saving Form Entries or Canceling Processes

As you can see in Figure 5.9, buttons take up the full screen width by default. Under **LAYOUT**, as with many other UI elements, you can change the width under **Width and Height** to **Fit Content**. This sets the width exactly to the length of the respective content. You can of course also assign a fixed size, but remember our earlier tip about relative sizes.

5.1 Developing a Basic App

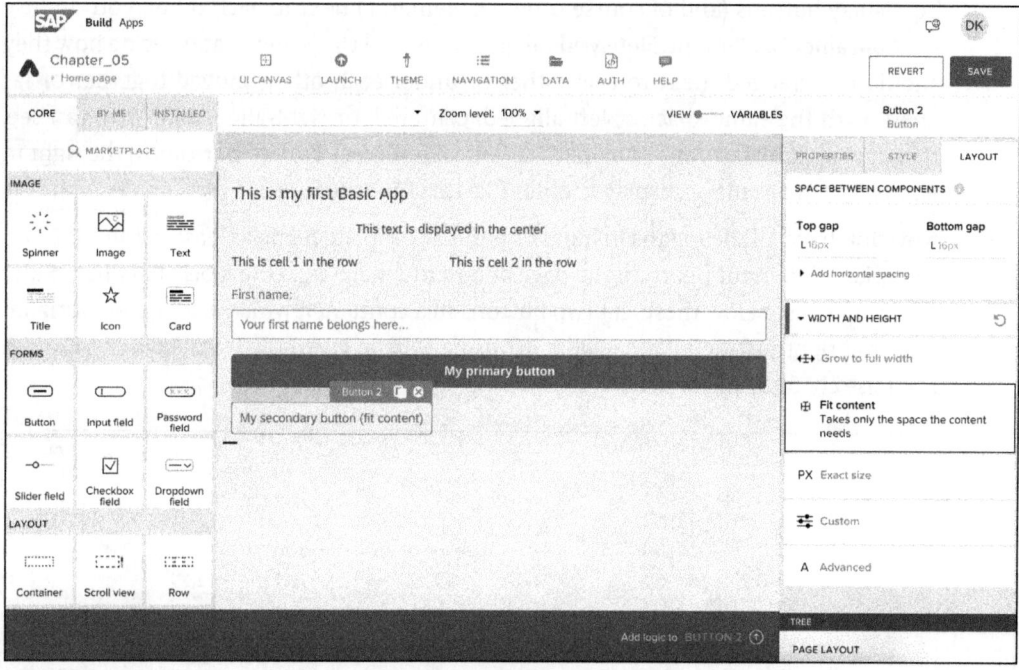

Figure 5.9 Different Layout Options to Change Appearance of Buttons

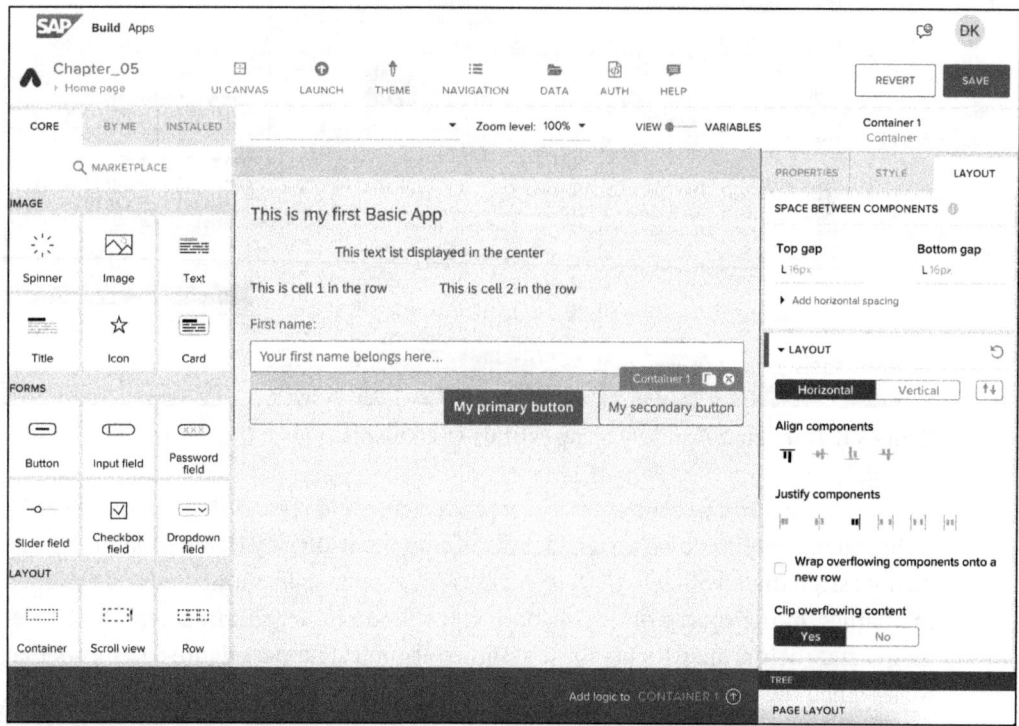

Figure 5.10 Containers to Group UI Elements and Define Styles and Alignments of Container Content

5 Developing Applications

To display buttons (and of course other UI elements) next to each other, you can use the **Container** UI element. Here you can group several UI elements and decide how they should be arranged: next to each other or under each other, moved together or far apart, with the same spacing, left-aligned, centered, or right-aligned. As you can see, there are no limits to your imagination. We have aligned the two buttons to the right in Figure 5.10 to be able to display a footer like this in the future.

Now you have all the means in hand to build a complete form, at least in the UI. To do this, place two input fields on the page and name them with the labels **First name** and **Last name**. Just below these, set two buttons in a container, which are aligned horizontally and right-justified. These two buttons have been given the texts **Save** and **Cancel**. All these changes are shown in Figure 5.11.

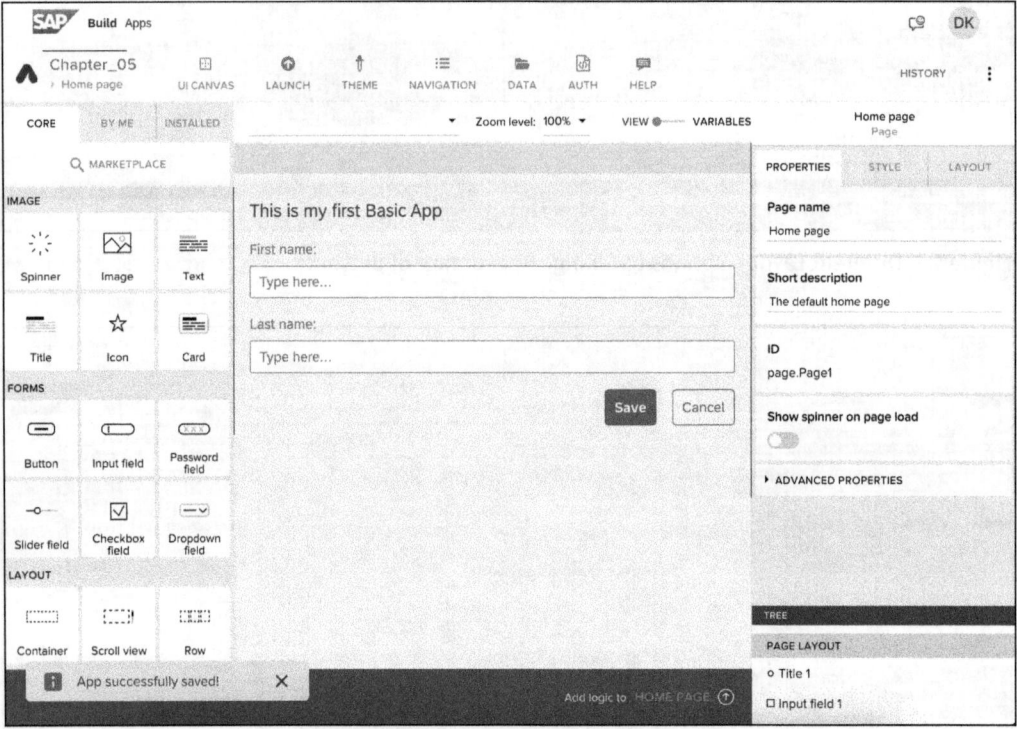

Figure 5.11 Form Built Manually Using Existing UI Elements

So that you can find UI elements in a technical sense and recognize them more easily, technical names should be assigned to the **Component Display Name** property. If you do not assign these yourself, all UI elements are automatically numbered and assigned the names. In the course of this section, you will see how you can access the UI elements and their properties via these assigned technical names. Figure 5.12 shows that we assigned the name "input_first_name" to the first input field.

5.2 Theming

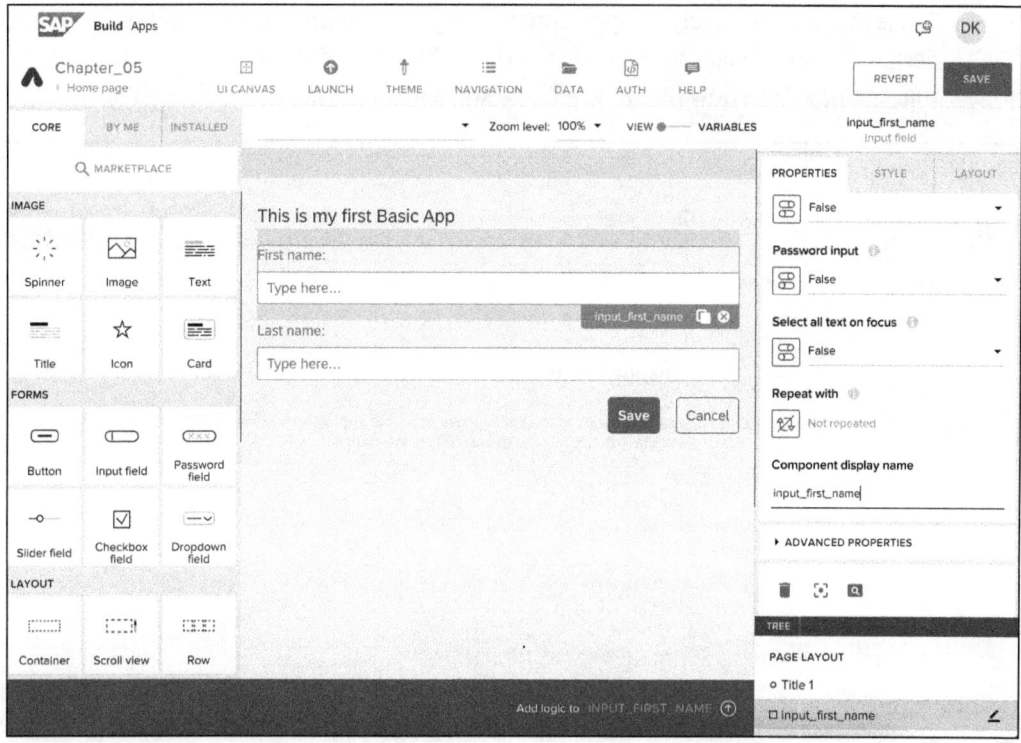

Figure 5.12 Stable IDs Help to Access UI Elements

5.2 Theming

Before we continue on to testing and application logic, let's look at how to adapt an app to your own corporate design. To do this, you first need to switch to the **THEME** tab, as shown in Figure 5.13.

As you can see in this tab, there are basically three themes that can be selected. Two of these themes correspond to SAP's own theme, with **Morning Horizon** being the latest theme delivered in the SAP Fiori design system. **SAP Quartz Light** was the standard theme for a long time and the predecessor of Horizon.

> **Horizon Theme**
>
> The Horizon theme was introduced as one of the last themes in the SAP Fiori design system and is now the standard theme for all SAP apps, both on-premise and in the cloud. Customer developments should also adhere to this design in the SAP environment so that the various applications are still coherent. You can learn more about this theme on the official website of the SAP Fiori design guidelines:
> *https://experience.sap.com/fiori-design-web/look-feel-and-wording/*

145

5 Developing Applications

These themes are merely templates and are equipped with a standard set of colors and design elements. You can select one of these themes and still make any color and textual changes. Let's now take a closer look at these options.

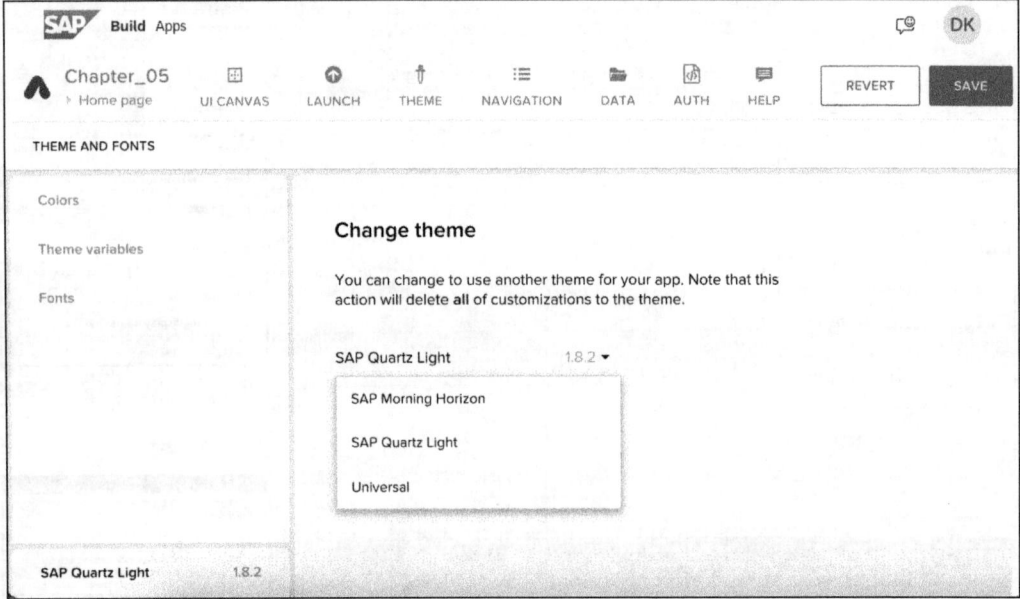

Figure 5.13 Select SAP Theme or Create Your Own

On the left-hand side, under **Colors**, you can assign the individual colors for the respective semantic colors. There are color groups such as **Primary**, **Secondary**, **Positive**, **Negative**, and many more, which should generally apply to a uniform color in the program. This is to ensure that if several UI elements use the color set for primary, they all change automatically when a change is made. Nevertheless, you could also assign a separate color for each UI element instead of using these general color groups. Such an adjustment is also shown in Figure 5.14, where the basic colors can be defined on the one hand, but the colors can also be adjusted again for each UI element and theme.

> **[+] Colors with Theme Variables**
>
> Instead of changing the colors for the UI elements individually, we recommend that you create global theme variables that save the corresponding colors. Access via the variable ensures that all UI elements in the app change when a change is made.

Technically speaking, we use Cascading Style Sheets (CSS) for changes and classes, which are the technical descriptions of the layout. As colors have their own codes in web development, we recommend that you use a CSS color picker, which is already integrated into SAP Build Apps. With such a tool, you can easily find the corresponding hex code for a color and then enter it.

5.2 Theming

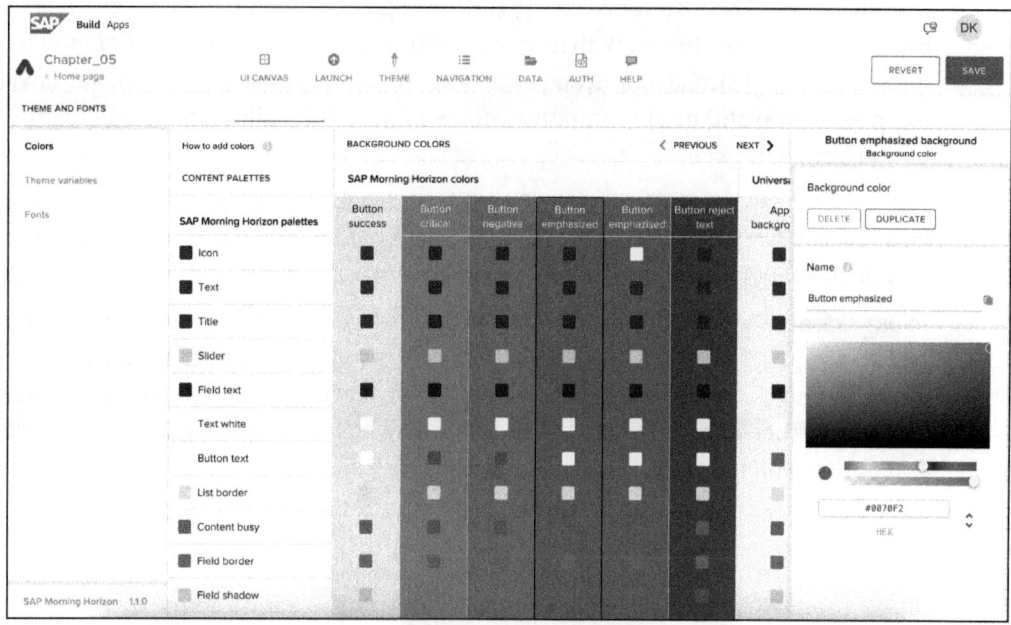

Figure 5.14 Change Colors or Fonts to Integrate Corporate Design

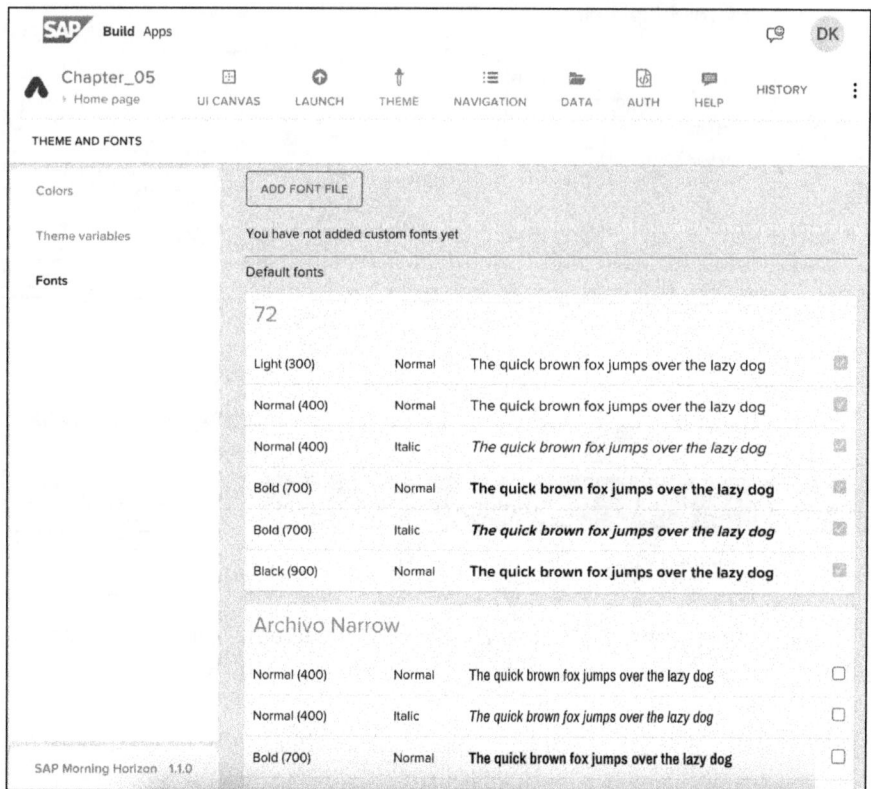

Figure 5.15 SAP Default Fonts

147

5   Developing Applications

Not only colors but also fonts can increase recognition potential. Many companies have their own fonts in use. With its newest theme, SAP has again put a lot of thought into the font, font size, shadows, and color ratios to ensure accessibility in the web environment. If you still need to import your own fonts, SAP Build Apps also offers this option (see Figure 5.15).

However, we recommend that you at least take a look behind the scenes and realize how much effort SAP has put into the right choice of font: *https://experience.sap.com/fiori-design-web/typography-horizon/*.

In our example, we have adapted the primary color in our app and used the main color of our company. Now all UI elements, such as buttons, are also set in this color by changing and globally affecting the semantic color. If you still want to change the color or make further adjustments to the style of a UI element, simply right-click the style you want to edit in the **STYLE** menu tab (see Figure 5.16).

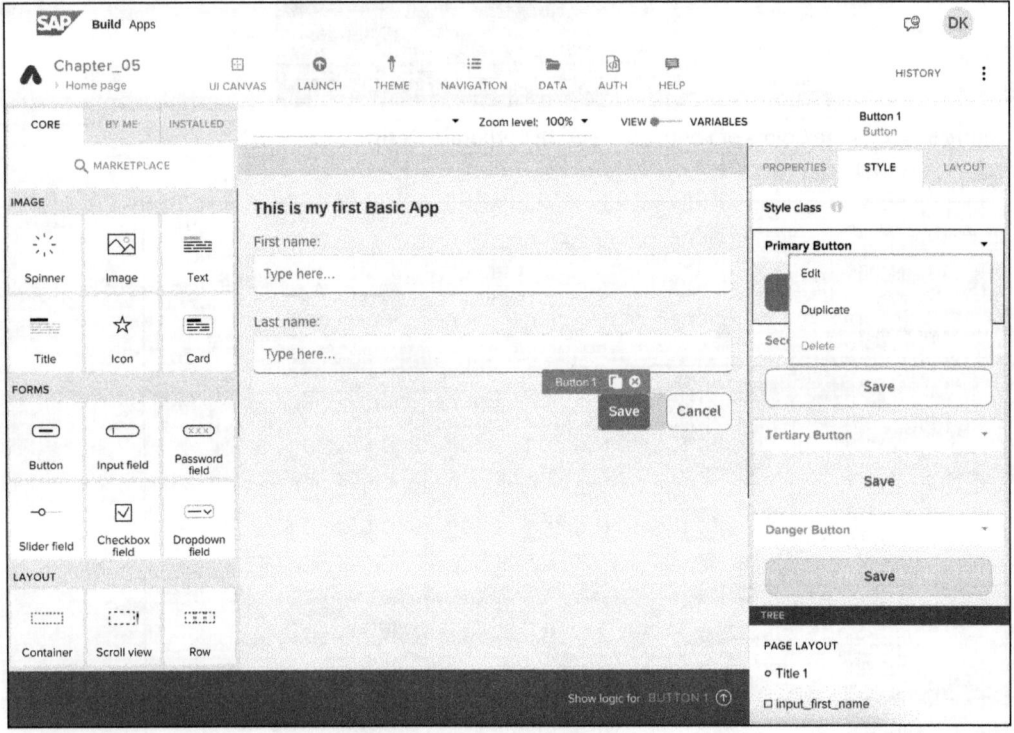

Figure 5.16 Changing Predefined Styles for Each UI Element

Edit takes you to the editing mode shown in Figure 5.17, where you can change the background color, color in general, font, border, special effects, width, height, padding, and margin of a style.

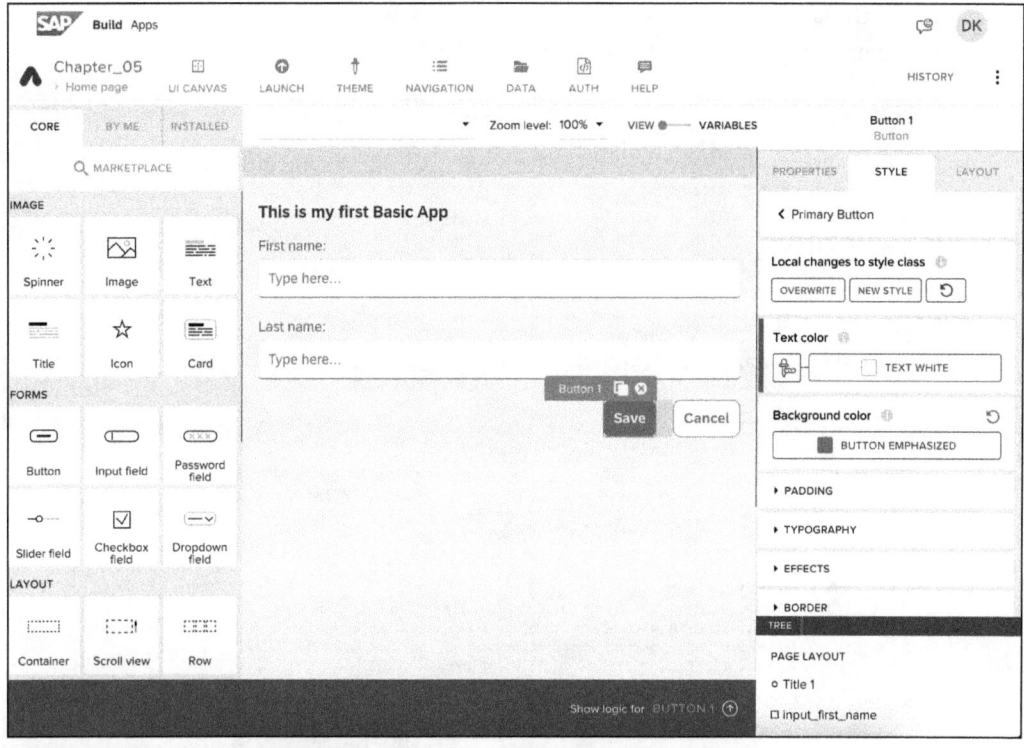

Figure 5.17 Local Style Class Changes to Change Colors, Font, Size, and More

Incidentally, if you have made unwanted changes or simply want to switch back to the standard theme, you can reset the changes you have made under **THEME**.

5.3 Building and Testing

What use is the development environment if we can't run the application and test it from the user's perspective? We'll show you how to do so in this section.

In the **LAUNCH** tab, you have the option to test the application on the one hand, but also to proceed to the build and then to the corresponding deployment. You'll see that you can test the application in various ways and start a preview. However, bear in mind that this is only ever a preview. The application cannot be executed by end users, nor is there any guarantee that the environment in which the application is later executed in real time will behave in the same way as in the preview. With that in mind, click the **Open Preview Portal** button shown in Figure 5.18 to begin your testing.

Figure 5.19 shows that you now have two important options for testing the app. The first option is **Open Web Preview**, which allows you to test the application directly in your current browser. However, if you've downloaded the native application from the

5   Developing Applications

respective app store, you can also connect it here with the code shown, which appears in the native app, and test the app on a mobile device.

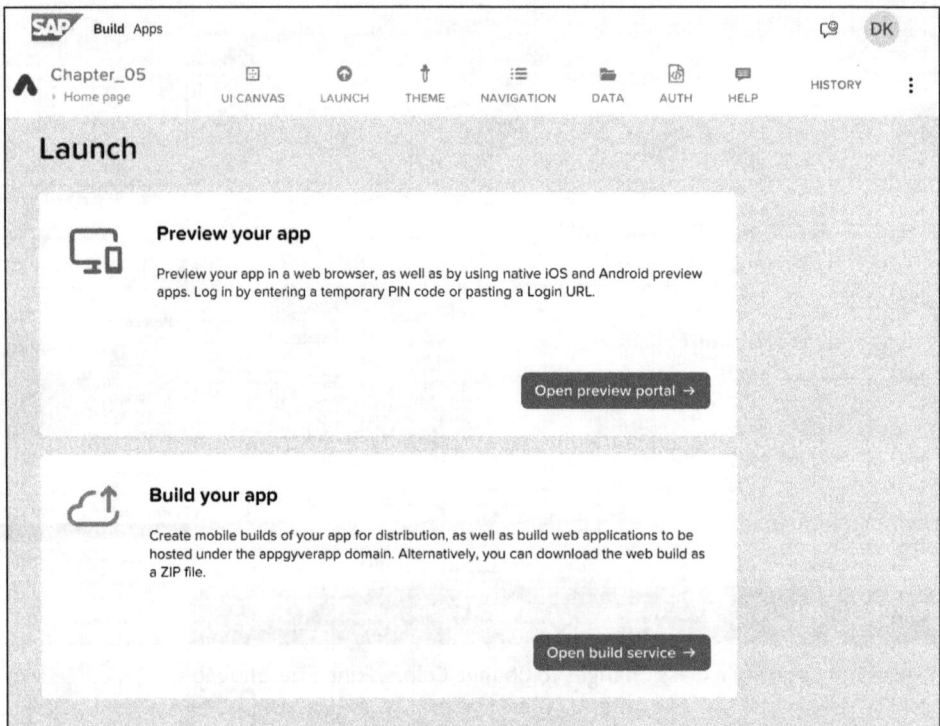

Figure 5.18 Open Preview Portal to Check App Usability

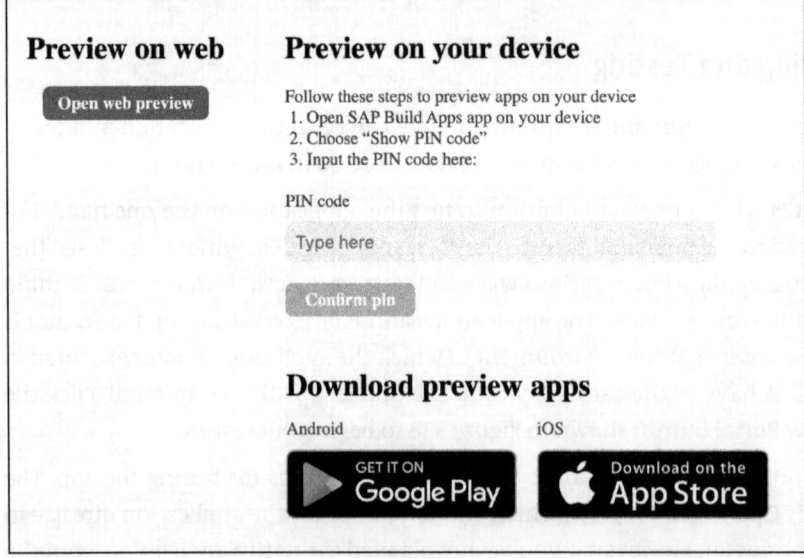

Figure 5.19 Different Preview Options for Displaying Apps in Browsers or on Mobile Devices

For now, select the variant for testing in the browser (**Open Web Preview**). You'll see all the SAP Build Apps applications available that you have built (see Figure 5.20). If you click **OPEN**, you'll start the preview for the respective project.

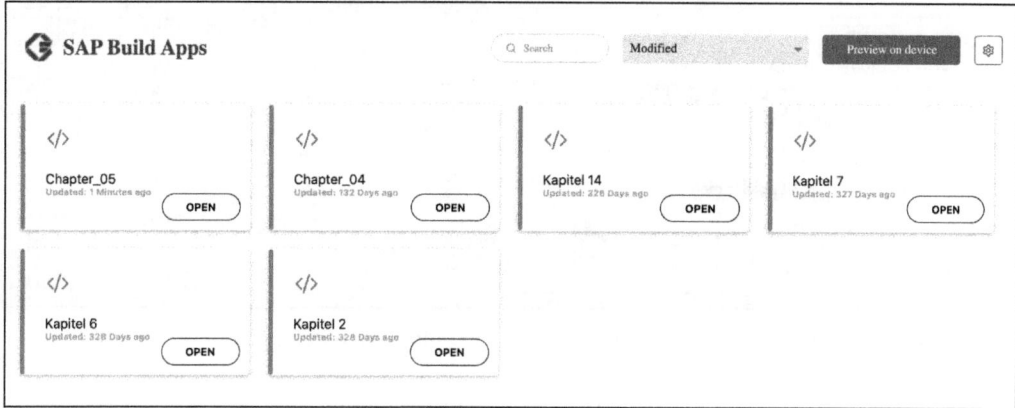

Figure 5.20 All Projects You Can Access Can Be Previewed

OPEN opens a new tab in which the application is opened in preview mode and can be tested, as shown in Figure 5.21. We'll do extensive testing in Section 5.4 and in Chapter 6.

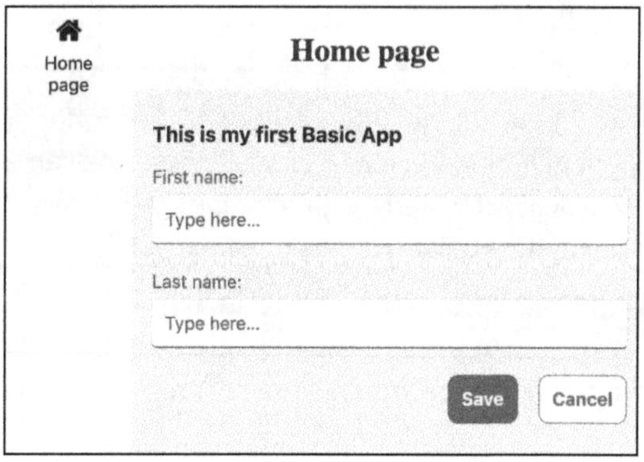

Figure 5.21 Preview to See How Applications Will Look to Users

You'll learn more about navigation in Chapter 6, but for now we want to show you how to implement a single-page application. *Single page* means that no other pages are used, so no navigation and therefore no navigation bar, menu, or anything else is required.

You can do this under the **NAVIGATION** tab by deactivating both the **Navigation Header Bar** and the **Navigation Menu** settings by selecting **No** under **Enabled?**. As shown in Figure 5.22, you can also see that by deactivating the navigation, a page must be set as the initial page. The initial page is displayed by default when the app is opened.

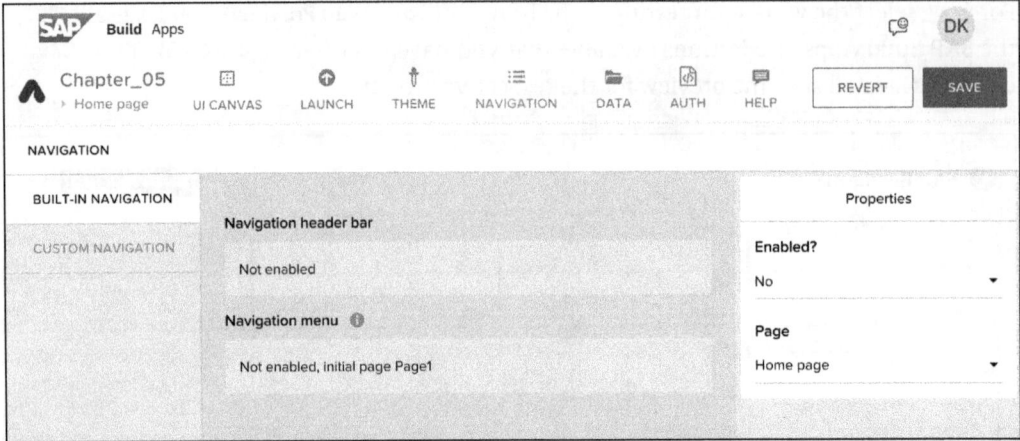

Figure 5.22 Disable Navigation Options for Single-Page Applications

To ensure that an application that you build in SAP Build Apps can also be accessed by end users, you need to make the application available either on a server or hosted somewhere in the cloud—for example, on SAP BTP. However, the application is not copied over 1:1, but must first be put into an executable state. This process is called the *build process* and ensures that the development files are converted into executable files and into a so-called minified version.

> **Deployment Process**
>
> In this section, we show you a short excerpt of how to build an executable version of an application. In Chapter 9, however, we will deal extensively with the topics of building and deployment. *Deployment* means making the executable version of the application available to end users.

If you click **Open Build Service**, as shown in Figure 5.18, you'll see a separate screen, as shown in Figure 5.23.

If you now scroll down this page, you'll see the options to bring an application into an executable form. Because different target systems require different types of such executable programs, there are also different build processes that can be carried out. So, for example, you can decide whether you want to run the application as a web application (hosted somewhere on a web server or on SAP BTP) or as a native application (on iOS or Android). Depending on your requirements, click **BUILD** to start the process, as shown in Figure 5.24. In **CONFIGURE**, further settings can be made for the corresponding target system, such as the file format of the build output, versioning, or naming.

5.3 Building and Testing

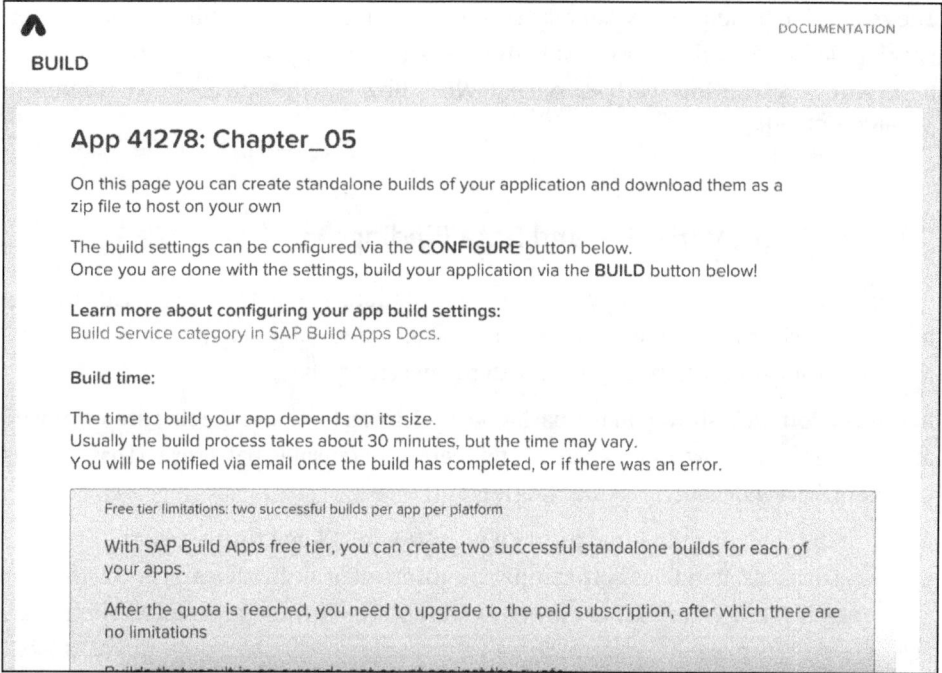

Figure 5.23 Build Information and History of Selected Application

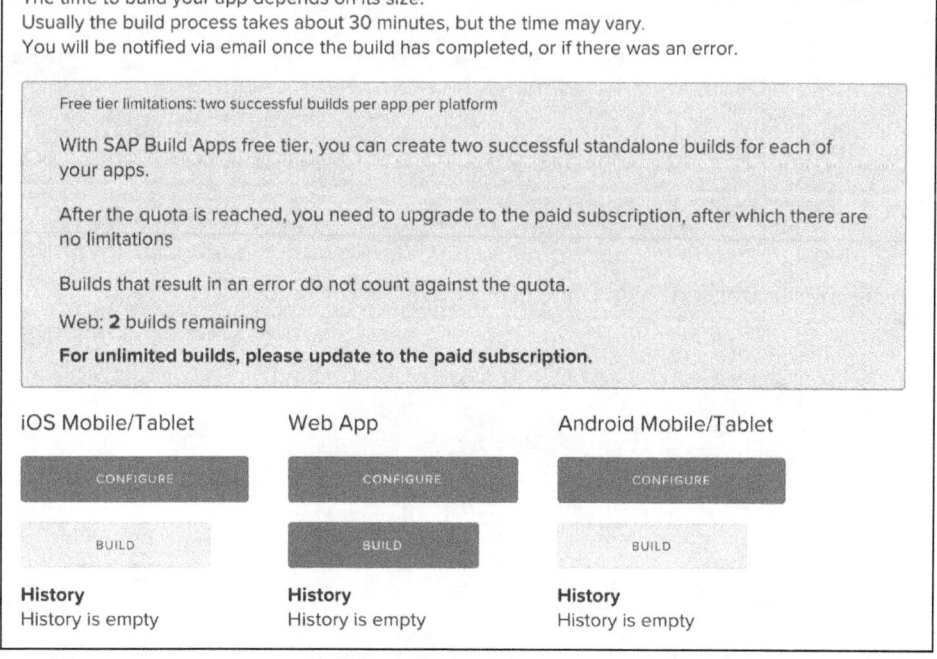

Figure 5.24 Start Build Process for Different Outputs

The respective build process can take some time, which is why it's first queued in the backlog. As soon as the process is complete, a button appears under the respective build request to continue with the output. We'll show you the options you have for this output in Chapter 9.

5.4 App Logic, Variables, and Data Binding

Building user interfaces is only part of the equation. There are also invisible forces behind it, such as the application logic and the use of variables so that both the UI and the application logic can be supplied with the necessary data.

In this section, we'll show you the basics of application logic and the first steps for variables. We'll again show you the basics and teach you reusable patterns so that you can use them for a wide variety of requirements.

Let's start by changing the slider from **VIEW** to **VARIABLES** and thus access the option of creating variables. Variables help temporarily store data and provide a central location for storing it. **PAGE VARIABLES** are only available on the current page and will help manage data on this page. In Figure 5.25, we've created two variables with the **ADD PAGE VARIABLE** button, named them "value_first_name" and "value_last_name", and set both to store a **Text** element.

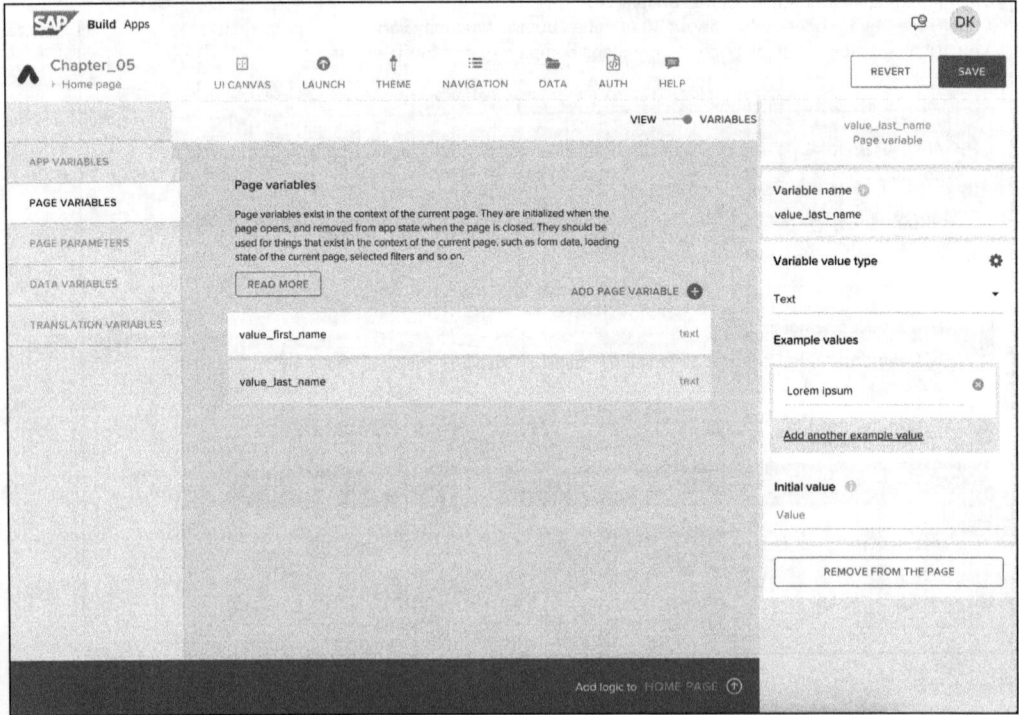

Figure 5.25 Page Variables to Store Data at Runtime

5.4 App Logic, Variables, and Data Binding

Now, of course, the question arises: How can these variables help you? One possibility is to bind one of these variables to a property of a UI element. This procedure is known as *data binding*. The advantage of doing so is that if a property to which the variable is bound changes because of user input, the new value is entered directly into the variable. Conversely, if the value of the variable changes due to the application logic, this has a direct effect on the user interface.

Let's consider an example of how the data binding between the **value_first_name** variable can be linked to the **Value** property of the input field of the first name. Start, as shown in Figure 5.26, by clicking the data binding symbol for the **Value** property.

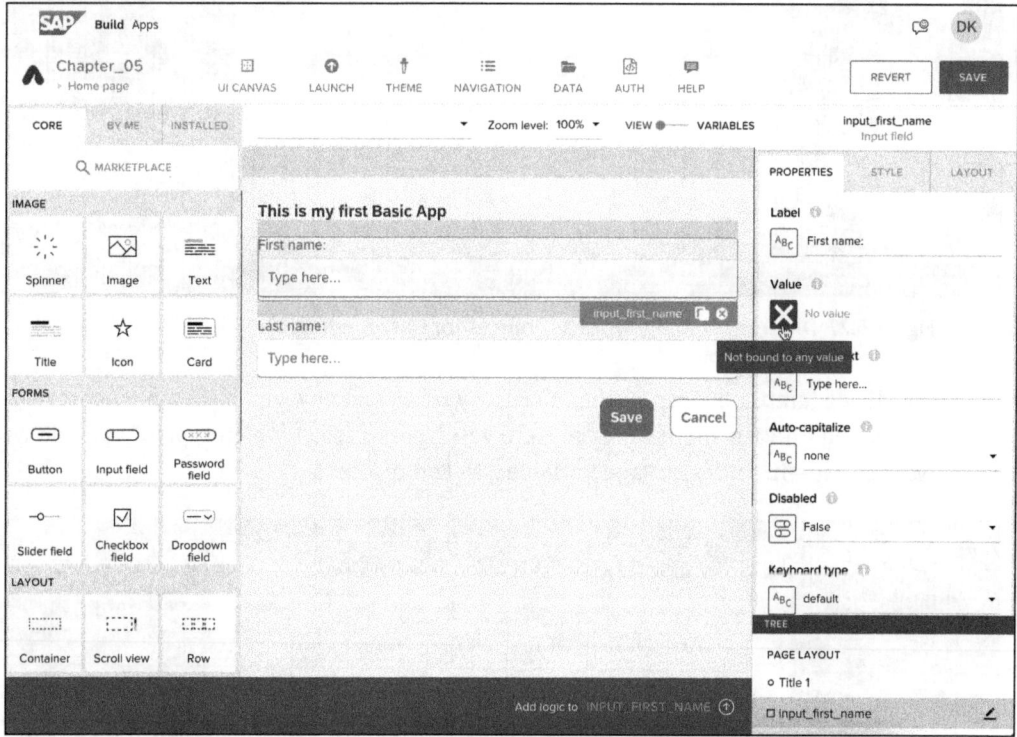

Figure 5.26 Data Binding to Interconnect UI Element Property and a Variable

Data binding can have different sources, but only one per property. We'll introduce a few more sources in this and the following chapters, but for now we'll start with variables. Clicking the data binding opens a dialog, in which you can select **Data and Variables** as the source in the first step (see Figure 5.27).

155

5   Developing Applications

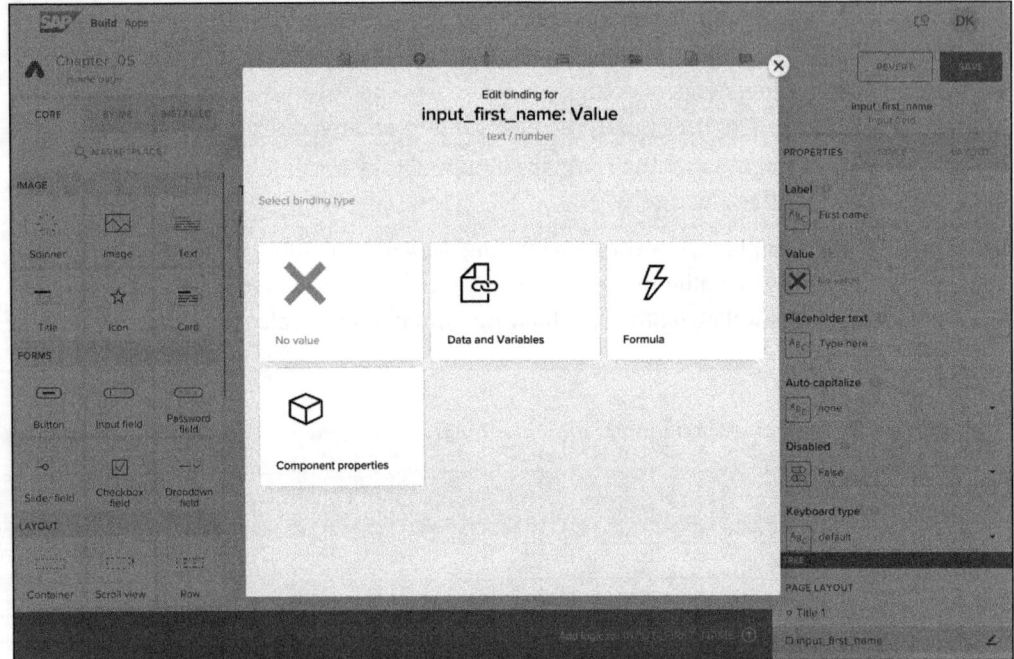

Figure 5.27 Different Binding Types as Sources for Data Binding

The dialog jumps one step further, like in a wizard, and now you'll see a further restriction based on your first selection. Here you have to select which type of variable we want to use—in this case, **Page Variable** (see Figure 5.28).

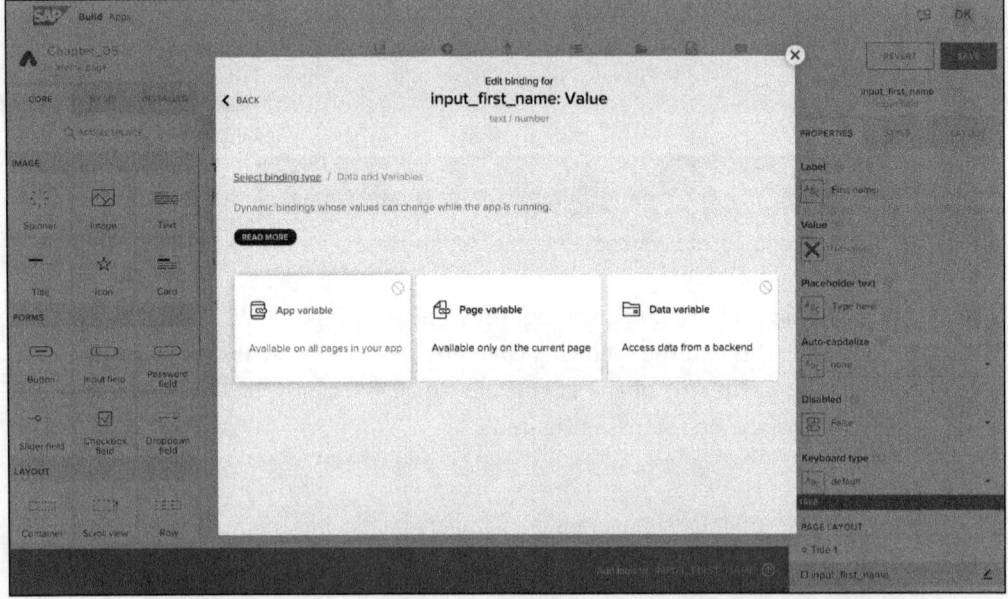

Figure 5.28 Different Types of Variables for Different Functionalities

In the next step, you'll see all the variables of the respective type that you have created. You can only select the variables that match the data type of the respective property. For example, you will not be able to bind a variable of the text type to the **Visible** property because **Visible** returns true/false values—that is, yes/no values. A text type can store much more than true/false values, so data binding is not suitable here.

In Figure 5.29, we've already thought about this and created the two variables with the corresponding data type. So, we selected the corresponding **value_first_input** variable for the input field. Carry out the same steps for the input field for the last name.

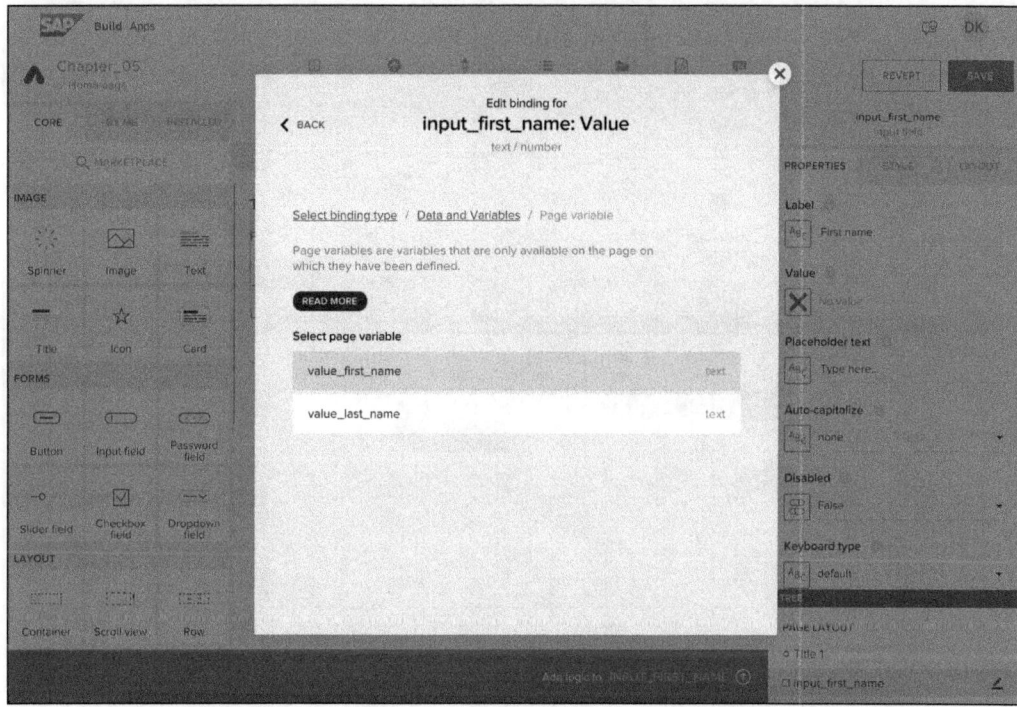

Figure 5.29 Preferred Variable to Store and Provide Data for Selected Property

To add application logic to UI elements, first select the corresponding UI element and click the black bar at the bottom with the **Add Logic To ...** title to open the logic canvas. The *UI canvas*, where we have been so far, is for building the user interface. The *logic canvas* is the counterpart to this and forms the corresponding application logic. Application logic can either be added to the entire view or to a specific UI element.

Select your **Save** button and open the logic canvas. This interface is a graphical workbench of concatenations of prefabricated logic blocks. These logic blocks can be dragged and dropped into the workspace from the left, and the sequence can be determined by connecting the blocks. The trigger is always an event, such as the component tap event.

Now drag in an **Alert** logic block and link it to the event. An *alert* is a dialog that shows a message to the user in the UI with a corresponding text. The displayed text can be stored in the **Dialog Title** property. Instead of storing a static text, click **Data Binding** again and assign the displayed text dynamically by using your two variables (see Figure 5.30).

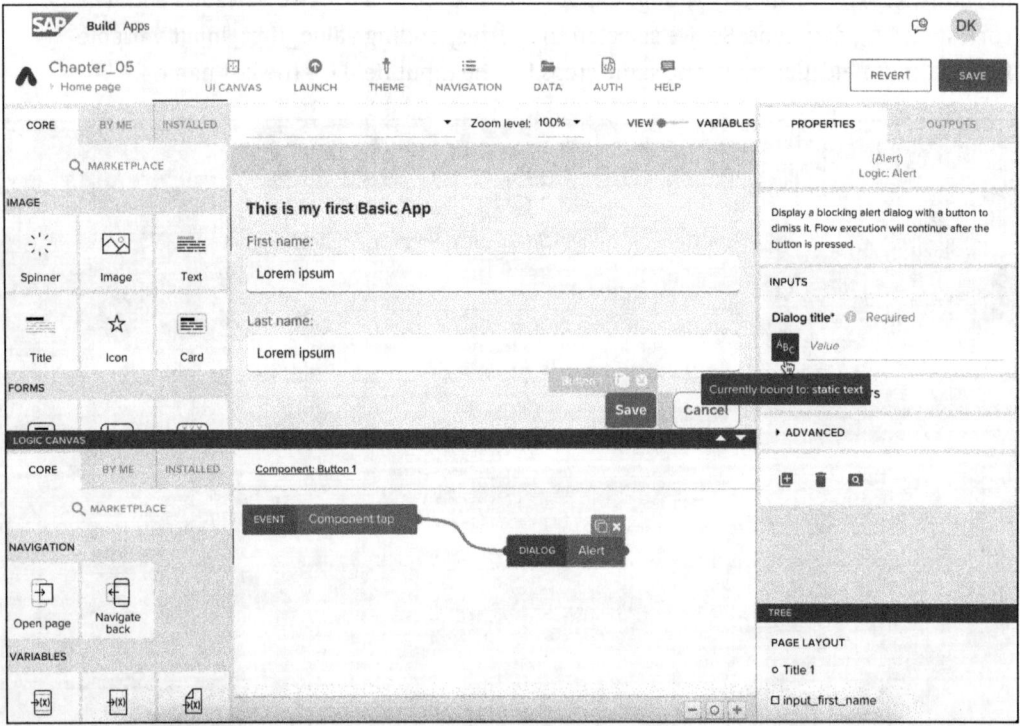

Figure 5.30 Data Binding in Logic Canvas for Properties of Logic Blocks

With this data binding, you can establish that a dialog will open when clicking **Save**, in which both the first name and last name are displayed concatenated as a text. In this case, you cannot bind the variable directly, as only one variable can be used in this case. So, you'll use a formula. Formulas help to build nonexistent data yourself based on calculations. Now go ahead and select **Formula**, as shown in Figure 5.31.

As a formula, store a calculation that does nothing other than concatenate a static text with the two variables. In our case, this formula is as follows:

"Your name: " + pageVars.value_first_name + " " + value_last_name

This formula says that when you click the **Save** button, a dialog opens, and the first name and last name from the two page variables are added to the **"Your name: "** text. These variables can be accessed with `pageVars.value_first_name` and `pageVars.value_last_name`, as shown in Figure 5.32.

5.4 App Logic, Variables, and Data Binding

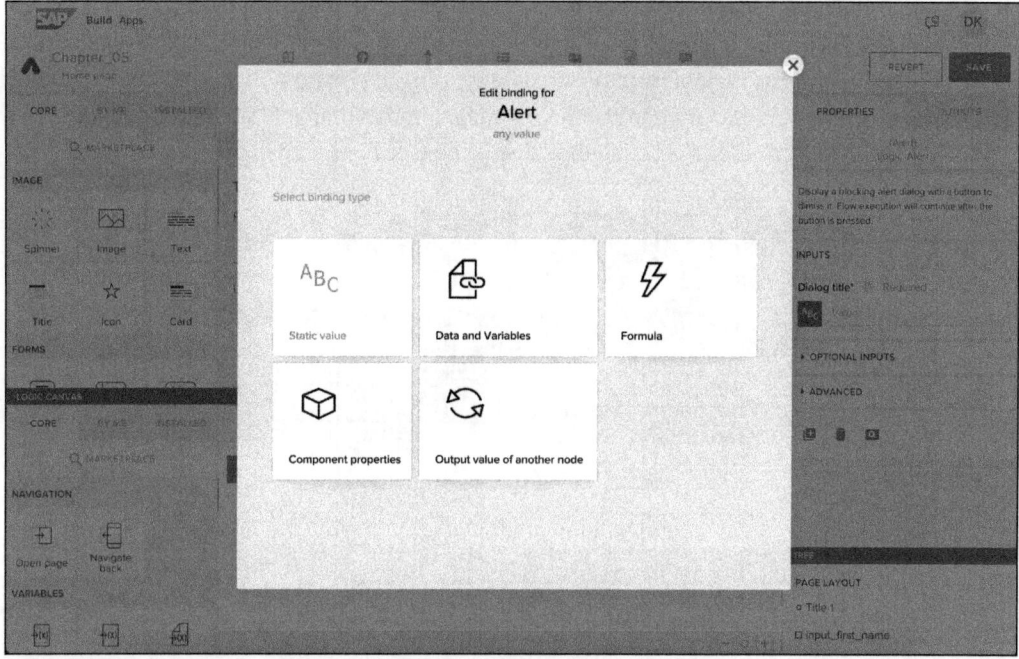

Figure 5.31 Select Formula for Calculations Done Directly at Runtime

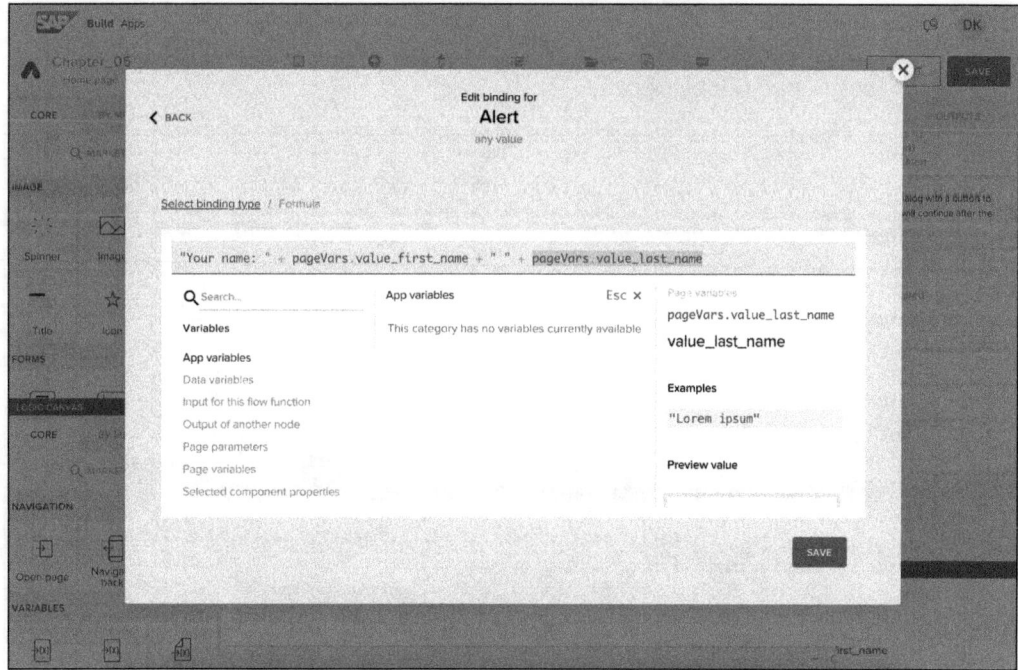

Figure 5.32 Creating Formulas to Define Calculations to Produce Desired Results

Once you've saved both the data binding and your SAP Build Apps project, you can test the app again. If you enter values in the input fields using keyboard input, these values are written to the variables thanks to data binding. If you now click **Save**, the app action logic behind it is executed and a dialog opens. In this dialog, a static text with the current values from the two variables is displayed (see Figure 5.33).

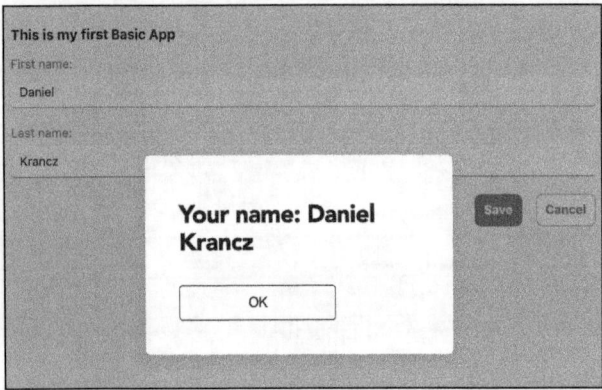

Figure 5.33 Alert Logic Block Opens Message Dialog at Runtime

What do you do now with your **Cancel** button? Well, let's get to know more logic blocks. With the **Set Page Variable** block, you can change the value of a page variable from the application logic. Just as a user can change the value using data binding by means of an input, you can also make this change via the application logic. As shown in Figure 5.34, we have inserted this block twice.

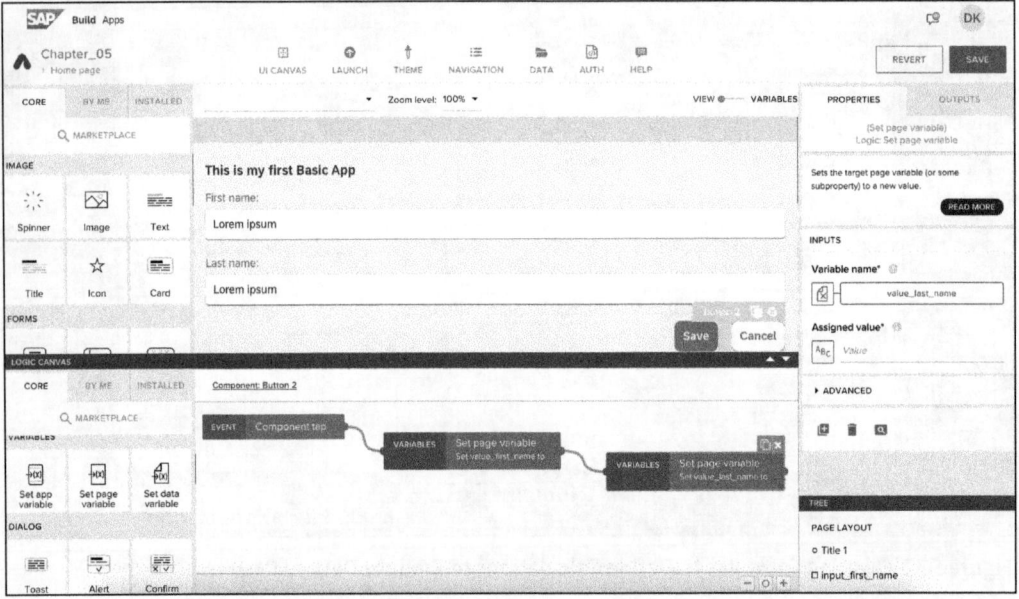

Figure 5.34 Data Binding to Ensure Input Fields Are Cleared if Variable Is Cleared

First, we've selected the **value_first_name** variable as the source for the data binding and the **Variable Name** property. We've left the **Assigned Value** property empty. This resets the variable and, as already mentioned, also clears the input field thanks to data binding. We've done the same for the input field for the last name. This means that both variables and both input fields are reset when the **Cancel** button is clicked.

5.5 Lifecycle Management and Team Collaboration

All these tasks are easier to complete if you share the work and implement the project with others. In this section, we'll look at available collaboration options. In this case too, we already have ready-made options that can be activated very easily.

As you know, someone had to take the first step and create the application locally in their own SAP Build lobby. This person is the first administrator of the project and can use the **Options • Manage Members** button to manage the members and invite new collaborators (see Figure 5.35).

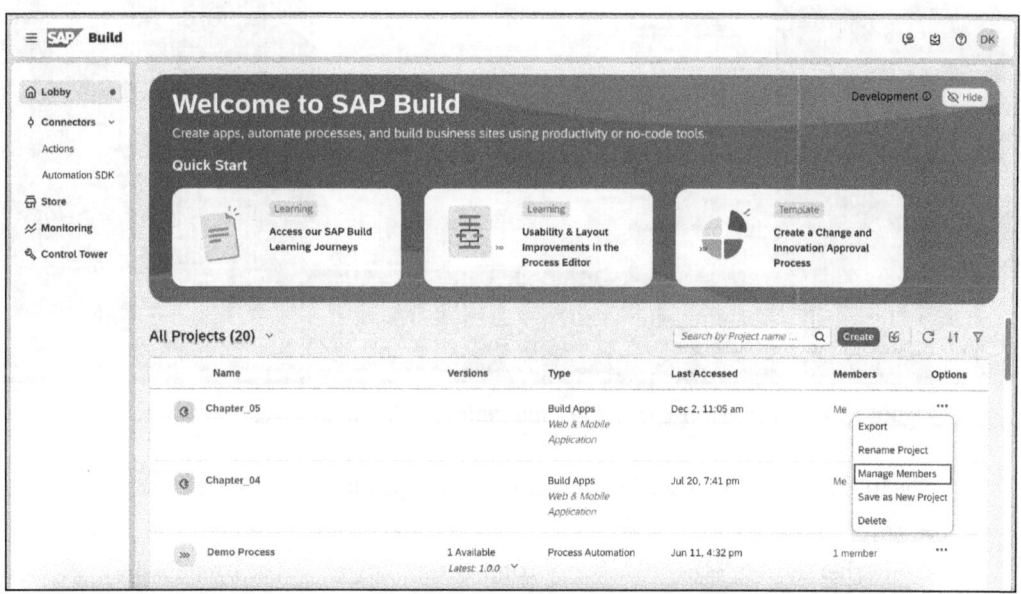

Figure 5.35 Manage Project Members to Invite Others to Participate

A dialog will now open for the corresponding project, in which you can add new project members in the upper section (see Figure 5.36). You must enter the members' emails yourself and also ensure that these members are maintained in SAP BTP with the corresponding authorization for SAP Build. You can choose from the following roles:

- **Administrator**
  Complete access, including editing, sharing, and deleting a project

5 Developing Applications

- **Developer**
  Edit, deploy, release, manage dependencies in a project, and publish to the library
- **Viewer**
  View and deploy a project

In the lower area, you can view the existing members, change the authorizations, and remove members. The only person who cannot be removed is the creator of a project.

From this point on, you can work together with all administrators and developers on the SAP Build Apps project and jointly edit the pages, application logic, theming, and variables and thus share the work.

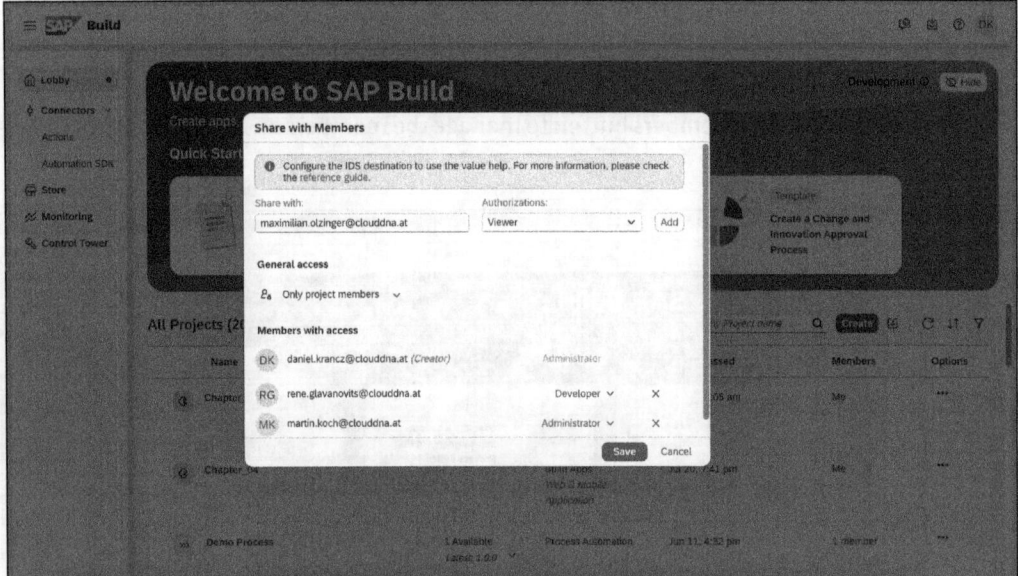

Figure 5.36 Add or Remove Members and Define Their Project Roles

The following rules are specified by SAP Build Apps for collaborative work:

- Unsaved changes are continuously applied on top of the latest saved version by another user. In other words, nonconflicting changes will be immediately reflected for all members, without interfering with their work.
- Editing the same parts of the project, such as the page name or component style class, will cause a conflict. Conflicts are resolved by reloading the page, whereupon conflicting changes are discarded. Note that conflicts are resolved in favor of the latest saved version.
- User profile coloring in the top bar reflects the current state of the user. Grey implies no unsaved changes, while orange denotes that the user is currently editing the project.

Especially when resolving conflicts, we unfortunately have no choice but to refresh the app and thus discard the conflicts that occur in comparison with the last actively saved version. A merge, as known to pro code developers from other source code–management tools such as Git, unfortunately does not (yet) exist here.

5.6 Summary

In this chapter, we looked at many different topics relating to application development. We started with the user interface, and you got to know the first UI elements. You've seen how the properties, style, and layout of these UI elements can be customized so that they behave according to your wishes. On the subject of style and layout, you saw that you can establish your own theme and integrate your corporate design. We then turned our attention to the application logic and events as triggers. With application logic, you can create all the magic behind the scenes, again using drag and drop. You could use this to, for example, display a message box, change the values of variables, or access UI elements. Finally, we looked at how to work together with others on a project and share responsibilities.

Chapter 6
Data Integration and Authentication

Let's now look at the magic behind SAP Build Apps. How does data get into the app? How can you manipulate or delete this data, which may be stored on another system? And how can you securely allow all these actions after appropriate authentication? We'll answer all these questions and many more in this chapter, and also engage in a number of digressions on other topics.

An application can rarely do much without data. In the modern world of web applications, the application is split into two parts, to put it simply. As builders, we work with SAP Build Apps on the frontend and need a remote station with which we have to communicate. This remote station is called the *backend* and offers the opportunity to communicate via certain interfaces, make requests, and wait to see how these requests are processed and what kind of response is given. This is where the request-reply principle comes into play. Everything we need from our backend has to be sent to the backend via a request, which more or less triggers business logic in the backend in order to ultimately get data back or manipulate it. We will provide all the necessary basic knowledge for this in this chapter, as well as the individual options for communicating with a backend.

Another important point is user authentication. Of course, you do not want unauthorized persons to gain access to your application, nor do you want them to be able to communicate with a backend without permission. For this reason, the use of remote systems is only permitted in SAP Build Apps in combination with an authentication concept. If no identity provider is connected, you will not be able to access an SAP system or a non-SAP system in SAP Build Apps.

In Section 6.1, we'll talk about user authentication and the advantages that SAP Build Apps in combination with SAP Business Technology Platform offers. In Section 6.2, we'll first explain the basics of HTTP, OData, and application programming interfaces (APIs) to make it easier for you to consume the services. Building on these basics, in Section 6.2.1, we'll look at how to consume data from an SAP system in the cloud. In this context, we will also talk about sorting options. Section 6.2.2 will be about consuming data from an SAP S/4HANA system, where you can read customer data and search through it using a search function. We will also have a digression here: we'll look at the

Marketplace and how to install additional UI elements and logic blocks from it. Section 6.3 will cover the consumption of non-SAP systems—that is, how to easily connect web services via a URL. In an example, we'll display products with their stock levels and filter the products by product category using a dropdown field. Finally, in Section 6.4, we'll use the local storage of a device to save data on the device itself. In this context, we'll also show you how to navigate between several pages and how to transfer data from one page to the next.

You'll see that this is not as complicated as it appears at first glance. In SAP Build Apps, a very nice option exists to be able to use the same variables and logic blocks in SAP Build Apps again and again, regardless of the technology of the backend. In other words, there is a facade that doesn't care whether it communicates with an SAP system, a non-SAP system, or local storage.

6.1 User Authentication

You will often face the challenge of having to protect your app from unauthorized access. It's particularly important for transactional business apps to ensure confidentiality and integrity of their data. The terms *authentication* and *authorization* play a major role here.

These two terms are often mistakenly used interchangeably. During *authentication*, the identity of the user is checked. This means that he proves that he is the user he claims to be. This can be done either via a user name and password or via client certificates. In many cases, *multifactor authentication* is used, in which, in addition to a user name and password, you must confirm your identity via another factor, such as a smartphone. The second factor, like the smartphone app, is uniquely assigned to the user. This ensures that only the person in possession of the second factor can confirm the first factor.

You can compare this to everyday life. When you visit a company, such as a business partner, you have to identify yourself at reception. This can be done with a passport or identity card, for example. This allows them to ensure that you are who you say you are. In this case, this would be the same as authentication. Once you have been authenticated, you will receive an access card and you will be allowed into the company building. *Authorization* is used to determine which rooms you are allowed to access.

Authentication must be explicitly activated in SAP Build Apps. In Figure 6.1, you can see that you are greeted with an **ENABLE AUTHENTICATION** button in the **AUTH** tab and that you have no further options until you have pressed this button.

6.1 User Authentication

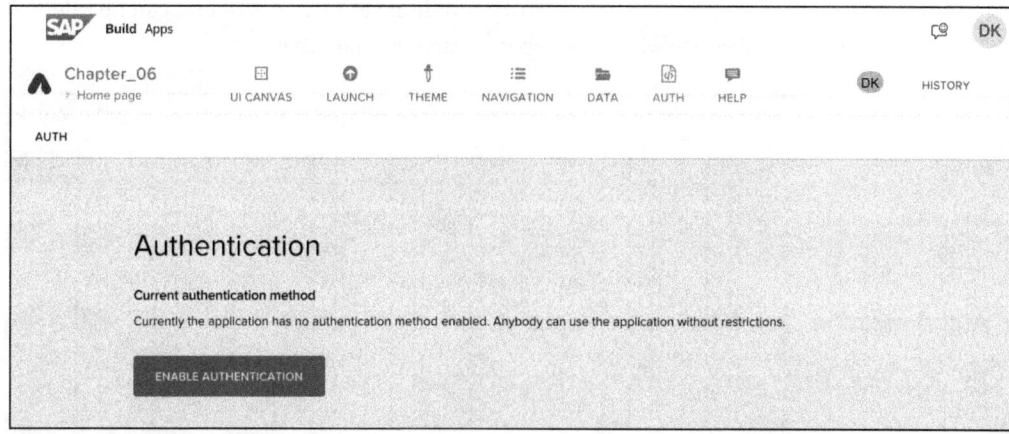

Figure 6.1 Enable Authentication in AUTH Menu Tab

You have three options for having authentication carried out by an identity provider (see Figure 6.2):

- Google Firebase email/password authentication
- SAP BTP authentication
- Direct third-party identity provider

The most common option is **SAP BTP Authentication**, as this can be seamlessly integrated into the existing infrastructure. With the other two methods, you will have to do the setup yourself. Nevertheless, with all three variants you get your own pages inserted into the project, which are equipped with standard content (application logic, variables, UI elements, etc.).

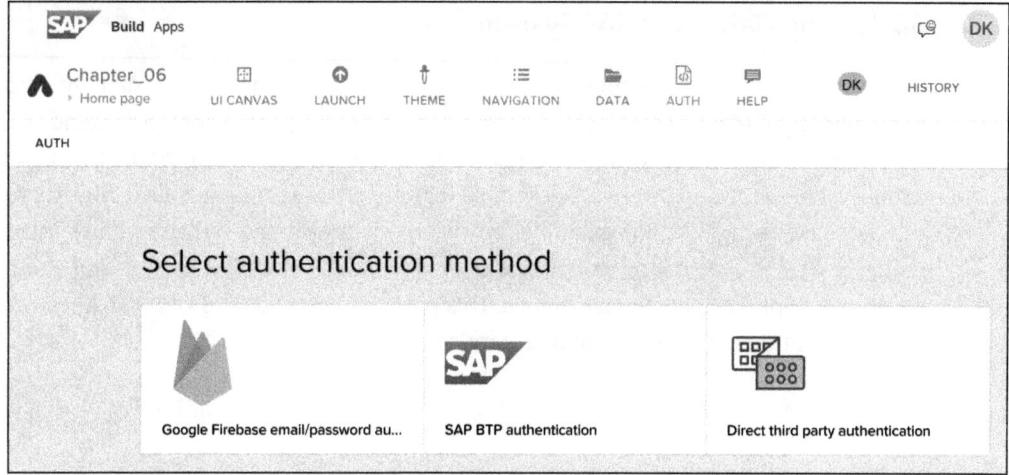

Figure 6.2 Select Desired Authentication Method and Identity Provider

SAP BTP authentication uses the identity provider set by SAP BTP and can be selected as the start page directly when the app is started (see Figure 6.3).

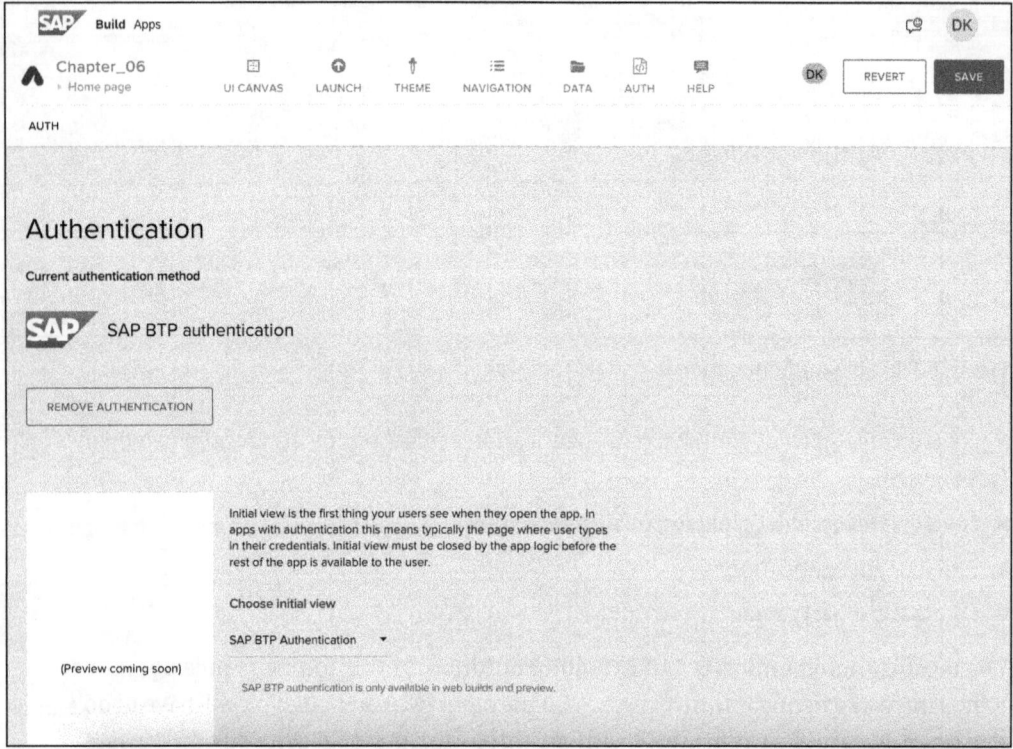

Figure 6.3 SAP BTP Authentication Activated

6.2 Using Data from SAP Systems

Now we come to one of the most interesting parts of app development—namely, data management. For web applications, the application is decoupled from the backend and all the data that is needed must be loaded into the app via an interface. The same applies when data is to be created, changed, or deleted. We no longer write directly to the database using SQL, for example, but need an intermediary that receives these requests, tries to execute the requests, and reports back whether it worked or not. From a web development perspective, we call this part the *backend*. In SAP Build Apps, we have four options for integrating such services:

- REST APIs
- OData APIs
- Google Firebase/Cloud Firestore
- Visual cloud functions

All four technologies are based on the HTTP/HTTPs protocol. This means that an open standard is used for communication with the service implementations. The modeling of the respective services or their connection within SAP Build Apps differs among the four technologies. For this reason, each technology is examined individually in the following sections.

All technologies have a common denominator—the HTTP protocol. *HTTP* is an extensible protocol that forms the basis of the internet as we know it. It is based on some basic concepts such as the notion of resources and Uniform Resource Identifiers (URIs) as well as a simple structure of messages. A client-server structure is used for the communication flow. Numerous extensions have been added to these basic concepts over the years, offering new semantics. URIs uniquely identify resources. You can think of the HTTP protocol as a mail carrier. Say that someone transmits messages from a sender to a recipient. In the case of the mail carrier, the message would be a letter. The letter is sent in an envelope, and the actual message is inside the envelope. The information about the sender and recipient is attached to the envelope itself. This is comparable to HTTP headers. This is control information that is required so that the message is delivered to the correct recipient. At this point, you can also specify the format and encoding via which your message is to be transmitted.

The HTTP protocol always returns an HTTP status code. You can compare this to a registered letter. You will also receive confirmation that the message has been delivered to the recipient or information if the recipient cannot be reached because they have moved and have a new address, for example. HTTP status codes are three-digit numerical values. It's important that all statuses in the range 200–299 correspond to successful processing on the service side. All statuses in the range 300–399 mean a redirect, which in turn means that the recipient was not reached at the original address, but you will be notified of the new address. Statuses in the range 400–499 mean a client-side error. This could be, for example, incorrect authentication because you have entered an incorrect password. However, it can also mean that you do not have sufficient authorization to access the service. A special status code, 404, means that the resource you have addressed does not exist. All statuses in the range 500–599 mean a server-side error. This may be the case, for example, if your request contains incorrect data or the format in which the request is made is not understood by the server. You should then be equipped to use the various service types.

However, we will not delve any further into the technical depths of the HTTP protocol here. It's used as the basis for the service connection when developing apps, but we will not have to use the protocol itself directly in programming or modeling. SAP Build Apps does this work for us in the background. A basic understanding of how it works, the HTTP methods provided, and the status codes used is nevertheless required, especially when it comes to REST and OData APIs.

We would also like to explain APIs at this point. APIs are mechanisms that enable two software components to communicate with each other. For example: Imagine you want to create an app that displays weather data. To do this, you need a system that provides this data. The weather app "talks" to this system via APIs and displays current weather data on your end device, such as a smartphone. In the context of application programming interfaces, the word *application* refers to any software with a specific function. The *interface* is a type of service contract between two applications. This contract defines how the two applications communicate with each other using requests and responses. The API documentation contains information about how the developers should structure and subsequently implement these requests and responses.

In this section, we have two main topics. Section 6.2.1 deals with the details of how to consume data from a cloud system (in our case, SAP S/4HANA Cloud). In Section 6.2.2, we will focus on an SAP S/4HANA on-premise system. You will see that the two approaches are very similar in SAP Build Apps, and we can build on our acquired knowledge in both subsections.

6.2.1 Cloud Systems

To be honest, we could spend another umpteen pages on the basics, but as the saying goes: learn by doing. For this reason, we start by consuming data from an SAP S/4HANA Cloud system. You will see that all this communication with the protocols is not that difficult: we use a graphical interface and a facade in SAP Build Apps, so we don't have to communicate directly with the interfaces. We will also gradually expand your knowledge so that we can apply what we have learned again and again. The integration of the data is so generic for us builders that it makes no real difference whether you're communicating with an SAP system or a standard web service.

Let's now switch to the **DATA** tab and look at the options for data integration. We will take a step-by-step look at some of the options for data and interface integration. Let's start with the goal we have set ourselves, the connection of a cloud system. First, click the **ADD INTEGRATION** button, as shown in Figure 6.4.

If you want to integrate a system, you have two options (see Figure 6.5):

- You'll learn more about the **Visual Cloud Functions** option in Chapter 7. This again involves development based on the no-code principle, but in the backend environment.

- Integration via the **SAP Systems** option means connecting any systems that are linked to SAP BTP. This can be SAP S/4HANA, SAP S/4HANA Cloud, SAP SuccessFactors, or another system.

[»] **Data Integration Only Possible with Authentication**
As you can also see in Figure 6.5, both options are always grayed out. In the case of SAP systems, authentication must be activated. With visual cloud functions, a visual cloud functions project must also be recognized for this option to be active.

6.2 Using Data from SAP Systems

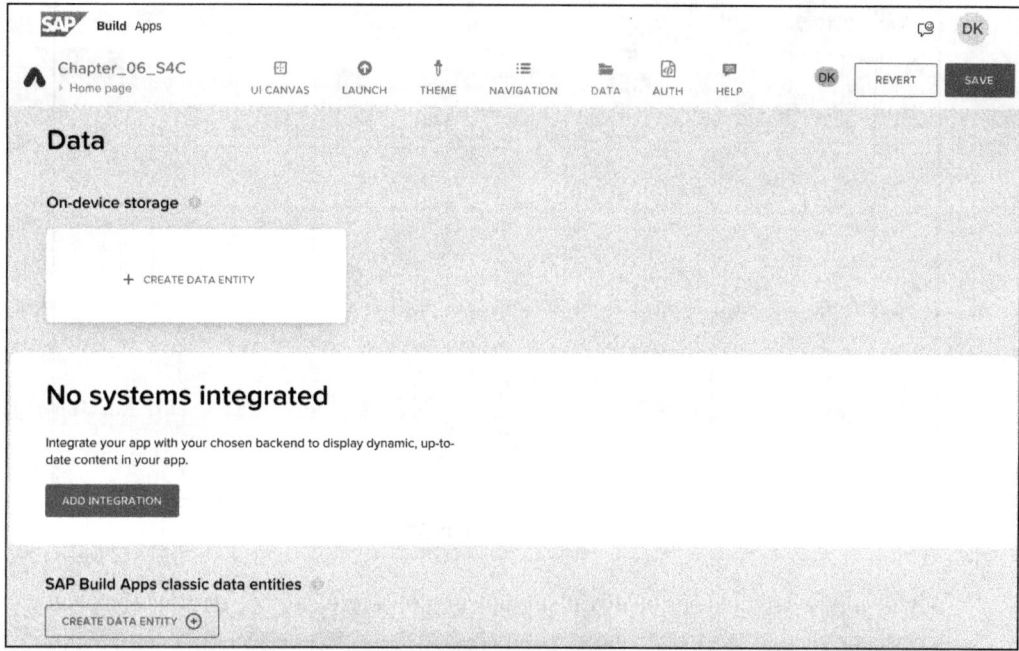

Figure 6.4 DATA Tab for Adding System Integrations

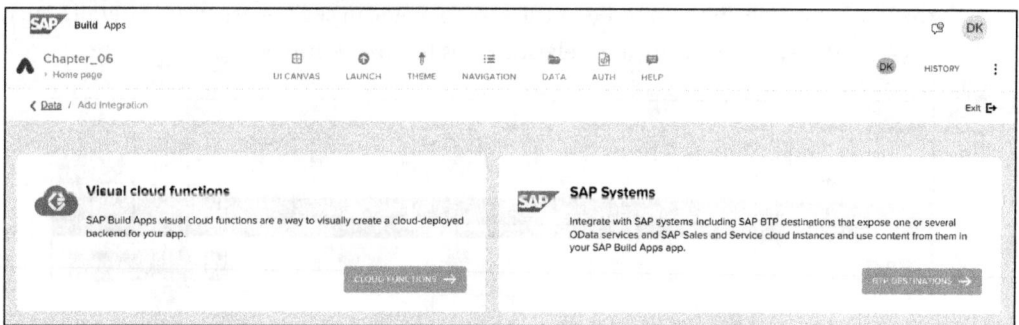

Figure 6.5 System Integration Not Possible Until Authentication Is Enabled

Once authentication is active, the option of integrating an SAP system is offered. Now click the **BTP DESTINATIONS** button to see a list of all destinations that are connected to your SAP BTP account.

As you can see in Figure 6.6, you're now offered all the destinations that you've maintained in your subaccount in SAP BTP. A *destination* is a technical tool in SAP BTP for accessing a host and port. Some of these entries can be expanded, which means that the corresponding destination is a *service collection*. This means that the destination does not only point to a single service, but offers several for use.

6   Data Integration and Authentication

Figure 6.6 SAP BTP Destinations for SAP and Non-SAP Systems

Once you've selected the relevant service, you're offered all the entities that you can consume via this service. An *entity* is a technical object in a service that can be accessed via an API. The service provider can decide whether these APIs are read-only or also writeable. As you can see in Figure 6.7, we have selected the **BookingCollection** entity on the left-hand side. In the middle of the screen, you can see the associated details, such as primary keys, properties, and relationships to other entities.

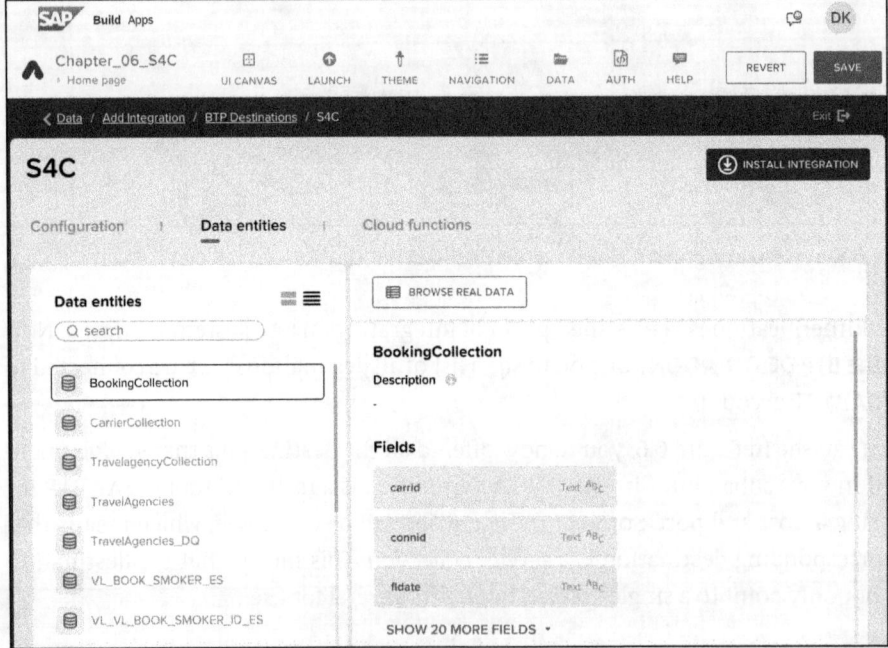

Figure 6.7 Service and Entity Selected

6.2 Using Data from SAP Systems

Before you can use this entity at all, you must first say that this service is intended to be used. You can do this with the **INSTALL INTEGRATION** button at the top right. This does not install anything in the classic sense, but simply activates this system for further use in SAP Build Apps via the destination. This installation does not introduce any advantages if you don't also activate the desired entities with the **ENABLE DATA ENTITY** button (see Figure 6.8).

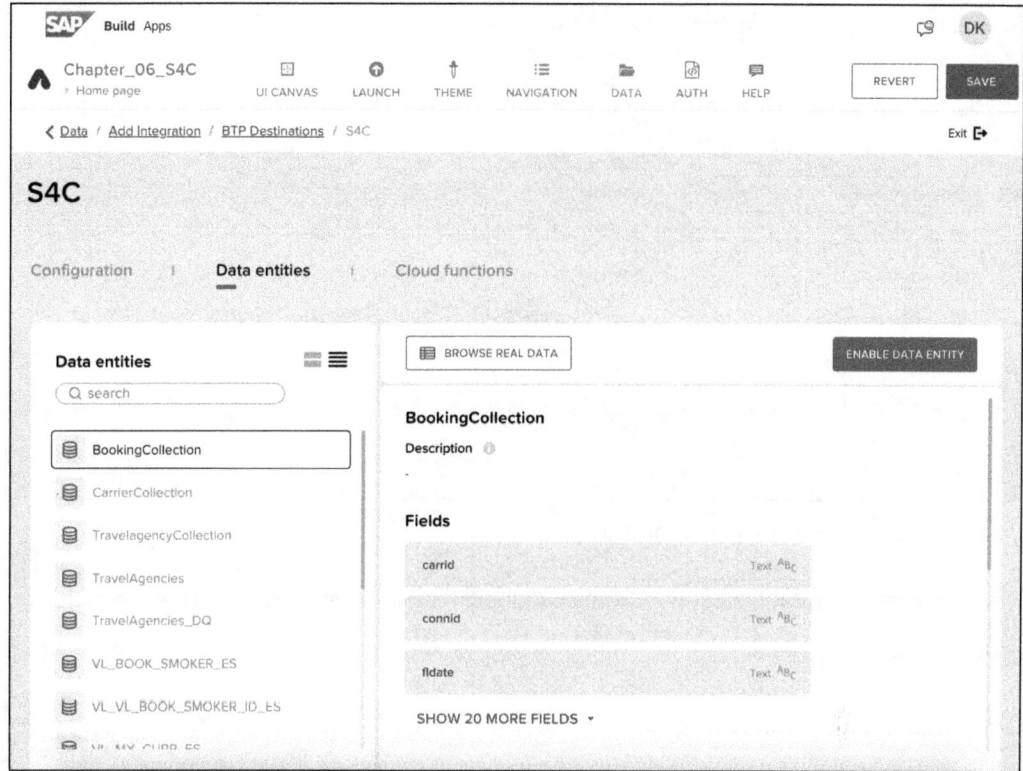

Figure 6.8 Data Integration Installed

Once an entity has been activated, the status in the list on the left also changes. Furthermore, the **BROWSE REAL DATA** and **BROWSE SAMPLE DATA** buttons can be used to view the real data supplied by the backend through the service and also to store sample data locally in the app for testing (see Figure 6.9).

Figure 6.10 shows what it might look like if you query an entity with the **BROWSE REAL DATA** button and have the data returned. This gives you a first impression of how the data comes from the backend and how you could use it for the user interface.

6 Data Integration and Authentication

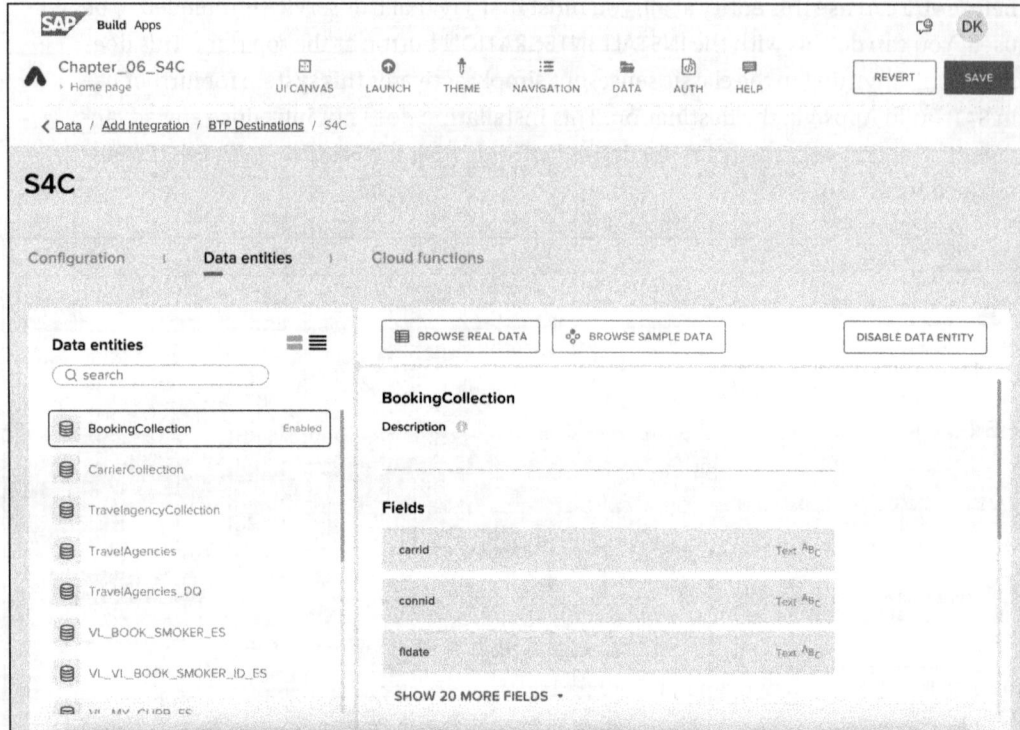

Figure 6.9 Entity Enabled for Use

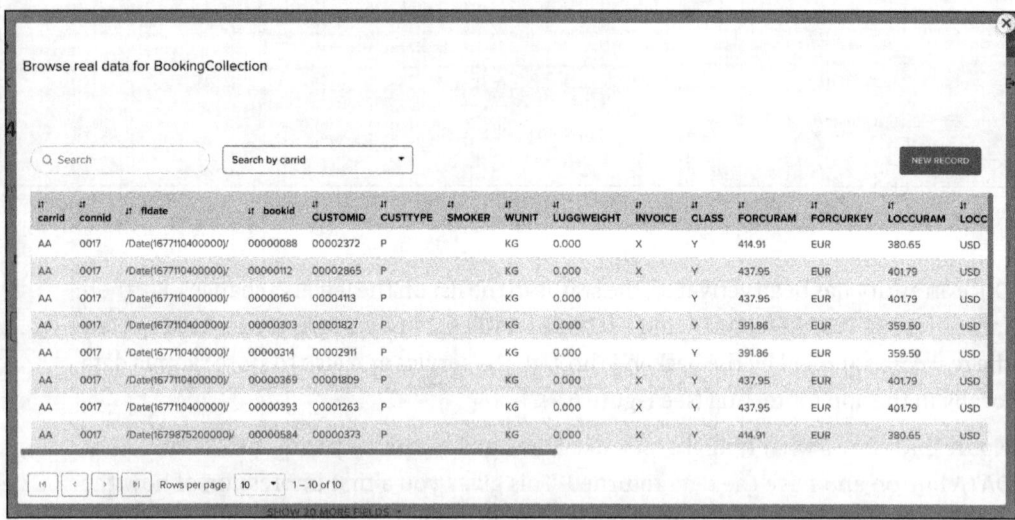

Figure 6.10 Preview of Data Sent Back Live from Backend Service

If you now go back to the **DATA** tab, you will see in the integrated systems that one or more entities are available via the corresponding destination. In this case, it is the

6.2 Using Data from SAP Systems

BookingCollection entity, which you can use in SAP Build Apps (see Figure 6.11). In the next steps, we'll look at how this entity can be used in development.

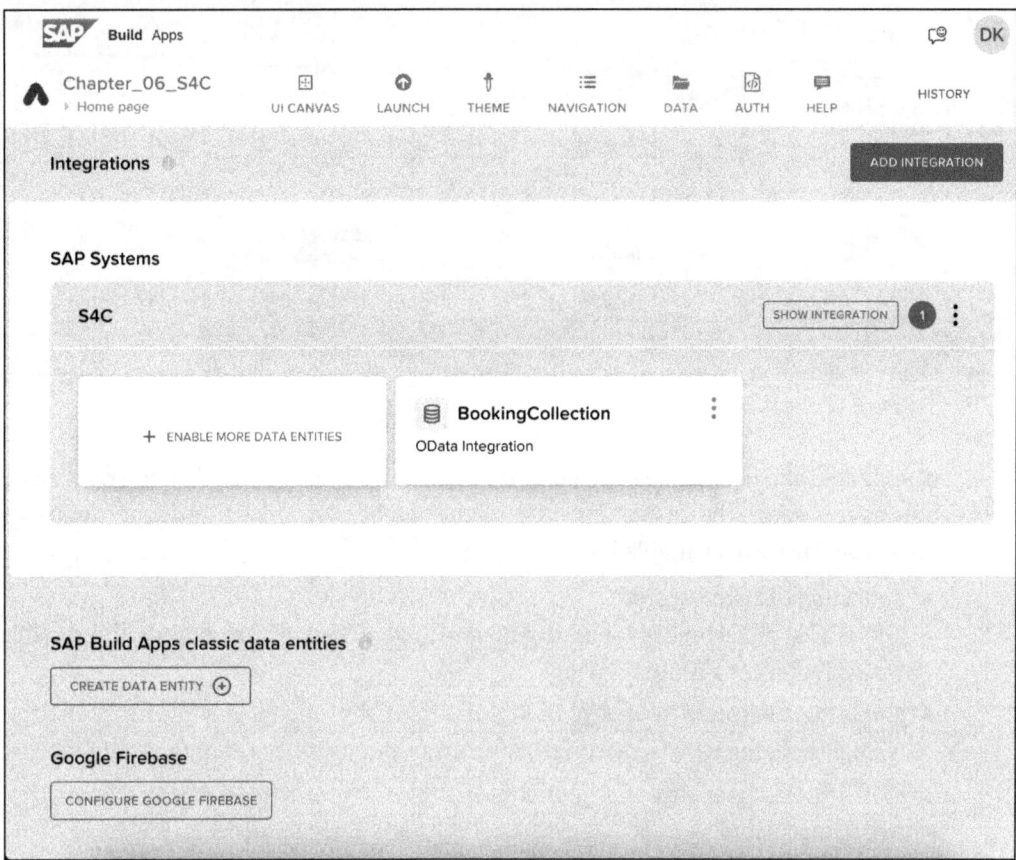

Figure 6.11 Entity Successfully Added

Unfortunately, these added data resources cannot yet be used directly for development. T be able to access the resources, you still need *data variables*. These enable you to communicate with the data resources and then further with the interfaces.

Let's switch to the UI canvas and to a page where we want to display data later. Now switch from **VIEW** to **VARIABLES** and go to **DATA VARIABLES** on the left-hand side. Here you can create a new variable using the **Add Data Variables** button (see Figure 6.12). All installed data resources and available entities are then displayed in a popover.

6 Data Integration and Authentication

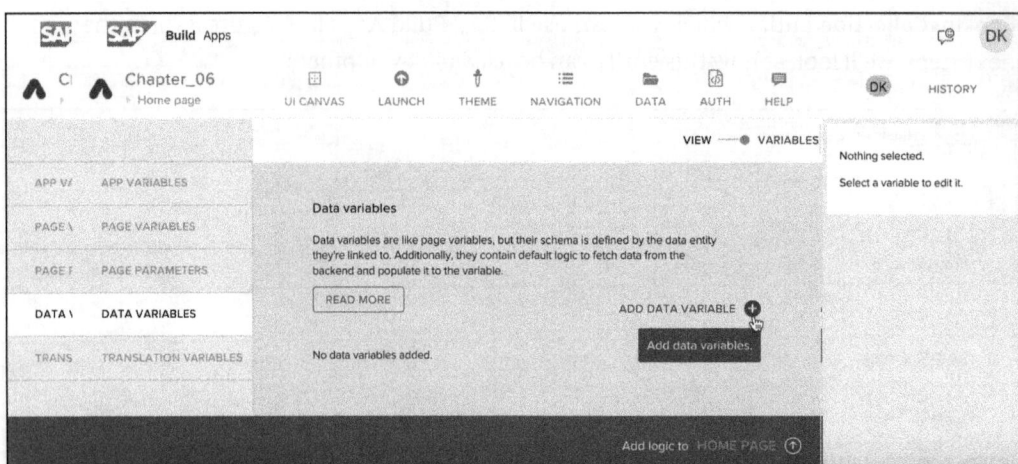

Figure 6.12 Creating Data Variables to Access Data Resources

The created variable can be selected, and the properties can be edited on the right-hand side as usual with the UI elements. The name can be customized, and you can decide what you'll use each variable for:

- **Collection of Data Records**
  This is a list or, technically speaking, an *array* of entities. In our example, the variable would represent a list of bookings.

- **Single Data Record**
  A single booking is not structured like an array, but technically only contains a single object.

- **New Data Record**
  If you were to use a variable to create new data records, this would be the way to go. This variable would be an empty shell that we could fill with data and then create via the application logic.

Filtering, sorting, and **Authentication** options could be set here for the collection. We'll come back to filtering and sorting as an example in this chapter. But first, let's create an ordinary data variable with the name "bookings", which represents a list of flight bookings (see Figure 6.13).

The aim now is to display all flight bookings supplied by the interface and therefore the backend in a list. Let's now go to the graphical editing of the user interface and insert the **List Item** UI element into the page, as shown in Figure 6.14. Also click the data binding symbol next to the **Repeat With** property in the **PROPERTIES** tab. With this property, you can define a rule for how often the list item should be displayed.

6.2 Using Data from SAP Systems

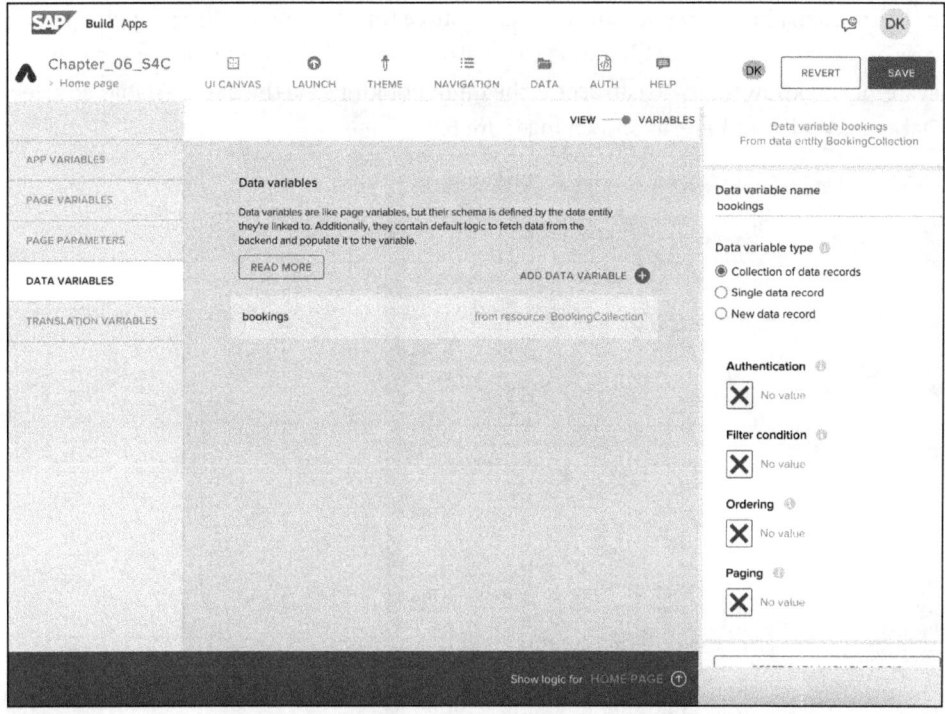

Figure 6.13 Data Variable Successfully Added

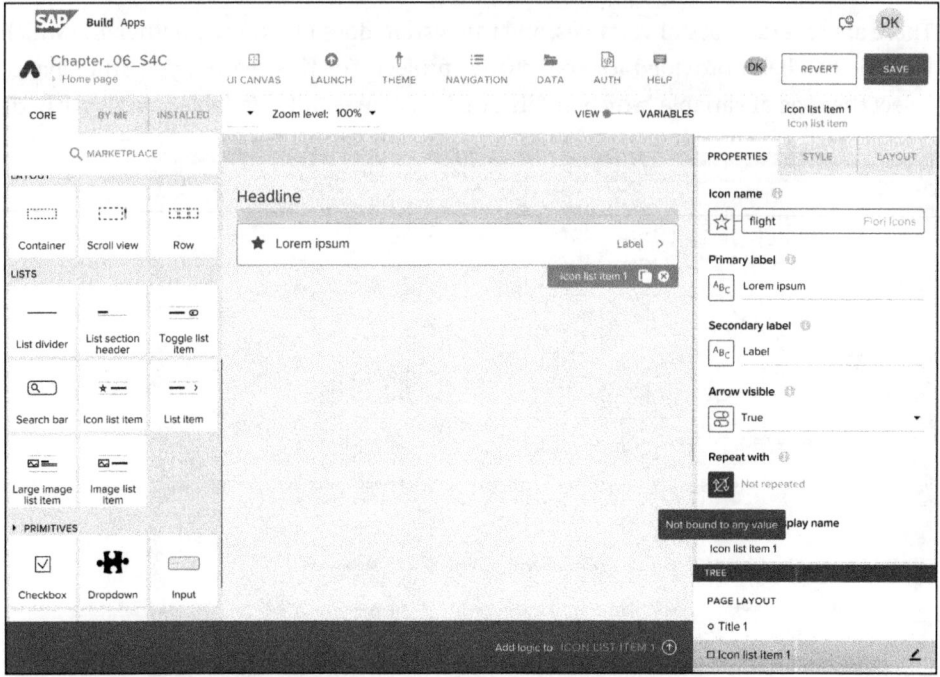

Figure 6.14 Use Repeat With Property

177

A dialog opens in which you can select the source for the **Repeat With** property. In this case, we want as many list elements to be displayed as there are flight bookings in the system. You know that you can access the flight bookings via the data variable, so select **Data and Variables** here, as shown in Figure 6.15.

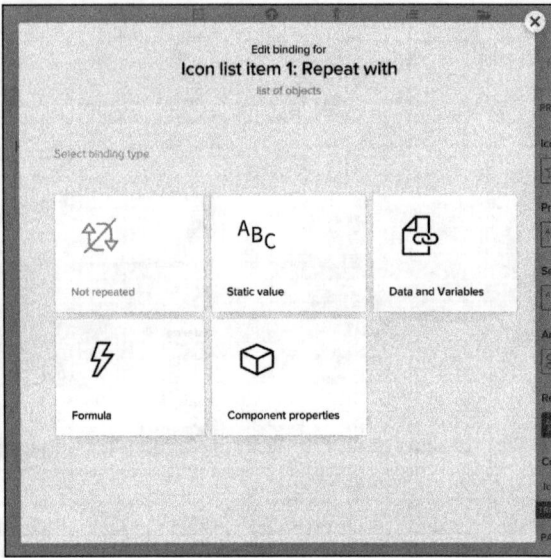

Figure 6.15 Data Binding Type for Repeat With Property

There are several types of variables, and the system does not know whether and which variable has been provided for the current project. For this reason, you still have to select the type of variable, which in this case is of course **Data Variable** (see Figure 6.16).

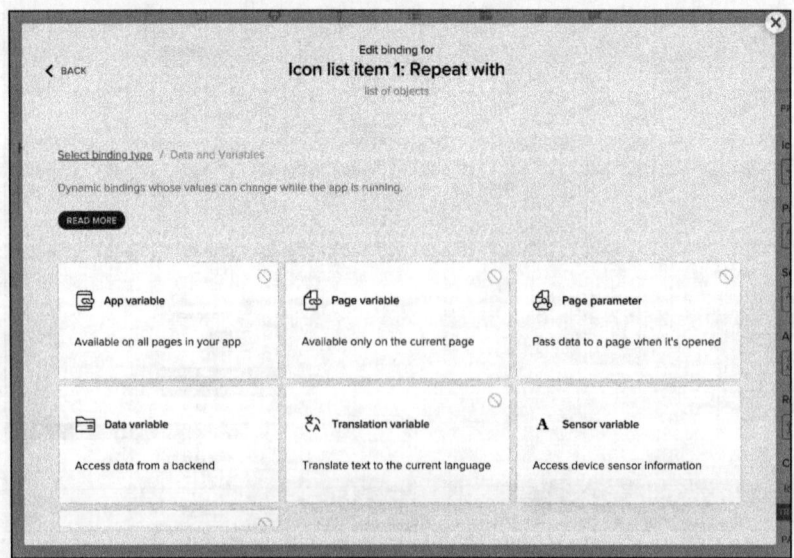

Figure 6.16 Data and Variables Available for Data Binding

Now you can select from the created data variables, which offer a list of objects. Select the **bookings** variable here, and then click **SAVE** (see Figure 6.17).

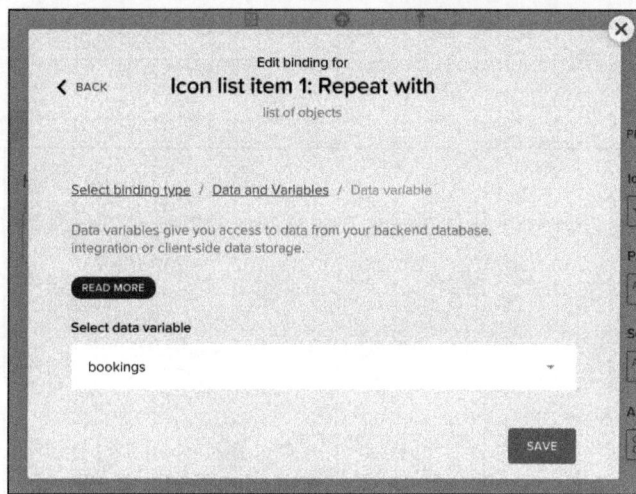

Figure 6.17 Available Data Variables for Data Binding

Now you can see as an example that the **Repeat With** property is set. You can recognize this by the repeating and fading list elements. Now at least the list elements are repeated based on the list of booking entries in the variable. However, this still does not help the user, as you haven't yet set the content of the list element. What you can easily change is the **Icon Name** property, where you can choose from a whole range of ready-made icons. Choose the **flight** icon for this property, which will display a plane as an icon at the beginning of each list element. Next, click the data binding for the **Primary Label** property, as shown in Figure 6.18. The primary label is the text that is displayed in every list element on the left.

> **Disclaimer to Repeat With**
>
> The fact that the elements fade with **Repeat Wth** in the workbench is only a given in this case. Of course, the real elements that are displayed on the productively executed app will not look like this. It's merely intended to show that these elements repeat depending on which data is bound behind the **Repeat With** property.
>
> It must also be said that the number here does not correspond to the truth either. The number of list elements is again fixed for this preview. In this case too, when the app is executed, the real number coming from the backend is used for the list elements.

You have already learned about data binding, but not yet about the possibility of accessing data as part of the **Repeat With** procedure. You have to imagine that **Repeat With** iterates over each element of the list like a loop and displays the UI elements one after the other. In this iteration, there is a pointer with the technical name **current** pointing to the respective object so that the attributes of the respective object can be accessed. In this way, you can dynamically specify that a list element should be displayed on the user interface based on a list of data and that a text should be output from the current element in each case. This may sound complicated, but based on the next practical use case, we are sure that you will quickly become familiar with this practice.

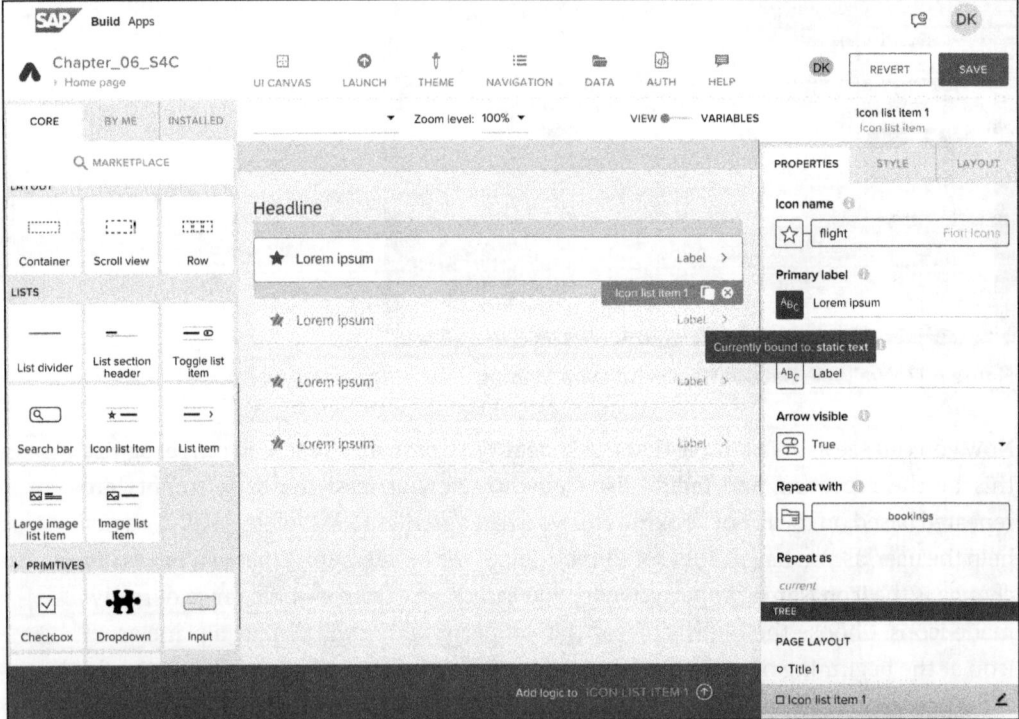

Figure 6.18 Repeat With Successfully Bound to List

As usual, a dialog opens for data binding, in which we have selected **Data Item in Repeat** as the binding type. After this selection, you are taken to the second step in the dialog, shown in Figure 6.19. Here, first select the pointer under **Select Repeat** that is to serve as the source for this data binding. Under **Select Repeat Data Property**, you must select the corresponding property from the object that is to be bound to the **Primary Label** property. The repeat data property is **current.PASSNAME**, which is the name of the passenger of a single flight booking. The last input, **Set Preview Value**, is not relevant for productive operation; it only serves as a preview text for the development interface and is only important for us builders. Finally, click **SAVE** and save the settings for data binding.

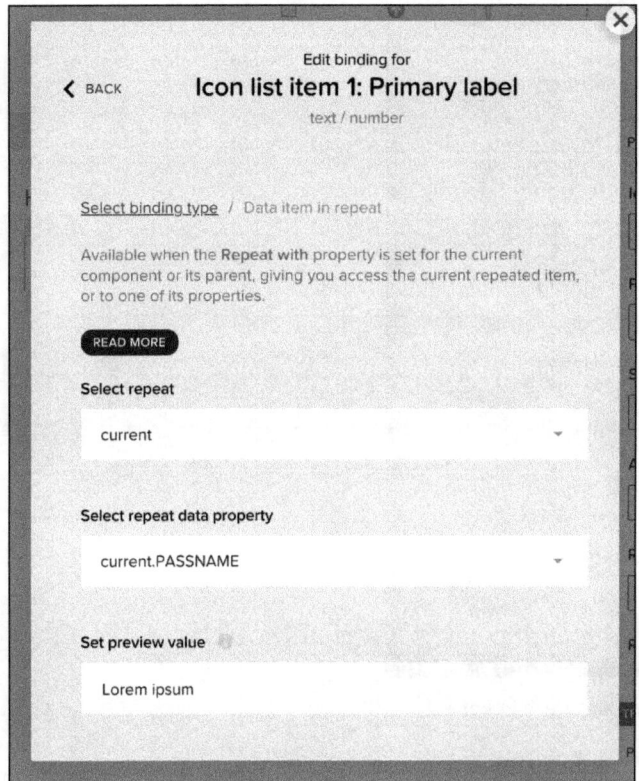

Figure 6.19 Using Data Binding of Data Item in Repeat

Follow the same process for **Secondary label**, which is a text displayed on the right in a list element. However, instead of directly using an attribute from the data item in repeat, here you can write a formula that displays two texts from the data item in repeat as one. To do this, select **Formula** as the binding type and store the following formula there:

repeated.current.carrid + " / " + repeated.current.bookid

You can see that the repeated.current keyword can be used to access the same pointer as the primary label. So you can access the carrid—that is, the ID of the carrier—and the bookid—that is, the ID of the flight booking—and concatenate them together with a slash (/) as a separator. As you can see in Figure 6.20, we have again assigned a text preview before clicking **SAVE**.

Before we take a look at the result of the data binding settings, we want to provide you with further helpful information and tips and tricks regarding data binding, data communication, and much more. In Figure 6.21, you can see that we have switched back to **DATA VARIABLES**. Say that you now want the data in your collection to be sorted accordingly and want to store the sorting criteria for the variable bookings on the right-hand side under **Ordering**.

6 Data Integration and Authentication

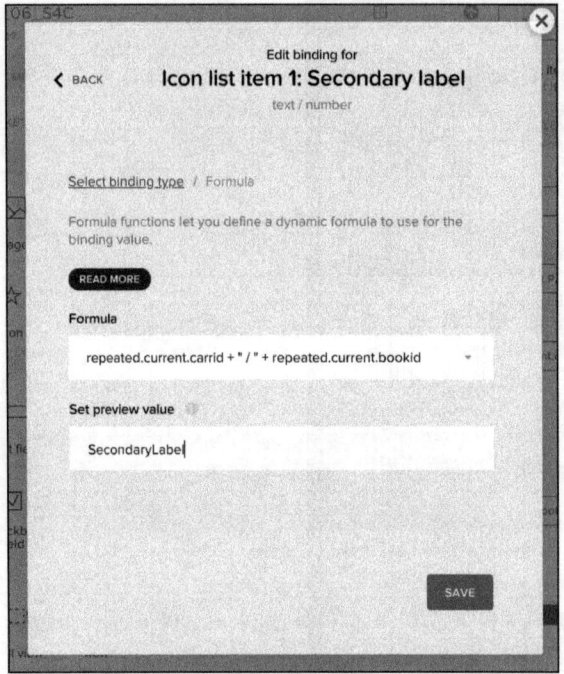

Figure 6.20 Using Formulas of Data Item in Repeat

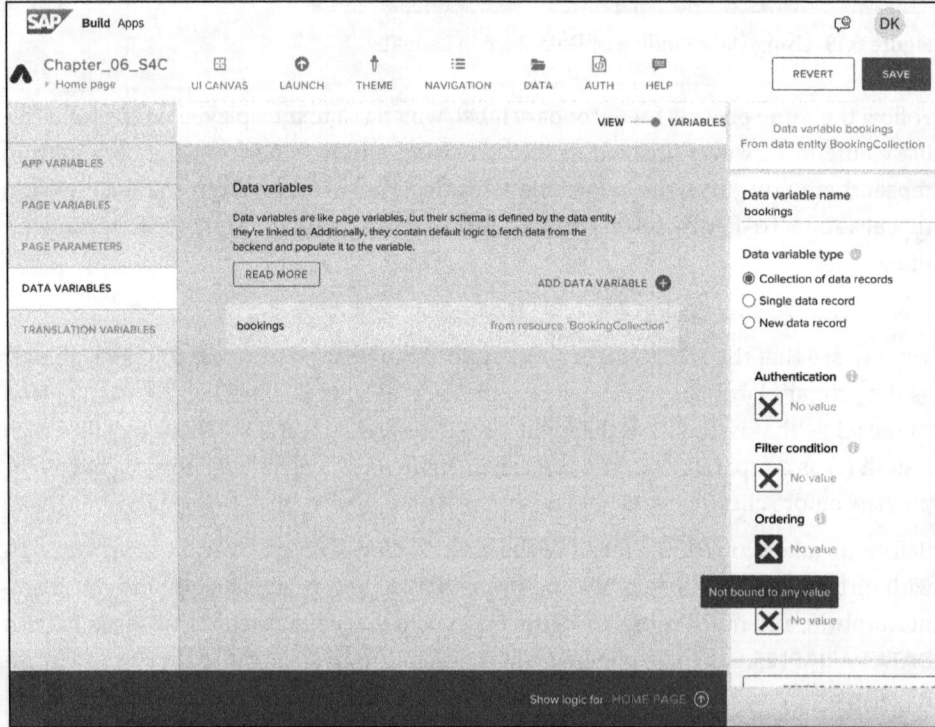

Figure 6.21 Sorting and Filtering Data within Data Variable

Ordering can also be assigned dynamically via data binding, but in this case, choose the **List of Values** binding type, which allows you to store static values. In Figure 6.22, the list is set to be sorted in descending order (**desc**) according to **fldate**—that is, the flight date—using a criterion. With **ADD SORT OPTION**, you can add further sort criteria and link them together.

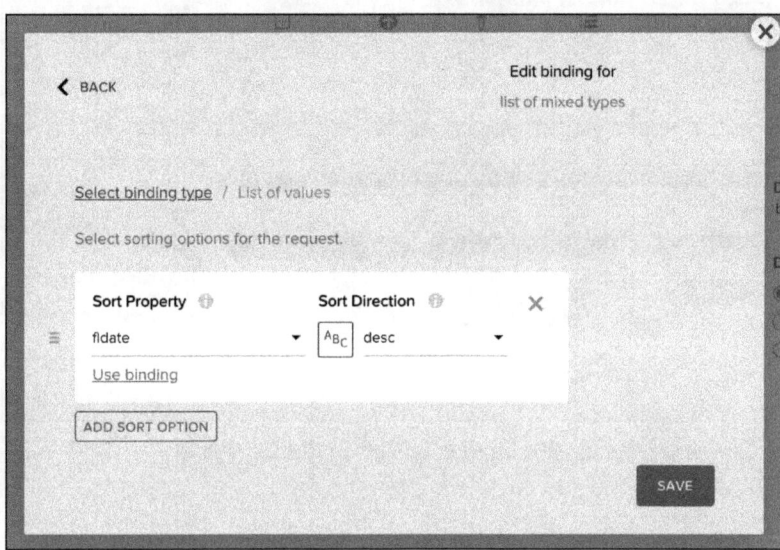

Figure 6.22 Sort Order Conditions for Collections

What we haven't yet told you is that variables can also have application logic and even have it by default. The default setting is that when the page is *mounted* (i.e., when the page is opened for the first time), the data is retrieved from the corresponding data resource and therefore the interface in the backend via the **Get Record Collection** logic block, and then the **Set Data Variable** logic block takes the output of the last logic block and sets the content of the data variable **bookings**.

You can also see what we have added in Figure 6.23. With the **Delay** logic block, we manage to ensure that precisely this process—namely, loading the data from the backend and saving it to the variable—is executed every few seconds. In this way, you can create a list for the end user that is automatically updated and thus always displays a dashboard with the latest flight bookings. Use this generic functionality to automatically present up-to-date information to end users (dashboards, inboxes with items to complete, news, notifications, and much more).

You've now achieved our first goal. You can test your first application, which delivers a list of flight bookings from the cloud system, sorted in descending order by flight date, and refreshes the list every few seconds so that the latest flight bookings are always reloaded and displayed. In Figure 6.24, you can see the result and also for the first time what the **Repeat With** property looks like from an end user's point of view.

6   Data Integration and Authentication

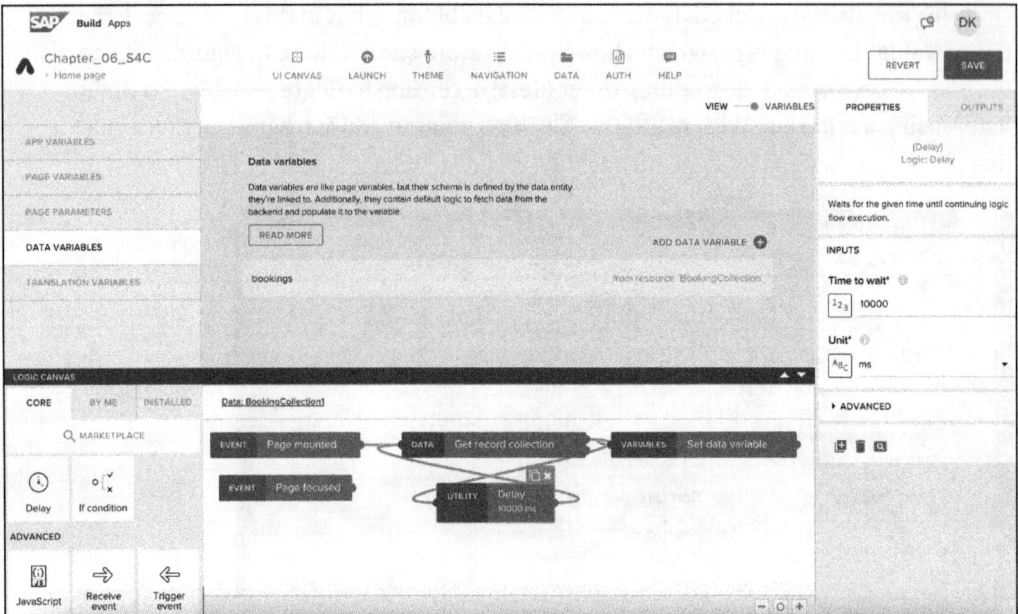

Figure 6.23 Logic Canvas of Data Variables and Possibility of Autorefresh

Figure 6.24 Preview of App Showing List of Flight Bookings

6.2.2 On-Premise Systems

In our next example, we want to consume data from an on-premise system—from an SAP S/4HANA system in this case. The goal will be to develop an application in which the customer data from a table in the SAP HANA database can be displayed in an SAP Build Apps web application and searched by name using a search dialog. To avoid repeating ourselves, we'll show you another way of displaying the data. In this example, we'll build a table in which the most important information about a customer is displayed in columns. To include other components, we'll show you how to use the Marketplace and install other UI elements or logic blocks that have been provided either by SAP or by the community.

> **Information from the Last Section**
>
> Although we dealt with other topics in the last section (consuming services from an SAP system), from the consumer's point of view (in this case, an SAP Build Apps application), the same approaches, practices regarding data binding and the structure of the user interface, and many other topics all come together. For this reason, we recommend that you also review the last section, as we are confident that the individual steps and basics are familiar and can be applied again.

You already know how to access, activate, and install data resources. For this reason, we just want to show you in Figure 6.25 that the interface here is the **ZHOUI5_CUSTOMERSET** entity. This entity is made available via an OData service developed with the ABAP RESTful application programming model. This entity provides all customers in the SAP S/4HANA system.

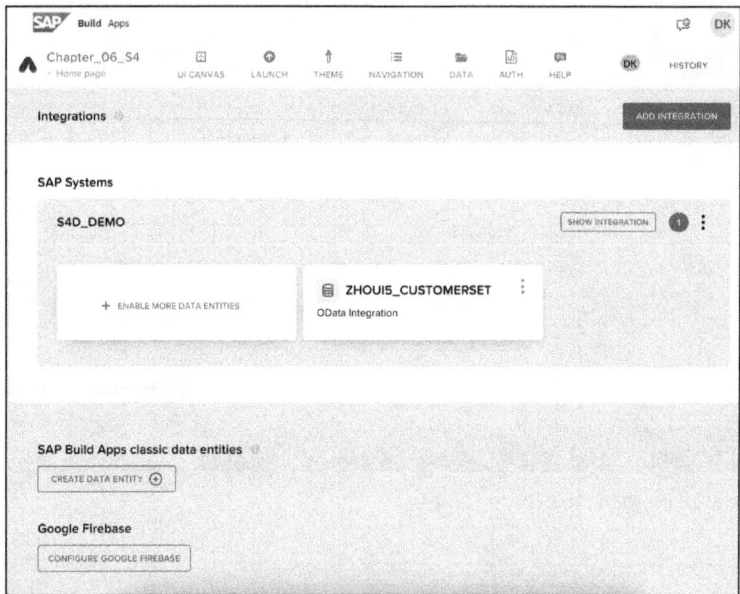

Figure 6.25 Entity Enabled from SAP S/4HANA System

6 Data Integration and Authentication

Figure 6.26 shows a preview of the live data from the SAP system. The customer data consists of the name, telephone number, email, and a linked website, among other things.

Figure 6.26 Data Preview of Corresponding OData Service

In Chapter 5, we explained that there is no such thing as a UI element for the table. Instead, we have to create a table using the **Row** UI element. To do this, insert a row that contains five cells. Then fill the individual cells with static texts using drag and drop (see Figure 6.27).

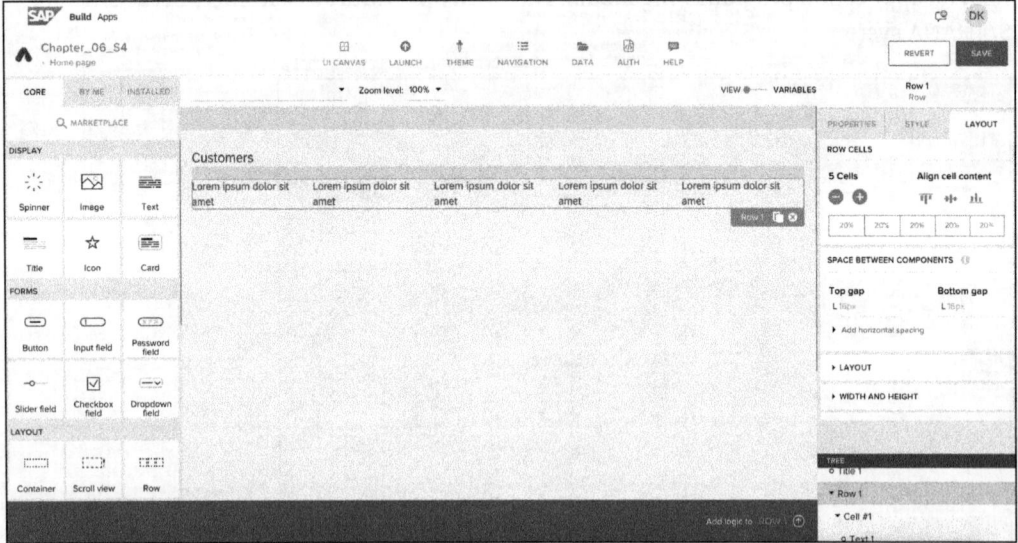

Figure 6.27 Using Row UI Element to Create Tables

In this case, this first row represents the column headings. For this reason, do not use data binding for the texts here, but simply static headings. You can see these headings in Figure 6.28.

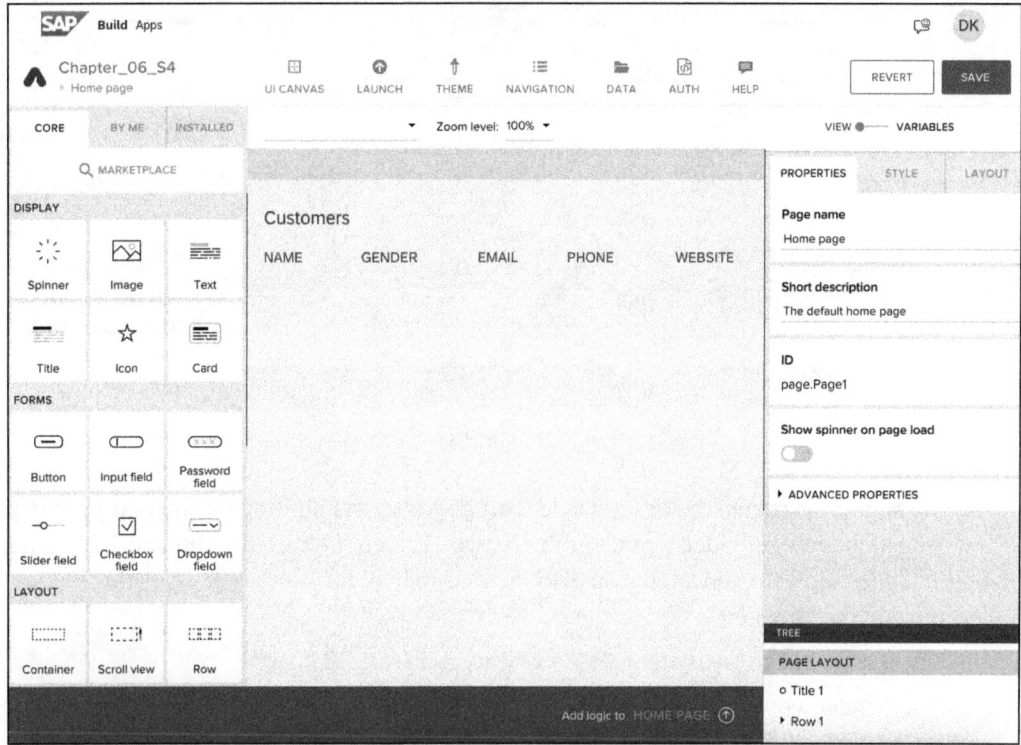

Figure 6.28 First Row Displays Column Headers

A second row will serve as a template, as was the case with the list element in the last section. Thanks to the **Repeat With** property, this row template is repeated as often as data is returned by customers in the collection.

We already created a data variable for the collection. We proceeded in exactly the same way as we did in the last section, but have not renamed the variable, leaving it set with the default name, **ZHOUI5_CUSTOMERSET1**. Now use this data variable for the **Repeat With** property so that the row is replicated based on the customer data coming from the SAP S/4HANA system (see Figure 6.29).

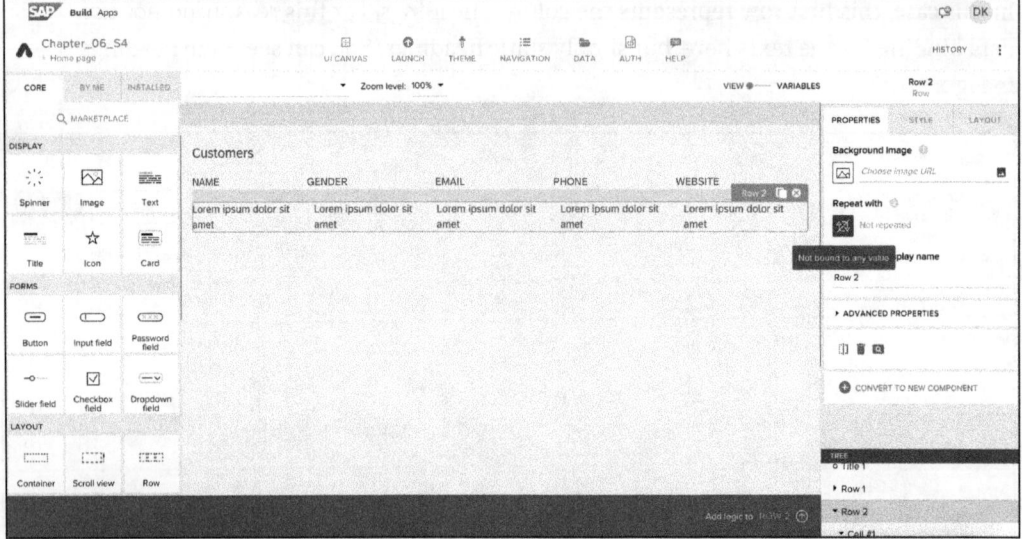

Figure 6.29 Second Row Is Template for Table Items

Now click the text in the first cell and then click the data binding for the **Text** property. The names are provided by the service in two different fields, **Lastname** and **Firstname**, but you can easily concatenate these two texts using a formula and display them as one field in the table:

`repeated.current.Lastname + " " + repeated.current.Firstname`

For **Gender**, you can also calculate the text to be displayed. Unfortunately, this is not entered as text in SAP Build Apps, but as technical keys, which are assigned in the backend by our business partner entity. These technical numbers, each of which stands for a corresponding gender, can be translated by a formula and output as text. With an IF statement, you can check whether `repeated.current.gender` is equal to 2, which corresponds to the technical key for Male, and thus it will output Male as text. Otherwise it will output Female as text. This is possible with the following formula (also shown in Figure 6.30):

`IF(repeated.current.Gender === "2", "Male", "Female")`

For all other texts, we can perform the data binding directly on the individual attributes from the entities and thus display email, phone, and website in addition to name and gender.

Now let's cover the requirement that end users should be able to click the website cell and thus open the corresponding website behind the URL. To do this, select the text that represents the website and open the logic canvas at the bottom. On the left-hand side, search for a corresponding logic block for this requirement—and this is the first time that you'll realize that this project cannot be covered by an element available in the standard system. This is where the Marketplace comes into play, which can be selected in the UI elements as well as opened in the logic blocks (see Figure 6.31).

6.2 Using Data from SAP Systems

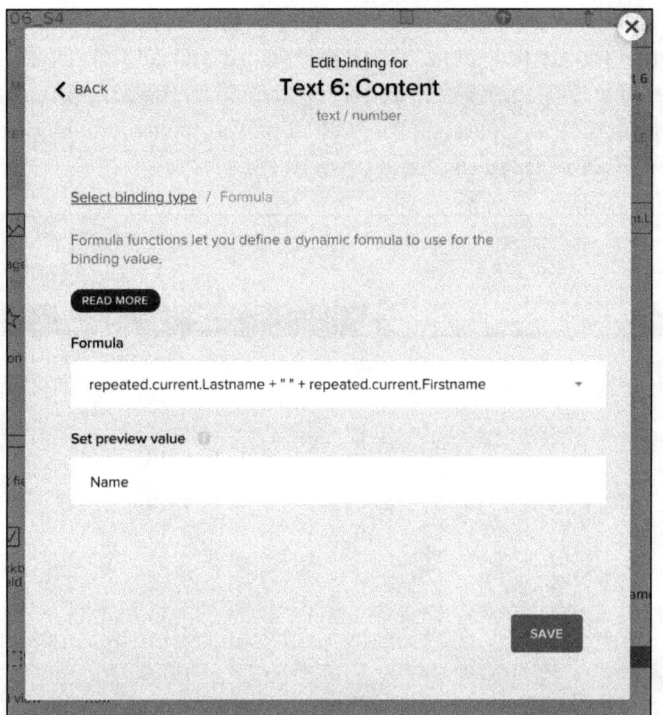

Figure 6.30 Showing Data in Cell Using Data Binding and Formulas of Data Item in Repeat

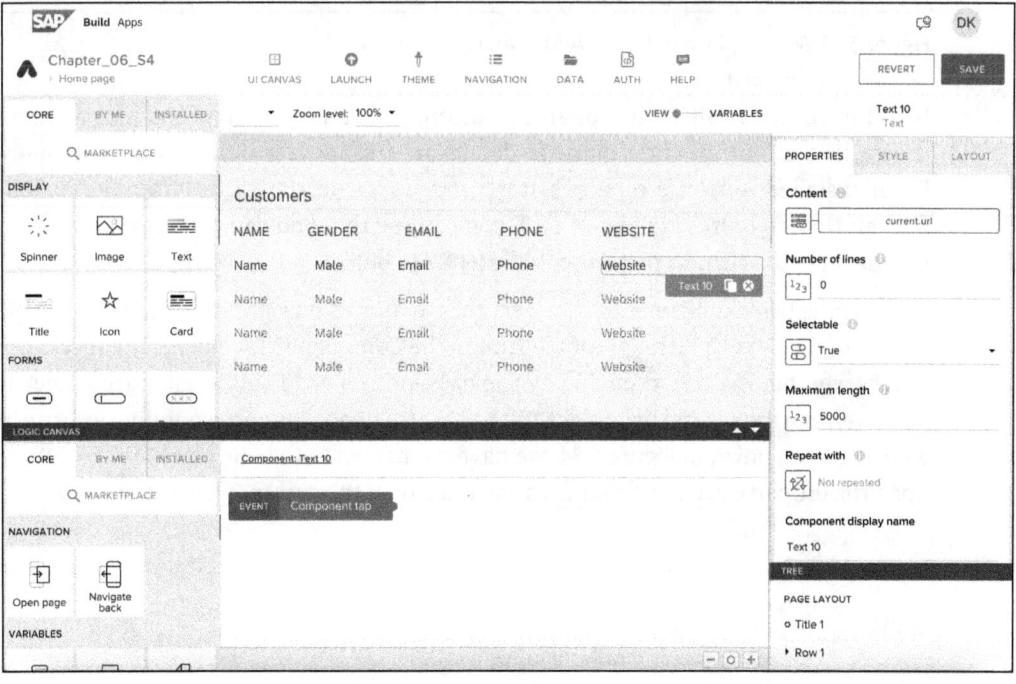

Figure 6.31 Table Prepared for Displaying All Customers

6 Data Integration and Authentication

In the Marketplace, you can search for additional UI elements as well as logic blocks that are not available in the standard system. These extensions and additional elements can either be delivered by SAP itself or created and provided by the community. In the search shown in Figure 6.32, we searched for "URL" and found two extensions that could fit the requirements: **Can Open URL?** and **Open URL**.

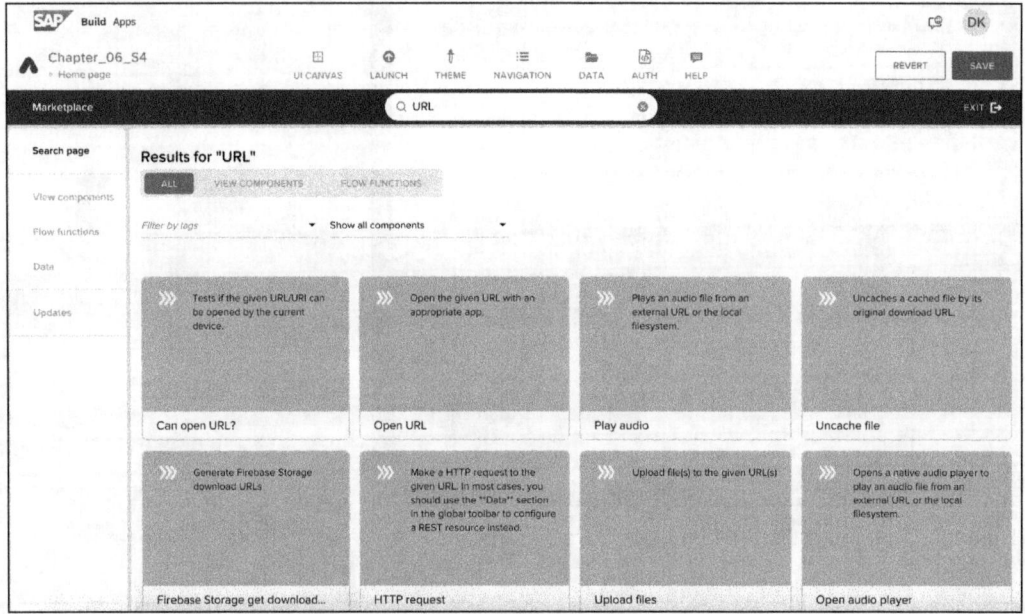

Figure 6.32 Marketplace to Install Additional Content

If you open the detailed view of an extension, you can see a short and also a longer description of how it works. Further details are included, such as release date, publisher, installed version, a version history, and, if your version differs from the current one, an **Update** button. In Figure 6.33, you can see that you can install or activate an extension in the current project with the **INSTALL** button at the top right.

Activate the two extensions and connect the two blocks one after the other in the logic canvas of the **Text** UI element for **website** in the **Component Tap** event. The first block checks whether the URL provided by the backend is a valid URL at all. If this check is positive, then the second logic block follows, which opens the corresponding website in a new tab. As shown in Figure 6.34, we have connected both logic blocks to the URL to open the property via data binding to the website of the current element.

6.2 Using Data from SAP Systems

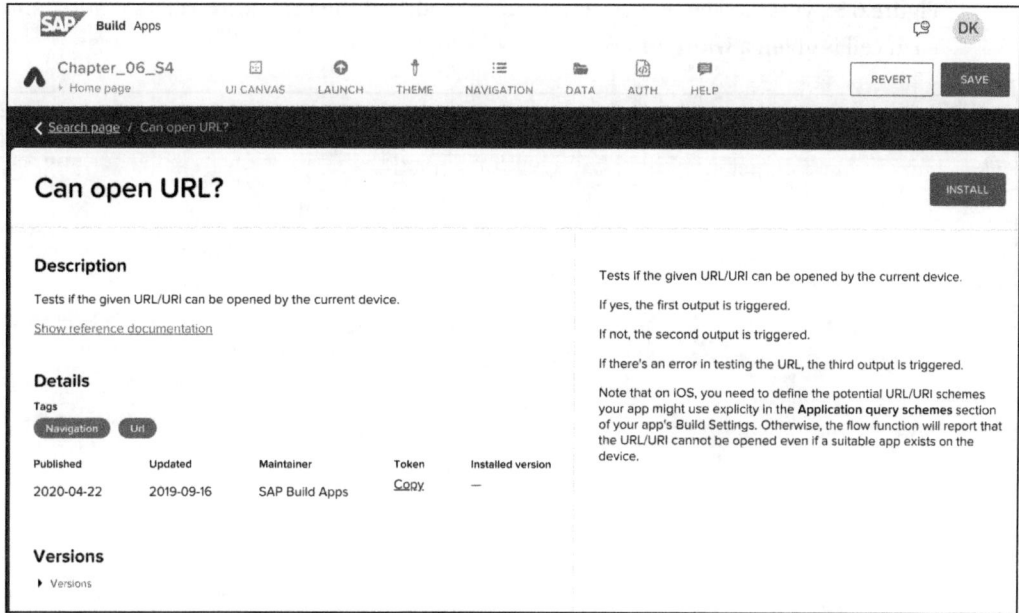

Figure 6.33 Extension Provided by SAP

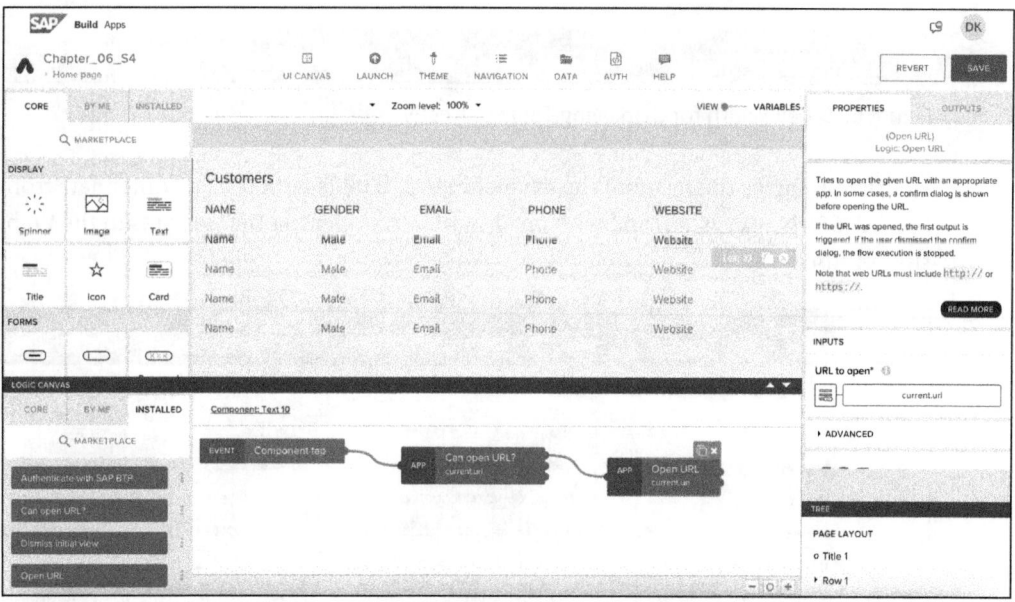

Figure 6.34 Logic Canvas for Opening Website when Text Is Clicked

Before we discuss the result, we have a tip to share. The width of the corresponding cells is more or less aligned with the width of the text elements. For this reason, it could happen that individual columns are not displayed neatly one below the other. In

Figure 6.35, you can see that we assigned the width of the individual cells as ratios, and each cell is given a width of 1:6.

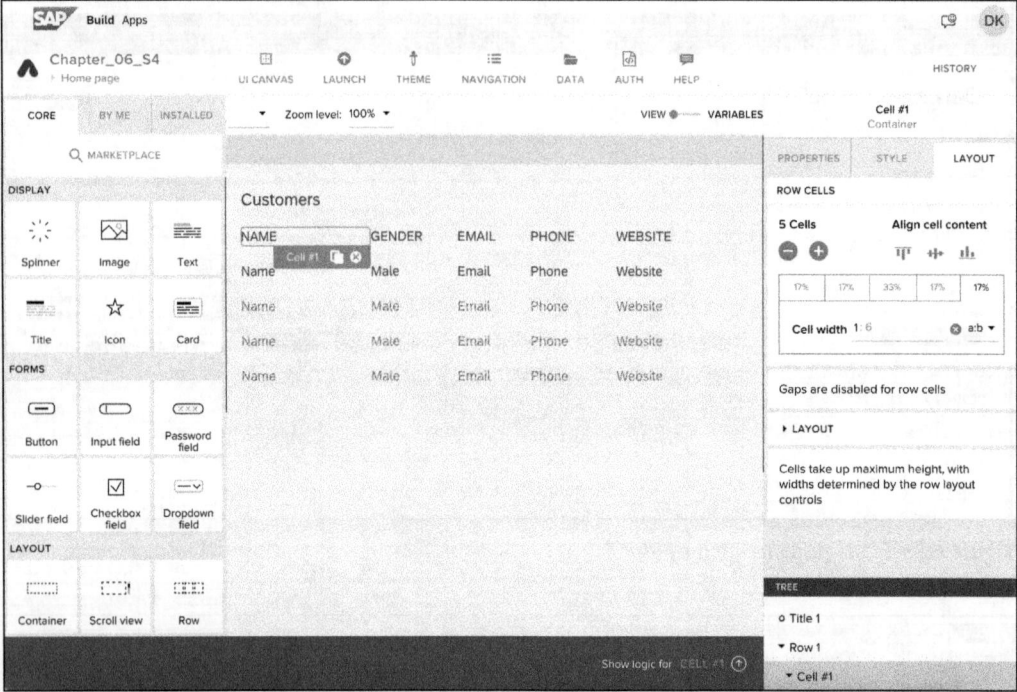

Figure 6.35 Cell Width for Displaying Table Correctly

Figure 6.36 shows the result of the developments: a table with the customer data from the SAP S/4HANA system and with an extension for opening the websites in the **Website** column.

Home page				
Customers				
NAME	GENDER	EMAIL	PHONE	WEBSITE
Koch Martin	Male	MARTIN.KOCH@CLOUDDNA.AT	+436641234567	https://clouddna.at
Krancz Daniel	Male	DANIEL.KRANCZ@CLOUDDNA.AT	+431234567892	https://clouddna.at
Glavanovits Rene	Male	RENE.GLAVANOVITS@CLOUDDNA.AT	+431111222233	https://clouddna.at
Olzinger Maximilian	Male	MAXIMILIAN.OLZINGER@CLOUDDNA.AT	+432222222222	https://clouddna.at
Mustermann Max	Male	MAX.MUSTERMANN@CLOUDDNA.AT	+431212121212	https://clouddna.at

Figure 6.36 Preview of Table Displaying Data of Customers

Once again, we don't want to leave the example application after following a simple use case; we'd prefer to extend it and show you further possibilities in application development with SAP Build Apps. So let's add a search dialog on the right above the table. In

6.2 Using Data from SAP Systems

Figure 6.37, you can see what we have already implemented. We'll briefly summarize these steps for you:

1. Create a **PAGE VARIABLE** of the **Text** data type and with the name **searchVar**, which will store the search text.
2. Insert a container above the table, which still contains the table heading, but now also contains the **Search Bar** UI element. The formatting means that the heading is left-aligned and the search bar is right-aligned.
3. Bind the **Value** property in the search bar to the created **searchVar** page variable via data binding.

You already know all these steps from the previous exercises. Let's focus on the business logic of the search bar, which ensures that the entered text is applied to the customer name as search text. To do this, change the event to **Property 'Value' Changed** in the logic canvas of the search bar and use **Get Record Collection** to query the customer data from the backend again. Save the result of the read operation again in the ZHOUI5_CUSTOMERSET1 data variable using the **Set Data Variable** block.

Now open the **Get Record Collection** logic block again and click the **Data Binding** button in the **Filter Condition** property on the right-hand side.

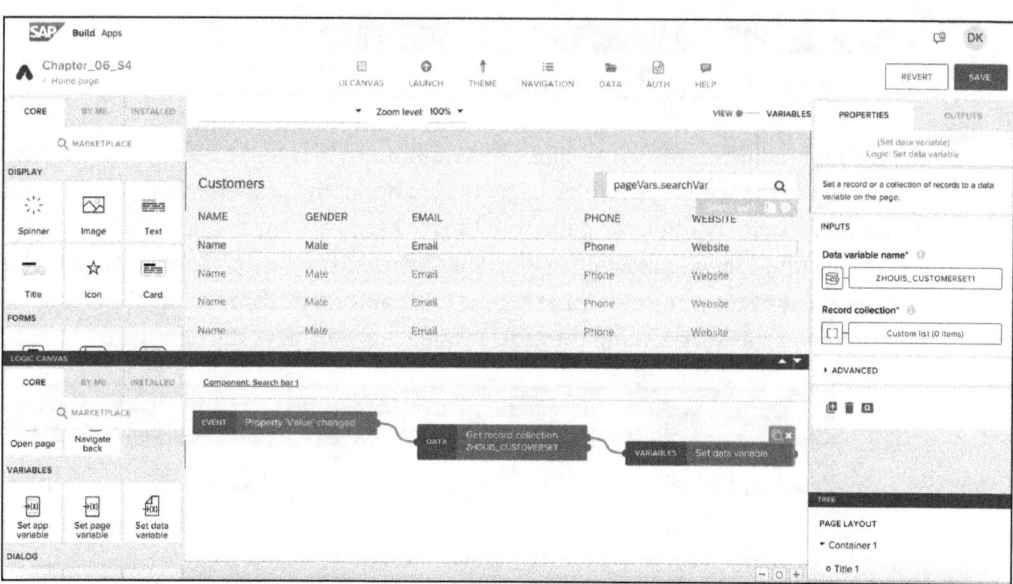

Figure 6.37 Search Field for Filtering Data

As shown in Figure 6.38, you now have the option of defining two filter criteria. The first filter is dedicated to **Firstname** and checks whether the content of the **searchVar** page variable matches. The same is also checked with the second criteria for **Lastname**. These two filter criteria are linked with **ANY**. This means that at least one of the filter criteria must match.

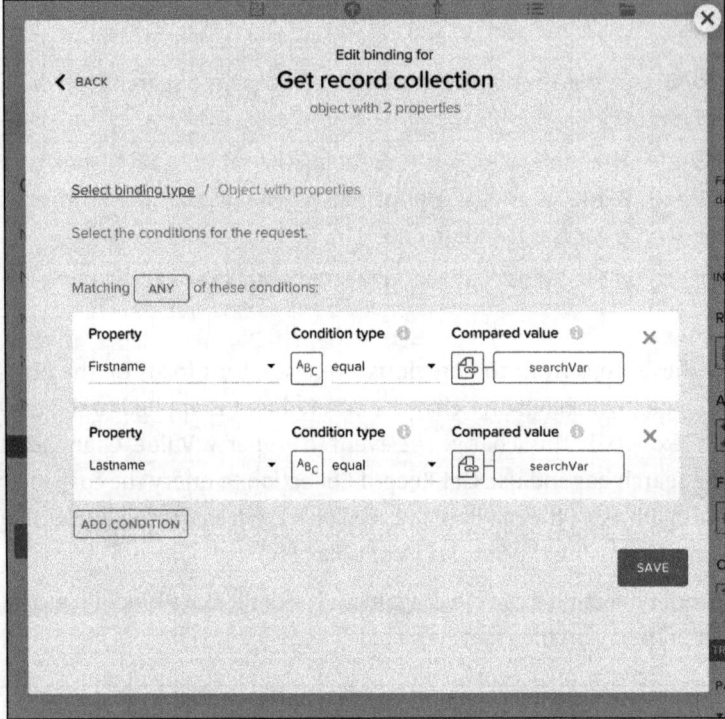

Figure 6.38 Filter Criteria Defined in Logic Block

These simple adjustments to the application are shown in Figure 6.39. Now the user can enter a text in the search bar, which triggers a change of the property value and triggers the logic blocks for reading the data from the backend with the text as a filter and then sets the result in the second step as content in the existing data variable. As the table displays the data from the data variable, the table automatically receives the change to the content of this variable and the table immediately displays the new data.

Figure 6.39 Filtering Works as Expected

6.3 Non-SAP Systems

We do not want to limit ourselves to SAP systems. You may well want to consume data from a web server via the familiar HTTP protocol, but this is neither an OData service

6.3 Non-SAP Systems

nor does it provide data from an SAP system. A classic example would be a service developed with Java and the Spring Framework, which in turn provides APIs for consuming and manipulating data.

> **Information from the Last Sections**
>
> Although we have dealt with other topics in the last sections (consuming SAP systems), from the consumer's point of view (in this case, for an SAP Build Apps application), the same approaches, practices regarding data binding and user interface structure, and many other topics play together. For this reason, we recommend that you also take a look at the other sections, as we are confident that the individual steps and basics are familiar and can be applied again.

In this section, we'll look at how to integrate a web server that is accessible via a URL and consume data. As is often the case, start in the **DATA** tab. There, we have already integrated two APIs or entities from a web server, which, as in the previous sections, was available as an SAP BTP destination. These two entities are **Products** and **Categories** (see Figure 6.40): several products can belong to a category, but each product can only ever belong to one category.

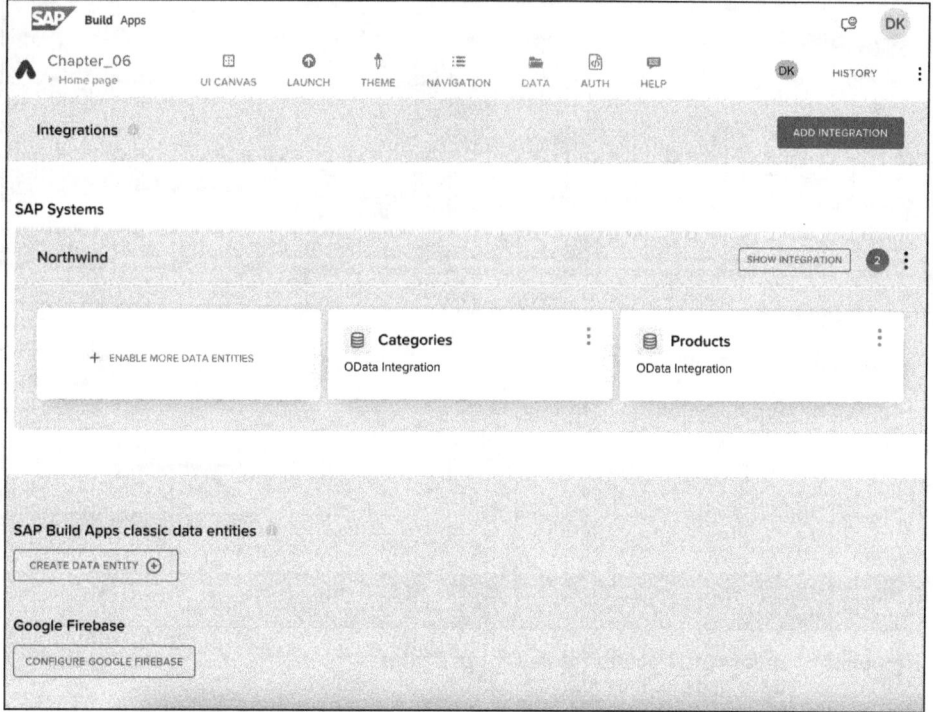

Figure 6.40 Entities for Products and Categories Enabled

First, let's display a list of all products, with the list title being the product name. On the right-hand side (as the secondary label), let's display the quantity currently available in the warehouse and also the quantity already ordered or reserved. To do this, carry out the following steps, which you've already seen in the previous sections (see Figure 6.41):

- Create a data variable for both the **Products** entity (**productsList**) and the **Categories** entity (**categoriesList**). Both contain a collection of the respective entities.
- Place a title with the value **List of Products** in the interface as a UI element.
- Place a **List Item** element in the view.
- The **Repeat With** property of the list item is bound to the **prdoctsList** data variable, which represents the list of products.
- The **Primary Label** property is bound to the **Data** item in repeat and displays the **ProductName** attribute of this entity.
- The **Secondary label** stores a formula which displays the **UnitsOnOrder** and **UnitsInStock** attributes of the data item in repeat. The formula looks like this:

  ```
  repeated.current.UnitsOnOrder + "/" + repeated.current.UnitsInStock
  ```

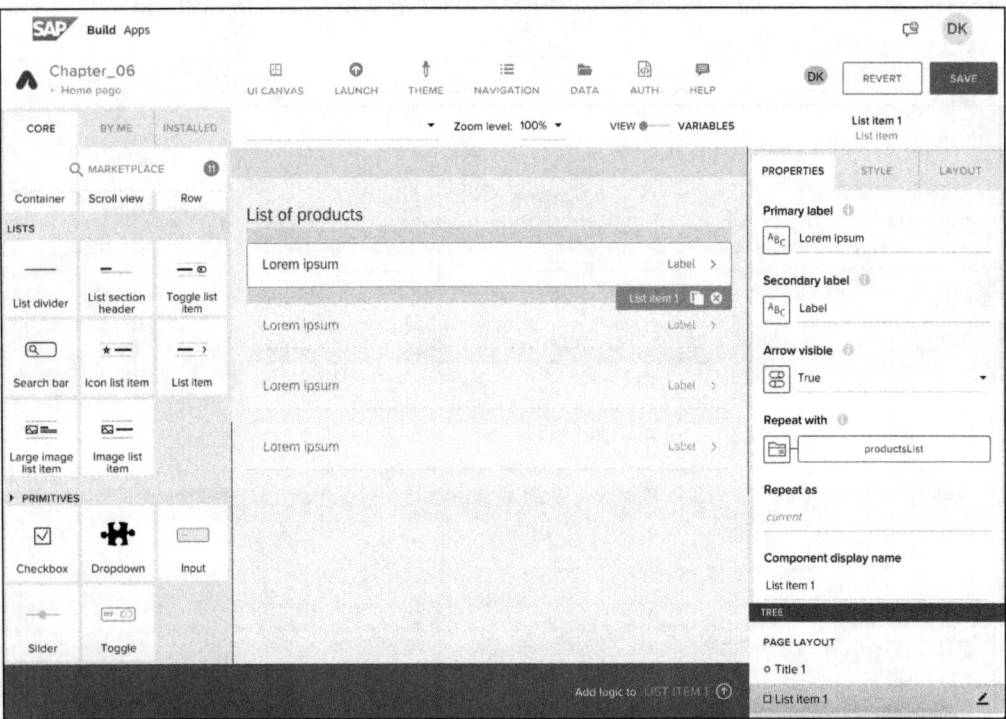

Figure 6.41 Repeat With Used to Display List of Products

Figure 6.42 shows the result of these steps: the list of products with the corresponding ordered quantities and the available stock.

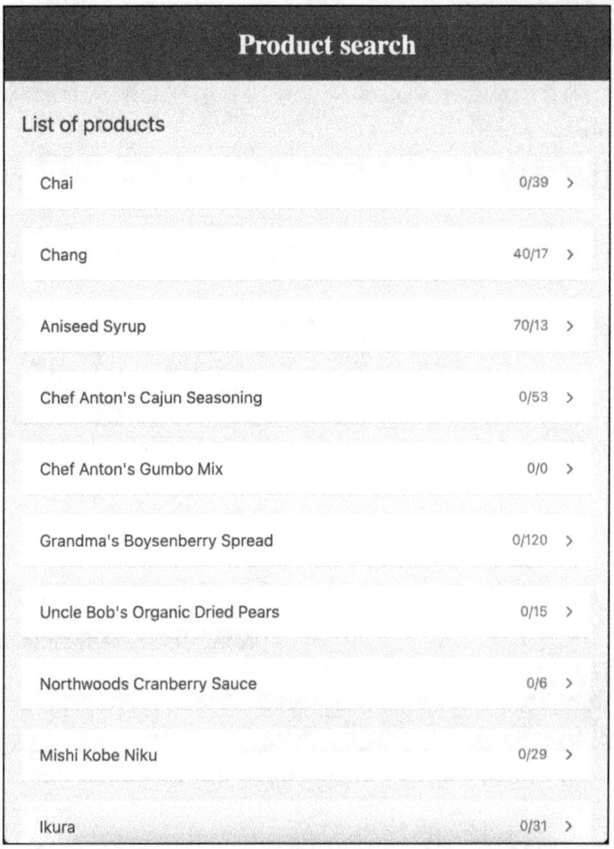

Product search	
List of products	
Chai	0/39 >
Chang	40/17 >
Aniseed Syrup	70/13 >
Chef Anton's Cajun Seasoning	0/53 >
Chef Anton's Gumbo Mix	0/0 >
Grandma's Boysenberry Spread	0/120 >
Uncle Bob's Organic Dried Pears	0/15 >
Northwoods Cranberry Sauce	0/6 >
Mishi Kobe Niku	0/29 >
Ikura	0/31 >

Figure 6.42 Products Displayed with Corresponding Quantities

Let's now expand the application by offering the list of categories using a dropdown field and filtering the list of products with the corresponding technical key of the category by selecting it. So that this technical key can be stored, create a page variable with the name **selectedCategory** and the data type **text**. Bind this variable to the **Selected** value property of the dropdown field.

To do this, insert the **Dropdown Field** UI element between the heading and the table and assign the **Select Category Filter** text to the **Label** text property (see Figure 6.43). Then click the data binding symbol for the **Option List** property, which represents the actual selection values in the dropdown.

The reason we want to show you this extension in this example is that the dropdown field has a rather strange behavior. An element in the dropdown field always consists of a label and the value behind it. The *label* is the text that the user also sees in the user interface. The *value*, on the other hand, is the technical ID or primary key of the individual object. This could be called a key-value pair relationship. As the structure of the **Category** entity does not correspond to this structure (which technically consists of a key-value pair), it's a little tricky to carry out the data binding at this point.

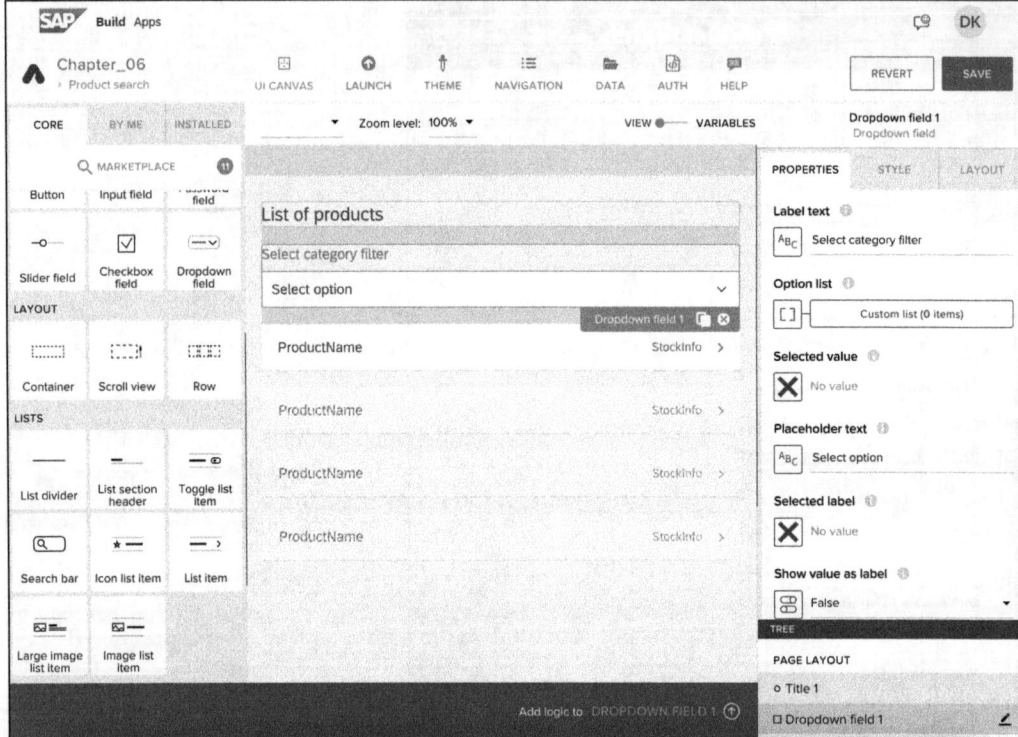

Figure 6.43 Dropdown Field Used for Filtering Products via Product Categories

For the data binding type, shown in Figure 6.44, select the **Mapping** type for the first time (as it isn't available for all data bindings either).

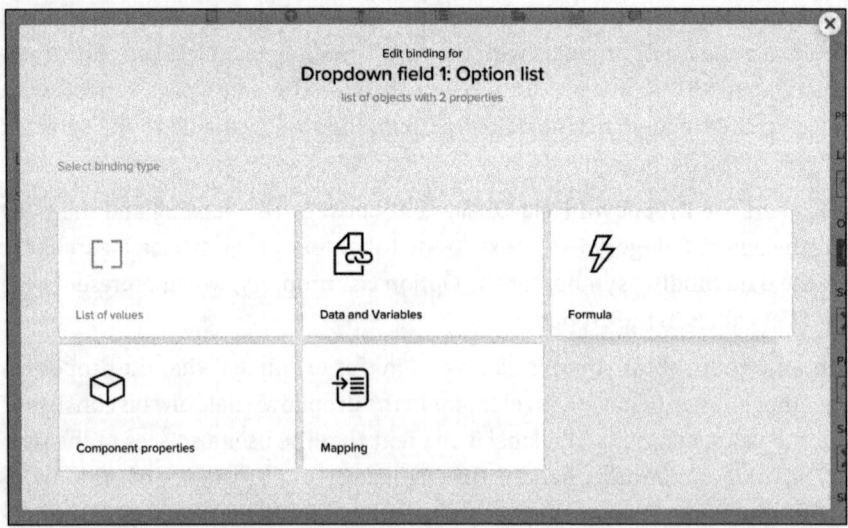

Figure 6.44 Using Data Binding for Items in Dropdown Field

Mapping is about selecting the two attributes from a larger and more extensive structure that represent this key-value pair—that is, the label and value (as shown in Figure 6.45). In this dialog, you can first select the source data (i.e., the **categoriesList** data variable in the case) and then assign the attributes to these two values using drag and drop.

> **Matching the Data Types**
>
> Both the label and value are of the **Text** data type for the list option. If the data type of the selected attribute does not match **Text**, you will not be able to assign it using drag and drop. We recommend (as we show in Figure 6.45) changing the data type via **Formula** to ensure that, for example, a true/false or number can still be assigned to the **Text** data type.

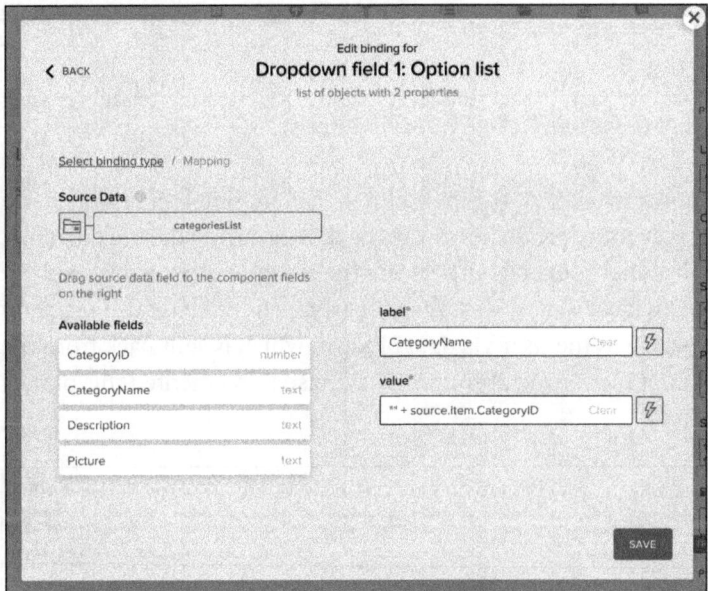

Figure 6.45 Using Data Binding Type Mapping

The last steps are only the correct application of the selection in the dropdown field and the setting of the filters. In Figure 6.46, you can see that we have opened the logic canvas for the dropdown field and carried out the following steps:

- We have changed the event that triggers the logic blocks one after the other to **Property 'Selected Value' Changed**.
- In the first subsequent logic block, **Get Record Collection**, we read the data from the **Products** data resource, sending the value of the **selectedCategory** page variable for the **CategoryID** attribute as a filter condition.
- In the last logic block after this, we use the output of **Get Record Collection** and write the returned values to the **productsList** data variable. As the list is also bound to this data variable, the list in the interface will also change automatically.

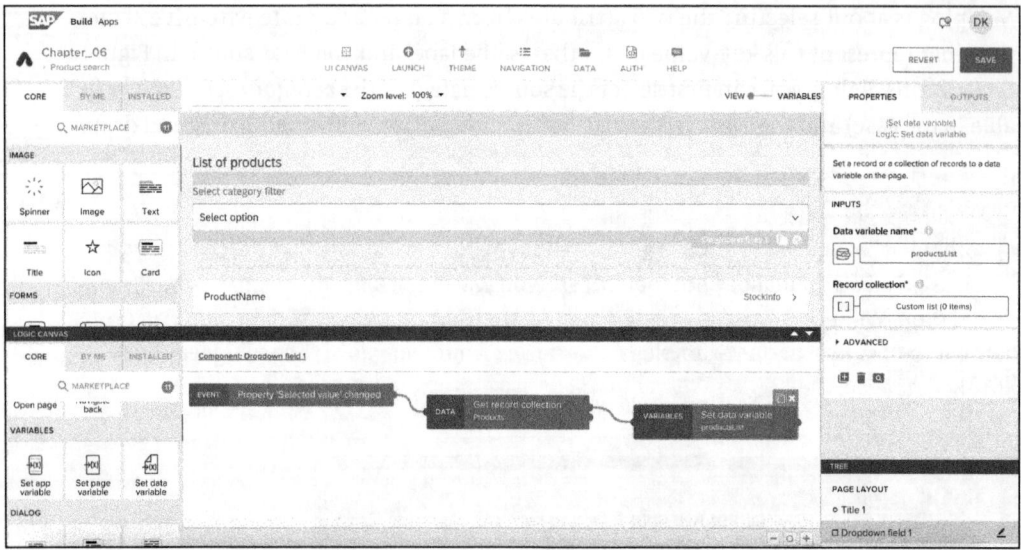

Figure 6.46 Logic Canvas of Dropdown Field to Filter Product List

What we haven't yet shown you is that the logic blocks that are sent to a remote backend using the request-reply principle do not always have to return a positive response. For this reason, such blocks have several outputs. For the second output, you could use the **Alert** logic block, which has a **Dialog Message** property. You could use this property to display the error message to the user via data binding, which is sent back from the backend and is the result of the second output. We have also implemented this project and shown it in Figure 6.47.

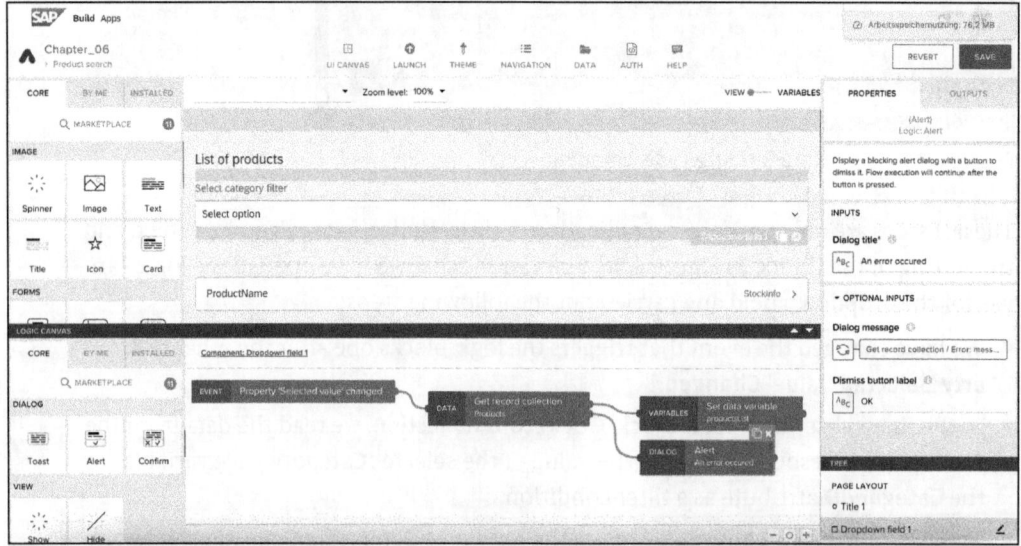

Figure 6.47 Using Error Callback and Showing Error in Alert Dialog Message

The application is now ready for use. In Figure 6.48, you can see that we have turned the application into a single-page application. Furthermore, the list below is automatically created based on the category selection in the dropdown.

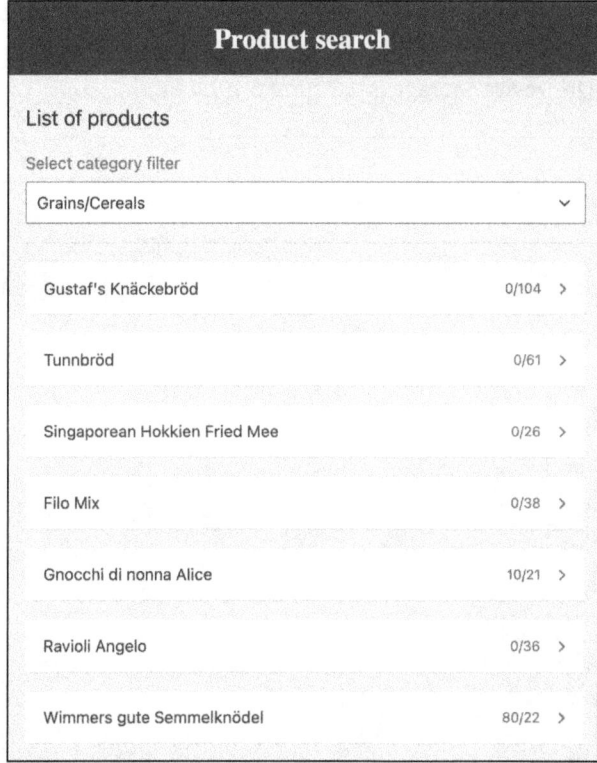

Figure 6.48 List of Products Filtered by Category

6.4 Local Data Storage

You have probably already worked with to-do lists yourself. Regardless of whether you use them on paper or in electronic form, the point is that you make a note of certain tasks so that you don't lose sight of them or forget them. Why not create a task list as an SAP Build Apps application? We limit ourselves here to local data storage.

It should be possible in the app to see an overview of all tasks, create new tasks, edit tasks, and mark a task as completed. In this section, you will also be confronted with navigation and transfer parameters during the navigation process.

> **Information from the Last Sections**
>
> Although we have dealt with other topics in the last sections (consuming SAP and non-SAP systems), from the consumer's point of view (in this case, for an SAP Build Apps application), the same approaches, practices regarding data binding and the structure

6   Data Integration and Authentication

> of the user interface, and many other topics play together. For this reason, we recommend that you also take a look at the other sections, as we are confident that the individual steps and basics are familiar and can be applied again.

As in the previous sections, start in the **DATA** tab. There, click for the first time the **CREATE DATA ENTITY** button under **On-Device Storage** (see Figure 6.49).

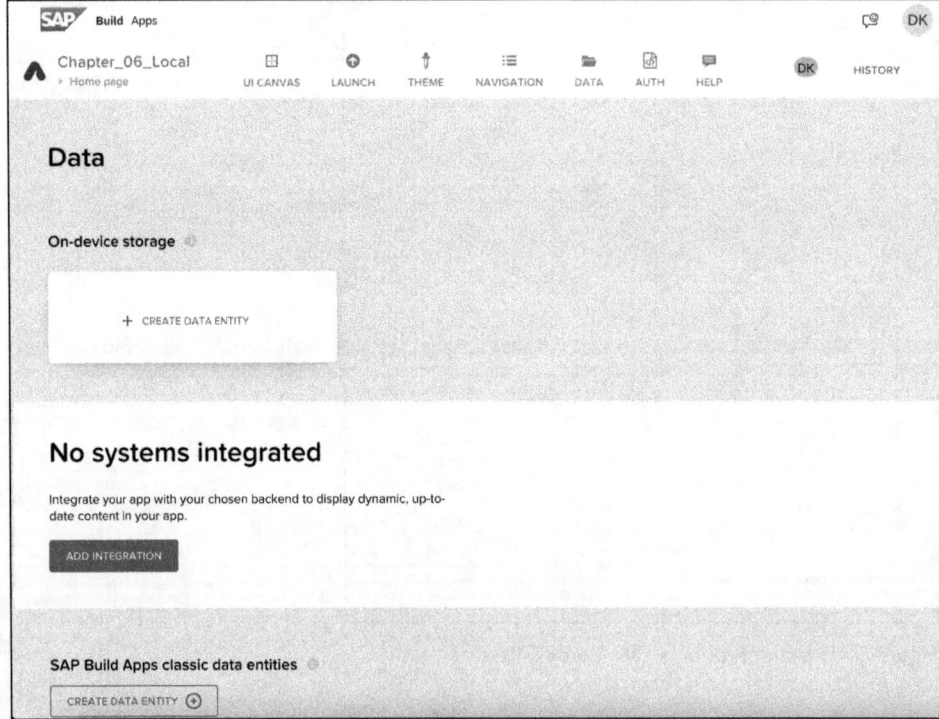

Figure 6.49 Using On-Device Storage for Saving Data in Your App

In this case, as with the last sections and a backend connection, the entity names are not predefined; you as the builder can assign a name. In Figure 6.50, you can see that we have assigned the name "task" to the new entity.

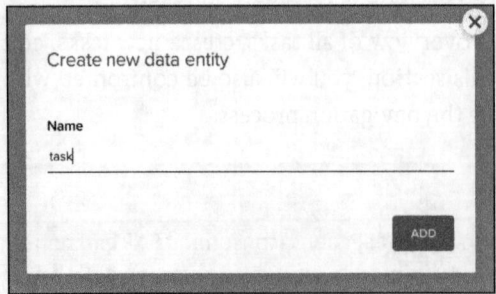

Figure 6.50 Create New Data Entity in Local Storage

6.4 Local Data Storage

As the structure of the entity is not predefined for a locally created data resource, you now have to perform these tasks in the application itself. An entity is always identified by an **ID**, whereby this attribute of the **text** data type is already predefined. You can now create further attributes in addition to the ID by clicking the **ADD NEW** button (see Figure 6.51).

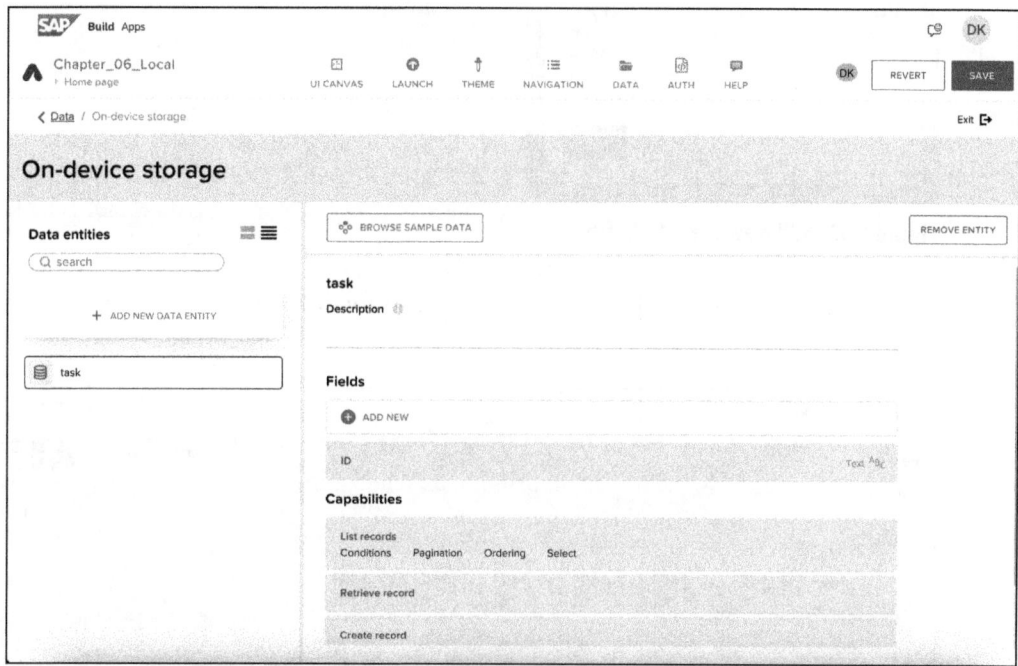

Figure 6.51 Add More Entities or Fields to Existing Entities

A new attribute first needs a name and the corresponding data type. Such attributes can, as shown in Figure 6.52, directly contain a primitive type (e.g., **text**, **number**, or **true/false**), but also a list or an object. For this example, create the additional attributes listed in Table 6.1.

Field Name	Field Type
"title"	text
"description"	text
"finished"	true/false

Table 6.1 Fields of Entity to Be Saved in Local Storage

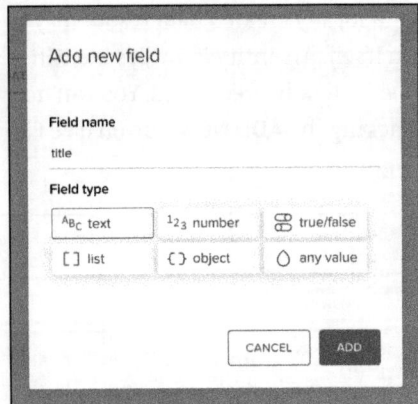

Figure 6.52 Add New Field to Entity

In Figure 6.53, you can see that you now have an entity in the on-device storage, which you can use with the created attributes to store data directly on the device.

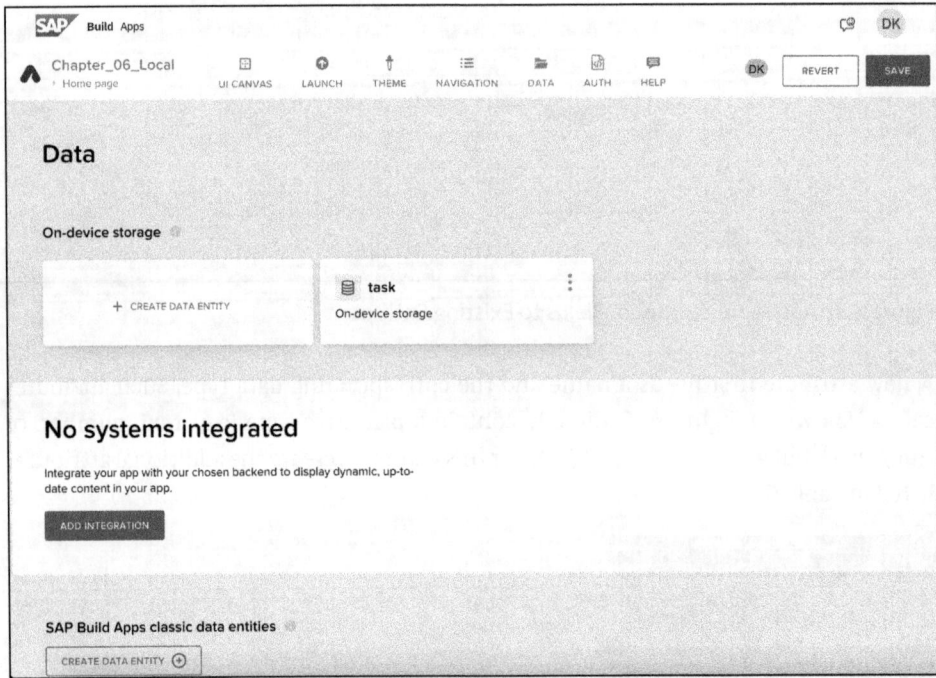

Figure 6.53 Entity in On-Device Storage Added

You will no doubt be familiar with the further development steps from the previous sections. For this reason, we will summarize what we have already done in Figure 6.54:

- We have created a **tasks** data variable, which should store a collection of the **task** entities.

- In the canvas UI, we have inserted a **Toggle List Item** element for the first time.
- We have bound the **tasks** data variable to the **Repeat With** property so that this item is repeated as often as there is data in our data variable.
- We have bound the **Content** property to the **title** of the data item in repeat.
- We have bound the **Toggle Value** property, which represents the toggle button on the right-hand side of the item, to the **finished** attribute of the data item in repeat.

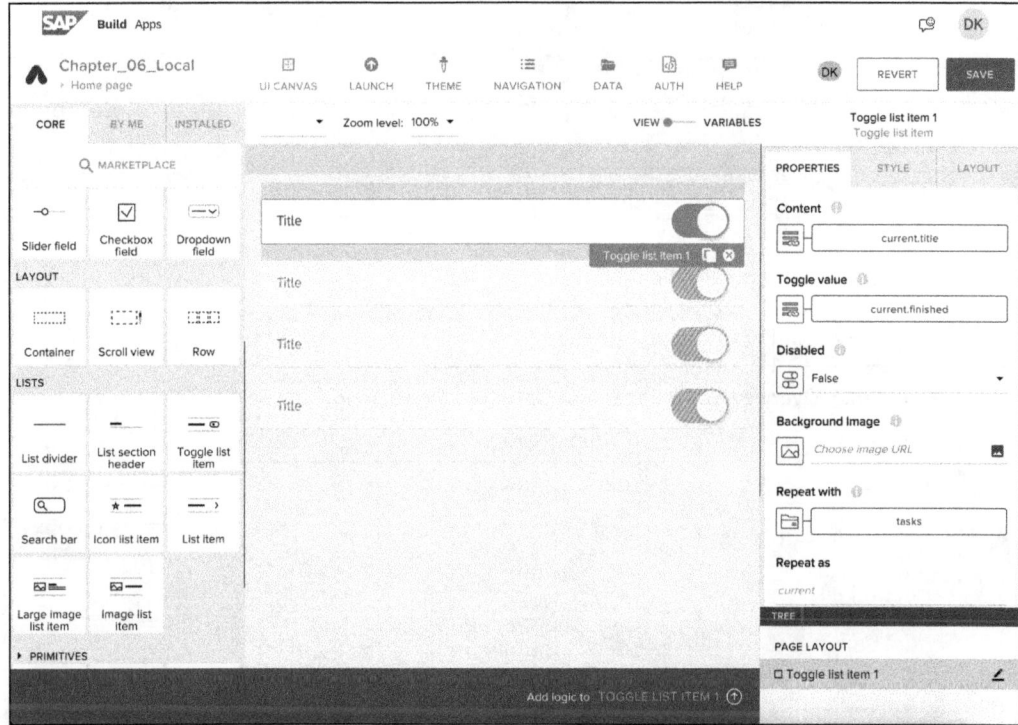

Figure 6.54 Creating List of Tasks Using Toggle List Item UI Element

Now let's create a second page for the first time, which displays the details of a task and/or allows a task to be edited or created.

Click the name of the page (in this case, **Home Page**) at the top left under the name of the project (in this case, **Chapter_06_Local**) to see all the pages that have been created (see Figure 6.55). Now click the **ADD NEW PAGE** button to create a new page. This will open a dialog with a single input field in which you must enter the name of the new page. We have assigned the name "Taskdetails".

On this new page, create a new parameter called **taskID** for **VARIABLES** on the left-hand side under **PAGE PARAMETERS** (see Figure 6.56). This parameter allows the ID of the task to be displayed to be specified when switching between pages. We'll show you how to define this behavior in a few steps.

205

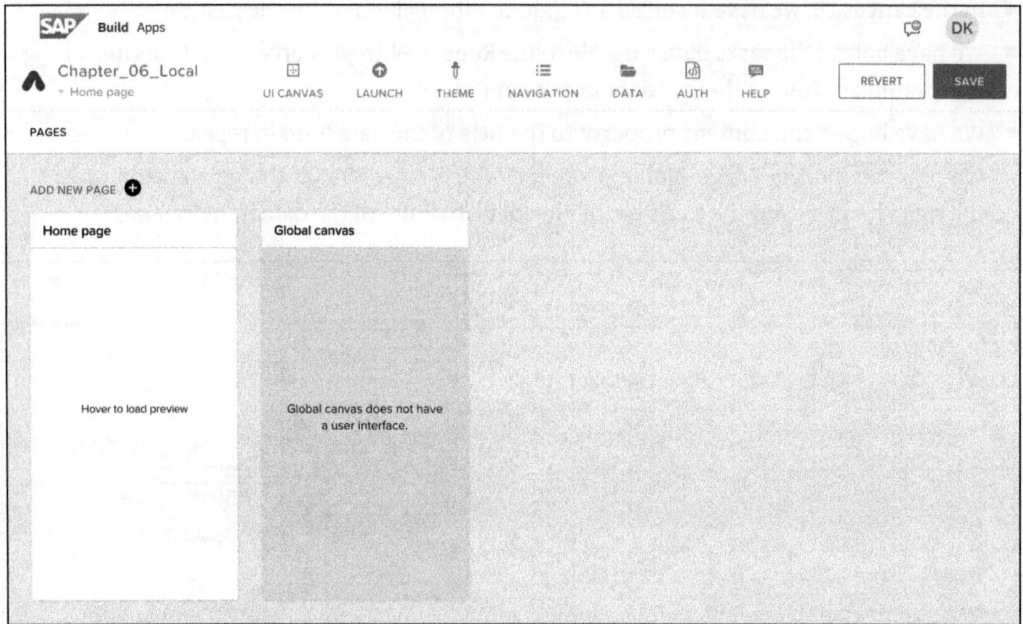

Figure 6.55 Overview of All Pages and Option to Create New Page

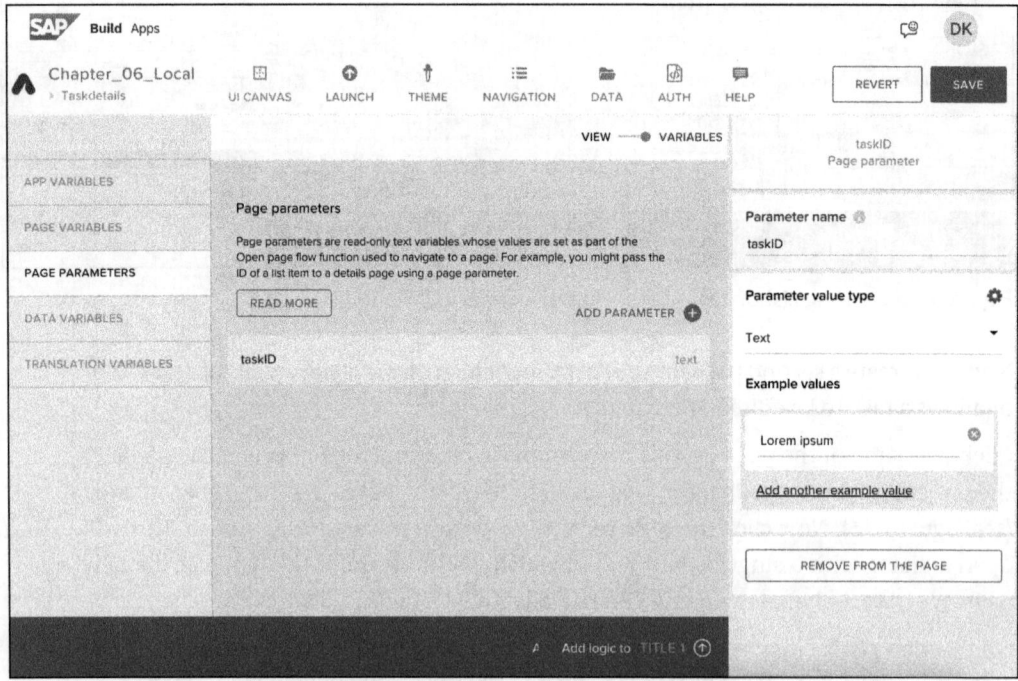

Figure 6.56 Add Page Parameters to Provide Option to Pass Data while Navigating

6.4 Local Data Storage

Let's also create **PAGE VARIABLES**, where we create an object called **variable1**. This object has exactly the same four attributes (**id**, **title**, **description**, and **finished**) that a task has (see Figure 6.57). Because we are using the same structure here, with the same names and data types, this will help us enormously later when reading and writing the data.

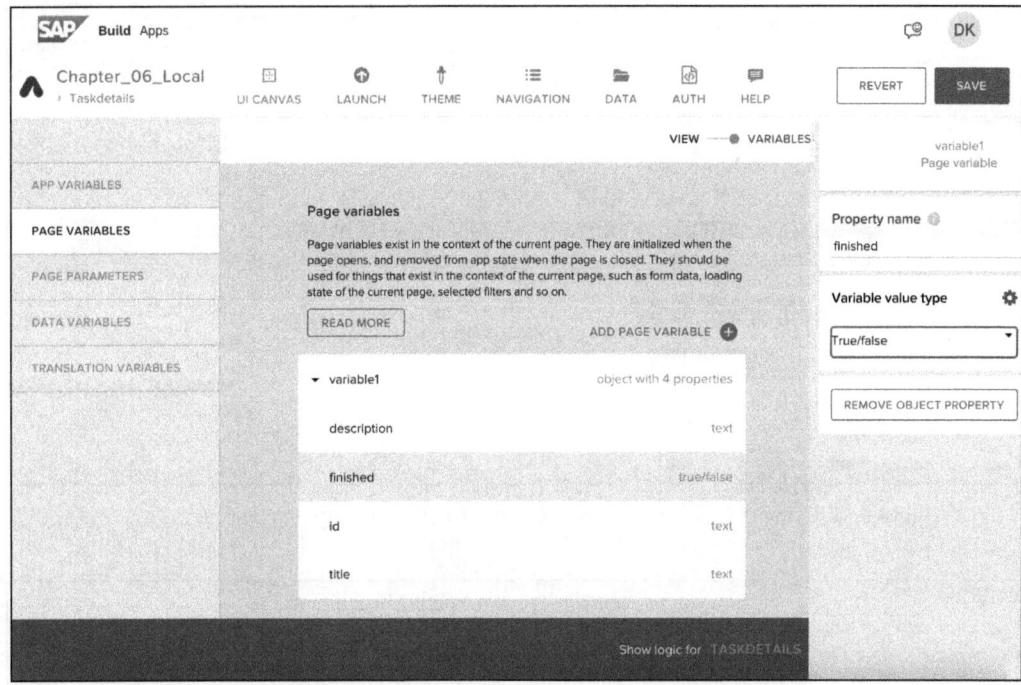

Figure 6.57 Creating Page Variable as Object/Structure with More than One Field

Now let's go back to the UI canvas and, without having selected a UI element, open the logic canvas. Here you can store logic blocks for the view and react, for example, when the view is opened or navigated to. You can see in Figure 6.58 that we react to exactly these two events and add the IF block immediately after them. In this IF, we use the following query to check whether a parameter has been specified or not:

IS_EMPTY(params.taskID)

In Figure 6.59, you can see which logic blocks we have provided for the detailed view. If the query was successful—that is, the parameter is empty—then we also set **variable1** to an empty object with the subsequent **Set Page Variable** logic block. Why are we doing this? We know that if no parameter has been provided, a new object will be created, and this is the easiest way to clear the form. If a parameter has been specified, we go the other way from the **IF** and first call up the details for a task with the specified **taskId** using **Get Record** and write the read values in the subsequent **Set Page Variable** logic block to the **variable1** variable and thus prefill the subsequent form with the current task details. The next step shows how **variable1** can be filled with the returned data.

207

6   Data Integration and Authentication

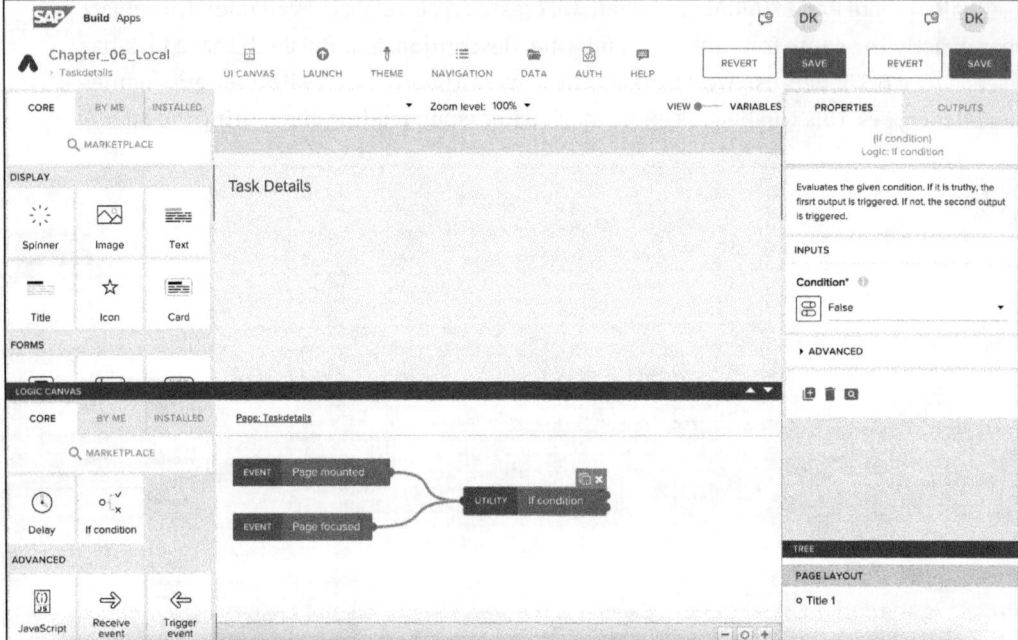

Figure 6.58 Using IF Condition in Logic Canvas to Check if Page Parameter Is Provided

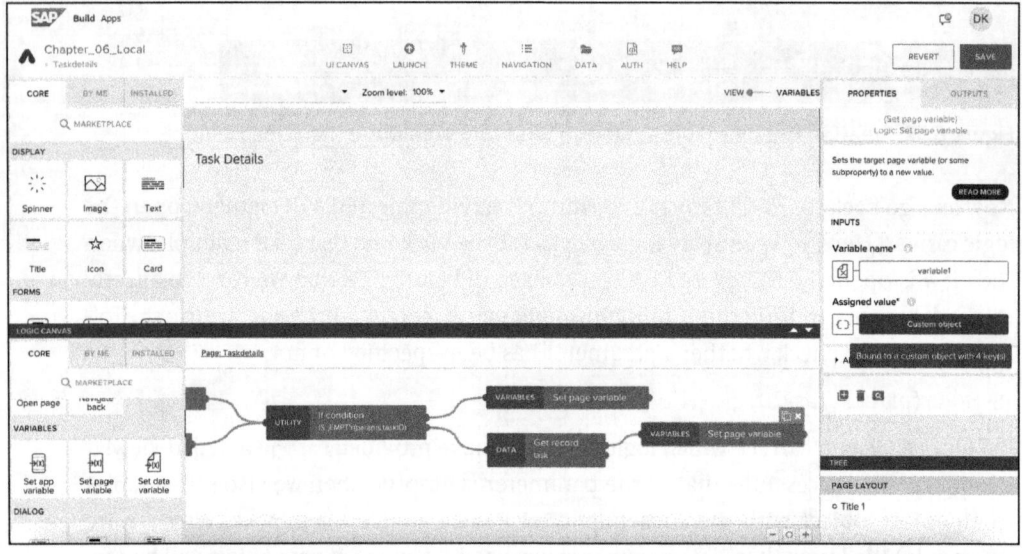

Figure 6.59 Resetting Page Variable or Reading Data from Storage and Setting Data as Content for Variable

We are now in the **Set Page Variable** logic block in the data binding of the **Assigned Value** property. Here we have selected **Object with Properties** as the binding type and set all attributes for each attribute in the **variable1** variable one after the other from the result of the **Get Record Collection**. This mapping or assignment of the data is shown in Figure 6.60.

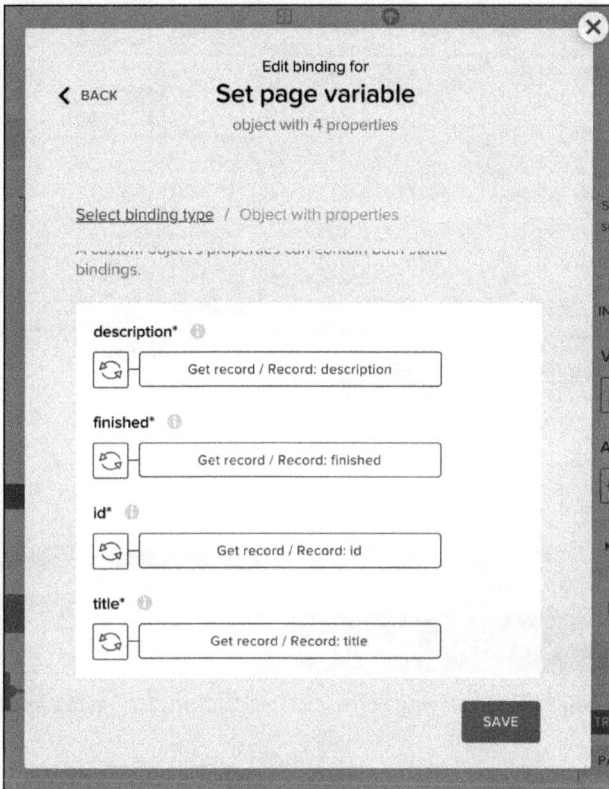

Figure 6.60 Mapping of Fields while Setting Page Variable

As the last steps for the detailed view, go to the corresponding form and perform data binding (see Figure 6.61):

1. Insert an **Input Field** for the task name. The **Value** property of the input field is bound to **variable1.title**.
2. Insert another **Input Field** for the task description. The **Value** property of the input field is bound to **variable1.description**.
3. Insert a **Checkbox Field** and bind the **Value** property of the checkbox to **variable1.finished**.
4. Insert two buttons with the names **Save** and **Cancel**. As the names suggest, you should be able to save tasks with **Save** and return to the list with **Cancel**.

The logic for the **Save** button is shown in Figure 6.62. Here you check in the first step with an **IF**, just as with the logic for the page, whether the page parameter is filled. If not, then you know that you need a **Create Record** logic block to be able to create a new task.

6 Data Integration and Authentication

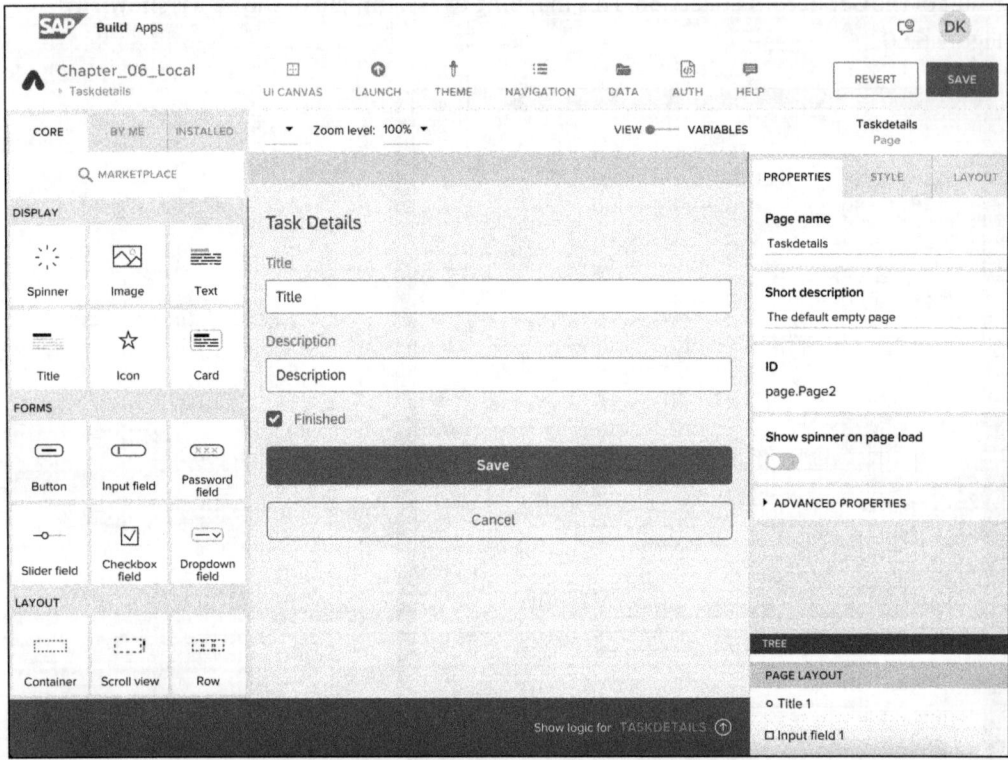

Figure 6.61 UI Canvas for Task Detail Page Displaying Form and Two Buttons for Saving and Canceling

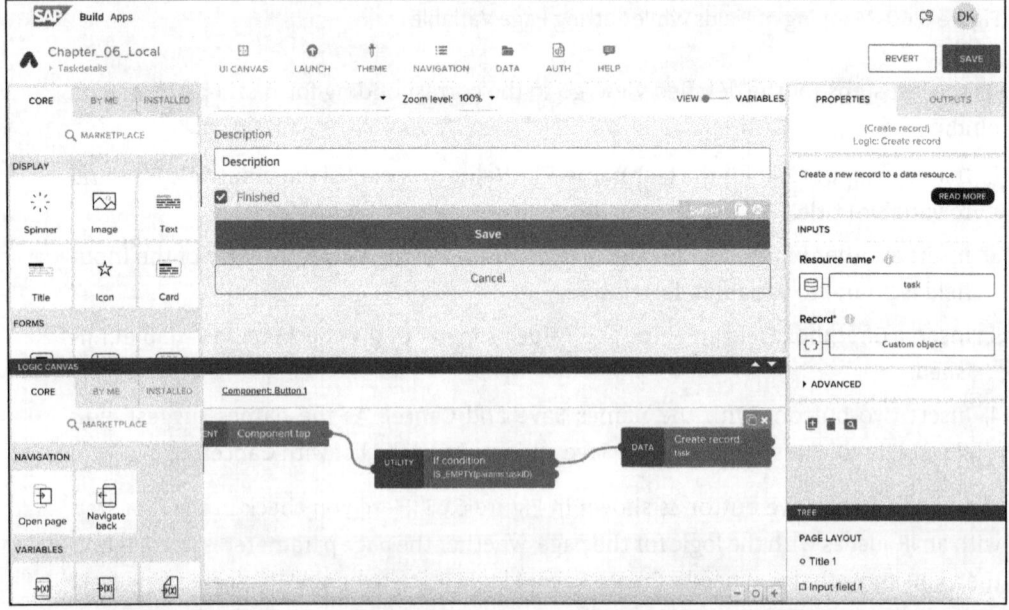

Figure 6.62 Logic Canvas of Save Button to Create New Record in Storage

In this **Create Record** logic, you must specify the data that must be created in the **Record** property. Because the form always binds to the attributes of **variable1**, you can take the data from it and send it along with the create action, as shown in Figure 6.63.

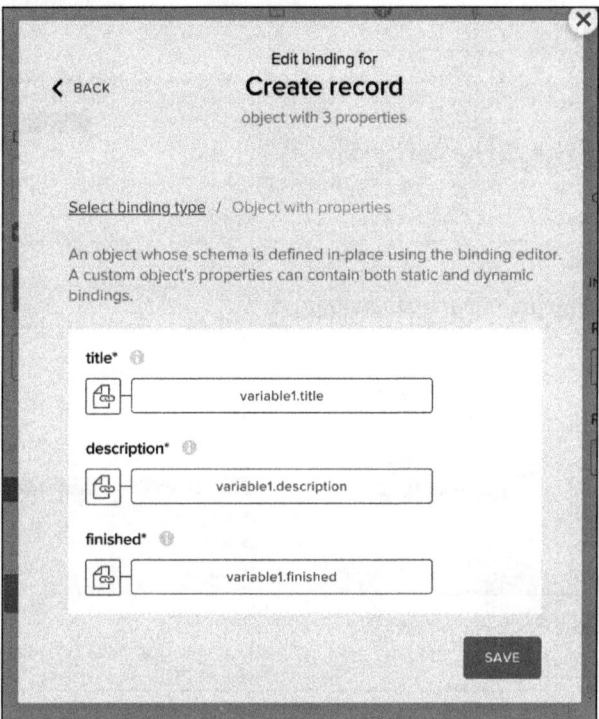

Figure 6.63 Creating New Record and Passing All Data Except ID, which Will Be Created Automatically

If a page parameter has been specified, you know that no new task needs to be created, but an existing one needs to be updated. For this reason, the **Update Record** logic block is executed in the second branch of the **IF** condition (see Figure 6.64). For the update record, in addition to the **Resource Name** and **Record** properties, the **ID** of the object must now also be specified, as was the case when the record was created. This ID is also stored in **variable1** under the **id** attribute and can therefore be specified directly via data binding.

In both cases, regardless of whether the creation worked or the update was successful, use the **Navigate Back** logic block to navigate back to the previous page, which in this case is the list (see Figure 6.65). Don't forget the **Cancel** button, which only needs to contain a **Navigate Back** in the logic canvas.

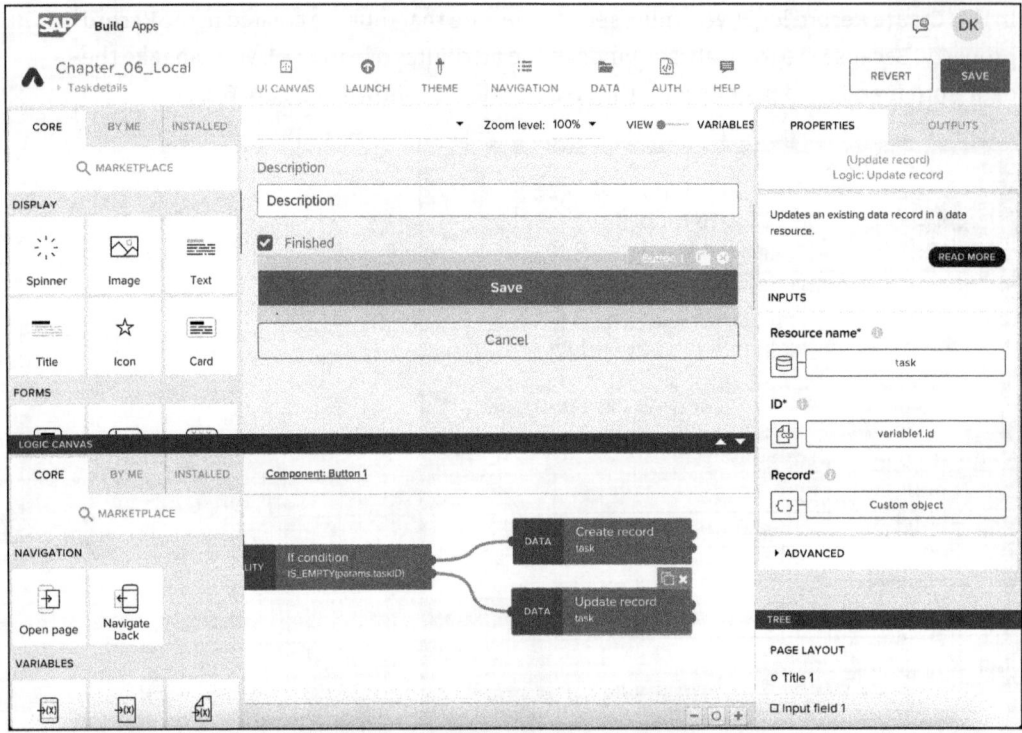

Figure 6.64 Update Existing Record if Page Parameter Was Provided

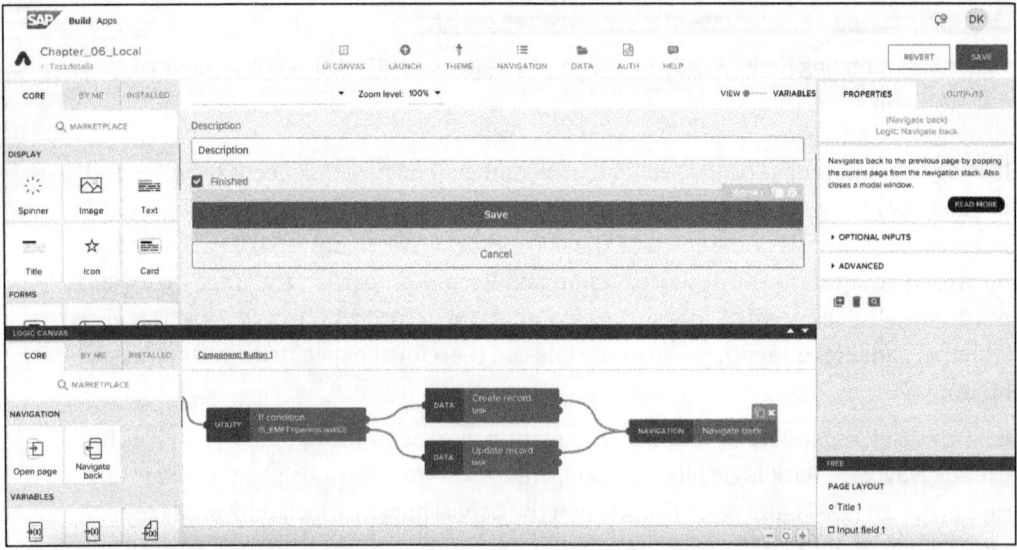

Figure 6.65 Navigate Back after Creation or Update Process Is Finished

Let's go back to the list and select a list element. The logic for navigating to the detail page is still missing for the list element. So, open the logic canvas from the list element and insert the **Open Page** logic block after the **Component Tap** event. For **Open Page**, the targeted page must be stored in the properties, as well as the ID of the current element, which can be accessed in the data binding with the data item in repeat.

We have also created a button and named it **Create Task**. When this button is clicked, you navigate to the detail page, but do not enter an ID, and so the detail page knows that you want to create a new task. You can see all these steps in Figure 6.66.

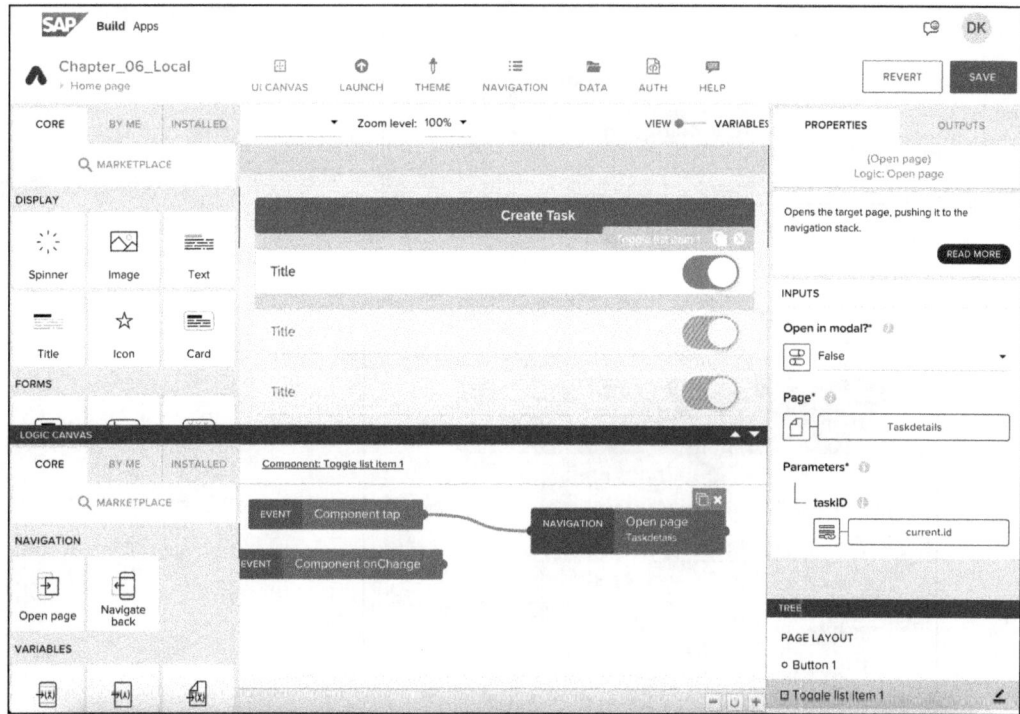

Figure 6.66 Open Page Task Details and Pass ID of Selected List Item as Page Parameter while Navigating

The time has finally come to test the application. In Figure 6.67, you can see that we have already created two tasks. The first task, **Test 1**, has already been completed, but the second has not. If you click **Create Task**, you are redirected to the details page.

The logic takes effect, and the details page knows that you want to create a new task, as no parameter was specified when navigating. You can enter the task details and then click **Save** to save the task in the device storage. If you click **Cancel**, you'll return to the list without saving. Figure 6.68 shows an example of the data that we have entered and will now save.

6 Data Integration and Authentication

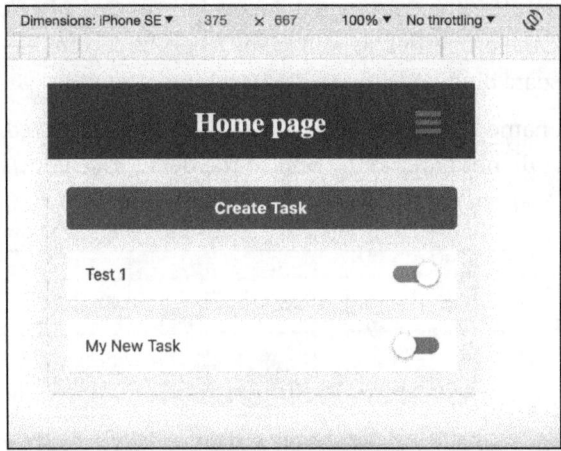

Figure 6.67 Preview of List of Tasks

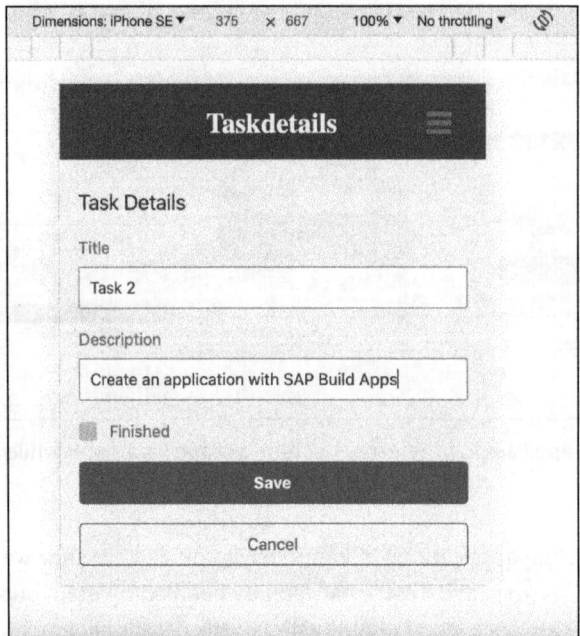

Figure 6.68 Creating New Task

6.5 Summary

In this chapter, we have shown you a lot about the topics of data integration and authentication, but also much more beyond that. You can now handle data variables and binding and display, as well as change and update data regardless of the source from which it is sent to the SAP Build Apps application. We also looked at other logic blocks in the logic canvas, looked at the Marketplace and installed extensions from it, talked about navigation, and tried out navigation with and without parameters and, at the end of the day, developed four different applications with filter and sorting options.

All in all, you have seen that, fortunately, it makes no difference to us builders whether the data resource we use is based on local storage, an SAP system, or a non-SAP system. We can always use the same logic blocks and data variables as facades for create, read, update, and delete processes.

Chapter 7
Visual Cloud Functions

Visual cloud functions are a backend service for no-code applications within SAP Build Apps. In this chapter, you'll learn about entity modeling, creating custom functions, and the deployment of visual cloud functions applications.

In this chapter, we'll focus on visual cloud functions. These basically provide a way to develop a backend application on a no-code basis. Everything is made possible, from the creation of entities and the addition of functions to the deployment of the application. This chapter teaches how to use visual cloud functions in SAP Build Apps, including entity modeling, functions, and deployment. The chapter provides a comprehensive overview of the tools and techniques used to design and deploy visual cloud functions, as well as best practices for managing these functions.

After a brief introduction in Section 7.1, we'll take a closer look at the modeling of entities in Section 7.2. Then in Section 7.3, we look at the creation of functions within visual cloud functions, before moving on to Section 7.4, which deals with the deployment of the application and thus the last step.

7.1 Introduction

Full-stack developers, as developers who work in both the frontend and backend areas, are able to develop applications in their entirety. In most cases, different skills and developer knowledge are required on the frontend and backend sides. This is because, on the one hand, the technologies differ on both sides, but the approach to implementation and the architecture is also completely different.

With the visual cloud functions within SAP Build Apps, SAP provides a way to implement a backend application on a visual basis for a frontend that was also created using SAP Build Apps. This backend application can be created by any no-code developer and then runs in the cloud.

Some use cases in which visual cloud functions are relevant could include the following:

- It's often necessary to perform complex calculations that are required not only in one application, but in several. Due to reusability and maintenance, it makes sense to outsource such logic.
- It's often necessary to carry out calculations based on certain criteria or to display certain data. This could also be based on authorizations, for example. If possible, such calculations and determinations should not be carried out in the client. Instead, this should be done in the backend so that only truly relevant data is made available to the user.
- It could also be that data based on different database tables needs to be read and combined. In this case, too, it's advisable to do this on the backend side in order to provide the user only with the data for which he is authorized.

To create visual cloud functions, proceed as follows:

1. First, SAP Build Apps must be started. Once this has been done, the lobby will appear and you can open the application creation wizard by clicking **Create**.
2. Now select **Build an Application**.
3. As shown in Figure 7.1, you must next select **Web & Mobile Application** or **Application Backend**. For visual cloud functions, you must select **Application Backend** here.

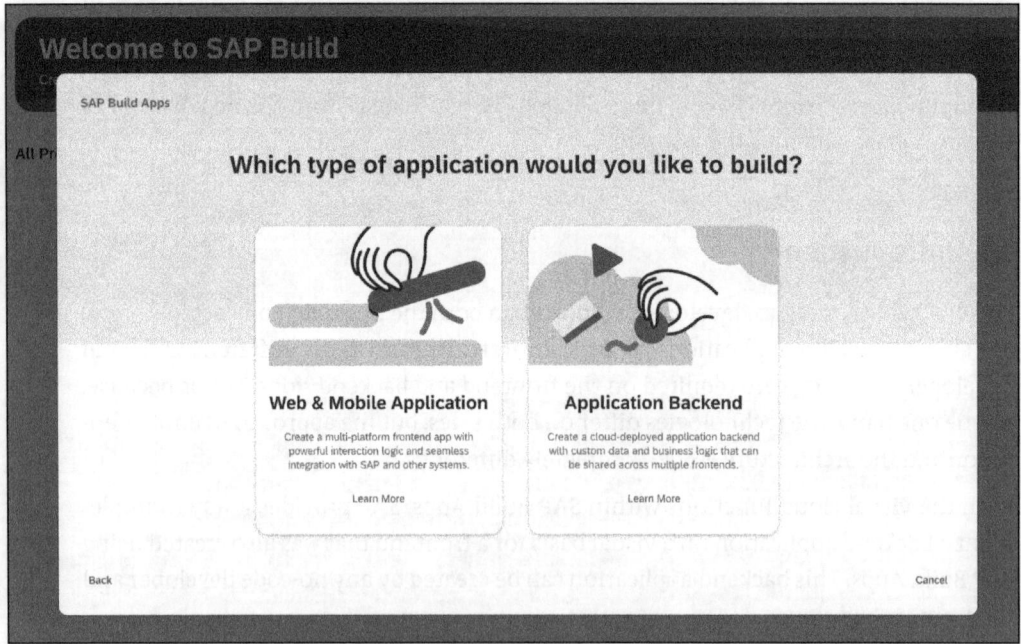

Figure 7.1 SAP Build Apps Creation Wizard: Application Selection

4. The next step is to enter data for the project. In this case, enter "Demo Project" as the project name and add a description if desired (see Figure 7.2).

7.1 Introduction

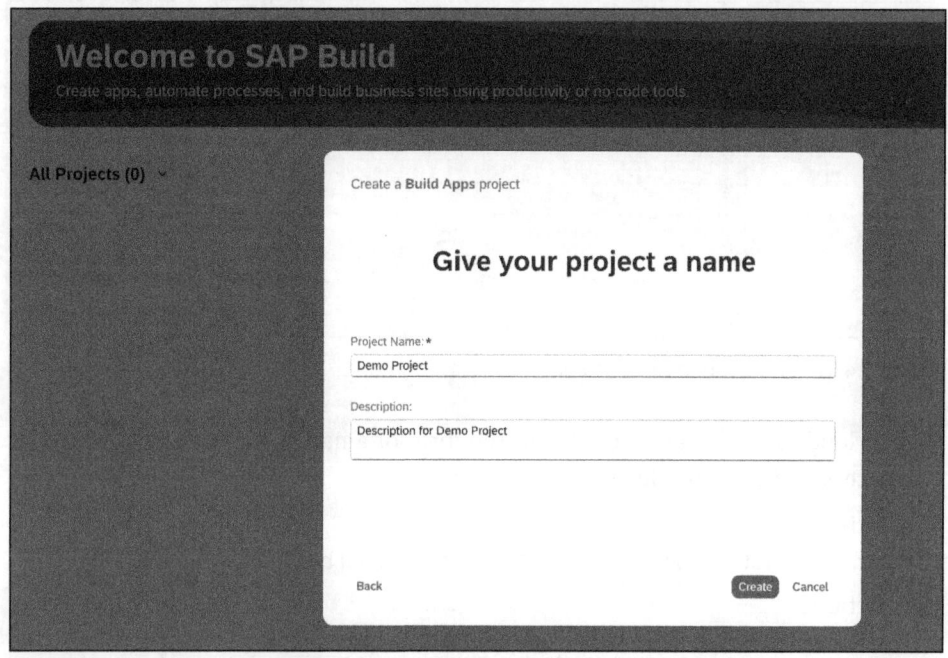

Figure 7.2 Enter Information for Project

5. Once this is done, continue by clicking **Create**. This creates the project under the specified name. It will also be visible in the SAP Build Apps lobby, as shown in Figure 7.3.

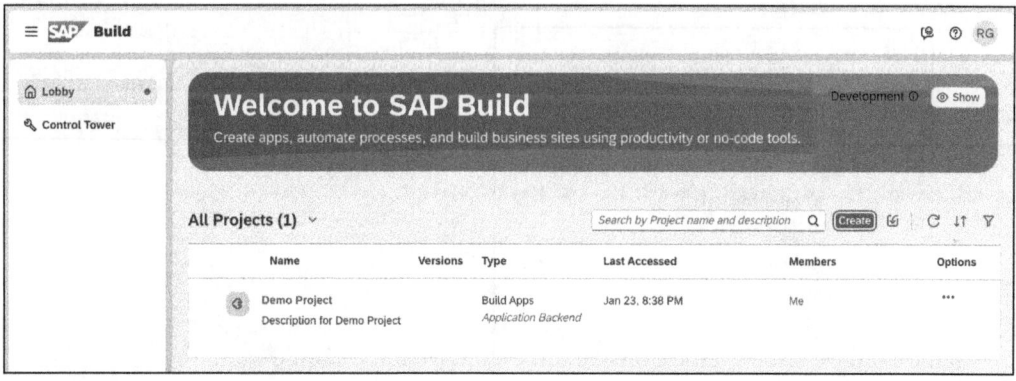

Figure 7.3 Project Overview in SAP Build Apps Lobby

6. Click the project to open it.

If you open a visual cloud functions project, the user interface looks as shown in Figure 7.4.

7 Visual Cloud Functions

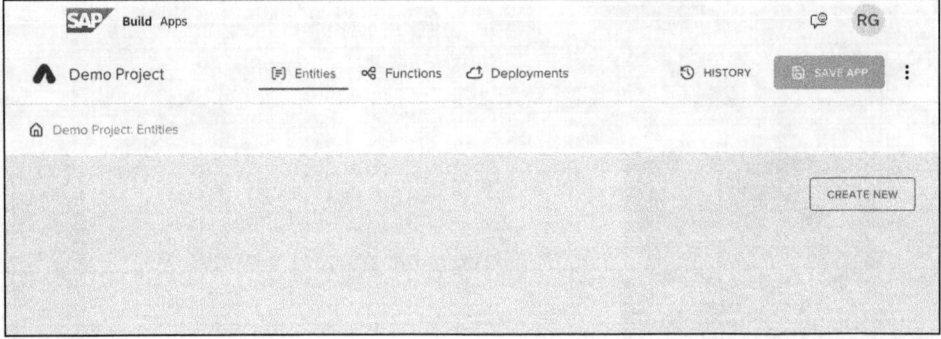

Figure 7.4 UI for SAP Build Apps Project

Visual cloud functions are provided in the form of a native user interface. This user interface essentially consists of three main views:

- **Entities**
  The data field structure can be developed in this area by creating entities.
- **Functions**
  Functions that have a specific visually defined sequence can be configured here.
- **Deployment**
  To use the project in the app builder at a later time, the building and deployment of the project can be done here.

The **History** button hides version management of the project. Clicking this button opens a dialog as shown in Figure 7.5.

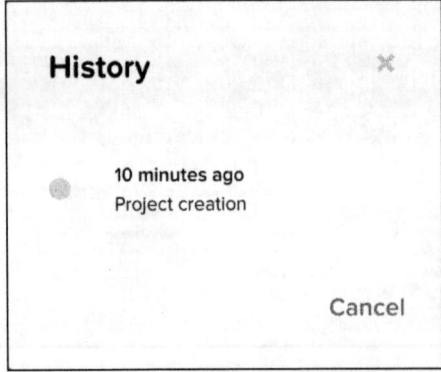

Figure 7.5 Version Control for Visual Cloud Functions Projects

7.2 Entity Modeling

To store data persistently in a backend application, be it a no-code app or any other application, database tables are required in which the corresponding data can be stored. The schema of the data to be stored is described by entities. Each entity can therefore be compared or equated with a database table. In SAP Build Apps, each definition of an entity basically consists of the following attributes:

- The *entity title*, a human-readable name
- The *entity description*, a summary
- The *entity ID*, a technical key

Furthermore, a distinction is made between two types of entities:

- Native entity
- Extended data entity

In the following sections, we'll first discuss these two entity types before moving on to discuss virtual fields and the data browser.

7.2.1 Native Entities

Native entities are the basic building blocks of a visual cloud functions project. These represent a single collection of data that is categorized according to fields. Let's look at how a native entity can be created:

1. First, after selecting the **Entities** tab, click the **Create New** button.
2. A dialog then opens in which the meta information for the entity must be specified (see Figure 7.6). Here, assign the following values:
   - **Entity Title**: "NativeEntityDemo"
   - **Entity Description**: "Demo Native Entity"
   - **Entity ID**: "NativeEntityDemo"

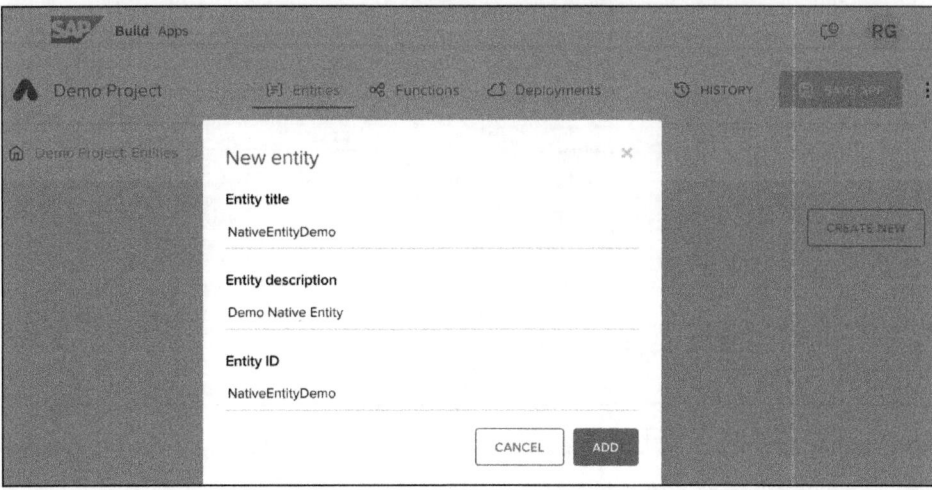

Figure 7.6 Creation of Native Entity

3. Confirm the entry by clicking **Add**, thus creating the entity.
4. Once this has been done, two tabs, **Configure Entity** and **Configure Fields**, are displayed:
   – In the **Configure Entity** tab, the title and ID of the entity can be changed or the entity deleted (see Figure 7.7).

Figure 7.7 Configure Entity Tab

   – The **Configure Fields** tab is used to add new data fields to the entity or to edit existing ones. You can also select fields to view their details (see Figure 7.8).

Figure 7.8 Configure Fields Tab

5. To add a new field to an entity, click **Add New** within the **Configure Fields** tab. A dialog then opens in which the field name and data type must be maintained. Here,

create a "Title" field of the **Text** data type (see Figure 7.9). In addition, the **Required** checkbox can be used to define whether the field must be filled—that is, whether it is nullable or not.

Figure 7.9 Create New Field for Entity

6. The field can be added to the entity by clicking **Add**. Once this has been done, the field is displayed accordingly in the **Configure fields** area (see Figure 7.10).

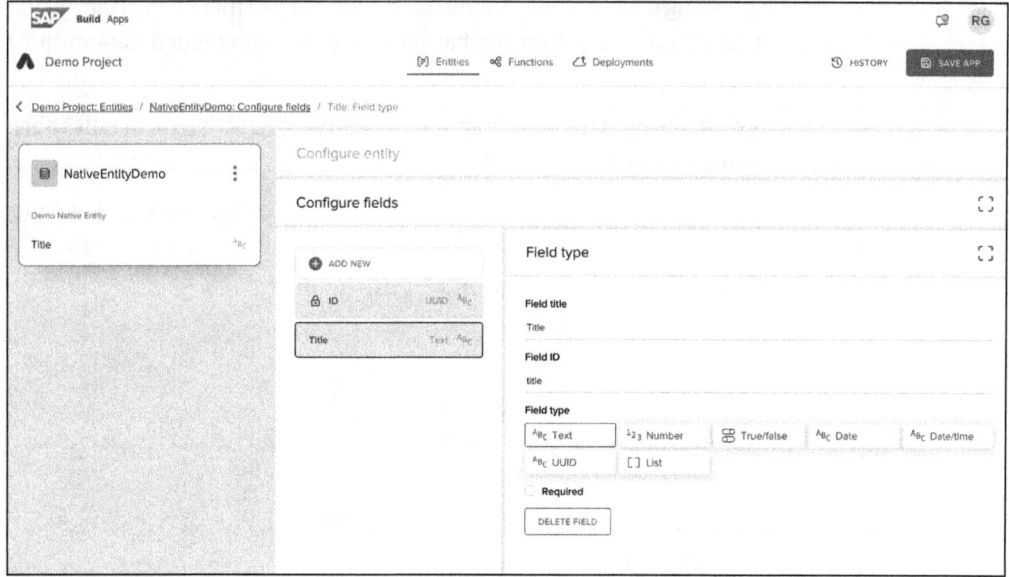

Figure 7.10 Entity after Adding Field

It's important to note that every added field can be edited. To change a field, it must be selected in the field list. However, changes to entities only become visible after the application has been deployed again. The data types in Table 7.1 are available for assigning the data type.

Display Name	PostgreSQL	JSON
Text	text	string
Number	numeric	number
Boolean	boolean	boolean
Date	date	string with format
Date/Time	timestamptz	string with format
UUID	uuid	string with format
List	array	array

Table 7.1 Datatypes for Fields in Entities

7.2.2 Extended Data Entities

Extended data entities are available to extend native entities. This happens to the extent that the extended data entities use formula expressions to change the data of the native entity. This makes it possible, for example, to hide desired fields and not make the corresponding data available to the user if desired. Furthermore, a data transformation can also be carried out on the backend side using extended data entities, which may simplify the process.

To create an extended entity, a native entity must already exist. Let's look at how to create an extended entity in detail:

1. First select a native entity that you want to extend, then click **Create Extended Entity** from the three-dot menu (see Figure 7.11).

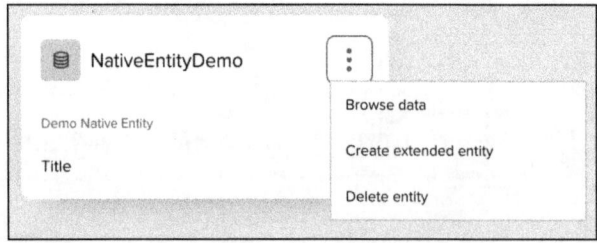

Figure 7.11 Three-Dot Menu for Native Entity

2. After clicking **Create Extended Entity**, a dialog opens in which the meta information of the entity to be created must be entered. For extended entities, this also involves

the **Entity Title**, **Entity Description**, and **Entity ID** parameters. After entering the following values, continue by clicking **Add** (see Figure 7.12):

- **Entity Title**: "ExtEntityDemo"
- **Entity Description**: "Demo Extended Entity"
- **Entity ID**: "ExtEntityDemo"

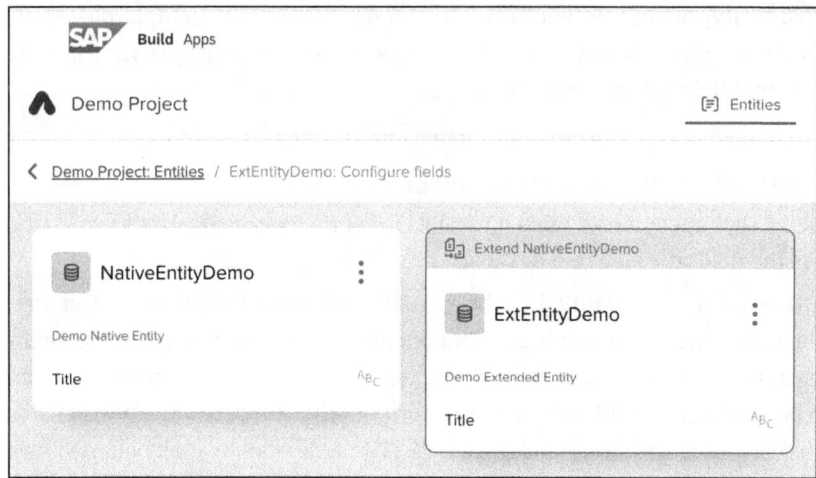

Figure 7.12 Creation of Extended Entity

3. Once the extended entity has been created, it's visible in the canvas. Furthermore, as shown in Figure 7.13, these entities stand out visually from native entities.

Figure 7.13 Available Entities

4. Extended entities also have **Configure entity** and **Configure fields** areas. With extended entities, the fields of the native entity that is being extended are adopted,

and you can only hide individual fields from the extended entity (see Figure 7.14). Technical properties of the fields cannot be changed. In addition, virtual fields can also be added to the extended entity.

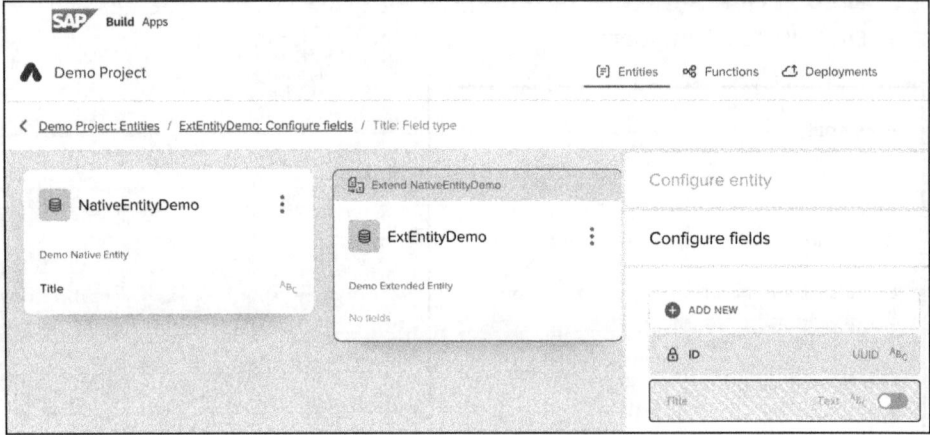

Figure 7.14 Remove Field from Extended Entity

By hiding fields in extended entities, these are not displayed in the app builder, which makes it possible to avoid making unwanted data visible to the user.

7.2.3 Virtual Fields

Virtual fields use a formula language with which data from the actual fields of the base entity (native entity) can be referenced and then transformed. For example, data could be masked in this way. All fields that are created in an extended entity are virtual fields. To create a virtual field, proceed as follows:

1. First an extended entity must first be created and selected.
2. Then, click the **Add New** button in the **Configure Fields** tab.
3. In the dialog that opens, enter meta information for the field again. Here, use "VirtualTitle" as the field title and **Text** as the data type (see Figure 7.15).
4. A formula must then be defined in the **Virtual Field Value** field for how this field should be filled. Clicking in the input field opens a dialog that helps you create the formula. In addition to attributes from the entity, constants or various functions can also be used here. In this case, define a formula that concatenates "Virtual" and the content of the "title" field (see Figure 7.16). Data fields can be easily queried here using the following syntax: `fields.fieldName`.

7.2 Entity Modeling

Figure 7.15 Create Virtual Field for Extended Entity

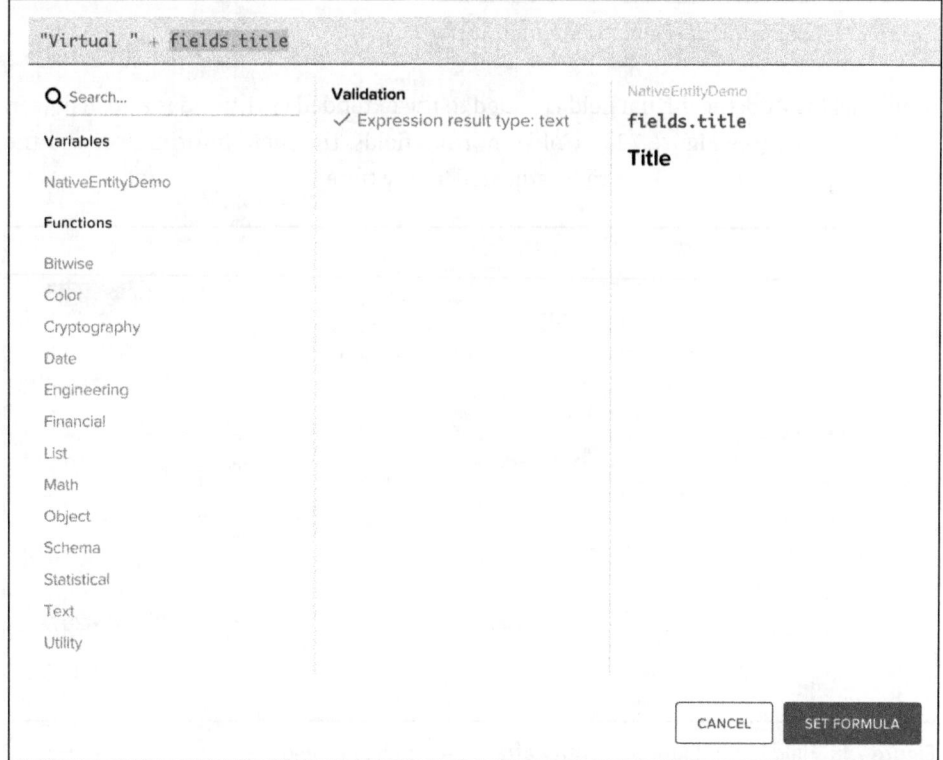

Figure 7.16 Virtual Field Formula

5. After clicking **Set Formula**, the entry is confirmed and transferred to the field creation screen (see Figure 7.17).

Figure 7.17 Adding Virtual Field to Extended Entity

6. By clicking **Add**, the virtual field is added to the extended entity and is also visible in the field list (see Figure 7.18). Unlike normal fields, the meta information and the data types of virtual fields can be adjusted at any time.

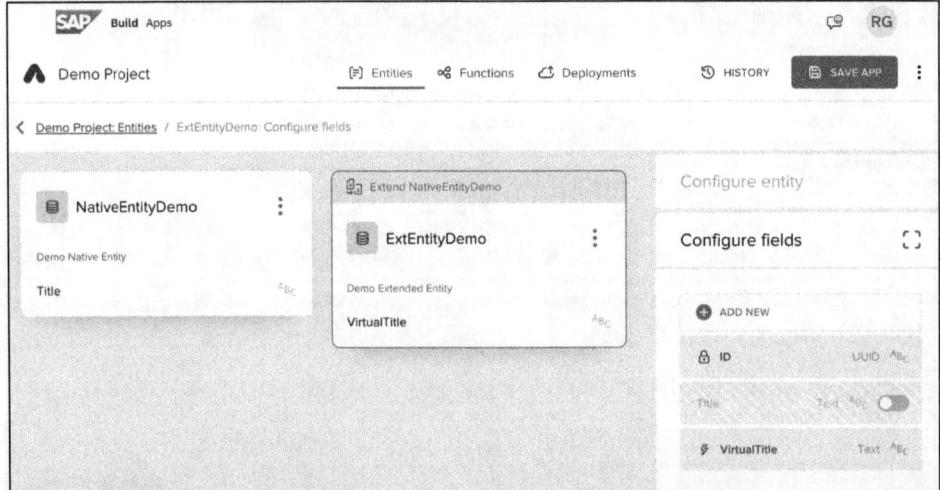

Figure 7.18 Field List of Extended Entity after Adding Virtual Field

7.2.4 Data Browser

All entities that are created in visual cloud functions can be searched visually. Furthermore, options for displaying, changing, and deleting data are offered directly in the data browser. To call up the data browser, select the **Browse Data** entry from the three-dot menu of an entity (see Figure 7.19).

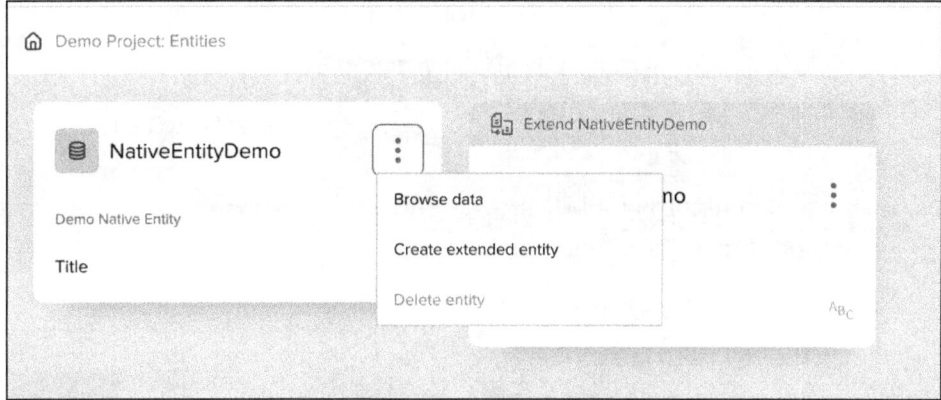

Figure 7.19 Browse Data Option in Three-Dot Menu

The data browser then opens for the selected entity. As shown in Figure 7.20, this entity is still empty.

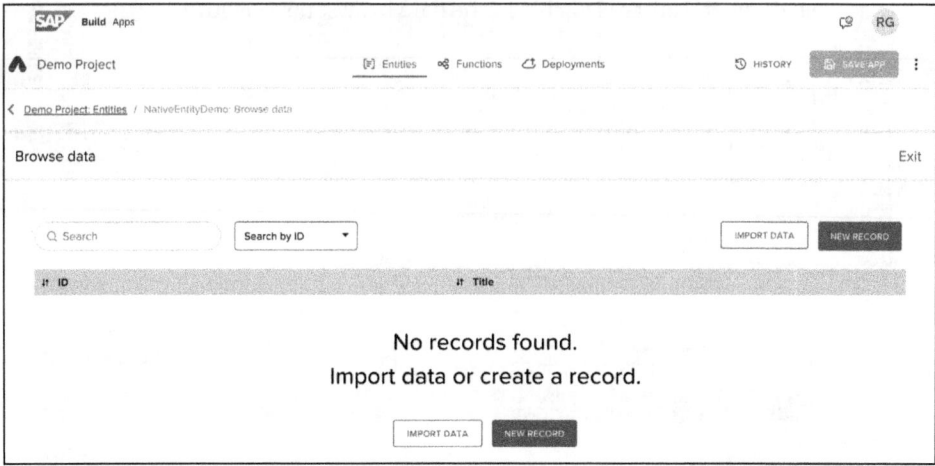

Figure 7.20 Data Browser for Empty Entity

To fill the entity with entries, either the import function can be used or a single data record can be created by clicking **New Record**. After clicking **New Record**, the attributes of the entry to be created can be maintained on the right-hand side (see Figure 7.21).

7   Visual Cloud Functions

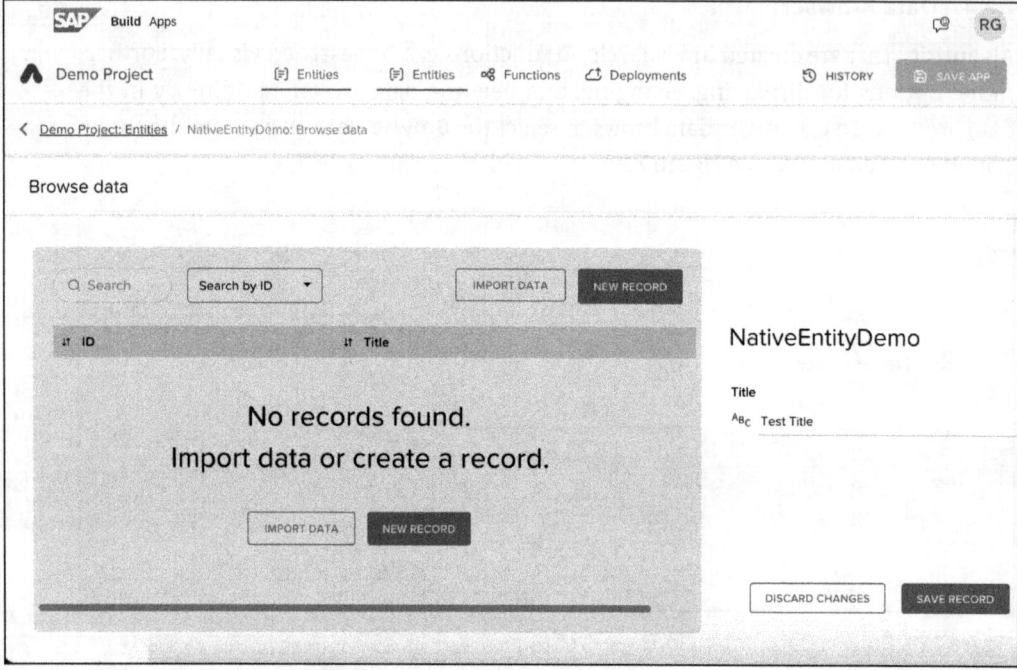

Figure 7.21 Create New Record for Entity

Once the fields have been filled in, the entry can be saved by clicking **Save Record**. Once this has been done, the entry is visible in the data browser (see Figure 7.22).

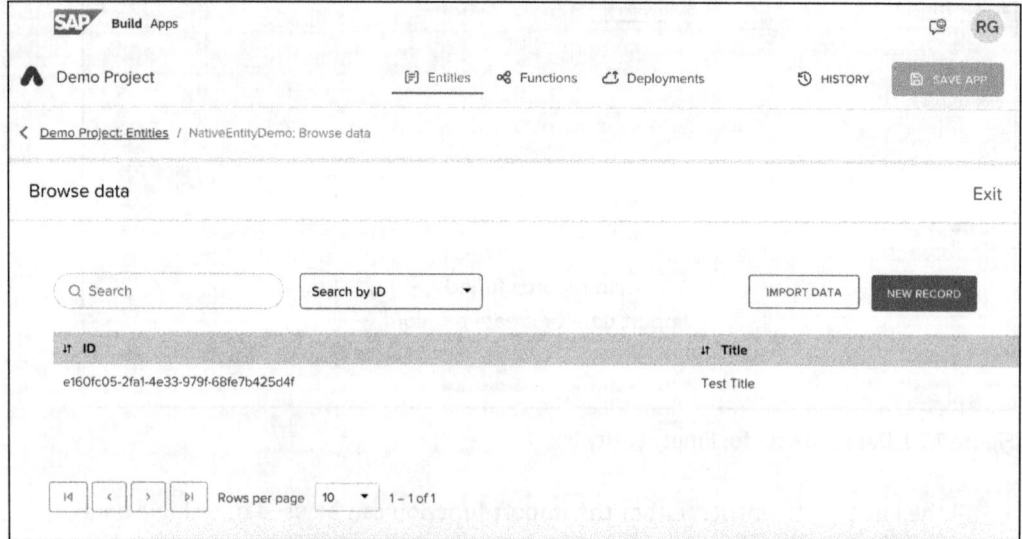

Figure 7.22 Data Browser after Adding Record

7.3 Functions

One of the main functions of visual cloud functions is the ability to define any functions that execute logical processes that are defined visually. In this section, we'll look at how functions can be created. Furthermore, you'll see which kinds of input and output parameters can be used and how to define a logic flow.

7.3.1 Create a New Function

To create, view, or edit functions, you have to go to the **Functions** tab. If you open the tab, it's still empty in this case, as shown in Figure 7.23. If functions already existed, they would be visible in the overview.

Figure 7.23 Functions Tab before Creating Function

To create a new function, proceed as follows:

1. First, click the **Create New** button. A dialog then opens in which meta information—name and description—of the function must be entered (see Figure 7.24). In this case, assign the following values here:
   - **Name**: "DemoFunction"
   - **Description**: "Demo Function"

Figure 7.24 Enter Meta Information for New Function

2. Once this has been done, continue by clicking **Create**. You'll then land in the logic canvas, in which the function can be created.

>
> **Logic Canvas**
>
> The *logic canvas* allows you to visually construct and configure the logical sequence of a function. The canvas always has a start/execution node, which is located on the left-hand side. The canvas can also contain several success and error nodes; these are located on the right-hand side. Additional logic nodes can also be inserted between the start and target nodes during the implementation of the function.

Figure 7.25 shows what the logic canvas initially looks like.

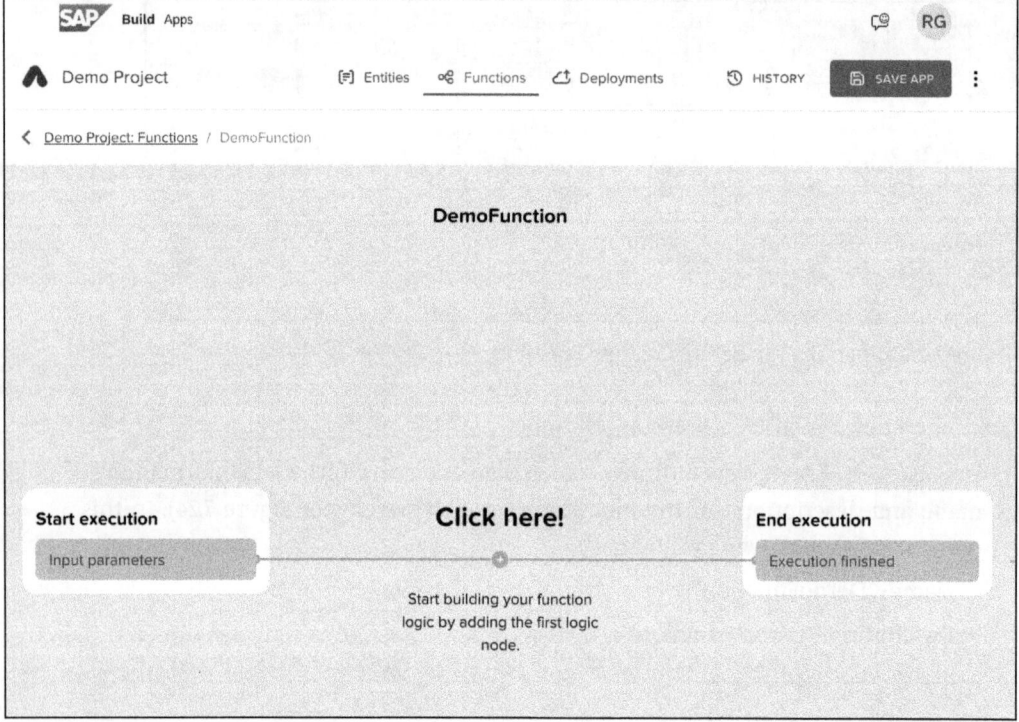

Figure 7.25 Initial State of Logic Canvas

7.3.2 Input Parameter

A function usually also requires incoming parameters. In visual cloud functions, input parameters are defined within the start execution node. Such a parameter essentially consists of the following attributes:

- **Name**
  Human-readable name of the input parameter

- **Description**
  Short description
- **Type**
  Determines if the parameter is of the text, number, or Boolean type
- **Required**
  Determines if the parameter is mandatory (if a mandatory parameter is missing, it causes an error)
- **Technical key**
  Machine-readable key used by formulas to access the parameter

Now that you know which attributes an input parameter has, let's take a look at how one is created:

1. First select the start-execution node, click **Input Parameters**, and then click **Add Parameter** (see Figure 7.26).

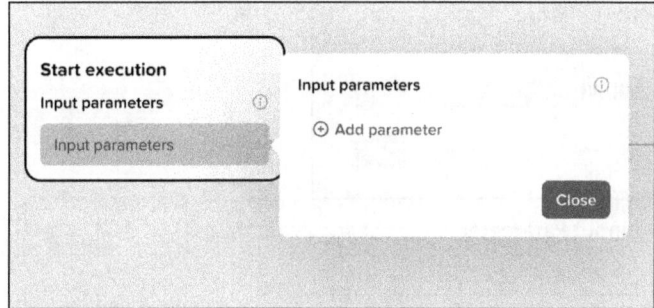

Figure 7.26 Creation of Input Parameter

2. After you click **Add Parameter**, another popover opens in which the previously named attributes of the parameter are to be maintained. Assign the following values (see Figure 7.27):
   - **Name**: "inputParam"
   - **Description**: "Demo Input Parameter"
   - **Type**: String
   - **Required**: Leave unchecked
   - **Default value**: Leave empty
   - **Technical key**: "inputParam"
3. Confirm the entries by clicking **Add**.
4. Once this has been done, the created parameter should be visible as shown in Figure 7.28.

7 Visual Cloud Functions

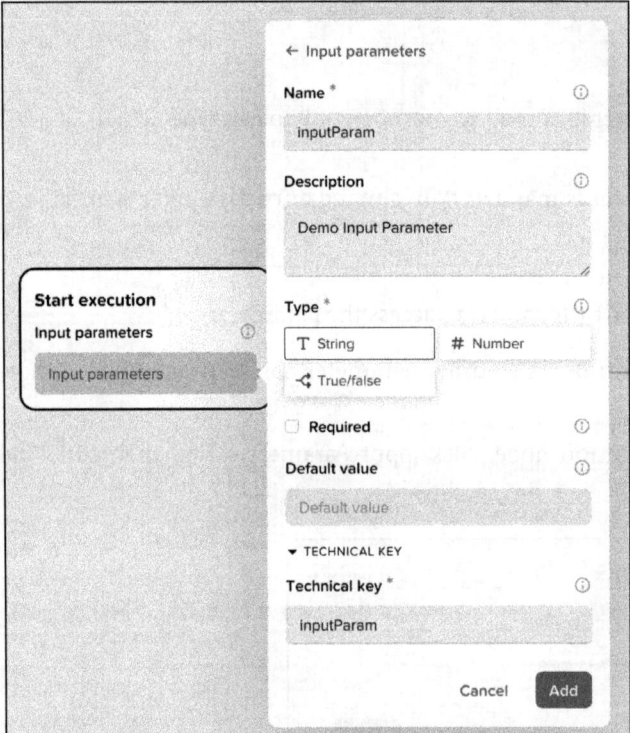

Figure 7.27 Enter Attributes of Input Parameter

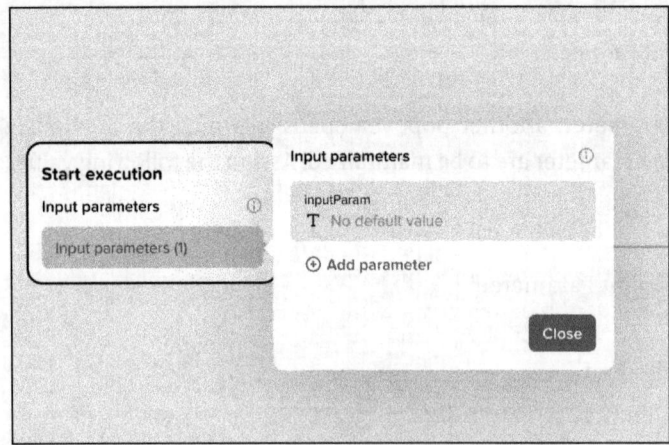

Figure 7.28 List of Input Parameters

7.3.3 Function Outcomes

Of course, it isn't important only to define input parameters. It's also much more important to define what the function should return. Here we speak of *outcomes*. We will take a closer look at these in this section.

The result of a function is defined by the success node on the right-hand side of the canvas. If this node is reached, the function execution is considered complete and the corresponding outcome is triggered. Such an outcome can be defined similarly to an input parameter and consists of the following attributes:

- **Name**
  Human-readable name
- **Description**
  Short description
- **Value**
  Defined as a formula that is evaluated when the success node is reached
- **Technical key**
  Machine-readable key used by formulas to access the parameter

To define an outcome, proceed as follows:

1. First select the success node and click **Execution Finished**, followed by clicking **Add Value** (see Figure 7.29).

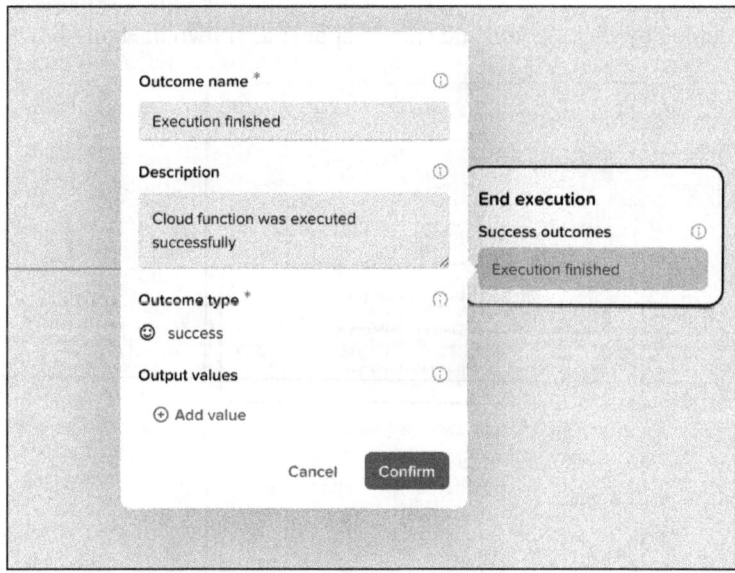

Figure 7.29 Open Dialog to Add Values to Outcome

2. Next, the attributes of the outcome parameter must be specified (see Figure 7.30). Use the following values in this case:
   - **Name**: "OutputValue"
   - **Description**: "Output Value Demo"
   - **Value**: ""Out: " + inputs.inputParam"
   - **Technical key**: "outputValue"

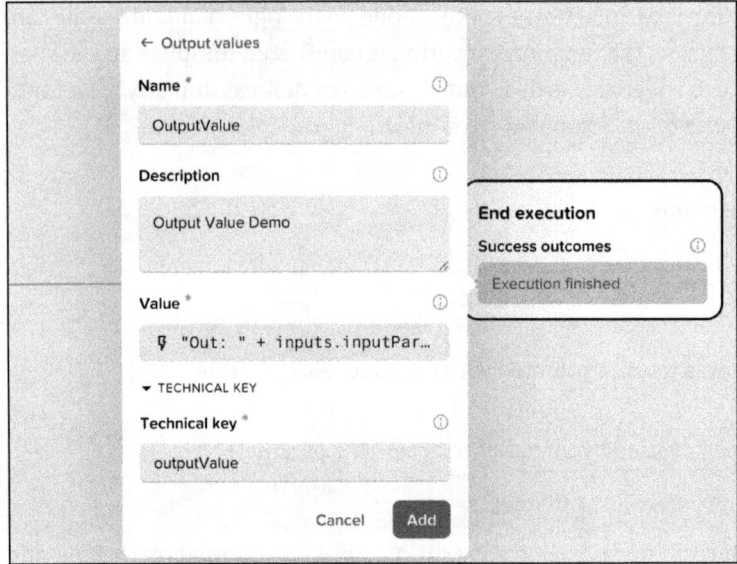

Figure 7.30 Enter Values for Output Value

3. The value can be added by clicking **Add** and should appear as shown in Figure 7.31.

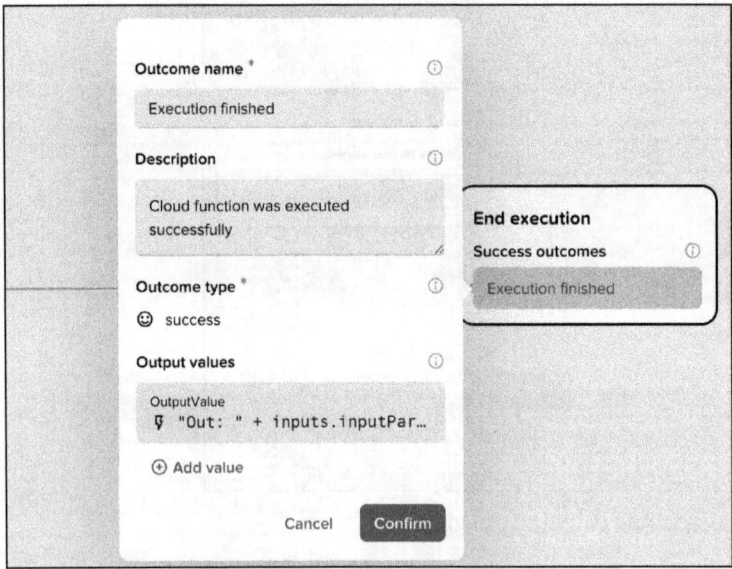

Figure 7.31 Overview of Outcome after Adding Value

4. The configuration can now be accepted by clicking **Confirm**.

7.3.4 Defining the Logic Flow

Now that you've defined an input parameter and an outcome, all that is missing is the logic between the start and end nodes. The **+** in the middle of the canvas can be used to add various records that are required to map the logic. The following are available:

- **Retrieve record**
- **List records**
- **Create record**
- **Update record**
- **Delete record**

After selecting a record, a new node appears in the canvas. You can then continue with the further configuration here. It should be noted that the configuration options differ depending on the type. While with **List record** you only specify the entity, with **Retrieve record**, for example, you also specify a formula based on which the data is read in addition to the entity. Let's look at how such a record can be added:

1. First, click the **+** in the middle of the canvas. This opens a popover (see Figure 7.32) in which you should select **List record**.

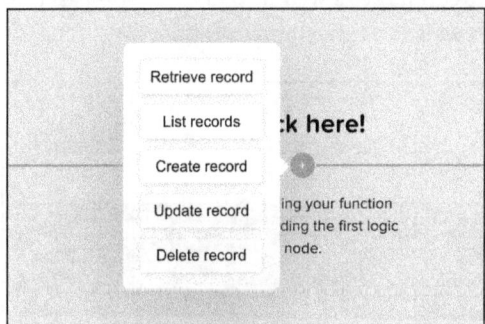

Figure 7.32 Select Record Type on Logic Flow Creation

2. The configuration must then be carried out within the new node. An entity must be specified here. In this case, select the **ExtEntityDemo** entity (see Figure 7.33).

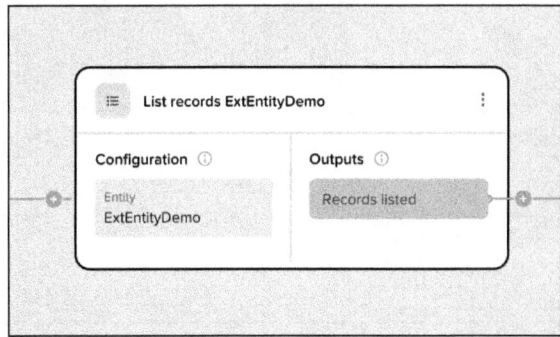

Figure 7.33 Node: List Records ExtEntityDemo

7 Visual Cloud Functions

As the **List record** node determines all entries of the specified entities and passes them on to the next node, you should create another outcome value in the end node at this point, which contains the results of the **List record** node. This value is created in the same way as in the previous chapter.

7.4 Deployment

At some point after development, you'll reach the point where the project needs to be deployed. In software development, *deployment* is the process of making a software product available for external use. In visual cloud functions, the deployment process publishes the latest version to be used in the app builder.

The version management of visual cloud functions documents every change that is made between versions. It's therefore important to ensure that the application is saved before deployment; otherwise, the last saved version will be deployed.

To get to the deployment view, switch to the **Deployment** tab in visual cloud functions. This tab consists of a single card, the deployment card (see Figure 7.34). All changes that have been made since the last deployment can be reviewed in this card. You can also see the status of the current deployment as well as a timestamp.

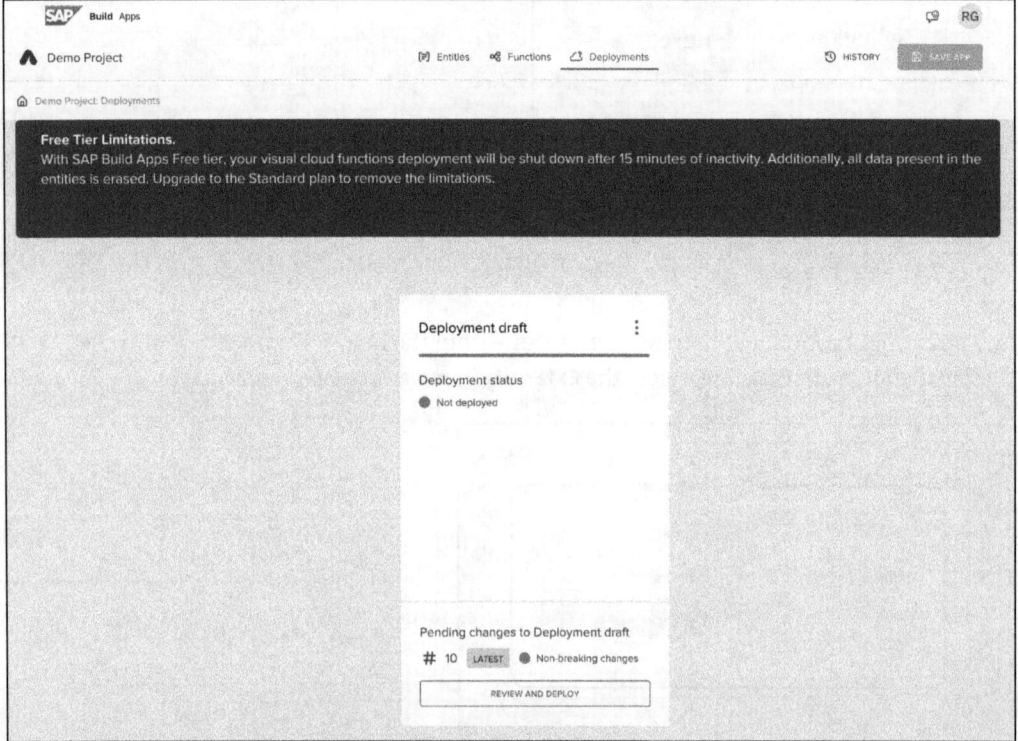

Figure 7.34 Deployment Tab of Visual Cloud Functions

238

In this section, we'll discuss the deployment card in detail. You'll see how a deployment can be triggered, paused, and stopped. Furthermore, you'll also learn about different types of changes and error handling.

7.4.1 Deployment Card

As already mentioned, the **Deployment** tab only contains the deployment card. In addition to the deployment status, a timestamp of the last deployment, and an overview of how many changes have been made since the last deployment, it also shows whether the current version contains breaking changes. *Breaking changes* are changes that could potentially affect the functionality of the application—such as the removal of an entity. As shown in Figure 7.35, we haven't yet carried out a deployment.

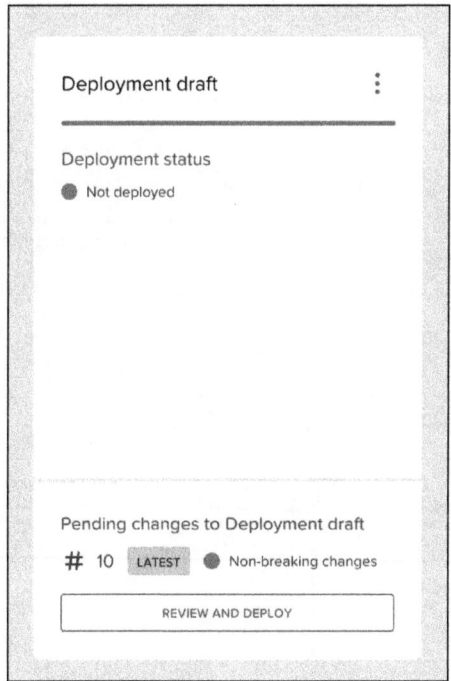

Figure 7.35 Deployment Card before First Deployment

Now let's go over the steps required to start a deployment:

1. First, click **Review and Deploy.** This opens a window like the one in Figure 7.36. Here you can see the main changes.
2. Deployment can now be initiated by clicking **Deploy to Deployment Draft**.
3. Once this has been done, the application should be deployed after a short wait.

7   Visual Cloud Functions

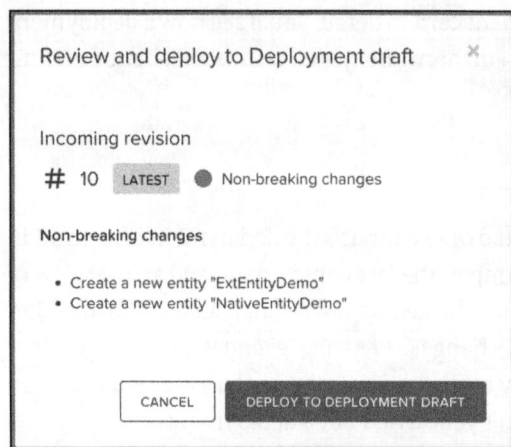

Figure 7.36 Review before Deployment

Once the deployment is complete, this is also reflected in the deployment card (see Figure 7.37).

Figure 7.37 Deployment Card after Deployment

7.4.2   Change Types

When making changes, visual cloud functions distinguishes between two categories with regard to deployment:

- Nonbreaking changes
- Breaking changes

Nonbreaking changes are changes that should not have a negative impact on the existing application. Such changes could include, for example, the following:

- Adding new entities
- Adding new data fields
- Adding new data browser entries

When we talk about *breaking changes*, we are referring to changes that could have a negative impact on the existing application or could result in the application no longer functioning correctly. Such changes could include the following:

- Deleting used fields
- Deleting used entities

7.4.3 Delete/Pause Deployment

It may well be that you want to permanently remove the currently active deployment. In this case, visual cloud functions offers the option of deleting the deployment. However, it should be noted that deleting a deployment also deletes all data that is already stored in entities.

If you do not want all data to be deleted but still want to remove the current deployment, there is also the option of pausing an active deployment. If a deployment is paused, it can be reactivated at a later point in time. To delete a deployment, click **Delete** in the three-dot menu of the deployment card (see Figure 7.38).

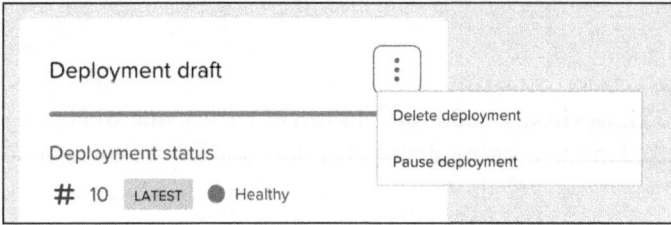

Figure 7.38 Three-Dot Menu of Deployment Card

You can then delete the deployment in the confirmation dialog by clicking **Delete** (or cancel it by clicking **Cancel**).

To pause a deployment, proceed in a similar way to deleting one. In this case, select the **Pause Deployment** button. The deployment is then paused, which can also be seen in the status shown on the deployment card (see Figure 7.39).

7   Visual Cloud Functions

Figure 7.39 Deployment Card of Paused Deployment

Click **Resume** to resume the deployment and make the application available again.

7.5 Summary

In this chapter, we showed how visual cloud functions can be created and what possibilities they offer. We walked through the modeling of entities and how to create functions within visual cloud functions. The final section dealt with the deployment of the application. In the following chapters, we'll see how extensions can be implemented in the context of SAP Build Apps, but also how applications can be deployed. These two topics are followed by examinations of SAP Build Work Zone and SAP Build Process Automation.

Chapter 8
Developing Extensions

There is no shame in outgrowing the standard if you do it right. Many requirements and processes are special and tailored to your own company. The applications used should therefore also support these processes in the best possible way. With SAP Build Apps, user interfaces can be created for these requirements according to the no-code principle.

An additional, specially developed web application is always used when the existing user interface delivered as standard does not meet your own requirements. This may be due to a specially required user interaction or used to be able to seamlessly cover the company's own processes with in-house development. In these cases, the corresponding existing system must be connected, the data consumed, and possibly also manipulated. Other requirements could be to provide the applications you have developed on a platform that comes close to the standard or even maps it 1:1 but was not intended in this way. An example of this would be a web application or mobile application, which is simply not provided for by the standard.

For this reason, in this chapter we look at how to develop extensions for existing systems, such as SAP SuccessFactors, SAP S/4HANA, and SAP S/4HANA Cloud. In Section 8.1, we will focus on SAP SuccessFactors and develop an application that makes the emergency contacts of employees available in a mobile application. Of course, this requires the system to be connected and the interface and application logic to be developed. To make this exercise easier for you, we'll call an example API from SAP Business Accelerator Hub (formerly known as SAP API Business Hub) in this case. The corresponding API is connected via API Management in SAP Integration Suite, which serves as a reverse proxy in our case. SAP Business Accelerator Hub can help you as an integration platform for other systems, but it can also provide API documentation and suitable interfaces for communication. SAP Integration Suite is more than just middleware in SAP BTP. In addition to integrations between two external systems, which may not be able to communicate with each other, you can also provide manually created APIs and offer them to consumers not only as a facade, but possibly also as monetization.

In Section 8.2, we'll connect an SAP S/4HANA system. In our example, the development will be so generic that you can use what you have learned for an SAP S/4HANA on-premise instance as well as for SAP S/4HANA Cloud. In this case, we'll build on the example already presented in Chapter 6 and will now completely set up the customer administration. It will be possible not only to search for customers, but also to create,

8 Developing Extensions

update, and delete customer data. SAP S/4HANA should already be familiar to all companies running SAP as the fourth generation of the SAP Business Suite, which is now also offered as a service in SAP BTP.

> **[+] Information from Other Chapters**
>
> Of course, this is a separate chapter in the book, but it nevertheless builds on the knowledge presented in the other chapters. So we rely on the fact that the terms *UI canvas*, *UI elements*, *properties*, *layout options*, *data resources*, *HTTP*, *APIs*, *data binding*, *variables*, *logic canvas*, and *routing* are not foreign to you. If you are puzzled by any of these terms, don't panic! Go back to the earlier chapters and come back to the current chapter later.

8.1 SAP SuccessFactors

Let's start in this section with the development of an application for SAP SuccessFactors. Imagine you want to be able to view the emergency contacts maintained by your employees. The employees should first be selectable via a list and then be able to be viewed in more detail. As this section is structured in several steps, we have also divided these individual steps into separate sections for the sake of clarity.

In Section 8.1.1, we show you how to use APIs from SAP Business Accelerator Hub. We use this service to access a sandbox API of an SAP SuccessFactor system without actually having to license a system yet. In Section 8.1.2, we start with the development of a list of persons from the SAP SuccessFactors sandbox system. This section is divided into the data integration first and then the actual development of the user interface. Section 8.1.3 then deals with the detail view and the display of emergency contacts. Because the person details and the associated emergency contacts come from a different API, we will create another data resource and entity before we build the user interface in the second part.

8.1.1 Consume API and Activate Authentication

In Figure 8.1, you can see that we are accessing the Sandbox API instead of a licensed SAP SuccessFactors instance. This API looks exactly the same as a productively used one would look, but you can test the communication with sample data free of charge. Go to the following website and log in with your SAP user: *https://api.sap.com/*.

Here, we selected the **/PerPersonal** API, which returns a list of people and their personal information located in SAP SuccessFactors Employee Central. With **Run**, you can test the current call and thus the API. Further down on this page, you can also view the response and thus draw conclusions about the structure of the data.

8.1   SAP SuccessFactors

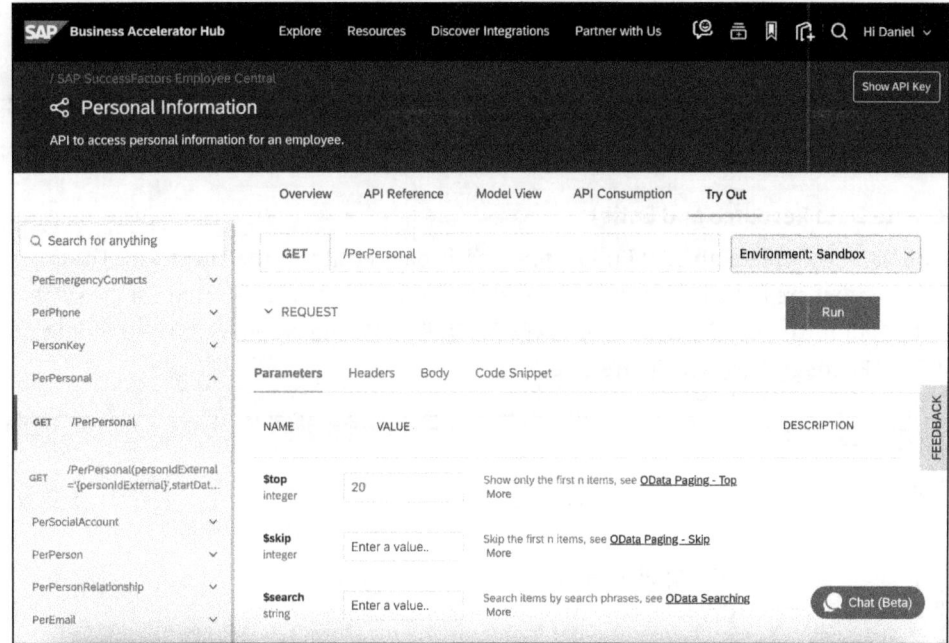

Figure 8.1 SAP Business Accelerator Hub for Discovering APIs of Many SAP Products

Once you've integrated this API in API Management of SAP Integration Suite and stored it as an SAP BTP destination, the actual work in SAP Build Apps begins. As you already know, we need to activate authentication under **AUTH** to connect data resources and a backend. In Figure 8.2, we have already activated SAP BTP authentication.

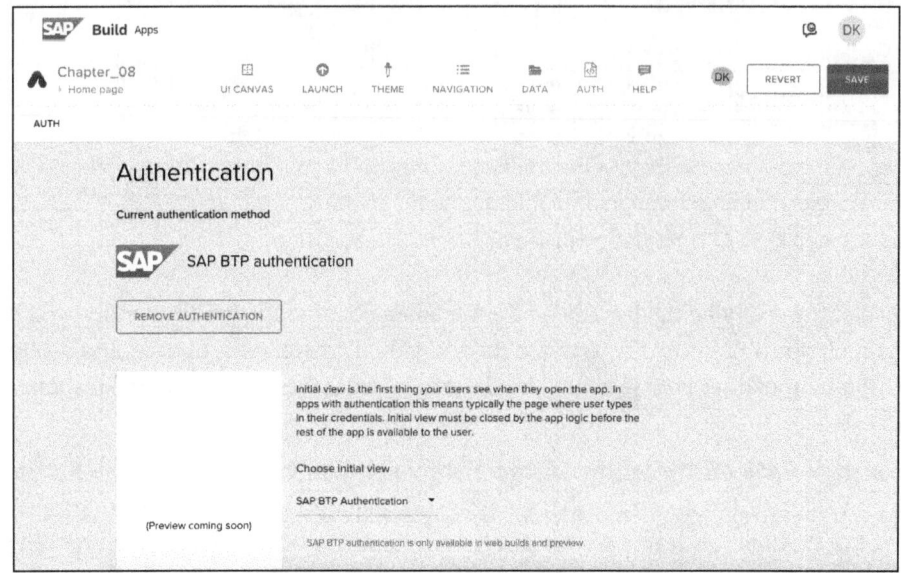

Figure 8.2 Enable SAP BTP Authentication to Use Data Integration

8 Developing Extensions

8.1.2 Build a List of Personal Data

In this section, we'll build the first page. The goal will be to display a list of all employees. To get the data into the app, you first need a data resource and the corresponding entity so that you can then build the UI.

Create Data Resource and Entity

You can now create an SAP Build Apps classic data entity in the **DATA** tab. These are entities that are not OData services of systems and cannot be integrated via the system integration. To do this, click on **CREATE DATA ENTITY** and select **SAP BTP Destination REST API Integration** (see Figure 8.3).

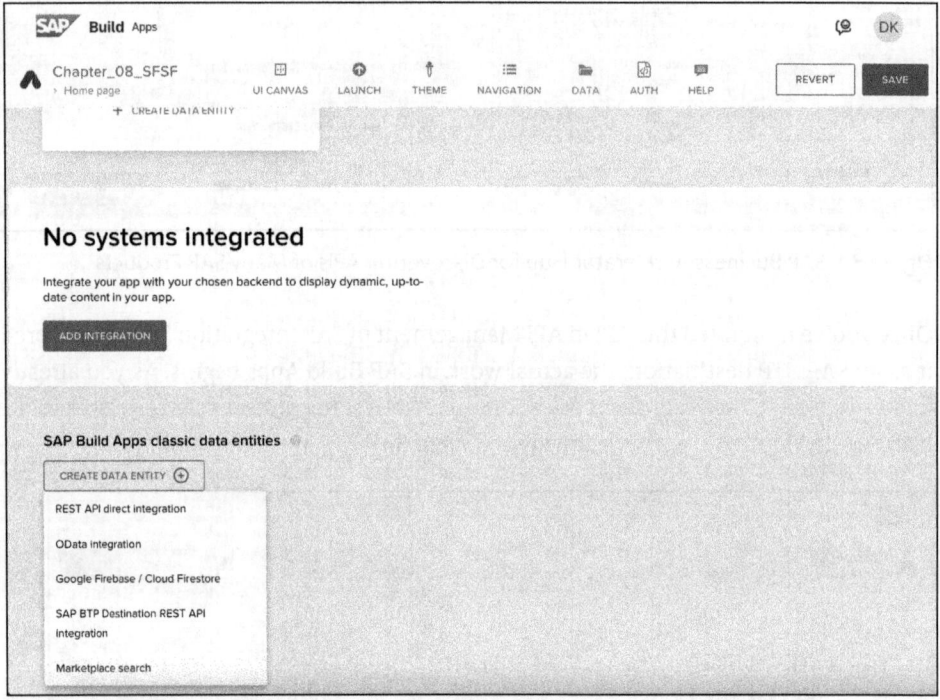

Figure 8.3 Add REST API Integration Based on SAP BTP Destination

Instead of the entity being described by a metadata document from the service, you'll integrate it yourself via a URL, activate the individual create, read, update, and delete (CRUD) functionalities yourself, and also determine what the structure of the response looks like.

Let's start at **BASE** on the left-hand side. Here you must first assign a name for the entity. In Figure 8.4, you can see that we assigned "Person" as the name. Furthermore, under **BTP Destination Name** we have selected the name of the destination that was created in SAP BTP and saved the URL to the corresponding API.

8.1 SAP SuccessFactors

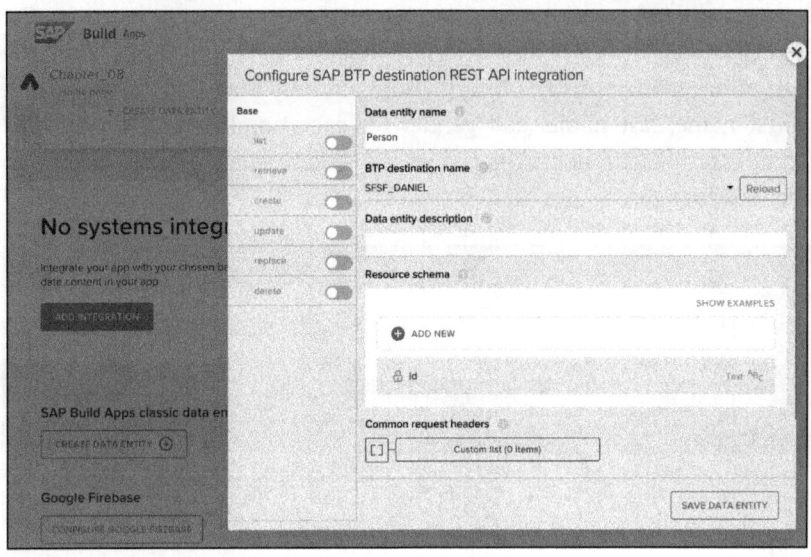

Figure 8.4 Configure Properties of REST API

In Figure 8.4, we clicked the **Data Binding** button in **Common Request Headers** and selected **List of Values**. These are header properties that should always be sent with the HTTP requests. As the system is a sandbox system, we only need to send an API key as authentication. In practice, you will of course send the authentication information of the currently logged in user per request instead of using one key for all users. You can find your API key for your sandbox instance in SAP Business Accelerator Hub.

In Figure 8.5, you can now see that we have assigned this **Header name** and will always send the API key as the **Header value**.

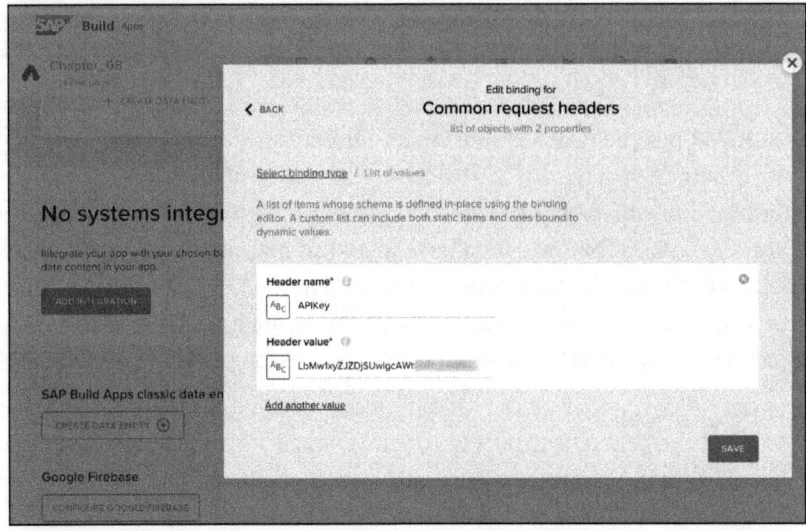

Figure 8.5 Add Common Request Headers to Be Applied on Every Request

247

8 Developing Extensions

To get a list of personal information back from the service, you must first activate **list** as a possible method on the left-hand side. Under **Relative Path and Query**, write the following path in order to access the information and only read out the person ID, nationality, first name, last name, and gender for each employee: "/PerPersonal?$select=personIdExternal,nationality,firstName,lastName,gender". Figure 8.6 shows where to make this change.

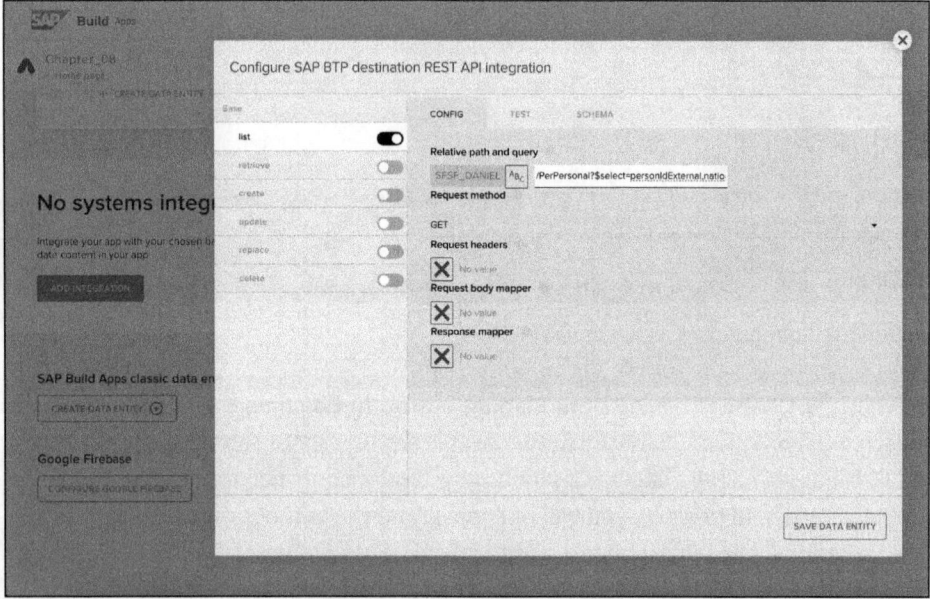

Figure 8.6 Enable List (HTTP GET) Functionality and Define Relative Path

In the second tab, **TEST**, in this dialog you can test the request of the relative path. You can see that the response is **OK (200)**, which in HTTP language means that everything went well. Below this, you can also see the response and the data that was returned (see Figure 8.7).

Switch to the **SCHEMA** tab and take a look at what is displayed. You won't see anything there for the time being. We have already told you that the structure of the response cannot be determined automatically, but that you actually have to build up which attributes are behind this API. So that you don't have to do this manually and step by step, click the **AUTODETECT SCHEMA FROM RESPONSE** button in the **TEST** tab. The structure of the response is then used to determine which attributes are actually available. You need the correct names of the attributes to be able to carry out a clean data binding later.

Figure 8.8 shows that the generation of the structure worked successfully and that the schema was created with the correct attribute names, data types, and sample data.

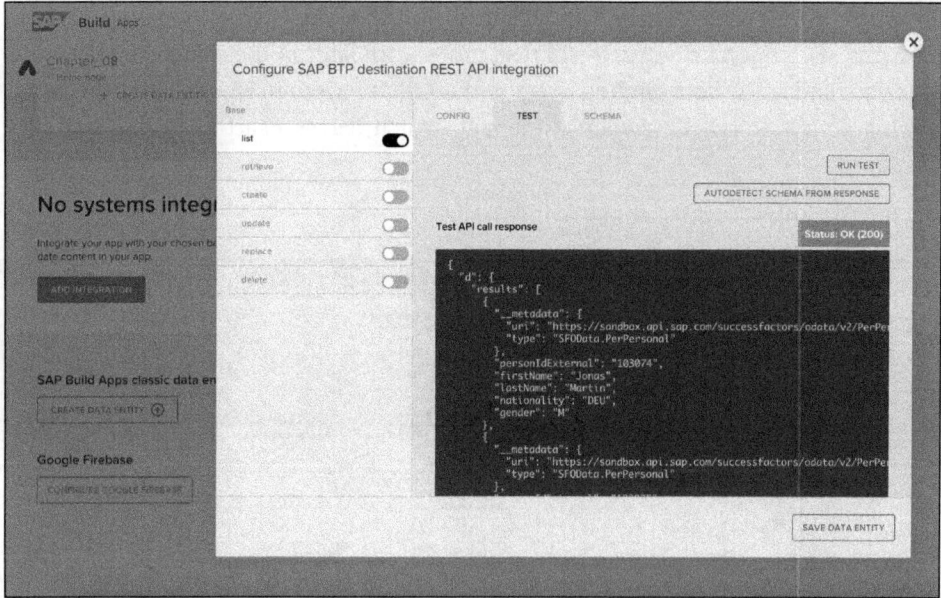

Figure 8.7 Test Request and Create Schema from Response

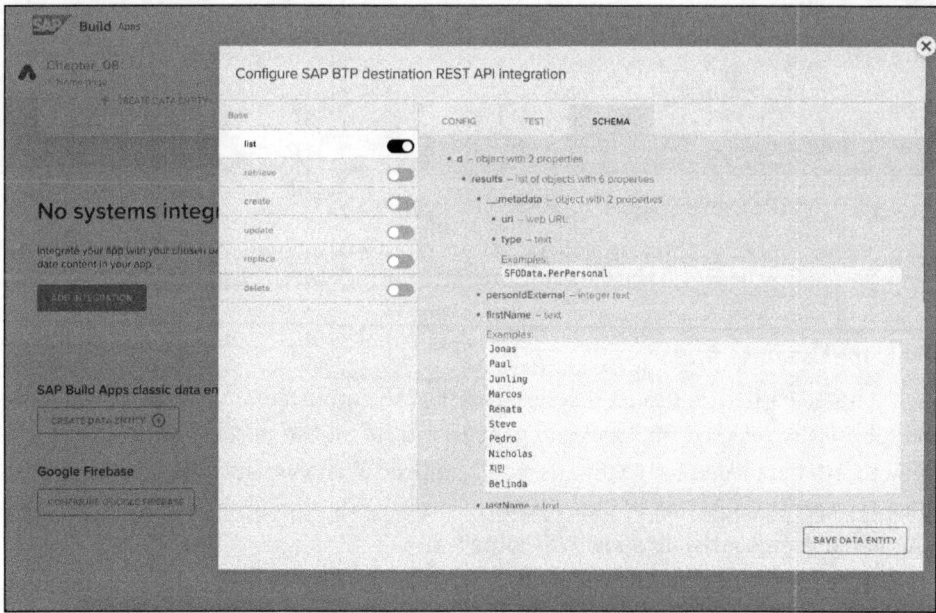

Figure 8.8 Schema of Response Sent Back for List

Now there's one more challenge to overcome before you can start developing the user interface. If you took a close look at the response from the API during the test, you will have noticed that the list or array is not returned directly as a response, but that the response is nested. For this reason, you have to define a mapper under **CONFIG** in the

8   Developing Extensions

last section, **Response Mapper**. This will break down the response so that only the array remains. If you don't do this, it will take more work later in the structure of the user interface and in the data binding to get to the list. As shown in Figure 8.9, store the following condition in the response mapper, which leaves exactly the list from the response:

```
{ records: response.data.d.results }
```

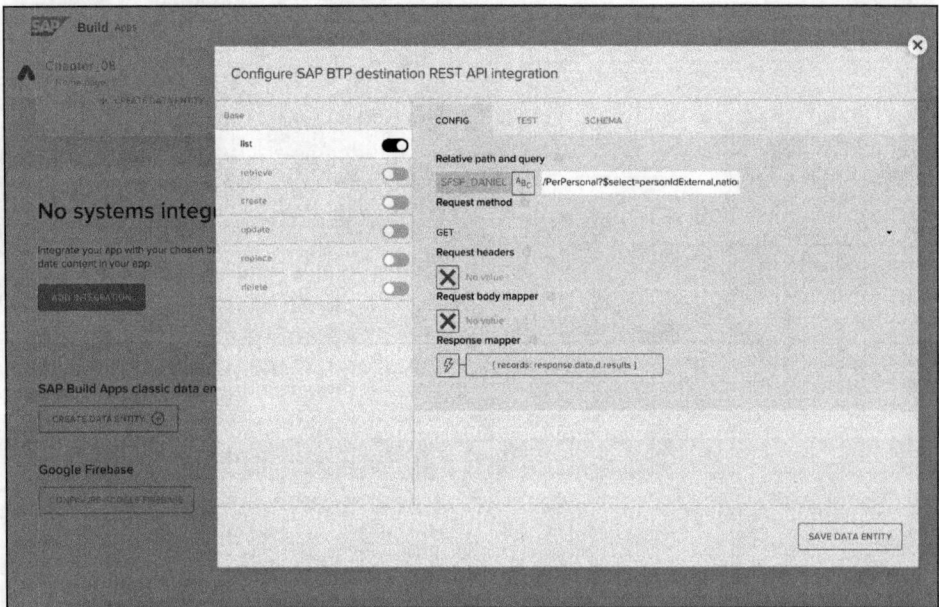

Figure 8.9 Add Response Mapper to Deserialize Reply Body

Now you're finished with the data resource and entities for the list.

Build the User Interface

As we noted earlier, you must first build a list of all employees. An employee can then be specifically selected, and the emergency contacts for the selected employee can be viewed in a detailed view. To do this, you first need a data variable with the **Data Variable Type** set to **Collection of Data Records**. This variable stores the list of people. Name this variable **personList**, as shown in Figure 8.10.

In the canvas UI, insert an **Icon List Item** element in **VIEW** and click the **Data Binding** button under **Repeat With** so that you can define the criteria according to which a list is to be displayed (see Figure 8.11).

8.1 SAP SuccessFactors

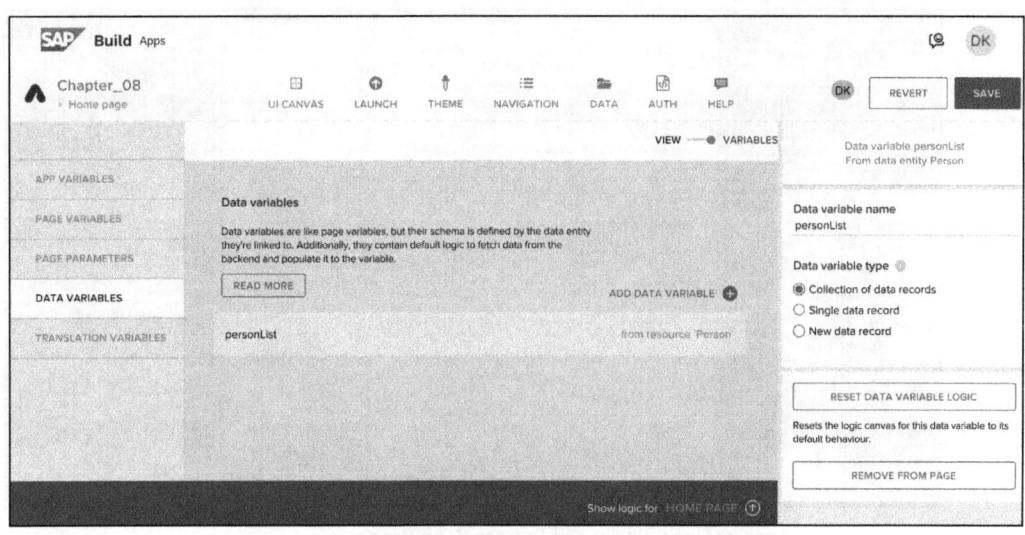

Figure 8.10 Create Data Variable to Store List of Persons (Employees in This Case)

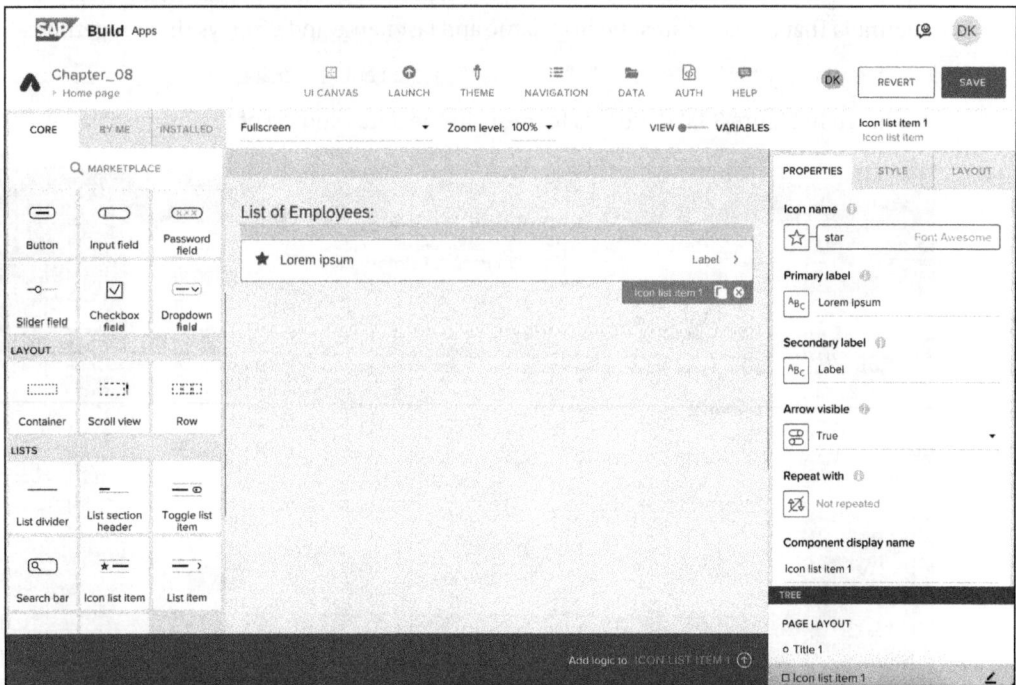

Figure 8.11 Insert Icon List Item in View

At **Repeat With**, bind the list element to the **personList** data variable because you want the icon list item to be inserted into the view as often as people are sent back from the backend and are stored in the variable (see Figure 8.12).

251

8 Developing Extensions

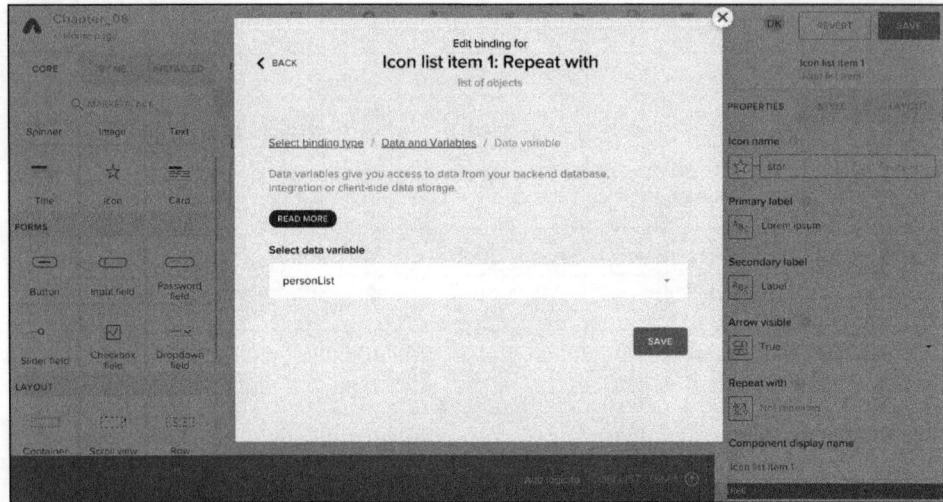

Figure 8.12 Customize Data Binding of Repeat With Property

Also perform data binding for the **Primary Label** property of the **Icon List** item. There, enter a **Formula** that concatenates the first name and last name and displays them together:

repeated.current.firstName + " " + repeated.current.lastName

As shown in Figure 8.13, store this formula in the data binding of the primary label.

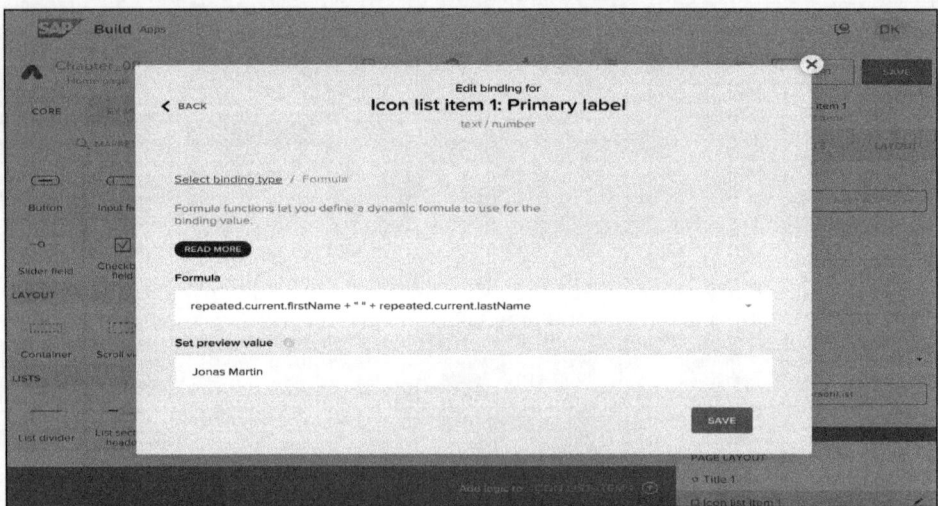

Figure 8.13 Data Binding for Primary Label of Icon List Item

With the **Secondary Label**, perform the data binding directly on the **Data Item in Repeat** and select **personIdExternal** as the data property. Figure 8.14 shows that you use the **current** pointer and can always access the current employee and output the personnel number when running through the list and inserting the list elements via this pointer (as in a *for-each* loop).

8.1 SAP SuccessFactors

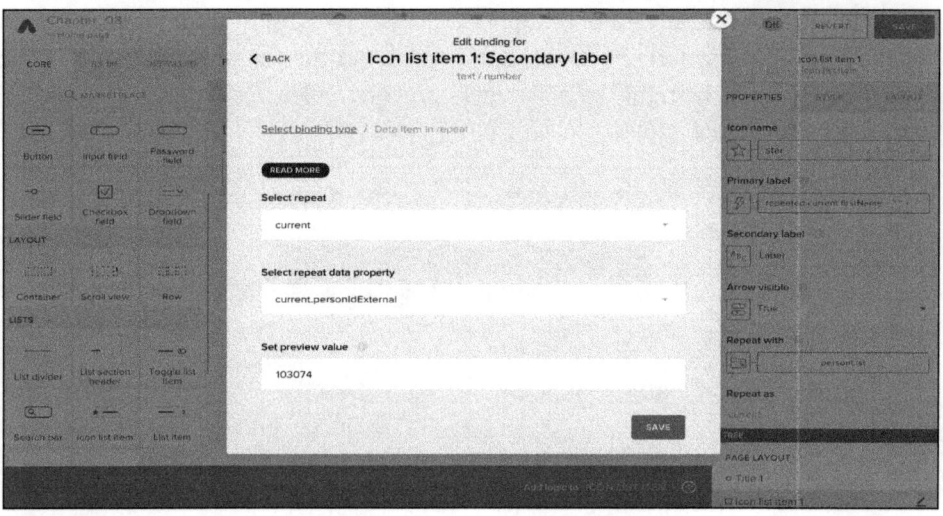

Figure 8.14 Data Binding for Secondary Label of Icon List Item

Last but not least, let's configure the **Icon Name** property of the **Icon List Item** accordingly and display a specific icon depending on gender. Enter a **Formula** to be able to define other conditions in the future:

```
repeated.current.gender === "F" ? "venus" : "mars"
```

In this formula, you check whether the current entry in the repeated.current.gender data property has saved the string "F" when iterating the list. If so, the "venus" icon should be displayed. Otherwise, the "mars" icon should be displayed.

Finally, let's deactivate the **Navigation Menu** in the **NAVIGATION** tab, but leave the header bar for the application header. You can statically change the header to **Search Emergency Contacts**, as shown in Figure 8.15.

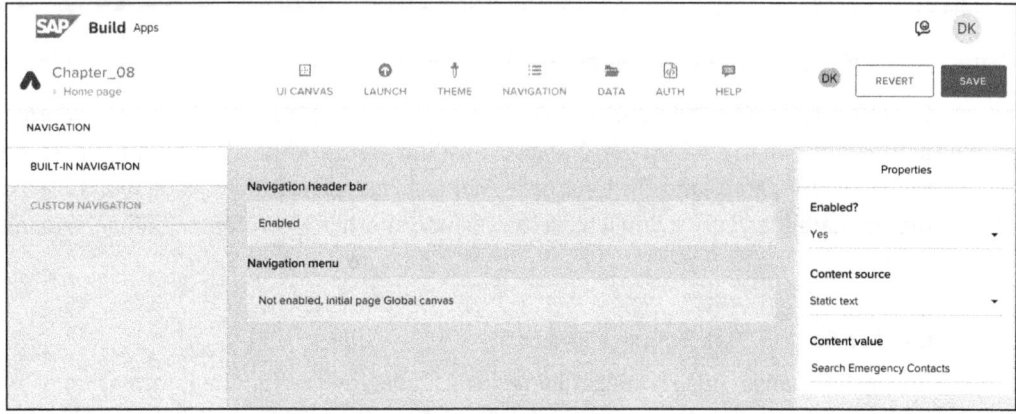

Figure 8.15 Changes for Navigation to Hide Navigation Menu

253

8   Developing Extensions

If you now look at the preview of the application, you see a list of employees. The icons for the items are formatted according to gender. On the left are the names of the employees, and on the right the personnel numbers, which are supplied by SAP SuccessFactors (see Figure 8.16).

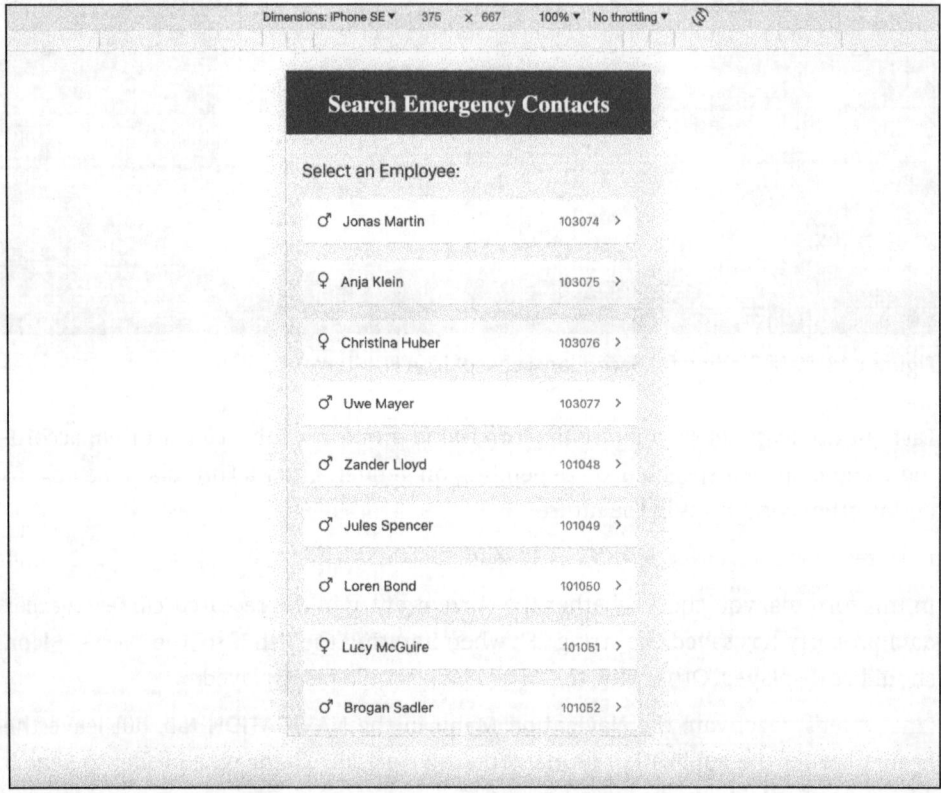

Figure 8.16 Preview of List of Employees

8.1.3   Build a Detail Page for Employee Details

To be able to navigate from the list to a detail page, you not only have to create another page, covered in this subsection, but also must integrate another data resource. Because the employee details and the emergency contacts are unfortunately not accessible via the already added entity, you'll have to consume another API before building the detail page.

Create a Second Data Resource and Entity

As the API that we integrated earlier is not suitable for reading the information of an individual employee, let's simply integrate a different interface. We'll proceed in the same way as we did with the **Person** entity. As shown in Figure 8.17, name the new entity "PersonDetail".

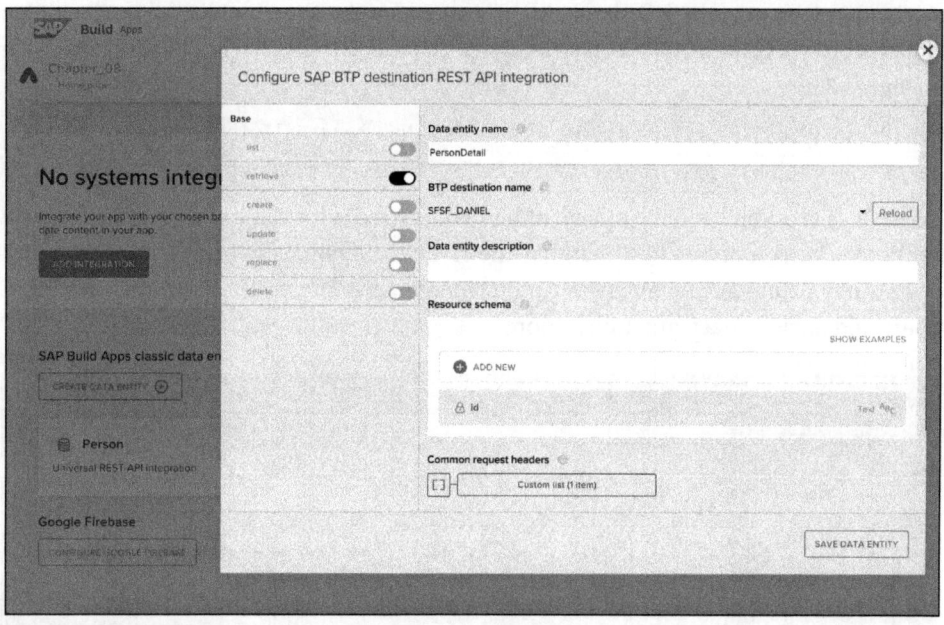

Figure 8.17 Configure Second REST API Integration for Person Details API

Stay within the **Base** menu and scroll all the way down until you reach the **Additional Inputs** section. Here, create a new **Additional Input** called **personIdExternal** (see Figure 8.18). This will later allow you to enter the ID of the employee you want to query the details of as a parameter when calling the API.

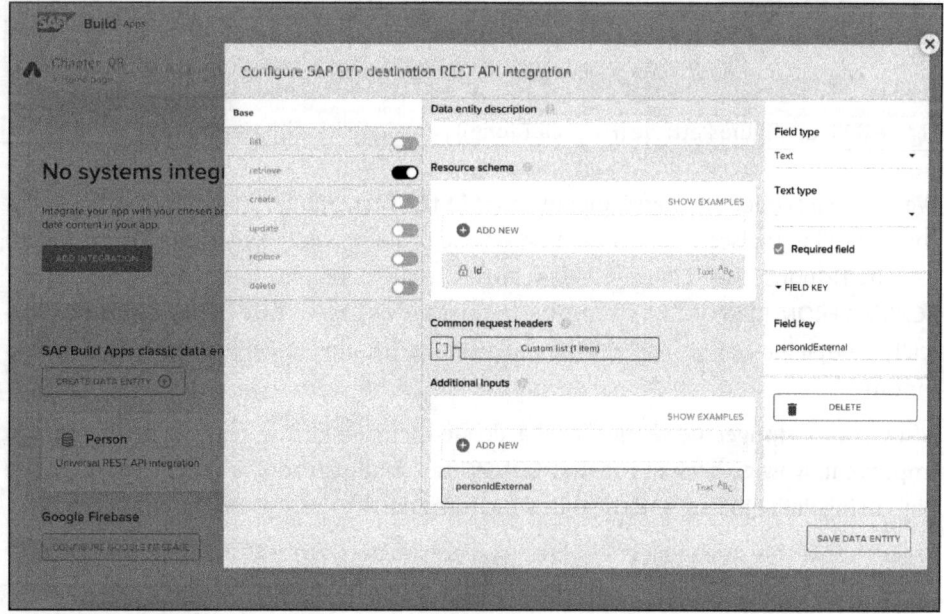

Figure 8.18 Add Additional Input to Be Used for Primary Key

8   Developing Extensions

Activate **retrieve** on the left-hand side. This enables the functionality to read a single data record via this entity. We have assigned the following as a relative path, as shown in Figure 8.19:

```
/PerPerson('108733')?$expand=emergencyContactNav,personalInfoNav,emergencyCon-
tactNav/relationshipNav,emergencyContactNav/relationshipNav/picklistLabels
```

With this path, you have the option of hardcoding the information from the employee with personnel number 108733 and at the same time requesting further information with query parameter $query, such as the emergency contacts, the relationship behind the emergency contact, and much more.

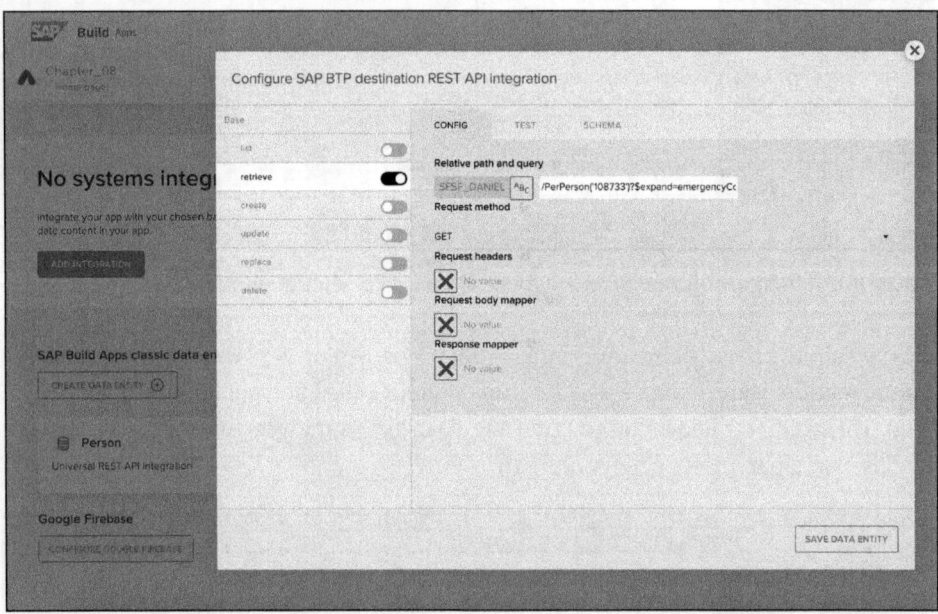

Figure 8.19 Configure Retrieve (HTTP GET [one]) Functionality with Hardcoded Path

We only hardcoded the personnel number in the first step for one reason: we wanted the test call in the second **TEST** tab to go through successfully and wanted to be able to use the response (see Figure 8.20), as with the previous entity, for the **AUTODETECT SCHEMA FROM RESPONSE** option. This in turn allows us to create the schema automatically so that we can also access the individual attributes in **PersonDetail** during data binding.

After you no longer need the first hardcoded version of the relative path, you can improve it. Now go back to the first **CONFIG** tab, and instead of a static text, build the path using data binding and the following formula:

```
"/PerPerson('" + query.additionalInputs.personIdExternal + "')?$expand=emergency-
ContactNav,personalInfoNav,emergencyContactNav/relationshipNav,emergencyContact-
Nav/relationshipNav/picklistLabels"
```

8.1 SAP SuccessFactors

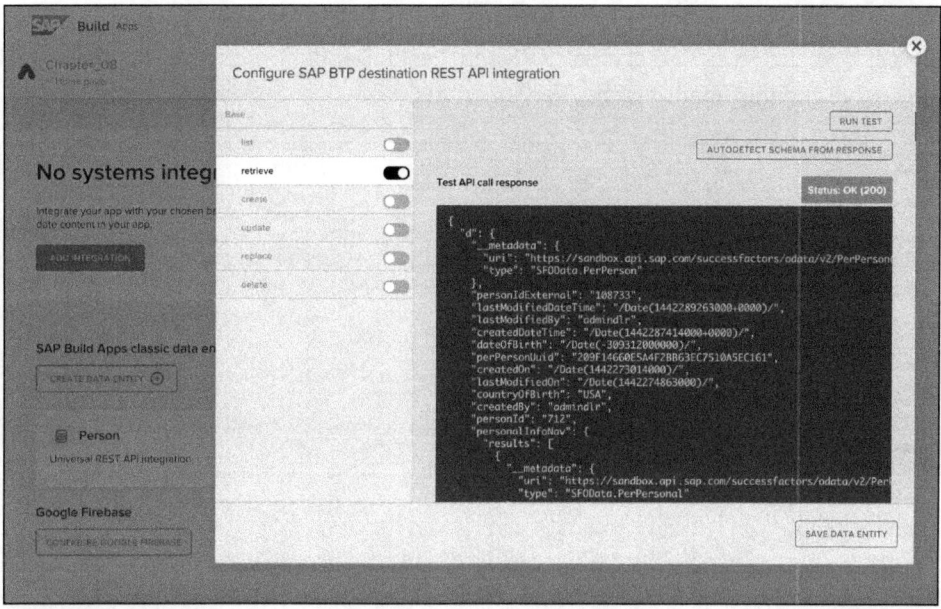

Figure 8.20 Hardcoded Path Used for Successful Response and Schema Mapping

With this formula, the call remains basically the same, but the personnel number is used dynamically. This will be the personnel number that you have already prepared in the **Base** menu under **Additional Inputs**. This allows you to enter the number later from the application and only get the information back from the requested employee. The stored formula is shown in Figure 8.21.

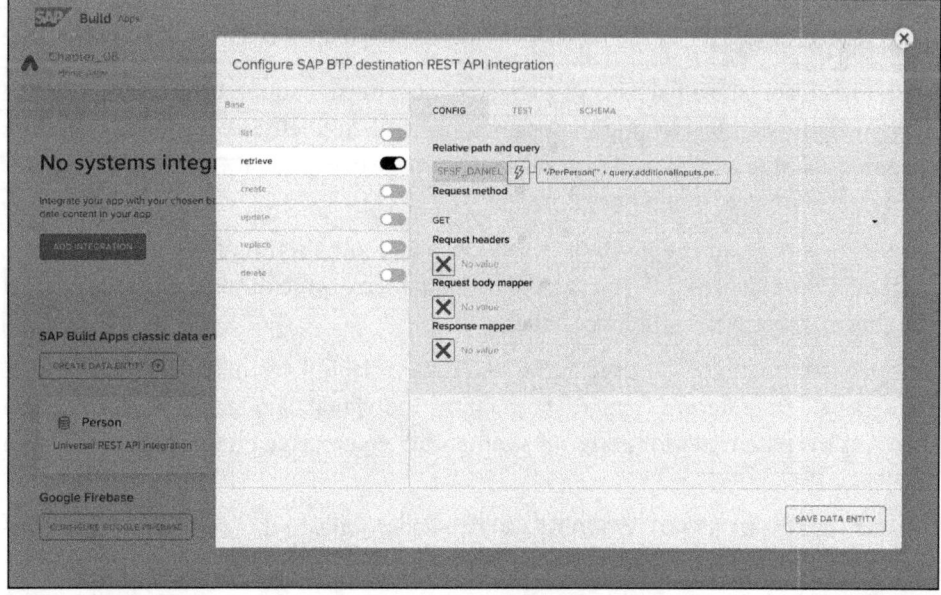

Figure 8.21 Change Path from Hardcoded to Relative Value

257

8   Developing Extensions

Figure 8.22 shows that there are now two entities available: **Person** for reading the list of persons via the /PerPersonal API and **PersonDetail** for reading an individual employee and the details via the /PerPerson API.

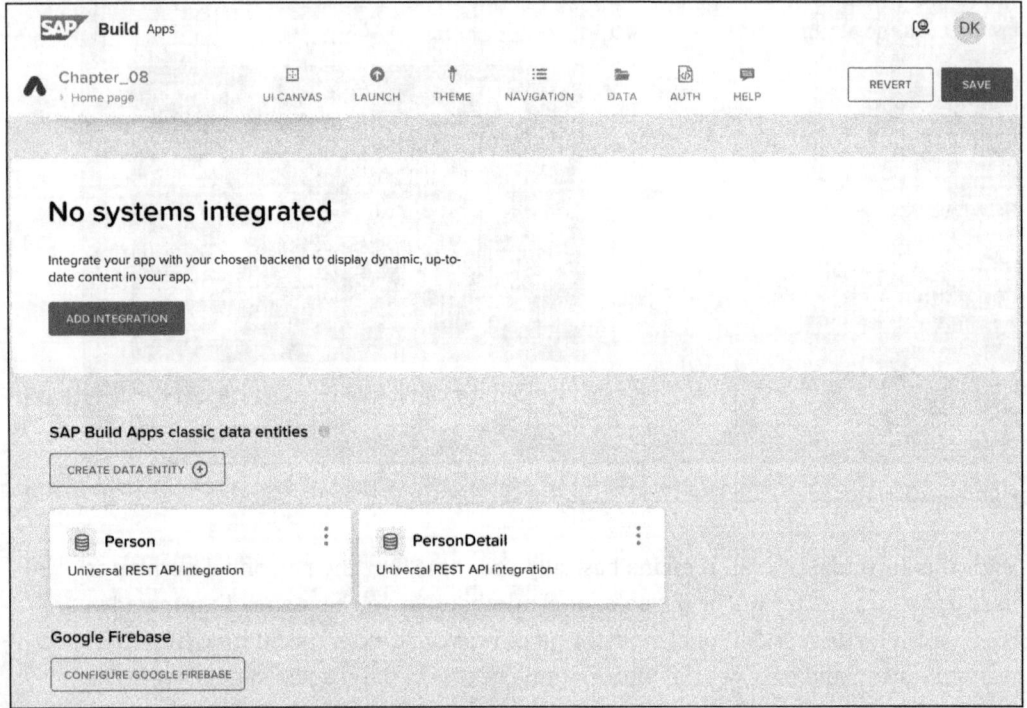

Figure 8.22 Two Data Resources Used for Displaying List and Details of Employees

Build the Detail Page

Let's now create a new page. As you already know from Chapter 4, you can access the view in Figure 8.23 by clicking the page name at the top left below the project name (**Chapter_08** in this case). You can create a new page using the **ADD NEW PAGE** button. Name the new page "Emergency Contacts".

On this new page, in the **VARIABLES** section, add a **PAGE PARAMETER** with the name "personid" (see Figure 8.24). You can use this parameter, which should be the personnel number, to request the employee details.

This page parameter will be used in the next data variable, **Person2Detail1**. This variable of the **Single Data Record** type will contain the individual data record of an employee and uses the parameter for **personIdExternal** that you receive on this page (see Figure 8.25).

You can also create a **PAGE VARIABLE** called **personName**. In this variable, you can display the name of the currently selected employee (see Figure 8.26). To ensure that this variable is filled automatically, open the logic canvas.

258

As the first block, drag and drop **Get Record** from the left-hand side into the workspace. Link this block to the **Page Mounted** and **Page Focused** events that exist by default. This means that the logic is triggered every time (whether for the first time or repeatedly) the page is opened. In **Get Record**, specify that you want to read the data from the **PersonDetail** data resource. As a parameter, this entity expects you to fill **Additional Input** with the name **personIdExternal**. Bind this value to the **personid** page parameter, which also stores the personnel number.

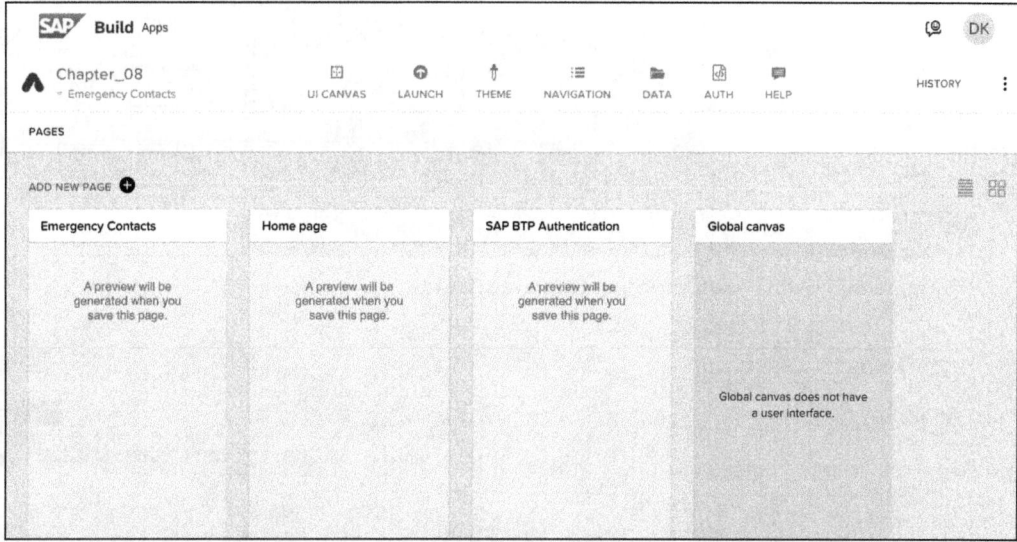

Figure 8.23 Page Overview and Option to Create New Pages

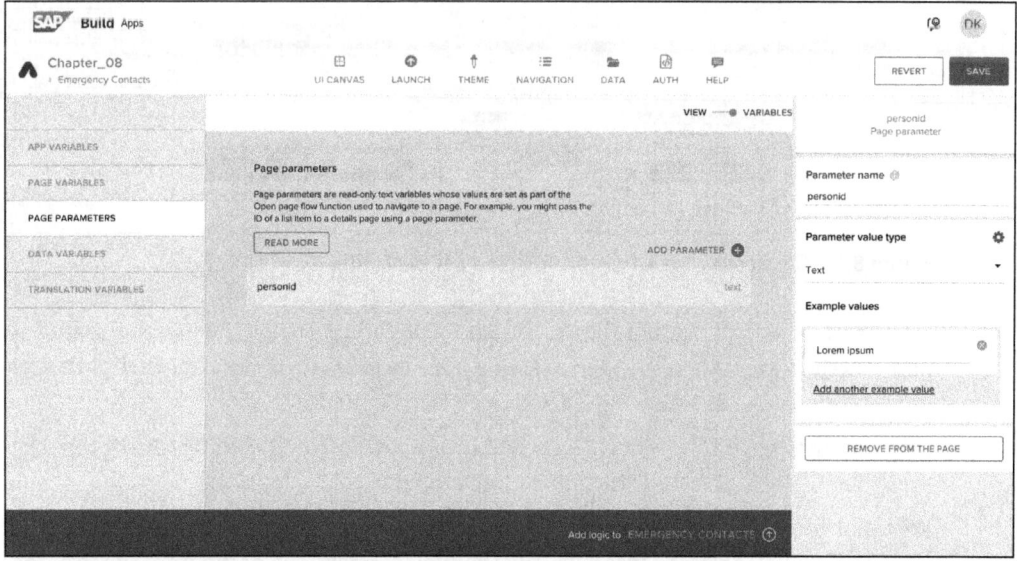

Figure 8.24 Page Parameter to Be Filled in when Navigating to This Page

8 Developing Extensions

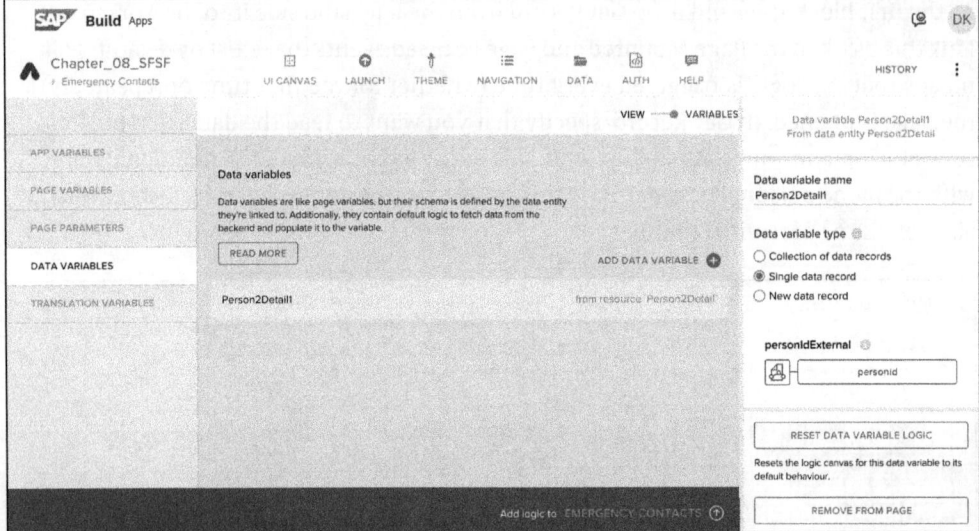

Figure 8.25 Data Variable to Store Employee Details

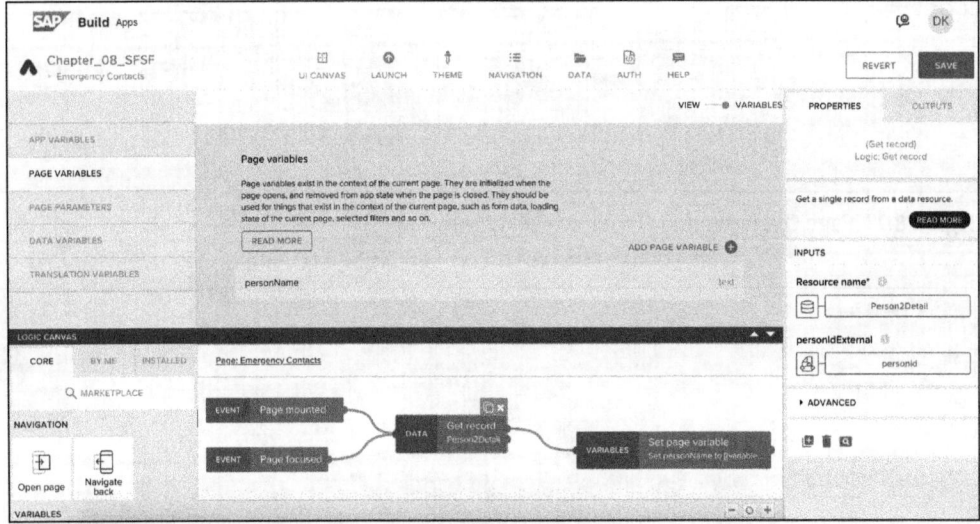

Figure 8.26 Get Record Logic to Read Selected Person Name

In the second step of the logic block, use **Set Page Variable** to set the first name and last name as the new value of the **personName** page variable from the result set of the previous block using the following formula (see also Figure 8.27):

```
outputs["Get record"].record.d.personalInfoNav.results[0].firstName + " " + outputs["Get record"].record.d.personalInfoNav.results[0].lastName
```

With this formula, you ensure that every time the page is opened, the page parameter is sent to the **PersonDetail** interface and that you can read the name from the response and display it in the view.

8.1 SAP SuccessFactors

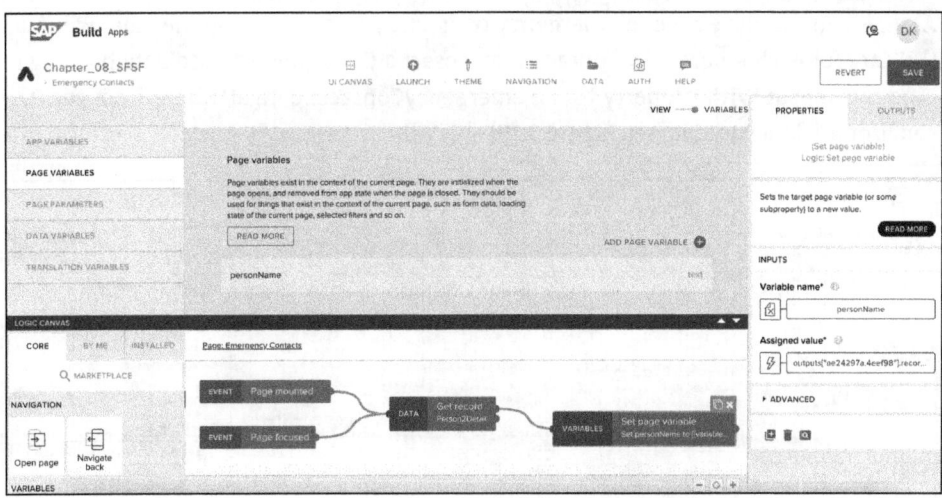

Figure 8.27 Set Page Variable Logic Block to Fill Page Variable with Data Returned by Get Record

Let's now switch to **VIEW** and change the **Content** property of the **Title** UI element via data binding to the following formula:

"Emergency Contacts of " + pageVars.personName + " (" + params.personid + "):"

With this formula, you can dynamically display the text "Emergency Contacts of" as the title with the person name from the variable **personName** and the personnel number from the page parameter **personid** in brackets behind it (see Figure 8.28).

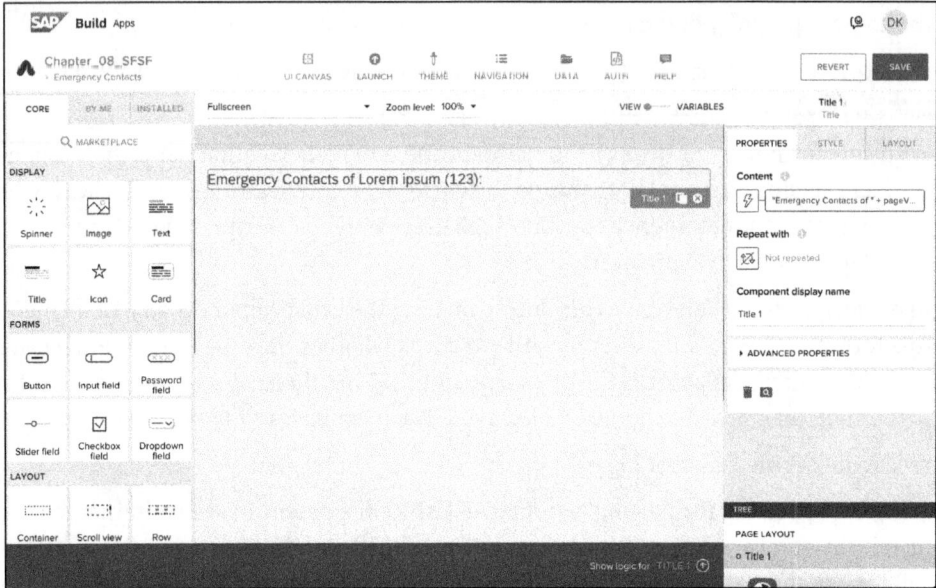

Figure 8.28 Data Binding for Title on This Page

261

8   Developing Extensions

As a person can have several emergency contacts, you need a container that will map the data for each emergency contact. Now insert a **Container** and use data binding to bind the **Repeat With** property to the **emergencyContacts** data attribute from the **Person2Detail1** data variable (see Figure 8.29).

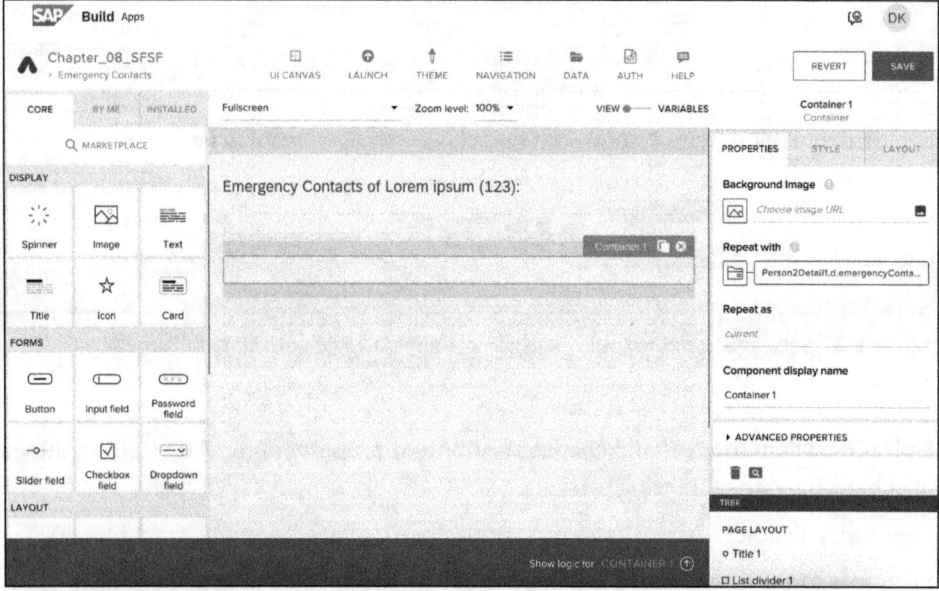

Figure 8.29 Container Used for Displaying Emergency Contacts One after Another

Now insert the **Title** UI element into this container, and assign the following formula to the content property of the title:

```
repeated.current.name + " (" + SELECT(repeated.current.relationshipNav.picklist-
Labels.results, item.locale === "en_US")[0].label + "):"
```

With this formula, you display the name of the emergency contact and the human-readable version of the relationship to the currently selected employee. This way, you know whether the emergency contact in the list is a friend, relative, or family member. This step is shown in Figure 8.30.

Insert an **Icon** to the left of the title and again bind the **icon** property of this UI element to a formula. Use the following formula to check whether the emergency contact the app is currently displaying is the primary contact. If yes, then the icon should show be star. If not, then no icon should be displayed. The formula is as follows:

```
repeated.current.primaryFlag === "Y" ? "star" : ""
```

For more appealing formatting, use an **Icon List Item** element in the container for the telephone number and email address (see Figure 8.31). For the first icon list item, statically assign the text "Phone" for **Primary Label**. Bind **Secondary Label** to the **phone** attribute from **Data Item in Repeat** via data binding. The **icon** property is again set statically and will represent a telephone symbol.

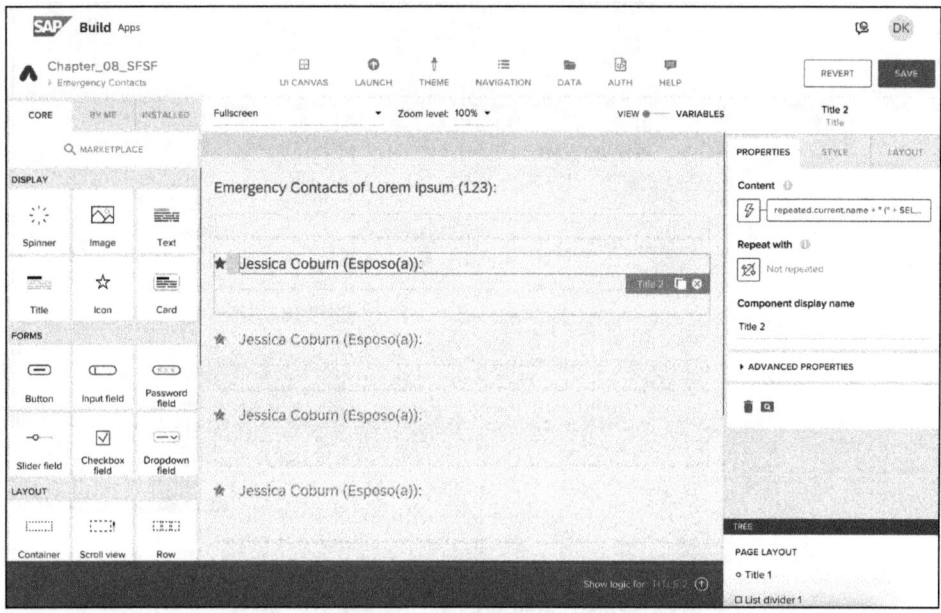

Figure 8.30 Title in Container Displaying Emergency Contact Name

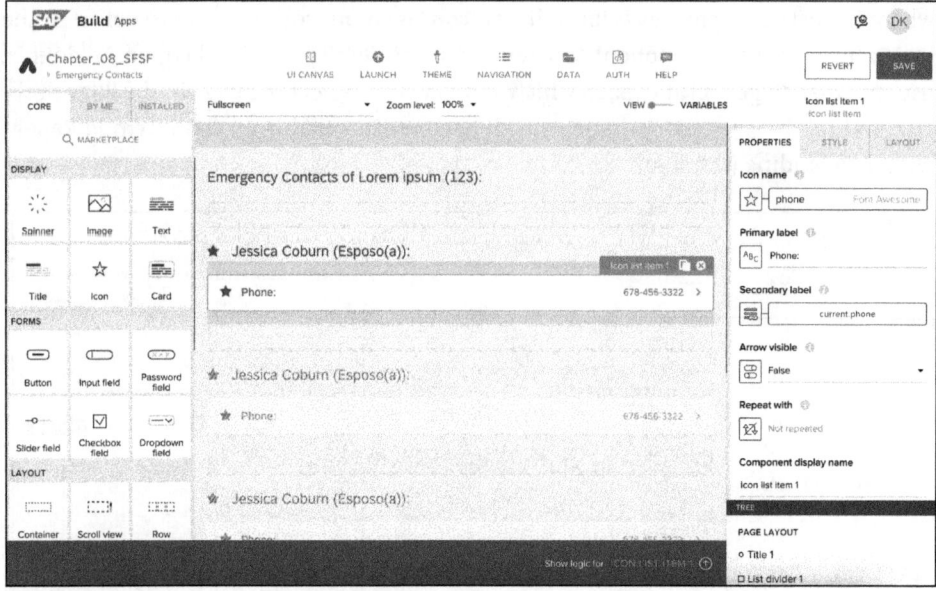

Figure 8.31 Icon List Item to Display Phone Number of Each Emergency Contact

For the email address, proceed in the same way as for the telephone number. However, here you can bind **Secondary Label** to the **email** data attribute from **Data Item in Repeat**. Of course, display an envelope as the icon. You can see these steps in Figure 8.32.

8   Developing Extensions

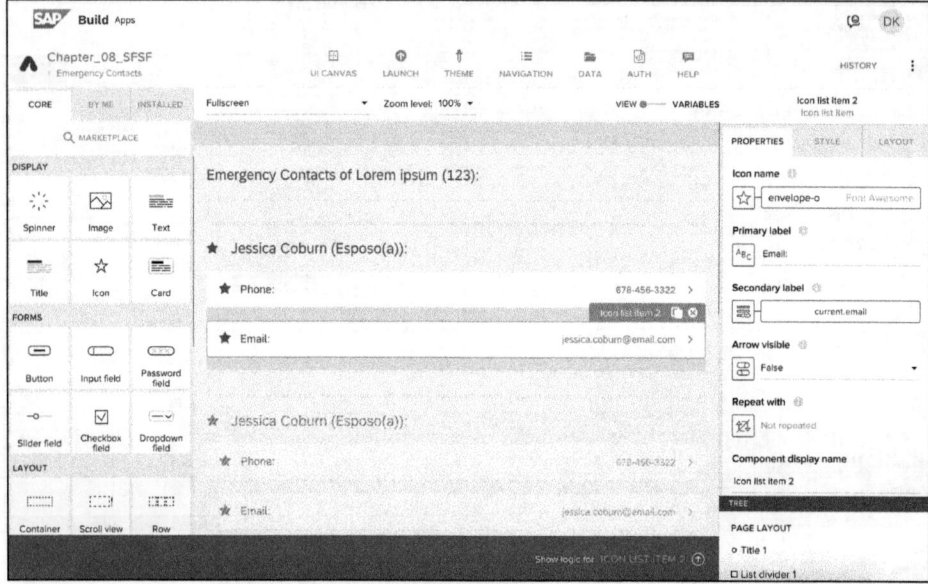

Figure 8.32 Icon List Item to Display Email of Each Emergency Contact

Now we're missing one last step to complete the development. Let's go back to the page where the list of employees is located. For **Icon List Item**, you still have to define in the logic canvas for the **Component Tap** event that the detail page should open. To do this, use the **Open Page** logic block, in which you naturally select the page in the **Page** property and fill the **personId** parameter with **personIdExternal** from **Data Item in Repeat** with data binding (see Figure 8.33).

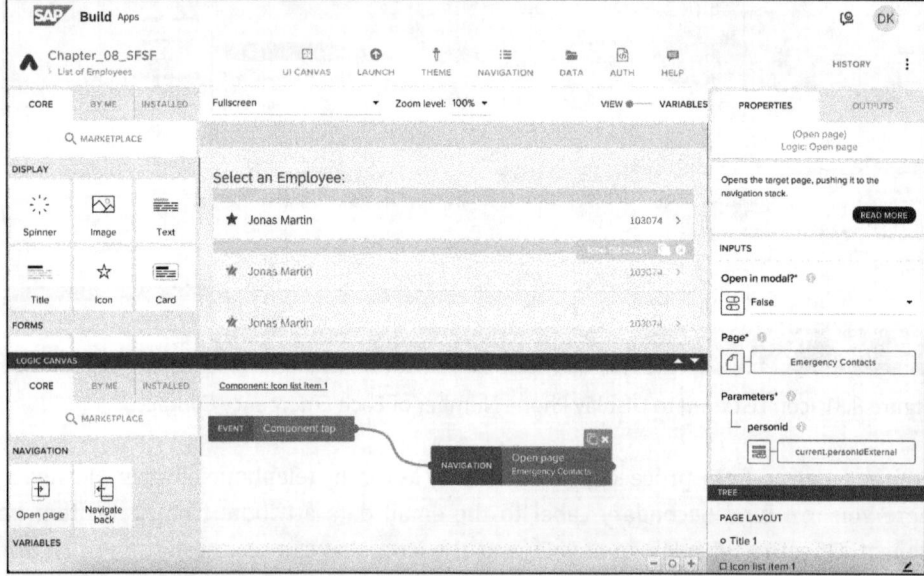

Figure 8.33 Logic to Navigate to Detail Page and Pass Parameter

264

When you test the app again, you'll now see that clicking an employee will navigate you to the details page. This details page contains the emergency contacts with their names, relationship to the employee, telephone numbers, and email addresses. The navigation includes the personnel number, which is used by the variables to communicate with the data resource and ultimately retrieve the data from SAP SuccessFactors. You can see the fully developed detailed view in the preview shown in Figure 8.34.

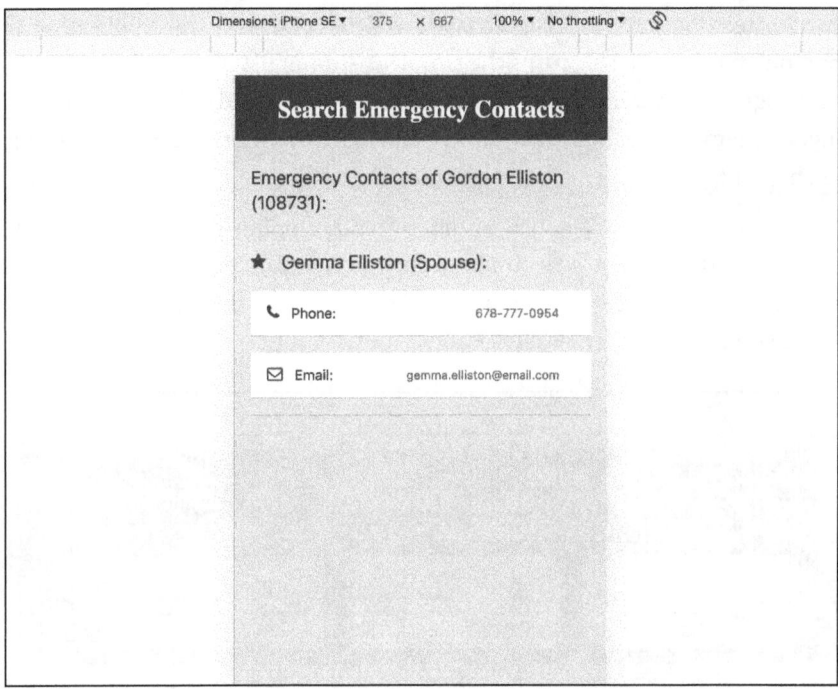

Figure 8.34 Preview of Detail Page Displaying Emergency Contacts of Selected Employee

8.2 SAP S/4HANA and SAP S/4HANA Cloud

You will certainly often be confronted with the requirement to develop a customer-specific web application for an SAP S/4HANA system. In addition to the SAP Fiori design system and the options for developing web applications with the SAPUI5 framework, SAP Build Apps is now also being added to the repertoire. SAP S/4HANA is the fourth generation of SAP Business Suite and is based on the SAP HANA database. Although SAP delivers many standard apps for SAP S/4HANA, you may well want to develop your own applications tailored to your own business processes. Now that the cloud has become an integral part of our professional lives, with SAP's new cloud-first ideology, you also have the possibility of licensing an SAP S/4HANA system in SAP BTP.

8   Developing Extensions

Whether you have an SAP S/4HANA or SAP S/4HANA Cloud system in front of you makes no difference in SAP Build Apps, as you already know from Chapter 6. Thanks to the way in which data resources can be consumed, you see a facade that works the same way for SAP Build Apps for a wide variety of data integrations.

In this section, we'll focus on the already familiar example from Chapter 6, in which we connected an on-premise SAP S/4HANA system and consumed customer data. As we only dealt with reading data in Chapter 6, we will now expand this application to include manipulative access. This turns the simple customer list into a fully functional customer management system, where you can create, update, and delete customers. If you no longer remember the final result from the chapter mentioned, it's shown in Figure 6.39 in Chapter 6.

To ensure that you neither influence an already functioning project nor have to develop everything again, there is an option to copy a project in the SAP Build lobby. As shown in Figure 8.35, select the **Save as New Project** option for this. A dialog will then open in which you can assign a new name.

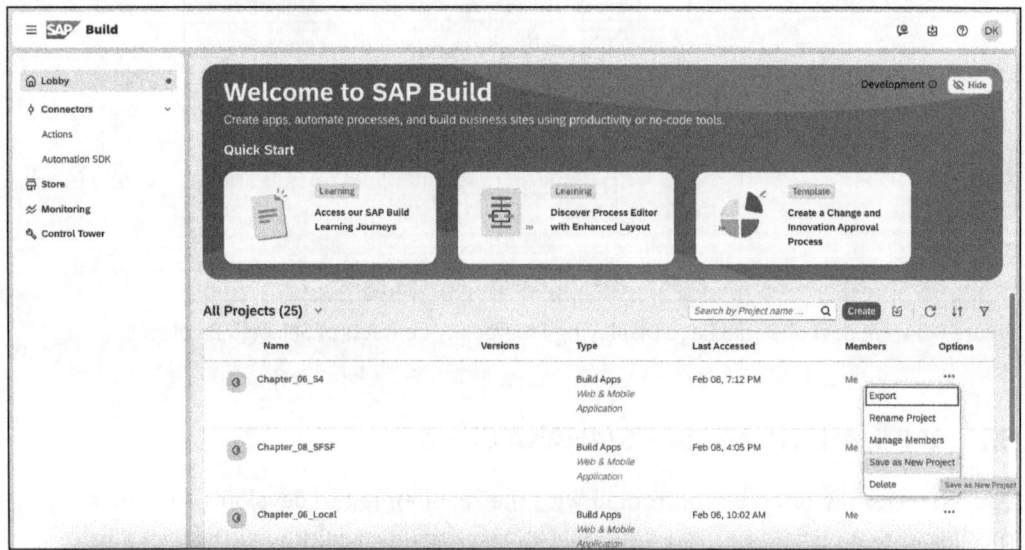

Figure 8.35 SAP Build Lobby: Copy Existing Project

In this new project, create a new page called **Create**. On this page, switch to **VARIABLES** and create a data variable called **ZHOUI5_CUSTOMERSET1**. This variable is of type **New Data Record** and will cache the data that we will send to the backend (see Figure 8.36).

8.2 SAP S/4HANA and SAP S/4HANA Cloud

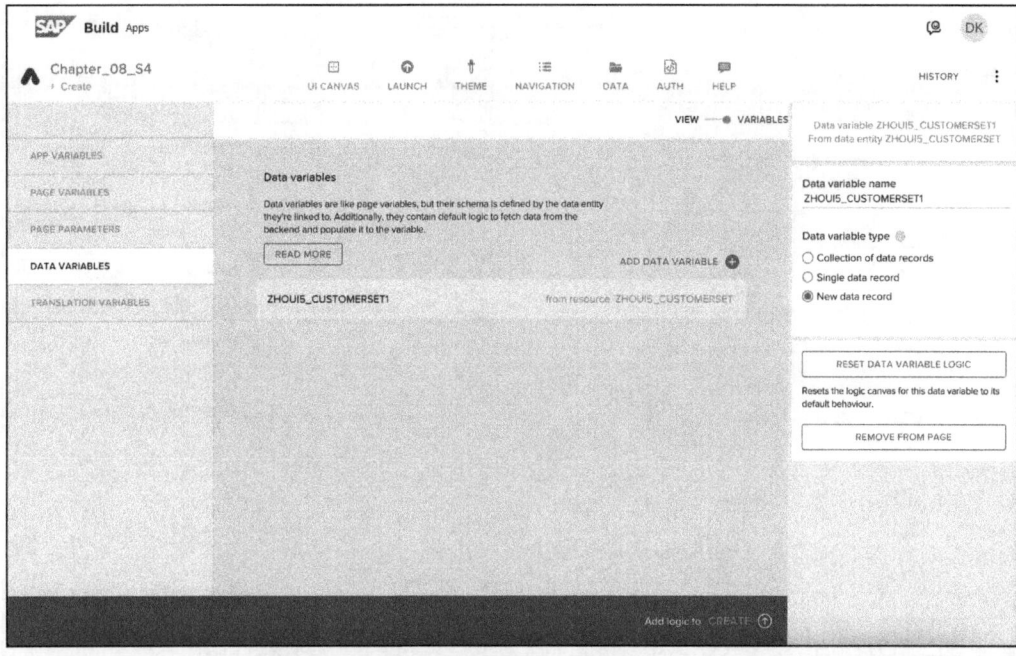

Figure 8.36 Creating Data Variable for Saving New Data Records

As you can see in Figure 8.37, we've already taken many steps in **VIEW** to prepare the page for creating a new customer:

- We have added the **Title** UI element and assigned the "Create Customer" text to the **Content** property.
- We have inserted an **Input Field** element for the first name. This input field has been given the label **Firstname**, and the **Value** property is bound to the **Firstname** field from the ZHOUI5_CUSTOMERSET1 data variable via data binding.
- We have repeated this step for **Lastname**, **Email**, **Phone**, and **Website**, with the data bindings here affecting the respective attributes.
- We have inserted a **Dropdown Field** element between **Lastname** and **Email**. This dropdown has been given the text "Gender" as the **Label Text**. The **Selected Value** property is bound to the **Gender** attributes from the ZHOUI5_CUSTOMERSET1 variable via data binding.

We have set the **Option List** property, which saves the possible selection values in the dropdown, statically as a list of values. In the backend, we defined that you can choose between "Female" and "Male" and that these are identified with the technical keys "1" and "2".

267

8   Developing Extensions

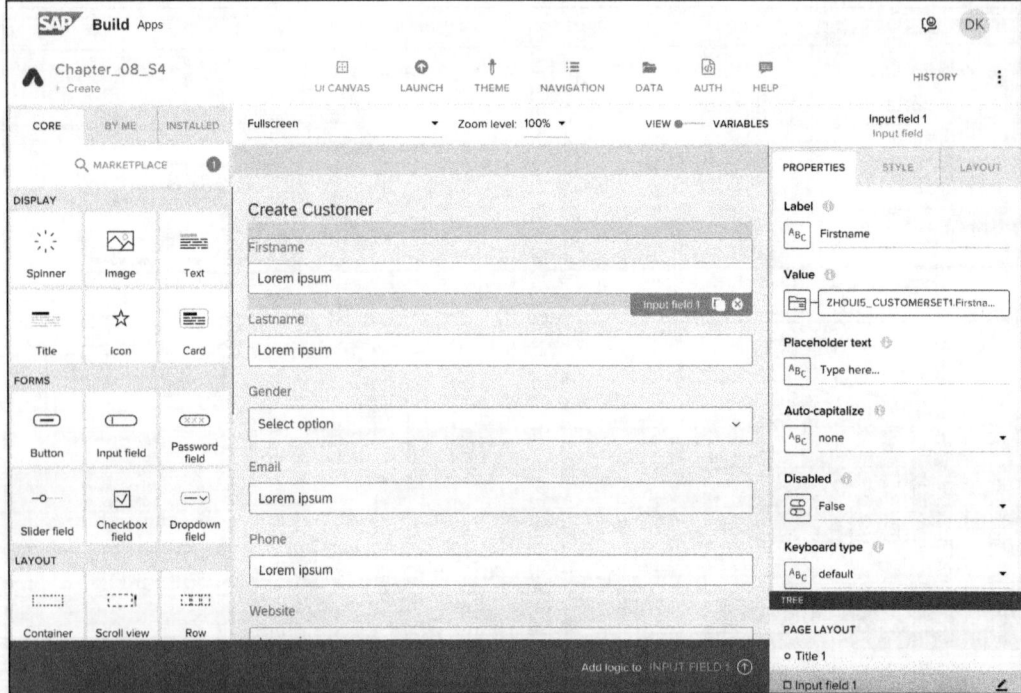

Figure 8.37 Developing View for Form

For a new data record to be saved, you still need corresponding action buttons so that users can also trigger this logic. Instead of ordinary buttons, we have installed the icon button extension from the Marketplace. We have packed the two buttons for **Save** and **Cancel** into a container and aligned them to the right. As shown in Figure 8.38, we have stored the following logic for the **Save** button:

- The logic is executed with the **Component Tap** event—that is, when the user clicks the button.
- With the **Create Record** logic block, the ZHOUI5_CUSTOMERSET data resource is called and the data that was temporarily stored in the ZHOUI5_CUSTOMERSET1 data variable is passed as a body. We will show you the easiest way to send the data in a moment.
- If the creation was successful (output 1 of **Create Record**), we issue a success message via **Toast**.
- After the toast event, we use **Navigate Back** to navigate back to the page from which we came.
- If the creation did not work (output 2 of **Create Record**), we use an **Alert** block to display the error message returned by the backend.

8.2 SAP S/4HANA and SAP S/4HANA Cloud

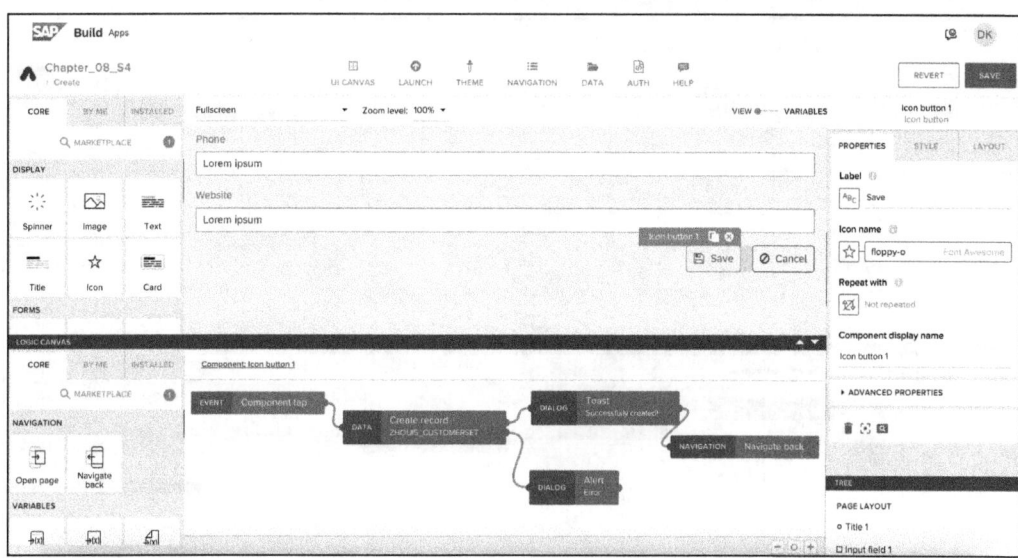

Figure 8.38 Logic of Save Button

In the **Create Record** logic block, define a mapping for the data to be sent via data binding. This mapping is shown in Figure 8.39. Select data variable **ZHOUI5_CUSTOMERSET1** as the source data and use drag and drop to assign the individual attributes from the variable to the body, which is sent to the backend for creation.

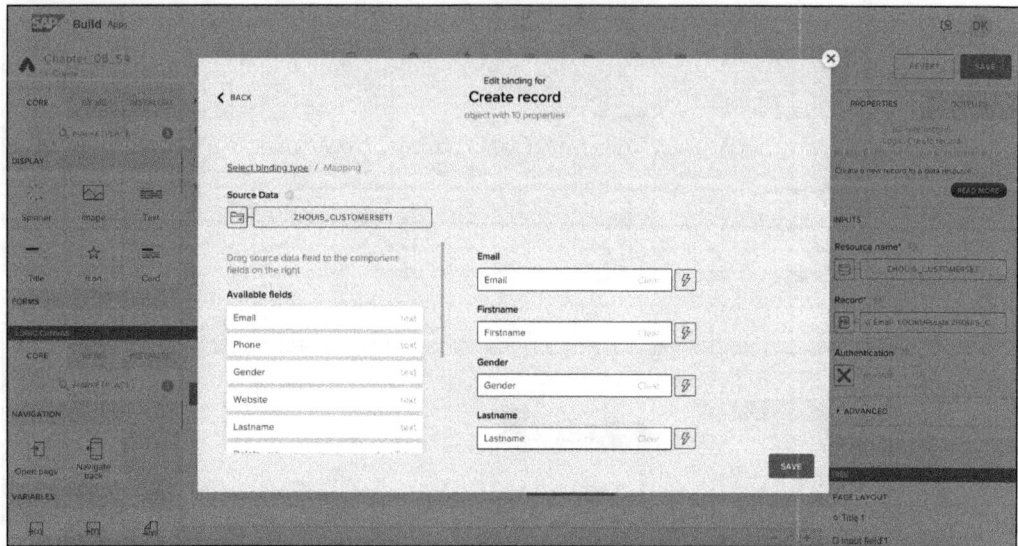

Figure 8.39 Data Mapping to Assign Values from Data Variable to Request Body

The logic behind the **Cancel** button is much simpler (see Figure 8.40). Here you let the user confirm with the **Confirm** dialog whether he really wants to cancel. If the user wants to (output 1 of **Confirm**), you navigate back again with **Navigate Back**.

269

8 Developing Extensions

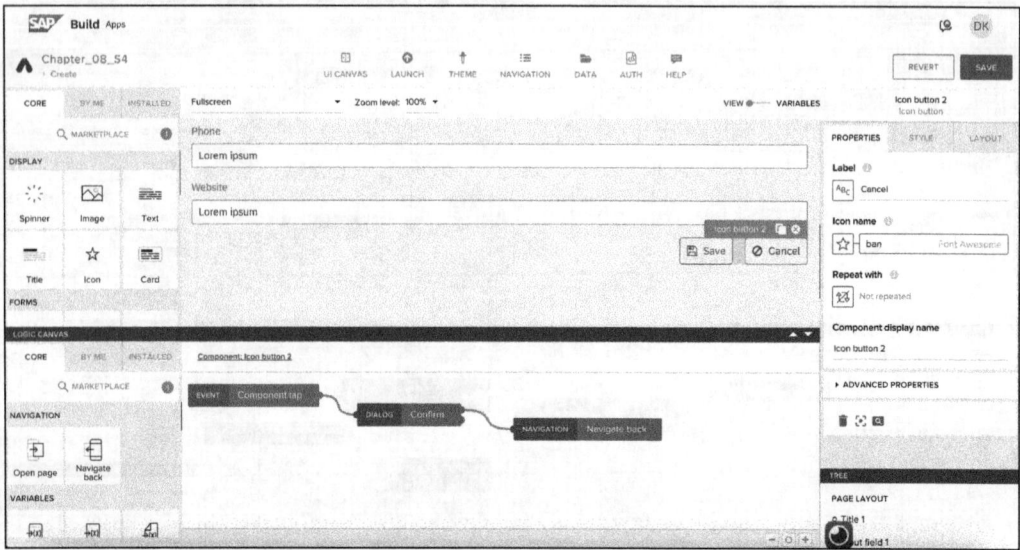

Figure 8.40 Logic Canvas of Cancel Button

To be able to access the page at all and create a new data record, you must either give the user the option of accessing the page via the menu navigation or via a separate button. In this case, we have inserted an **Icon Button** element above the table. If this button is clicked, the **Create** page is opened with **Open Page** (see Figure 8.41).

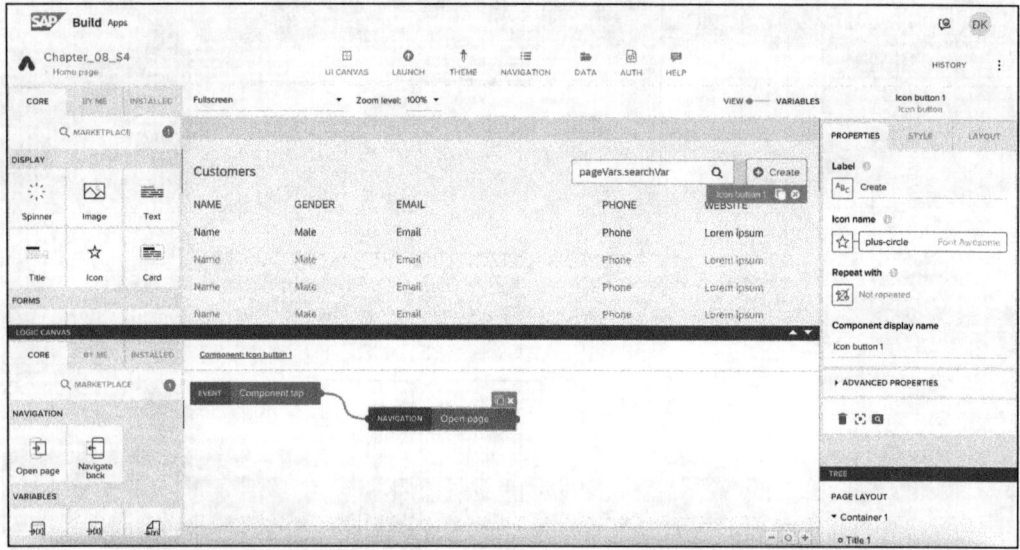

Figure 8.41 Logic Canvas of Create Button to Navigate to Create Form

Figure 8.42 shows that there is a button next to the search field above the table. Now, click this button, and expect to be taken to the **Create** page.

8.2 SAP S/4HANA and SAP S/4HANA Cloud

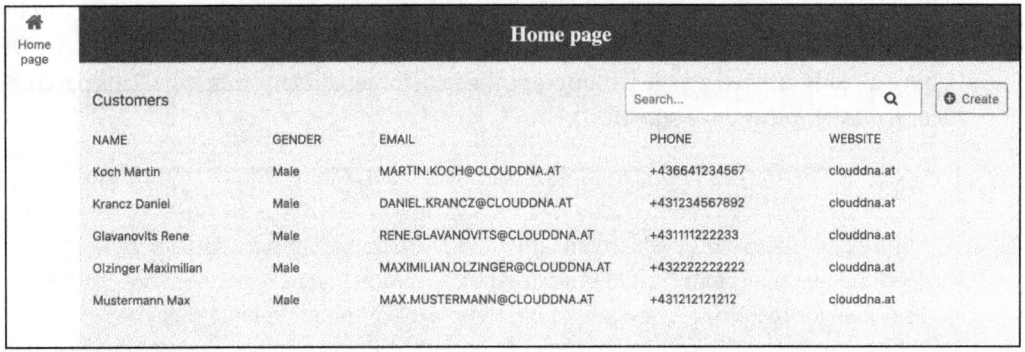

Figure 8.42 Preview of Table and Newly Added Create Button

Figure 8.43 shows the creation page. If you navigate to this page, you can make entries via the individual input fields and the dropdown field. These entries are stored temporarily in the data variable. As soon as the user clicks **Save**, the data is taken from the data variable and sent to the backend with an HTTP POST request. If everything goes well, the user is notified with a success message and returns to the table. By clicking **Cancel** or the **Home Page** menu item on the left, the user can cancel this process and return to the table without creating the customer.

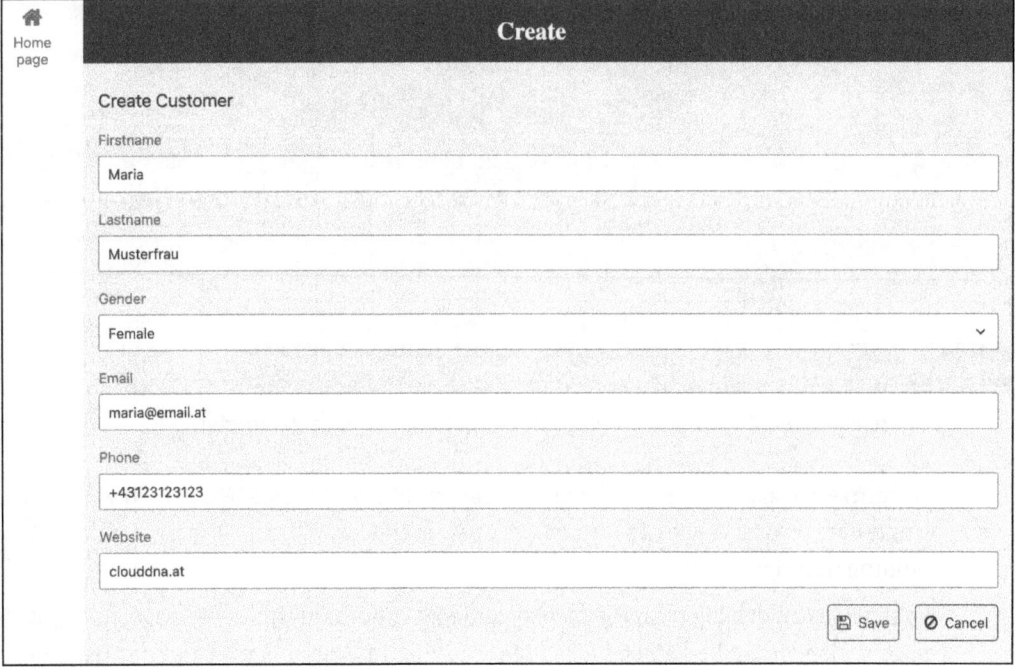

Figure 8.43 Form to Create New Customer

We have created a new page called **Detail** so that not only can new customers can be created, but also existing ones can be edited or deleted. As shown in Figure 8.44, we have already created a page parameter called **customerUuid** for this page. This parameter has been set as data type **UUID**.

> **What Is a UUID or a GUID?**
> In modern software development, Universally Unique Identifiers (UUIDs) or Globally Unique Identifiers (GUIDs) are often used to generate unique keys. These are naturally suitable for the primary keys of the database tables. The probability that the underlying algorithm will randomly generate two keys is very low. These keys only have one major disadvantage: they are not intended for human memory. Here is an example of what a GUID might look like: 936DA01F-9ABD-4D9D-80C7-02AF85C822A8.

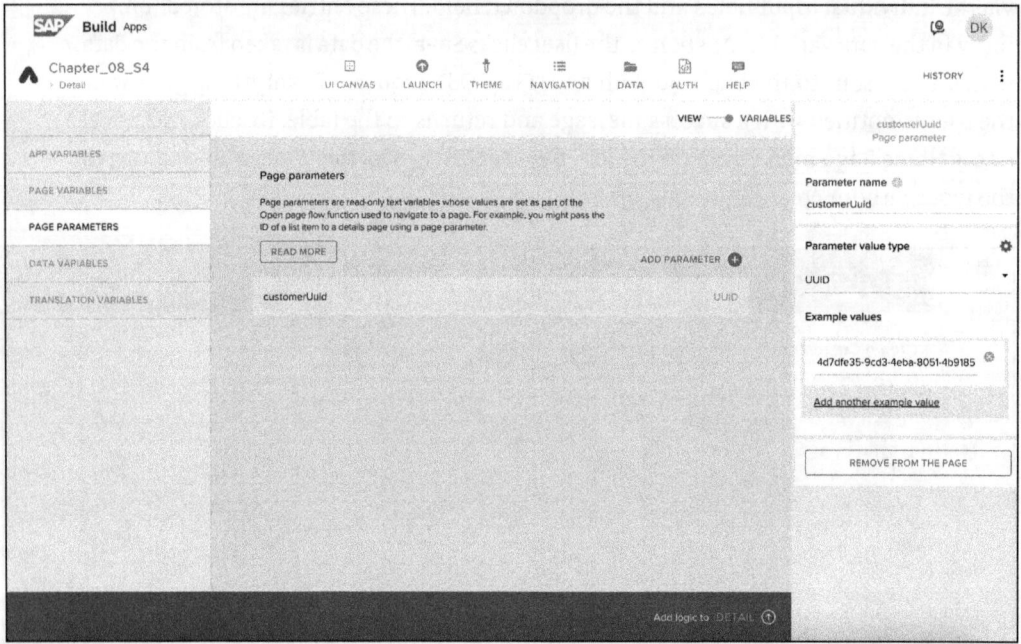

Figure 8.44 Detail View and Primary Key of Customer as Page Parameter

Of course, we also create a data variable called ZHOUI5_CUSTOMERSET1, which will store a single data record. We have assigned the **CustomerId** key to the page parameter via data binding (see Figure 8.45).

So that we can display both the display and edit modes on this detail page, we create a page variable called **editMode** (see Figure 8.46). This variable of data type **True/False** will help in the **VIEW** area to make the input fields ready for input or not.

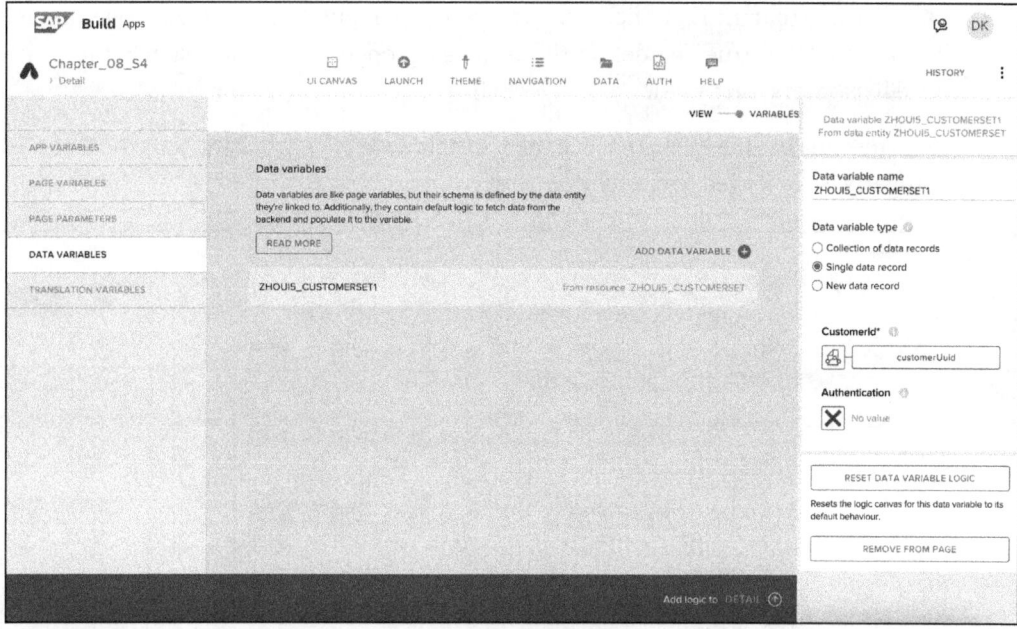

Figure 8.45 Data Variable to Save Data of Customer to Be Displayed

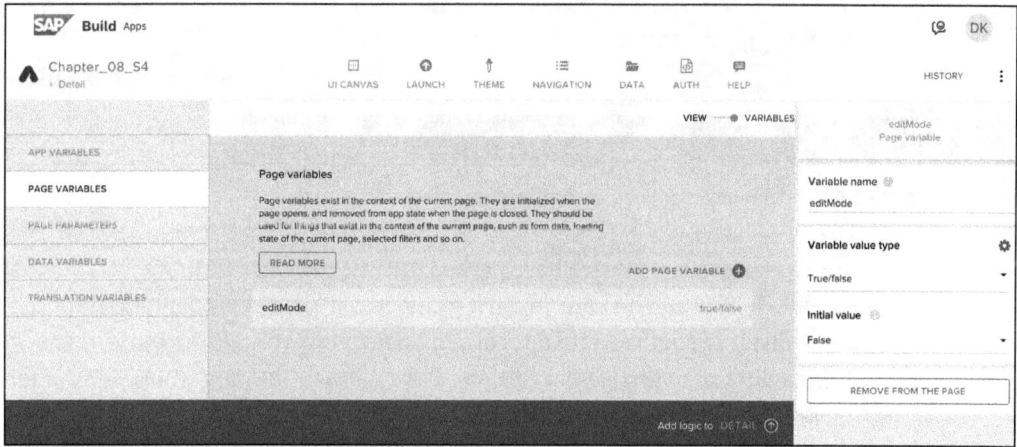

Figure 8.46 Page Variable to Help Switch between Display and Edit Mode

In Figure 8.47, you can see that we have already taken several steps, but these should not be new to you:

- We have changed the default title on this page to **Customer Details**.
- We have inserted several input fields and bound the **Value** properties of the input fields to the attributes from the data variable (**Firstname**, **Lastname**, **Email**, **Phone**, **Website**).
- We have inserted a dropdown field for **Gender** between **Lastname** and **Email**. This dropdown field has the value list set statically and the user can choose between Female and Male.

8 Developing Extensions

- To ensure that the input fields are only ready for input if the **editMode** page variable also contains **True**, we define the **Disabled** property with the following formula: NOT(pageVars.editMode).

This formula has the effect that the value in **editMode** is negated and thus the **Disabled** property is controlled correctly.

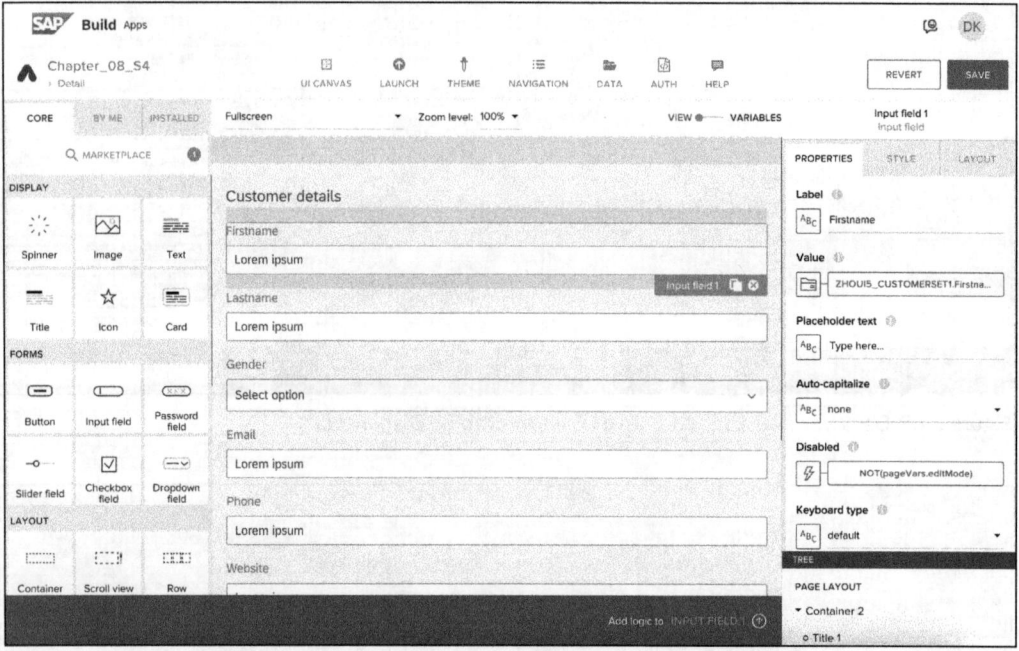

Figure 8.47 Detail View for Displaying Data of Selected Customer

In Figure 8.48, you can see that we have packed the title together with two icon buttons into a container. The buttons are in turn grouped via their own container and can therefore be displayed right-aligned together. The container with the buttons is only visible if the **editMode** variable is **False**. We have defined this condition using a formula. The **Delete** button has been given the following logic:

- The logic is triggered in the **Component Tap** event, when the button is clicked.
- In the first step, we use a **Confirm** dialog to confirm whether the user really wants to delete this customer.
- If the user confirms this dialog (output 1 from **Confirm**), we move on to the next logic block, **Delete Record**. This has been given the resource name ZHOUI5_CUSTOMER-SET1. The entity also expects us to specify the primary key of the customer to be deleted. As we already store the UUID of the customer in the page parameter anyway, we enter this in **CustomerId**.
- If the deletion was successful (output 1 from **Delete Record**), we navigate back to the table with **Navigate Back**.

8.2 SAP S/4HANA and SAP S/4HANA Cloud

- If, for whatever reason, the deletion ran into an error (output 2 from the **Delete Record**), we display the error message via **Alert**.

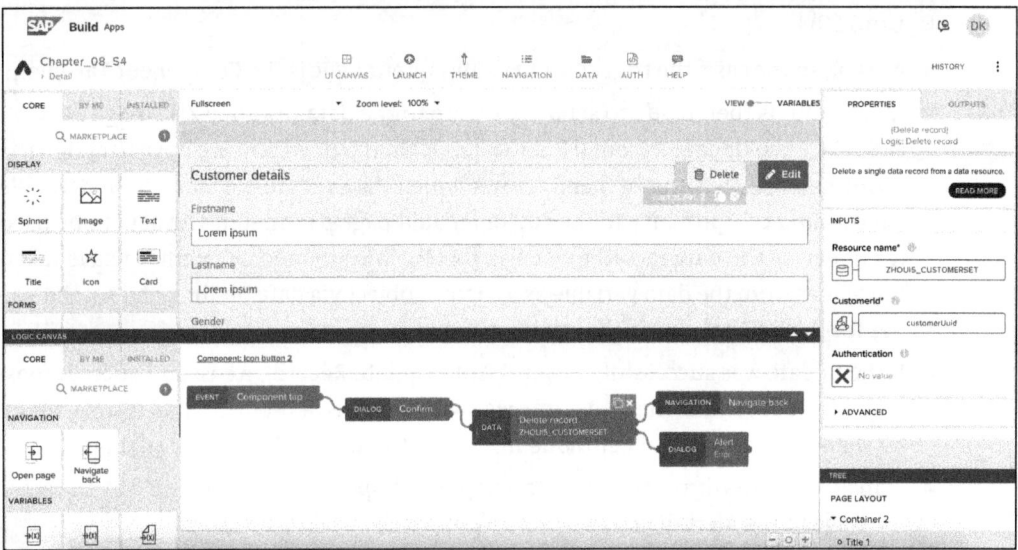

Figure 8.48 Logic Canvas of Delete Button to Perform Delete Request

When editing, the logic is surprisingly simple. If you remember, we have already made the fields ready for input via the data binding of **editMode**. As soon as the value of **editMode** changes from **False** to **True**, the input fields will also open for input. In Figure 8.49, you can see that all we have to do now is to set the value of **editMode** to **True** when clicking the **Edit** button with **Set Page Variable**.

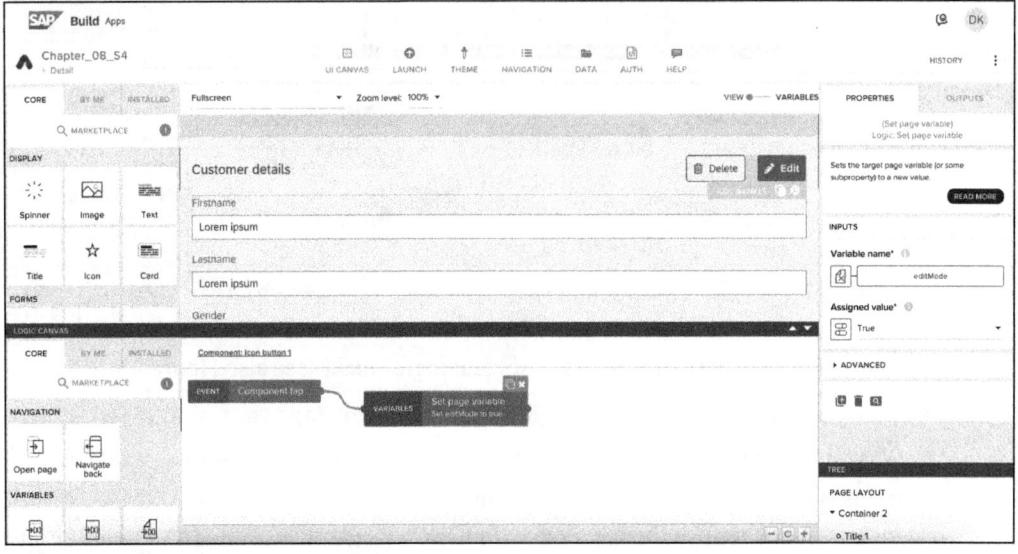

Figure 8.49 Logic Canvas of Edit Button

8 Developing Extensions

Below the last input field, we have inserted another **Container** element, which in turn contains two **Icon Button** elements. This container is only visible if the value of the page variable **editMode** is **True**. First, let's explain the logic behind the **Save** button (see Figure 8.50):

- As is often the case, the trigger for the subsequent logic is the **Component Tap** event.
- An attempt is then made to send an update to the data resource **ZHOUI5_CUSTOMERSET1** via the **Update Record** (HTTP MERGE). On the one hand, we have to assign the **CustomerId** so that the backend knows which data record we want to update. We have bound this property to the **customerUuid** page parameter. The data to be used for updating the data record must also be sent with the record. Here we have taken the values from the data variable as a custom object via data binding and set the values for the HTTP body via the known mapping.
- If the update was successful (output 1 from **Update Record**), we issue a success message as a **Toast** and at the same time we switch from edit mode back to display mode by changing the value of **editMode** from **True** to **False** via **Set Page Variable**.
- If an error occurs during the update (output 2 of **Update Record**), we display the error message in the **Alert** dialog.

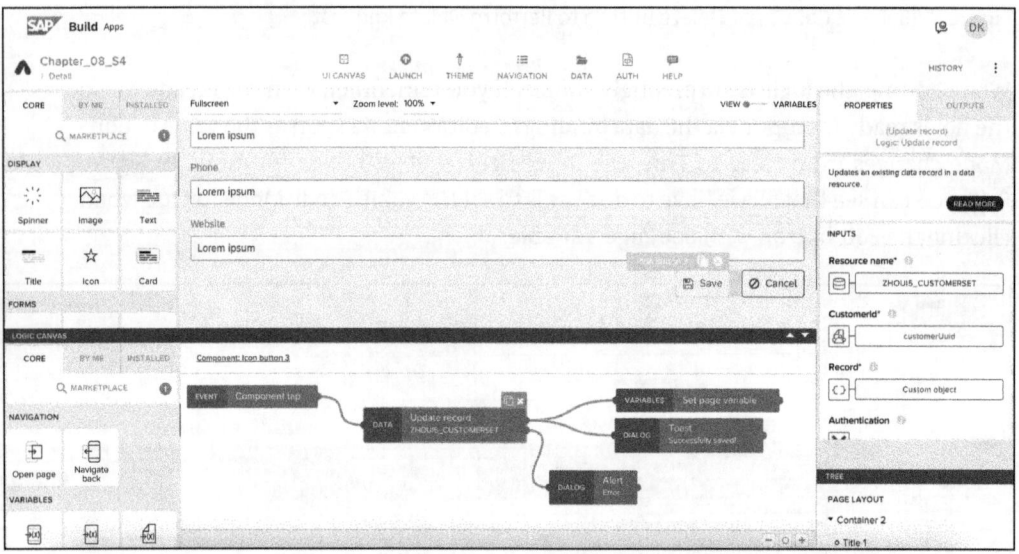

Figure 8.50 Logic Canvas of Save Button to Send POST Request to Backend

The **Cancel** button in turn consists of several logic blocks (see also Figure 8.51):

- First, you want the logic to start with the **Component Tap** event.
- We use **Confirm** to confirm once again whether the user really wants to cancel.
- If this is the case (output 1 from **Confirm**), we read in the data again with **Get Record**. In **Get Record**, we use the resource name **ZHOUI5_CUSTOMERSET1** and bind the **customerUuid** page parameter to **CustomerId**.

- After reading (output 1 from **Get Record**), we put the returned data into the data variable. In this way, we have very simply reset the values from the user. Furthermore, we must not forget to switch from edit mode to display mode by changing the value of the **editMode** page variable from **True** to **False**.

As a final step, don't forget to go back to the table and assign the logic for the row of the table so that users are navigated to the detail page when they click a row.

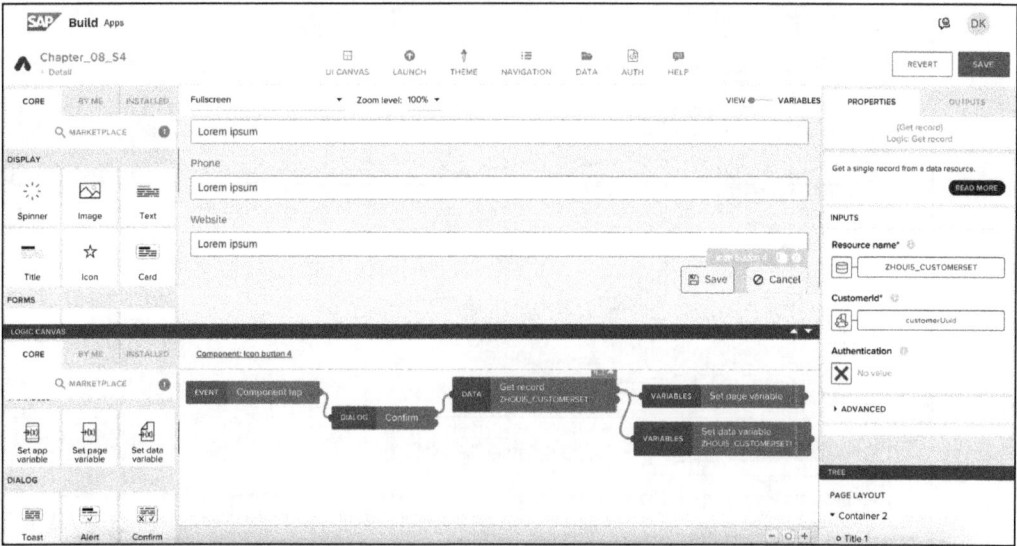

Figure 8.51 Logic Canvas of Cancel Button

The result is shown in Figure 8.52. Now you can click a line to go to the detailed view. There, the input fields are not ready for input so long as you do not click the **Edit** button. The **Delete** button would open a dialog that allows the user to delete after confirming again.

Figure 8.52 Preview of Detail Page Displaying Data of Customer

As soon as you are in edit mode, the input fields are ready for input. The entries can be saved with the **Save** button and discarded with the **Cancel** button (see Figure 8.53). Saving displays a **Successfully saved!** message and switches from edit mode back to display mode.

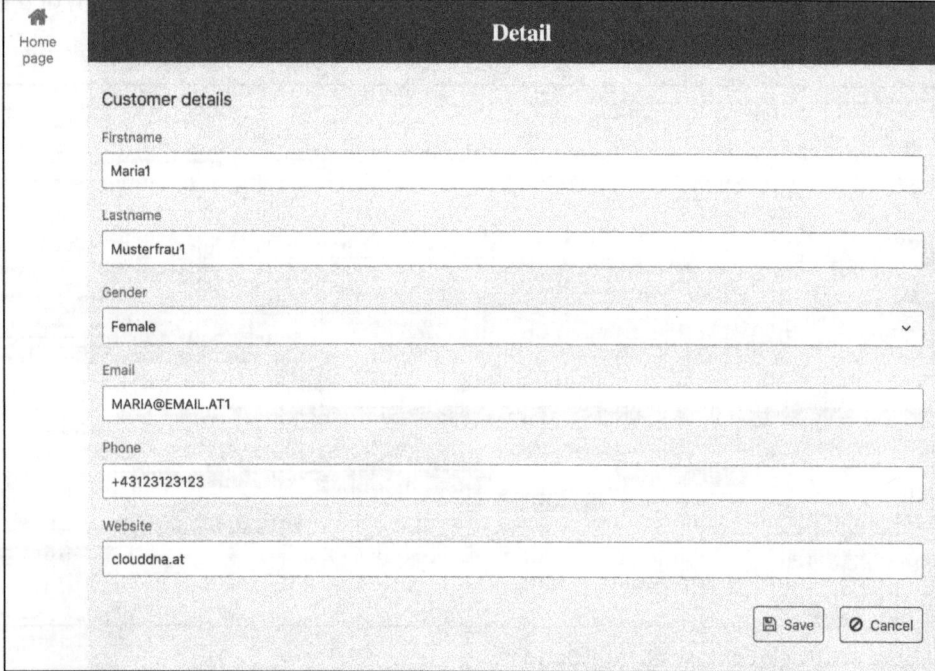

Figure 8.53 Preview of Form in Edit Mode Where Data Can Be Changed

In Figure 8.54, you can see that we have not only created a new customer record, but also changed it. You can also see these changes in the table as soon as you go back to the home page.

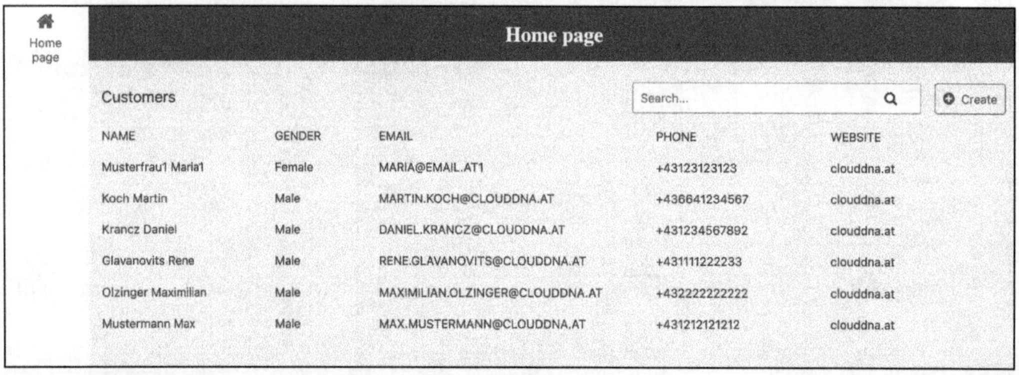

Figure 8.54 Customer Created Now Displayed in Table

8.3 Summary

In this chapter, we applied a number of things you learned in the previous chapters. For example, you consumed data from an SAP SuccessFactors sandbox instance via more than one APIs. Data variables and data binding for the correct display of data coming from the system were important key elements of the development. On a second page, you were able to display details of a selected dataset using navigation and a parameter. In addition to the display, you used an SAP S/4HANA system to create, change, and delete customer data. The logic blocks used here can be used for any data integration as they work generically.

Chapter 9
Deploying Applications

Developing apps with SAP Build Apps usually results in making the app accessible to a broader audience. This step is called deployment and can be done with SAP BTP as the runtime environment. The deployment can be either done manually using command line tools or automated directly out of the SAP Build Apps IDE.

Once you've developed your app, you'll also want to make it available to users. With SAP Build Apps, you can choose from different options for publishing your app depending on your requirements.

Development with SAP Build Apps was not originally aimed at enterprise applications in the SAP environment. This becomes clear when you look at the available deployment options. The focus was on providing apps within the SAP Build Apps platform in the form of web apps or creating corresponding mobile apps that run on iOS and Android. Deployment in SAP BTP was only possible in a roundabout way for a long time, but for many SAP customers it's a mandatory requirement for the deployment of apps. Therefore, SAP added a deployment option to SAP BTP in 2023.

In this chapter, we'll show how to deploy an app in SAP BTP in Section 9.2. This is the simplest option for making your app available to the wider world. In addition to making apps available on SAP BTP, you also have the option of simply performing a build within SAP Build Apps and exporting the build result in the form of a web application. You can then run this in any web server. Prominent examples of this are Apache Web Server or the NGINX web server software. From SAP's perspective, SAP BTP is the strategic platform for the development and provision of applications in a cloud environment.

Finally, in Section 9.3, we show how mobile apps can be created and deployed. This is a common use case that you will encounter again and again. This functionality was missing when SAP App Gyver was renamed to SAP Build Apps and moved to SAP BTP. It was made available in SAP Build Apps in late 2023 again. At this point, SAP is focusing on the iOS and Android platforms. As both platforms have a combined market share of almost 100 percent, you can use them to serve almost all mobile devices.

All types of distribution presented in this chapter have one thing in common: the build must be configured once for each platform, as you'll see in Section 9.1.

9 Deploying Applications

9.1 Build Configuration

The creation of the app is also referred to as a *build*. During the build, the application is prepared in such a way that it can be made available on the desired platform in the appropriate form.

Apps are deployed in a dedicated area of SAP Build Apps. If you are working in the UI canvas area, click the **LAUNCH** button to navigate to the build entry point (see Figure 9.1).

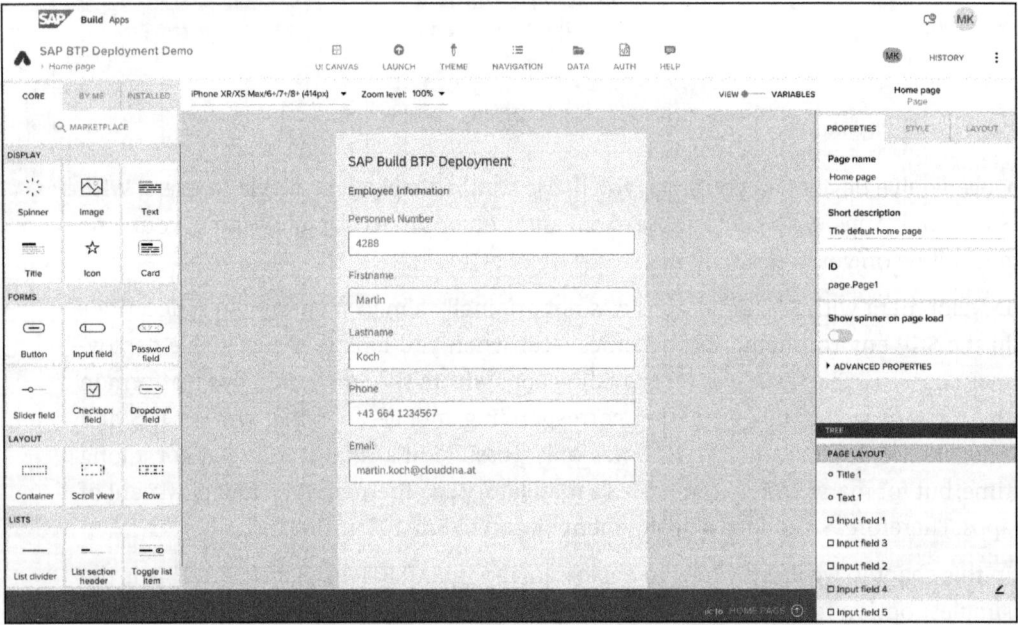

Figure 9.1 SAP Build Apps: UI Canvas

There, in the launch area, you have the option of viewing an application in preview mode, but you can also create and deploy the app directly in the build area (see Figure 9.2). Click the **Open Build Service** button to configure the build.

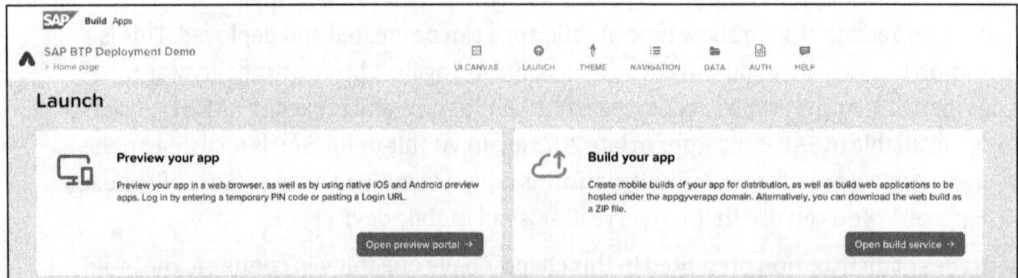

Figure 9.2 SAP Build Apps: Launch Area

In the build service, you must configure the desired build target to trigger a build. There is a configuration for iOS builds and Android builds, both for mobile and tablet, as well as web app builds. iOS and Android will be covered in Section 9.3. The entry point for configuring the respective target platform is shown in Figure 9.3.

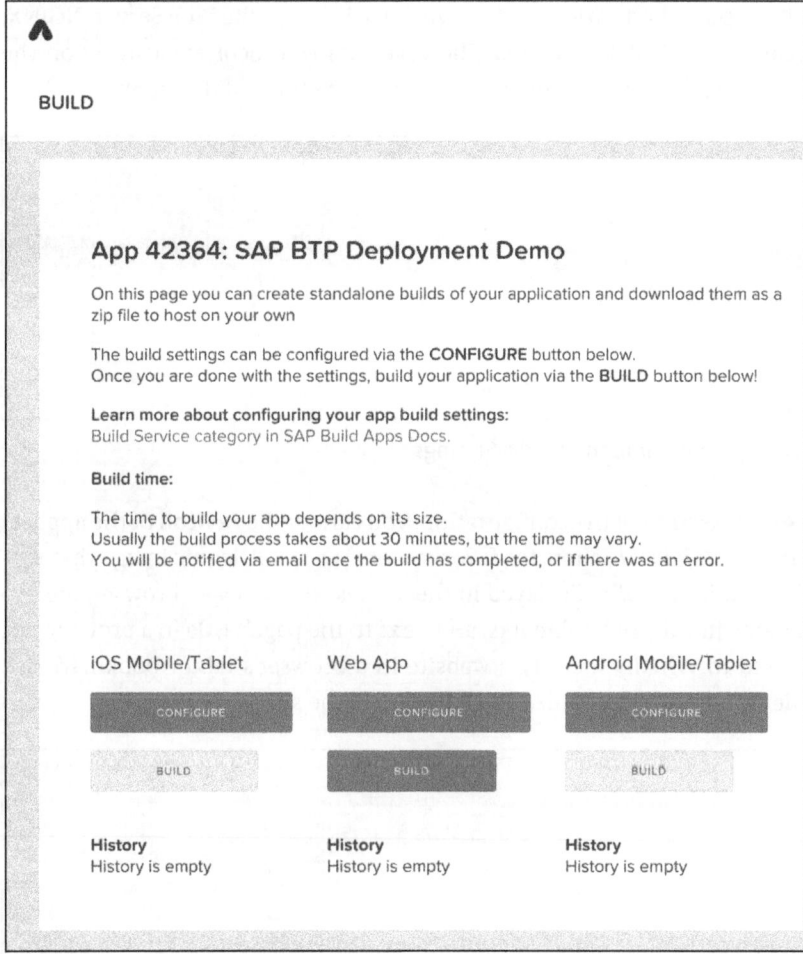

Figure 9.3 Build Configuration

9.2 SAP Business Technology Platform

An app is provided on SAP BTP in the form of a web application. The format required for this is MTAR (*multitarget archive*). Within SAP Build Apps, a build can be carried out with the *MTAR* target. In addition, the app can also be deployed directly from SAP Build Apps to a subaccount of SAP BTP, as explained in detail in this section. To access the configuration for the build of a web app, you must click the **Configure** button in the **Web App** area, as shown in Figure 9.3.

9   Deploying Applications

The configuration settings for the build process, as shown in Figure 9.4, comprise three main areas: **Bundle Settings**, **Image Settings**, and **Plugins**. Our explanation begins with the bundle settings, which allow you to define the schema for the build. Here you can select the output format in which your project is to be created. You can choose from two formats: **MTAR** and **ZIP**. The ZIP format allows you to export a classic web application that can be executed on various web services such as Apache Web Server, NGINX, or SAP NetWeaver AS ABAP. In this book, however, we will focus exclusively on the MTAR format, as the other options are not relevant to the topic of this book.

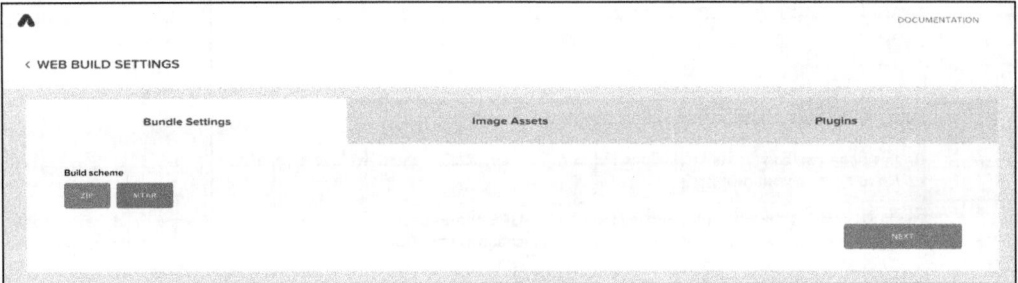

Figure 9.4 Web App Configuration: Bundle Settings

In the **Image Assets** section of the configuration, you can set the **Favicon** of the application (see Figure 9.5). A *favicon*, short for *favorite icon*, is a small, iconic image that represents a website. It is typically displayed in the address bar of a web browser, next to the website's name in a list of bookmarks, and next to the page's title in a browser tab. Favicons are used to visually identify a website in a browser's tab or bookmark list, making it easier for users to recognize and navigate to the site.

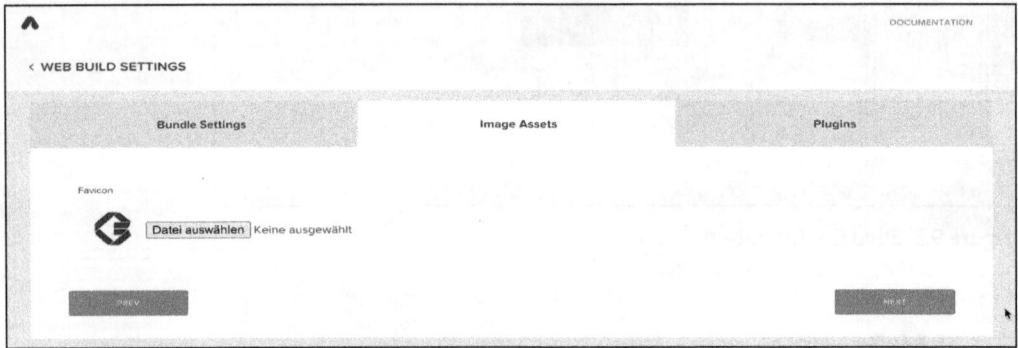

Figure 9.5 Web App Configuration: Image Assets

Finally, in the **Plugins** section, you can find a list of all required, installed, and available plugins (see Figure 9.6). As you can see from the screenshot, the plugins are shipped as *react-native* plugins, which means that SAP Build Apps is based on the React framework, and not on SAPUI5, as you might have expected. You cannot make any changes in this section. Its purpose is to give you an overview of what's going on behind the scenes.

284

9.2 SAP Business Technology Platform

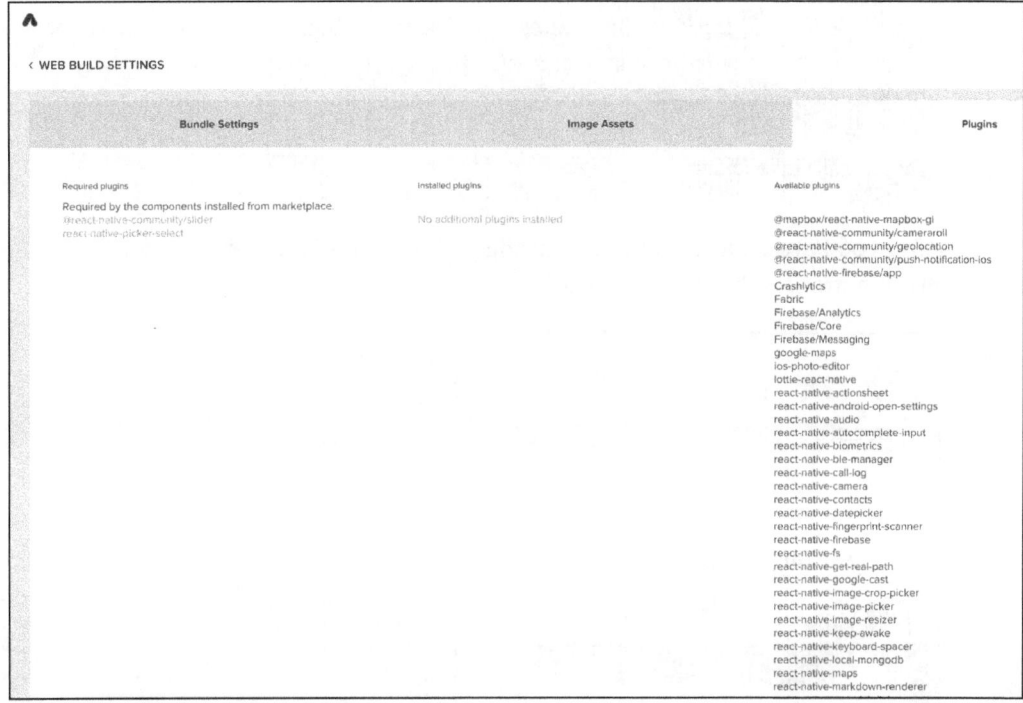

Figure 9.6 Web App Configuration: Plugins

After you have completed the configuration, you can trigger a new build. Click the **BUILD** button to do so, as shown in Figure 9.3.

Next you must configure a few things before the build can start, as shown in Figure 9.7. First, you can select **ZIP** or **MTAR** for **File Type**. To be honest, we don't know why you could set this in the web app configuration if it must still be selected here. The **Client Runtime Version** is also shown as information, but it cannot be changed. You must also set a value for **Version**, and you are free to choose whatever version number you want.

Application Versioning

We recommend using the widely adopted three-digit versioning system in application development. It is characterized by the following sequence: major.minor.patch. This systematic approach is crucial for maintaining consistency, clarity, and predictability in software evolution. The first digit, the major version, signifies substantial changes that often include backward-incompatible alterations, reflecting a significant evolution of the application's capabilities or architecture. The second digit, the minor version, denotes incremental enhancements and additions that are backward-compatible, indicating an expansion in functionality without fundamental changes to the existing system. Finally, the third digit, known as the patch level, is reserved for minor bug fixes and improvements that address specific issues without altering the application's overall functionality or features. This structured versioning framework not only aids developers

285

9 Deploying Applications

> in managing and tracking the progression of software but also helps users in understanding the scope of changes and the stability of each release. By adopting this approach, developers can effectively communicate the nature and impact of each update, ensuring a clear understanding of the application's development lifecycle among all stakeholders.

After you have set a version, you can trigger the build of the web app by clicking the **BUILD** button, as shown in Figure 9.7.

Figure 9.7 Trigger MTA Build

After triggering the build, you will be redirected to the build settings. There you can see the **Status** of the build. In Figure 9.8, you can see that the build is **queued**. The build may take up to 30 minutes, and you will get a notification via email when the build finishes. Be aware that you can only trigger two builds per app and target platform when using the SAP Build Apps free plan.

After the build has finished, you can download the build artifact—the MTAR file—by clicking the **DOWNLOAD** button (see Figure 9.9). SAP Build Apps also gives you the option to deploy the MTAR file to an existing SAP BTP subaccount. Therefore, you must click the **DEPLOY MTA** button. We'll focus on this option in the subsequent explanation.

9.2 SAP Business Technology Platform

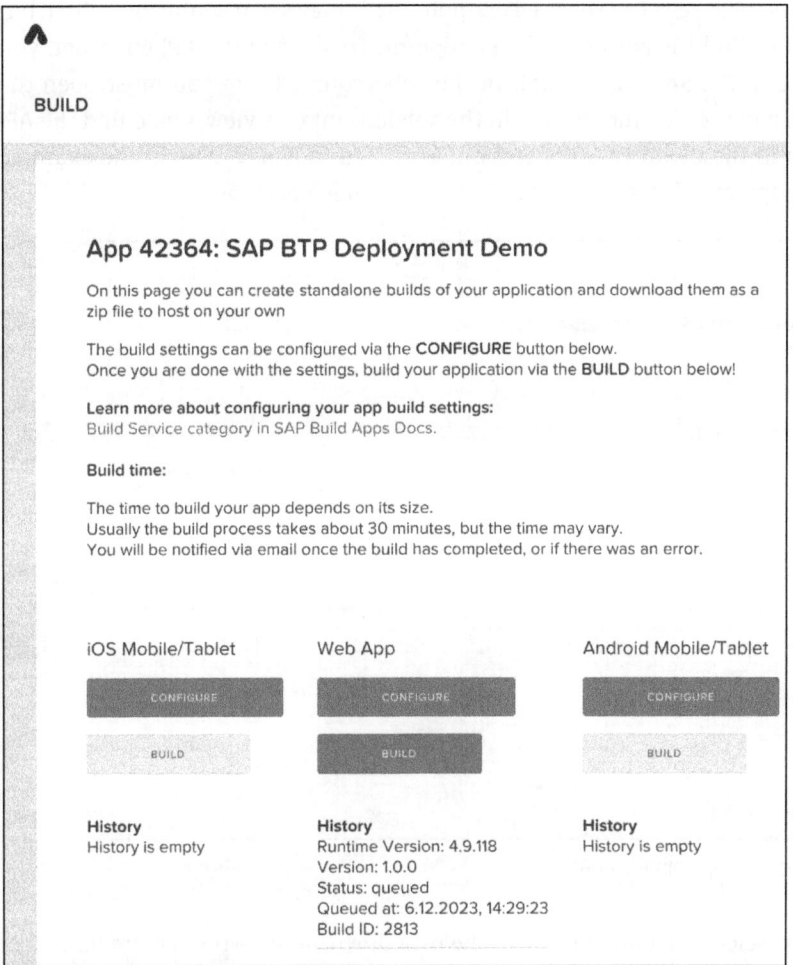

Figure 9.8 MTA Build Status

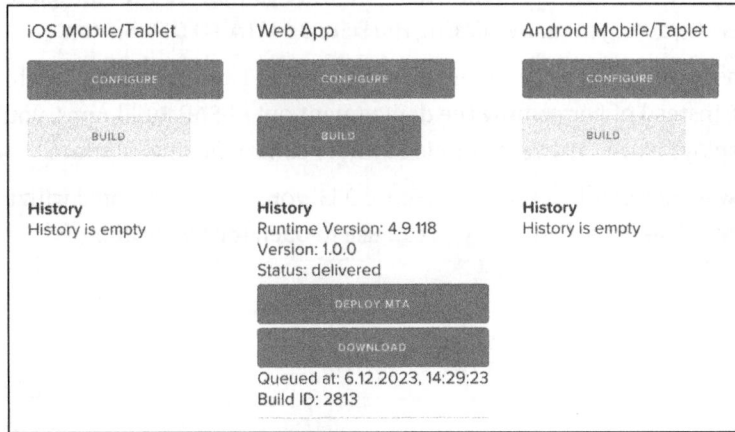

Figure 9.9 Deploy Generated MTA to SAP BTP

First, it's necessary to select the desired endpoint. This may seem confusing at first, but it's quite easy to find the corresponding endpoint. To identify the API endpoint, you must navigate to the SAP BTP cockpit of the subaccount. There you must open the **Overview** entry in the side menu. Within the subaccount overview, you'll find the **API Endpoint** info in the **Cloud Foundry** section, as shown in Figure 9.10. In the example shown, the endpoint is *https://api.cf.us10-001.hana.ondemand.com*.

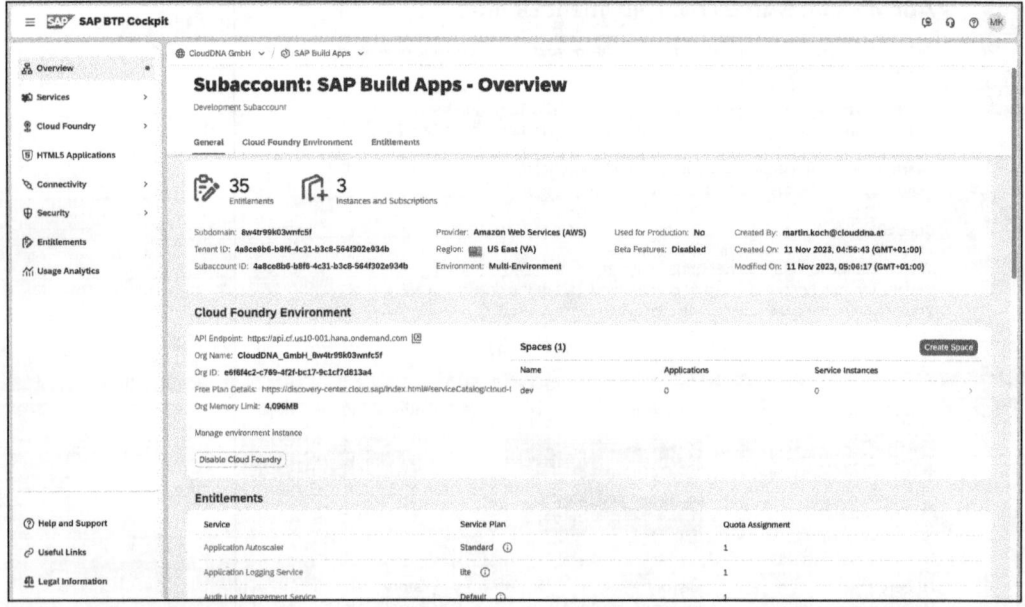

Figure 9.10 Determine API Endpoint

You must now select the endpoint from the dropdown, as shown in Figure 9.11. After you have selected the endpoint, the Cloud Foundry **Organization** linked to this endpoint is automatically selected. Finally, you must select a Cloud Foundry **Space** that belongs to the Cloud Foundry organization. When you have completed the configuration, you can trigger the deployment by clicking the **DEPLOY MTA TO DEV** button.

You will then see the status of the deployment (see Figure 9.12). The deployment might take a few minutes. Instead of performing the deployment out of SAP Build Apps, you can do a manual deployment by using the Cloud Foundry command line interface.

If the deployment was successful, as shown in Figure 9.13, you will be presented with a link to the application. This is the URL that you can use to open the application.

9.2 SAP Business Technology Platform

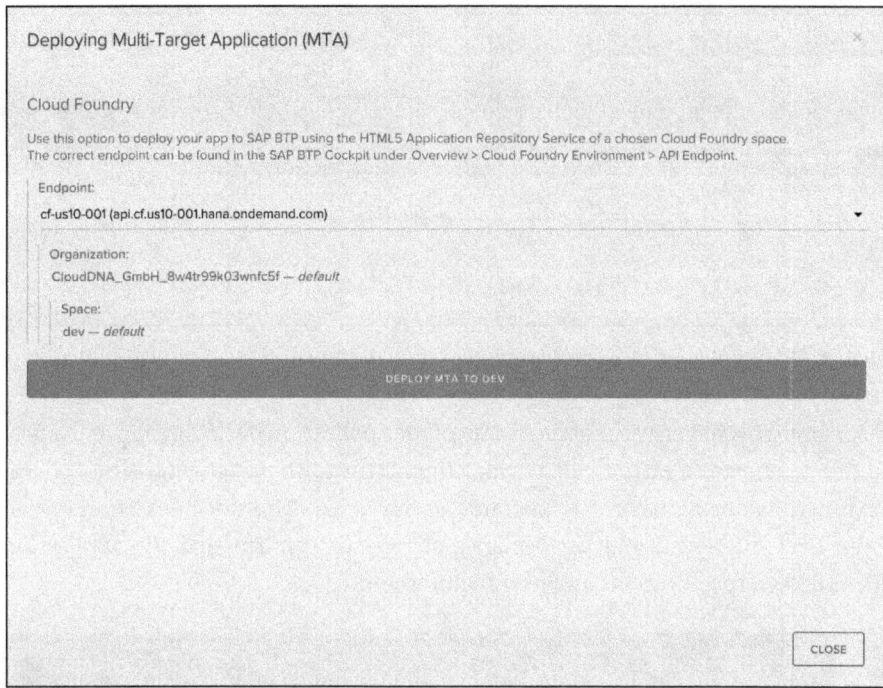

Figure 9.11 Trigger MTA Deployment

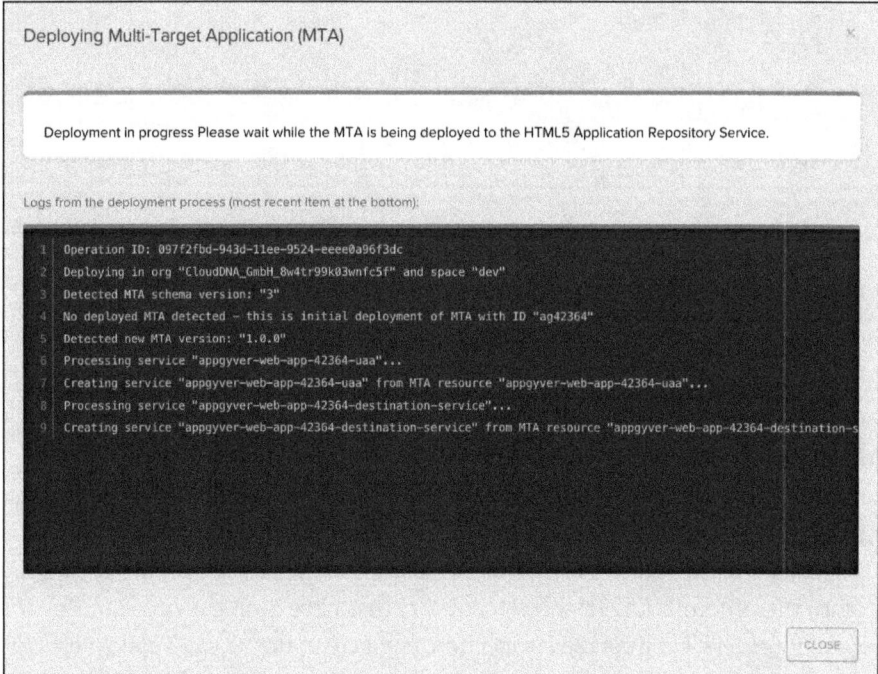

Figure 9.12 MTA Deployment Status

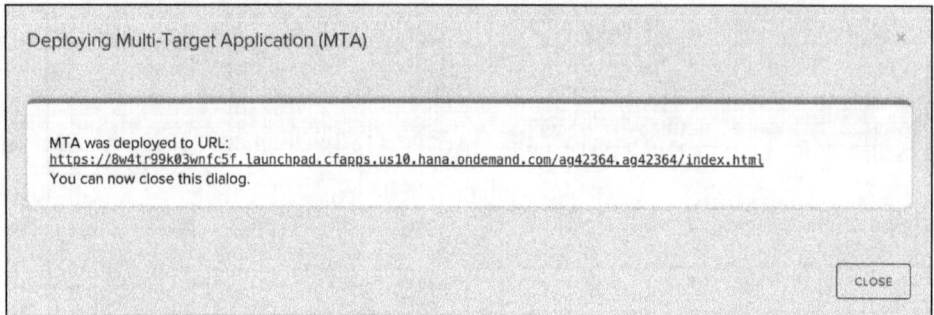

Figure 9.13 MTA Deployment Success Message

Next, it's advisable to verify the accessibility of the application by navigating to the provided link as shown in Figure 9.14. It's also important to thoroughly test the application's functionality to ensure it operates as intended. This includes checking all features, user interfaces, and performance aspects to confirm that the application meets its design specifications and user requirements.

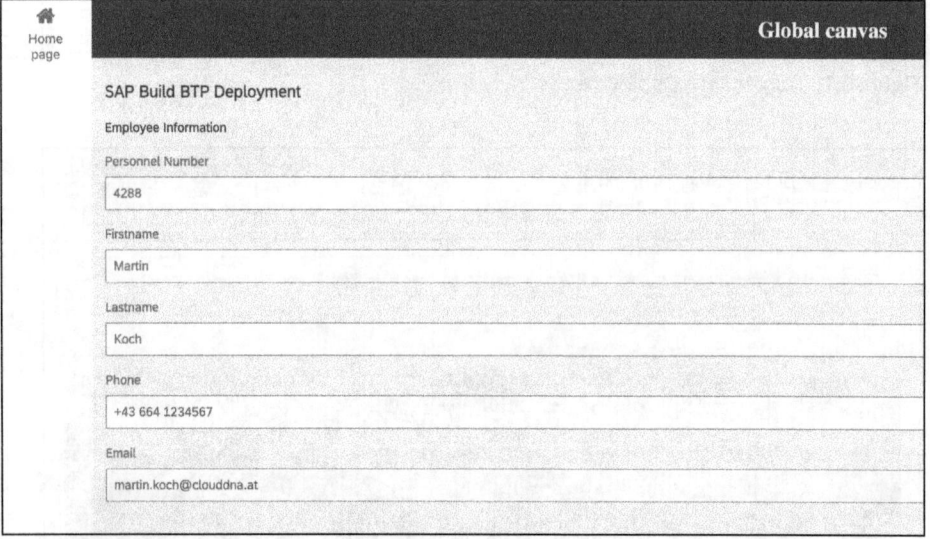

Figure 9.14 Running Deployed Application

Numerous activities occur behind the scenes during deployment at both the subaccount and space levels. To gain a comprehensive understanding of these actions, it is highly recommended to access the SAP BTP cockpit of your subaccount (see Figure 9.15). Once there, you should review the artifacts that have been generated during the deployment process. For a detailed inspection, navigate to the **HTML5 Applications** section located in the side menu of the cockpit. This will provide a clear view of the deployed applications and their respective statuses, ensuring you are fully informed about the deployment outcomes and the current state of your applications within the SAP BTP environment. There you will see that an HTML5 application was created. The name was randomly chosen by the deployment process.

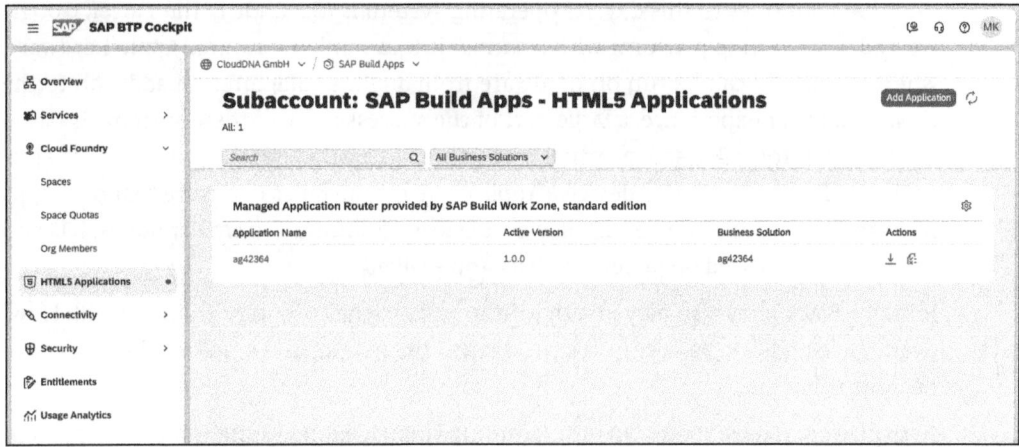

Figure 9.15 Subaccount HTML5 Applications

Next, we recommend checking the Cloud Foundry space. In the example shown in Figure 9.16, the space is named *dev*. Inside this space, navigate to **Services • Instances** in the side menu. There you will find three instances: one SAP HTML5 Application Repository service for SAP BTP, one SAP Destination service, and one SAP Authorization and Trust Management service.

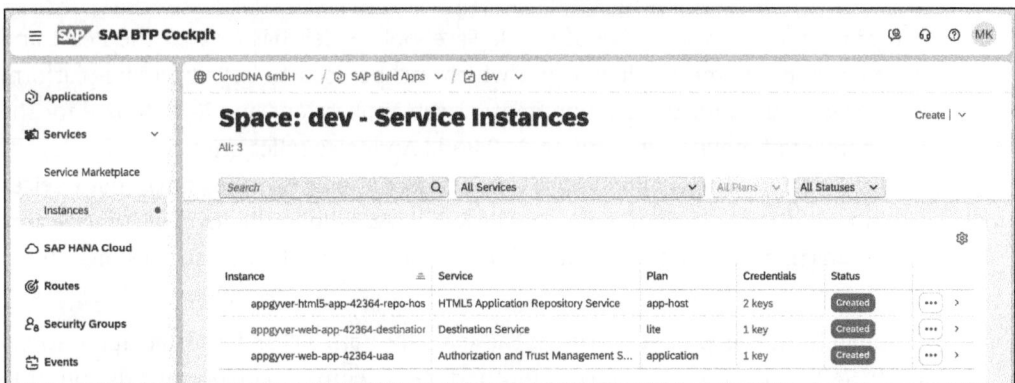

Figure 9.16 Cloud Foundry Space Instances

With the successful deployment of the app, you now have the option of calling it up directly. You have also created an important basis for integrating the app into the SAP Build Work Zone. This makes it possible to make the app accessible to end users via an SAP Fiori launchpad. This integration increases user-friendliness, as the app is seamlessly embedded in the existing work environment and offers users a familiar interface.

9.3 Mobile Deployment

Mobile applications can look back on an incredible success story. In 2007, Apple launched the iPhone, the first serious smartphone on the market. This started a revolution that continues to this day. Its operating system is *iOS*. Shortly thereafter, Google created a competing platform with *Android*. iOS can only be run on Apple hardware, whereas Android can be run on hardware from any manufacturer. In addition to the design and user experience, a large part of the success story of these systems is based on their *app stores*. These are virtual marketplaces through which apps can be offered, sold, and distributed to end devices. Apple even goes so far as to ensure that only apps from the Apple Store can be installed on devices. In addition to smartphones, iOS and Android are also used on tablets, such as Apple's iPad.

Initially, BlackBerry also played a significant role in the corporate environment. However, Apple and Google very quickly took over the market and made BlackBerry disappear into oblivion.

Both platforms have their own programming languages and tools for the development of apps that must be used. For Apple, this originally included the Objective-C programming language and later Swift. For Android, the programming languages C++, Java, and Kotlin are used for development. As you can already deduce from this, it isn't possible to develop native apps that run on both platforms. As a developer, you will therefore have to develop two independent apps if you want to offer them on both platforms. The development and maintenance of both apps is therefore very time-consuming. Some companies have recognized this and taken it as an opportunity to develop tools that can be used for cross-platform development.

One prominent representative of this is Facebook, which has created the React and ReactNative programming languages, which can be used to develop cross-platform apps. *React* is used for the development of web applications and *ReactNative* for the development of mobile applications. SAP Build Apps also relies on these technologies. The apps created with SAP Build Apps can therefore also be executed on mobile devices without additional development using ReactNative, even in the form of a native app. This means that you can access native functions of the end devices, such as the camera or GPS.

Although the development process is the same for both iOS and Android applications within SAP Build Apps, there are significant differences in the creation and distribution of the apps. We therefore dedicate a separate section to each platform. We start in Section 9.3.1 with the creation and deployment of iOS apps, followed by the creation and deployment of Android apps in Section 9.3.2.

Both providers want to prevent their app stores from being circumvented. The applications must therefore be signed before being installed on the end devices. The respective procedure for signing differs on both platforms.

There are exceptions for the distribution of apps within a company. However, the prerequisite for this is that the end devices are managed with a device-management system. This gives you as a company the opportunity to control your employees' end devices with the help of guidelines. You can specify, for example, which apps may be installed, whether roaming may be used, or what should happen if a password is entered incorrectly too often. You also have the option of remotely wiping the device if it is stolen or lost, for example. The apps must also be signed when using managed devices.

9.3.1 iOS

To distribute apps via the iOS operating system, you need a *developer account* with Apple. This is a paid account that must be renewed annually. The developer account can either be issued directly to developers or to a company to which various developers are assigned. The developer account gives you access to the *Apple Developer Portal*, where you can find basic and essential resources for developing for the iOS platform.

To create iOS apps on the SAP Build Apps platform, you first need a *distribution certificate*, which you can generate in the Apple Developer Portal. Import this certificate into the keychain, called the *key store*. From there, you can export the certificate in the form of a P12 file. You must assign a password for this. Therefore, you must open the Keystore app on your Mac and locate the appropriate certificate (see Figure 9.17). From there you can trigger the export of the certificate as a P12 file by using the context menu of the corresponding entry.

> **P12**
>
> A P12 file, also known as a *PKCS#12* file, is a file format used to store private keys and public certificate chains securely. PKCS#12 is an abbreviation for Public-Key Cryptography Standards #12, a standard developed by RSA Laboratories.

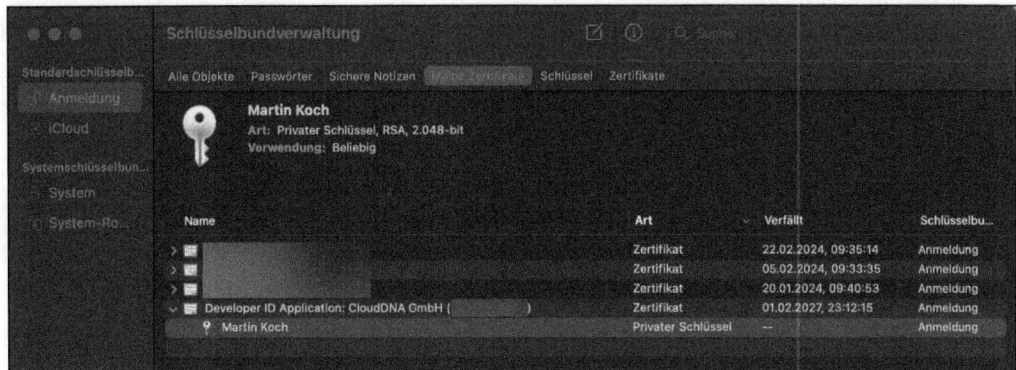

Figure 9.17 Export Certificate

9 Deploying Applications

In addition to the distribution certificate, which can be used independently of the app, you need a provisioning profile linked to the app. This provisioning profile allows you to distribute the app via the App Store or your Enterprise App Store. Before you can create the provisioning profile, you need the bundle identifier of the app. The easiest way to obtain this is in the iOS build setting of the app (see Figure 9.18).

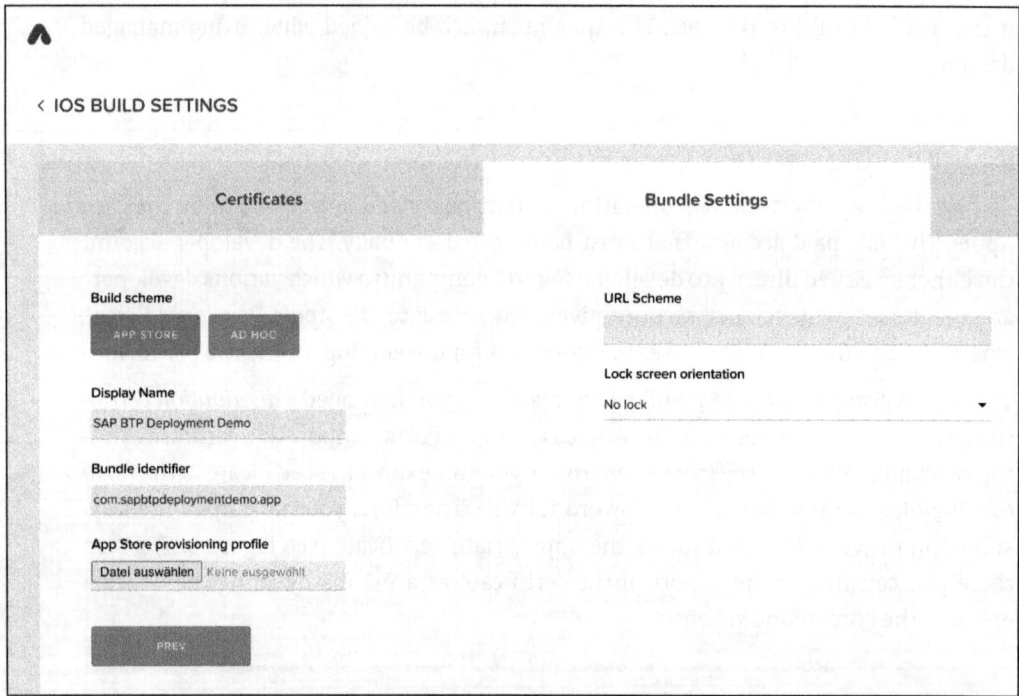

Figure 9.18 Get Bundle Identifier

You then need to perform a few manual steps in the Apple Developer Portal. This includes registering an app ID. Click on **Register an App ID** as shown in Figure 9.19.

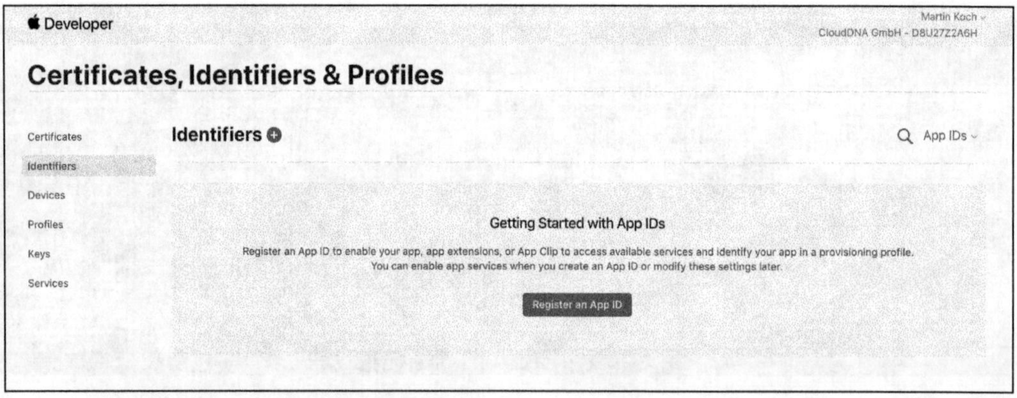

Figure 9.19 Register App ID

Then select the **App IDs** option, as shown in Figure 9.20, and click **Continue**.

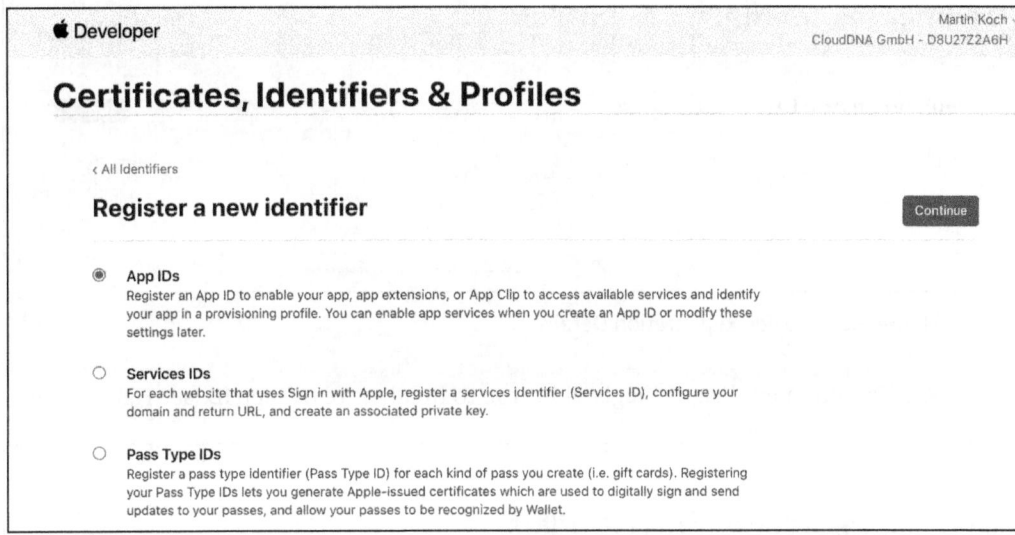

Figure 9.20 Register New Identifier

You must then select the app type. Click the tile with the name **App** and then click the **Continue** button (see Figure 9.21).

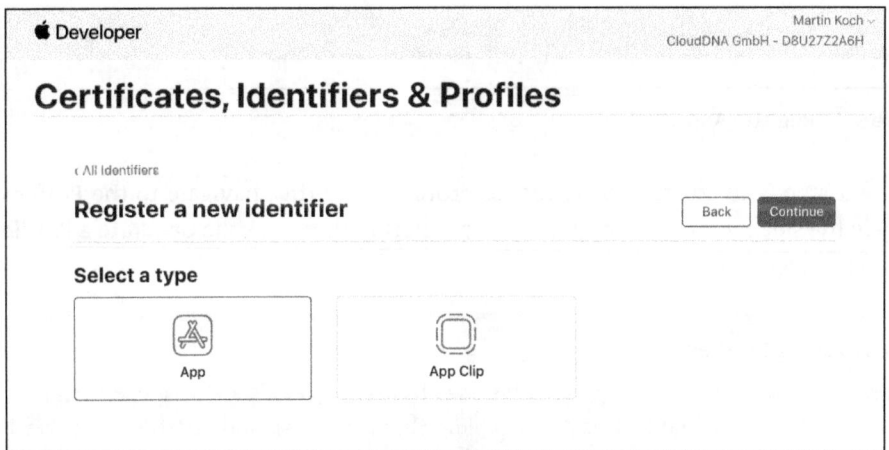

Figure 9.21 Select App Type

Now enter a descriptive name for the **Description** field and copy the bundle ID determined previously into the **Bundle ID** field. Then click **Continue** (see Figure 9.22).

9   Deploying Applications

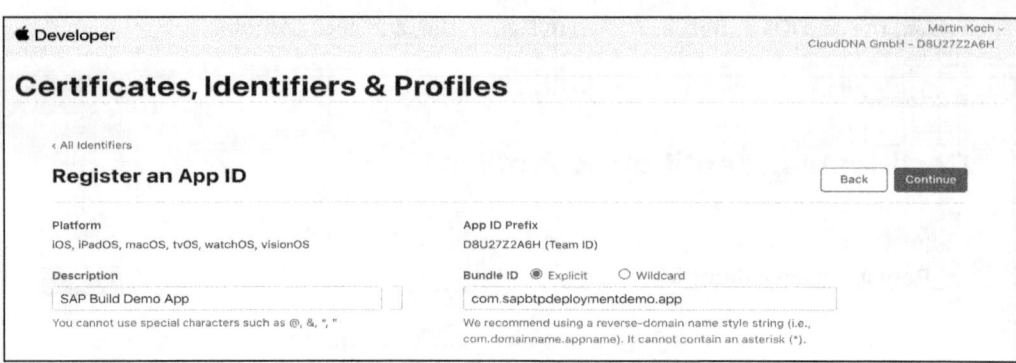

Figure 9.22 Provide Registration Details

Finally, you must click the **Register** button to finally save the app ID (see Figure 9.23).

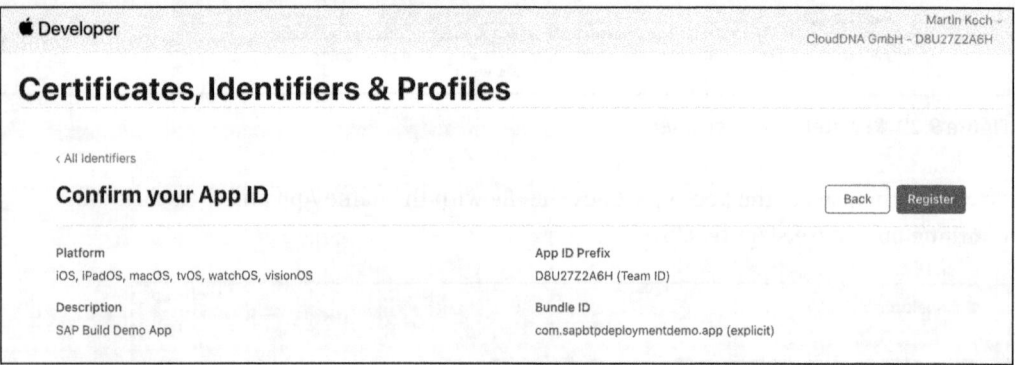

Figure 9.23 Register App ID

The next step is to create a provisioning profile. To do this, navigate to the **Profiles** entry in the side menu in the Apple Developer Portal. Then click the **Generate a Profile** button (see Figure 9.24).

> **Provisioning Profiles**
>
> A *provisioning profile* for iOS apps is a file used by Apple to associate developers' registered devices, apps, and certificates, enabling them to install and test their apps on physical devices. It acts as a link between the device and the Apple Developer account, ensuring that the app is being developed and tested by a registered developer on authorized devices. This profile contains information about the app's ID, the development certificate, and device identifiers, and it ensures that the app adheres to the guidelines set by Apple. Provisioning profiles are crucial for testing apps on actual devices before their final submission to the App Store and are also used for distributing apps internally within an organization.

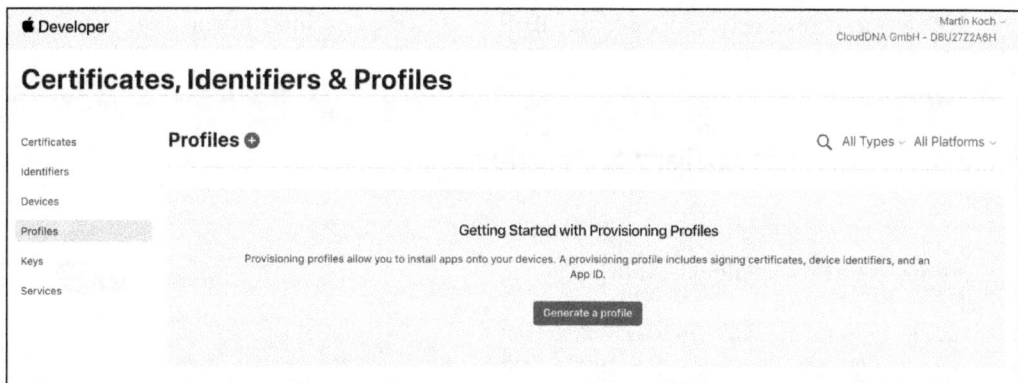

Figure 9.24 Generate Provisioning Profile

Select the **App Store** option in the **Distribution** section, as shown in Figure 9.25. Then click the **Continue** button.

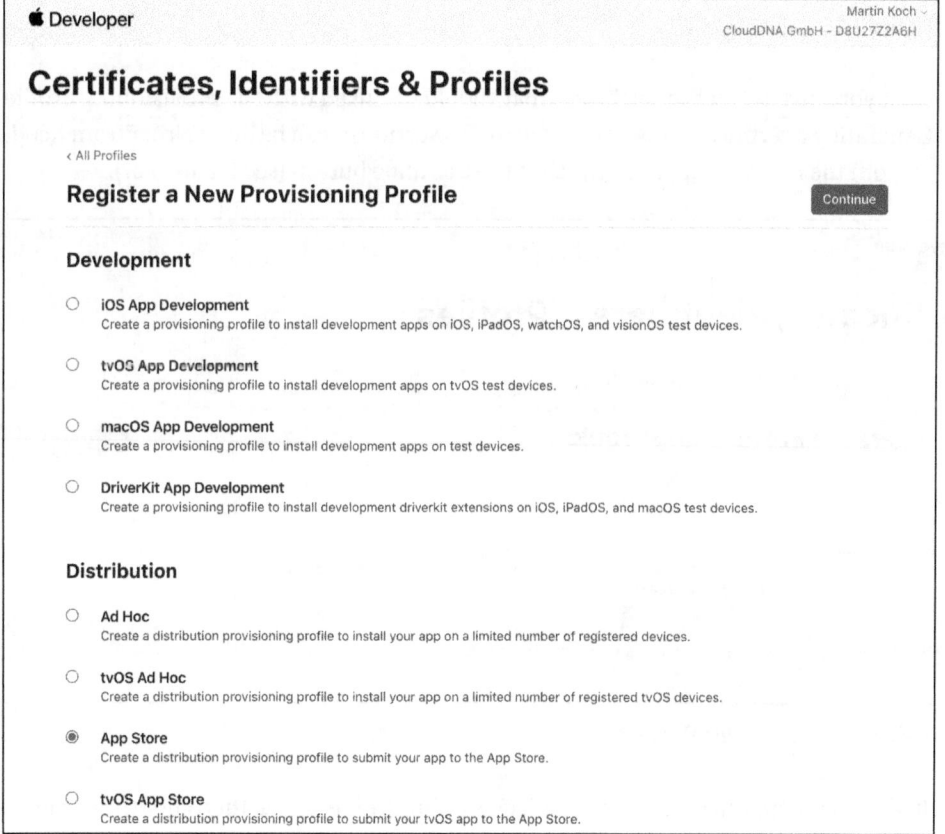

Figure 9.25 Select Provisioning Type

9 Deploying Applications

Select the previously registered **App ID**, then click the **Continue** button (see Figure 9.26).

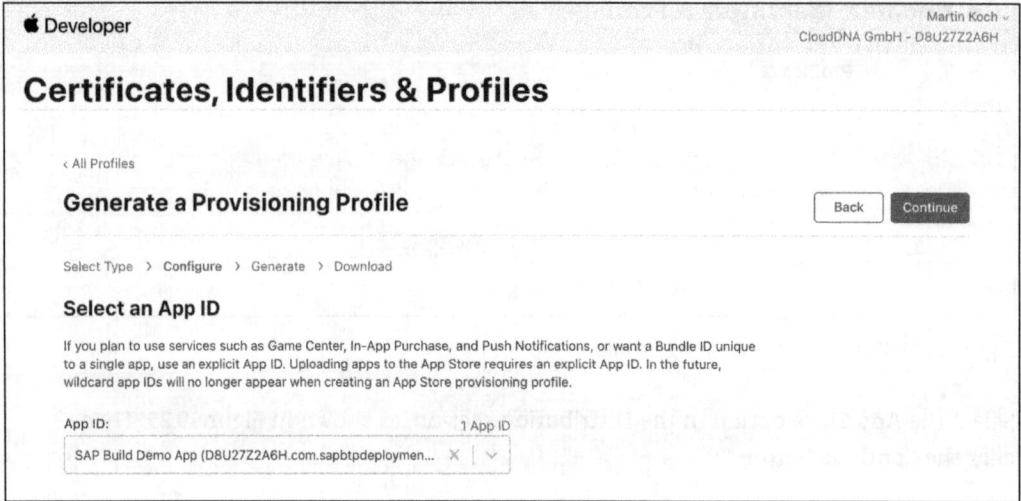

Figure 9.26 Select App ID

Next you must select the certificate that should be linked with the provisioning profile. Generating a certificate was not shown in this section. It can be done either from Xcode or from the Keystore app. Finally, click the **Continue** button (see Figure 9.27).

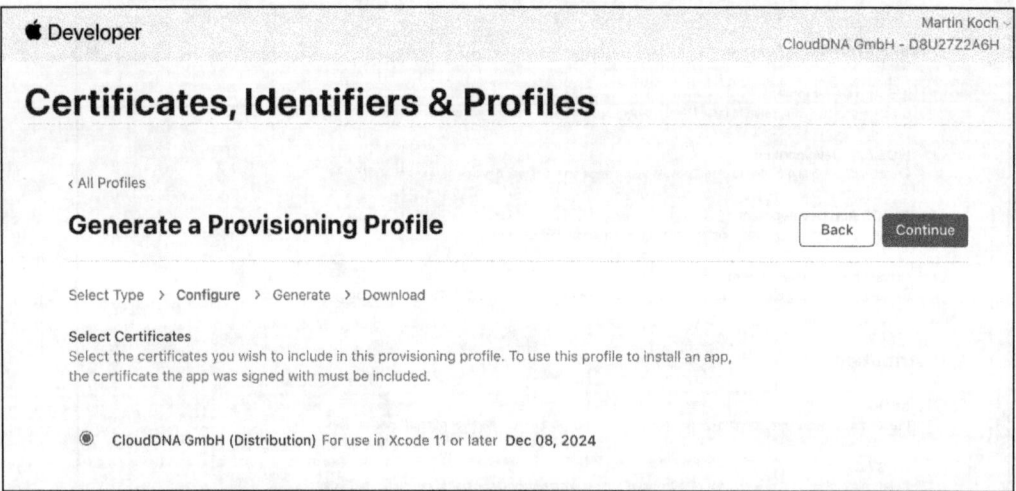

Figure 9.27 Select Certificate

In the final step, you must provide a **Provisioning Profile Name**, then click the **Generate** button (see Figure 9.28).

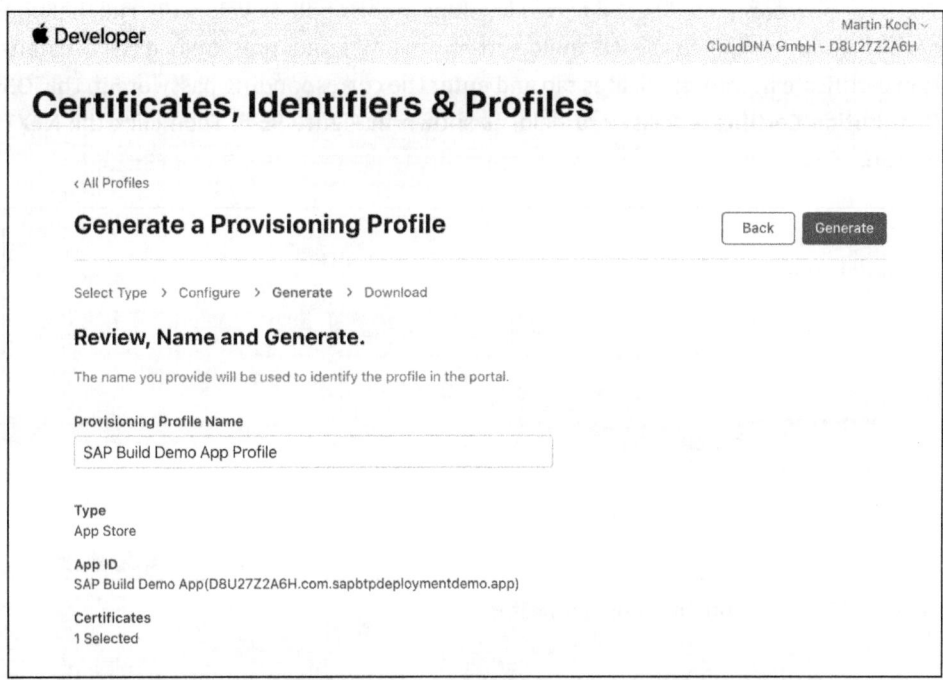

Figure 9.28 Generate Certificate

You can now download the provisioning profile by clicking the **Download** button (see Figure 9.29).

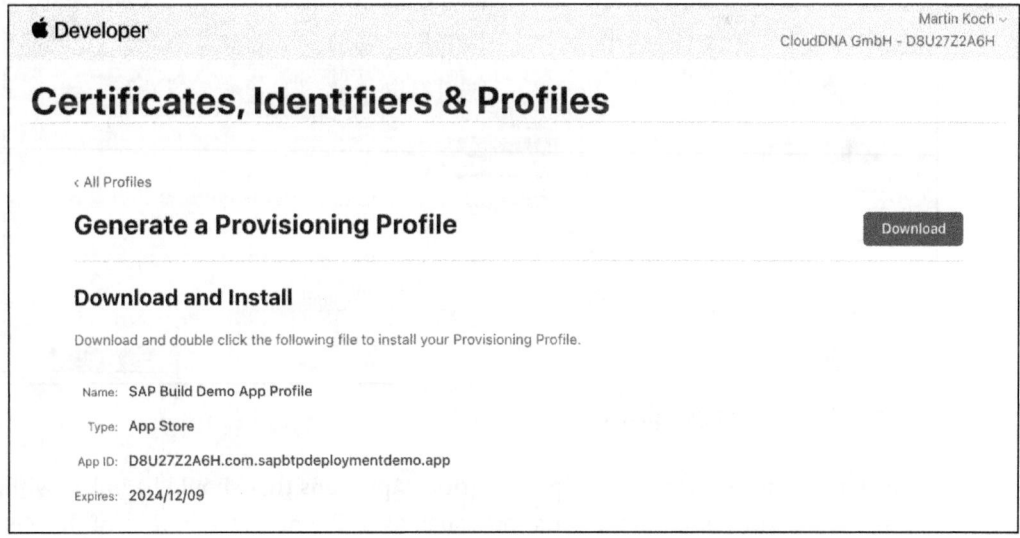

Figure 9.29 Download Profile

You have now completed all the necessary steps in the Apple Developer Portal that you need to build the app. In the **iOS Build Settings** in SAP Build Apps, upload the distribution certificate in the **Certificates** tab and enter the corresponding password in the **iOS Distribution Certificate Password** field, as shown in Figure 9.30. Then click the **NEXT** button.

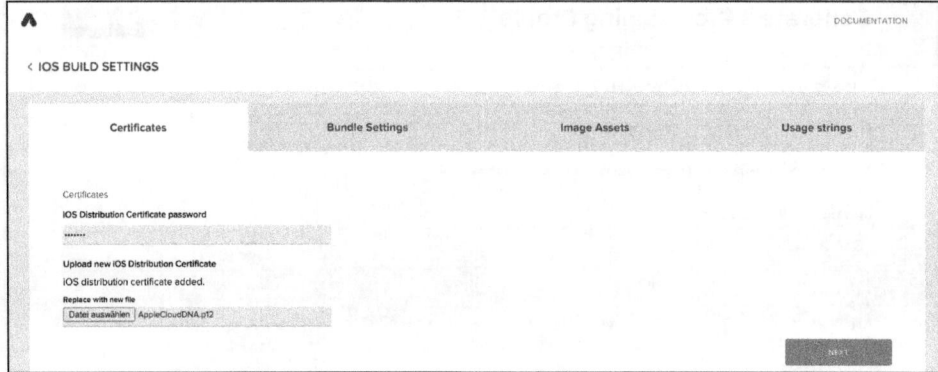

Figure 9.30 Build Configuration: Certificates

In the **Bundle Settings** tab, you must provide a **Display Name**. The **Bundle Identifier** is already prefilled. You must upload the previously generated **App Store Provisioning Profile** (see Figure 9.31). Finally, click the **NEXT** button.

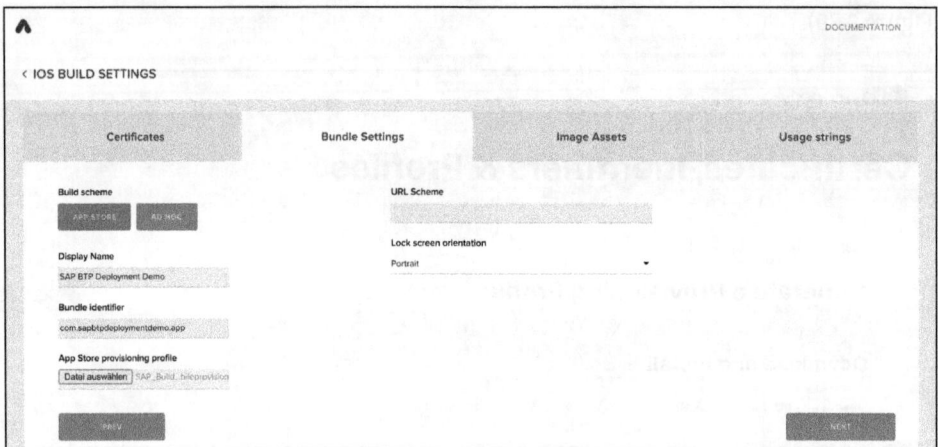

Figure 9.31 Build Configuration: Bundle Settings

In the **Image Assets** tab, you can upload various **App Icons** that should be linked with the app and shown in different situations, such as in the notification area of the iOS device (see Figure 9.32).

Finally, you can override the **Usage Strings** of the app as shown in Figure 9.33. *Usage strings* are short descriptions provided by developers that explain why their app needs access to certain features or data on a user's device, like the camera, photo library, or location services. These strings are displayed to the user in a dialog box when the app first requests access to the feature, ensuring transparency and informed consent. They are essential for privacy compliance and user trust as they clarify the intent behind the access request, helping users make informed decisions about their data. When you're done, click the **NEXT** button.

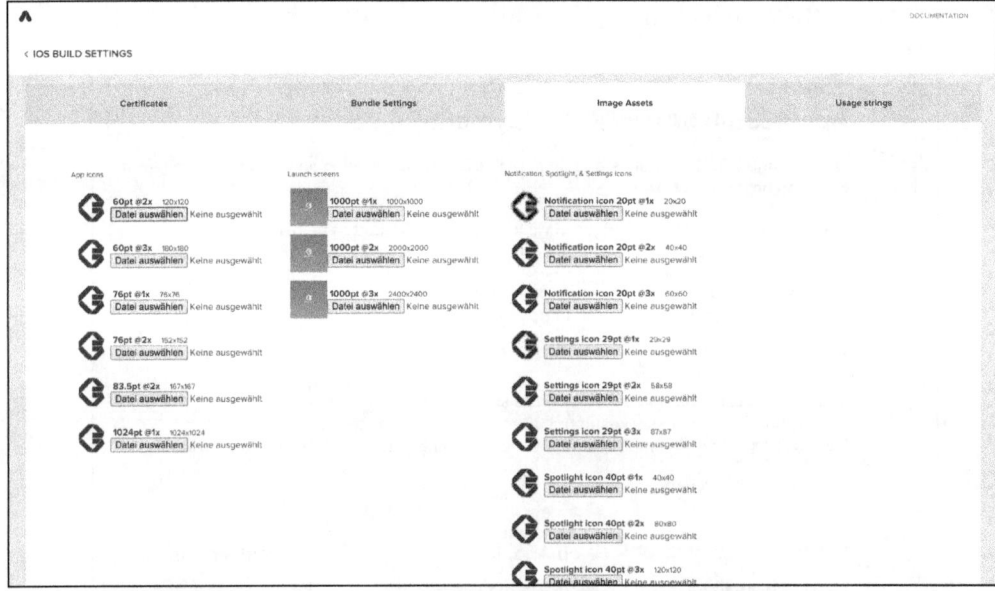

Figure 9.32 Build Configuration: Image Assets

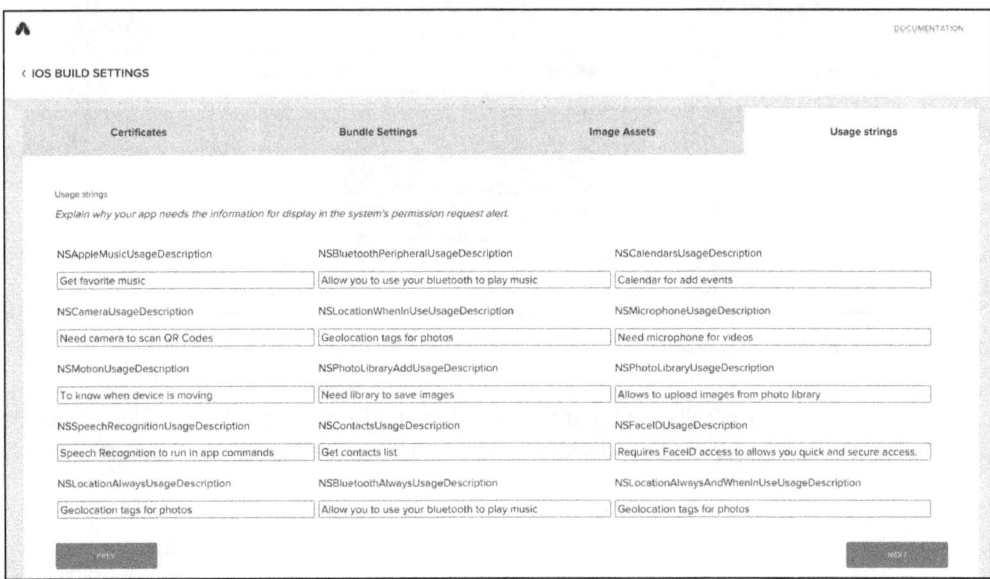

Figure 9.33 Build Configuration: Usage Strings

9   Deploying Applications

You are now ready to start the build of the app. Therefore, click the **BUILD** button in the **iOS Mobile/Tablet** section, as shown in Figure 9.34.

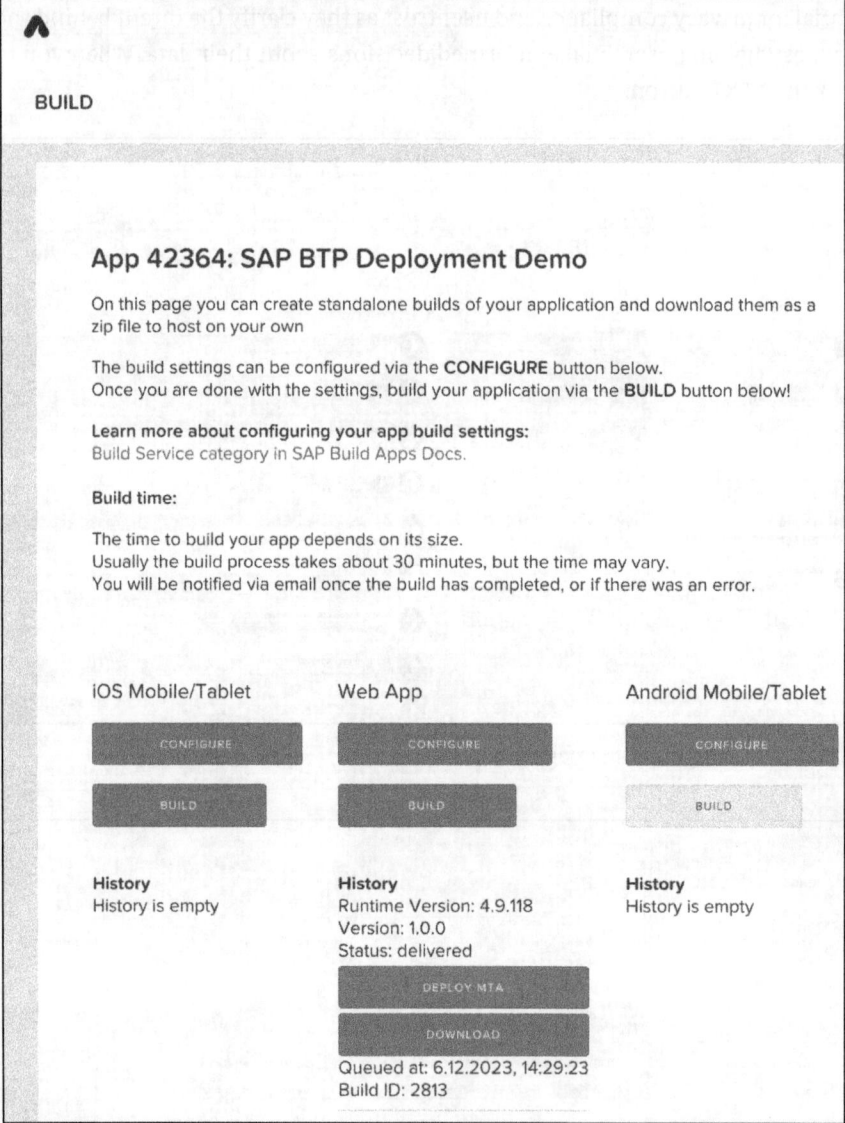

Figure 9.34 Trigger iOS Build

As shown in Figure 9.35, you must now select a **Build Type** and assign a **Client Runtime Version** as well as a **Version** and a **Short Version**. Then click the **BUILD** button to start the build.

9.3 Mobile Deployment

Figure 9.35 Provide Build Details

The build may take a few minutes. After the successful build, you can download the file with the .ipa extension and distribute it to the users' devices.

9.3.2 Android

The configuration of Android applications differs from that of iOS applications. Android works with a *keystore* instead of certificates (see Figure 9.36). You must therefore upload the keystore in the **Keystore** area and enter its password in the **Keystore Password** input field. When the keystore is created, an alias including the associated password is assigned. This alias must be entered in the **Keystore Alias** input field and the password in the **Keystore Alias Password** input field.

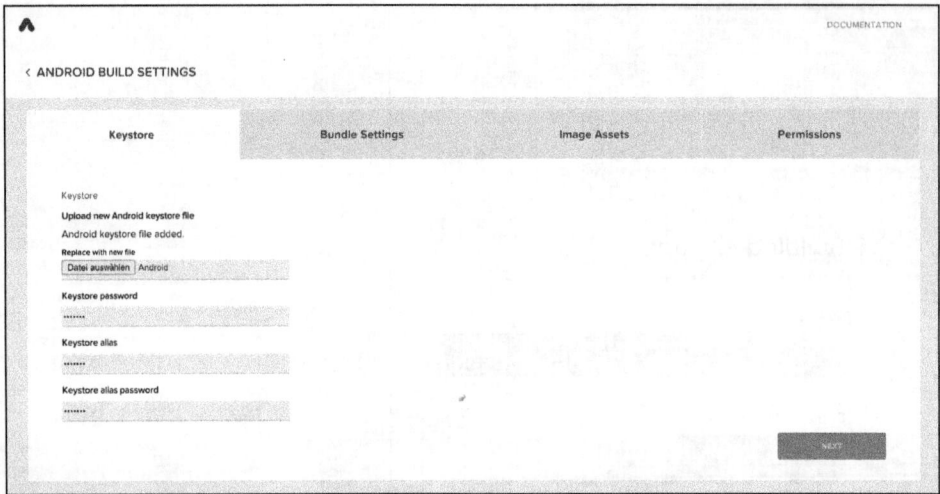

Figure 9.36 Android Keystore

In the **Bundle Settings** area, you can assign a package identifier and a display name (see Figure 9.37). The package identifier is unique for your app. You also have the option of assigning a **URL Scheme** here. This allows other apps and web applications to communicate with your app and call it up. You can later choose between two build schemes: APK and AAB. The APK format is the installable, executable format for Android apps. It's best used for testing. AAB is the Android App Bundle. It is only used for publication and contains the entire program code of the Android app. However, it cannot be used to install apps on end devices. It's used to optimize the size of a build.

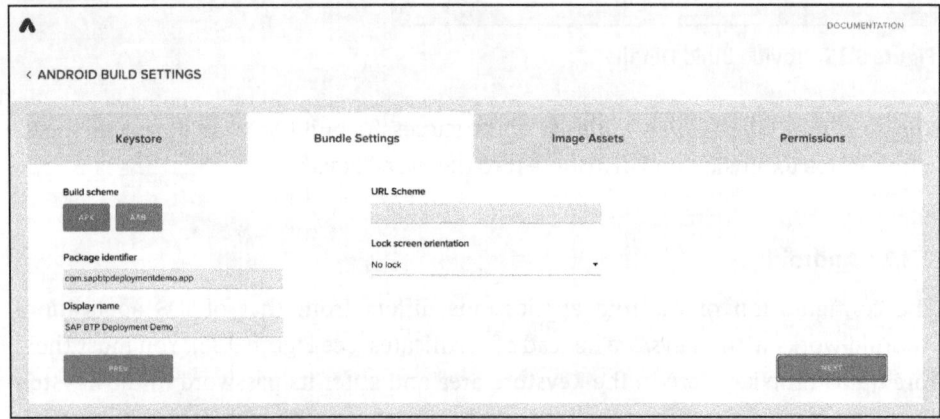

Figure 9.37 Android Bundle Settings

You must also provide the icons in the **Image Assets** area on the Android platform (see Figure 9.38). On Android, a distinction is made between icons, splash screens, and notification icons.

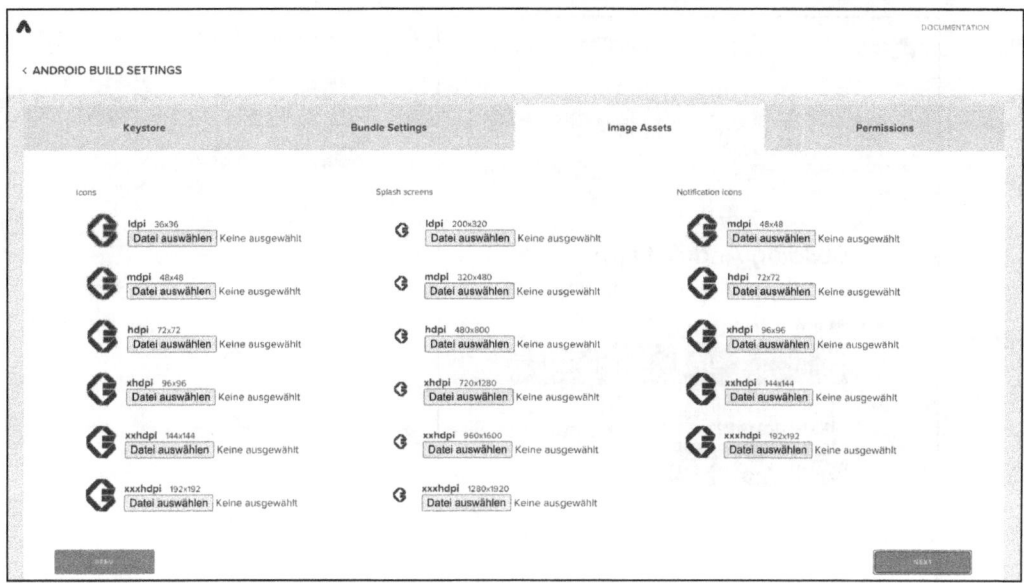

Figure 9.38 Android Image Assets

In contrast to iOS, you must also explicitly define the required permissions for Android. This is done in the **Permissions** area (see Figure 9.39). There you will see **Included Permissions** and **Not Chosen**. By clicking the name of a permission, you can either accept it as a required permission or remove it.

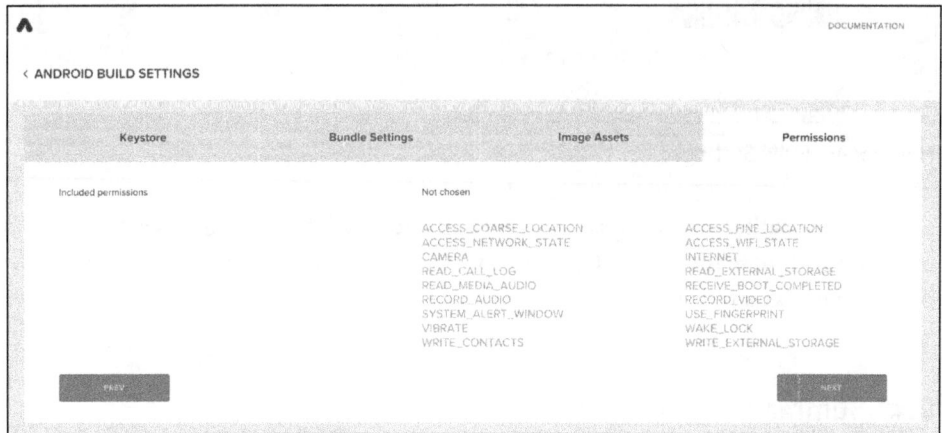

Figure 9.39 Android Permissions

Once you have completed the configuration, you can start the build. Therefore, click the **BUILD** button under **Android Mobile/Tablet**. The build is identical to the build of the web application or the iOS app. You must choose the APK or AAB file type. And you must also enter the **Version Code**, which represents the internal version number, and the **Version Name**, which represents the semantic version number (see Figure 9.40).

9   Deploying Applications

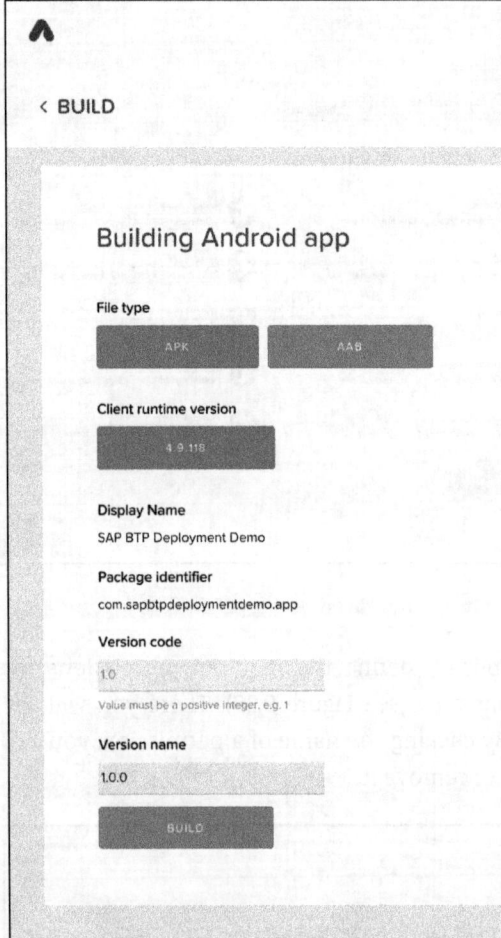

Figure 9.40 Build Settings

As explained earlier, the result is provided in the form of an APK or an AAB file. You can either upload these files to the Google Play Store or to the enterprise app store of your device-management solution.

9.4   Summary

We started the chapter with the configuration of the build showing you how you can influence the build process for various platforms This chapter also covered the various strategies and methodologies for deploying applications across different platforms and environments. Section 9.2 discussed the specific processes and best practices for deploying applications to SAP BTP. This section included details of how to prepare an application and ensure the application meets SAP BTP standards for deployment.

Section 9.3 focused on the deployment of mobile applications, outlining the unique requirements and steps necessary to successfully launch mobile apps. Section 9.3.1 and Section 9.3.2 delved into platform-specific guidelines for iOS and Android, respectively. These covered the specifics of packaging, provisioning, and distributing applications through platforms such as the Apple App Store for iOS devices and the Google Play Store for Android devices. This included managing certificates, provisioning profiles, and adhering to platform-specific guidelines and requirements to ensure a smooth deployment process for mobile users.

PART III
SAP Build Work Zone

Chapter 10
Introduction to SAP Build Work Zone

This chapter explores the two main editions of SAP Build Work Zone: the standard edition and the advanced edition. The chapter will examine the features and capabilities of each edition.

SAP Build Work Zone is one of the three services included in *SAP Build*, along with SAP Build Apps and SAP Build Process Automation. This service makes it possible to create clear business pages quickly and easily, which can also be personalized. In concrete terms, a distinction is made between two different editions:

- Standard edition
- Advanced edition

First, in Section 10.1, we'll look at the basics of SAP Build Work Zone, including which former services gave rise to it. In Section 10.2, we'll look at SAP Build Work Zone, standard edition in detail; this offering replaced the SAP Launchpad service. Then, in Section 10.3, we'll look at SAP Build Work Zone, advanced edition, which replaced SAP Work Zone. In addition to introducing the two services, we'll also look at the differences between them in the following section.

10.1 SAP Build Work Zone

Many people are probably familiar with the SAP Launchpad service and SAP Work Zone. The two editions of SAP Build Work Zone are basically just rebrandings of these former services. These services were replaced by SAP Build Work Zone in November 2022. It's important to note that this is only a rebranding. This means that pages created using the now-obsolete services can still be used.

The other two services included in SAP Build are also not completely new services. These already existed before and were brought together under a different name in SAP Build (see Figure 10.1). SAP Build thus expands SAP's low-code offering, and in addition to applications, processes or business pages now can also be created by drag and drop, so less technical know-how is required for the development of various applications.

10   Introduction to SAP Build Work Zone

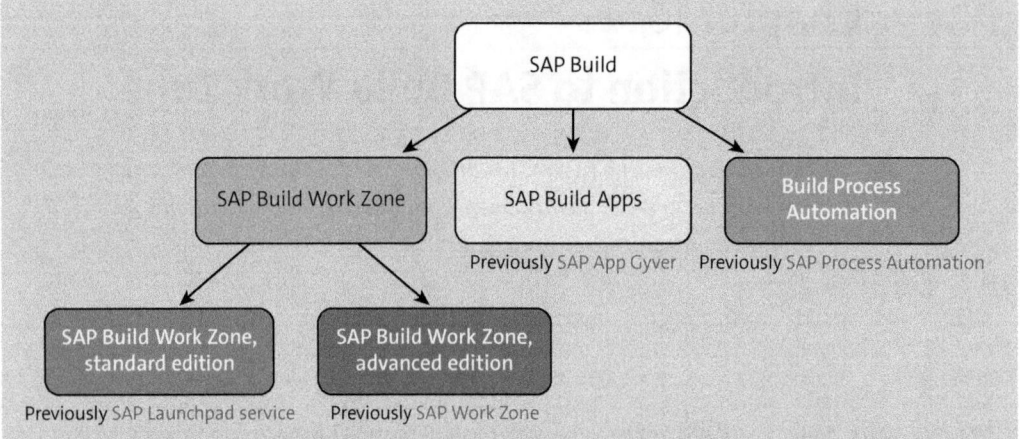

Figure 10.1 SAP Build Services

The question now arises as to how the former services differ. The *SAP Launchpad service* and *SAP Work Zone* were two different services from SAP, each with specific purposes and functionalities. The SAP Launchpad service served as a central platform for accessing various applications and business processes. It provided a customizable user interface through which users could access information and tools relevant to them. The focus here was on providing uniform and user-friendly access to applications. Furthermore, the SAP Launchpad service provided a central entry point through which users could access various applications.

In contrast, SAP Work Zone was a comprehensive collaboration platform that enabled users to collaborate; create, share, and manage content; and coordinate tasks and projects. The functionalities of the SAP Launchpad service included customizable dashboards and tiles to simplify access to applications and information, as well as integration between SAP and non-SAP applications. Users could design their personal workspaces and organize applications as needed. SAP Work Zone, on the other hand, offered comprehensive collaboration features such as collaborative document editing, project management, task management, calendar integration, and discussion forums. The focus here was on collaboration and the exchange of information in a digital working environment.

The SAP Launchpad service was aimed at companies and organizations that wanted to optimize their internal work processes and simplify access to various applications and information. It was ideal for companies that wanted to create a consistent user experience across different systems. SAP Work Zone, on the other hand, was aimed at teams and project groups that need efficient collaboration and a central place to share information. It was particularly suitable for companies that want to improve collaboration between employees at different locations or in distributed teams. Although the two services differed in their functions and purposes, they could also be used in combination. For example, companies could use the SAP Launchpad service to facilitate access to different applications while using SAP Work Zone as a platform for collaboration and information sharing. In this way, companies could benefit from the advantages of both services and create a comprehensive digital working environment.

Today, a large number of companies have rather complex enterprise landscapes. These landscapes are characterized by heterogeneity and fragmentation because they span a large set of content types, UI tools, content repositories, hybrid IT systems, channels, and applications.

The variety of content types can include, for example, text, images, videos, documents, and more. Each of these types requires specific tools and technologies to manage them effectively and integrate them into the organization's workflows.

Furthermore, companies often use a wide range of UI tools that enable them to implement intuitive and appealing user interfaces. These tools vary greatly in complexity. They range from simple drag-and-drop editors to complex design platforms, but always with the purpose of visually designing applications.

If you think about the IT systems in companies, they can be very diverse and include different software solutions for various purposes. Just to name a few, these solutions could include the following:

- Enterprise resource management
- Customer relationship management
- Content management

It's important that these systems can be integrated into each other without any problems and that they enable a smooth exchange of data between each other.

It should also not be forgotten that companies want to reach their customers on different channels. Each channel has its own characteristics and thus requires special adjustments to ensure interaction with the customer. Channels include, for example:

- Web applications
- Mobile applications
- Social media

Business users often have problems finding their way around in such corporate landscapes and accessing the right information. Furthermore, they often experience difficulties when it comes to switching between different applications. This leads to users despairing when it comes to finding the right application (see Figure 10.2).

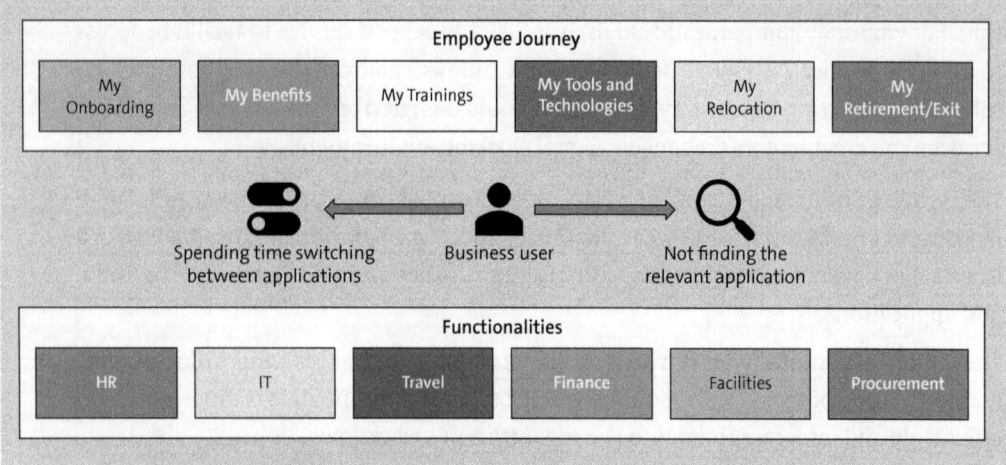

Figure 10.2 Employee Journey

Staff experience is often fragmented. Decisions are often made and actions taken without a complete overview. As a result, recommendations and insights cannot be based on the full context.

To counteract this, users want a unified, intelligent work environment that not only increases their productivity but also their efficiency and engagement.

SAP Build Work Zone is a service designed to solve these problems. It enables companies, IT departments, and employees to improve the fragmented experience by bringing everything together in one central place and making it easily accessible. Companies are provided with a central platform where all content, be it information, tools, or resources, can be found. This makes it easier for users to access the information they need, which leads to a more efficient way of working. It's also important to note that this creates a consistent and engaging user experience across business processes and applications. This makes it easier for users to find their way around, even when they are in different applications. This has a positive impact on productivity and subsequently reduces the need for training.

Let's now briefly look at the functionalities that SAP Build Work Zone offers. At a high level, SAP Build Work Zone, standard edition aims to allow users to work independently on the system without being aware of and interacting with other users on the same system. In contrast, the advanced edition focuses more on enabling user interaction in a structured way. Users are given the opportunity to interact and collaborate directly with each other on the system (see Figure 10.3).

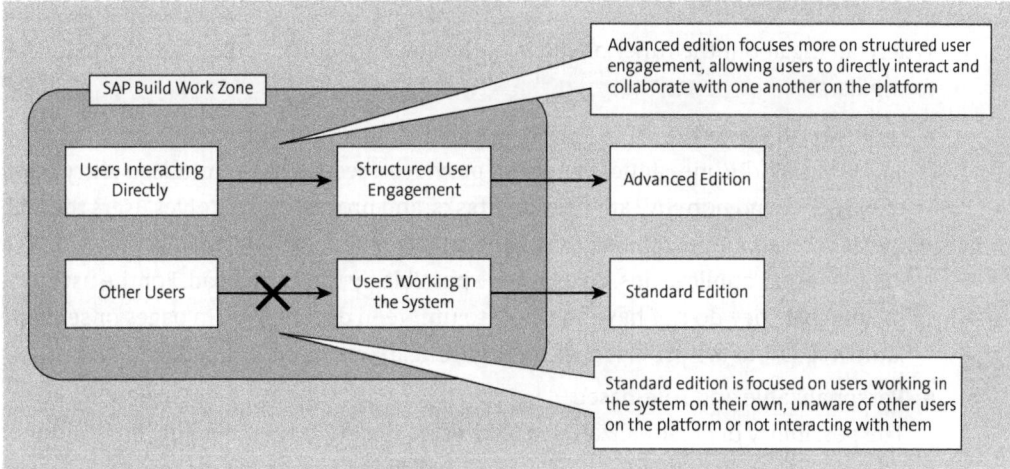

Figure 10.3 Major Features of SAP Build Work Zone

Let's now go into detail about the two editions, starting with the standard edition.

10.2 Standard Edition

As already mentioned, the standard edition is a possible variant of the SAP Build Work Zone. This edition is the successor to the previously existing SAP Launchpad service. In principle, nothing has changed in its function; only the name of the service was changed as of November 2022. The service still has the same range of functions. It's also important to note that the rebranding has no effect on existing content. This means that no actions have to be taken on the customer side—for example, to transfer pages to the new service. All content created in the SAP Launchpad service also works seamlessly with SAP Build Work Zone, standard edition. This service can also be used free of charge with RISE with SAP.

> **RISE with SAP**
>
> *RISE with SAP* is not a new product, but a package designed to support companies that want to switch to SAP S/4HANA. This makes it easier to optimize business processes in the cloud. SAP takes care of the necessary steps such as analysis, operation, support, and the selection of relevant hyperscalers.

In addition, existing licenses can continue to be used. Thus, nothing has to be considered on the customer side in this regard either.

In the following, we'll take a look at the functionality, working environment, and installation of SAP Build Work Zone, standard edition.

10.2.1 Functionality

When talking about the functionalities of the standard edition, there are the following four pillars:

- **Central entry point**
  SAP Build Work Zone, standard edition provides a secure and centralized entry point to both SAP and non-SAP applications, tasks, and processes. This gives users the ability to access various applications, whether they are SAP S/4HANA apps, SAP BTP apps, or other applications. This is made possible by the launchpad. For the user, this means that they do not have to resort to umpteen different login pages or separate launchpads, but are provided with everything in one place, in a uniform design.

- **Personalizable and role-based**
  The possibility of personalization allows users to design their working environment according to their own needs and, in addition, layouts and variants can also be saved as such. For example, elements can thus be changed in their arrangement to achieve the most ideal user experience. Furthermore, functions and information are adapted to the roles and responsibilities of the respective user in the company through the role-based setup. Users are usually only given access to specific functions and information.

- **Adaptable and expandable**
  Administrators have the option of customizing the platform according to their requirements. For example, adjustments can be made in the direction of the corporate design of the respective company in order to ensure a uniform appearance. In addition to visual adjustments such as font, color scheme, and layout, an individual logo can also be inserted and displayed. Administrators also have the option of expanding the platform. For example, additional modules or integrations can be added.

- **Content integration**
  Another important factor is the integration of content. With this in mind, other systems can be connected and integrated to ensure collaboration with other systems. Users can thus retrieve data and information from different systems without jumping around between applications. Standardized interfaces such as APIs and preconfigured integration scenarios are used for this purpose. In addition, third-party systems can thus be connected alongside SAP systems.

There are a number of advantages to using SAP Build Work Zone, standard edition. SAP Build Work Zone, standard edition provides the ability to do the following:

- Connect to applications with integrated access to both applications and data
- Utilize an enterprise-wide search feature to easily locate the information and resources you require
- Access important business updates and information to remain informed about the latest developments

- Use unified workflows
- Leverage engagement services for communication and information sharing
- Gain the flexibility to work anytime and anywhere
- Access embedded reports and alerts that are tailored to the specific context of your work
- Enjoy a personalized experience with tailored recommendations to help you become more self-sufficient and work efficiently

Figure 10.4 shows what a business page created with the standard edition might look like. This is only an example. In the course of this chapter, you will create your own page in the standard edition.

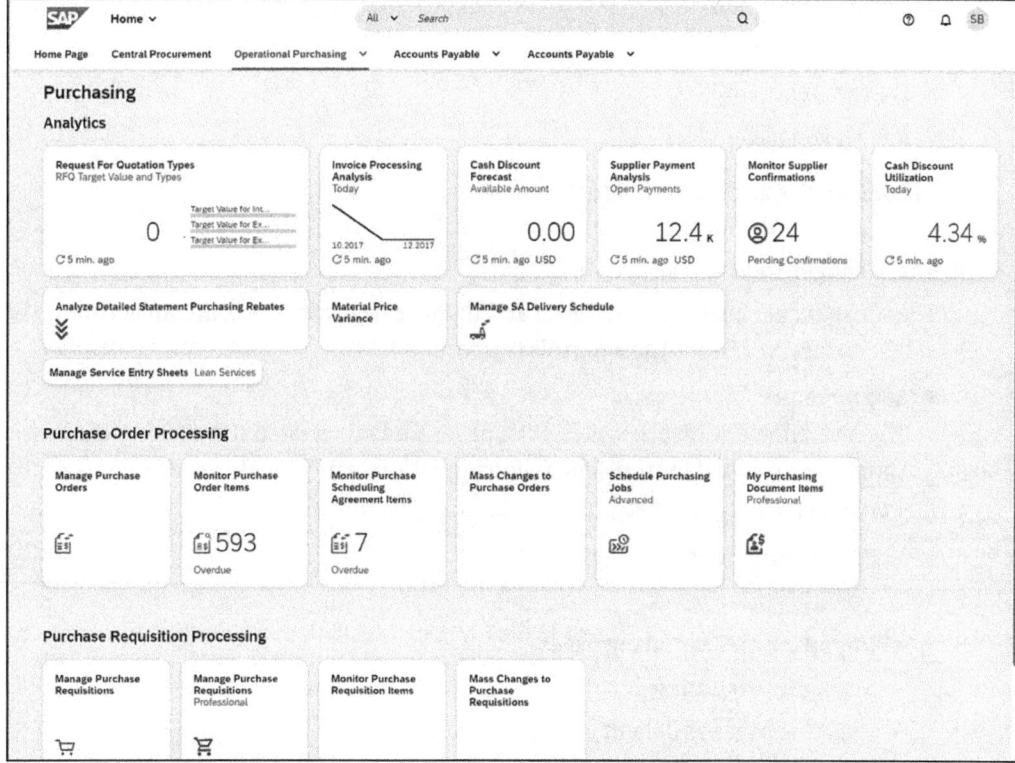

Figure 10.4 Business Page Created with SAP Build Work Zone, Standard Edition

For users, SAP Build Work Zone, standard edition offers a range of use cases:

- Central access platform
- Personalised dashboard
- Task management
- Collaboration and interaction
- Mobile accessibility

> **Availability Zones and Multitenancy**
>
> The standard edition also supports the concept of availability zones in the AWS and Azure data centers. These enable higher availability, fault tolerance, and scalability of services than if they were only running in a single data center. In addition, multitenancy is also supported.

10.2.2 Working Environment

SAP Build Work Zone, standard edition provides administrators with a range of tools and functions to create and manage business pages. These include the following:

- Site manager tools
- Content manager
- Business content
- Spaces and pages

Ahead, we look at these individual tools in more detail.

Site Manager Tools

The *site manager* is used to configure and manage pages for a specific subaccount. The sidebar contains various tools for this purpose:

- **Site directory**

  The site directory displays tiles with pages already created. By clicking the corresponding icons and buttons, the following actions can be performed:
  - Create new pages
  - Edit existing pages
  - Delete existing pages
  - Importing and exporting pages
  - Managing page aliases
  - Selecting pages as default
  - Opening pages

- **Content manager**

  The content manager can be used to manage the business content elements of the subaccount. These include applications, catalogs, groups, roles, and shell plugins. Business content elements can be used in every page of the subaccount. There are two different ways to add such elements to the subaccount:
  - Integrating content from content providers
  - Manually integrating content created with content editors

- **Channel manager**
  Using the business content, it's possible to manage content providers. These content providers make content available that can be integrated into the pages. A distinction is made between remote content providers and SAP BTP content providers. Remote content providers include the following:
  - SAP Integrated Business Planning for Supply Chain (SAP IBP)
  - SAP S/4HANA Cloud
  - SAP S/4HANA
  - SAP Business Suite
  - SAP Enterprise Portal

  SAP BTP content providers basically provide HTML5 applications.

You should also be aware that settings for the subaccount can be adjusted. Settings made here affect all pages created in the respective subaccount. Adjustments can be made in the following areas:

- **Error Log**
  Errors that have occurred in the Approuter of the system in the last 24 hours can be troubleshot here.

- **Notifications**
  There are two elements to be configured in this area. On the one hand, you can determine which identifier is to be used for authentication: the email or the user ID.

  On the other hand, credentials can also be generated here. This is relevant if a developer wants to publish notifications in an application. In this case, credentials are required, which must be stored in the destination. However, it's important that the relevant roles are assigned to a role collection beforehand.

- **Alias Mapping**
  If a backend system has system aliases, these are to be mapped here. The mapping is done on the basis of the respective destination and allows successful navigation in the applications. This point is only relevant when applications are integrated manually, not when federated content is involved.

- **Security Headers**
  To introduce another layer of security, special HTTP headers can be added here. These make it possible to restrict actions that can be carried out by the browser and server in order to provide more security. One or more headers can be defined for this purpose.

- **Neo Site Conversion**
  Of course, it should also be possible to transfer pages created in Neo environments to Cloud Foundry. This can be done via this menu item. After a page has been exported from the Neo environment and the switch for the conversion has been

activated here, the resulting ZIP file can be converted into the correct format for the Cloud Foundry environment. Afterward, the page can also be imported into Cloud Foundry and added.

- **Identity Provisioning**
  In this area, the subaccount can be linked to Identity Provisioning in SAP Cloud Identity Services. This takes care of the provision of authorizations for the various cloud and on-premise business applications.

- **Identity Authentication**
  Identity Authentication in SAP Cloud Identity Services is a cloud service that is responsible for authentication, single sign-on, and user management. Before it can be used, a trust between the service and SAP BTP must be configured.

Content Manager

The content manager contains the following subtabs:

- **My Content**
  In this tab (see Figure 10.5), all contents of the subaccount are displayed that have been integrated either by content providers or manually. By using content editors, elements from providers can be displayed, and new manually integrated content can be displayed, edited, and configured.

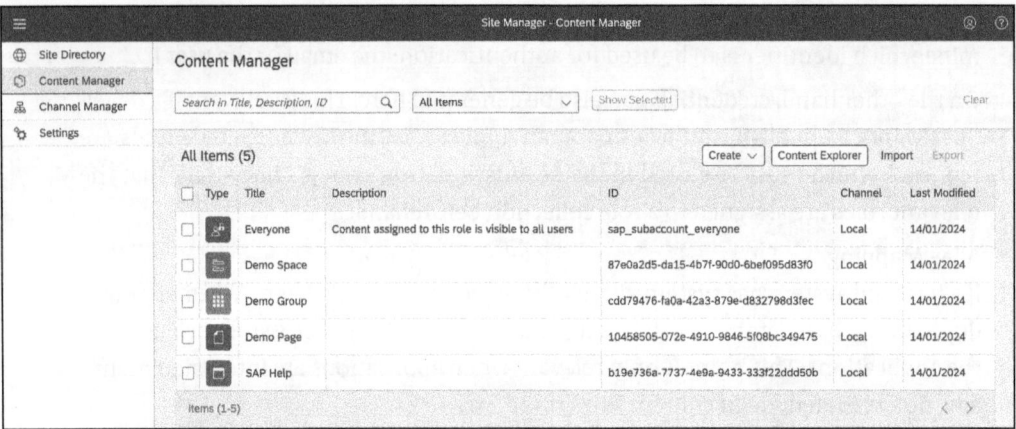

Figure 10.5 My Content Tab in Content Manager

- **Content Explorer**
  In the **Content Explorer** tab (see Figure 10.6), on the other hand, all available content providers are displayed. These can then be searched for content, and this content can ultimately also be added to the subaccount. If content is selected and added here, it's also visible under **My Content**.

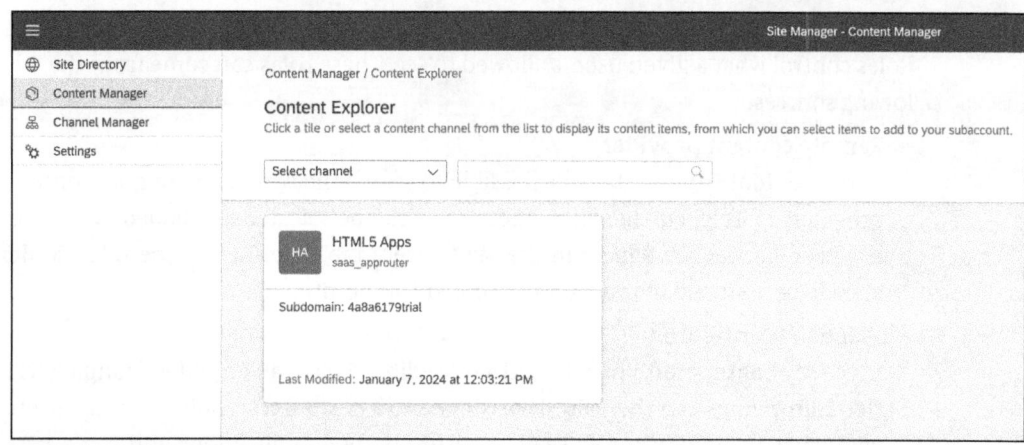

Figure 10.6 Content Explorer Tab in Content Manager

Business Content

With the help of the content manager, content such as applications, catalogs, groups, roles, and shell plugins can also be added. Let's discuss what exactly this means:

- **Apps**
  Applications can be in two different areas: SAP BTP provider–based or manually configured. To add an SAP BTP provider–based application, you must switch to the **Content Explorer** tab. Here, content providers can be searched, applications selected, and finally added to the subaccount. These then become visible under **My Content**.

  The situation is somewhat different for manually configured applications. Here, the app editor can be used in the **My Content** tab to make manual configurations. The following UI technologies can be integrated:
  - URL
  - Dynamic URL
  - SAPUI5
  - SAP GUI for HTML
  - Web Dynpro ABAP

  At runtime, a tile is usually used to display the app. It should be noted that different tiles can also call up the same application with different configurations.

- **Catalogs**
  You can display several applications bundled under one catalog. The catalogs can be found in the App Finder depending on their assignment.

- **Groups**
  Similar to catalogs, groups also allow several apps to be displayed in a bundle. Normally, applications that are related to each other are brought together within a group. If an application is not assigned to a group, it is not displayed.

- **Roles**
  Roles control what a given user is allowed to see. These roles can come from the following sources:
  - **Remote content provider**
    In the **Content Explorer** tab, it's possible to search the connected remote content providers by role. Furthermore, these roles can be selected and added. Once this has been done, they appear in the **My Content** tab. In principle, the role should already be assigned to the relevant apps at this point.
  - **Manually configured**
    In the **My Content** tab, you will find a role editor. Roles can be added manually via this editor. Apps can then also be assigned to a created role. Subsequently, users who have assigned the role see the corresponding applications and can access them.
  - **Everyone role**
    In addition to the two variants mentioned previously, there is also a special *Everyone* role provided out of the box. If apps are assigned to this role, they are visible to all users.

  To determine which pages can be accessed by which users, a role must be assigned to one or more pages.

- **Shell plugins**
  Shell plugins make it possible to influence the functionality and behavior of a page's user interface. These plugins are automatically loaded, and once the page is loaded, they are automatically initialized at runtime. Basically, these are HTML5 applications. These plugins can be integrated either by SAP BTP or by remote content providers.

Spaces and Pages

In SAP S/4HANA on-premise, it's possible to organize content not only in groups but also in spaces and pages. This gives users the opportunity to organize content in an even more structured and clear way. The concept of spaces and pages basically involves four elements:

- **Space**
  A space can be compared to a group. Such a group consists of one or more pages.
- **Page**
  A page in turn contains one or more sections.
- **Section**
  Within a section, you'll find one or more tiles.
- **Tile**
  Tiles are used to display and open applications.

10.2 Standard Edition

> [!] **Browser Compatibility**
>
> With regard to the browser compatibility of SAP Build Work Zone, standard edition, you can orientate yourself with the compatibility of the SAPUI5 framework. Basically, all common browsers such as Google Chrome, Microsoft Edge, and Mozilla Firefox are supported with regard to the runtime.
>
> When talking about design time, only those browsers are supported that are also supported for SAPUI5 on desktop devices.

10.2.3 Installing SAP Build Work Zone, Standard Edition

To install SAP Build Work Zone, standard edition, you first need an account in SAP BTP. In our case, we used a trial account, which can be created at *https://hanatrial.ondemand.com*. Once this account is created, a subaccount must be created. Then you can continue with the next steps:

1. First, log into SAP BTP and navigate to your subaccount.
2. Now select **Services • Service Marketplace** from the left menu bar (see Figure 10.7).

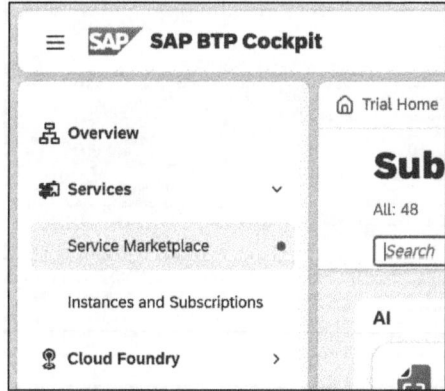

Figure 10.7 Select Service Marketplace in SAP BTP Subaccount

3. You can then search for "SAP Build Work Zone" using the search field (see Figure 10.8).
4. After the service has been found, it can be selected with a click. If this is done, a detailed view of this service appears on the right-hand side (see Figure 10.9). Click **Create** to initiate the creation of the service.
5. After clicking the **Create** button, a dialog opens in which the following data must be selected (see Figure 10.10):
   - **Service**: SAP Build Work Zone, standard edition
   - **Plan**: standard

10 Introduction to SAP Build Work Zone

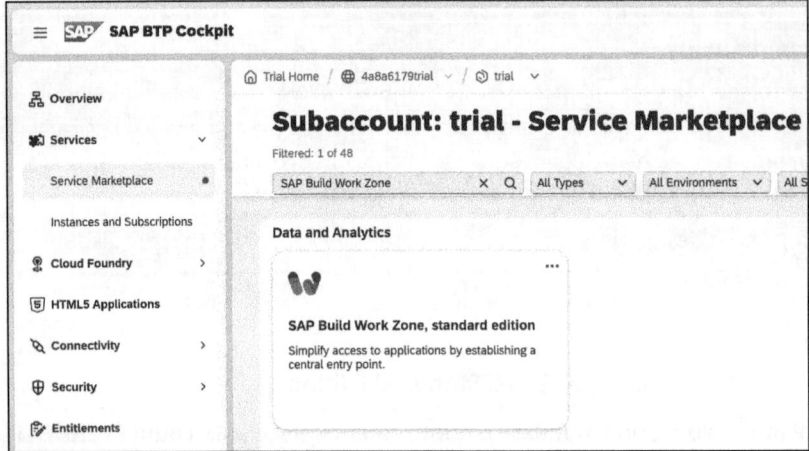

Figure 10.8 Search for SAP Build Work Zone in List of Services

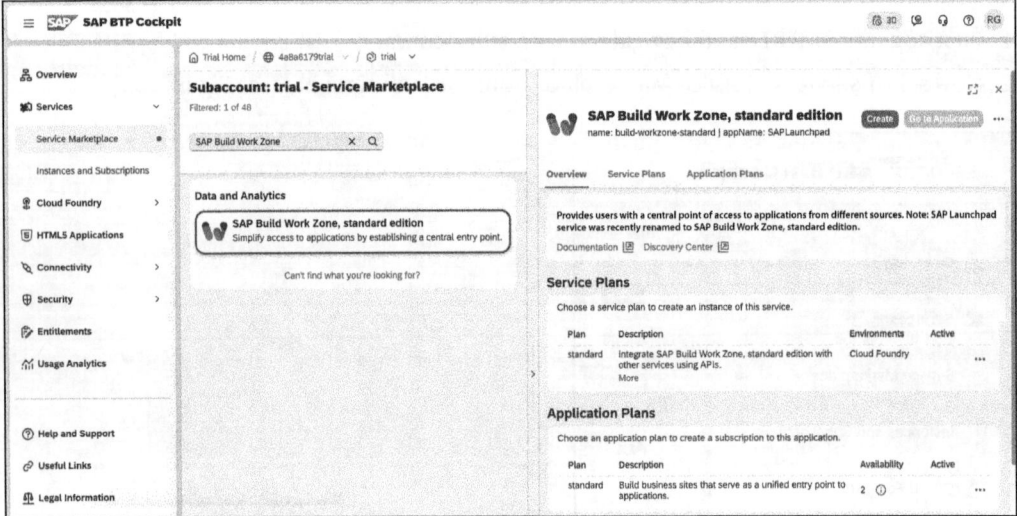

Figure 10.9 Details of SAP Build Work Zone, Standard Edition Service

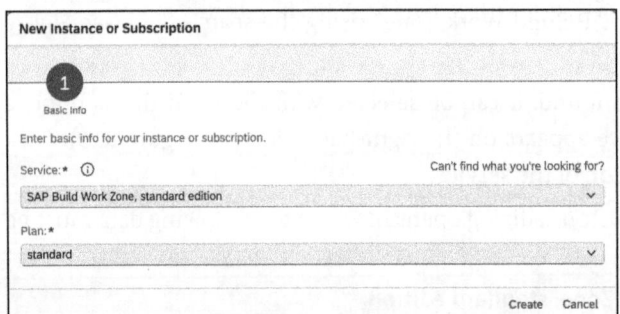

Figure 10.10 Subscribe to SAP Build Work Zone, Standard Edition

6. Once the selection has been made, the creation can be completed with **Create**. After clicking **Create**, it takes a little while for the creation to complete. Afterward, the previously created entry will appear in the list of instances under **Services • Instances and Subscriptions** (see Figure 10.11).

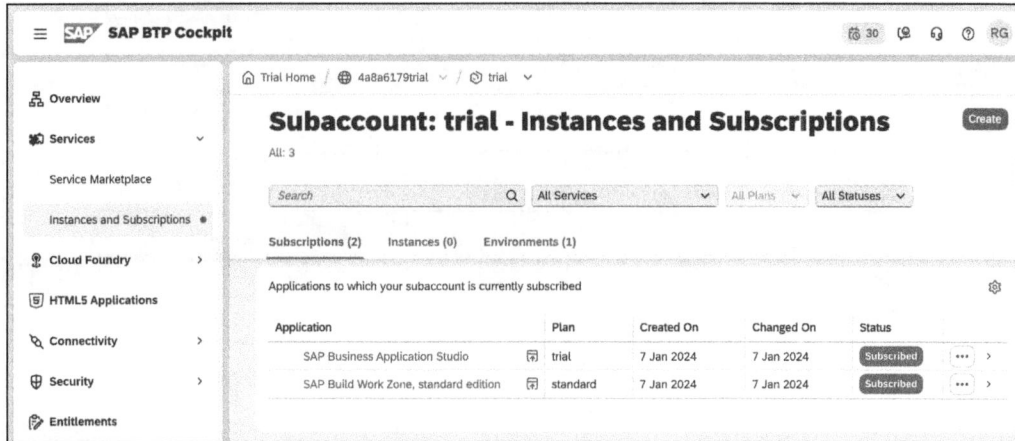

Figure 10.11 List of Subscriptions in SAP BTP Subaccount

7. By selecting the SAP Build Work Zone, a detail dialog opens, via which it's possible to jump to SAP Build Work Zone with **Go to Application** (see Figure 10.12).

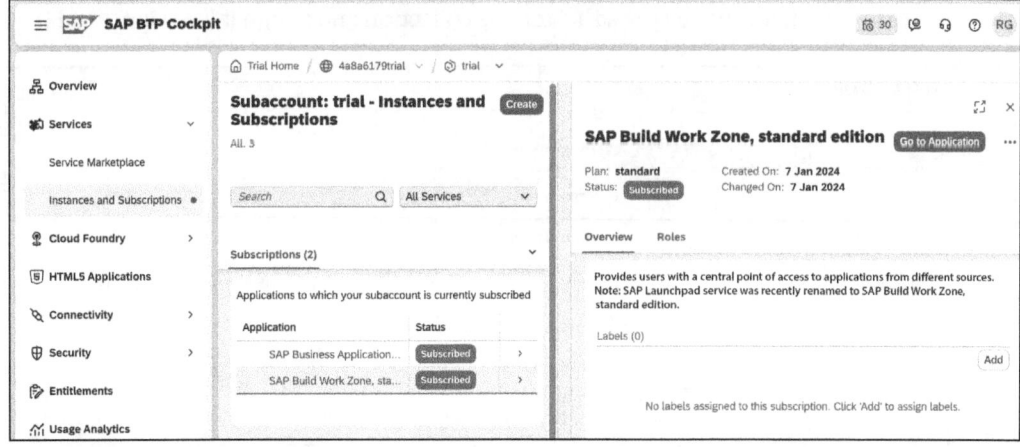

Figure 10.12 Details of Subscribed Service: SAP Build Work Zone, Standard Edition

If the application is now started, an authorization error is currently displayed. This is because the role collection is not assigned to the user. To remedy this, the Launchpad_Admin role collection must be assigned to the calling user in SAP BTP. This can be done at the subaccount level under **Security • Users**, as follows:

1. Navigate to the **Users** page and select the calling user from the list of users.

2. Afterward, a detailed view of the user opens on the right-hand side, in which, among other things, the assigned role collections can be seen (see Figure 10.13).

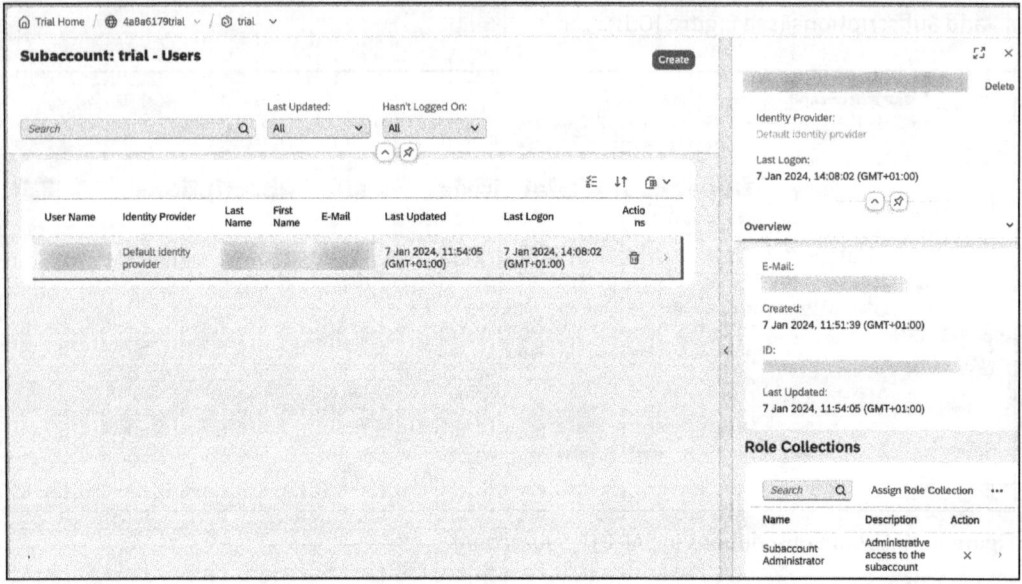

Figure 10.13 Details of User in SAP BTP

3. Click **Assign Role Collection** to assign one or more role collections to the user. In this case, search for the Launchpad_Admin role collection and assign it (see Figure 10.14).

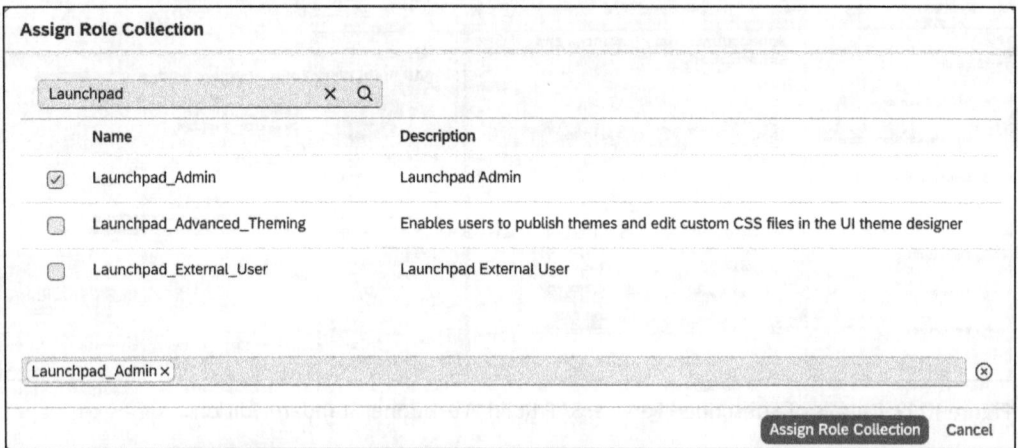

Figure 10.14 Assign Role Collection to User in SAP BTP

4. After clicking **Assign Role Collection**, the assignment is made and can be seen in the user's details (see Figure 10.15).

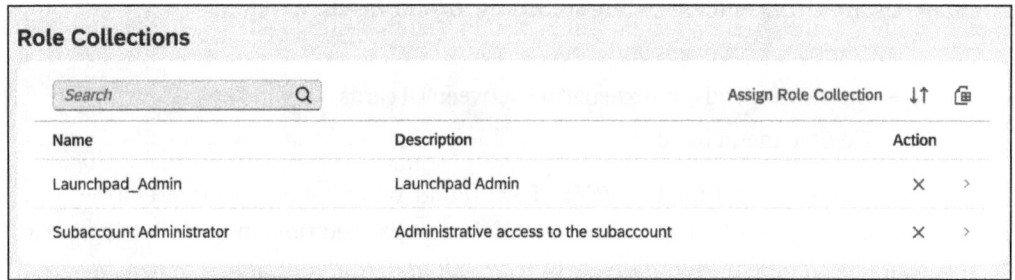

Figure 10.15 Role Collections of User in SAP BTP after Assigning New Role Collection

5. After this assignment, it may be necessary to log in again or clear the browser cache. SAP Build Work Zone, standard edition can then be called up.

10.3 Advanced Edition

In this section, we'll take a closer look at SAP Build Work Zone, advanced edition. This edition can be used especially when you want to build solutions for the digital workplace that have a positive impact on employee productivity. Thus, by implementing SAP Build Work Zone, advanced edition, access to relevant business applications, processes, information, and communication is centralized. As already mentioned, SAP Build Work Zone, advanced edition is the successor to the SAP Work Zone service, which was replaced in November 2022. With regard to its functionality, however, nothing has changed, as it is merely a rebranding.

If you compare the functionalities of the advanced edition with those of the standard edition, four categories stand out in the advanced edition—namely:

- **Empowering key users from business units**
  - Decentralized content creation and authoring
  - Expanded list of widgets for flexible page building
  - Advanced business content packages
- **Unifying business data and unstructured content**
  - Business data side by side with rich content
  - Multimedia content (images, videos, audios)
  - Integration with Microsoft Teams
- **Increasing user engagement**
  - Interactive workspaces (communities)
  - Document access and repository
  - Knowledge base with articles

- **Customizing and extending according to your needs**
  - Templates for pages and workspaces
  - Interaction and context sharing between UI cards
  - Flexible menu builder

If you look at the functionalities of SAP Build Work Zone, advanced edition, you'll notice one or two overlaps with those of the standard edition. In the following, we will take a look at the most important functionalities of this edition:

- Modern and consistent user interface for a unified, intelligent working environment
- Unified access to business and third-party applications
- Ability to share and interact with colleagues
- Secure access
- Integration of on-premise and cloud applications
- Global access on multiple channels
- Support for the multitenancy concept

As most people are probably aware, SAP BTP has a variety of different services that can be used as needed. Most of these services have a simple onboarding procedure. With SAP Build Work Zone, advanced edition, however, this is somewhat different. This is because it consists of several architecture components. While some of these components are optional—such as SAP Business Application Studio—others are included as standard or are required for the onboarding process.

Standard components of SAP Build Work Zone, advanced edition are as follows:

- **Digital workplace service**
  One of the main components is the digital workplace service, often referred to as the SAP Build Work Zone core component. This service offers features that can be used to create a kind of intranet. For this purpose, home pages and workspaces are created in a grid-based layout using a page editor.
- **Launchpad**
  Another standard component is the launchpad, which creates a central entry point to SAP and non-SAP applications. The question here is to what extent this launchpad differs from SAP Build Work Zone, standard edition. It should be mentioned in advance that it is not necessary to activate both services (standard edition and advanced edition).

In addition to the functions of the standard edition, the advanced edition also offers functions that make it possible to create and operate your intranet with homepages and workspaces. The following are some further differences between the two editions:

10.3 Advanced Edition

- SAP Build Work Zone, standard edition allows multiple pages to be created, while SAP Build Work Zone, advanced edition only allows a single default page to be created.
- Furthermore, SAP Build Work Zone, advanced edition has a work zone configurator, which is used in the onboarding process. SAP Build Work Zone, standard edition, on the other hand, has various other settings, such as **Notification Settings**.
- Finally, one more point is relevant for the end users. When an end user opens SAP Build Work Zone, standard edition, they immediately see the assigned SAP Fiori apps and groups. In contrast, SAP Build Work Zone, advanced edition initially shows a home page with a grid-based layout.

The differences mentioned here are also illustrated in Figure 10.16.

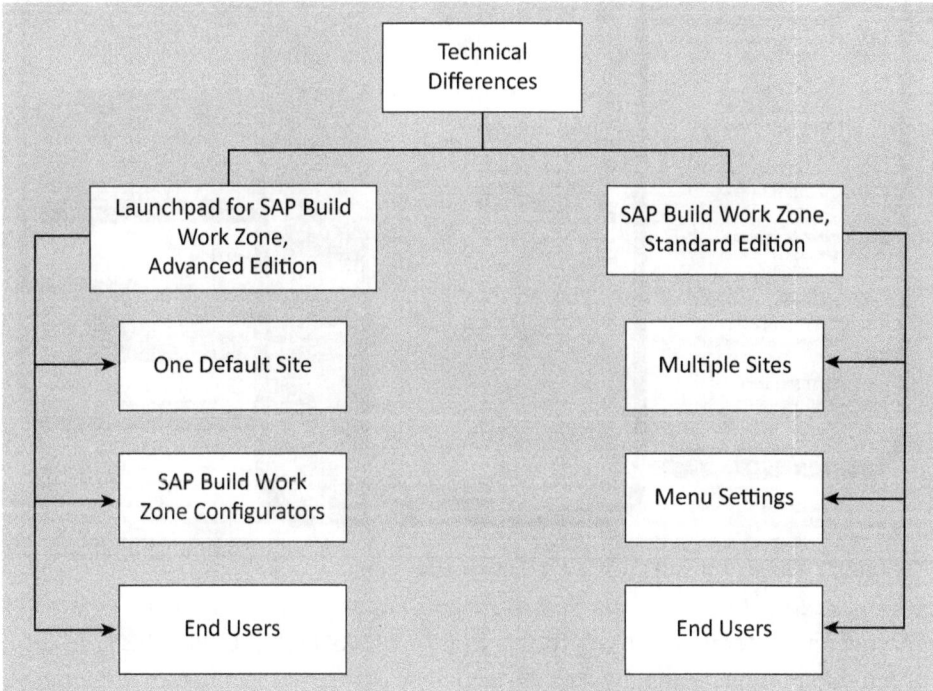

Figure 10.16 Key Differences between SAP Build Work Zone, Standard and Advanced Editions

Because SAP Build Work Zone uses the SAPUI5 framework and also follows the SAP Fiori design system, it basically looks and works great on all platforms. The only requirement is a web browser with internet access. No further settings are necessary to enable mobile use.

In addition, it's also possible to use native mobile applications; currently, the iOS and Android mobile platforms are supported. Once the mobile application is installed, the user can access the key functions of SAP Build Work Zone. For example, personal and team workspO10_OO1aces can be visited, colleagues can be invited to workspaces, and

329

content can be shared and viewed. Figure 10.17 shows how an application might look schematically.

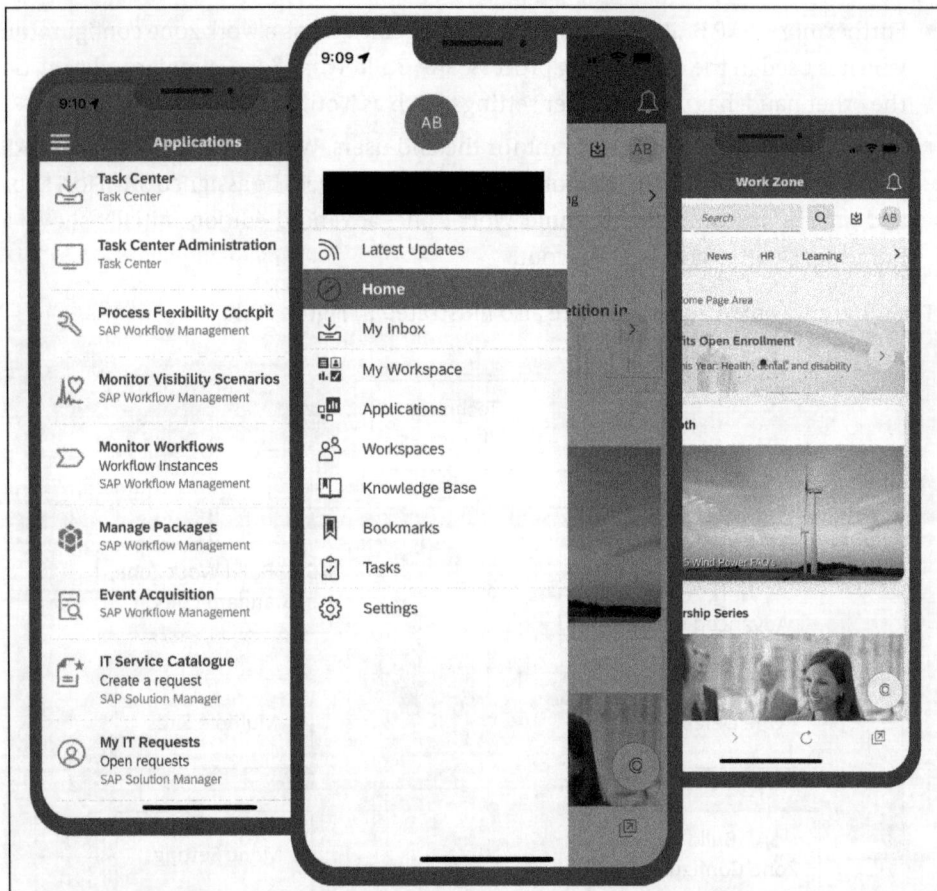

Figure 10.17 Preview of SAP Build Work Zone Mobile App

Another option for mobile use is the application of SAP Mobile Cards. This is an application that presents company data as microapps. Figure 10.18 shows what the mobile cards can look like.

With the help of the UI theme designer, it's possible to create your own themes in order to adapt an app's appearance. For example, the corporate design of your company can be applied to the pages. Technically, this designer is a browser-based tool with a WYSIWYG editor. The use of this editor significantly simplifies the creation and maintenance of themes.

10.3 Advanced Edition

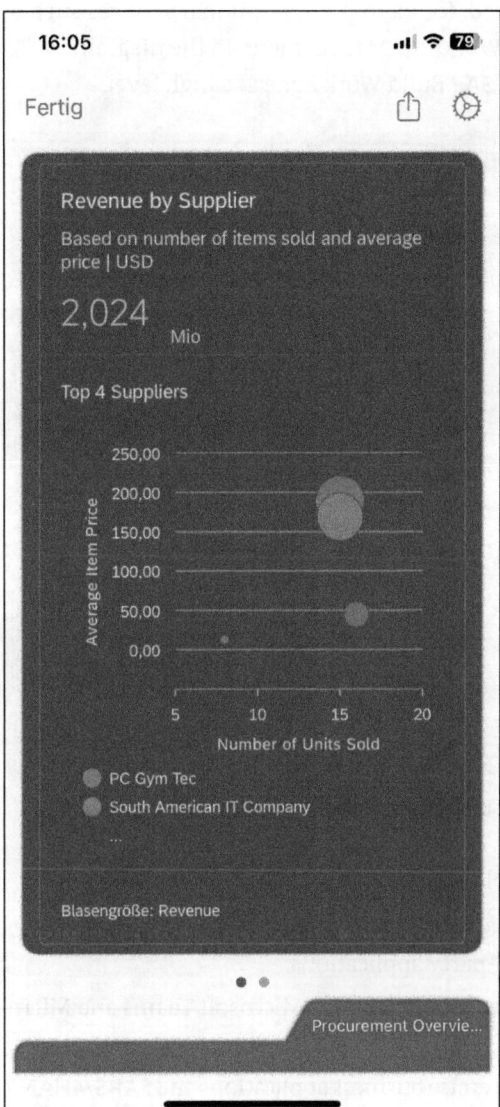

Figure 10.18 SAP Mobile Cards: Preview

> **Installing the Advanced Edition**
> Installation of SAP Build Work Zone, advanced edition, will be covered in Chapter 11.

10   Introduction to SAP Build Work Zone

These standard services are supplemented, for example, by optional services such as SAP Business Application Studio or SAP Workflow Management. In the diagram in Figure 10.19, you can see the architecture of SAP Build Work Zone at a high level.

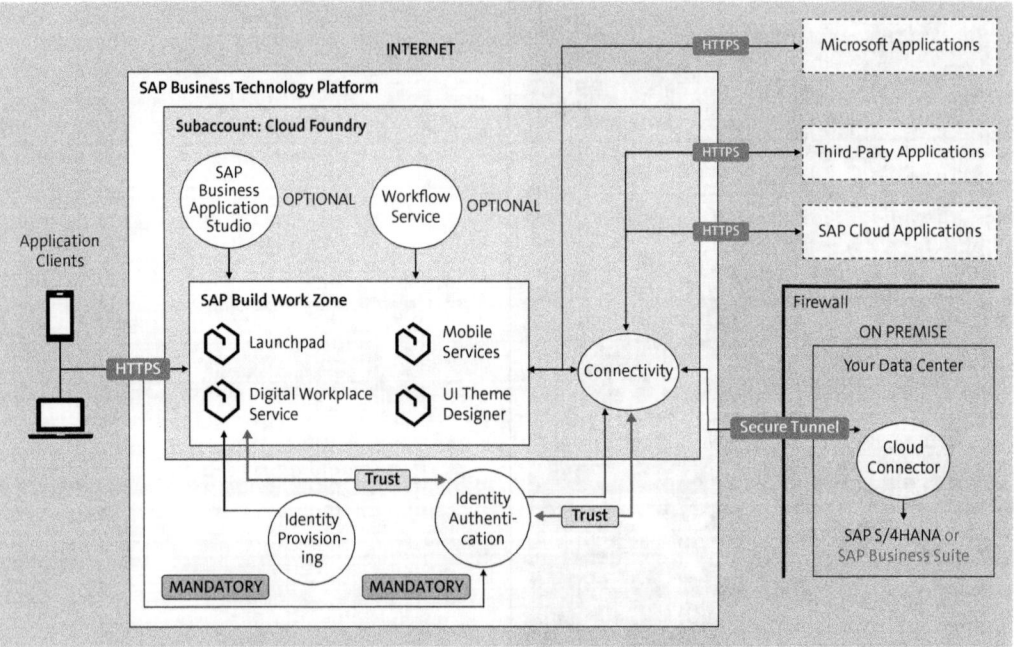

Figure 10.19 High-Level Architecture of SAP Build Work Zone

Key features of this architecture are as follows:

- As you can see, SAP Build Work Zone acts as a central entry point and enables integration with SAP cloud apps and third-party applications.
- It also has out-of-the-box integration with tools such as Microsoft Teams and Microsoft SharePoint.
- It can also be integrated with both on-premise business applications and SAP S/4HANA. If SAP S/4HANA is used, the cloud connector opens a secure tunnel through which communication with the underlying system takes place.
- Because security also plays an important role, the SAP Connectivity service covers authentication when it comes to integrating SAP cloud, SAP on-premise, or third-party applications.
- In addition, SAP Cloud Identity Services—especially Identity Authentication and Identity Provision—are used to ensure secure login and user lifecycle management.

> **Why Is SAP Cloud Identity Services Mandatory for the Advanced Edition?** [Ex]
>
> SAP Cloud Identity Services is a mandatory requirement for SAP Build Work Zone, advanced edition. Nowadays, it is not uncommon for software products to require information about their users. This is also the case with SAP Build Work Zone, advanced edition. In addition to the information about the logged-in user, which is obtained via the authentication process (via SAML assertion or JSON web token), this service also requires information about other users so that profile information can be displayed in blog posts or comments, for example. This is where Identity Provisioning comes into play, ensuring that user information is correctly extracted from the source system.
>
> Because various other services are also used seamlessly with SAP Build Work Zone, the Identity Authentication and Identity Provisioning services are indispensable to ensure that these other services can be accessed without logging on again and without manually creating users.

Finally, we should discuss the fact that SAP Build Work Zone, advanced edition distinguishes among different types of administrators. These administrators are particularly useful when it comes to administrative matters, such as setting up SAP Build Work Zone. There are the following types of administrators:

- **Company administrator**
  Full access within the administration console and can customize pages as required by the company.
- **Area administrator**
  Manages the content of an area, works within a specific area, and has more limited rights compared to the company administrator.
- **Support administrator**
  Has access to a subset of the functionalities of the company administrator and can support them.
- **Page content administrator**
  These are key users who have limited admin rights that have been granted to them by an administrator. They have the option of changing the content of a custom home page.
- **Workspace administrator**
  These are key users who create a workspace and thus become the administrator of that workspace. These users can also be added to another workspace as an administrator.

Table 10.1 lists the authorizations in the administration console for the company, area, and support admin. The page content admin and workspace admin only have end user authorizations and are therefore not listed here.

Screen in Administration Console	Company Admin	Area Admin	Support Admin
Change admin area	X	X	X
Overview Screen			
Overview screen for the company	X	X	X
Overview screen for an area in the company	X	X	
General			
View general information of a site	X		X
Users			
Users (viewing and adding new users)	X	X (area-specific)	X
User lists	X	X (area-specific)	X
External users (management)	X		X
Alias accounts	X		X
Authentication and Authorization			
SAML trusted IDPs (adding)	X	X	
SAML local identity provider	X	X	
Theming and Branding			
Theme manager (configuration)	X	X (area-specific)	X
Local theme designer (configuration)	X	X (area-specific)	X
Email templates (configuration)	X		
Area and Workspace Configuration			
Administrative areas (configuration)	X		
Home page (customization)	X	X (area-specific)	X
Profiles (administration)	X		X
Content templates	X		X
Workspaces (management)			
Workspace templates (administration)	X	X (area-specific)	X

Table 10.1 Administrator Permissions in SAP Build Work Zone

Screen in Administration Console	Company Admin	Area Admin	Support Admin
UI Integration			
Cards	X	X	
Widget builders	X		
External Integrations			
Business content	X		
Microsoft Teams	X		
External solutions	X	X (area-specific)	
OAuth clients (adding)	X	X (area-specific)	
Bots	X		
Extensions catalog	X	X	
Feature Enablement			
Kudos (configuration)	X		
Hashtags (administration)	X		
Knowledge base categories	X		
Features (configuration)	X		
Compliance and Security			
External user terms of service (configuration)	X		
Custom terms of service (configuration)	X		
Content administration (configuration)	X		
Compliance (monitoring)	X		
Security (configuration)	X		
Analytics			
Reports	X	X (area-specific)	
Third-party analytics	X		

Table 10.1 Administrator Permissions in SAP Build Work Zone (Cont.)

10.4 Summary

In this chapter, you received an introduction to the two editions of SAP Build Work Zone: standard edition and advanced edition. In addition to the functions offered by these two products, you also learned how SAP Build Work Zone, standard edition can be installed. In the following chapter, we'll look at the installation of SAP Build Work Zone, advanced edition.

Chapter 11
Installing and Configuring SAP Build Work Zone

SAP Build Work Zone is a comprehensive digital workplace solution that enables businesses to create engaging and connected experiences for their employees, partners, and customers. The installation process involves setting up the necessary infrastructure, configuring the system according to the organization's specific needs, and integrating it with existing business applications and data sources.

This chapter will walk you through the critical steps of installing and configuring SAP Build Work Zone. Section 11.1 delves into the step-by-step process of installing SAP Build Work Zone, advanced edition. We aim to provide clear and detailed instructions based on the SAP BTP booster. Whether you're a beginner or a seasoned user, our guidelines are crafted to cater to all levels of technical expertise.

Once the installation is complete, the next phase is configuration. Section 11.2 is dedicated to helping you customize and configure the system to meet your specific needs and preferences. We'll guide you through various settings and options, offering tips and best practices.

11.1 Installation

In this section, we'll show you step by step how to install and configure SAP Build Work Zone, advanced edition. Here too, we start by creating a subaccount in SAP BTP, as shown in Figure 11.1.

Next, navigate to the **Boosters** area in the side menu, as shown in Figure 11.2. There you will find a total of four boosters in connection with SAP Build Work Zone. Open the booster named **Get Started with SAP Build Work Zone, Advanced Edition**.

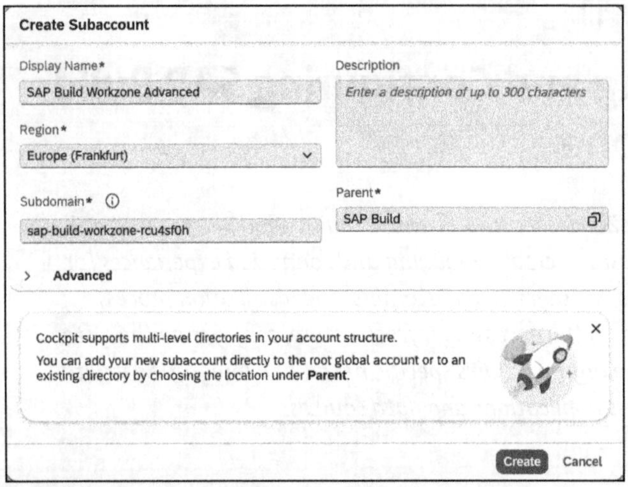

11   Installing and Configuring SAP Build Work Zone

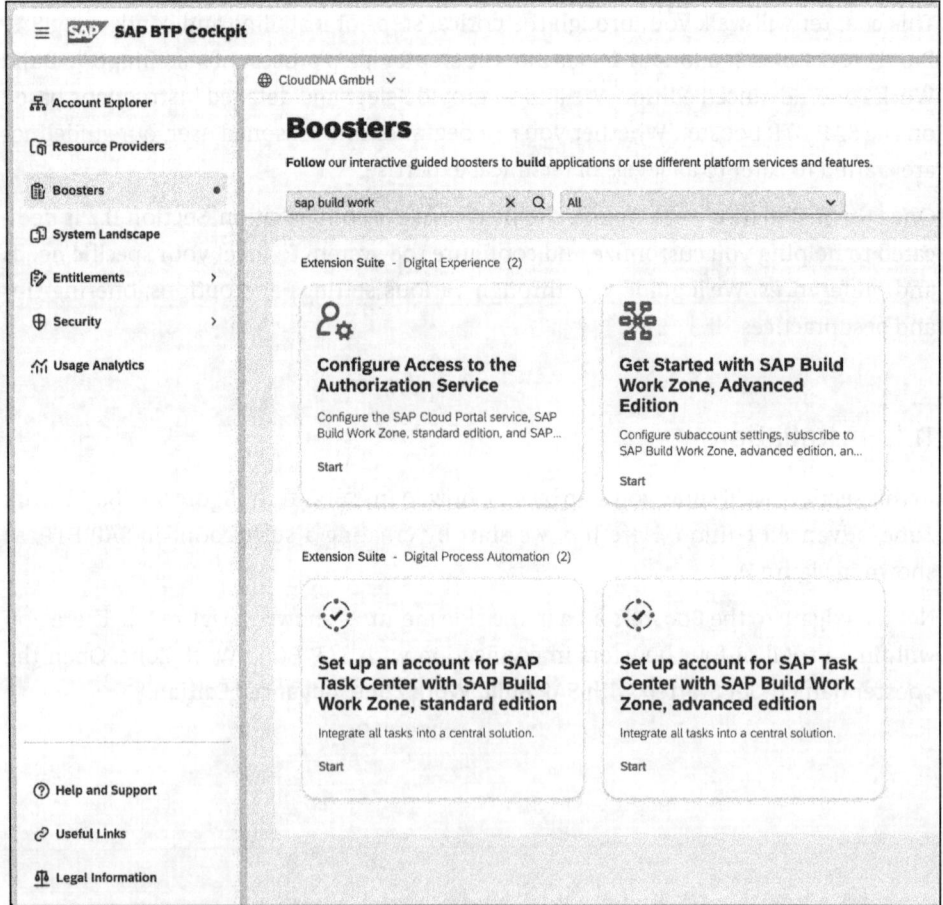

Figure 11.1 Create Subaccount

Figure 11.2 SAP BTP Booster for SAP Build Work Zone

Now open the **Components** tab as shown in Figure 11.3 for an overview of the required services and subscriptions.

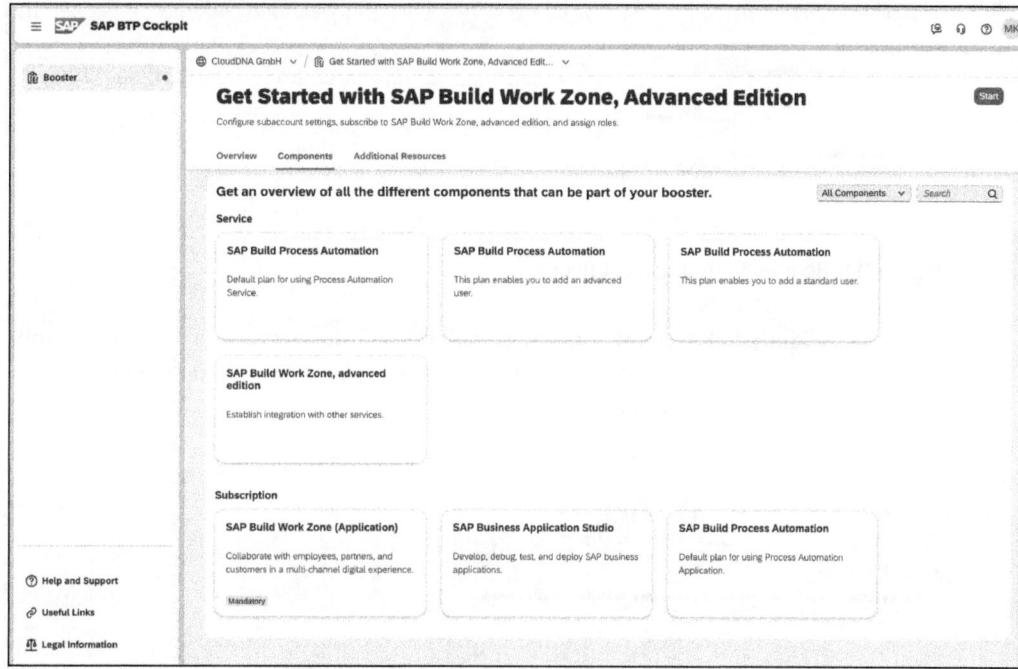

Figure 11.3 SAP Build Work Zone Component Overview

Next, click **Start** to start the booster. First, you must confirm that you want to run the booster for SAP Build Work Zone, and not for SAP SuccessFactors Work Zone, by clicking **Continue** (see Figure 11.4).

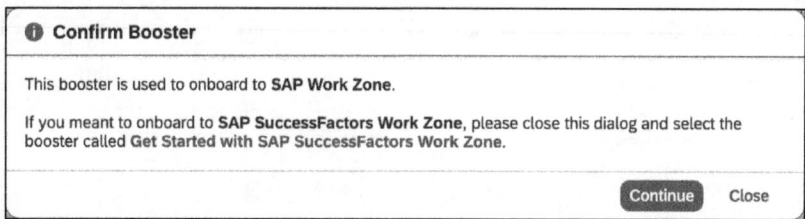

Figure 11.4 Confirm Start of Correct Booster

First, the requirements are checked as usual (see Figure 11.5). The user's authorizations are checked as well as whether the required entitlements are available in the global account. You cannot make any adjustments at this point. If an error occurs, an error message is displayed. If both checks are successful, you can proceed to the next step of the configuration by clicking **Next**.

11 Installing and Configuring SAP Build Work Zone

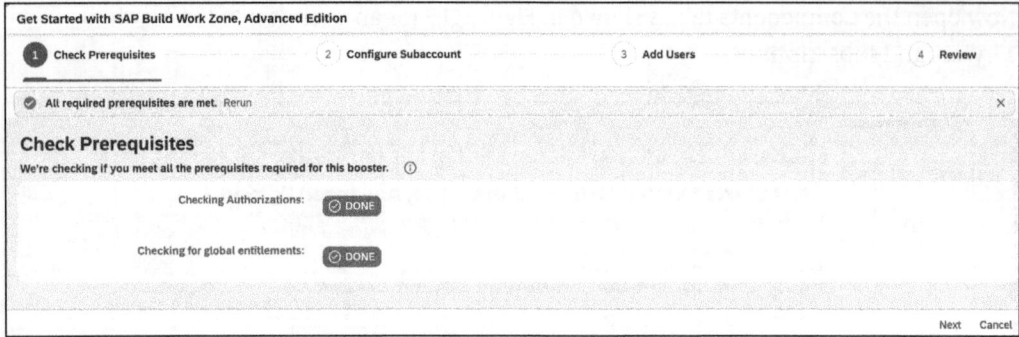

Figure 11.5 Booster Prerequisites Check

In the **Configure Subaccount** step, you can see all the services and subscriptions required for SAP Build Work Zone, advanced edition in the **Entitlements** area (see Figure 11.6). This also contains all optional services and subscriptions.

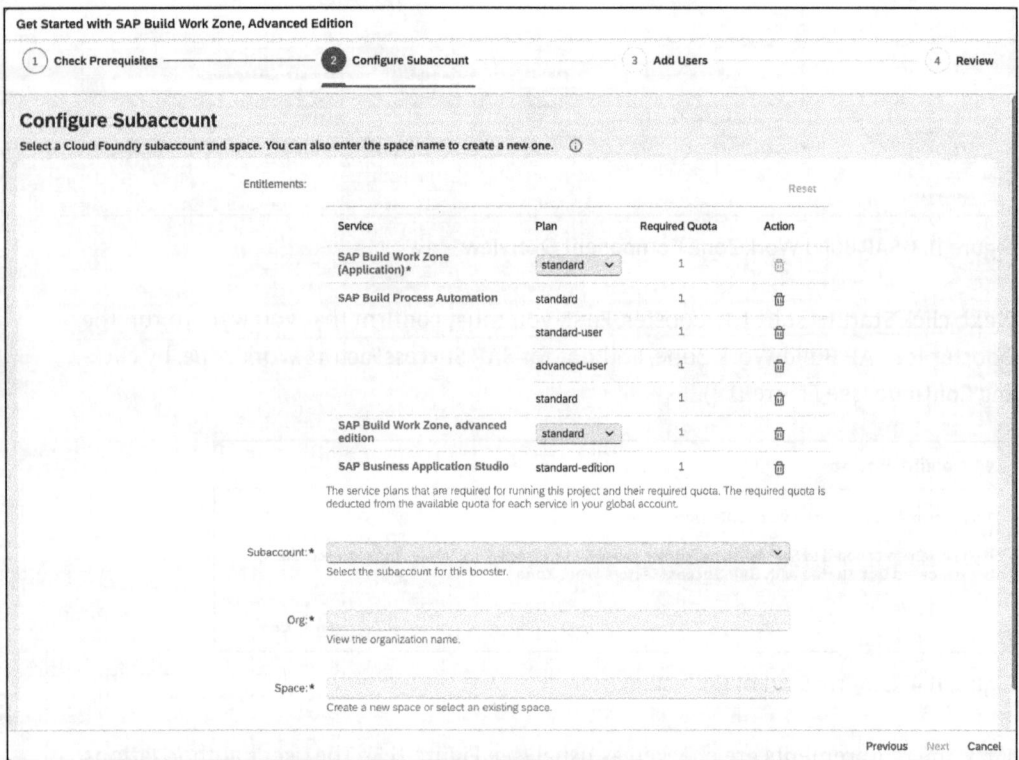

Figure 11.6 Booster Subaccount Configuration

As shown in Figure 11.7, you can delete the optional components from the list of **Entitlements**. Then select the previously created subaccount in the **Subaccount** field. The **Org** and **Space** fields are then filled in automatically. You can adjust these if necessary. When you're ready, click **Next**.

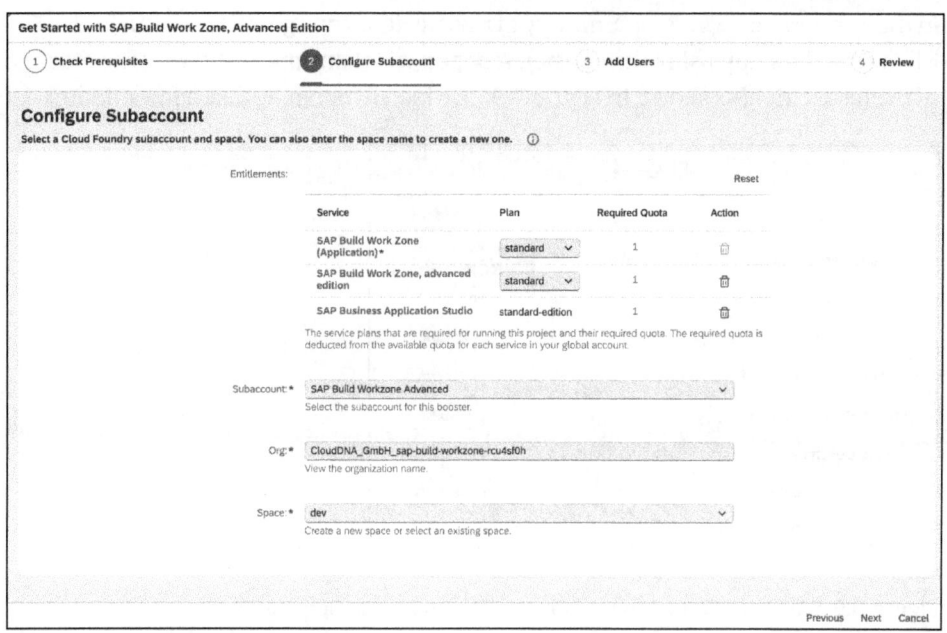

Figure 11.7 Booster Subaccount Selection

You must now carry out the initial assignment of users in the **Add Users** step (see Figure 11.8). You must select which identity provider is to be used for the platform users and application users. In the **Administrators** field, you can also maintain the users who are initially to be assigned administrator authorizations. In the **Developers** field, you can maintain the users who are to be initially authorized as developers. Then click **Next**.

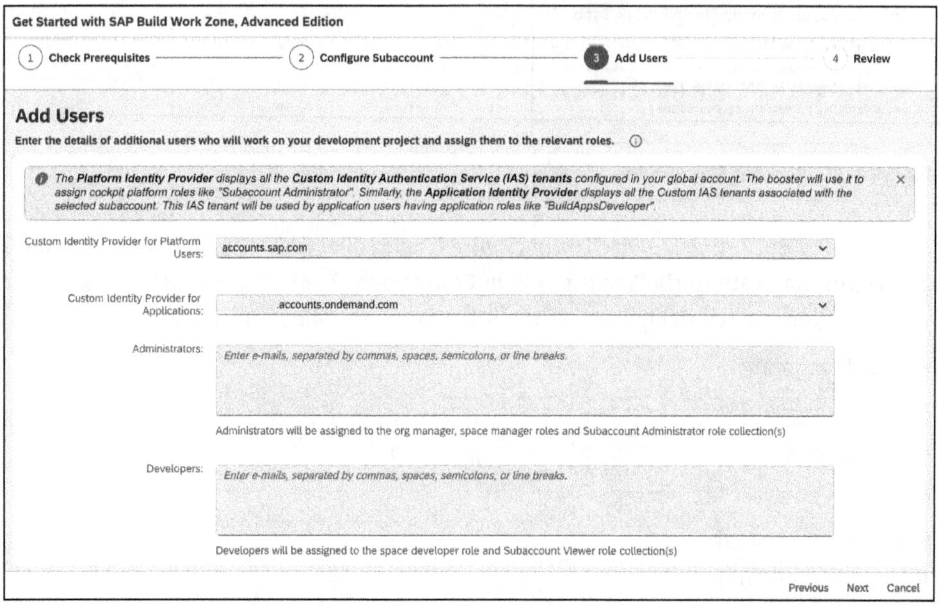

Figure 11.8 Booster User Management

In the last step, you will see a summary of the previous configuration as usual. Click **Finish** to start the setup using the booster. A popup will then appear showing the progress of the installation, as shown in Figure 11.9. The installation may take a few minutes.

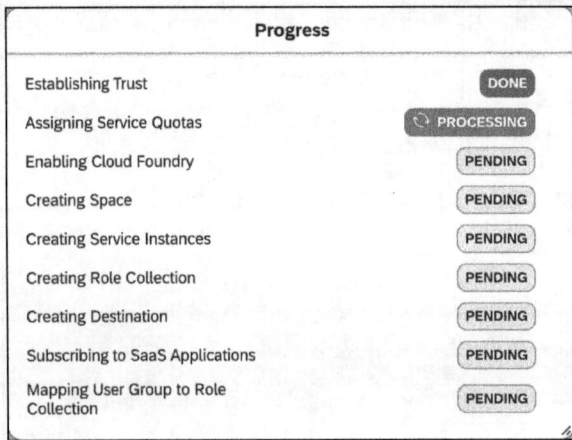

Figure 11.9 SAP Build Work Zone, Advanced Edition Installation Status

After the installation and configuration have been successfully completed, you will see a success message as shown in Figure 11.10. Here, click **Navigate to Subaccount** to jump to the SAP BTP cockpit of the subaccounts.

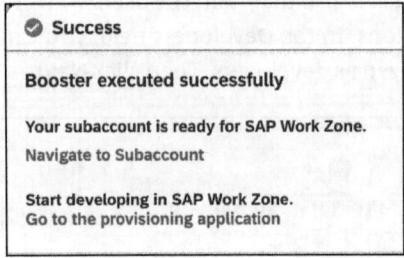

Figure 11.10 Booster Success Message

In the next step, it makes sense to check which role collections the booster has created. To do this, navigate to the **Security • Role Collections** area in the SAP BTP cockpit side menu. As you can see in Figure 11.11, the following role collections have been created:

- Workflow_Admin
- Workflow_End_User
- Workzone_Admin
- Workzone_Advanced_Theming
- Workzone_Area_Admin
- Workzone_End_User
- Workzone_XSUAA_Access

11.1 Installation

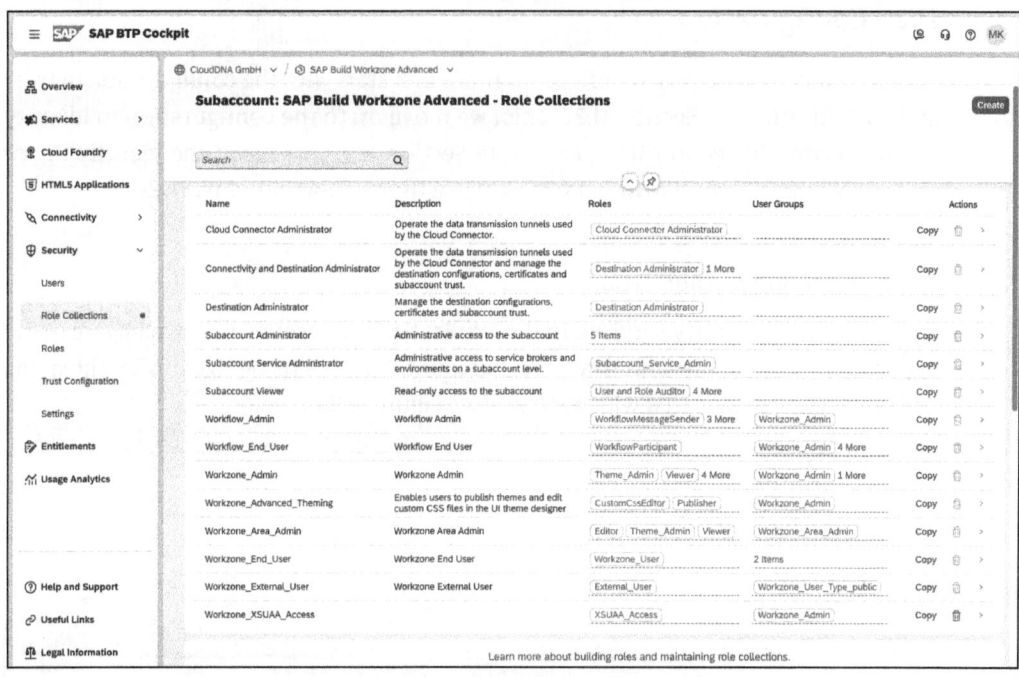

Figure 11.11 SAP Build Work Zone, Advanced Edition Role Collections

Now navigate to the **Security • Trust Configuration** area in the side menu (see Figure 11.12). There you will see that the trust relationship with Identity Authentication has been successfully established using OpenID Connect as the **Custom Identity Provider for Applications**.

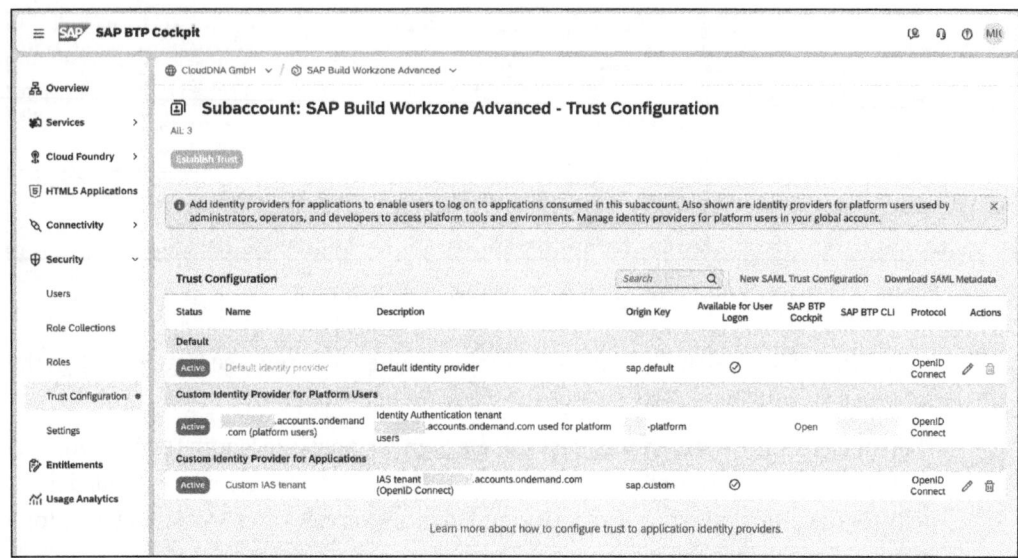

Figure 11.12 Trust Configuration

11.2 Configuration

Now it's time to move on to configuration. We first start with the configuration of trust and authorizations in Section 11.2.1. Then we move on to the configuration in Identity Authentication in Section 11.2.2. Finally, in Section 11.2.3, we cover the Identity Provisioning configuration that allows you to create users in SAP Build Work Zone.

11.2.1 Trust and Authorizations

Open SAP Cloud Identity Services. In the **Applications & Resources** area, navigate to the application that corresponds to your subaccount. You can recognize this by the name of the subaccount prefaced with *XSUAA_* (see Figure 11.13).

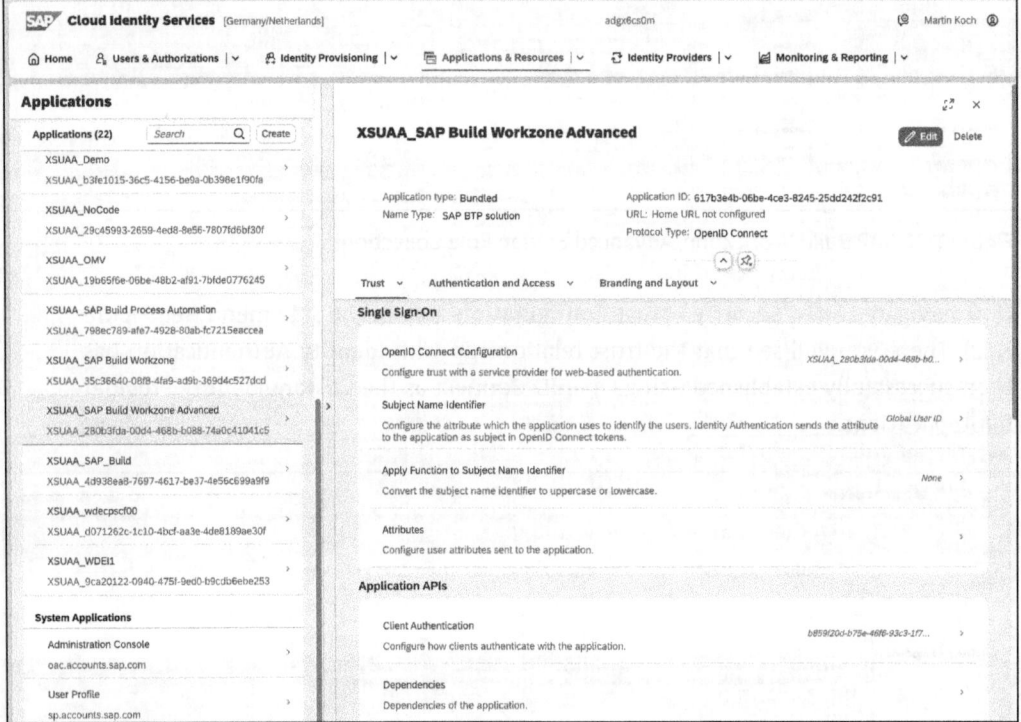

Figure 11.13 Identity Authentication Service Provider Adjustments

In the **Trust • Single Sign-On** area, go to **Subject Name Identifier** (see Figure 11.14). In the **Select a Basic Attribute** field, change the value to **Global User ID**.

Then open the attributes in the **Trust • Single Sign-On** area (see Figure 11.15). As you can see from the figure, six attributes are currently available. These must be extended manually for the interaction between SAP Cloud Identity Services and SAP Build Work Zone to function correctly.

11.2 Configuration

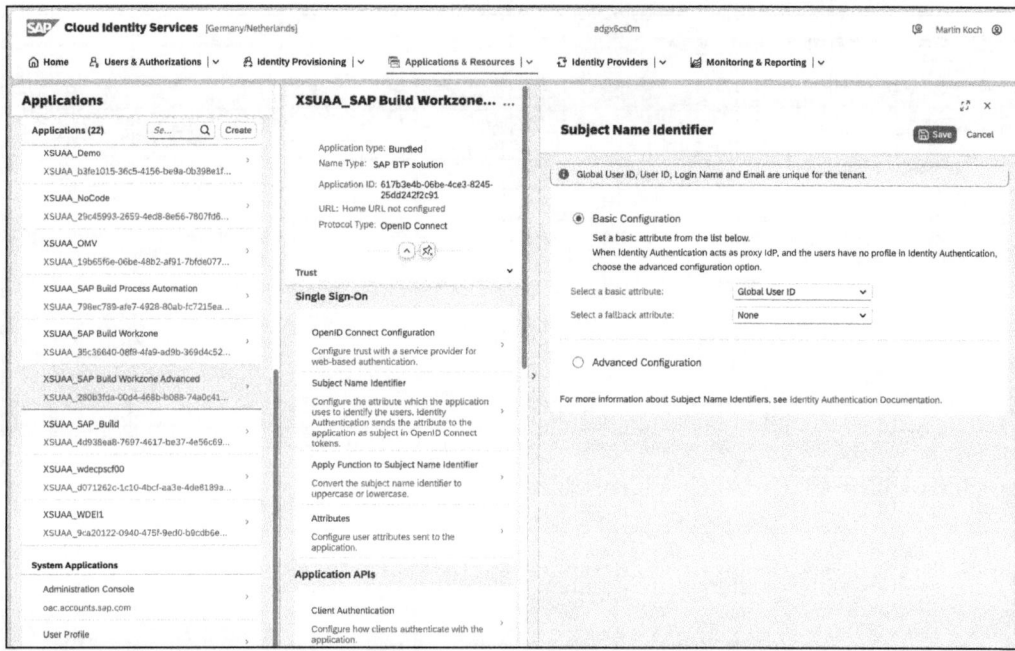

Figure 11.14 Adjust Subject Name Identifier

Figure 11.15 Insert Attributes

Click on **Add** as shown in Figure 11.16 to insert a new attribute in the table. Insert the attributes listed in Table 11.1, then click **Save**.

11 Installing and Configuring SAP Build Work Zone

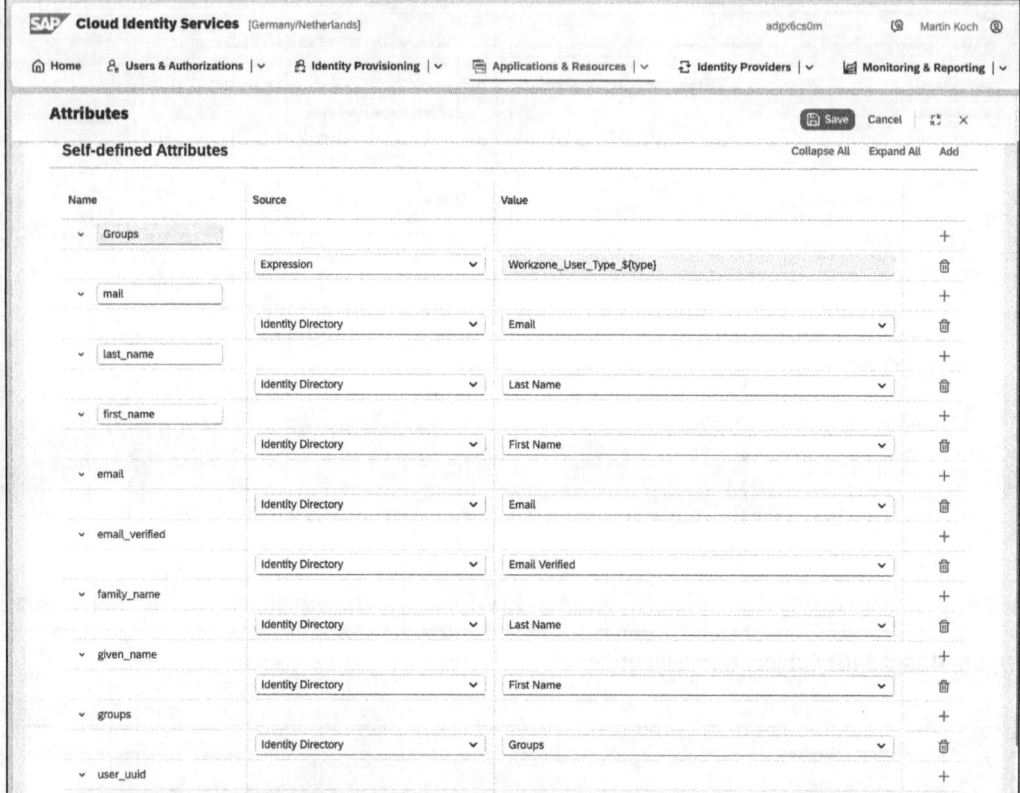

Figure 11.16 Insert Attributes

Name	Source	Value
"first_name"	Identity Directory	First Name
"last_name"	Identity Directory	Last Name
"mail"	Identity Directory	Email
"Groups"	Expression	"Workzone_User_Type_${type}"

Table 11.1 User Attribute Configuration

Now jump to the **Users & Authorizations • Groups** area as shown in Figure 11.17. There you must manually create new groups that are mapped to the role collections in the subaccount.

11.2 Configuration

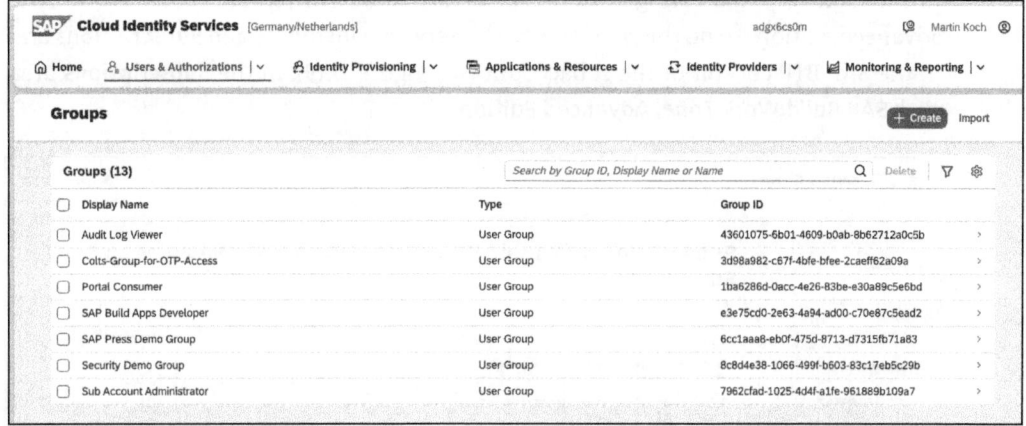

Figure 11.17 Create Groups

Click **Create** to create a new group (see Figure 11.18). Create the following groups in this way:

- `Workzone_Admin`
- `Workzone_Area_Admin`
- `Workzone_End_User`
- `Workzone_Page_Content_Admin`
- `Workzone_Support_Admin`
- `Workzone_User_Type_public`

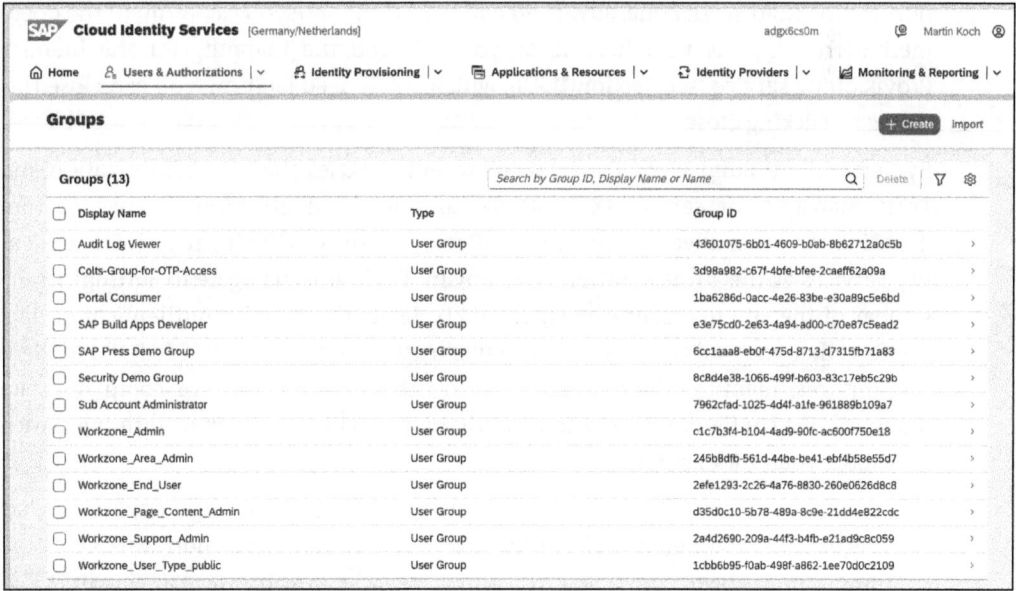

Figure 11.18 SAP Build Work Zone Groups

In the next step, the configuration can be carried out within SAP Build Work Zone, advanced edition. To do this, navigate to the **Services • Instances and Subscriptions** area in the SAP BTP cockpit of the subaccount (see Figure 11.19). In the **Subscriptions** area, click **SAP Build Work Zone, Advanced Edition**.

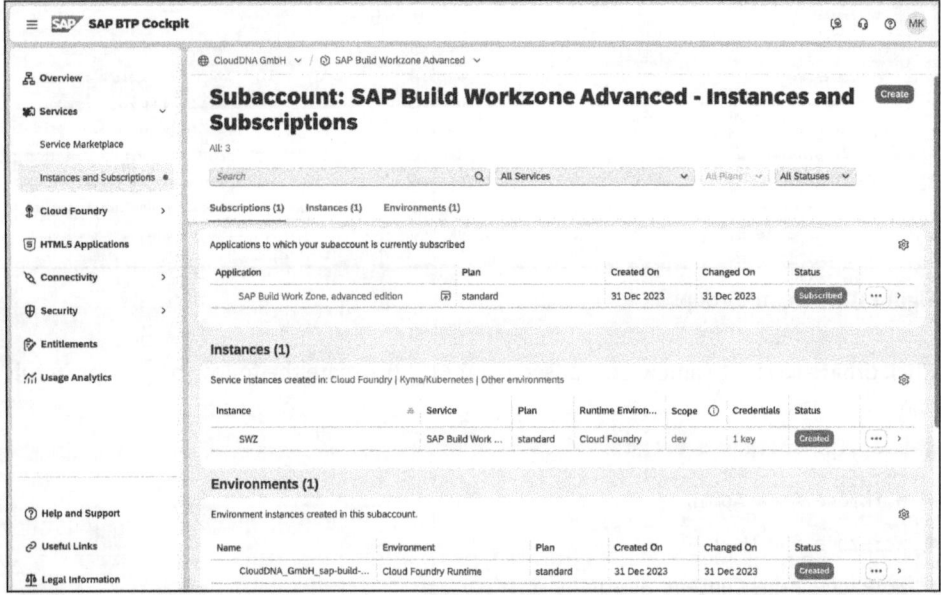

Figure 11.19 Open SAP Build Work Zone, Advanced Edition

As shown in Figure 11.20, you will now receive a note on the installation and configuration of SAP Build Work Zone, advanced edition. From there you can jump to the documentation, which is very helpful, as you will find the mapping for the Identity Provisioning service (see Section 11.2.3), which you'll need later on. You can close this dialog by clicking **Close**.

By default, SAP Build Work Zone, advanced edition uses the SAP Authorization and Trust Management service (XSUAA) for identity authentication. To use custom domains and other advanced features, SAP recommends switching to SAP Cloud Identity Services as the authentication mechanism. To do this, navigate to **Settings** in the sidebar of the site manager (see Figure 11.21). Select the **Identity Authentication** tab. Confirm that your subaccount has an active trust configuration with SAP Cloud Identity Services, Identity Authentication. To do this, activate the **I've completed the required trust configuration and confirm that I would like to continue with the switch** checkbox. Then click **Enable**.

This process can take up to 15 minutes. If you see an error message, experience has shown that it is advisable to refresh the browser window after a short wait. Once the configuration has been completed, you should see a success message as shown in Figure 11.22.

11.2 Configuration

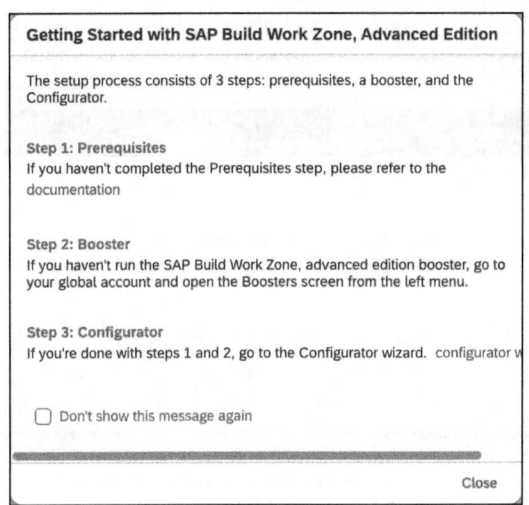

Figure 11.20 SAP Build Work Zone: Note on Initial Configuration

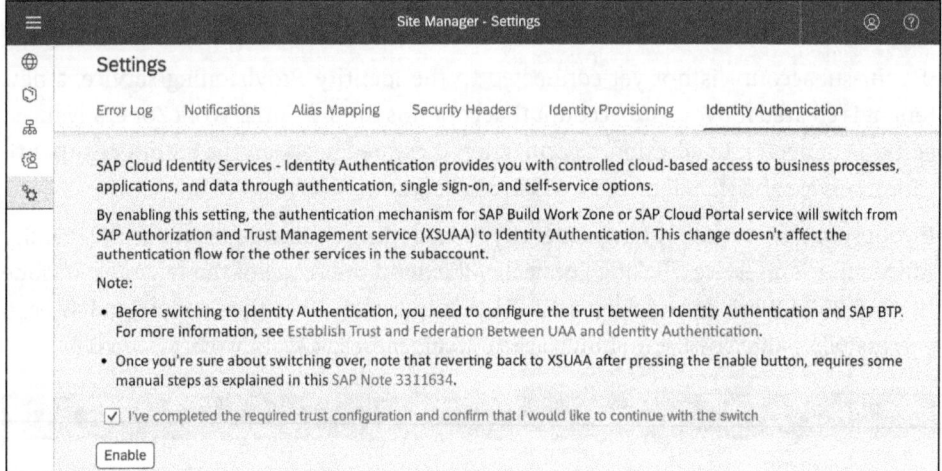

Figure 11.21 Identity Authentication Connection

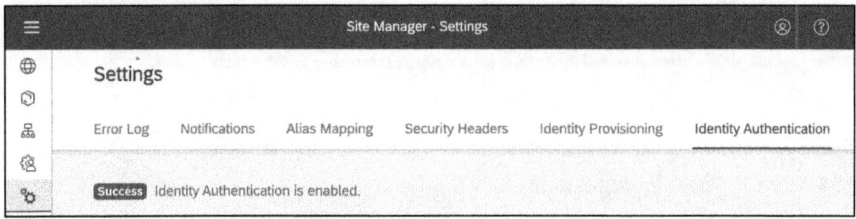

Figure 11.22 Identity Authentication Successful Connection

SAP Build Work Zone, advanced edition uses SAP Cloud Identity Services, Identity Authentication as the user management system and SAP Cloud Identity Services,

Identity Provisioning as the user provisioning system. Switch to the **Identity Provisioning** tab as shown in Figure 11.23 and click **Connect**.

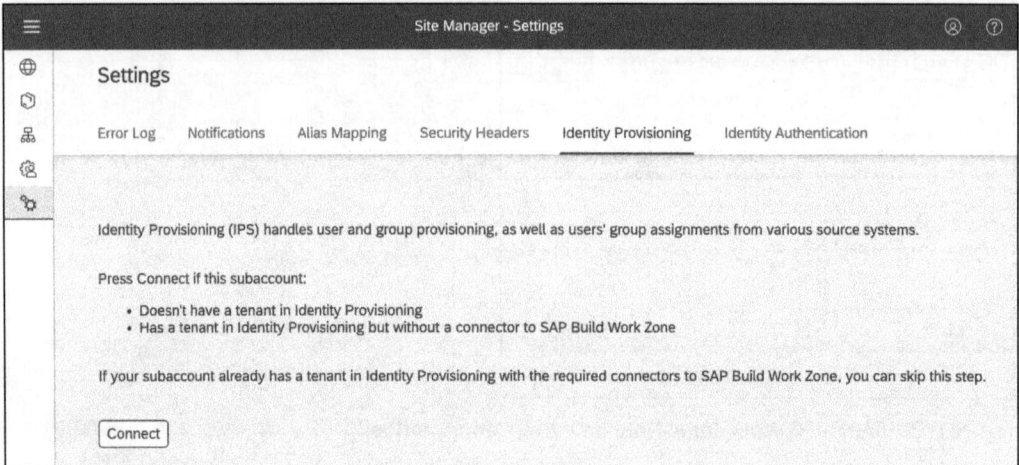

Figure 11.23 Establish Identity Provisioning Connection

If your subaccount is not yet connected to the Identity Provisioning service, a new tenant is created for your subaccount that contains the SAP Build Work Zone, advanced edition connector. In addition, a connection is created between the Identity Authentication service and the Identity Provisioning service.

If your subaccount already has an Identity Provisioning tenant connected to the Identity Authentication service, clicking **Connect** will extend the scope of the tenant to include the SAP Build Work Zone, advanced edition connectors. Once the connection has been successfully established, you should see a success message as shown in Figure 11.24.

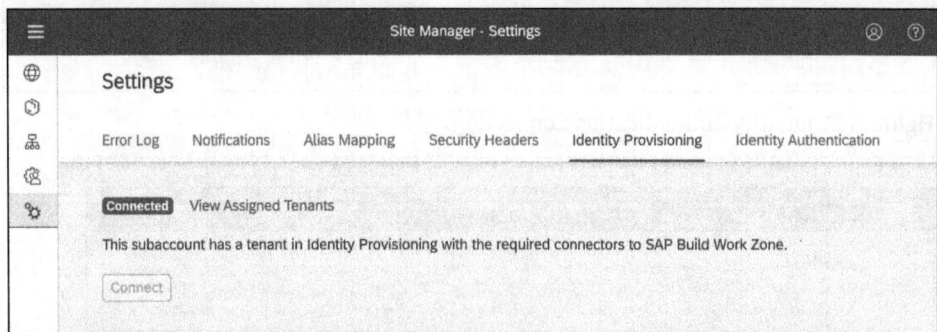

Figure 11.24 Identity Provisioning Successfully Connected

Once you have completed all the previous steps, follow the configuration wizard to complete the implementation process. You can access this via the configurator side menu (see Figure 11.25). In the first step, select the **I want to create a new service instance** option. Then click **Next**.

11.2 Configuration

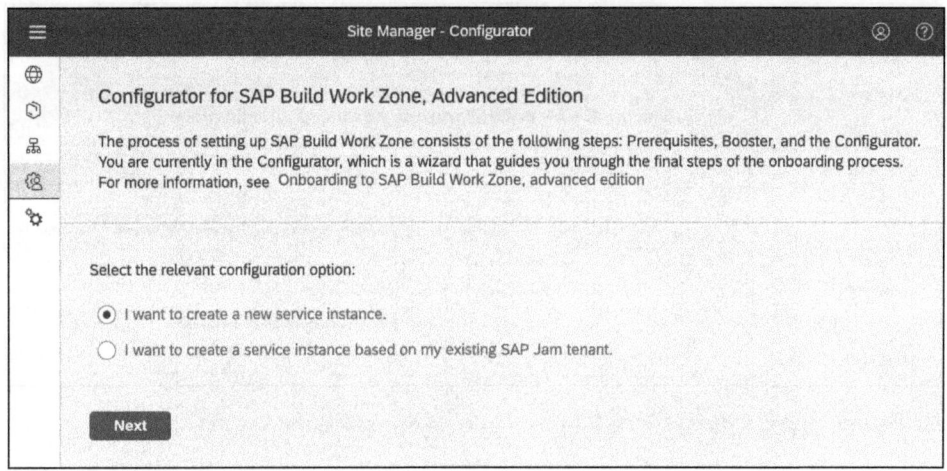

Figure 11.25 SAP Build Work Zone, Advanced Edition Configurator

You can now optionally create a destination that establishes the connection to SAP SuccessFactors (see Figure 11.26). This is particularly interesting if you want to use the HR business content packages. We will skip this step at this point, so click **Skip Step** to go to the next step.

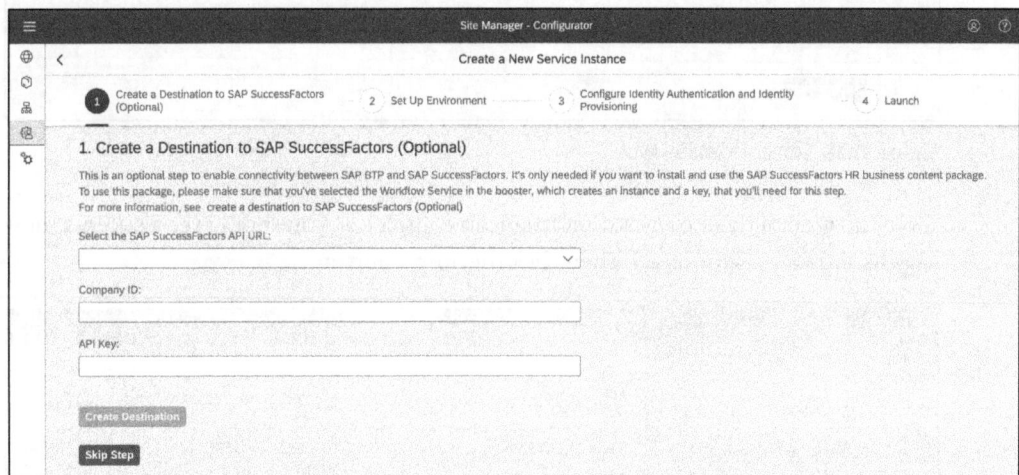

Figure 11.26 Create SAP SuccessFactors Destination

Before you start the setup, open the **Connectivity • Destinations** area in the SAP BTP cockpit of the subaccount and click the **Download Trust** button above the table of destinations (see Figure 11.27). If you do not perform this step, the next step in the configuration will run into an error.

Now select the **Default** option in the **Select Domain Type** area (see Figure 11.28). If you are using a custom domain, you must of course select the other option. Then click **Trigger Setup**.

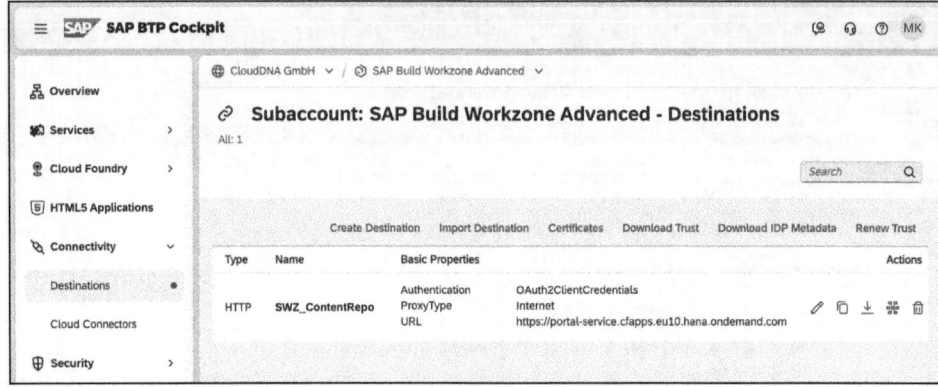

Figure 11.27 Download Trust

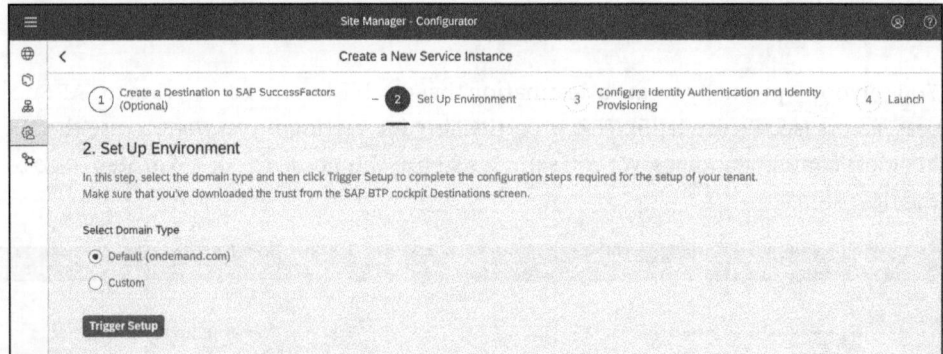

Figure 11.28 Select Domain Type

Once this step of the configuration has been completed, you should see a success message as shown in Figure 11.29. Then click the **Step 3** button.

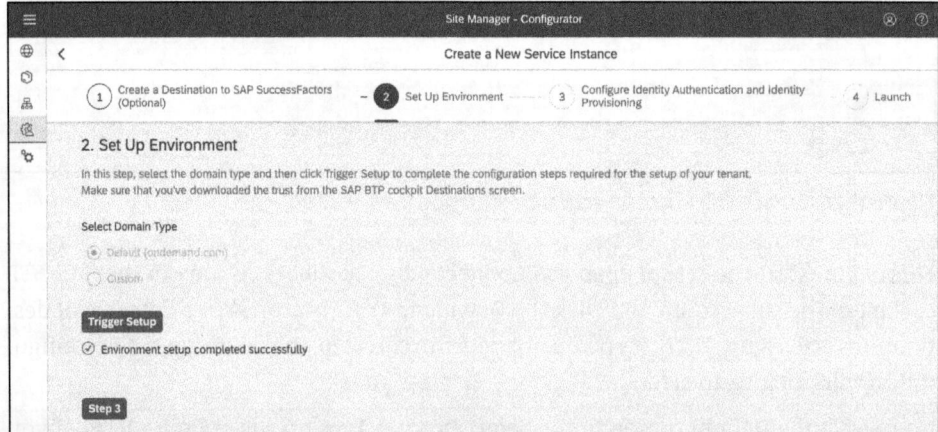

Figure 11.29 Environment Setup Success Message

11.2.2 Identity Authentication

You must now manually configure the Identity Authentication service and the Identity Provisioning service to activate user authentication and user provisioning. In SAP Cloud Identity Services, go to **Applications & Resources • Applications** and click **Create** to create a new application (see Figure 11.30).

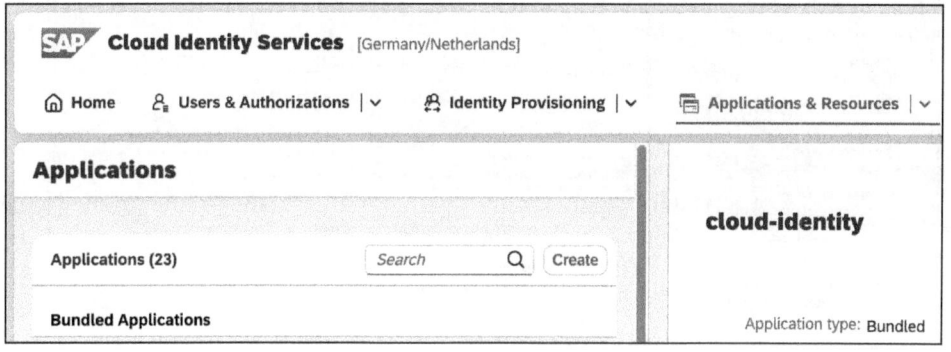

Figure 11.30 Create Application

Enter a meaningful **Display Name** as shown in Figure 11.31, such as "SAP Build Workzone Advanced". Select the **SAP BTP Solution** option as the **Type** and the **SAML 2.0** option for the **Protocol Type**. Then click **+ Create**.

Figure 11.31 Configure Application

Now open the **SAML 2.0** configuration of the application you have created, as shown in Figure 11.32.

11   Installing and Configuring SAP Build Work Zone

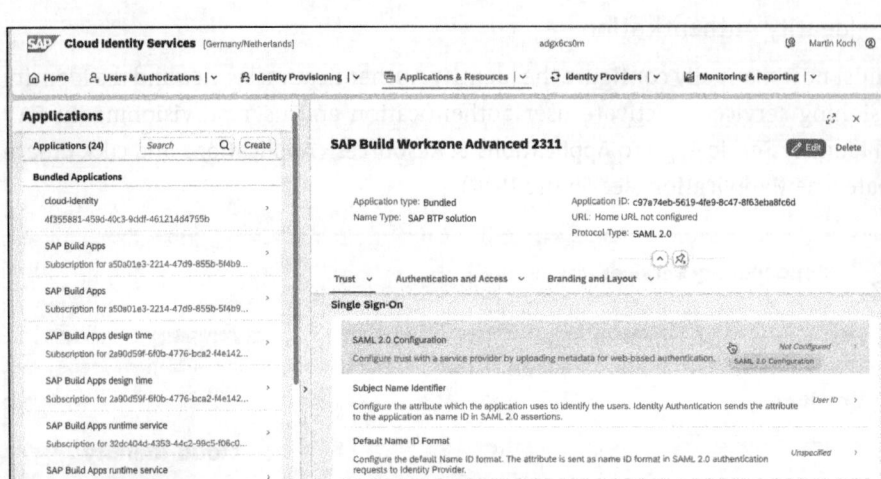

Figure 11.32 Service Provider Details

Download the SAML 2.0 metadata of the service provider from the configurator by clicking the **Download Metadata** link (Figure 11.48 shows this later in Section 11.2.3). Insert the metadata of the service provider by clicking **Browse** (see Figure 11.33). Save your changes by clicking **Save**.

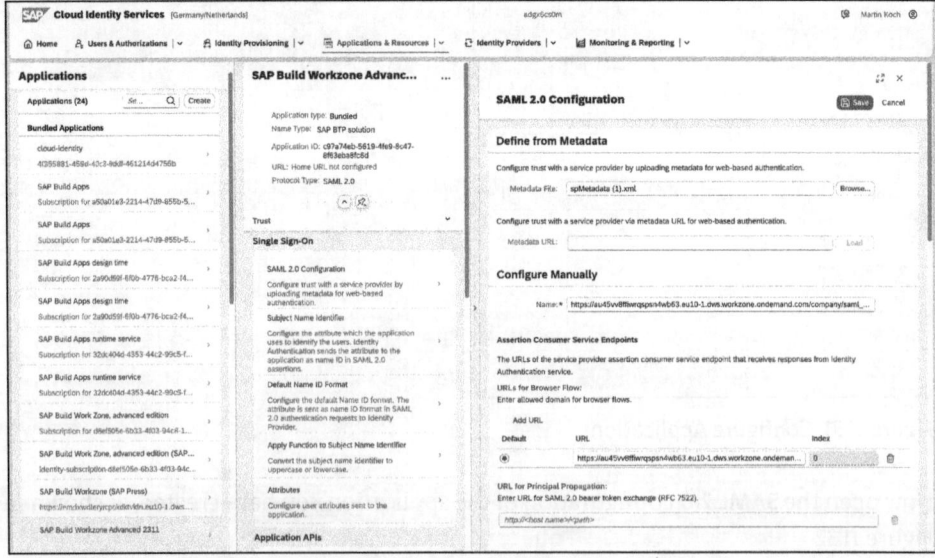

Figure 11.33 Upload SAML 2.0 Service Provider Metadata

11.2 Configuration

You must now open the **Subject Name Identifier** editor of the application, as shown in Figure 11.34. Change the **Basic Configuration** attribute from **User ID** to **Global User ID**. Then click **Save**.

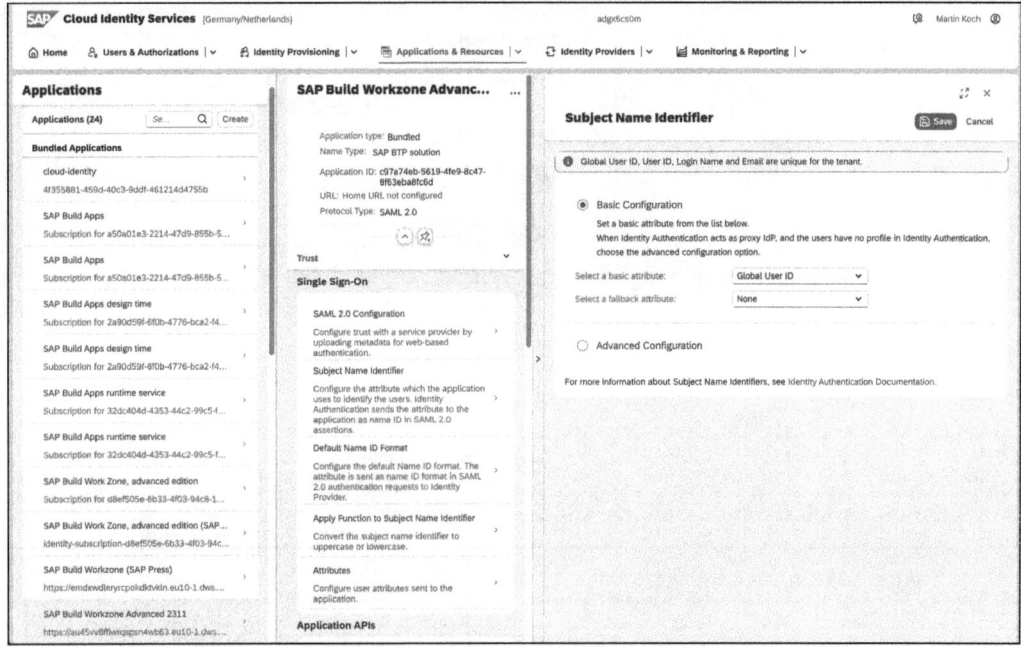

Figure 11.34 Adjust Subject Name Identifier

11.2.3 Identity Provisioning

You must now create a system as an administrator that is required for the Identity Provisioning service. To do this, navigate to the **User & Authorizations** • **Administrators** area. Click **Add**, and create an administrator of the **System** type as shown in Figure 11.35. In the **Configure Authorizations** area, assign the authorizations shown in Figure 11.35 and then click **Save**.

Now click **Secrets** to access the administration of the client secrets (see Figure 11.36). In the **Secrets** area, click **+ Add**.

You can now enter a meaningful description for the secret in the **Description** field, as shown in Figure 11.37. Then click **Save**.

A client ID and a client secret are now created. You can only view the client secret at this time (see Figure 11.38). Make a note of the client ID and the client secret for subsequent use.

355

11 Installing and Configuring SAP Build Work Zone

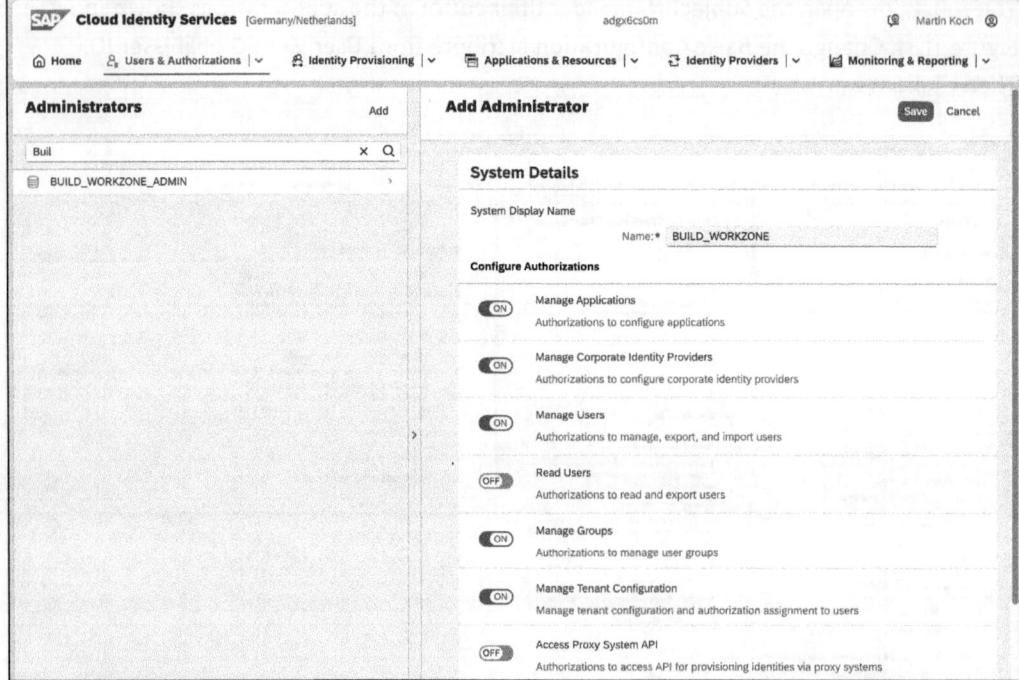

Figure 11.35 Create Administrator

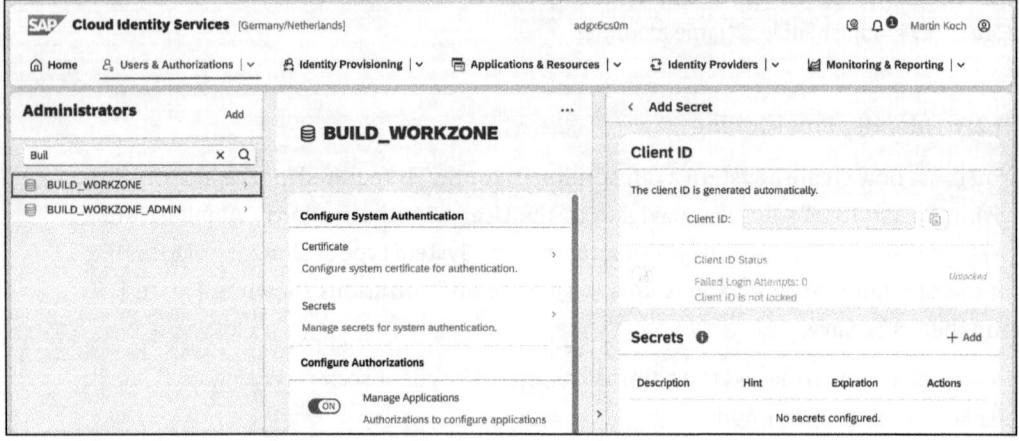

Figure 11.36 Create Secret

Figure 11.37 Secret Configuration

11.2 Configuration

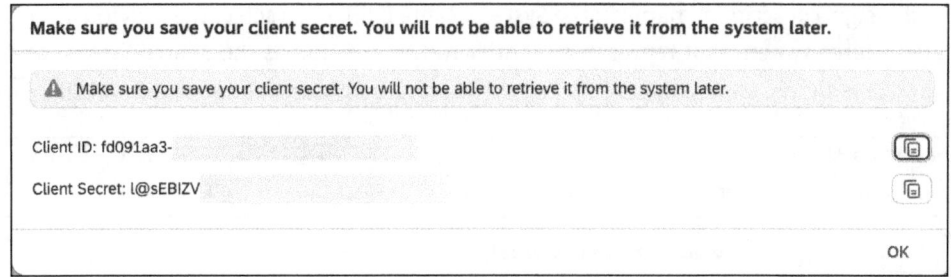

Figure 11.38 Copy Client Secret

You can now continue with the configuration of Identity Provisioning. To do this, jump to the **Identity Provisioning • Source Systems** area (see Figure 11.39). Then click **+ Add** in the footer area of the table. Assign a descriptive name and select the **Identity Authentication** type. Use the suggested default values for the properties. Enter the previously created client ID as the value for the **User** property and the corresponding client secret for **Password**.

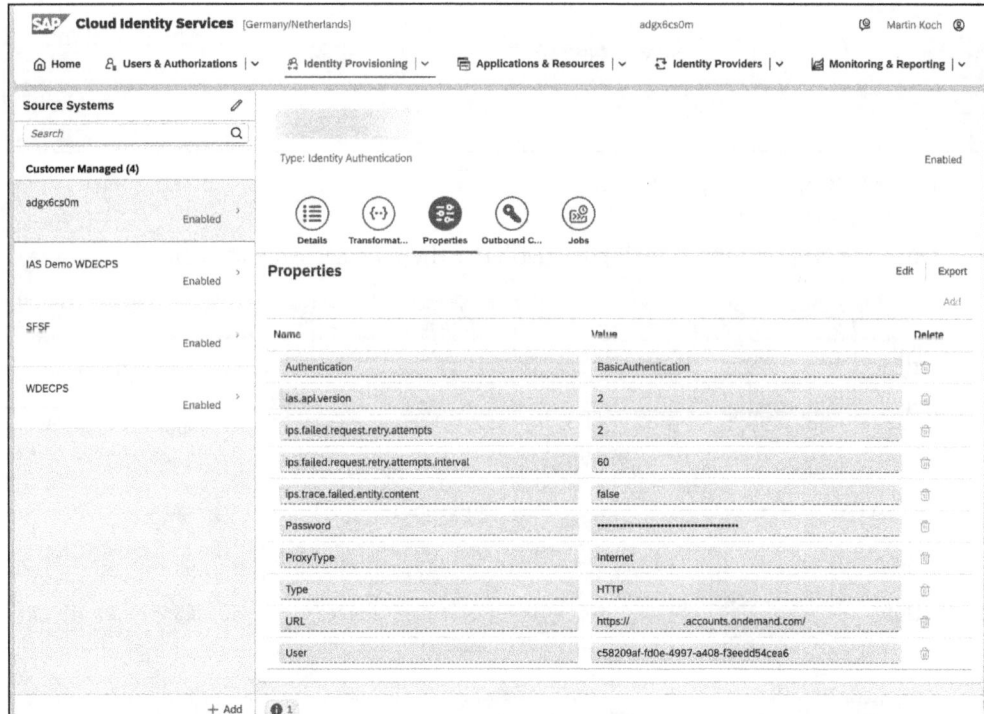

Figure 11.39 Create Identity Provisioning Source System

In the next step, you must create an SAP Build Work Zone, advanced edition target system. To do this, navigate to the **Identity Provisioning • Target Systems** area, as shown in Figure 11.40, and click **+ Add**. Select **SAP Build Work Zone, Advanced Edition** as the **Type**

357

11 Installing and Configuring SAP Build Work Zone

and assign a descriptive name in the **System Name** field. Then select the previously created source system that represents Identity Authentication as the **Source System** and click **Save**.

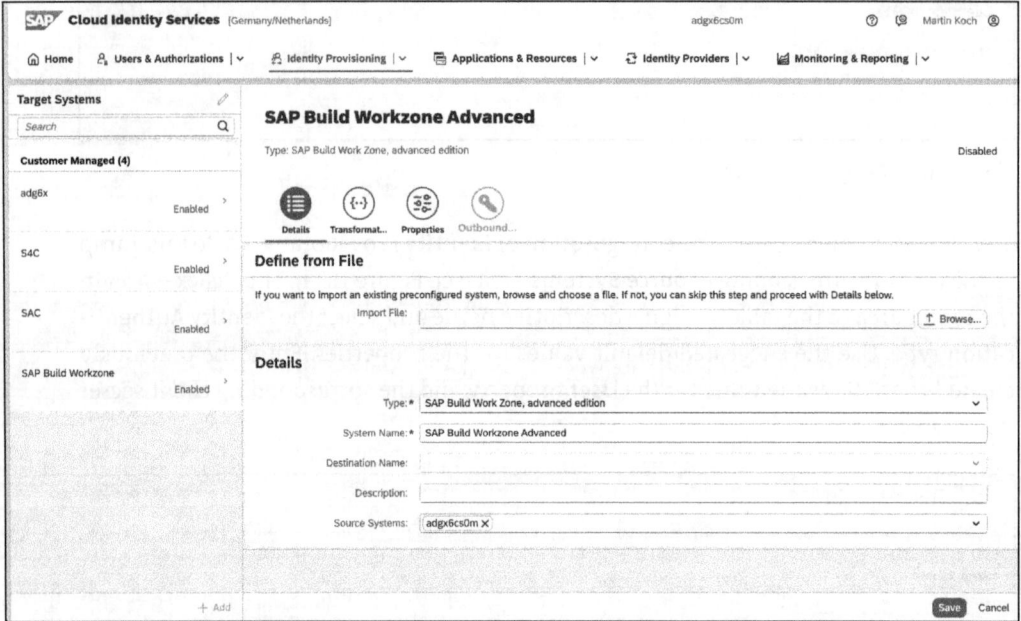

Figure 11.40 Create Target System

Open the **Properties** tab (see Figure 11.41). Use the settings listed in Table 11.2.

Identity Provisioning Property Name	Value
Type	"HTTP"
URL	Copy this value from the wizard, in the **Integration URL** field
ProxyType	"Internet"
Authentication	"BasicAuthentication"
User	Copy this value from the wizard, in the **OAuth Client Key** field
Password	Copy this value from the wizard, in the **OAuth Client Secret** field
OAuth2TokenServiceURL	Copy this value from the wizard, in the **Token Service URL** field

Table 11.2 Target System Configuration Properties

Identity Provisioning Property Name	Value
ips.failed.request.retry.attempts	"3"
ips.failed.request.retry.attempts.interval	"60"
ips.delete.existedbefore.entities	"true"
ips.trace.failed.entity.content	"true"

Table 11.2 Target System Configuration Properties (Cont.)

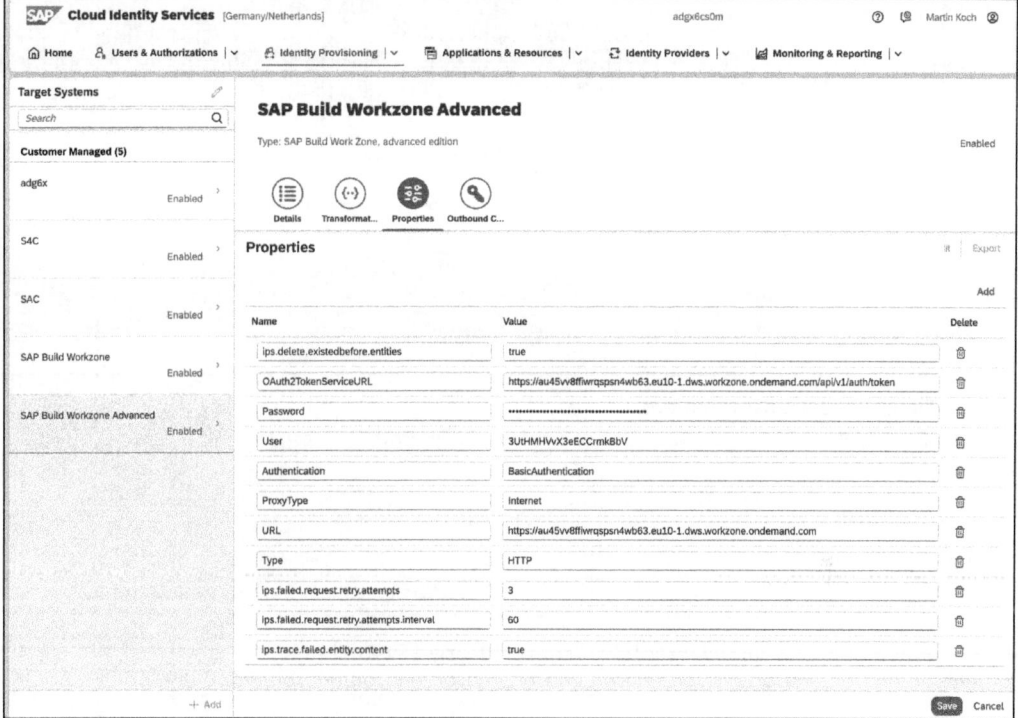

Figure 11.41 Configure Target System Properties

You must now adjust the standard mappings. To do this, open the **Transformations** tab, as shown in Figure 11.42, and switch to the code view. Copy the transformation shown in Listing 11.1 into the editor. You can also find the current transformation in the SAP Build Work Zone, advanced edition documentation.

11   Installing and Configuring SAP Build Work Zone

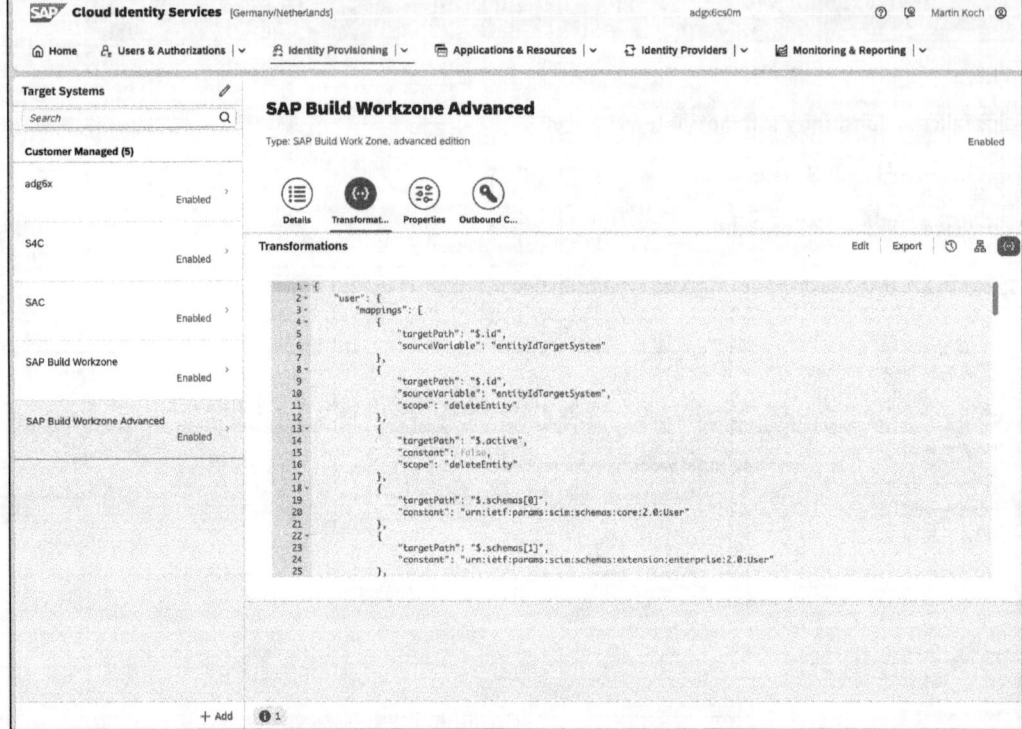

Figure 11.42 Transformations

```
{
    "user":
    {
        "mappings": [
        {
            "sourceVariable": "entityIdTargetSystem",
            "targetPath": "$.id"
        },
        {
            "sourceVariable": "entityIdTargetSystem",
            "targetPath": "$.id",
            "scope": "deleteEntity"
        },
        {
            "constant": false,
            "targetPath": "$.active",
            "scope": "deleteEntity"
        },
        {
            "sourcePath": "$.userName",
```

```
            "optional": true,
            "targetPath": "$.userName"
        },
        {
            "sourcePath": "$[
'urn:ietf:params:scim:schemas:extension:sap:2.0:User']['userUuid']",
            "targetPath": "$.userName"
        },
        {
            "sourcePath": "$[
'urn:ietf:params:scim:schemas:extension:sap:2.0:User']['userId']",
            "targetPath": "$.externalId"
        },
        {
            "sourcePath": "$[
'urn:ietf:params:scim:schemas:extension:sap:2.0:User']['userUuid']",
            "targetPath": "$[
'urn:ietf:params:scim:schemas:extension:sap:2.0:User']['userUuid']"
        },
        {
            "constant": "employee",
            "targetPath": "$.userType"
        },
        {
            "condition": "$.groups[?(@.display == 'Workzone_User_Type_public')]
EMPTY false",
            "constant": "public",
            "targetPath": "$.userType"
        },
        {
            "sourcePath": "$.name.givenName",
            "targetPath": "$.name.givenName",
            "optional": true
        },
        {
            "sourcePath": "$.name.familyName",
            "targetPath": "$.name.familyName",
            "optional": true
        },
        {
            "sourcePath": "$.name.honorificPrefix",
            "targetPath": "$.name.honorificPrefix",
            "optional": true
        },
```

```
        {
            "sourcePath": "$.timezone",
            "targetPath": "$.timezone",
            "optional": true
        },
        {
            "condition": "($.locale EMPTY false) && ($.addresses[?(@.type ==
'work')].country EMPTY false)",
            "sourcePath": "$.locale",
            "targetPath": "$.locale",
            "functions": [
            {
                "function": "toLowerCaseString"
            },
            {
                "function": "concatString",
                "suffix": "_"
            },
            {
                "function": "concatString",
                "suffix": "$.addresses[?(@.type == 'work')].country"
            }]
        },
        {
            "sourcePath": "$.emails[0].value",
            "targetPath": "$.emails[0].value"
        },
        {
            "condition": "$.emails[0].length() > 0",
            "constant": true,
            "targetPath": "$.emails[0].primary"
        },
        {
            "sourcePath": "$.phoneNumbers",
            "targetPath": "$.phoneNumbers",
            "preserveArrayWithSingleElement": true,
            "optional": true
        },
        {
            "sourcePath": "$.addresses",
            "targetPath": "$.addresses",
            "preserveArrayWithSingleElement": true,
            "optional": true
        },
```

```
        {
            "condition": "$.[
'urn:sap:cloud:scim:schemas:extension:custom:2.0:User']['attributes'][?(@.name =
= 'customAttribute1')]['value'] EMPTY false",
            "sourcePath": "$[
'urn:sap:cloud:scim:schemas:extension:custom:2.0:User']['attributes'][?(@.name =
= 'customAttribute1')]['value']",
            "targetPath": "$[
'urn:sap:cloud:scim:schemas:extension:custom:2.0:JamCustomUser'][
'customAttribute1']['value']"
        },
        {
            "condition": "$.[
'urn:sap:cloud:scim:schemas:extension:custom:2.0:User']['attributes'][?(@.name =
= 'customAttribute1')]['value'] EMPTY false",
            "constant": "Custom 1",
            "targetPath": "$[
'urn:sap:cloud:scim:schemas:extension:custom:2.0:JamCustomUser'][
'customAttribute1']['name']"
        },
        {
            "condition": "$.[
'urn:sap:cloud:scim:schemas:extension:custom:2.0:User']['attributes'][?(@.name =
= 'customAttribute2')]['value'] EMPTY false",
            "sourcePath": "$[
'urn:sap:cloud:scim:schemas:extension:custom:2.0:User']['attributes'][?(@.name =
= 'customAttribute2')]['value']",
            "targetPath": "$[
'urn:sap:cloud:scim:schemas:extension:custom:2.0:JamCustomUser'][
'customAttribute2']['value']"
        },
        {
            "condition": "$.[
'urn:sap:cloud:scim:schemas:extension:custom:2.0:User']['attributes'][?(@.name =
= 'customAttribute2')]['value'] EMPTY false",
            "constant": "Custom 2",
            "targetPath": "$[
'urn:sap:cloud:scim:schemas:extension:custom:2.0:JamCustomUser'][
'customAttribute2']['name']"
        },
        {
            "condition": "$.[
'urn:sap:cloud:scim:schemas:extension:custom:2.0:User']['attributes'][?(@.name =
= 'customAttribute3')]['value'] EMPTY false",
```

```
            "sourcePath": "$[
'urn:sap:cloud:scim:schemas:extension:custom:2.0:User']['attributes'][?(@.name =
= 'customAttribute3')]['value']",
            "targetPath": "$[
'urn:sap:cloud:scim:schemas:extension:custom:2.0:JamCustomUser'][
'customAttribute3']['value']"
        },
        {
            "condition": "$.[
'urn:sap:cloud:scim:schemas:extension:custom:2.0:User']['attributes'][?(@.name =
= 'customAttribute3')]['value'] EMPTY false",
            "constant": "Custom 3",
            "targetPath": "$[
'urn:sap:cloud:scim:schemas:extension:custom:2.0:JamCustomUser'][
'customAttribute3']['name']"
        },
        {
            "condition": "$.[
'urn:sap:cloud:scim:schemas:extension:custom:2.0:User']['attributes'][?(@.name =
= 'customAttribute4')]['value'] EMPTY false",
            "sourcePath": "$[
'urn:sap:cloud:scim:schemas:extension:custom:2.0:User']['attributes'][?(@.name =
= 'customAttribute4')]['value']",
            "targetPath": "$[
'urn:sap:cloud:scim:schemas:extension:custom:2.0:JamCustomUser'][
'customAttribute4']['value']"
        },
        {
            "condition": "$.[
'urn:sap:cloud:scim:schemas:extension:custom:2.0:User']['attributes'][?(@.name =
= 'customAttribute4')]['value'] EMPTY false",
            "constant": "Custom 4",
            "targetPath": "$[
'urn:sap:cloud:scim:schemas:extension:custom:2.0:JamCustomUser'][
'customAttribute4']['name']"
        },
        {
            "condition": "$.[
'urn:sap:cloud:scim:schemas:extension:custom:2.0:User']['attributes'][?(@.name =
= 'customAttribute5')]['value'] EMPTY false",
            "sourcePath": "$[
'urn:sap:cloud:scim:schemas:extension:custom:2.0:User']['attributes'][?(@.name =
= 'customAttribute5')]['value']",
            "targetPath": "$[
```

```
'urn:sap:cloud:scim:schemas:extension:custom:2.0:JamCustomUser'][
'customAttribute5']['value']"
        },
        {
            "condition": "$.[
'urn:sap:cloud:scim:schemas:extension:custom:2.0:User']['attributes'][?(@.name =
= 'customAttribute5')]['value'] EMPTY false",
            "constant": "Custom 5",
            "targetPath": "$[
'urn:sap:cloud:scim:schemas:extension:custom:2.0:JamCustomUser'][
'customAttribute5']['name']"
        },
        {
            "sourcePath": "$.title",
            "optional": true,
            "targetPath": "$.title"
        },
        {
            "condition": "$.groups[?(@.display == 'Workzone_Admin')] EMPTY
false",
            "constant": "Administrator",
            "targetPath": "$.roles[0].value"
        },
        {
            "condition": "$.addresses[?(@.type == 'work')].country EMPTY false",
            "constant": true,
            "targetPath": "$.addresses[1].primary"
        },
        {
            "type": "remove",
            "targetPath": "$.groups"
        },
        {
            "sourcePath": "$[
'urn:ietf:params:scim:schemas:extension:enterprise:2.0:User']['department']",
            "targetPath": "$[
'urn:ietf:params:scim:schemas:extension:enterprise:2.0:User']['department']",
            "optional": true
        },
        {
            "sourcePath": "$[
'urn:ietf:params:scim:schemas:extension:enterprise:2.0:User']['organization']",
            "targetPath": "$[
'urn:ietf:params:scim:schemas:extension:enterprise:2.0:User']['organization']",
```

```
            "optional": true
        },
        {
            "sourcePath": "$[
'urn:ietf:params:scim:schemas:extension:enterprise:2.0:User']['costCenter']",
            "targetPath": "$[
'urn:ietf:params:scim:schemas:extension:enterprise:2.0:User']['costCenter']",
            "optional": true
        },
        {
            "sourcePath": "$[
'urn:ietf:params:scim:schemas:extension:enterprise:2.0:User'][
'employeeNumber']",
            "targetPath": "$[
'urn:ietf:params:scim:schemas:extension:enterprise:2.0:User'][
'employeeNumber']",
            "optional": true
        },
        {
            "sourcePath": "$[
'urn:ietf:params:scim:schemas:extension:enterprise:2.0:User']['division']",
            "targetPath": "$[
'urn:ietf:params:scim:schemas:extension:enterprise:2.0:User']['division']",
            "optional": true
        },
        {
            "sourcePath": "$[
'urn:ietf:params:scim:schemas:extension:enterprise:2.0:User']['manager'][
'value']",
            "targetPath": "$[
'urn:ietf:params:scim:schemas:extension:enterprise:2.0:User']['manager'][
'value']",
            "optional": true,
            "functions": [
            {
                "function": "resolveEntityIds"
            }]
        },
        {
            "sourcePath": "$[
'urn:ietf:params:scim:schemas:extension:enterprise:2.0:User']['manager'][
'displayName']",
```

```
                "targetPath": "$[
'urn:ietf:params:scim:schemas:extension:enterprise:2.0:User']['manager'][
'displayName']",
                "optional": true
            }]
    },
    "group":
    {
        "ignore": true,
        "mappings": [
        {
            "sourceVariable": "entityIdTargetSystem",
            "targetPath": "$.id"
        },
        {
            "constant": "urn:ietf:params:scim:schemas:core:2.0:Group",
            "targetPath": "$.schemas[0]"
        },
        {
            "sourcePath": "$[
'urn:sap:cloud:scim:schemas:extension:custom:2.0:Group']['name']",
            "targetPath": "$.displayName"
        },
        {
            "sourcePath": "$.members[*].value",
            "preserveArrayWithSingleElement": true,
            "optional": true,
            "targetPath": "$.members[?(@.value)]",
            "functions": [
            {
                "type": "resolveEntityIds"
            }]
        }]
    }
}
```

Listing 11.1 Target System Mapping

Then save the transformation by clicking **Save** (see Figure 11.43). You do not have the option to test the transformation directly at this point. This is only possible when you carry out a provisioning run. You should therefore exercise particular caution when adapting mappings or attributes.

11 Installing and Configuring SAP Build Work Zone

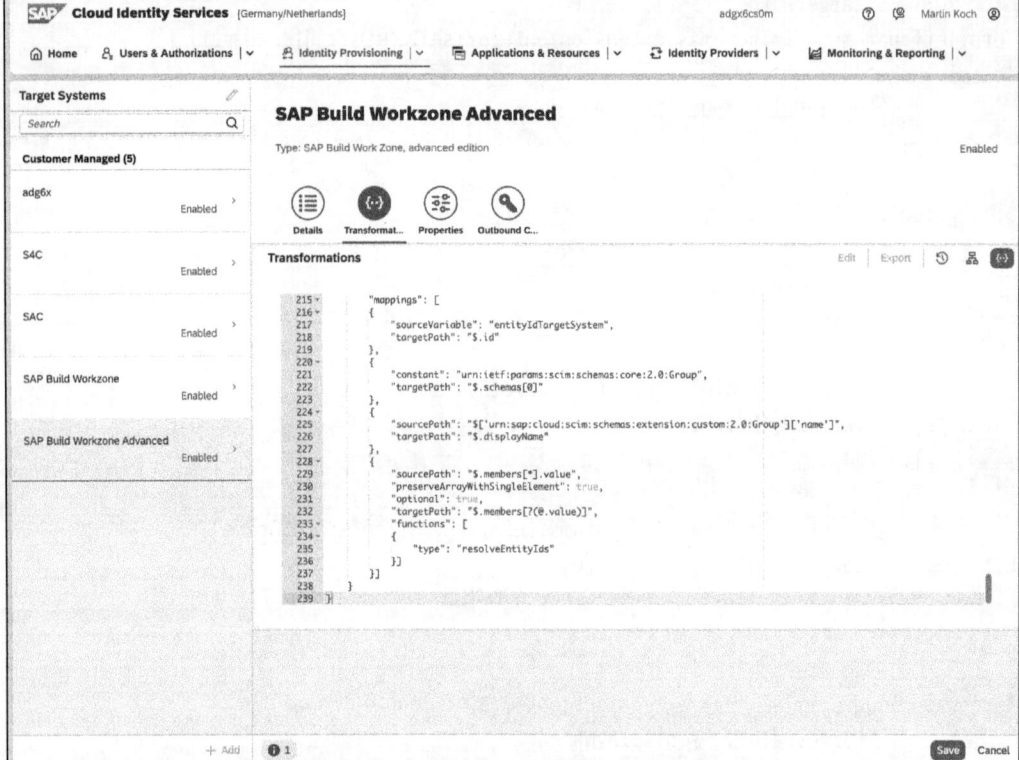

Figure 11.43 Save Transformation Configuration

When saving the transformation, you must maintain a description of the changes as shown in Figure 11.44. Then click **Confirm** to apply the changes.

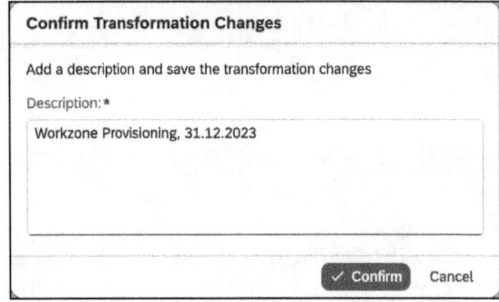

Figure 11.44 Confirm Changes to Transformation

Once all the necessary settings have been made, you can start the initial provisioning. This is always done from the source system. To do this, navigate to the previously created source system (see Figure 11.45). Open the **Jobs** tab and click **Run Now** for the **Read Job**. You should schedule a job for productive operation that carries out provisioning at regular intervals.

11.2 Configuration

Creating New Users for SAP Build Work Zone, Advanced Edition
If you want to create new users, this is done within SAP Cloud Identity Services. You must also assign the authorizations there.

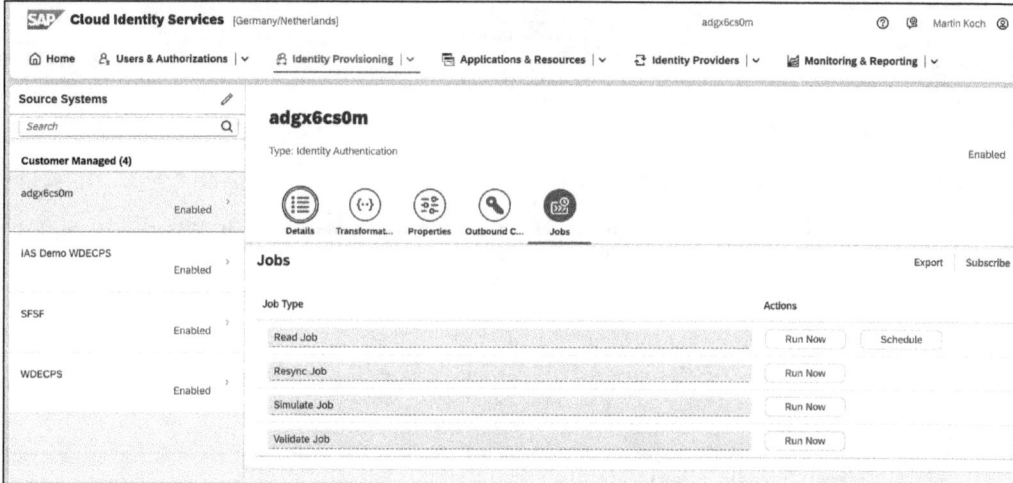

Figure 11.45 Execute Job

Then open the provisioning logs. To do this, navigate to the **Identity Provisioning • Provisioning Logs** area as shown in Figure 11.46. Open the **Job Logs** tab. As you can see in the figure, the job has been completed with the status **Finished Successfully**. Click the corresponding entry in the table to jump to the details.

Figure 11.46 Open Provisioning Log

In the **Job Execution Details** area, you will find all relevant information about the provisioning run (see Figure 11.47). There you can see, among other things, how many users and groups were read from the source system and adjusted in the target systems. At this point, you will also find error messages if the provision encounters an error. This can be the case, for example, if a transformation is incorrect.

369

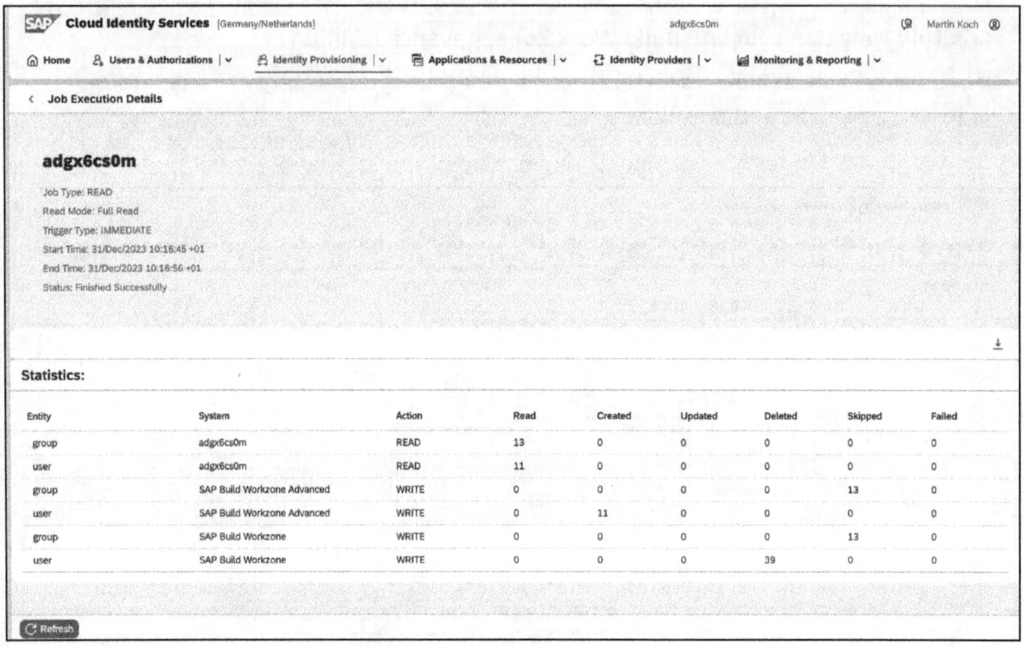

11   Installing and Configuring SAP Build Work Zone

Figure 11.47 Job Execution Details

Now that you have fully configured SAP Cloud Identity Services, you can continue with the configuration in SAP Build Work Zone, advanced edition. Check the relevant boxes for all tasks and then click **Step 4** (see Figure 11.48).

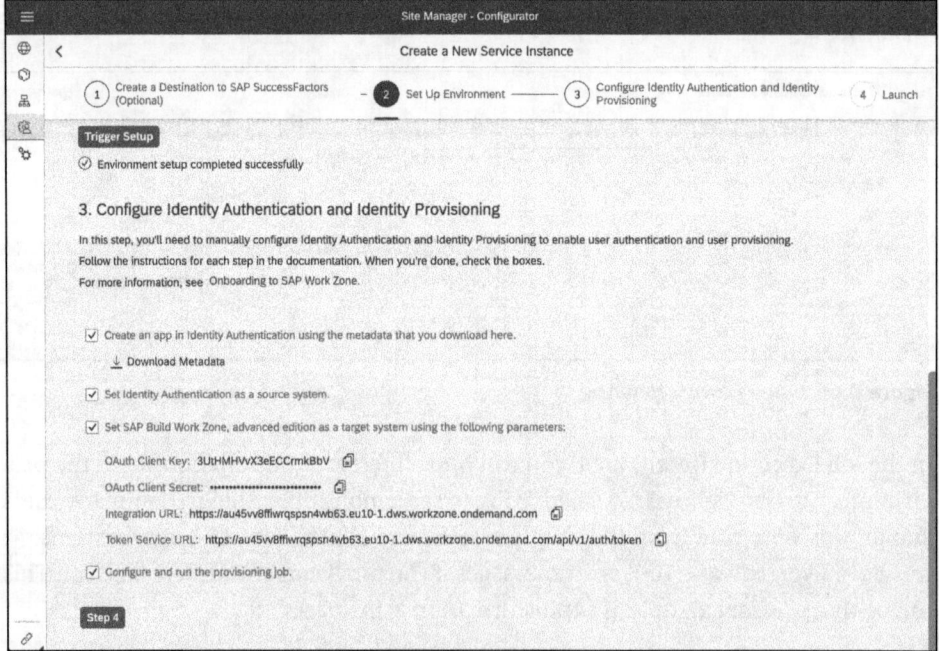

Figure 11.48 Identity Authentication Setup

You should now see that the onboarding has been successfully completed. By clicking the **Open SAP Build Wok Zone, Advanced Edition** link, you can jump to the application for end users.

Once you have jumped to the end user view, you should see the screen shown in Figure 11.49 after successful authentication.

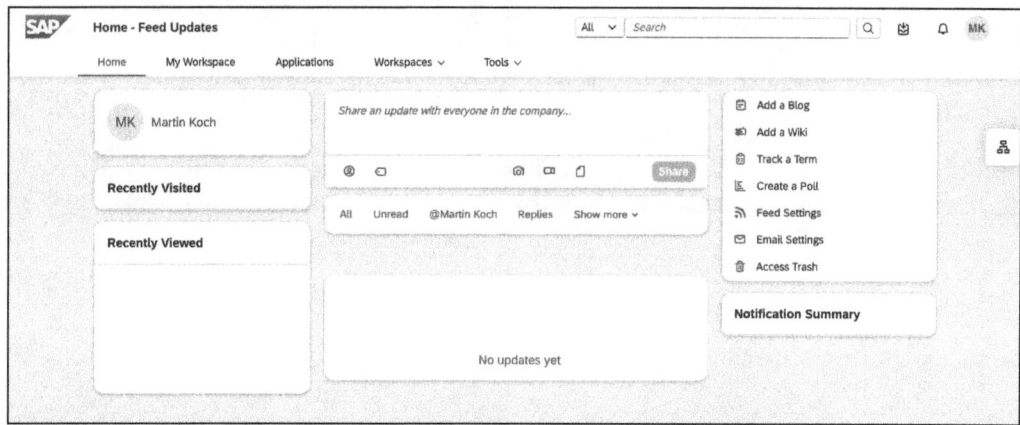

Figure 11.49 SAP Build Work Zone, Advanced Edition: End User Home Page

11.3 Summary

In this chapter, we navigated through the critical stages of installing and configuring *SAP Build Work Zone, advanced edition*, a robust platform designed for creating engaging digital workplace experiences. Section 11.1 began by guiding you through the installation process for SAP Build Work Zone, advanced edition. Emphasis was placed on using the SAP BTP booster–based installation. Following the installation, we delved into the configuration of SAP Build Work Zone, advanced edition in Section 11.2.

Chapter 12
UI Integration

SAP Build Work Zone offers various UI elements that can be used to create work pages. This allows intuitive and clear pages to be created, and these pages make daily work easier and more enjoyable for users.

SAP Build Work Zone provides a variety of UI elements that developers can configure, arrange, and deploy in work pages. In this chapter, you'll learn how to develop and install these different UI elements.

We'll begin by looking at content packages in Section 12.1. These are groups of several different artifacts. In Section 12.2, we'll look at UI integration cards. These exist in various forms to represent lists, diagrams, forms, and other UI elements. UI integration cards can be part of a content package, for example.

Finally, in Section 12.3, we'll look at widgets. Widgets are basically containers in which artifacts can be included. In addition, the widget builder can be used to create your own widgets for embedding in other websites.

12.1 Content Packages

With the help of content packages, artifacts that are built can be delivered in the form of a ZIP file. For example, content packages can contain UI integration cards. Content packages can be used for internal use—such as for testing or illustration purposes—but also for the delivery of content. Content packages can contain various artifacts, such as the following:

- UI integration cards
- Workspace templates
- Exported workspaces and work pages

A distinction is made between two types of content packages for users:

- **Centrally provided content packages**
  These are content packages that are available to all customers in SAP Build Work Zone. They do not require manual uploading. Furthermore, these content packages cannot be downloaded or adapted, but must be installed by the administrator before they can be used.

12 UI Integration

- **Custom content packages**
  These are content packages that have been developed by the customer and uploaded to the subaccount by the administrator. These content packages must also be installed by the administrator before they can be used.

In the following sections, we'll discuss how to create and deploy a content package.

12.1.1 Creating a Content Package

In this section, we'll look at how to create a content package. SAP Business Application Studio is used for this purpose. First, a dev space must be created here. It is important here that the development tools for SAP Build Work Zone are included. This can be specified when creating the dev space by selecting the checkmark next to **Development Tools for SAP Build Work Zone**, as shown in Figure 12.1.

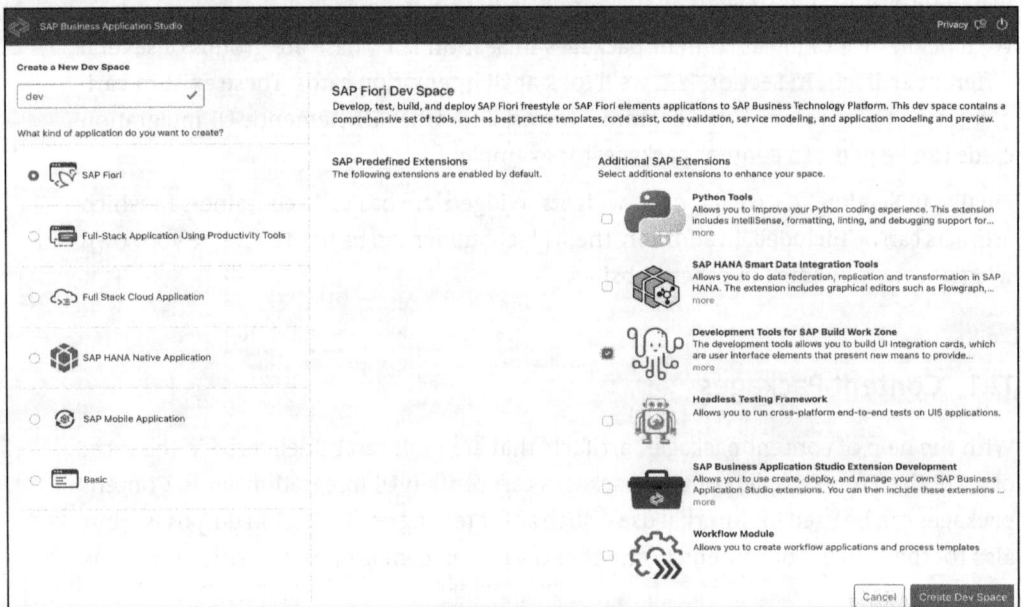

Figure 12.1 Creating Dev Space in SAP Business Application Studio

After entering a name for the dev space and selecting the required tools, the dev space can be created using the **Create Dev Space** button. Once this has been done, the dev space can be started and then opened. After opening the dev space, SAP Business Application Studio opens. You can then continue with the creation of a content package.

In SAP Business Application Studio, select **File • New Project from Template** in the context menu (see Figure 12.2).

12.1 Content Packages

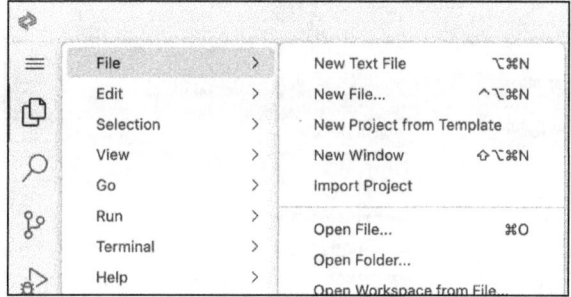

Figure 12.2 Create Project from Template

In the wizard that opens, the template must first be selected. Here, select **Content Package** (see Figure 12.3).

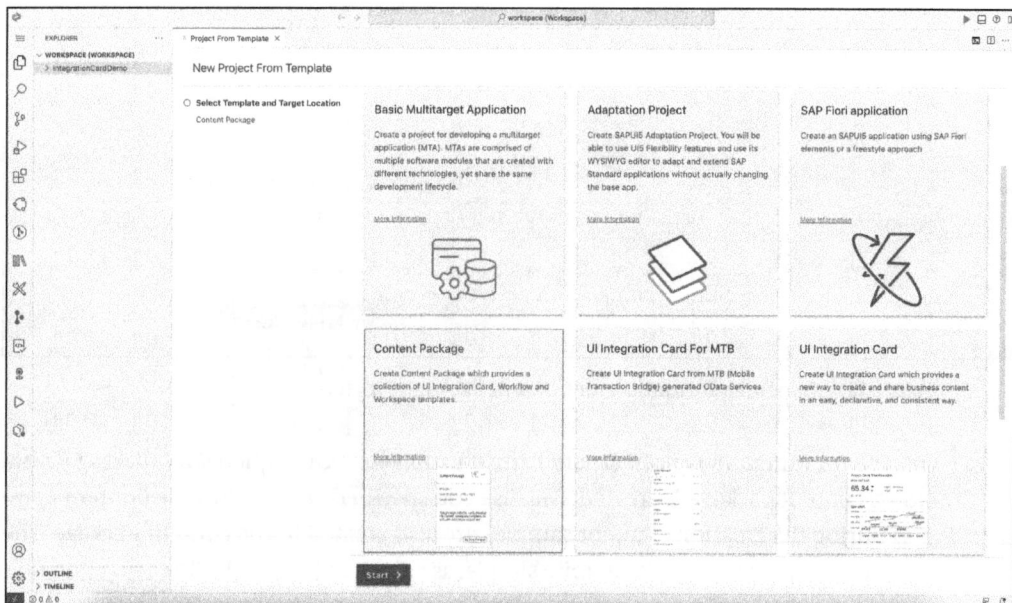

Figure 12.3 Select Content Package as Template for Project

Click **Start** to jump to the next step. Here, general information about the project must be entered. For this, assign the following values (see Figure 12.4):

- **Project Name**: "contentPackageDemo"
- **Namespace**: "at.clouddna"
- **Title**: "Demo Content Package Title"
- **Subtitle**: "Demo Content Package Subtitle"
- **Include Content Samples**: True

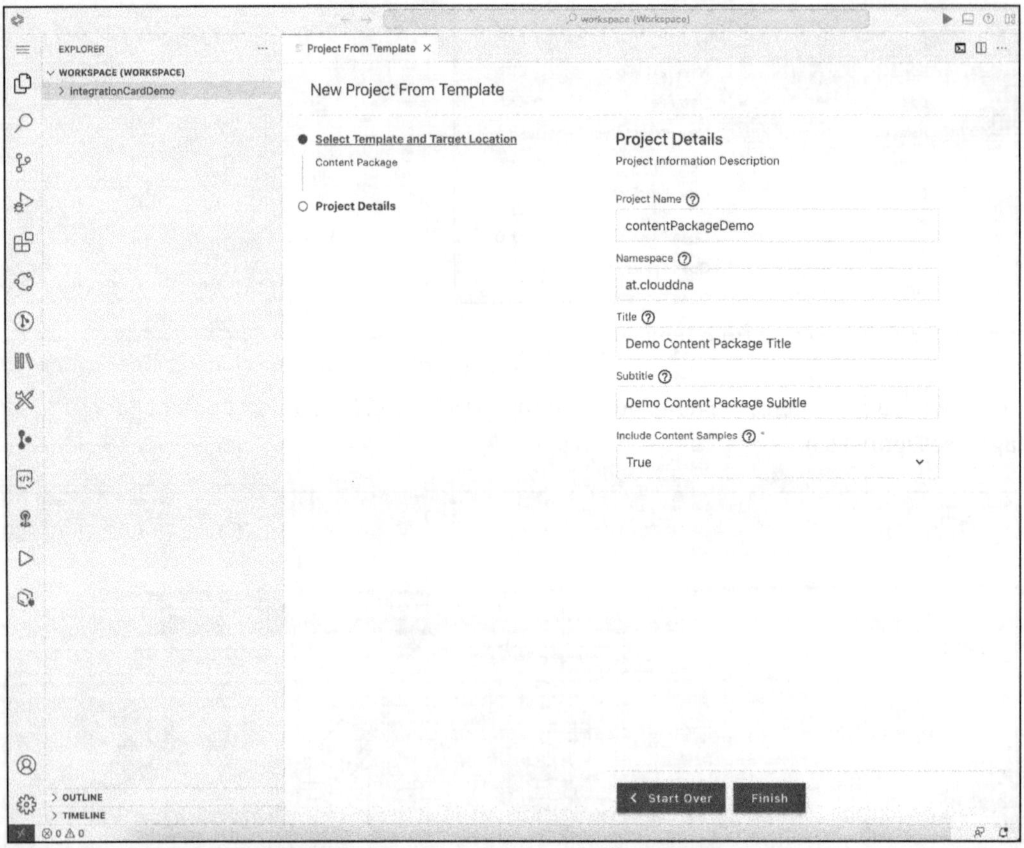

12 UI Integration

[Screenshot of "New Project From Template" dialog in an IDE workspace showing Project Details with fields: Project Name (contentPackageDemo), Namespace (at.clouddna), Title (Demo Content Package Title), Subtitle (Demo Content Package Subitle), Include Content Samples (True).]

Figure 12.4 General Information for Content Package Project

After these values have been entered, the creation can be completed by clicking **Finish**. Now a content package is created. Because you specified **True** for **Include Content Samples** during the creation, content samples are also created in the content package. The structure of this project is kept quite simple, as shown in Figure 12.5. The two main files are *manifest.json* and *content.json*.

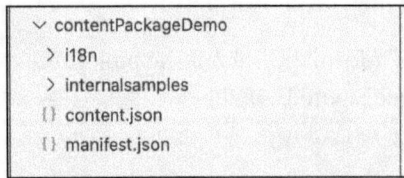

Figure 12.5 Project Structure of Content Package

There are now basically three actions that can be carried out for the content package that has been created:

12.1 Content Packages

- **Delete the content package**
  This deletes the project.

- **Editing the content of the content package**
  By right-clicking the *content.json* file and selecting **Content Package: Content Edit**, the content of the content package can be adjusted. Among other things, existing artifacts can be edited, and new artifacts can be added (see Figure 12.6). Alternatively, this can also be entered manually directly in the JSON file.

- **Deploying the content package**
  The content package can be deployed either directly via SAP Business Application Studio or manually via import and export. We will use the latter variant later.

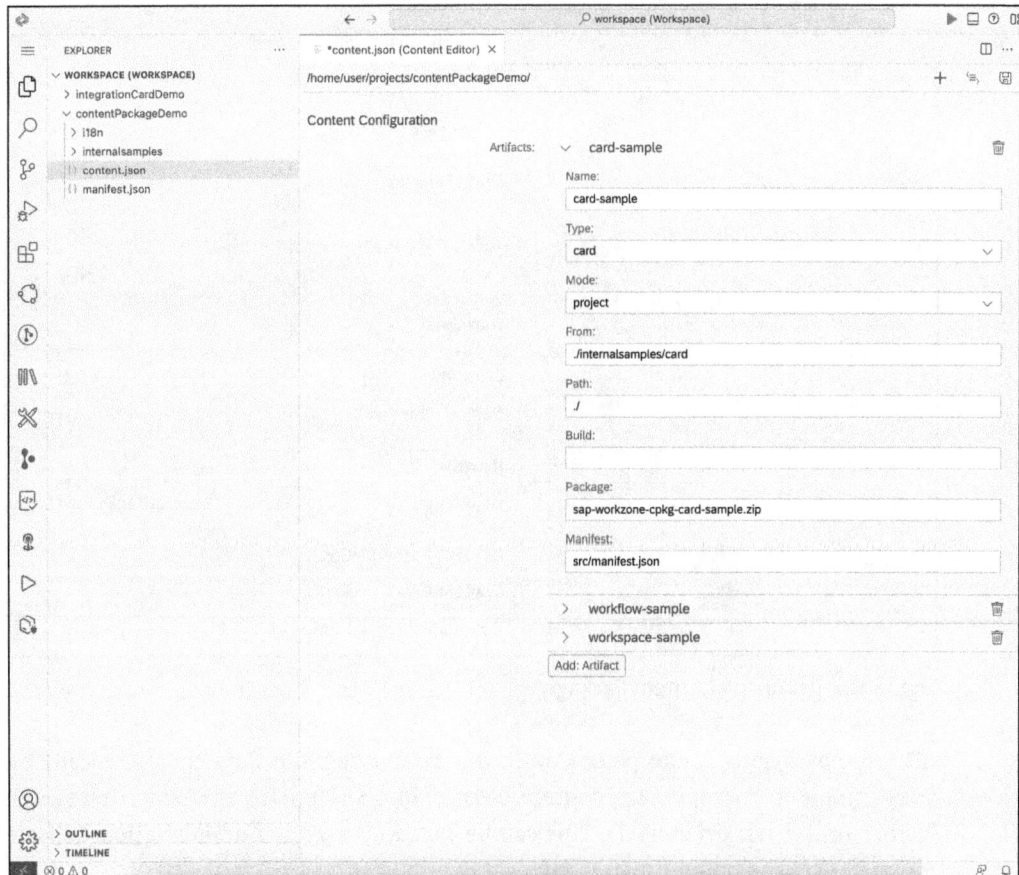

Figure 12.6 Content Editor

12.1.2 Deploying a Content Package

Because you've already created the content package with sample artifacts, you can now export the content package so that it can be imported again into SAP Build Work Zone.

To export the content package, first right-click **manifest.json** and select **Content Package: Package** (see Figure 12.7). This creates a ZIP file containing the content package, which you can then download and store anywhere.

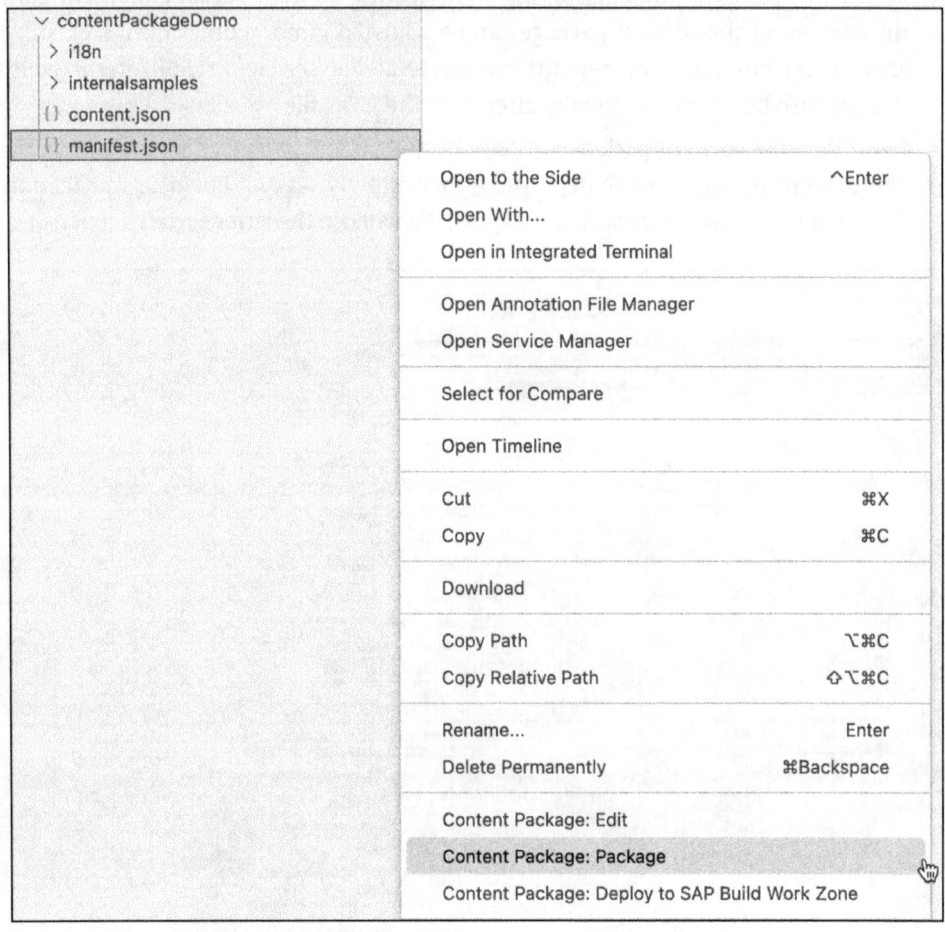

Figure 12.7 Packaging Content Package

After the packaging is complete, you have a *package.zip* file. Save this file locally on your computer. To import the content package into SAP Build Work Zone, first switch to the administration console. This can be called up via the **Administration Console** menu item found under the user icon (see Figure 12.8).

Here you'll find a subitem called **Content Packages** in the left-hand bar under **UI Integration**. By selecting this entry, you get an overview of the currently available content packages (see Figure 12.9). In addition, further content packages can be uploaded via an **Upload** button.

12.1 Content Packages

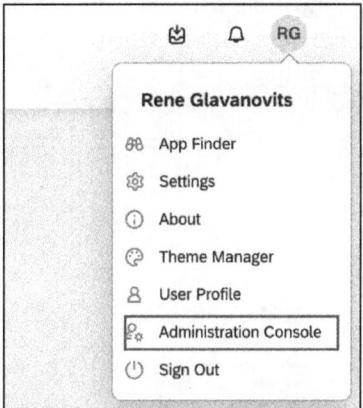

Figure 12.8 Open SAP Build Work Zone Administration Console

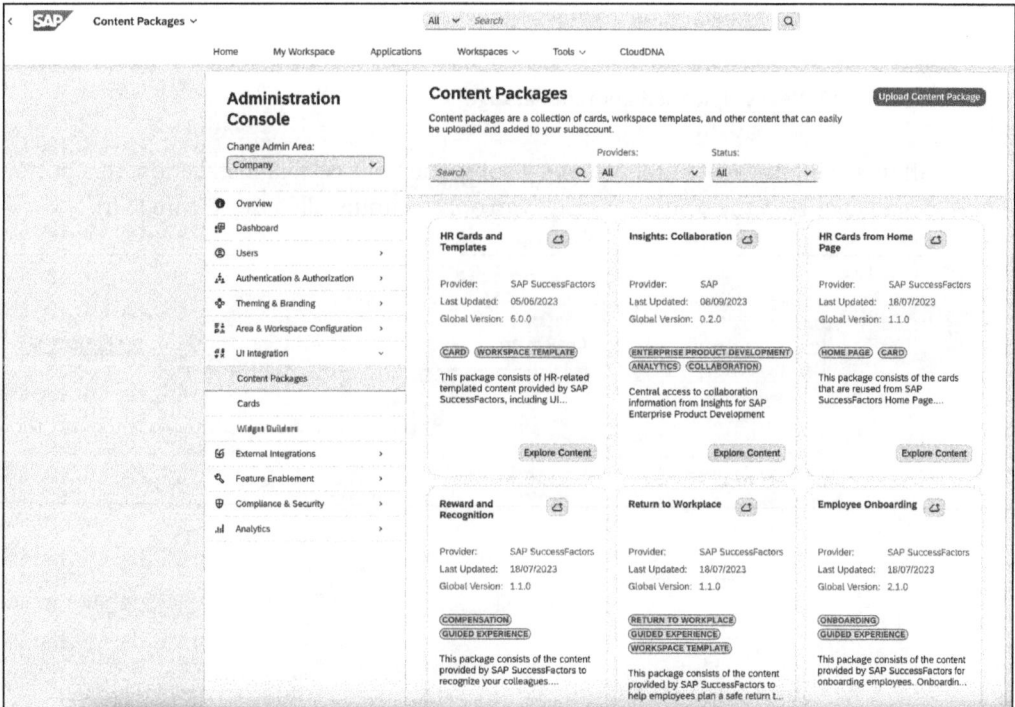

Figure 12.9 Overview of Content Packages

To upload the previously exported content package, click the **Upload Content Package** button and then select the exported ZIP file in File Explorer and confirm this selection. After confirming the selection, the content package is uploaded and processed. Once this is complete, the overview can be updated and the uploaded content package should be visible in it. To ensure that the contents of this content package can actually be used in SAP Build Work Zone, the content package must be installed. This can be done via the **Install** button in the header area of the package (see Figure 12.10).

Figure 12.10 Newly Uploaded Content Package

After clicking the **Install** button and a subsequent successful installation, this button disappears. Now the content package can only be uninstalled (see Figure 12.11).

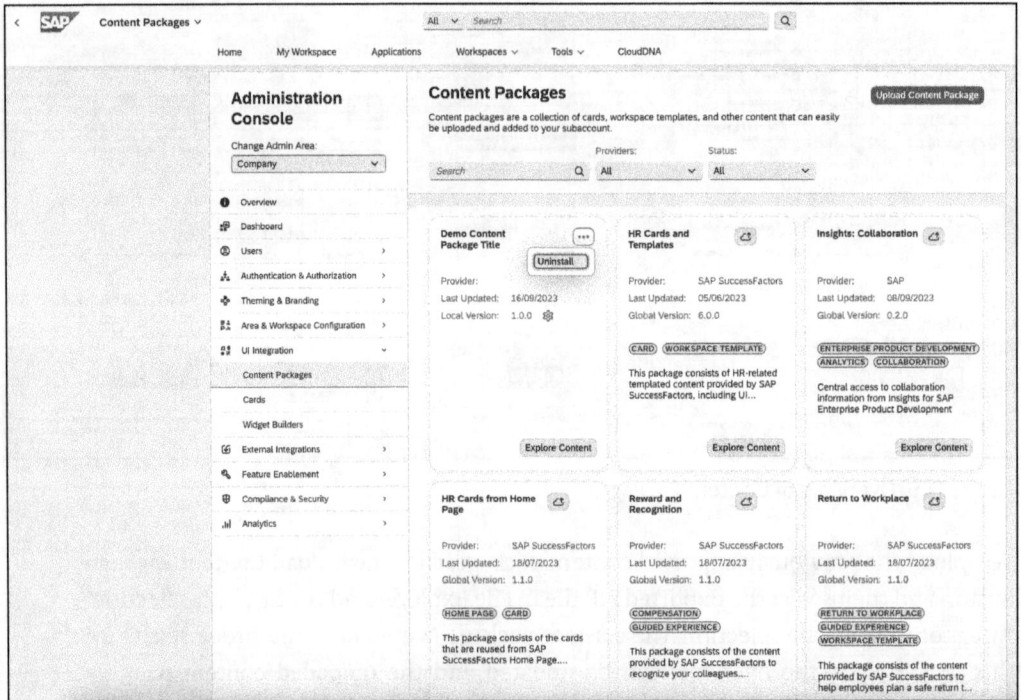

Figure 12.11 Overview of Content Packages after Installation

Because the content package also contains a UI integration card, you'll now find this card in the **UI Integration • Cards** tab after installing the package, with a note that it comes from a content package (see Figure 12.12).

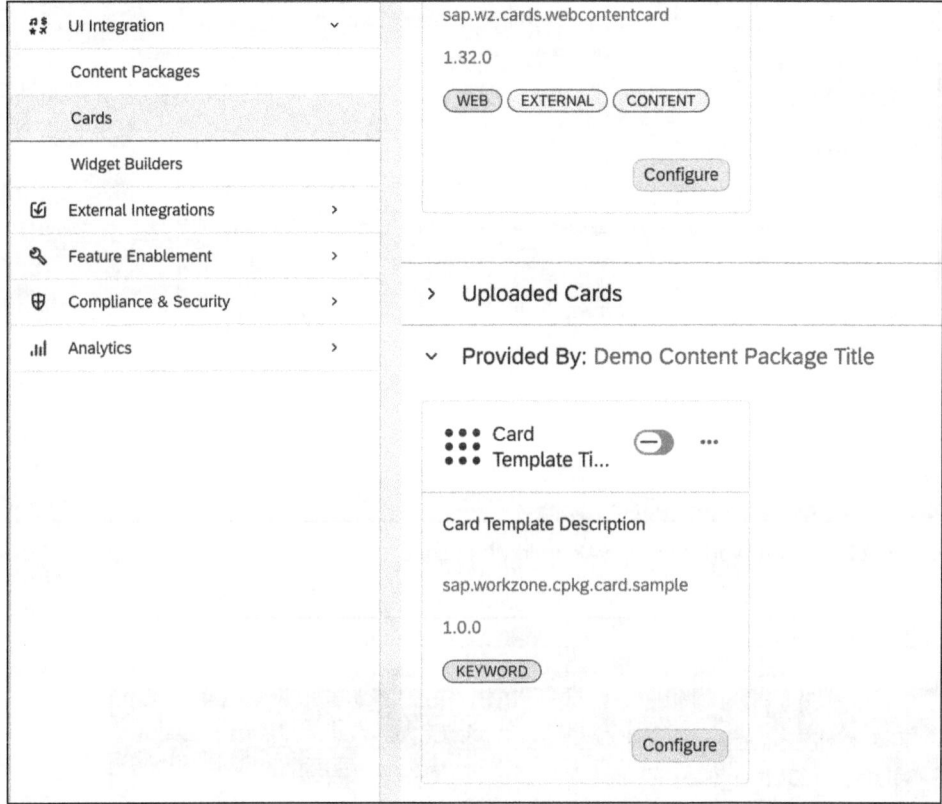

Figure 12.12 Card Imported from Content Package

For the card to be visible and selectable for the user, it must be activated via a switch. To display the contents of the content package—in this case, the integration card—in a work page, proceed as follows:

1. Open a workspace in SAP Build Work Zone. To do this, navigate to **Workspaces • View All Workspaces** and select a workspace. In this case, open the workspace named **Demo_Workspace** (see Figure 12.13).
2. After selecting a workspace, you'll see all the work pages contained in it. In this case, select the **Demo_Workpage** work page (see Figure 12.14). This work page then opens.
3. After opening the work page, you can switch to edit mode using the **Edit** button on the right-hand side. If you do so, you'll see a screen like that shown in Figure 12.15.

12   UI Integration

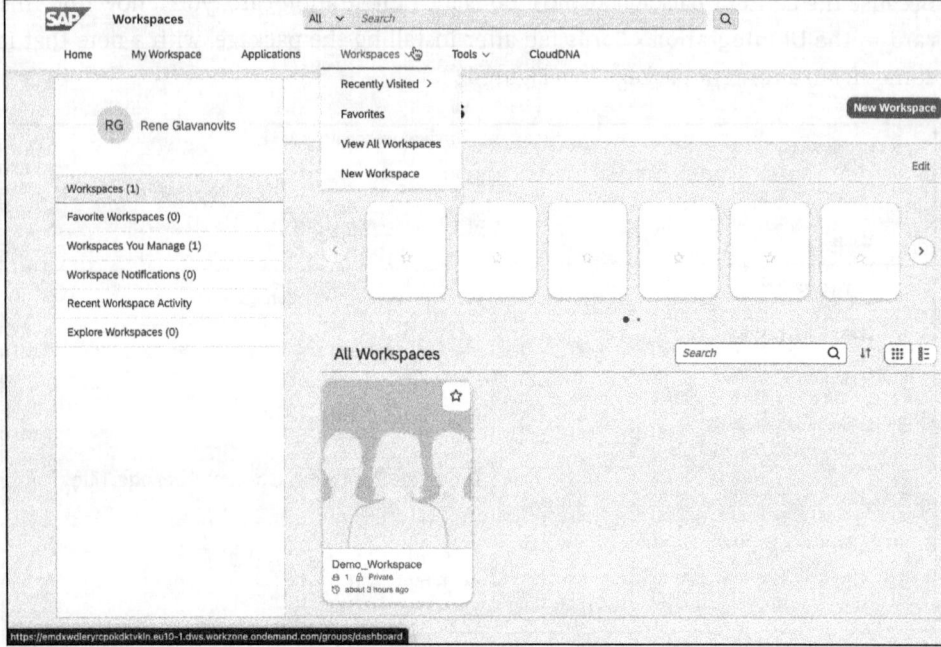

Figure 12.13 Open Workspace in SAP Build Work Zone

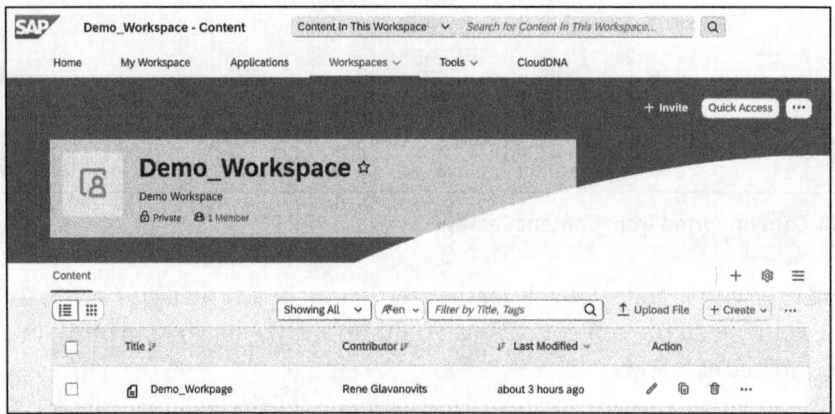

Figure 12.14 Overview of Work Pages in Workspace

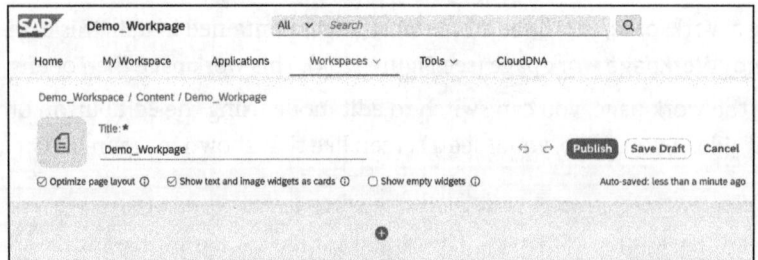

Figure 12.15 Empty Work Page

Once you're in edit mode, the card can be added to the work page. A detailed description of the steps required to add a card to the work page is given in Section 12.2.5. For now, just select the card template, which you'll find under **Add Widget • Cards** (see Figure 12.16).

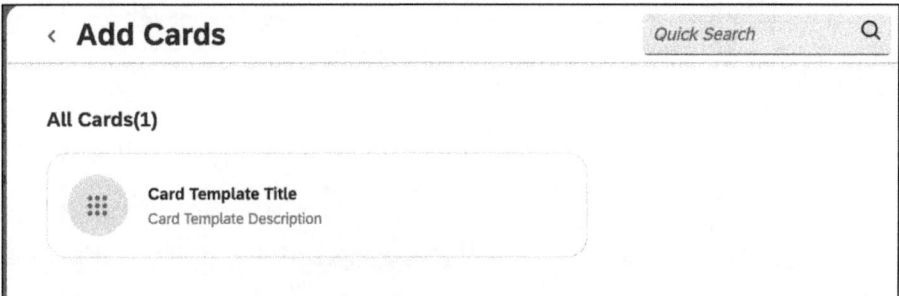

Figure 12.16 Select Card from Content Package

After selecting this card, a dialogue opens in which configuration parameters can be defined. Here, leave the default values and confirm this choice with **Save** (see Figure 12.17).

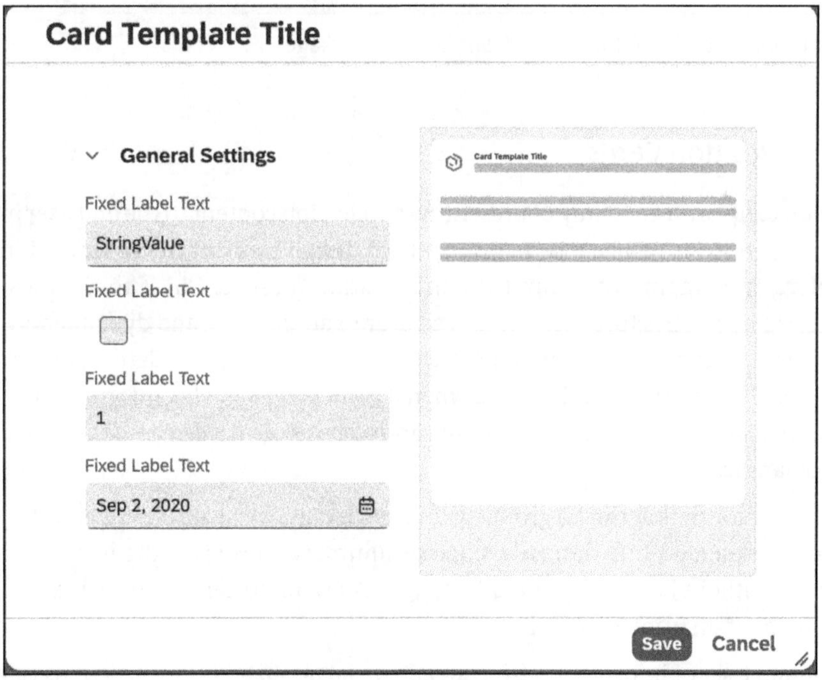

Figure 12.17 Provide Values for Card Template

After clicking **Save** and then publishing with **Publish**, the work page can be called up, and the added card is displayed accordingly (see Figure 12.18).

12   UI Integration

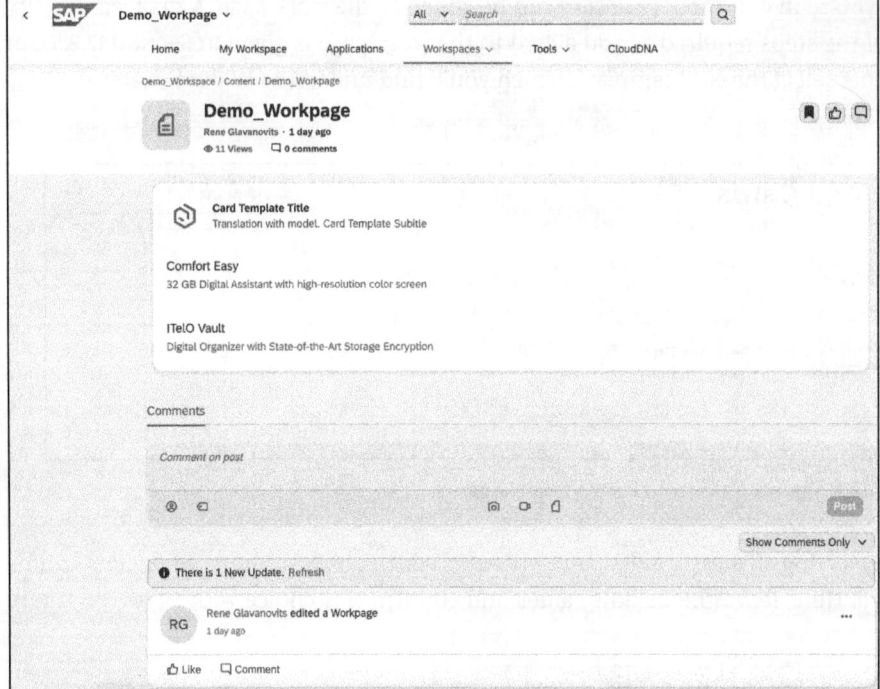

Figure 12.18 Work Page after Adding Card from Content Package

12.2 UI Integration Cards

UI integration cards are a good way of presenting application content to the end user in a pleasant and, above all, consistent way. A *card* is a design element that is limited in size. Therefore, it is important to limit the information presented in it to the most essential. It also helps structure content for the user in an intuitive and dynamic way, similar to tiles. However, the layout and structure of the cards are predetermined by the framework. Cards are most widely used when it comes to displaying information in flexible layouts. This ensures that the content can still be recognized even if the size of the display changes.

By using cards, information can be grouped, links to details can be displayed, or information can be presented in a short and concise summary. This gives the end user a good overview without having to open other pages. A few important UX concepts you should know about up front are as follows:

- Cards contain a set of information from an app or a page. In addition, functions can also be offered.
- Cards are representations of an app or page and provide content to the end user in a specific context.
- Cards can be a mash-up of several different apps (e.g., a list card with links).

- The structure of the cards is predefined.
- Individual apps or pages can be represented by multiple cards, each representing different aspects.
- End users get an overview at a glance without having to navigate to other pages. In addition, cards can also offer navigation options.

In the following sections, we'll look at the structure of cards and their types, and how to develop one, upload it, and add it to a work page.

12.2.1 Structure of Cards

Cards basically consist of three parts:

- **Card container**
  The card container defines the background and the frame of the card. The frame of the card contains the header and the content.
- **Header**
  The header contains, among other things, a title, a subtitle, an icon, and a status. If the header is of the numeric type, it can also contain other attributes that describe numeric indicators, such as KPIs. The header should also make clear what information is illustrated with this card. In addition, the header can also be used for navigation. If a list is displayed in the content of the card, the header is also often used to show how many entries are available in this list.
- **Content**
  The content area of a card is used to display the data of the respective data source. Depending on the type of card, different UI elements are embedded here. The respective UI elements also determine which type of interaction and visualization is used.

Figure 12.19 shows that the forms of representation can be used for cards; the clear structure of the cards can also be seen.

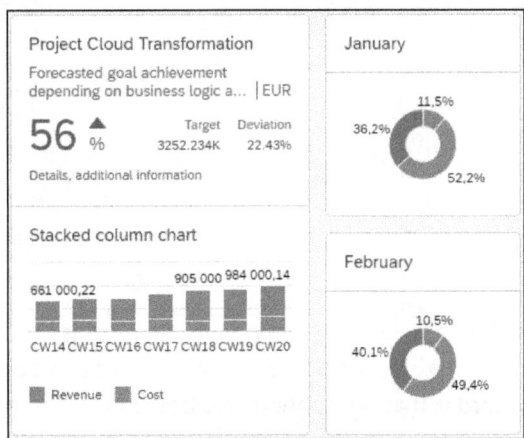

Figure 12.19 Sample Integration Cards

Integration cards are defined in a declarative way by defining a *manifest.json* file. This type of definition makes it relatively easy to integrate and reuse a card. At the same time, the card control must reference the manifest file.

In this *manifest.json* file, the developer must define the header, the content, the data source from which the information is read, and, if available, the possible actions. If the created card is to be integrated into an application at a later time, the height and width of the card and the behavior for the defined actions can also be defined.

12.2.2 Card Types

Now let's look at the different types of cards and their characteristics. It should be noted that each card is designed in a different style and supports different content formats. Basically, we distinguish between two card types:

- Transactional
- Analytical

While analytical cards are usually used to display key figures or diagrams, transactional cards are usually used when information needs to be displayed in list or table form.

Adaptive Card

Adaptive cards can be used to display and reuse cards created using Microsoft's Adaptive Cards specification and a manifest file.

Analytical Card

Analytical cards are used for data visualization. As a rule, a numerical header and analytical data content are defined for these. The following analytical card contents can be used for this purpose:

- Line chart
- Donut chart
- (Stacked) column chart
- (Stacked) bar chart
- Bubble chart
- Scatter plot chart

Component Card

This type of card allows you to display different SAPUI5 components and is thus used on an individual basis for use cases for which there are no other suitable card types or structures. A peculiarity of this type of card is that the content area of the card can be moved upward.

Web Page Card

The web page card offers the possibility to embed HTML content into the content area of a card.

List Card

If you want to display information in list form, you can use the list card. This allows you to display several list items of different types. In addition, this type of card also offers the possibility to display aggregated information in the line items. For displaying the items in list form, this card type uses the sap.m.List UI element.

When using this card type, however, it should be noted that the counter in the header area is mandatory here. You can see what such a card might look in Figure 12.20. In addition, the counter indicator in the header area and actions in the individual list items can also be seen here.

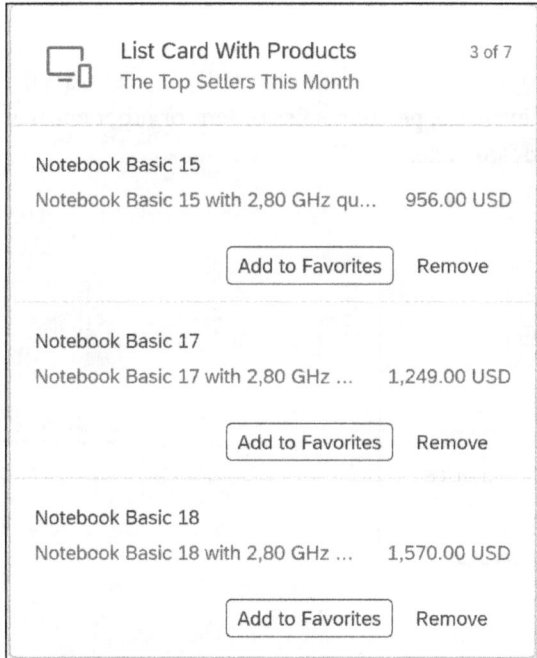

Figure 12.20 Sample for List Card

Table Card

Similar to list cards, table cards are also used to display multiple entries of an entity. Unlike lists, tables contain columns so that information can be displayed in tabular form. Within the card content, the sap.m.Table UI element is used for this card type. Clicking a table entry would open its detail page. You can see an example of a table card in Figure 12.21.

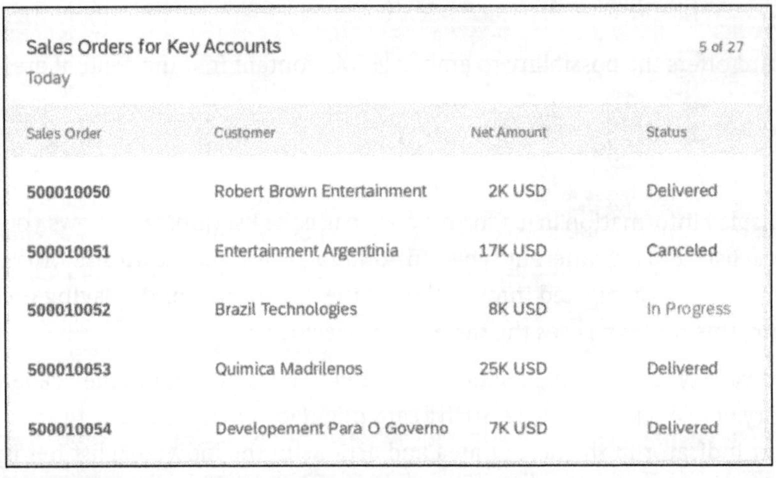

Sales Orders for Key Accounts
Today
5 of 27
Sales Order
Customer
Net Amount
Status
500010050
Robert Brown Entertainment
2K USD
Delivered
500010051
Entertainment Argentinia
17K USD
Canceled
500010052
Brazil Technologies
8K USD
In Progress
500010053
Quimica Madrilenos
25K USD
Delivered
500010054
Developement Para O Governo
7K USD
Delivered

Figure 12.21 Sample for Table Card

Object Card

If you want to display detailed information about a specific object, the object card is the best choice. These objects can be, for example, persons, sales orders, or other entities. Figure 12.22 shows what an object card looks like.

Donna Moore
Sales Executive
Contact Details
First Name:
Donna
Last Name:
Moore
Phone:
+1 202 555 5555
Company Details
Company Name:
Company A
Address:
481 West Street, Anytown OH 45066, USA
Website:
www.company_a.example.com
Organizational Details
Direct Manager:
John Miller

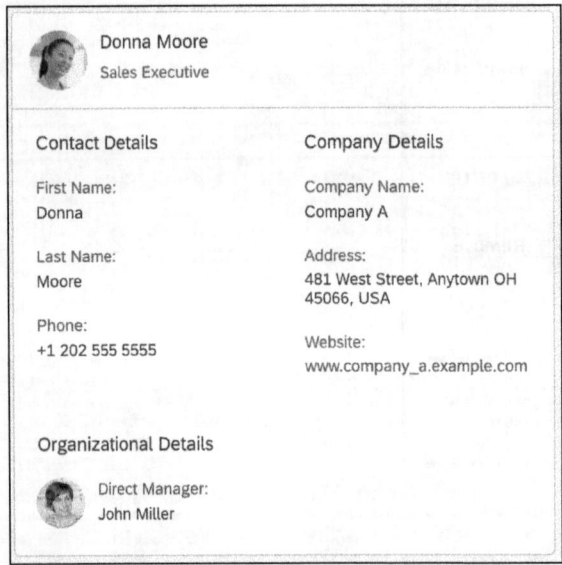

Figure 12.22 Sample for Object Card

Timeline Card

A timeline card contains, as the name suggests, a timeline. This type of card is particularly useful when you want to map a newsfeed or comment history, for example. You can see an example of this type of card in Figure 12.23.

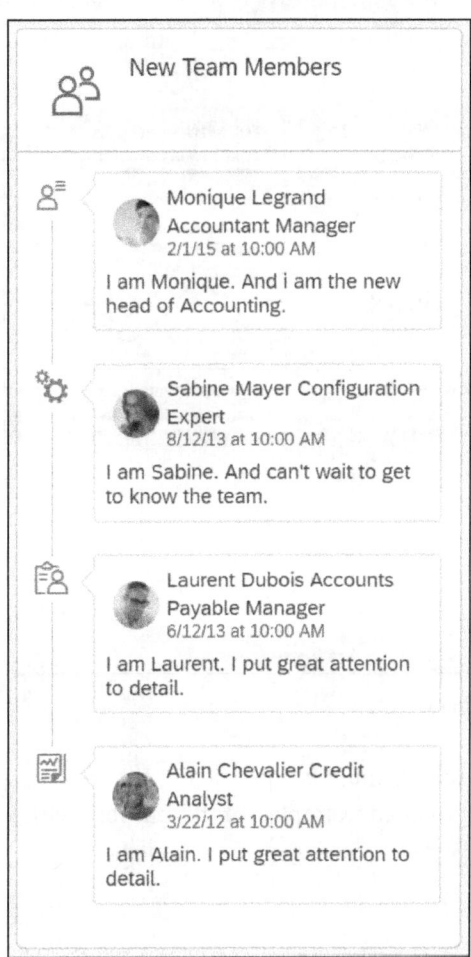

Figure 12.23 Sample for Timeline Card

12.2.3 Developing a UI Card

In this section, we'll look at how such a UI card can be developed. As with the development of content packages, SAP Business Application Studio is used here. Here too, the development tools for SAP Build Work Zone must be activated when creating a dev space.

To create a UI integration card, first select **File • New Project from Template** in the context menu. A wizard opens in which the template must be selected. **UI Integration Card** must be selected here (see Figure 12.24). The wizard can then be continued by clicking **Next**.

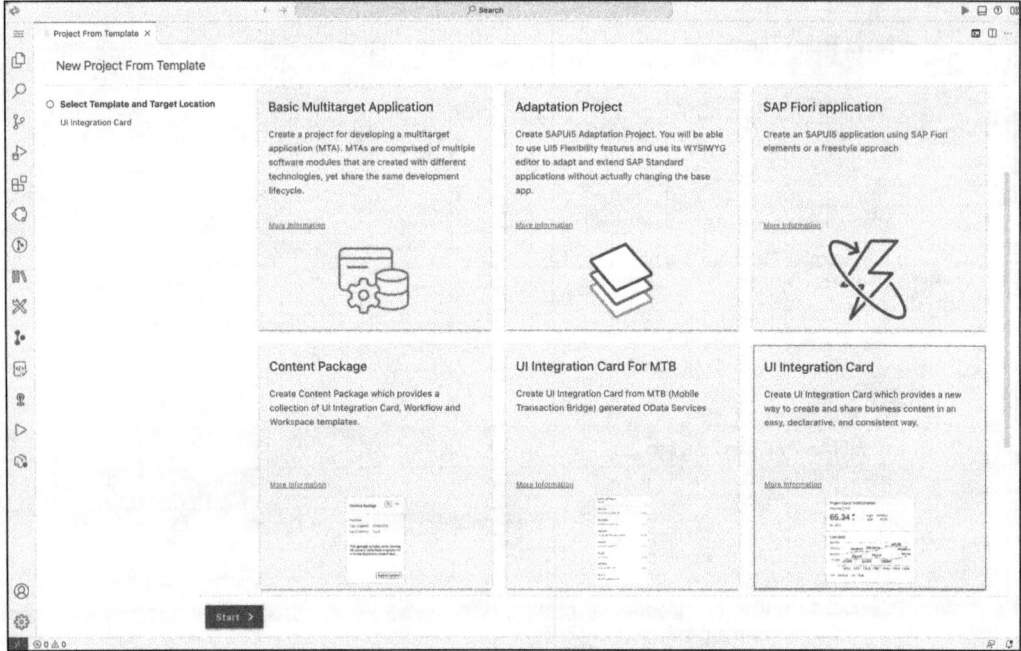

Figure 12.24 Select UI Integration Card as Template

After selecting the template, general information must be specified in the following wizard step. In this case, let's create a table card as an example. For this purpose, assign the following values, as shown in Figure 12.25:

- **Project Name:** "integrationCardDemo"
- **Namespace:** "at.clouddna"
- **Select a Card Sample:** Table Card
- **Title:** "Demo Card Title"
- **Subtitle:** "Demo Card Subtitle"
- **Compatible with SAP Mobile Cards:** False

Then you can finish the creation with **Finish**. After the card has been successfully created, the folder structure should look like that shown in Figure 12.26.

The configuration of the card can be found in the *manifest.json* file. Besides the general information about the card, the card type is also defined here. Also defined is what the header contains and what content is displayed in the card. The main properties that determine the layout and content of the card are essentially type, header, and content in the manifest file (see Figure 12.27).

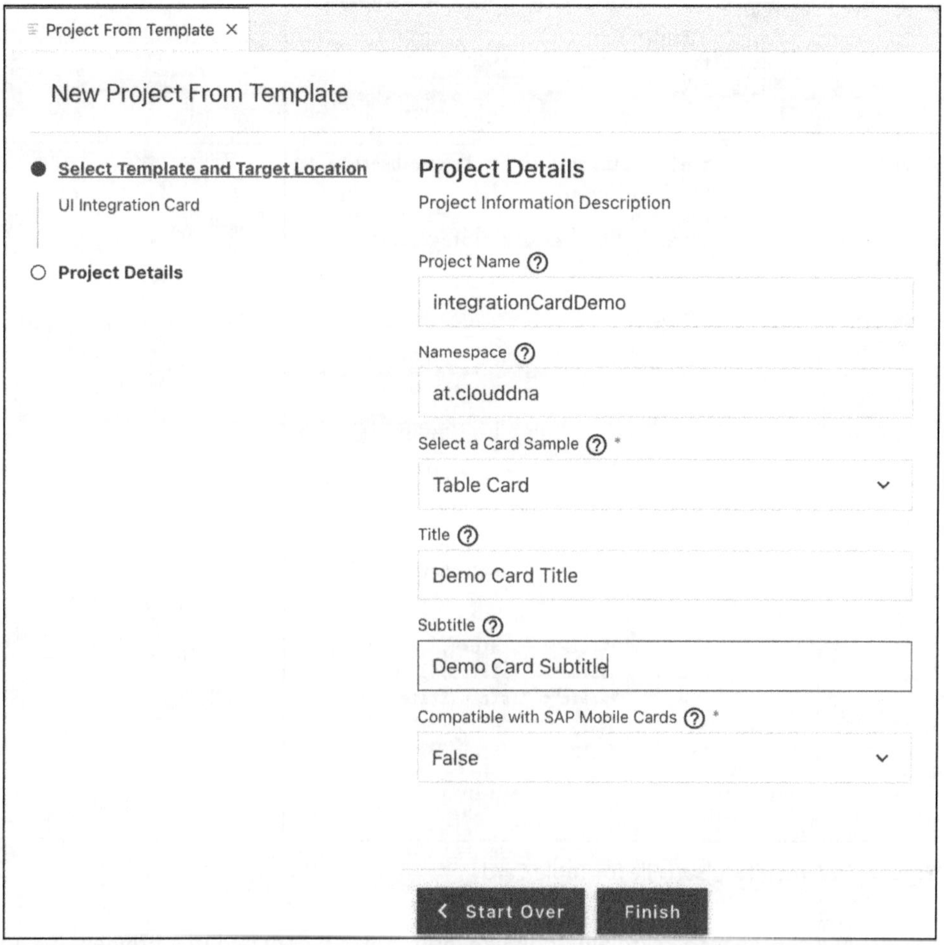
Project From Template
New Project From Template
Select Template and Target Location
UI Integration Card
Project Details
Project Details
Project Information Description
Project Name
integrationCardDemo
Namespace
at.clouddna
Select a Card Sample *
Table Card
Title
Demo Card Title
Subtitle
Demo Card Subtitle
Compatible with SAP Mobile Cards *
False
Start Over
Finish

Figure 12.25 General Information for Integration Card

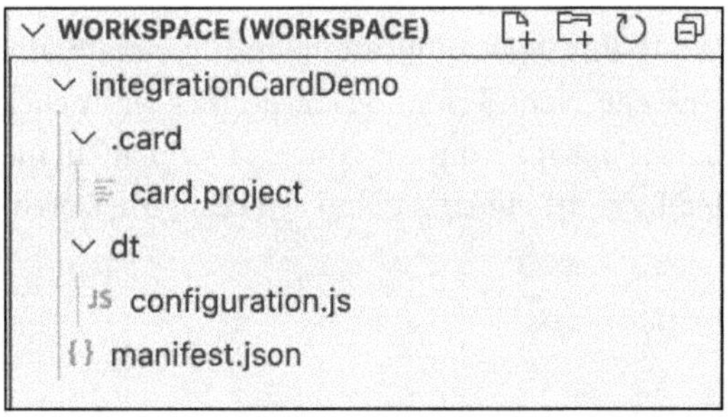
WORKSPACE (WORKSPACE)
integrationCardDemo
.card
card.project
dt
configuration.js
manifest.json

Figure 12.26 Folder Structure of Integration Card

```
19          "sap.card": {
20              "type": "Table",
21              "designtime": "dt/configuration",
22 >            "data": {…
87              },
88              "header": {
89                  "title": "Sales Orders for Key Accounts",
90                  "subTitle": "Today",
91                  "status": {
92                      "text": "{headerData/statusText}"
93                  }
94              },
95              "content": {
96                  "row": {
97 >                    "columns": [{…
101                     },
102                     {
103                         "title": "Customer",
104                         "value": "{customerName}"
105                     },
106                     {
107                         "title": "Net Amount",
108                         "value": "{netAmount}"
109                     },
110                     {
111                         "title": "Status",
112                         "value": "{status}",
113                         "state": "{statusState}"
114                     }
115                 ]
116             }
117         }
118     },
```

Figure 12.27 manifest.json File for Table Card

To visually see what the card will look like later, it can be opened in a preview. To do this, right-click **manifest.json** and select **UI Integration Card: Preview**. The card will then open in a separate tab (see Figure 12.28).

A card can be deployed either directly via SAP Business Application Studio or manually via export and import. In this case, choose the second method. To do this, right-click **manifest.json** and select **UI Integration Card: Package** from the context menu. A ZIP file is then created, which can be downloaded by right-clicking it and selecting the **Download** menu item.

12.2 UI Integration Cards

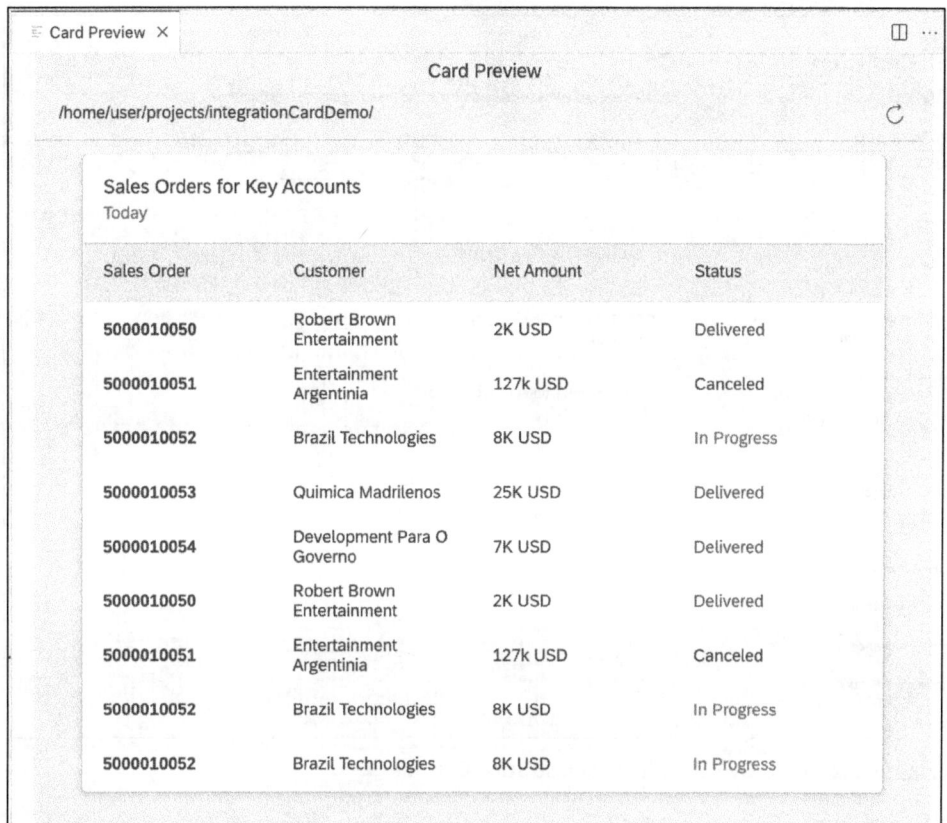

Figure 12.28 Preview of UI Integration Card

12.2.4 Uploading a UI Integration Card

To upload the previously zipped and downloaded card to SAP Build Work Zone, you have to jump to the admin console. This can be accessed by clicking the user icon and then selecting the **Administration Console** entry if you are in SAP Build Work Zone.

In the administration console, go to the **UI Integration** option in the left bar, and then the **Cards** suboption. This provides an overview of already available UI integration cards, but you can also upload new ones by clicking the **Upload Card** button (see Figure 12.29).

To upload a new card, click the **Upload Card** button. An explorer window opens in which the previously exported ZIP file can be selected. After selecting the card, an info dialog opens that shows, among other things, the title, subtitle, and ID of the card (see Figure 12.30). In this dialog, the upload can be initiated by clicking **Upload**.

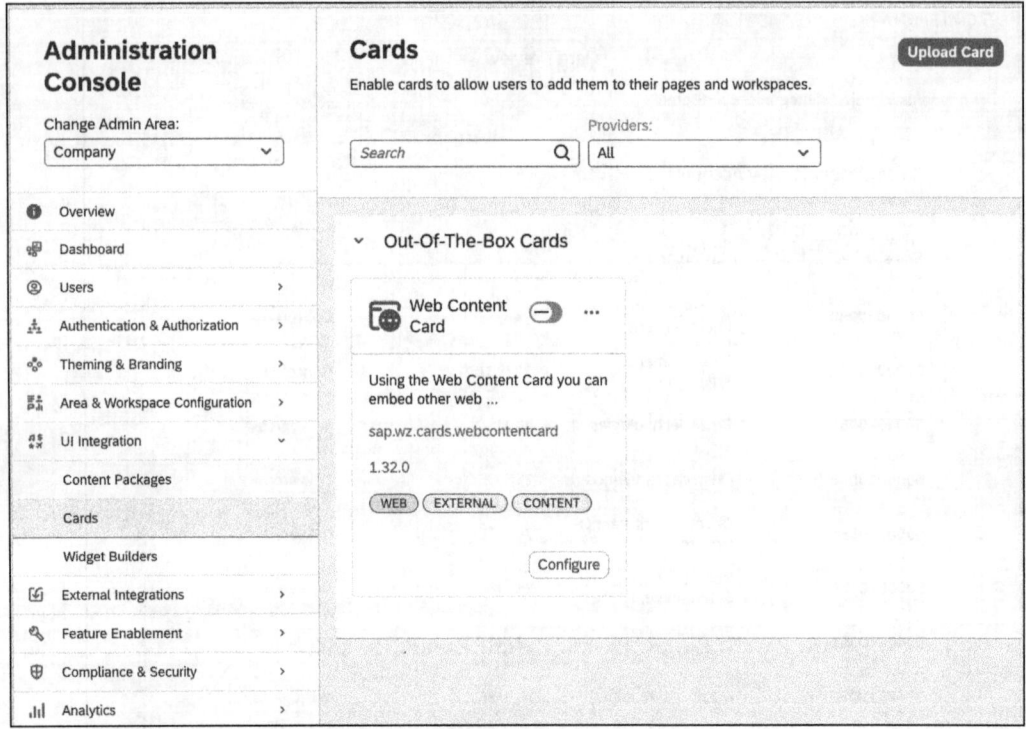

Figure 12.29 Card Overview in Administration Console

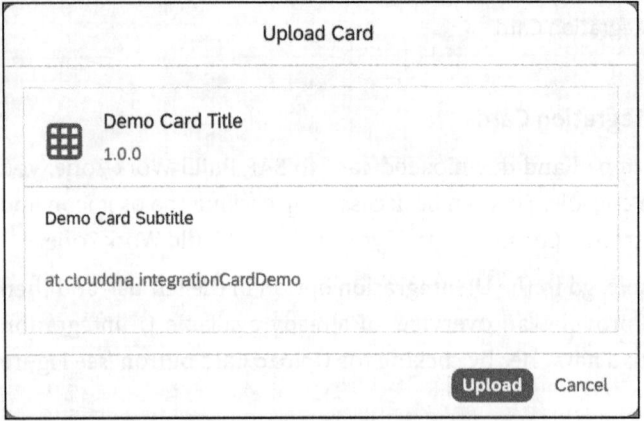

Figure 12.30 Upload Dialog for UI Integration Card

After the upload is complete, this card is displayed in the overview under **Uploaded Cards**. However, for this card to also be selected in SAP Build Work Zone, it must still be activated. This can be done via the slider button in the title bar of the card. After the card has been activated, the status of the button in the header changes (see Figure 12.31).

12.2 UI Integration Cards

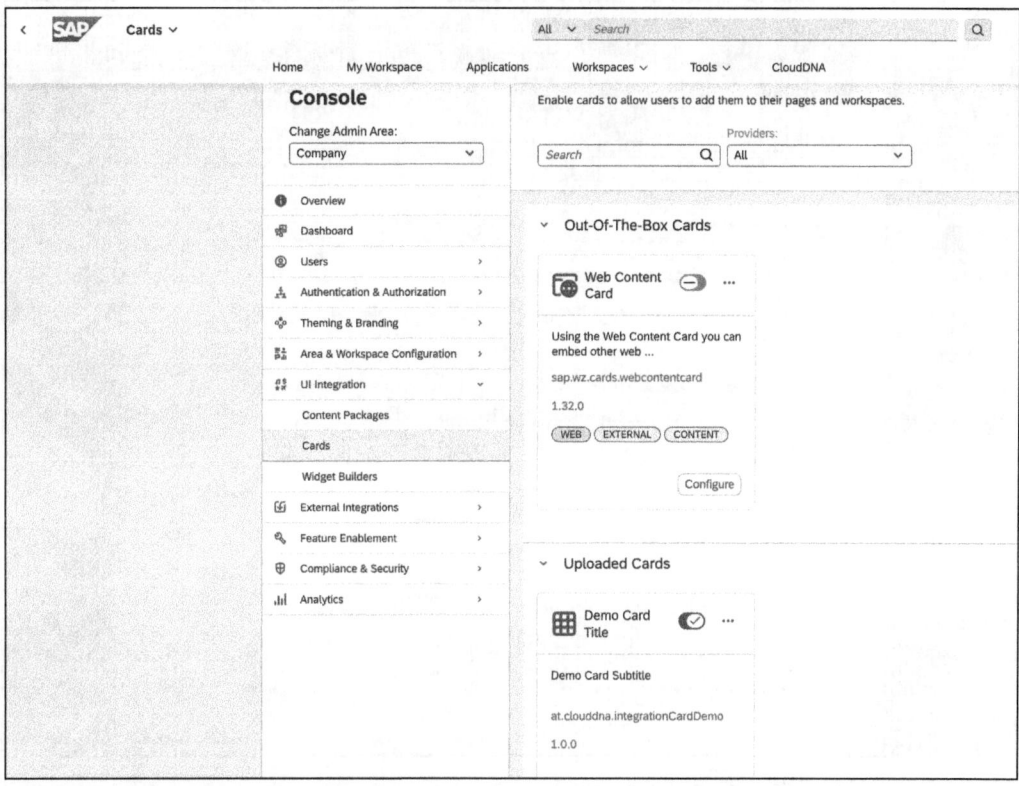

Figure 12.31 Overview of Available Cards

12.2.5 Adding a UI Integration Card to a Work Page

Now that the card has been added and activated in the administration console, it can be added to a work page in SAP Build Work Zone as follows:

1. Open a workspace and a work page within it.
2. Click the **Edit** button to change to edit mode for the work page.
3. To add new elements to the work page, click the **+** symbol. This will create a new section in which a widget can be inserted by clicking the **Add Widget** button.
4. After clicking the **Add Widget** button, a dialog opens in which the type of content can be selected. In this case, select **Cards** (see Figure 12.32). You'll see an overview of all cards that are available for selection. In this case, select the previously imported demo card (see Figure 12.33).

12 UI Integration

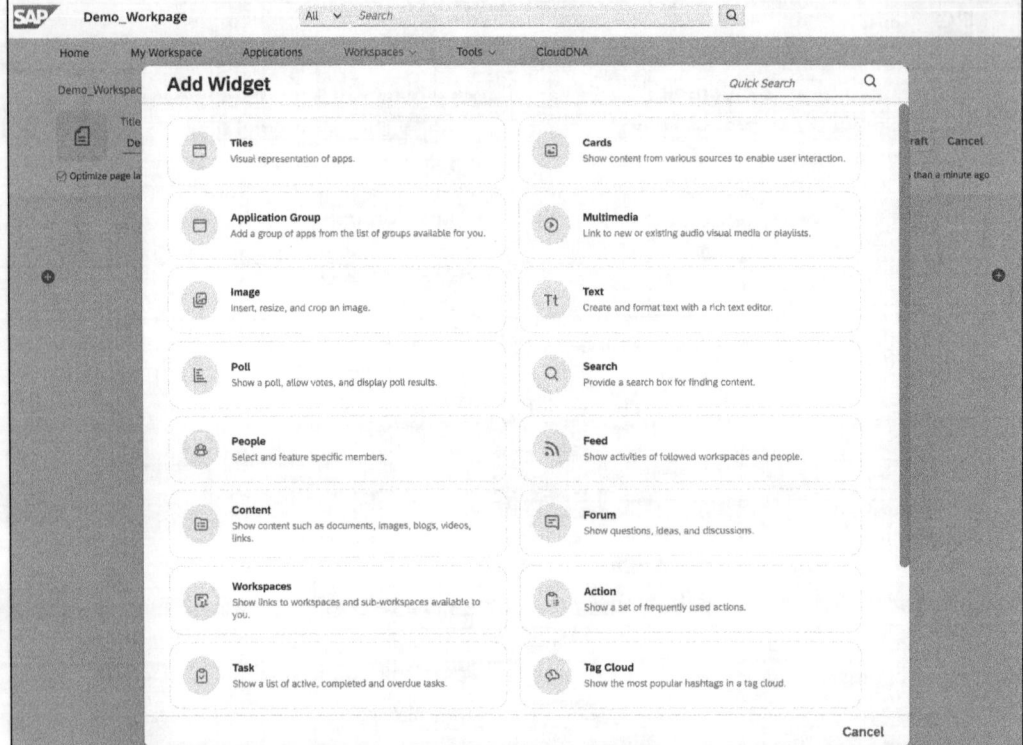

Figure 12.32 Add Widget Dialog

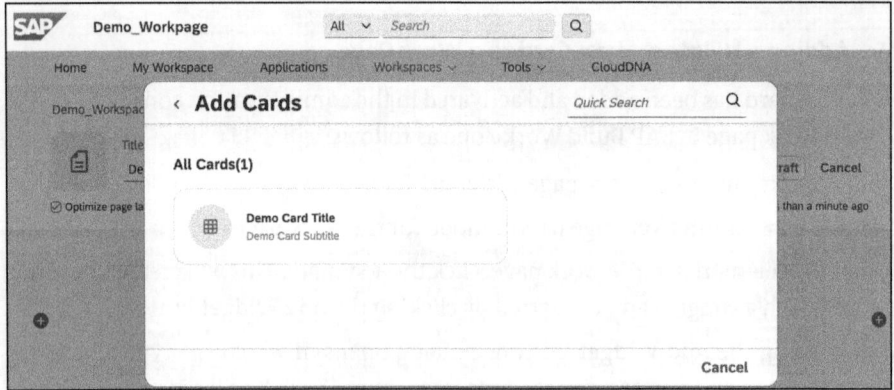

Figure 12.33 Select Card in Cards Overview

5. After selecting the card, it's displayed in the previously added section (see Figure 12.34).
6. Click **Publish** to save the changes and publish the work page. In addition, an optional comment and the visibility of the work page can be specified in a subsequent dialog (see Figure 12.35).

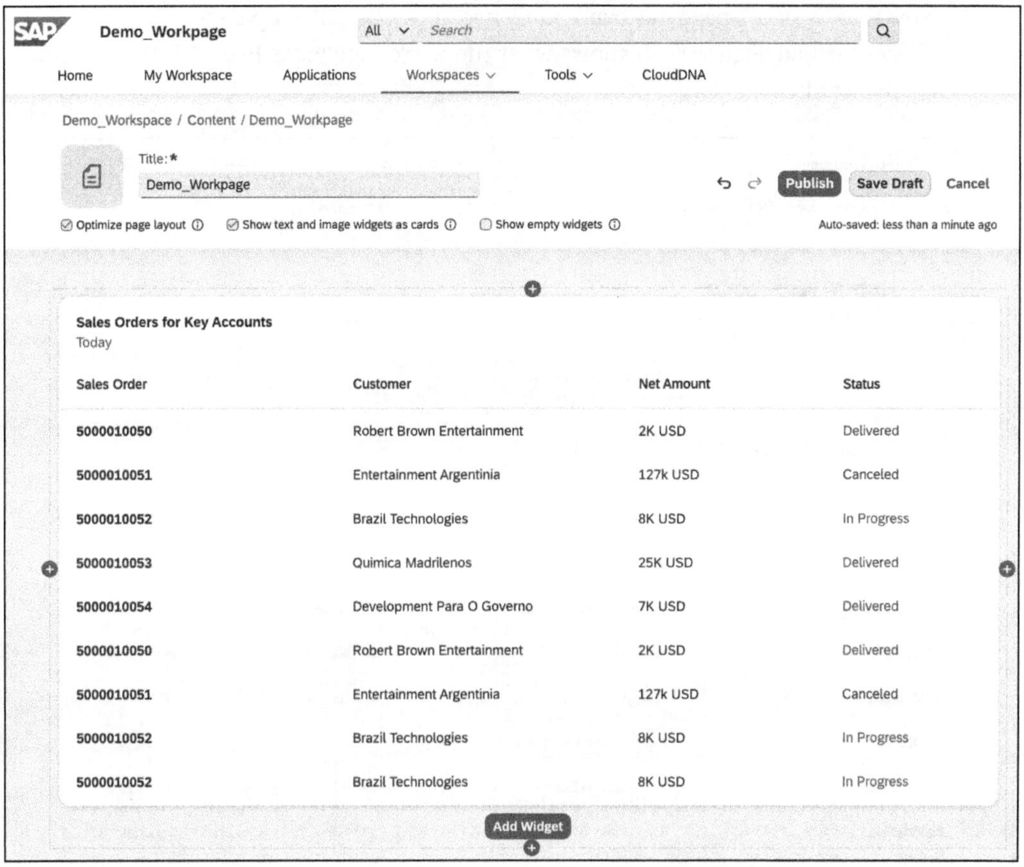

12.2 UI Integration Cards

Figure 12.34 Preview of Recently Added Card

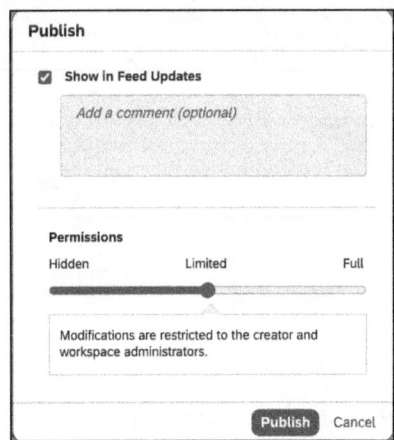

Figure 12.35 Publish Work Page

Now that the work page has been published, it can be called up depending on the visibility specified. Figure 12.36 shows what the work page looks like with the previously created card.

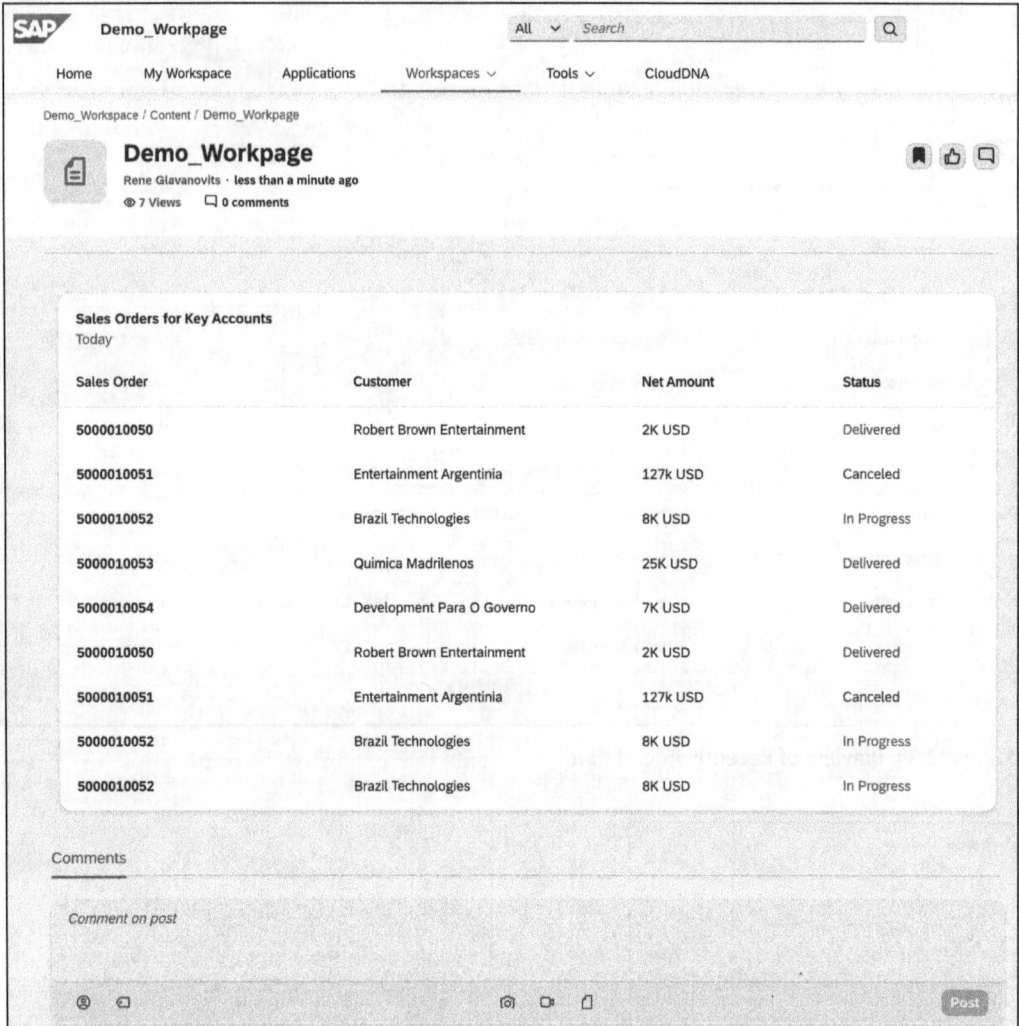

Figure 12.36 Work Page after Adding UI Integration Card

12.3 Widgets

In SAP Build Work Zone, widgets are used to display various content in a work page. Basically, widgets are containers for concrete UI elements. If you create a work page in SAP Build Work Zone, you can insert sections into it. Within a section, any number of widgets can be displayed. A widget, in turn, can contain a UI element. Table 12.1 offers an overview of the widgets available by default.

12.3 Widgets

Widget Type	Purpose
Tiles	Visual representation of apps
Cards	Show content from various sources to enable user interaction
Application group	Add a group of apps from the list of groups available for you
Multimedia	Link to new or existing audiovisual media or playlists
Image	Insert, resize, and crop an image
Text	Create and format text with a rich text editor
Poll	Show a poll, allow votes, and display poll results
Search	Provide a search box for finding content
People	Select and feature specific members
Feed	Show activities of followed workspaces and people
Content	Show content such as documents, images, blogs, videos, and links
Forum	Show questions, ideas, and discussions
Workspaces	Show links to workspaces and subworkspaces available to you
Action	Show a set of frequently used actions
Task	Show a list of active, completed, and overdue tasks
Tag cloud	Show the most popular hashtags in a tag cloud
Slideshow	Select a presentation (PDF, PPT, or PPTX files)
Event	Show upcoming or recent events
Rotating banner	Show a carousel with up to 10 slides of headlines or news
Tool content	Show content from various tools
Knowledge base	Show featured, last updated, most viewed, or most liked articles

Table 12.1 Widgets Available Out of the Box

As you can see, a large number of widgets for various use cases are already delivered by default. In addition, the widget builder can be used to create your own widgets, which can be integrated into any web page.

The widget builder can be found in the Administration Console under the item **UI Integration • Widget Builder**. Here you can choose among three different builders:

- **Feed Widget Builder**
  This builder is used to create feed widgets. Any feeds can be displayed in them, such

12 UI Integration

as the following: company feed, my follows feed, workspace feed, external object feed, external wall feed, mentions feed, and content item feed.

- **Recommendations Widget Builder**
  Using this widget builder, a widget can be created in which recommendations are displayed. These recommendations can include content, people, or workspaces. In addition, you can also define how many recommendations should be displayed.

- **Share Widget Builder**
  This builder allows you to create your own share icon with any link behind it and to display it later. For example, you can link to your own company website or any other URL. In addition, you can also specify whether specific branding should be used.

To use the widget builder, proceed as follows:

1. Select **UI Integration • Widget Builders** from within the administration console (see Figure 12.37).

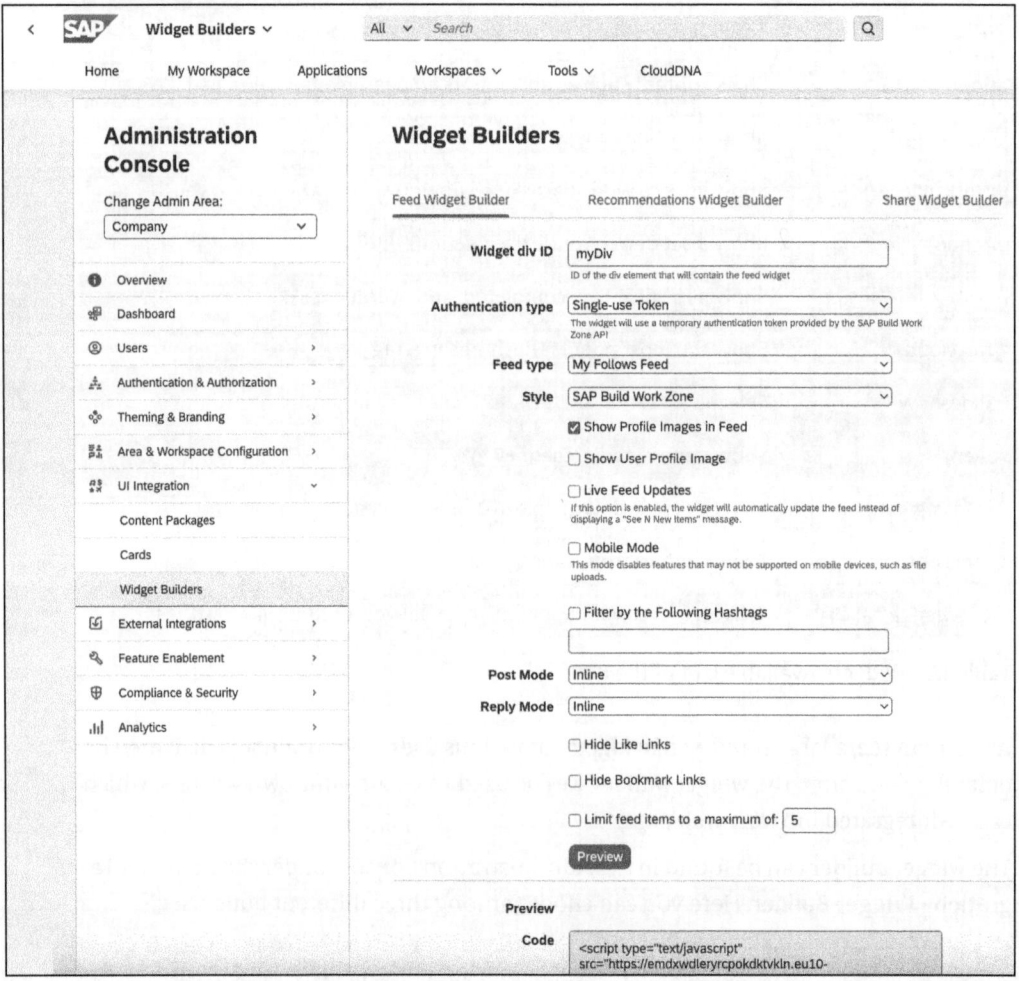

Figure 12.37 Widget Builder in Administration Console

2. Select the type of widget to be created.
3. Depending on the type, different properties must be assigned. By clicking the **Preview** button, the widget can be examined at any time. The JavaScript code that defines the widget is also displayed.
4. Once you've finished configuring the widget, you can copy the JavaScript code and paste it into any web page to display the widget.

Figure 12.38 shows what the preview of such a widget might look like.

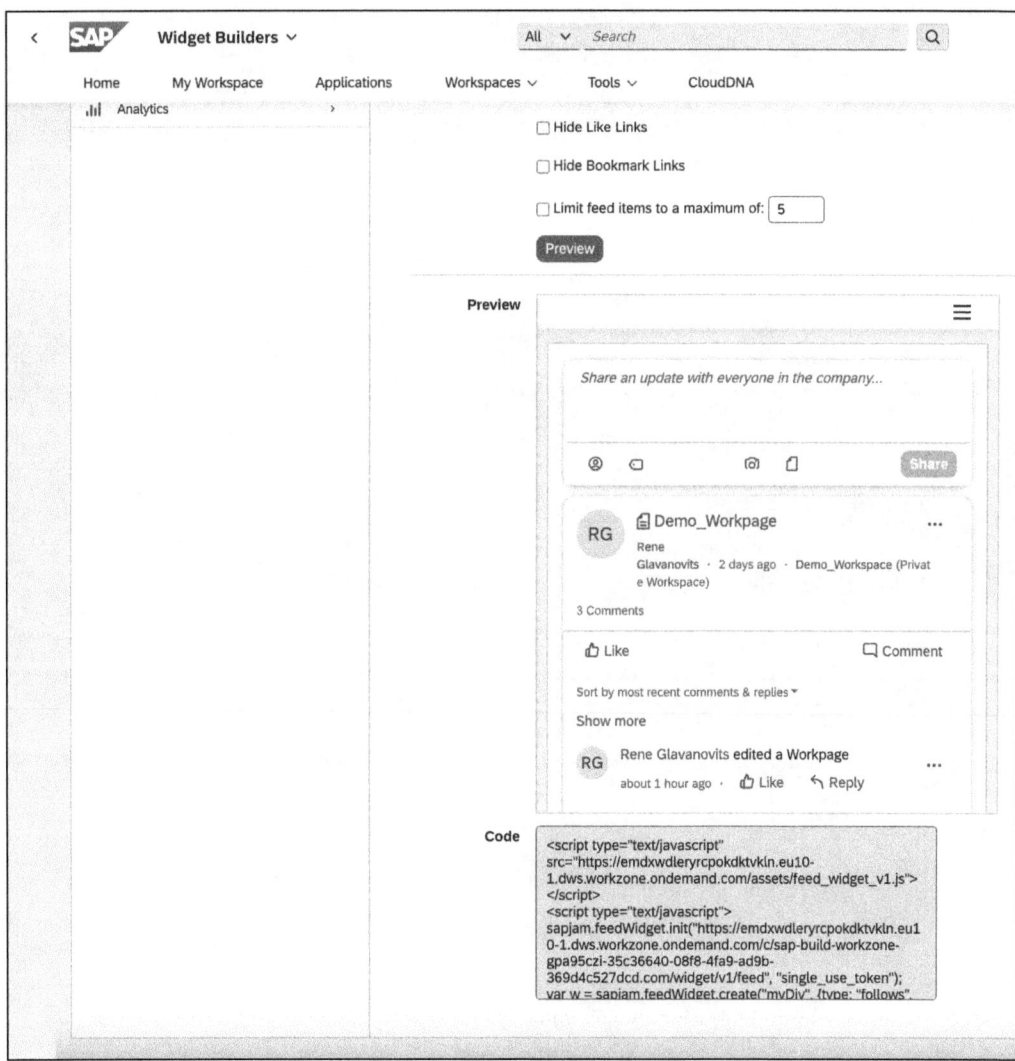

Figure 12.38 Preview of Widget Created with Widget Builder

12.4 Summary

In this chapter, we took a comprehensive look at the possibilities of widgets. In addition to how we can use widgets, we also looked at how to create new ones using the widget builder. We also looked at the creation, deployment, and activation of content packages and UI integration cards, respectively, in SAP Build Work Zone. In the following chapter, we'll look at the integration of external content—such as from Microsoft Teams—and business content. You'll learn how external services can be integrated into SAP Build Work Zone.

Chapter 13
External Integrations and Content Federation

In this chapter, we'll discuss integration of external content into SAP Build Work Zone. You have the capability to establish connections with diverse external applications and systems, forming a centralized access hub that fosters an efficient working environment for your users within SAP Build Work Zone.

Serving as a central nexus, SAP Build Work Zone assumes the role of an entryway, casting a wide net for the integration of diverse applications. Going beyond My Inbox and the many applications that belong to SAP Build Process Automation, this platform accommodates the assimilation of external content. For example, it includes SAP Fiori apps, sourced directly from an on-premise SAP S/4HANA system, a phenomenon referred to as *business content integration*. This integration not only showcases the platform's flexibility but also underscores its capacity to synergize disparate elements into a cohesive ecosystem. By leveraging preexisting assets, you can expedite the integration process, leading to the efficient deployment of content within SAP Build Work Zone. Importantly, this integration is equally attainable across both editions of SAP Build Work Zone: the standard and advanced editions. The necessary configuration is presented in detail in Section 13.1. In this section, you'll get a wide-ranging view of the integration process and an in-depth understanding of the functions at play.

Microsoft Teams is a collaboration platform that encompasses chat exchanges, meeting orchestration, file dissemination, and app integration. Intertwining Microsoft Teams with SAP Build Work Zone presents an avenue where the strengths of both platforms can coexist, culminating in an ecosystem that maximizes efficiency, amplifies collaboration, and accelerates task accomplishment. This integration is discussed in Section 13.2. By diving into the intricacies, you'll learn about the mechanisms that form the foundation of this integration.

13.1 Business Content

The content manager tool provides the capability to integrate applications from various origins into the work pages of your site via content providers. By utilizing content providers, you gain the ability to merge business applications and content originating from diverse sources. This procedure, known as *content federation*, facilitates the integration of this content into the **Applications** page of SAP Build Work Zone, advanced edition. These content providers serve as bridges, allowing content to be exposed from both cloud-based and on-premise systems. Presently, the supported content providers are as follows, with each offering distinct content exposure capabilities:

- For *SAP BTP content providers*, the content explorer displays the apps exposed by the content provider.
- For *remote content providers*, the content explorer displays the roles exposed by the content provider. Clicking a role displays the list of apps assigned to it.

The contents of an SAP BTP content provider are integrated as follows:

1. In the content manager, select a content provider and fetch its latest content.
2. Explore the available apps in the selected content provider and add them to **My Content**.
3. Assign the app to a group.
4. Assign your app to at least one role, to determine which users can access it.
5. To access the apps in the **Applications** page, assign the same role assigned to your apps to your site.
6. Add apps to the work pages in your site.

Remote content providers make on-premise and cloud applications from external origins accessible. The integration of the content made available occurs at the level of roles. Furthermore, all associated content elements linked to these roles, encompassing applications, groups, and catalogs, are seamlessly merged, culminating in comprehensive visibility within the operational runtime.

Presently, the following cloud solutions are provided with support as content providers:

- SAP Integrated Business Planning for Supply Chain (SAP IBP)
- SAP S/4HANA Cloud
- SAP BTP, ABAP environment

Currently, the following on-premise solutions are endorsed and operational as content providers:

- SAP S/4HANA
- SAP Business Suite
- SAP Enterprise Portal

The overarching process at a higher level can be outlined as follows:

1. Select a content provider and fetch its latest content.
2. Explore the roles in the selected content provider and add them to **My Content**.
3. Click a role to display the list of apps assigned to it.

4. To access the apps in the **Applications** page, assign the same role assigned to your apps to your site.
5. Add apps to the work pages in your site.

13.1.1 Create Destinations

To use SAP Fiori apps from on-premise systems, specific destinations are required in the SAP BTP subaccount. In the SAP BTP cockpit, configure a design-time destination and one or more runtime destinations. You must create a design-time destination to define the location from which to fetch the exposed content. In addition, you must configure the runtime destinations for launching that app. Therefore, navigate to **Connectivity • Destinations** in the side menu as shown in Figure 13.1 and click the **New Destination** button.

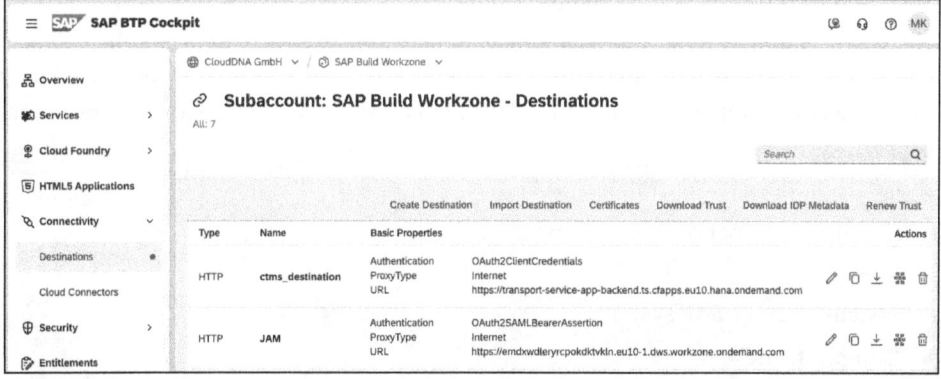

Figure 13.1 Destination Overview

The design-time destination establishes the source from which to retrieve the content made available by the content provider. Within the channel manager, you utilize the design-time destination to configure the content provider, enabling the retrieval and presentation of the provider's content through the content explorer.

The name of the destination can be chosen freely, but we recommend using the suffix "dt" in the name. As shown in Figure 13.2, we have chosen the name "s4hanadt". Select **HTTP** as **Type** and specify the virtual host in the **URL** field. This must have the path "/sap/bc/ui2/cdm3/entities" appended to it. Type in "OnPremise" for **Proxy Type** and "BasicAuthentication" for **Authentication**. Enter a valid SAP user in the **User** field and the corresponding password in the **Password** field. Create the additional "sap-client" property, to which the client of the SAP system is assigned. Finally, click **Save** and perform a connection check.

You must also create a runtime destination. The runtime destination defines the location from which to obtain the resources needed to run the federated apps during runtime. When federating applications from SAP S/4HANA or SAP Business Suite, there may be more than one runtime destination.

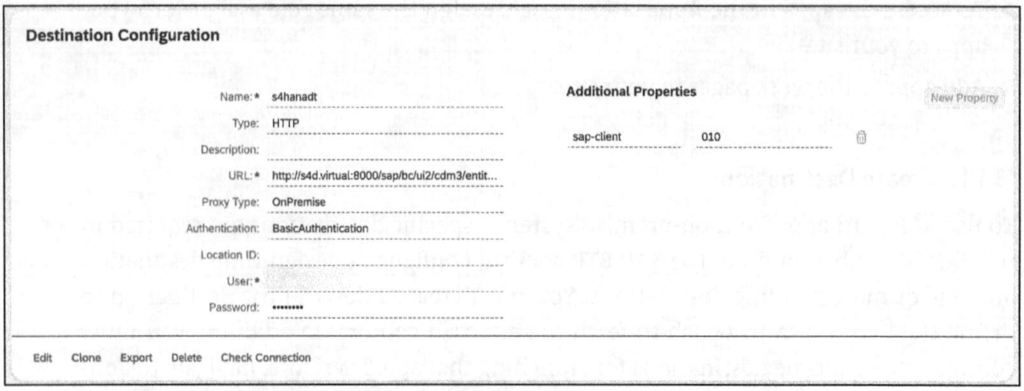

Figure 13.2 Design-Time Destination

The name of the destination can be chosen freely, but we recommend using the suffix "rt" in the name. As shown in Figure 13.3, we have chosen the name "s4hanart". Select **HTTP** for **Type** and specify the virtual host in the **URL** field. Select **OnPremise** for **Proxy Type**. Select **Principal Propagation** for **Authentication**. Create the following additional properties:

- "sap-client"
  The client number of the ABAP system.
- "sap-sysid"
  System ID of the SAP system.
- "sap-platform"
  Set to "ABAP".
- "sap-service"
  A concatenated string that contains four characters: the first two characters are 32, and the last two characters are the instance number of the ABAP application server (or the SAP system number).

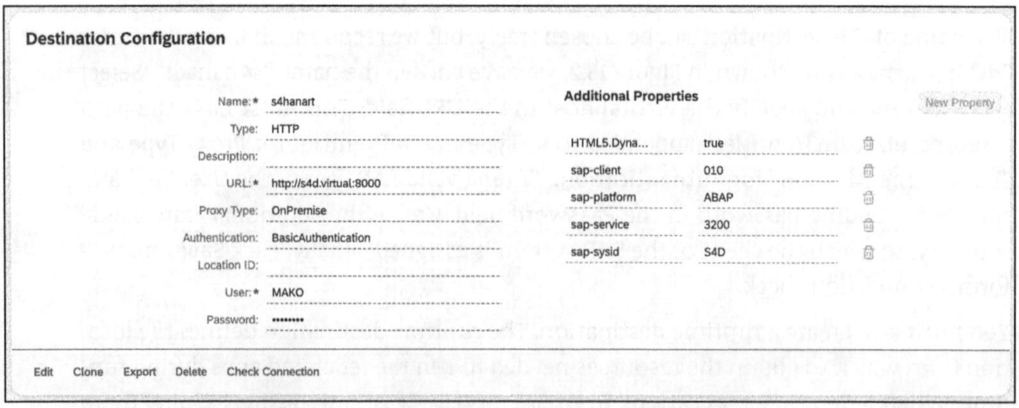

Figure 13.3 Runtime Destination

- **"HTML5.DynamicDestination"**
  Set to "true".

- **"sap-provider-label"**
  This is an optional property. Add this property to provide a user-friendly display name for this destination. This label is used in various runtime features inside SAP Build Work Zone.

Finally, click **Save** and perform a connection check. The connection check should show a success message.

13.1.2 Expose Content

In the next step, you must expose the desired SAP Fiori launchpad content to SAP BTP by choosing specific roles. As a result, the related content, such as groups, catalogs, pages, or spaces, can then be consumed in SAP BTP. One prerequisite for successful content exposure is the activation of the */sap/bc/ui2/cdm3* ICF node. Therefore, launch Transaction SICF using SAP GUI and navigate to the **sap • bc • ui2 • cdm3** path. If the node is not active, as shown in Figure 13.4, open the context menu and click **Activate Service**.

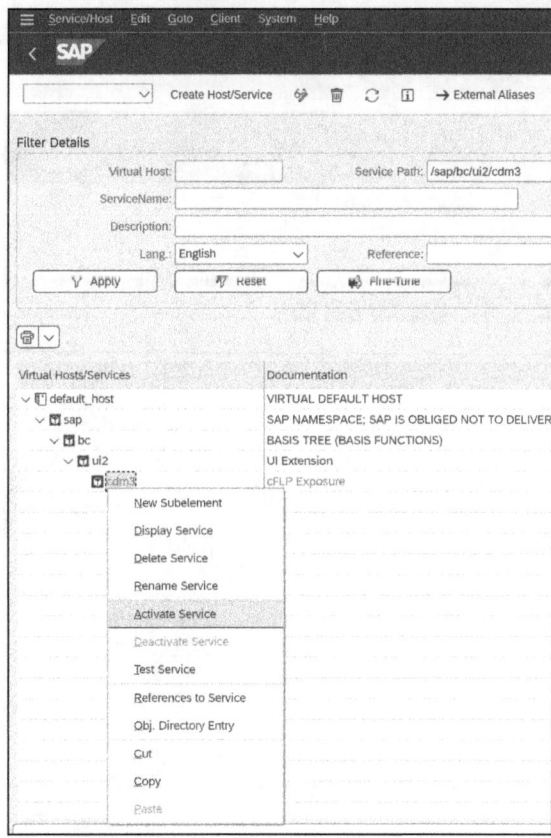

Figure 13.4 Activate ICF Node

13 External Integrations and Content Federation

To display, edit, or expose roles, you need to launch Transaction /UI2/CDM3_EXP_SCOPE. Therefore, the SAP_FLP_ADMIN role must be assigned to your user. With this role, the administrator can expose content to the launchpads on SAP BTP. Inside the transaction, you can choose either to expose all roles with launchpad content or to select specific roles that you want to expose (see Figure 13.5). To start the exposure, click the **Expose Content** button.

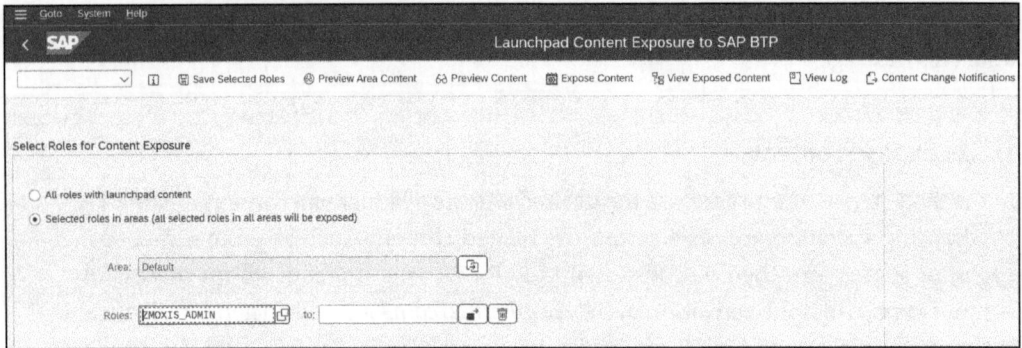

Figure 13.5 Select Roles for Exposure

After the execution of the exposure has completed, the **Exposure Log** opens (see Figure 13.6). Check to make sure that there are no errors in the log.

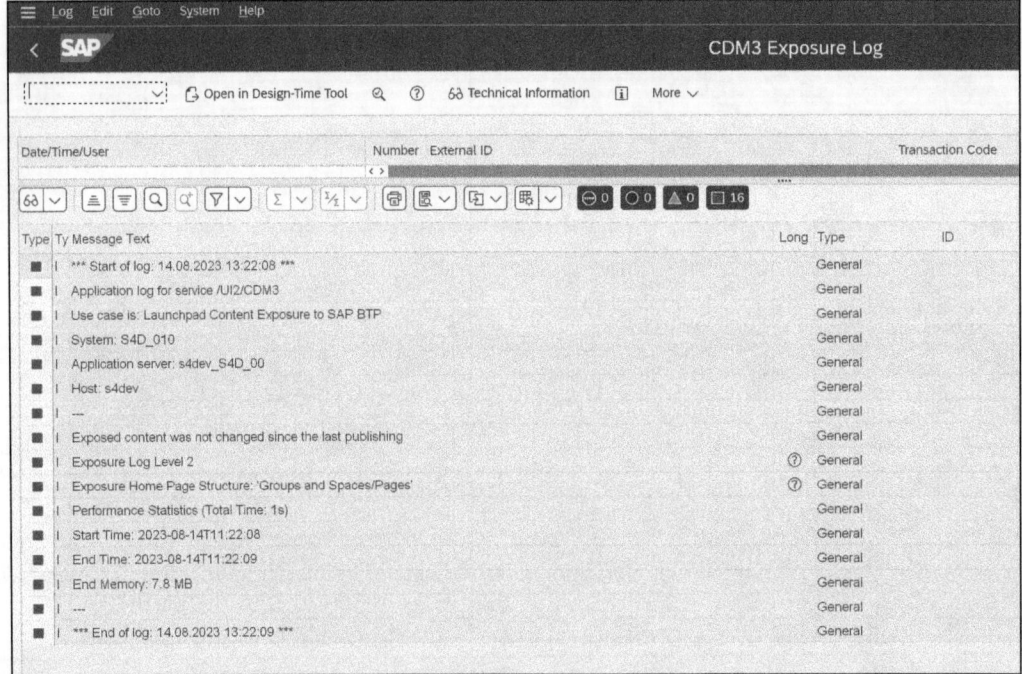

Figure 13.6 Exposure Log

13.1.3 Add Content Channel

That's all you need to do in the SAP S/4HANA system. The next step is to add a new content channel in SAP Build Work Zone. To do this, open SAP Build Work Zone and navigate to **Channel Manager** in the side menu (see Figure 13.7). Then click the **+ New** button.

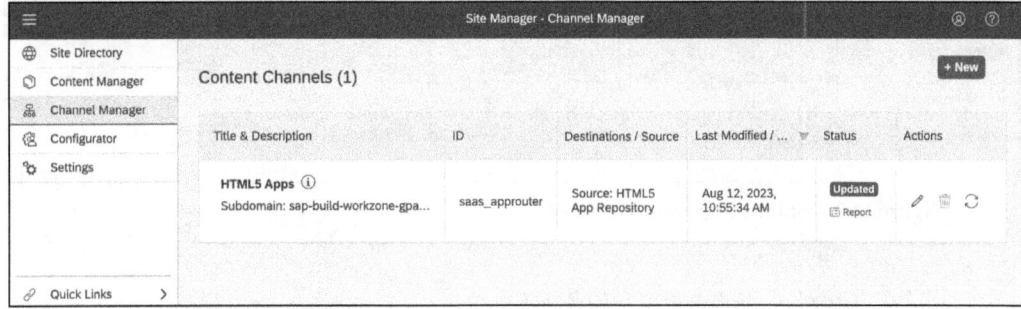

Figure 13.7 Add Content Provider

A dialog for creating a new content provider opens (see Figure 13.8). Maintain the **Title**, **Description**, and **ID** attributes. Select the previously created **Design-Time Destination** and **Runtime Destination**. For the remaining parameters, you can leave the default settings. Then click **Save**.

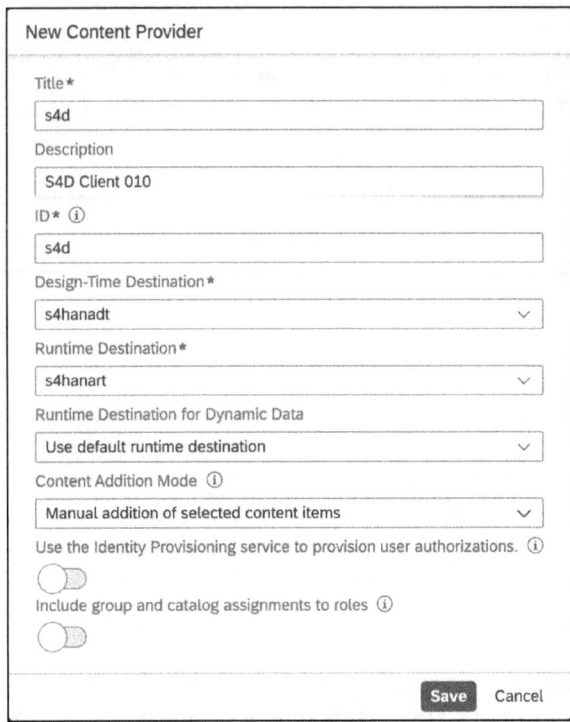

Figure 13.8 Content Provider Configuration

13 External Integrations and Content Federation

The new content channel should now be displayed as shown in Figure 13.9. In the **Status** field, you should see the **Updated** flag.

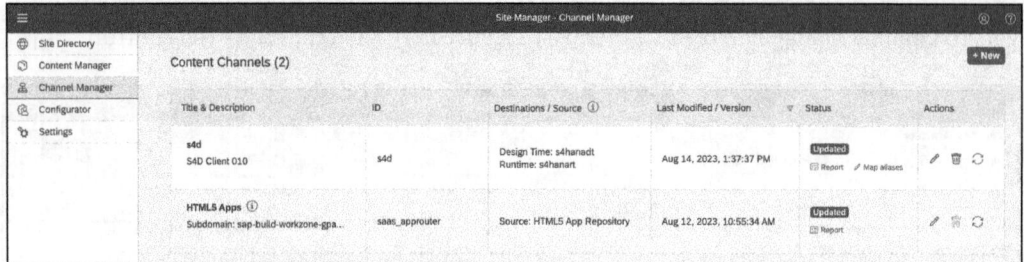

Figure 13.9 Content Channel Overview

13.1.4 Check Content and Add Roles

In the next step, you can check the published content. To do this, navigate to the **Content Manager** area in the side menu. The previously added content channel should now be visible there as a separate tile (see Figure 13.10). Click the corresponding tile to jump to the content.

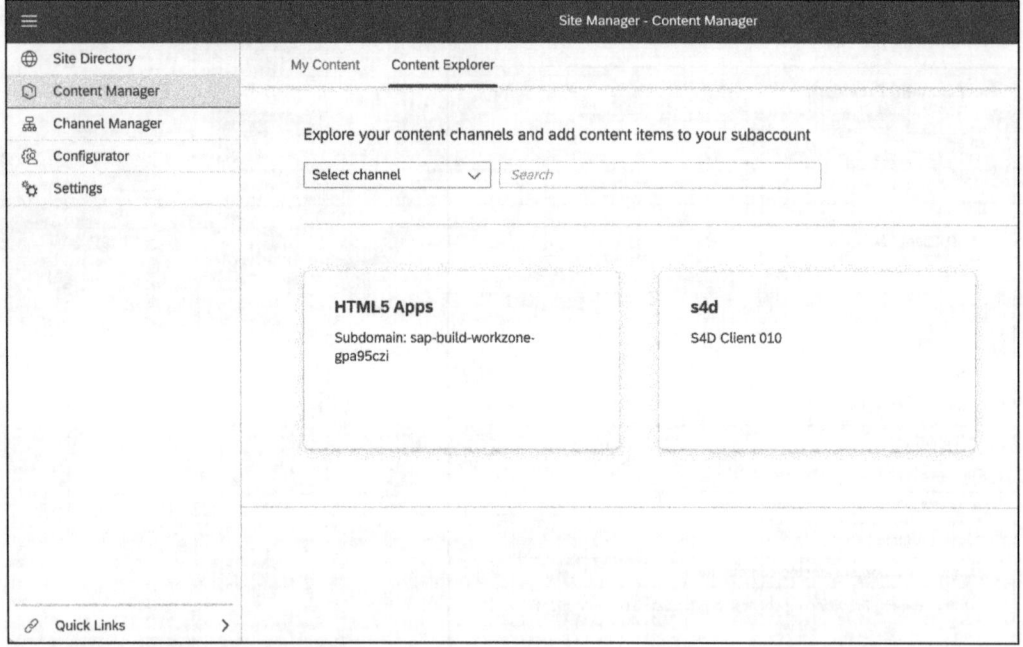

Figure 13.10 Explore Content

You'll now see a list of all roles published in the SAP system. The example in Figure 13.11 shows the **Moxis Administrator Role** and the **Moxis User Role**. You can add these roles to SAP Build Work Zone by selecting the desired roles in the **Items** list and then clicking **Add to My Content**.

13.1 Business Content

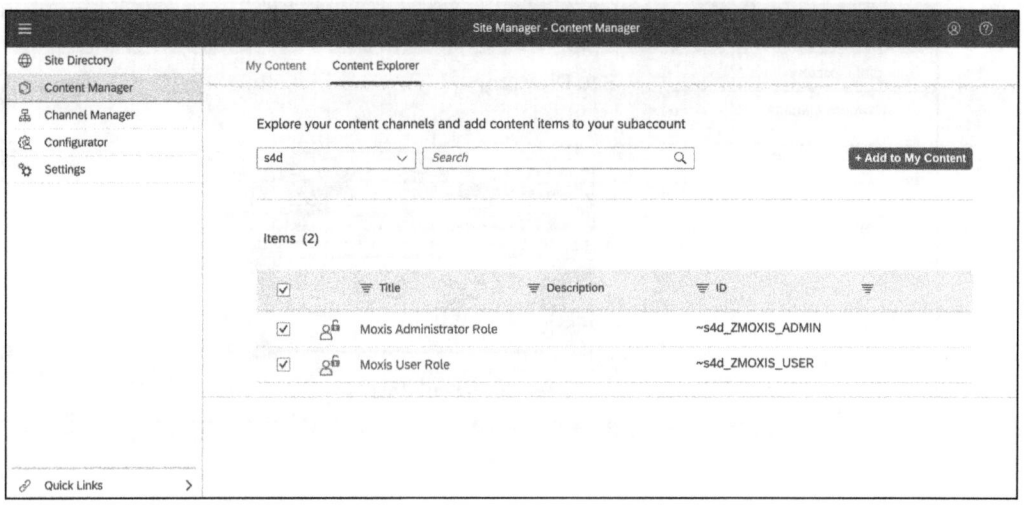

Figure 13.11 Add Roles to Content

Then, as shown in Figure 13.12, open the **My Content** tab to check whether the roles have actually been transferred. You have thus created the basis for using the roles from the on-premise SAP S/4HANA system in a site.

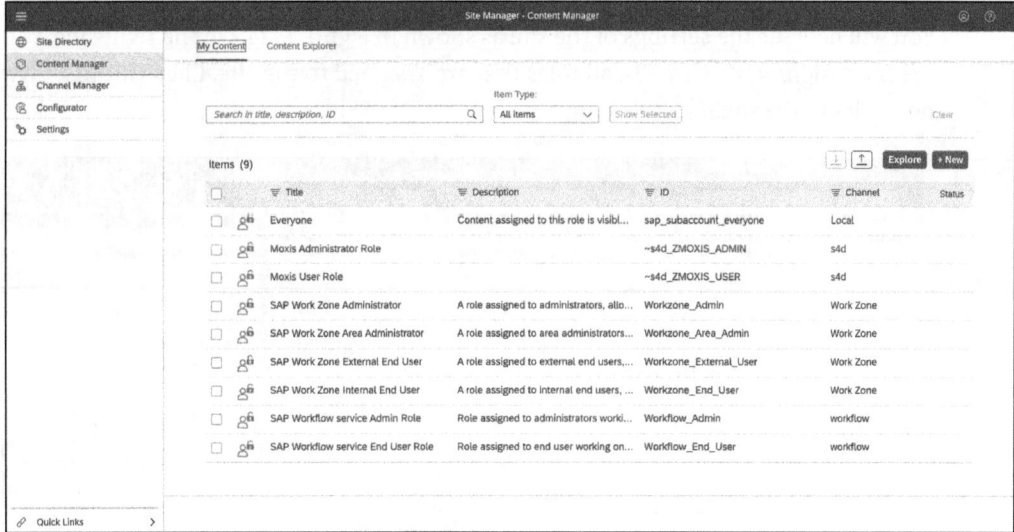

Figure 13.12 Check My Content

13.1.5 Configure Site

You can now proceed with the configuration of the site. To do so, navigate to the **Site Directory** section in the side menu as shown in Figure 13.13. Click the gear icon ⚙ to open the settings.

13 External Integrations and Content Federation

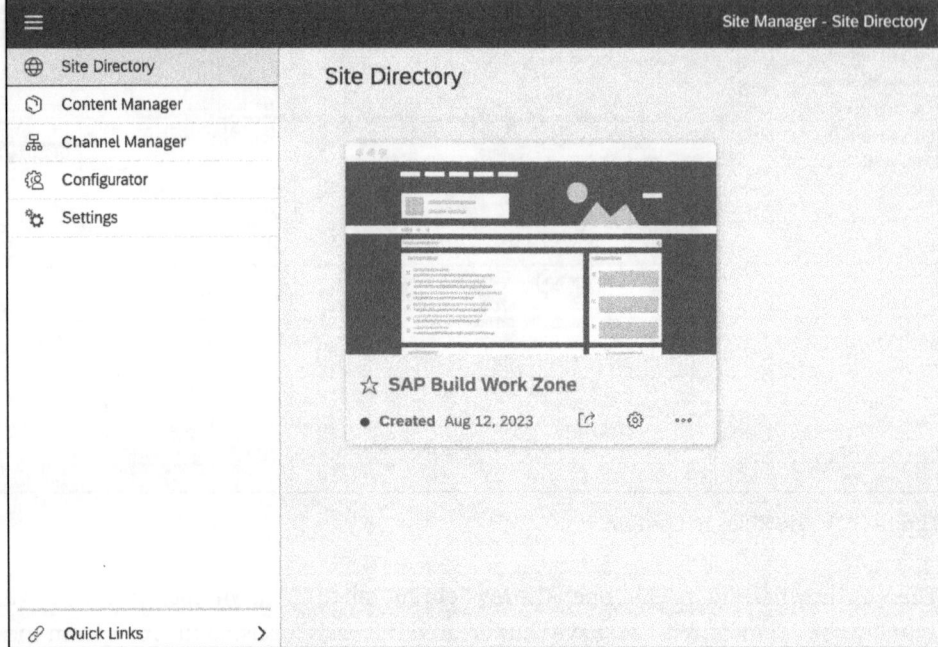

Figure 13.13 Site Directory

You will now see the settings of the site as shown in Figure 13.14. On the right side, you see the assignments—that is, all roles that are assigned to the site. Click **Edit** to assign new roles to the site.

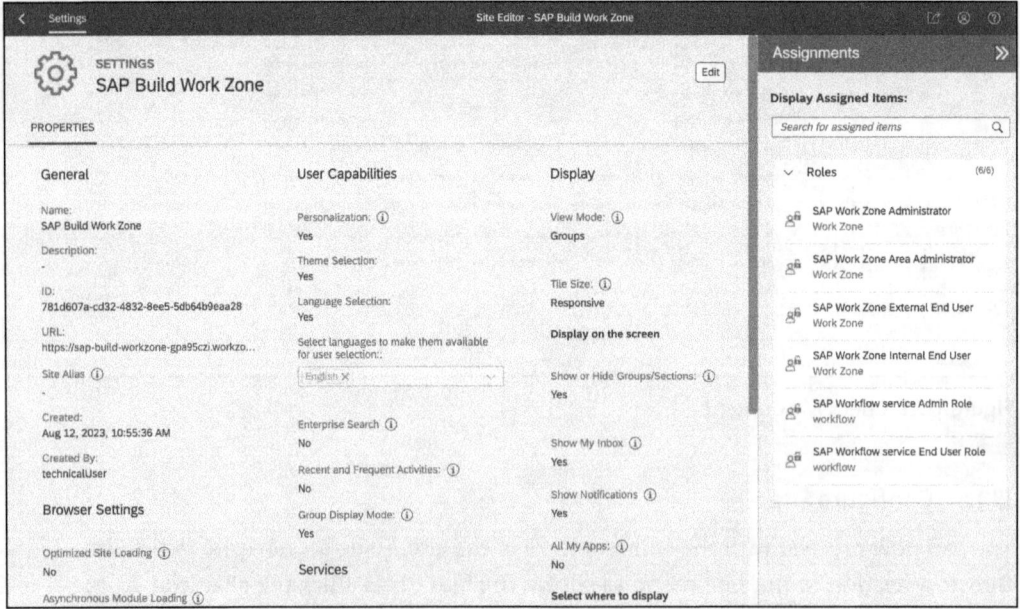

Figure 13.14 Site Settings

13.1 Business Content

Those roles that are not yet assigned to the site can be recognized by the **+** button (see Figure 13.15). By clicking this button, you can assign the role to the site. Finally, you must save the assignment by clicking the **Save** button.

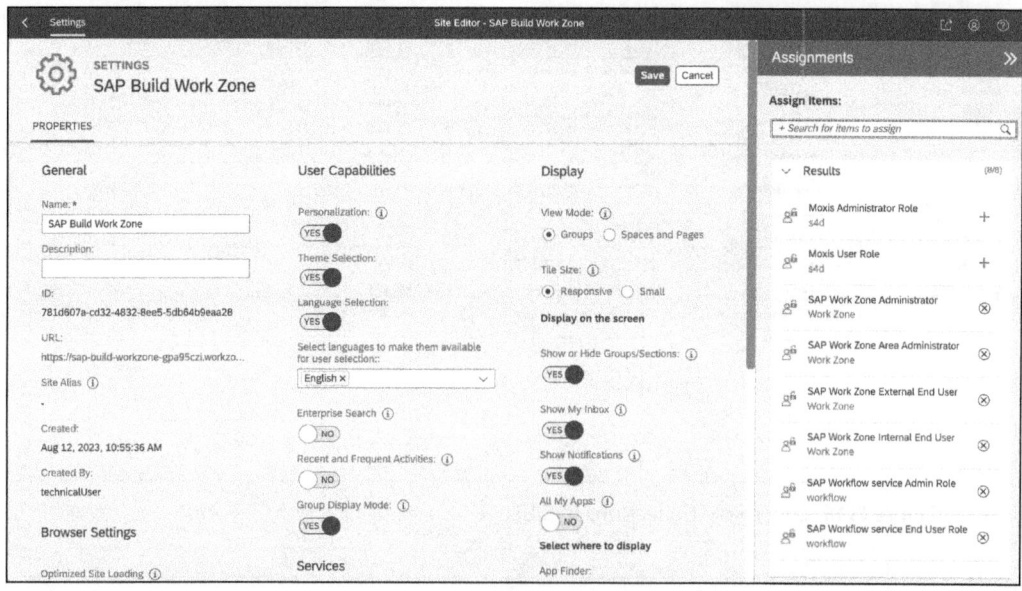

Figure 13.15 Add Roles to Site

13.1.6 Assign Roles

For users to access the launchpad, they still need to be assigned the appropriate role. Contrary to expectations, this is not done within SAP Build Work Zone, but in the subaccount in which the SAP Build Work Zone subscription is provided. To do this, navigate to the **Security • Role Collections** section in the side menu of the SAP BTP cockpit, as shown in Figure 13.16. Then click the desired role to open the details.

The assignment of role collections can be done either by assigning the users directly to the role collection or by mapping user groups from the identity provider. In both cases, it's necessary to switch to edit mode by clicking the **Edit** button (see Figure 13.17). Now enter the email address of the desired user in the **ID** field in the **Users** section. Then click the **Save** button.

You can now test whether the assignment of the roles has worked as desired. To do this, open SAP Build Work Zone. You should see the home screen, as shown in Figure 13.18. Click the **Applications** tab to jump to the app overview.

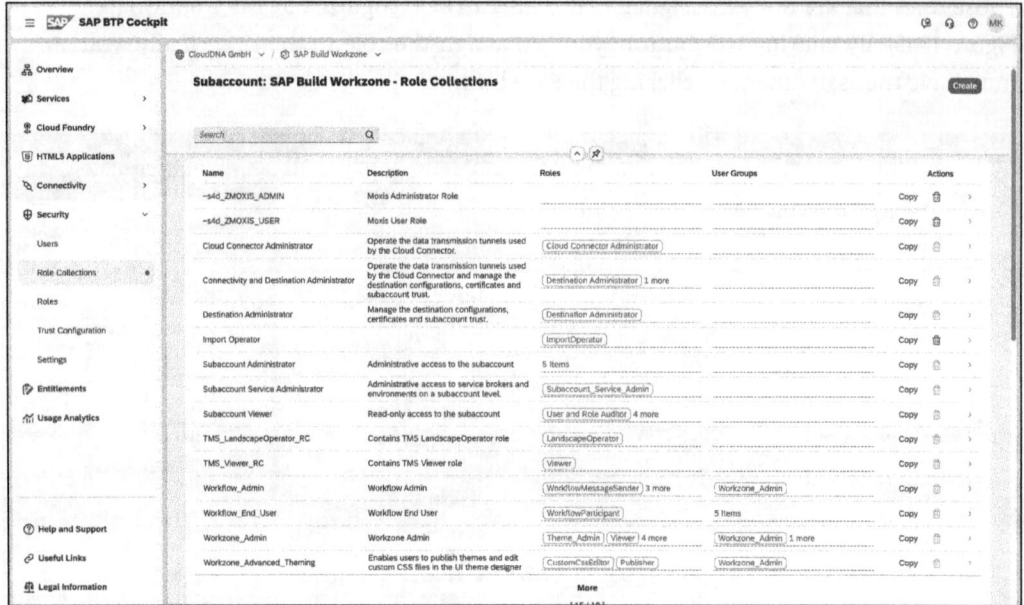

Figure 13.16 Check Role Collections in SAP BTP

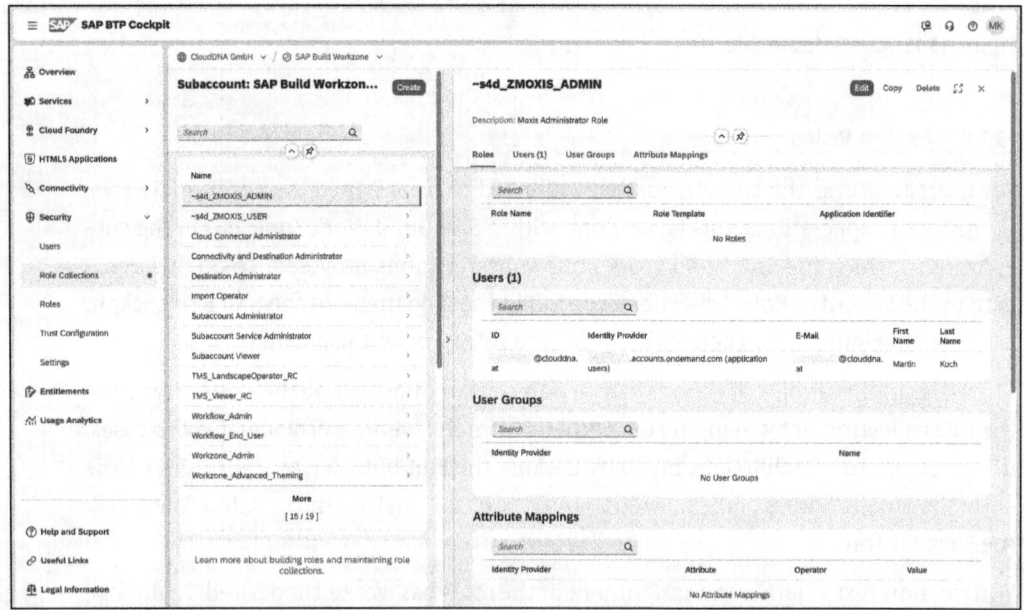

Figure 13.17 Add User to Role Collection

You should now see the apps contained in the roles (see Figure 13.19).

You can jump into an app by clicking, identical to clicking a tile representing the desired app in an on-premise SAP Fiori launchpad. After that, you should see the app's launch screen, as shown in Figure 13.20.

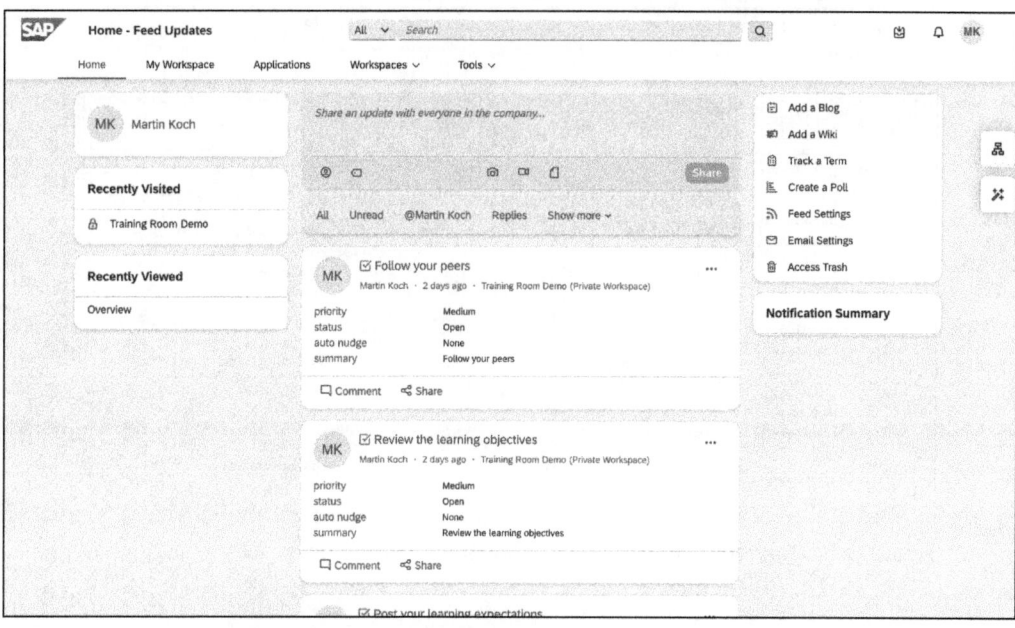

Figure 13.18 Check Results in SAP Build Work Zone

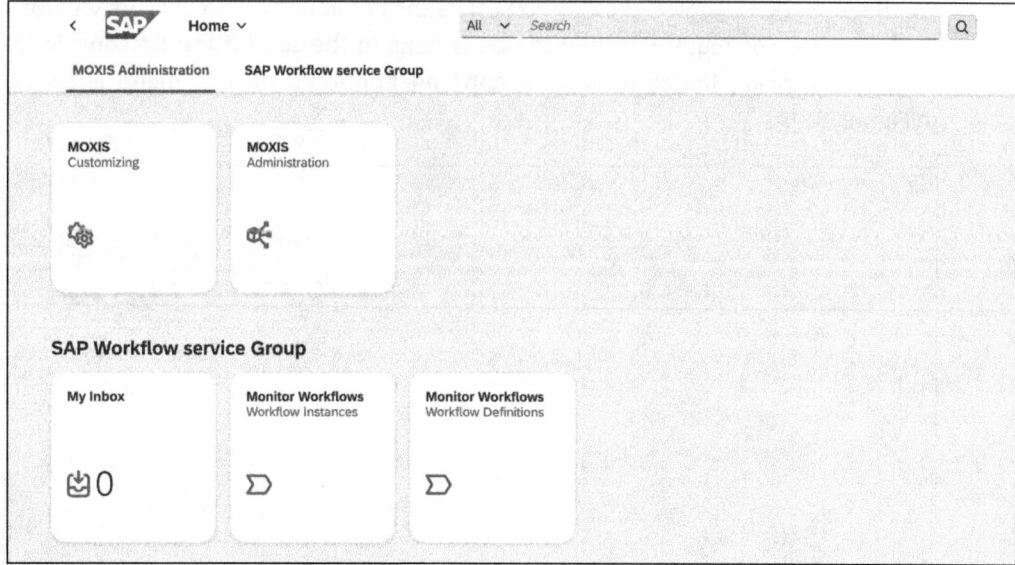

Figure 13.19 Check App Availability

This means you have successfully completed the configuration and performed a positive test. You can also perform a negative test by opening the launchpad with a user who lacks the required permissions. You should not see any apps in this case.

13　External Integrations and Content Federation

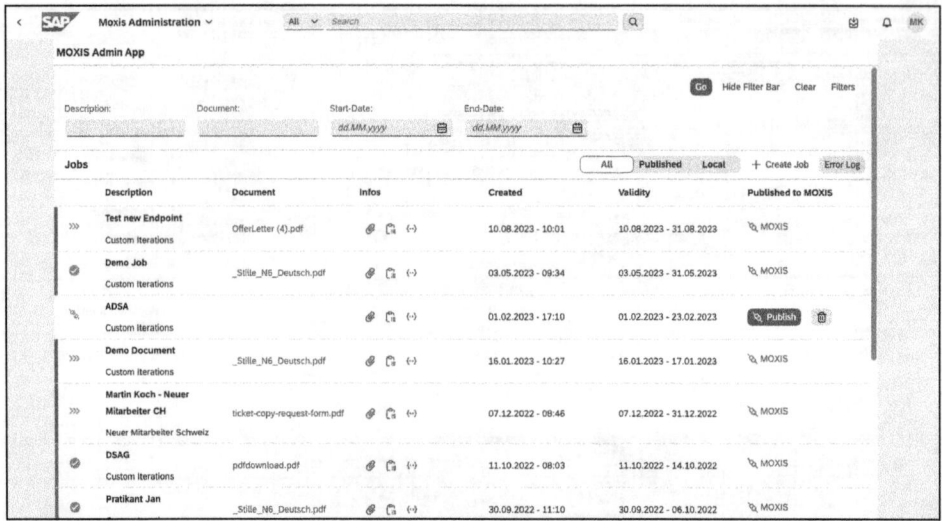

Figure 13.20 On-Premise App Running in SAP Build Work Zone

13.2　Microsoft Teams

SAP Build Work Zone, advanced edition offers the possibility of integration with Microsoft Teams. The required configuration is done in the administration console. To begin, open the app finder through the **App Finder** option in the user menu, as shown in Figure 13.21.

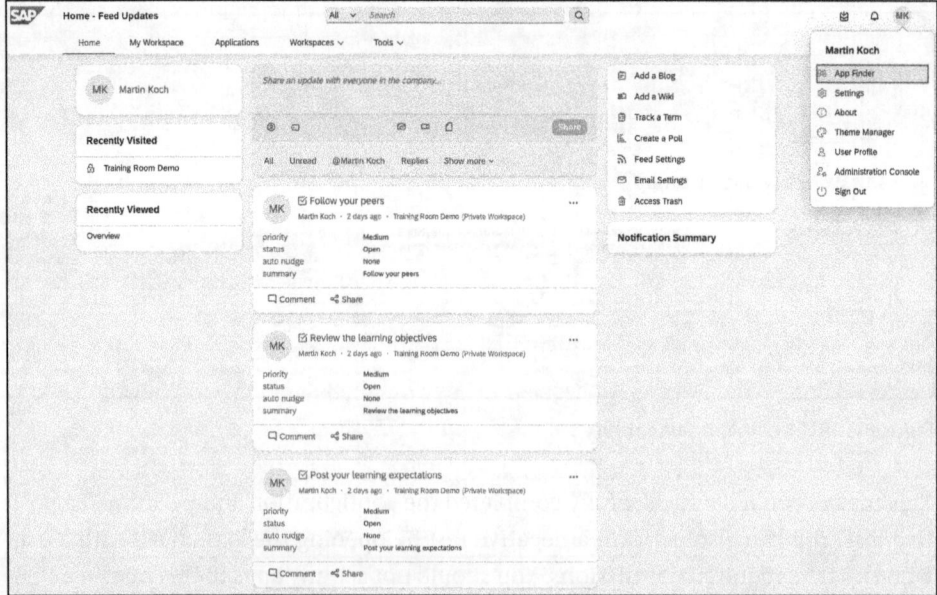

Figure 13.21 Open App Finder

13.2 Microsoft Teams

In the **SAP Work Zone** group, look for the tile named **Administration Console** (see Figure 13.22).

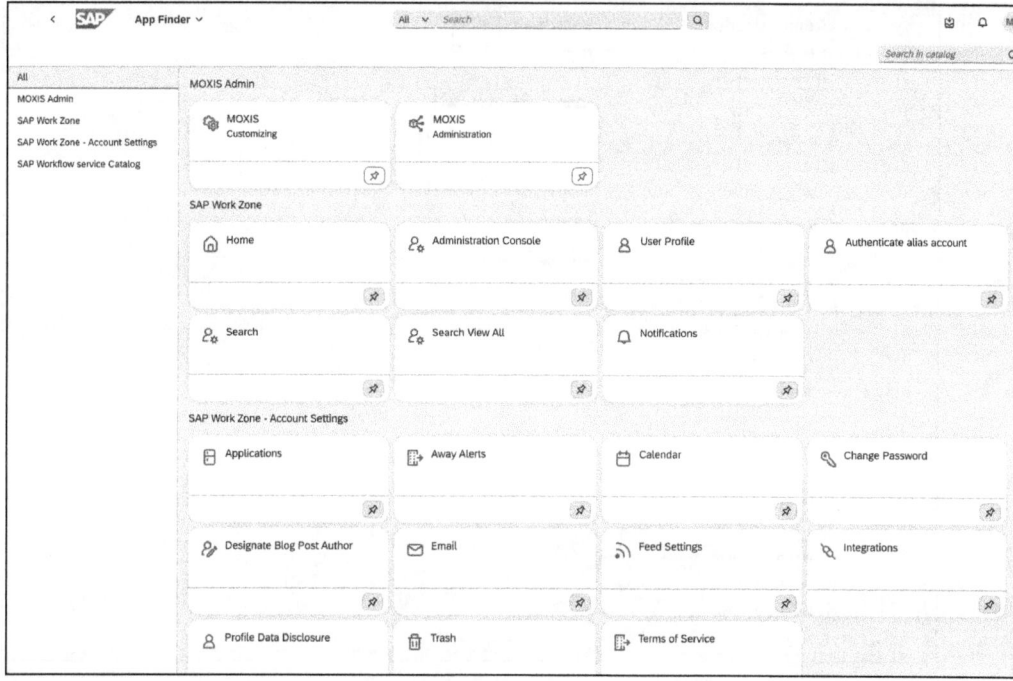

Figure 13.22 Search for Administration Console

Click this tile to launch the administration console (see Figure 13.23). Because you will need this app constantly in your daily work as an administrator, we recommend that you add the tile to your launchpad by clicking the pin icon.

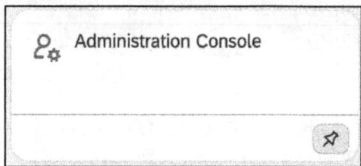

Figure 13.23 Launch Administration Console

In the admin console side menu, navigate to the **External Integrations • Microsoft Teams** section (see Figure 13.24). Activate the switch for **Chats**, and then click the **Download ZIP** button to generate the content for import into Microsoft Teams.

Now open Microsoft Teams, then navigate to the **Apps** section in the side menu as shown in Figure 13.25. From there, click **Manage Your Apps** in the menu.

13 External Integrations and Content Federation

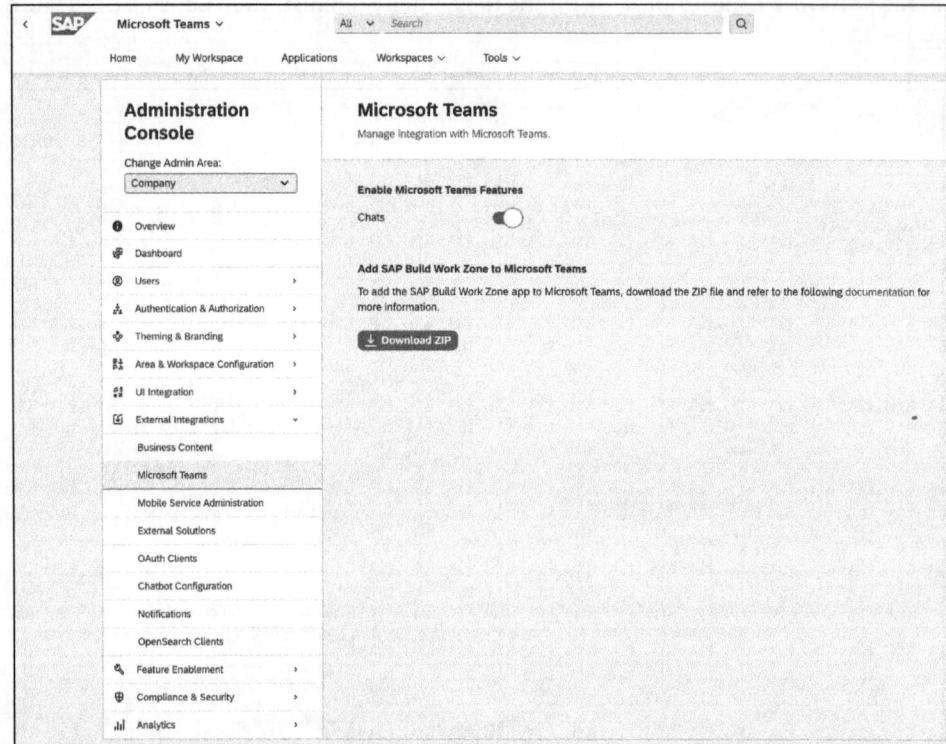

Figure 13.24 Download Configuration

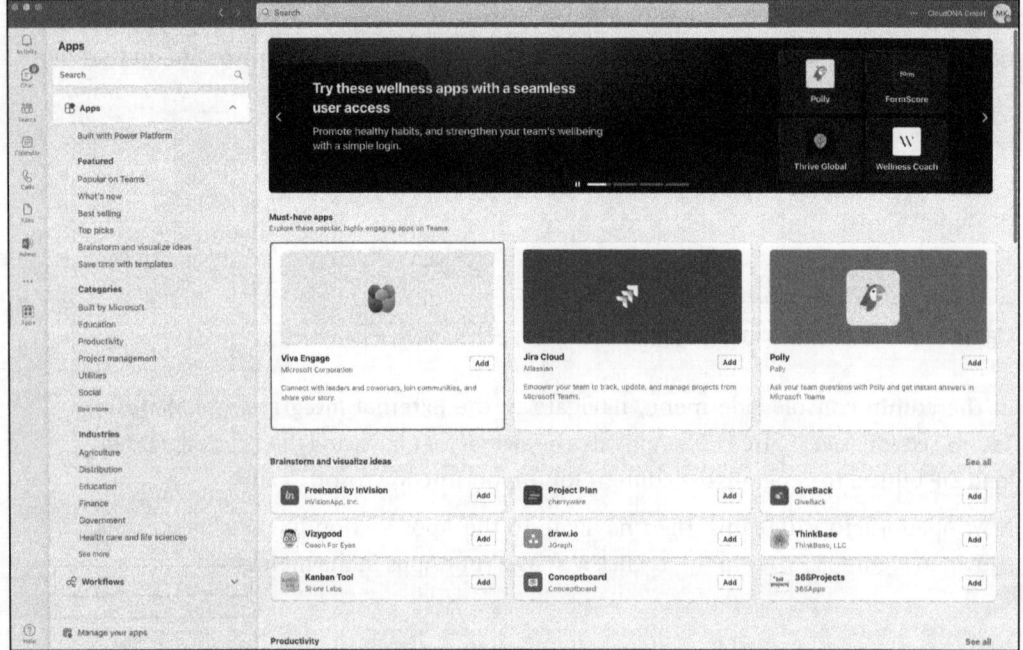

Figure 13.25 Open Microsoft Teams Apps

13.2 Microsoft Teams

Within the **Apps** tab, click the **Upload an App** button (see Figure 13.26).

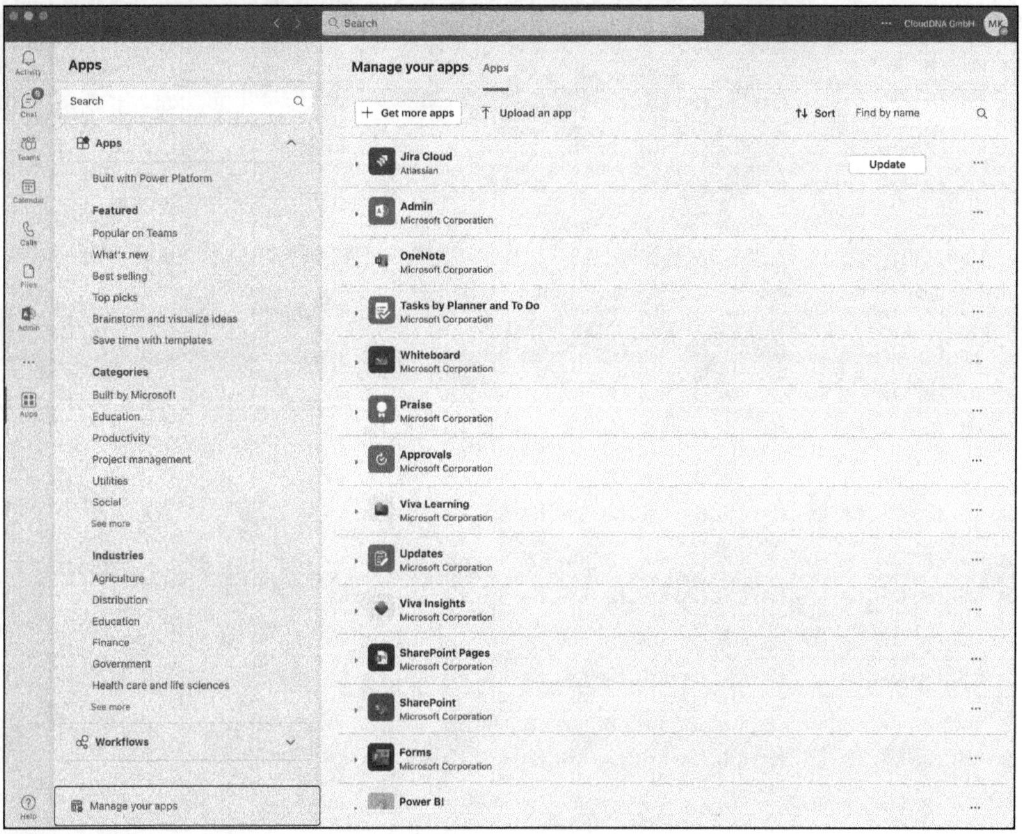

Figure 13.26 Manage Apps

After that, a dialog opens, as shown in Figure 13.27. Click **Upload an App to Your Org's App Catalog**. Now the **File Upload** dialog opens, in which you can select the previously exported SAP Build Work Zone content.

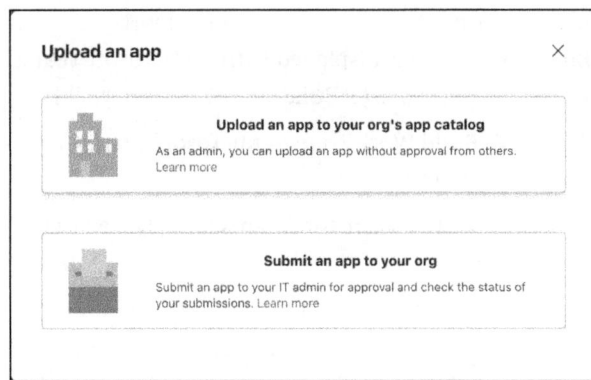

Figure 13.27 Upload an App

The app should now be displayed, as shown in Figure 13.28. Click the tile.

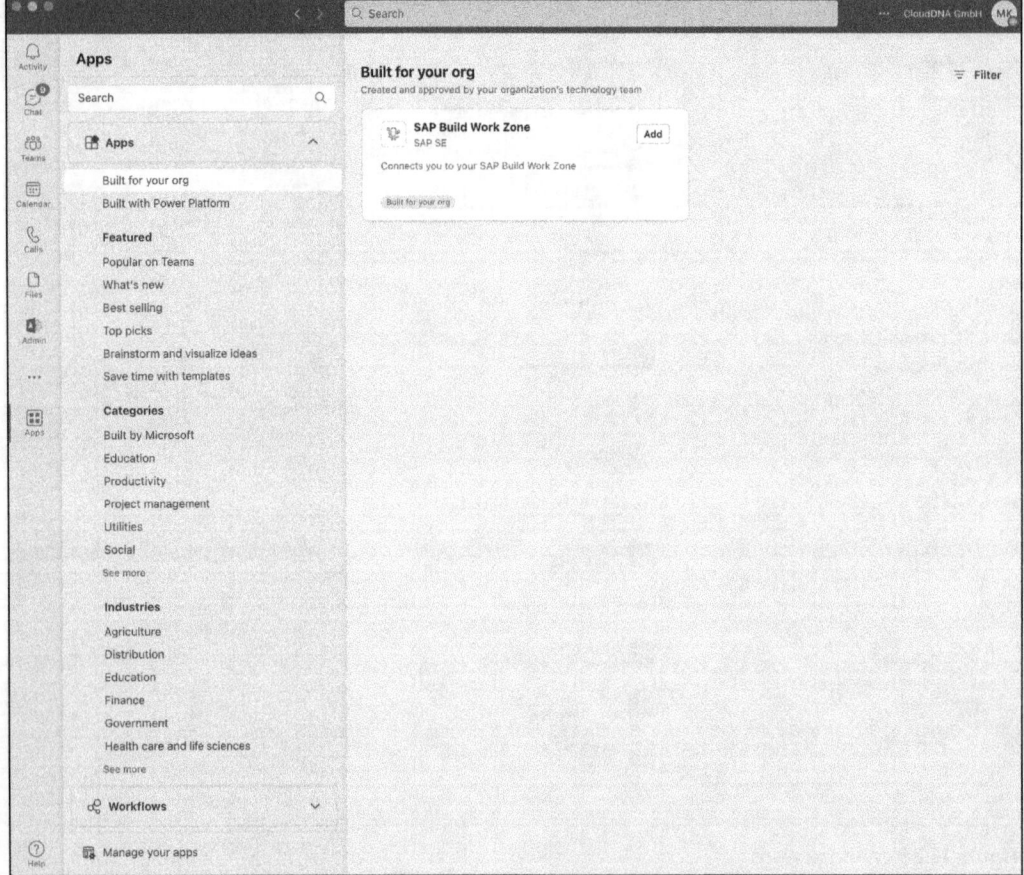

Figure 13.28 Check Result of Upload

You will now see the details of the app. Here you will find the **Overview and Permissions** tabs. Click the **Add** button (see Figure 13.29) after carefully reading the contents.

You must now log into SAP Build Work Zone. To do this, use the credentials of your user. After successful login, SAP Build Work Zone is displayed within Microsoft Teams. As you can see in Figure 13.30, only certain areas of SAP Build Work Zone are visible.

You can now also share content from SAP Build Work Zone in Microsoft Teams. To do this, open the **User Profile**, as shown in Figure 13.31.

You can click 🗨 in the user profile to jump off to Microsoft Teams (see Figure 13.32).

13.2 Microsoft Teams

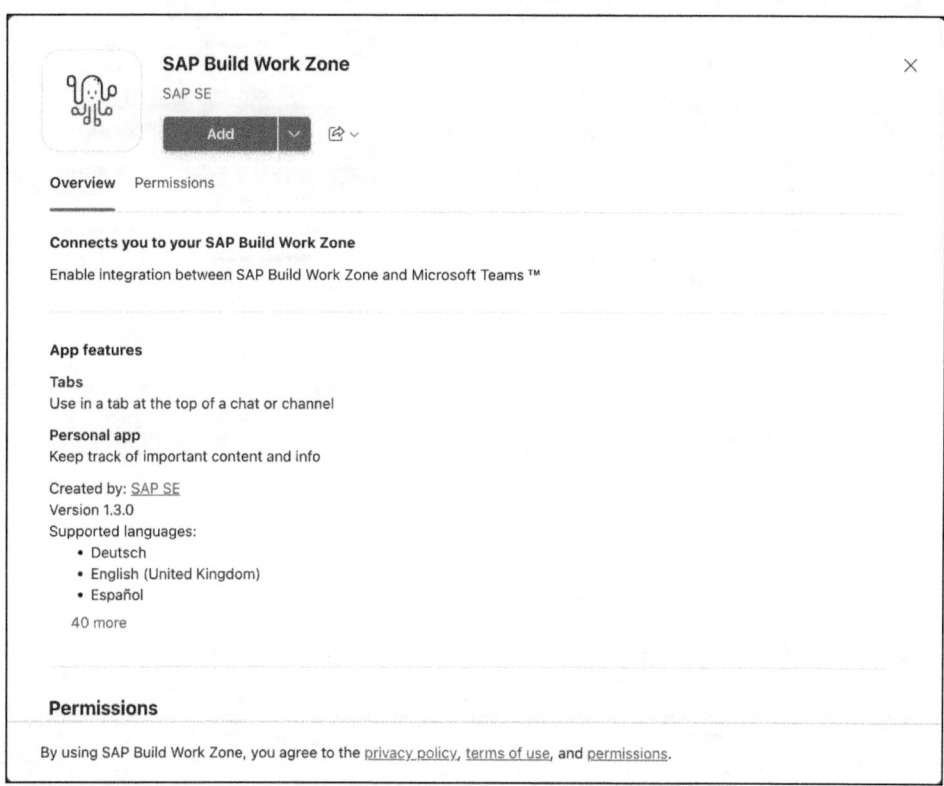

Figure 13.29 Add App

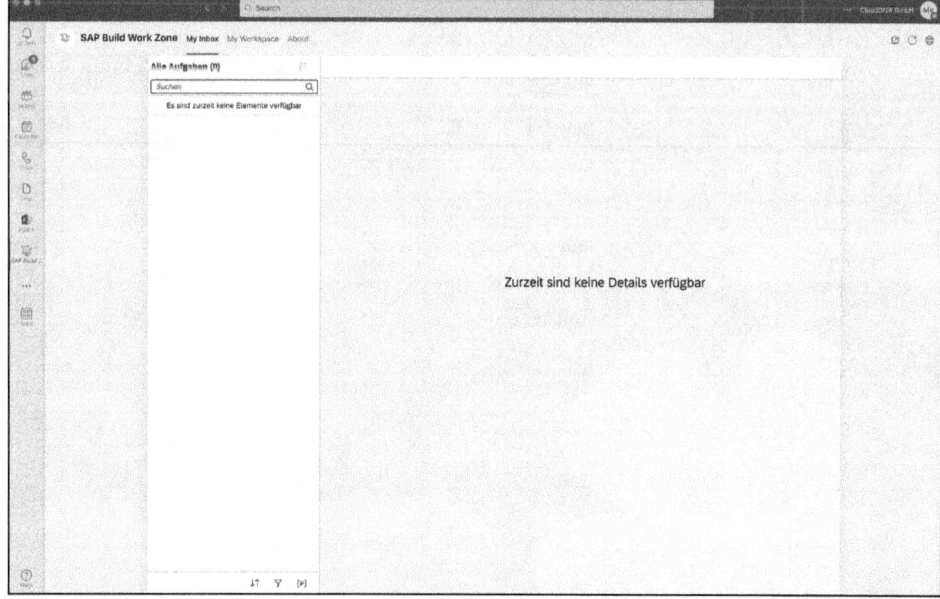

Figure 13.30 Check SAP Build Work Zone Apps inside Microsoft Teams

13　External Integrations and Content Federation

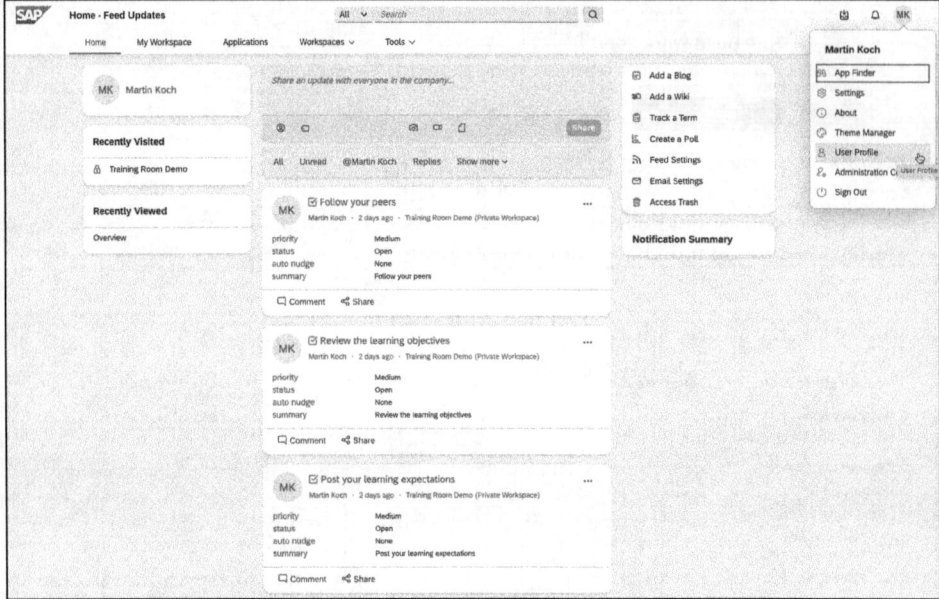

Figure 13.31 Open User Profile

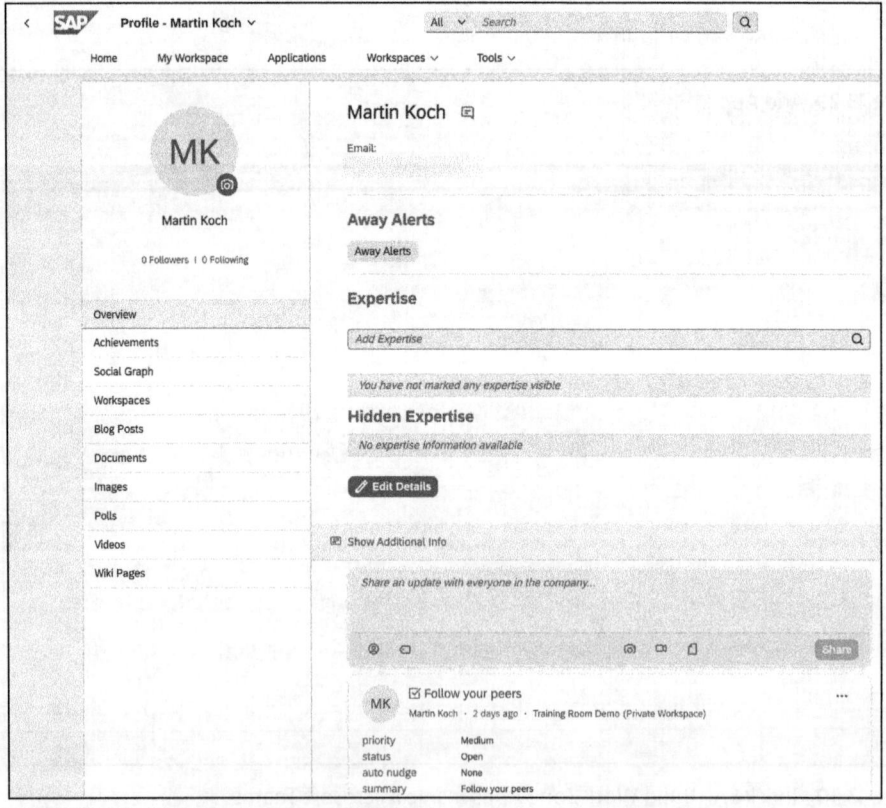

Figure 13.32 Start Chat

422

After that, the Microsoft Teams app should open (see Figure 13.33).

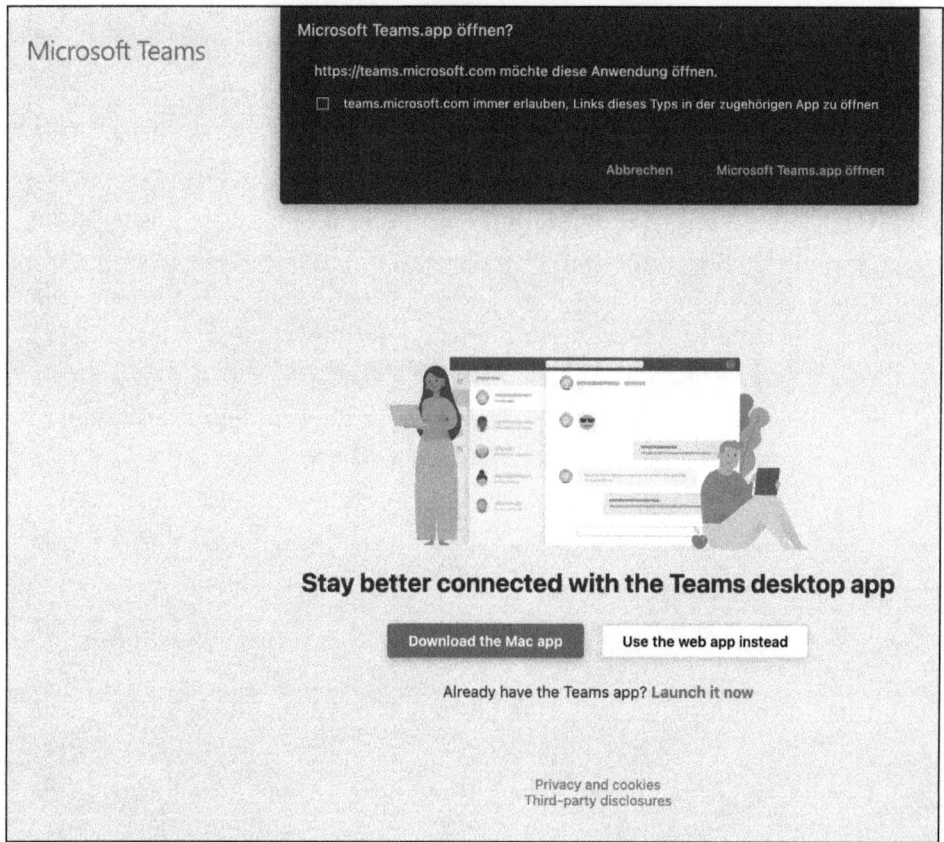

Figure 13.33 Microsoft Teams Launch

13.3 Summary

In this chapter, you have learned about some of the integration possibilities of SAP Build Work Zone. In Section 13.1, we integrated business content from an on-premise system in the form of SAP Fiori apps. This gives you the option to deploy an SAP Fiori launchpad on SAP BTP. The integration into the on-premise landscape is done via the cloud connector.

Another common use case is integration with Microsoft Teams, an application that has established itself as the de facto standard for enterprise collaboration. We discussed the required minimal configuration in Section 13.2.

Chapter 14
Content Transport

Running SAP Build Work Zone in a staged environment makes it necessary to transport content between different stages. This can be done either manually via file export and import or by using SAP Cloud Transport Service. We'll discuss both options in this chapter.

SAP Build Work Zone, advanced edition provides functionality for the transfer of business content between various services and environments. Building upon SAP's legacy of transport management in the on-premise world, this capability extends its reach not only to moving content between different services but also to facilitating the transfer of specific site content items between subaccounts and data centers. In SAP Build Work Zone, advanced edition, the transport process is remarkably like that in the on-premise world: exporting content from the source system and effortlessly importing the exported file into the target system, ensuring that content migration remains a straightforward operation. The following use cases are supported:

- **Streamlining environment transitions**
  One of the core applications of content transport is when you need to migrate content between distinct environments. For example, you might be working on a project that requires you to transfer content from the development (dev) environment to the testing (test) and eventually the production (prod) environments. This ensures that changes and updates made in one environment can be seamlessly deployed and rigorously tested in others, contributing to a more agile and efficient workflow.

- **Interservice content exchange**
  SAP Build Work Zone, advanced edition exists within a broader ecosystem that encompasses various services, such as SAP Cloud Portal; SAP Build Work Zone, standard edition; and SAP SuccessFactors Work Zone. With content transport capabilities, you gain the ability to smoothly transport content between these diverse services. This opens new horizons for collaborative efforts, allowing cross-functional teams to work harmoniously by sharing and updating essential content, regardless of the service they utilize.

- **Facilitating content migration across subaccounts**
  Beyond services, SAP Build Work Zone caters to the nuanced requirements of subaccounts. You can transport a wide spectrum of content items, spanning from home pages and workspaces to workspace templates, from one subaccount to another.

This level of flexibility empowers organizations to efficiently distribute content across different subaccounts, fostering a more organized and cohesive work structure.

To harness the full potential of content mobility, it's crucial to understand what precisely can be transported between services when SAP Build Work Zone, advanced edition serves as the content source. First, let's look at the target system, as follows:

- **SAP Cloud Portal service**
  The supported content includes specific business content items like apps, roles, groups, catalogs, and shell plugins. Notably, the content that gets filtered out during this transport process is the out-of-the-box content from SAP Build Work Zone, advanced edition as it doesn't align with the requirements of the SAP Cloud Portal service.

- **SAP Build Work Zone, standard edition**
  The supported content remains consistent with the selected business content items, encompassing apps, roles, groups, catalogs, and shell plugins. Like the previous case, the content to be filtered out comprises the out-of-the-box content from SAP Build Work Zone, advanced edition as it holds no relevance within the context of SAP Build Work Zone, standard edition. This ensures that the transported content remains finely tuned to the specific needs and functionalities of the chosen system.

When you shift your focus to the scenario in which SAP Build Work Zone, advanced edition, serves as the target system for transporting selected content items, the dynamics change yet again. Given the advanced edition's elevated functionality, it becomes crucial to discern what elements from the source system are directly applicable and relevant to its enhanced features. In the context of content transportation, different source systems present unique considerations, as follows:

- **SAP Cloud Portal**
  The content earmarked for transportation includes selected business content items. These encompass apps, roles, groups, catalogs, and shell plugins, representing the essence of the digital workspace. However, there's a deliberate exclusion during the transport process. Freestyle site entities, such as pages, menus, and widgets, are filtered out as they don't align with the transport requirements for this scenario.

- **SAP Build Work Zone, standard edition**
  The transported content remains consistent with the selected business content items, mirroring the essentials of apps, roles, groups, catalogs, and shell plugins. Notably in this case, there's no filtered content, signifying that everything selected for transport finds its place seamlessly within SAP Build Work Zone, standard edition.

- **SAP Build Work Zone, advanced edition**
  The transported content encompasses selected business content items, including apps, roles, groups, catalogs, and shell plugins. What sets this scenario apart is the inclusion of SAP Build Work Zone, advanced edition content items, such as home

pages, workspaces, and workspace templates, which enrich the digital workspace's functionality. Importantly, the filtration process targets the exclusion of SAP Build Work Zone, advanced edition's out-of-the-box content. This content doesn't necessitate transportation as it's inherently delivered with any site, streamlining the transport process for maximum efficiency.

Content transfer offers various options, depending on your preferences and requirements. Manual content transfer, as explained in detail in Section 14.1, gives you full control over the transfer process. Here you can select content and move it to the desired destination. This is especially useful if you have specific customizations or preferences that require individual handling. Alternatively, SAP Cloud Transport Management on SAP BTP offers a structured and logged approach, as explained in Section 14.2. We generally recommend this latter approach because it is an efficient and transparent method for content transfer. Here you can perform the transport process in a systematic framework, which greatly facilitates traceability and management.

14.1 Manual Transport

Manual content transport is the simplest form of transport. The content to be transported is first downloaded to the administrator's computer, then the content is uploaded and imported into the target system by means of a manual upload. This form of transport has the major disadvantage that it is error-prone. For example, you might unintentionally capture an older version of the content during the upload. The following content types can be transported in this way:

- Business content (Section 14.1.1)
- Home page (Section 14.1.2)
- Workspaces (Section 14.1.3)
- Workspace templates (Section 14.1.4)

14.1.1 Business Content

Business content is transported using the content manager. To do this, open the administration console as shown in Figure 14.1 and navigate to the **External Integrations • Business Content** area in the side menu. From there, jump to the content manager by clicking **Content Manager**.

In the content manager, select the objects you want to transport (see Figure 14.2). You can transport apps and plugins, catalogs, groups, and roles. Then click on the **Download** button [⬇].

14 Content Transport

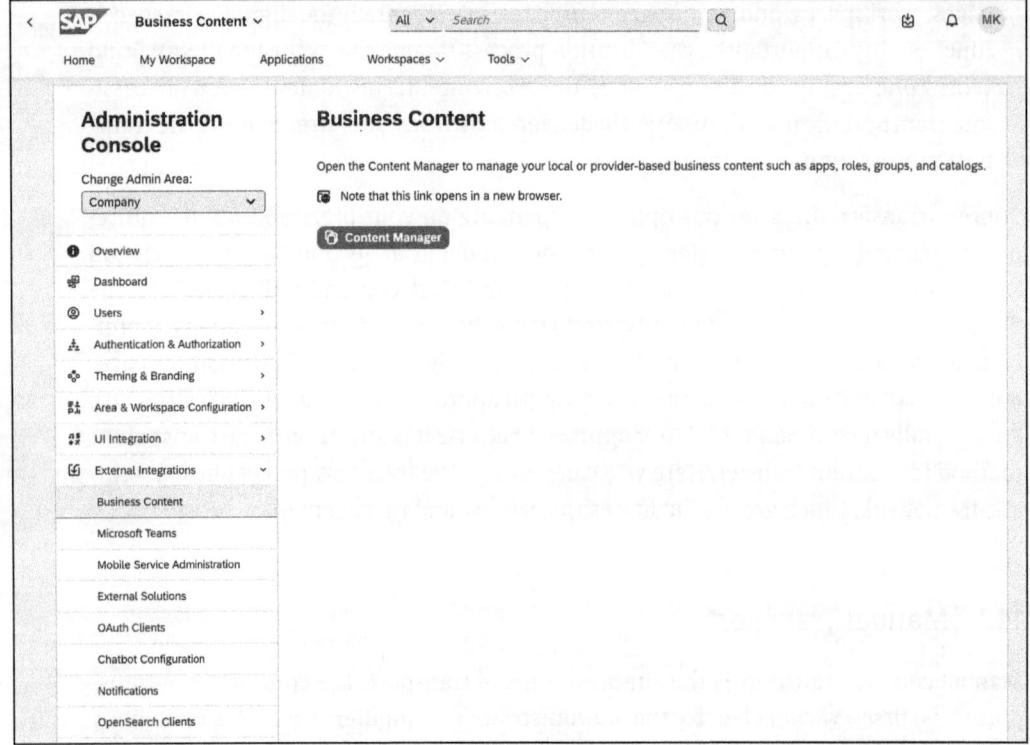

Figure 14.1 Content Manager

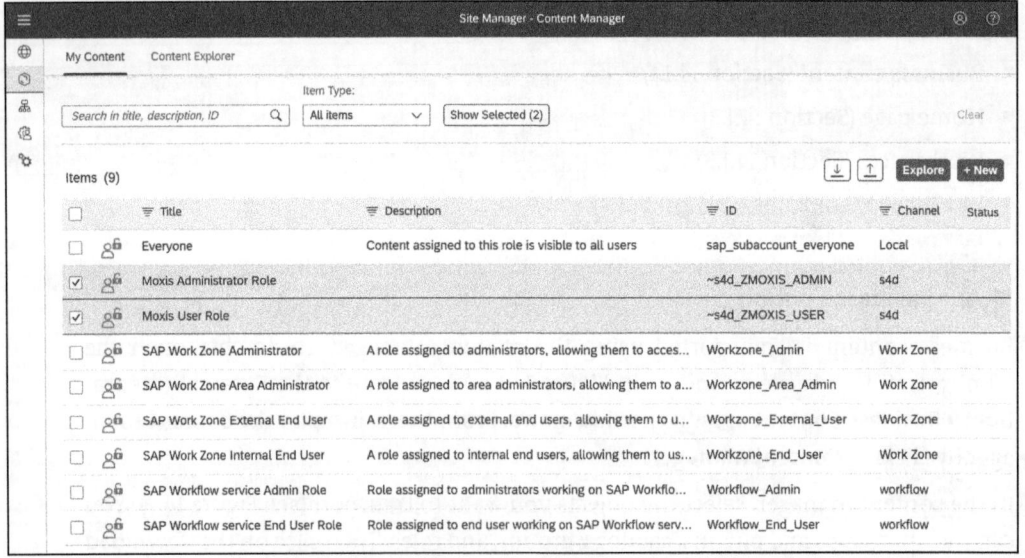

Figure 14.2 Select Objects for Transport

You will now see a popup in which the selected objects are displayed (see Figure 14.3). At this point, check whether all objects are included. Then click the **Export** button.

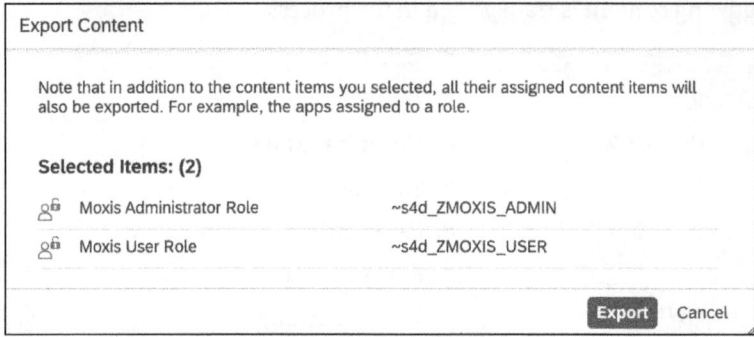

Figure 14.3 Confirm Content Export

The content is then downloaded to your computer in the form of a ZIP file. In the next step, the content must be imported into the target system—for example, into the productive SAP Build Work Zone system. To do this, log onto the target system and open the content manager. Click on the **Upload** button ⬆ to start the import wizard.

A popup will now open where you have to select and import the previously exported content (see Figure 14.4). First click the **Browse** button and then select the ZIP file to be imported. Then click **Agree and Import**.

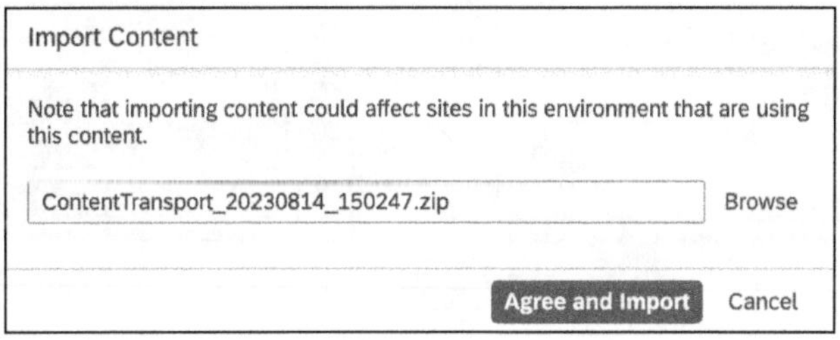

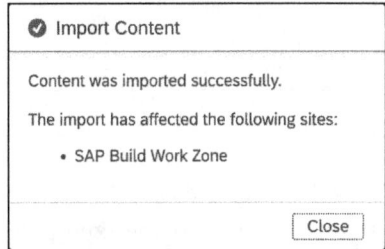

Figure 14.4 Select Content for Import

After the import has been performed, you will see a summary of the import, as shown in Figure 14.5. You can also see which sites are affected by the import.

Figure 14.5 Import Confirmation

14.1.2 Home Page

SAP Build Work Zone, advanced edition offers the possibility to transport home pages manually. The following content is transported in the process:

- Folders
- Blog posts
- Documents (e.g., TXT, DOCX, or XLSX files)
- Links
- Images
- Polls (without voting results)
- Videos (MP4 files only)
- Wiki pages
- Tags
- Home page tabs
- Subtabs
- Page settings
- Row settings
- Widget settings
- Widget tile translations

To do this, in the administration console, navigate to the **Area & Workspace Configuration • Home Page** section in the side menu. Then open the **Export** tab (see Figure 14.6). There, select the home pages you want to transport, and then click **Export Home Page**.

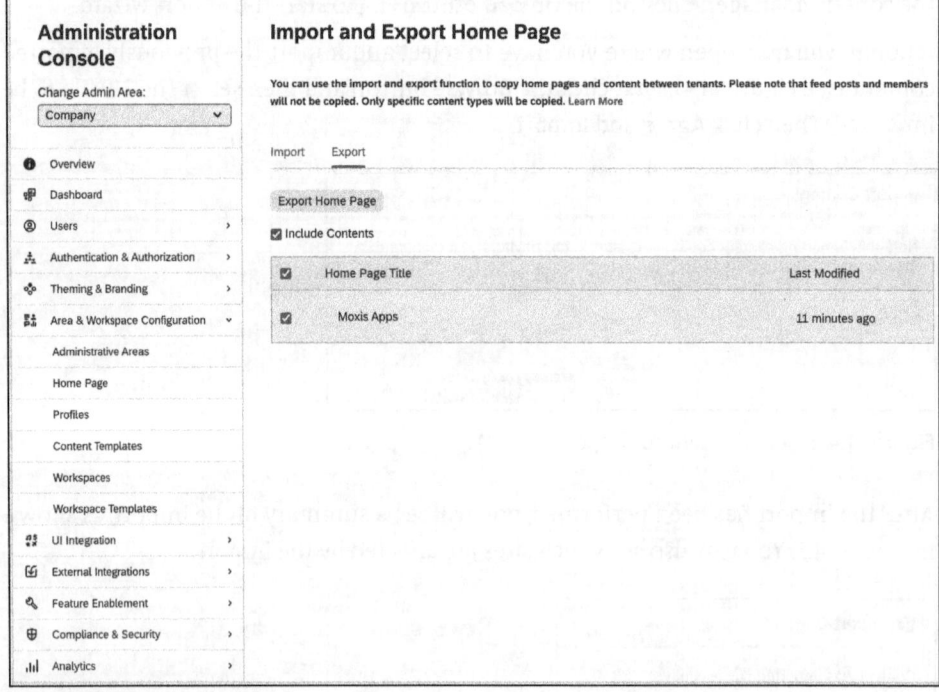

Figure 14.6 Export Home Page

You will be informed again in a popup that the content will be exported as a ZIP file and that the export may take some time. Click **Export** to start the creation of the export file. The file will be downloaded to your computer.

Now open the target system into which the content is to be imported. As before, jump to the **Import and Export Home Page** area. This time, however, open the **Import** tab, as shown in Figure 14.7. Then click **Import Home Page**.

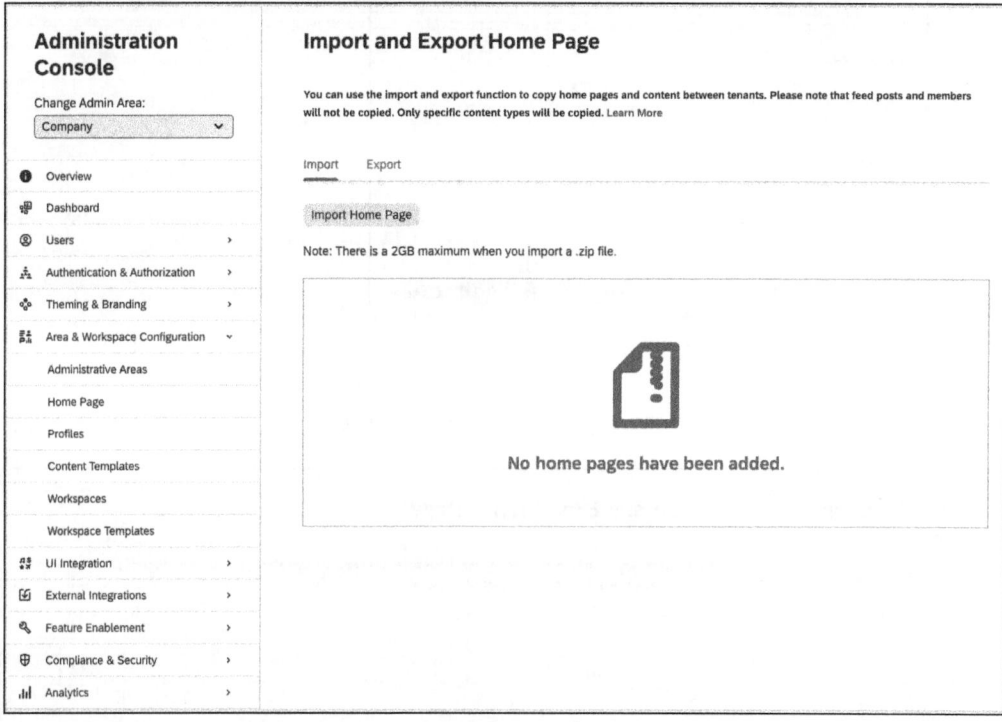

Figure 14.7 Import Home Page

You will then receive some important notes on the import, as shown in Figure 14.8. Read them carefully, and then click **Import**.

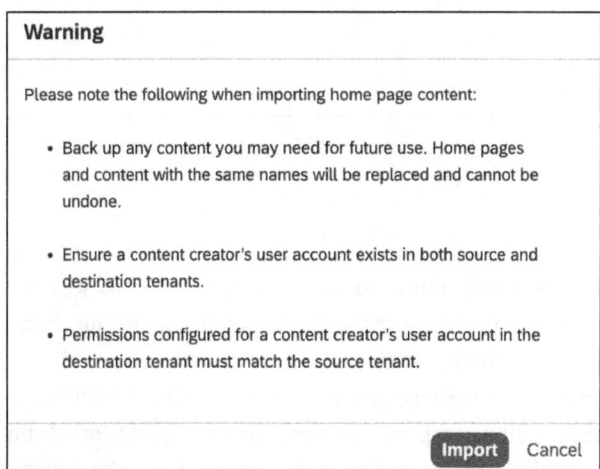

Figure 14.8 Import Notes

14   Content Transport

After that, select the ZIP file with the content you want to import and click the **Import** button (see Figure 14.9).

Figure 14.9 Select Content for Import

Now check the status of the import, as shown in Figure 14.10.

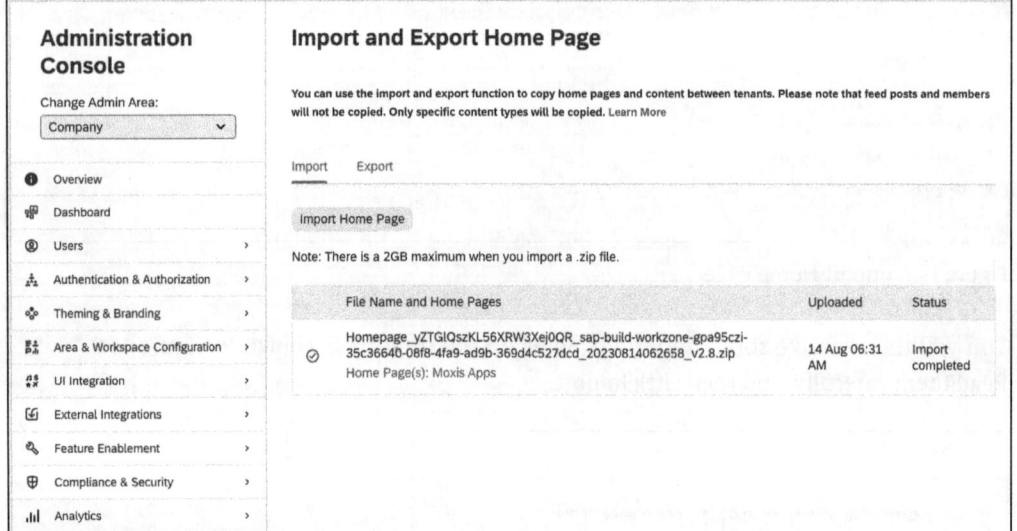

Figure 14.10 Import Status

14.1.3   Workspaces

As mentioned earlier, you also have the option to export workspaces. To do so, open the administration console and navigate to the **Area & Workspace Configuration • Workspaces** section in the side menu. There, in the **Workspaces You Manage** tab, you can view and export the workspaces you manage (see Figure 14.11). Or in the **All Workspaces** tab, you can view and export all workspaces available in SAP Build Work Zone. Click **Actions** in the desired workspace, then select **Export Workspace** from the menu.

14.1 Manual Transport

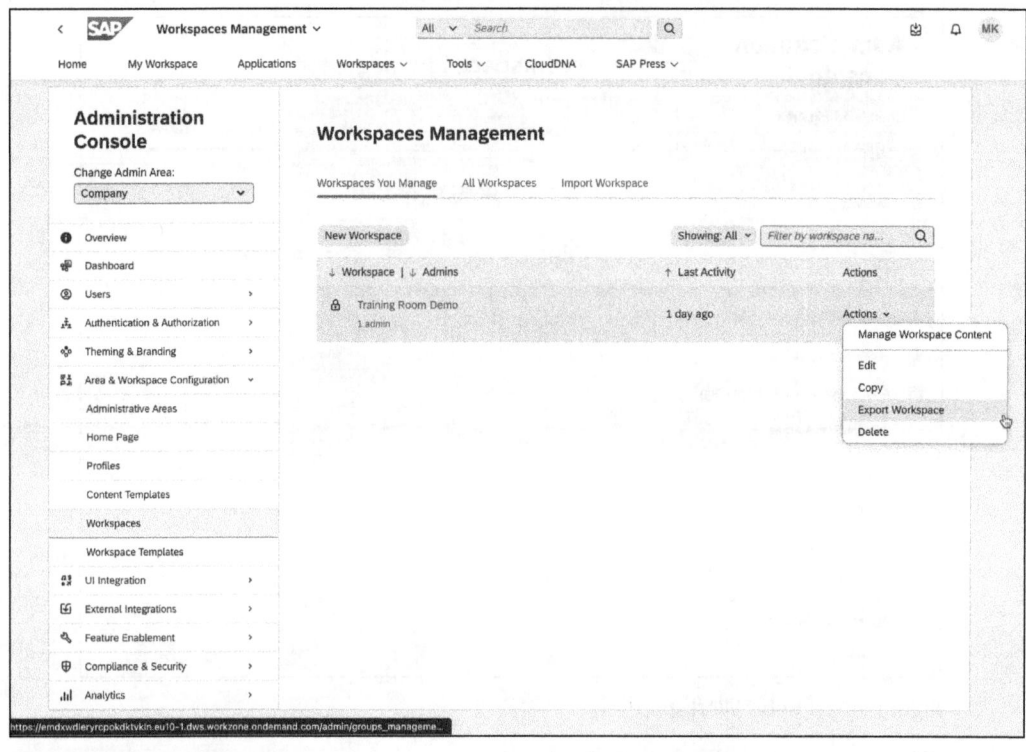

Figure 14.11 Export Workspace

You will see a note regarding the export, and you will be informed that the workspace will be exported in the form of a ZIP file and that the export may take longer depending on the number and size of the contained objects. Click the **Export** button to trigger the export.

The import of a workspace is done in the target system in the **Workspaces Management** area within the **Import Workspace** tab (see Figure 14.12). Click **Import and Export Workspace** to open the import functionality.

In the **Export** tab, you also have the option to start the export (see Figure 14.13).

Open the **Import** tab as shown in Figure 14.14 to initiate the import. Click the **Import Workspace** button in it.

In the next step, you will be shown some warnings and hints that you must pay attention to so that the import can be performed without any problems. In the popup, click the **Import** button. Now you need to select the ZIP file you want to import and then click the **Import** button.

14 Content Transport

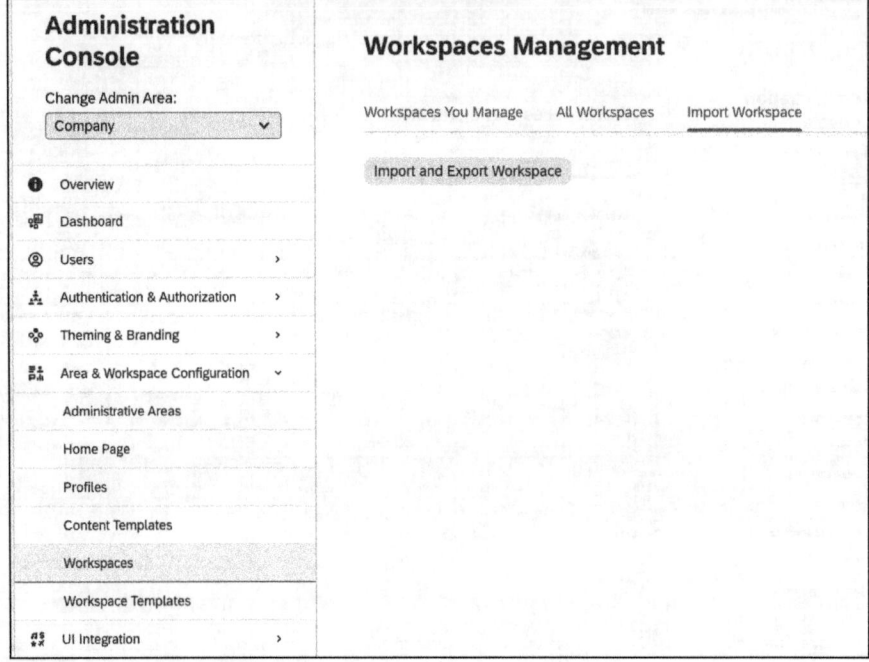

Figure 14.12 Import Workspace

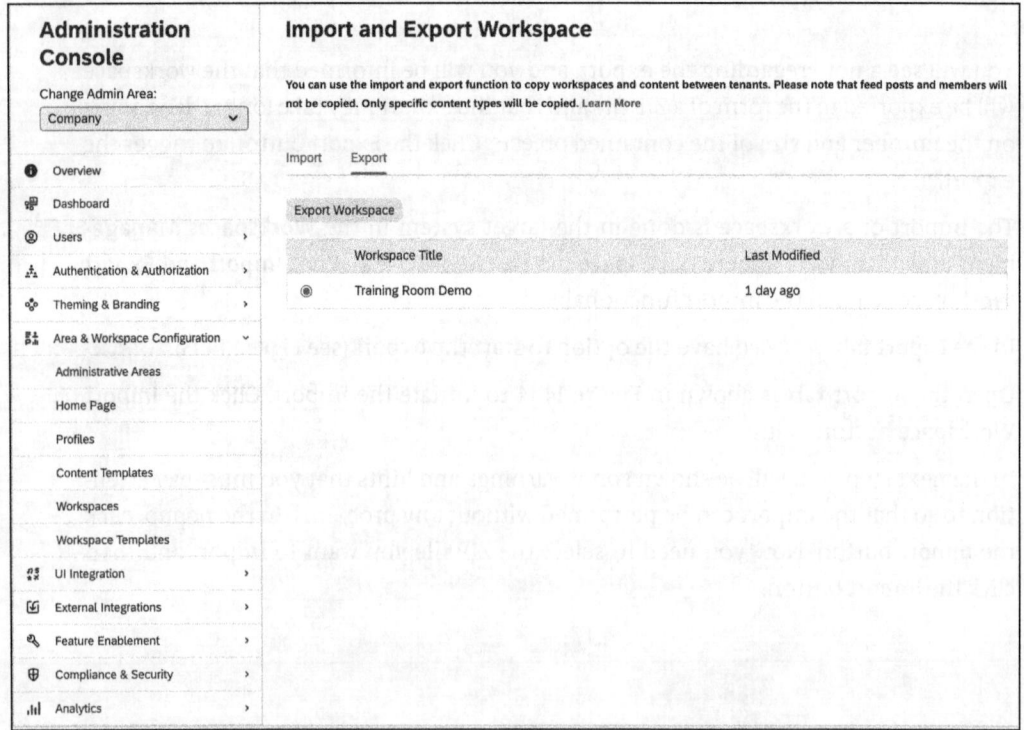

Figure 14.13 Workspace Selection

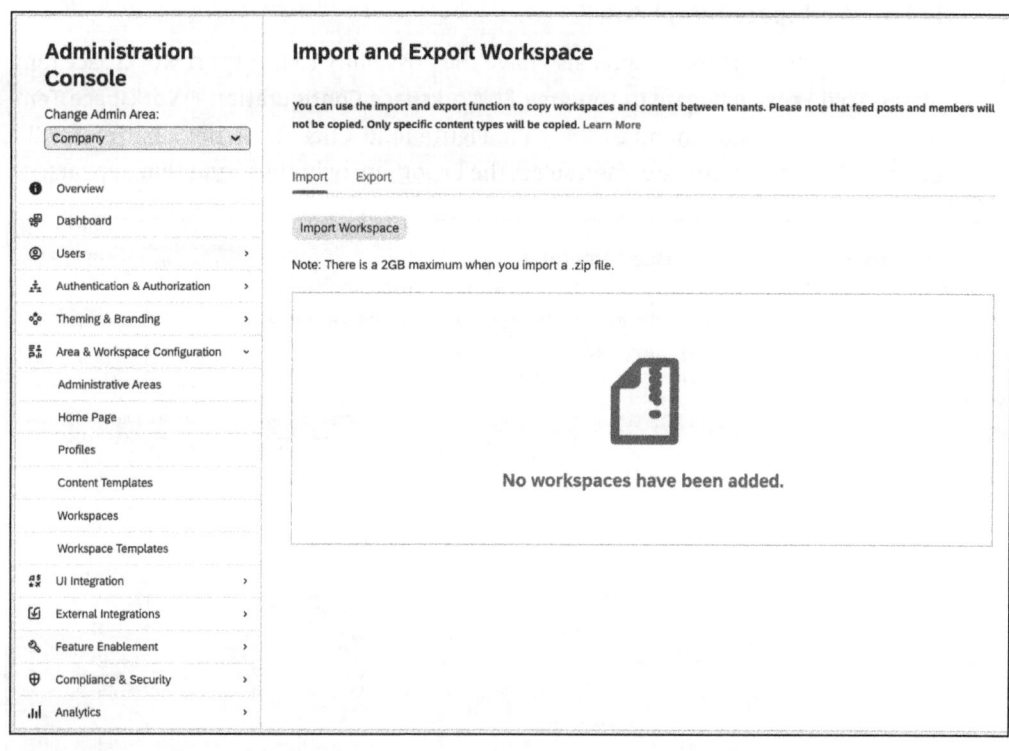

Figure 14.14 Import Workspace

You will then see the status of the import. As shown in Figure 14.15, an error may occur under certain circumstances, such as if a workspace with the same name already exists in the target system.

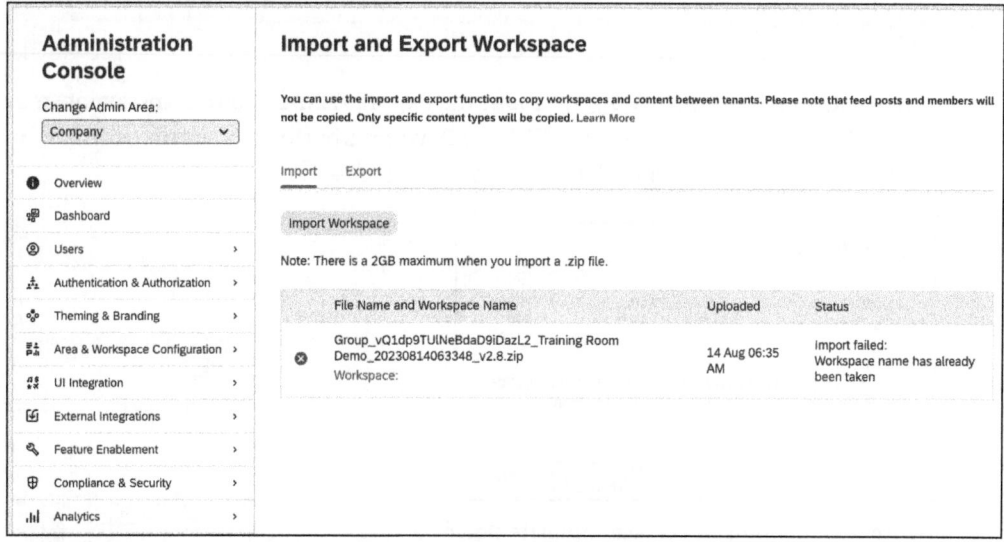

Figure 14.15 Check Import Status

14.1.4 Workspace Templates

Similarly to the workspaces, you also have the possibility to transport workspace templates. To do this, navigate to the **Area & Workspace Configuration • Workspace Templates** area in the side menu as shown in Figure 14.16. Click the **Actions** button for the desired workspace template, then select the **Export** entry in the menu that appears.

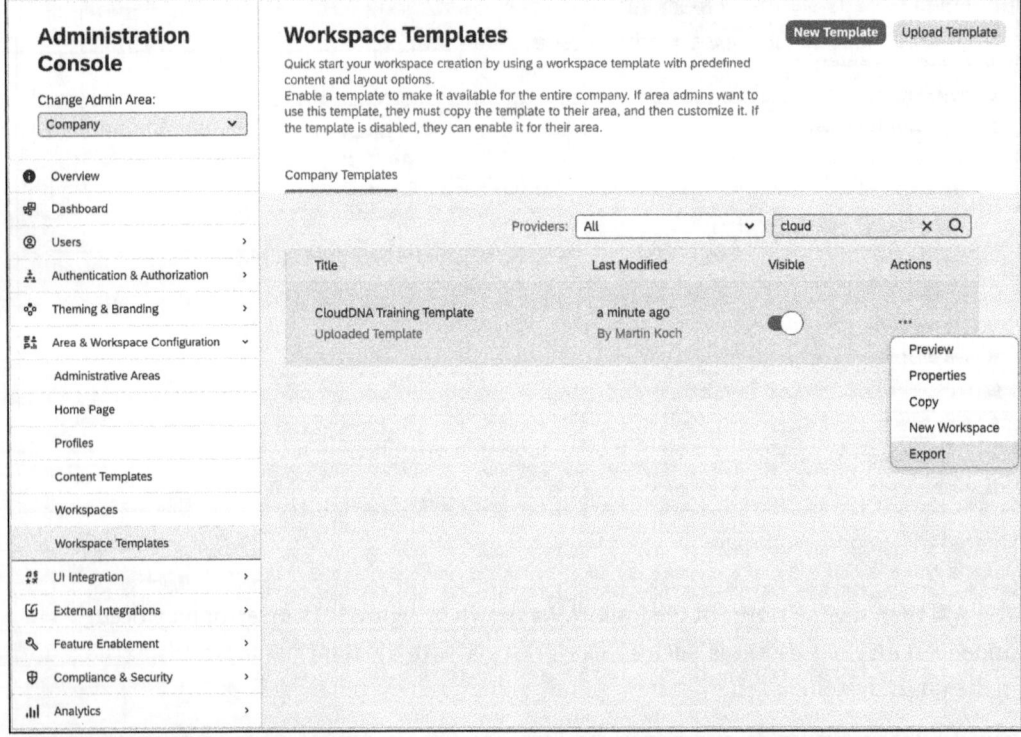

Figure 14.16 Trigger Export

After that, you again will see a note about the export, telling you that the export may take a little longer (see Figure 14.17). Click the **Download** button to start the download of the template. The download will be saved on your computer.

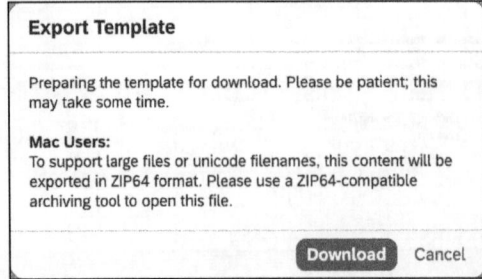

Figure 14.17 Confirm Workspace Template Download

14.1 Manual Transport

Now log into the target system and navigate to the **Workspace Templates** area as shown before. Click the **Upload Template** button (see Figure 14.18) to open the **Upload** dialog.

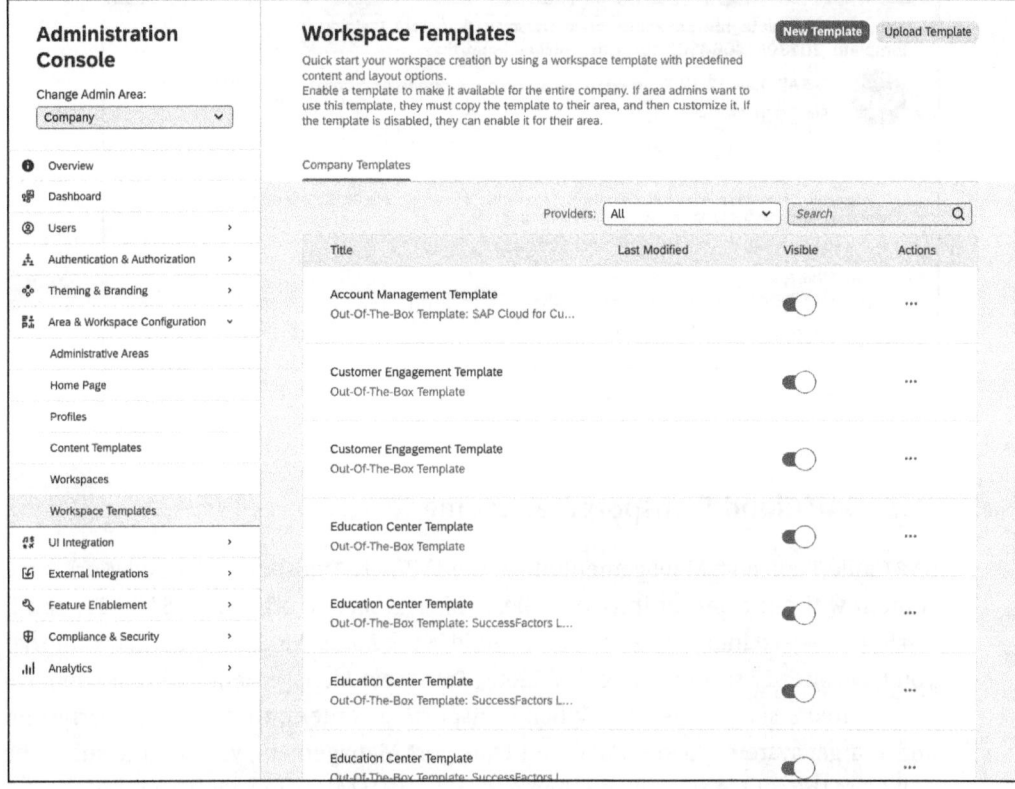

Figure 14.18 Trigger Import

You must now select the file to be imported, as shown in Figure 14.19. Then click the **Upload** button to import the workspace template into the target system.

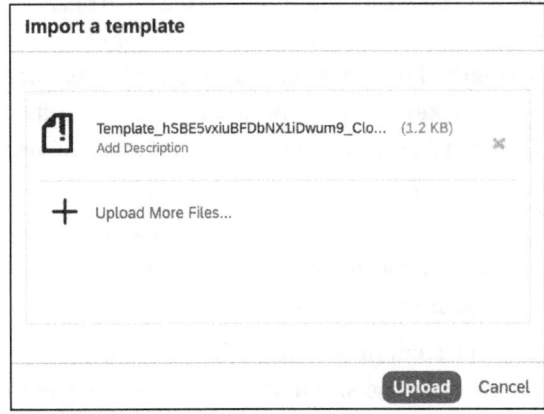

Figure 14.19 Select Template for Import

14 Content Transport

You will see a message that the import will take some time and that you will be notified by email when the import is successful. Press **OK** to close the message. Figure 14.20 shows an example of a notification email about the successful import.

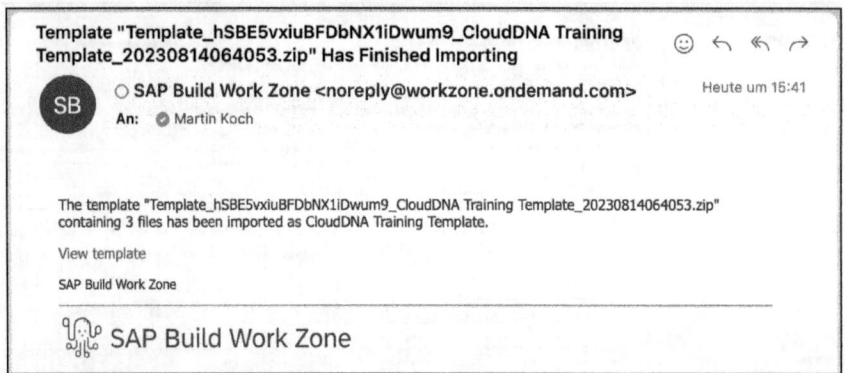

Figure 14.20 Check Import Status

14.2 SAP Cloud Transport Management

SAP Cloud Transport Management offers the ability to transport SAP Build Work Zone content without a detour into local files. This is a service offered on SAP BTP, and it must be licensed independently of SAP Build Work Zone. We assume at this point that you have already subscribed to SAP Cloud Transport Management and that you have also created a service instance. When transporting content, there is a source system and a target system. To use SAP Cloud Transport Management, you need to subscribe to it from the source system and then perform initial setup and configuration.

14.2.1 Initial Setup

In the corresponding subaccount, navigate to the **Services • Instances and Subscription** area. In the **Subscriptions** section, check if the subscription for the **Cloud Transport Management** application exists in the **Application** list, and if an instance for the **Cloud Transport Management** service with a **standard** plan has been created in the **Instance** list (see Figure 14.21). Now you need a service key for the application. To get this, click the **Actions** button for the service, as shown in Figure 14.21, and select the **Create Service Key** menu item from the popover.

Assign a name for the service key as shown in Figure 14.22, then click **Create**. No further configuration is required at this point. Now load the created service key onto your local computer. It contains all the information required for the destination created.

To see the transport site or transport selected content options in the site manager, you need to configure a destination. Therefore, open the SAP BTP cockpit and navigate to the subaccount. Open **Connectivity • Destinations** from the side menu.

14.2 SAP Cloud Transport Management

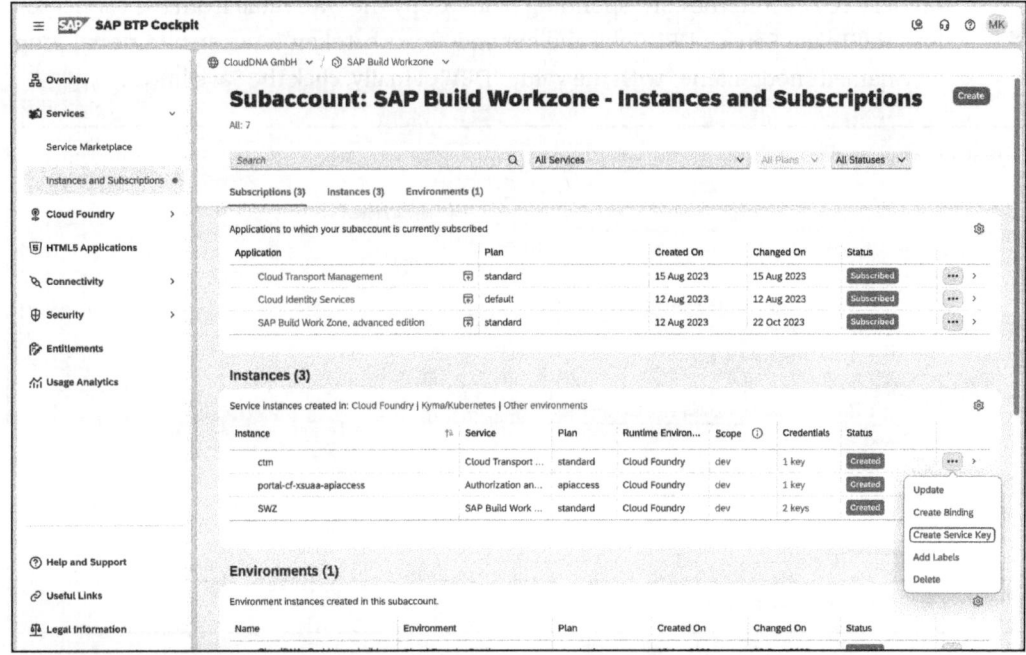

Figure 14.21 Subaccount Instances

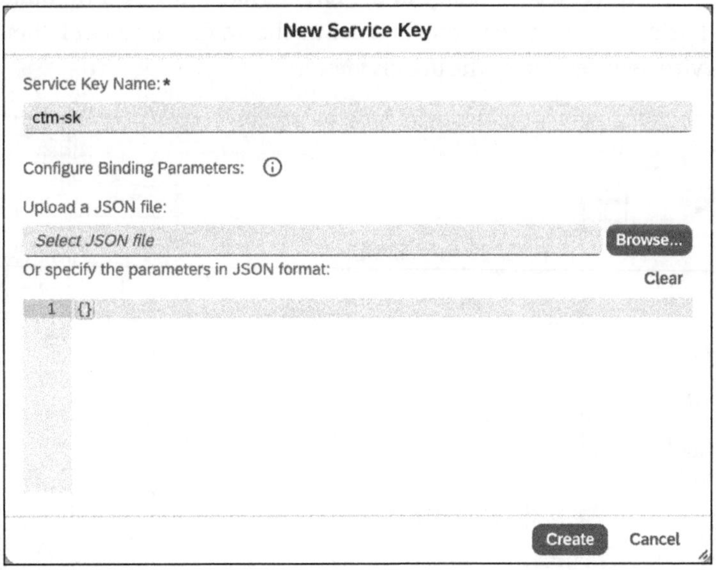

Figure 14.22 Create Service Key

Now, create a new destination. The configuration of the destination is shown in Figure 14.23. The name of the destination can be freely chosen. Select **HTTP** for **Type**. Copy the URI from the service key into the **URL** field. Select **Internet** for **Proxy Type** and **OAuth2-ClientCredentials** for the **Authentication** field. Copy the **Client ID** and **Client Secret** from

14   Content Transport

the service key into the corresponding fields. Copy the URL value from the service key into the **Token Service URL** field, and append "/oauth/token". Add an additional property named "node-name" with the value "DEV". Finally, click the **Save** button.

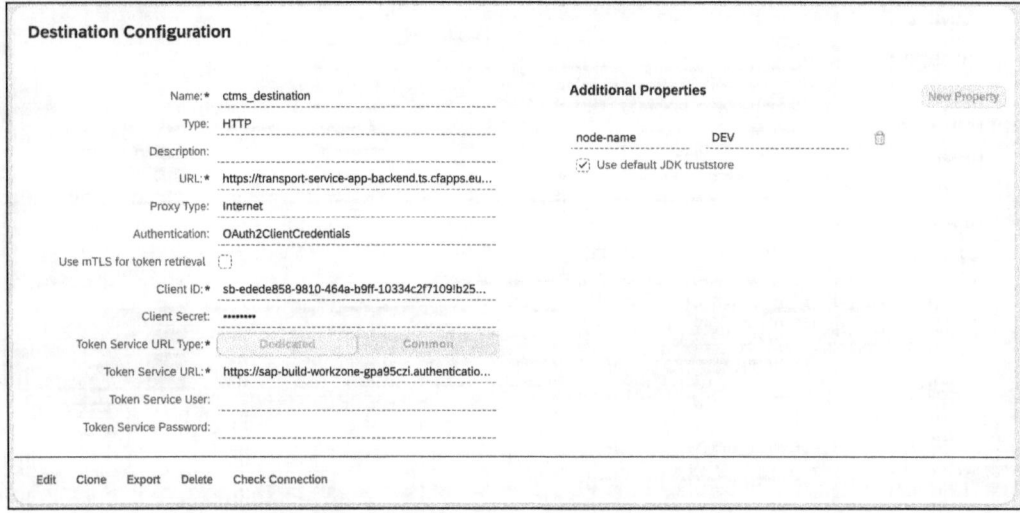

Figure 14.23 Create Transport Management Destination

In the next step, you must create a service key in the target system for the SAP Build Work Zone instance. To do this, navigate to the instance as shown in Figure 14.24, then click **Create** in the **Service Keys** section of the details view.

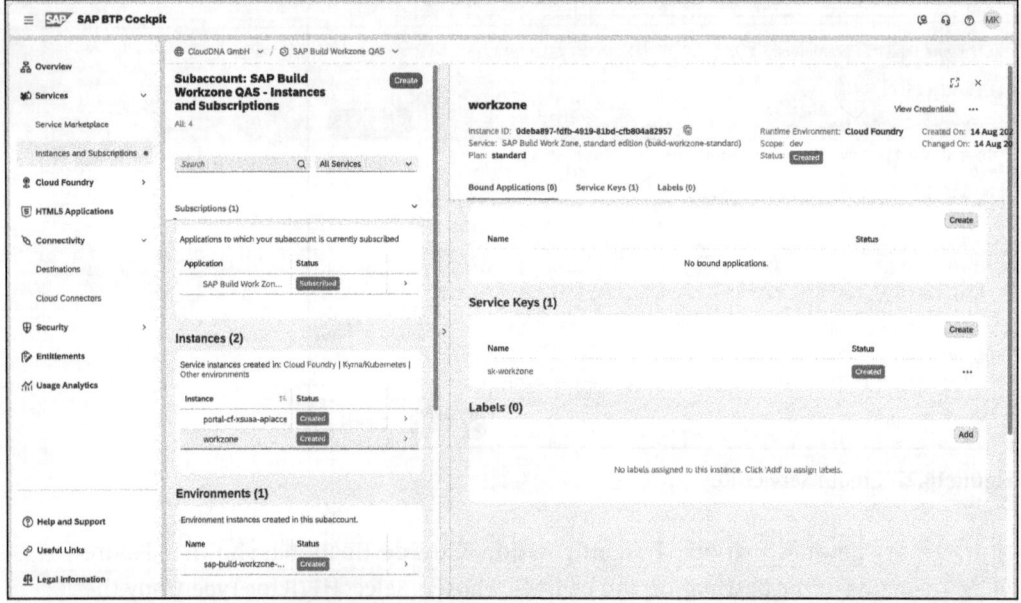

Figure 14.24 Create Service Key

14.2 SAP Cloud Transport Management

You can freely select the name for the service key (see Figure 14.25). No further configuration is required. Click **Create** to create the service key.

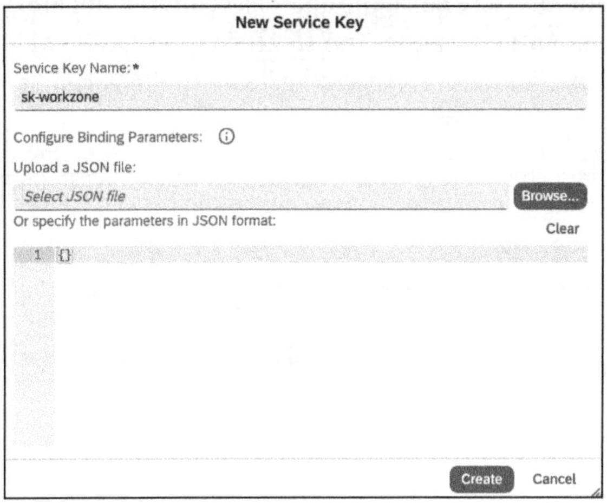

Figure 14.25 Service Key Configuration

Now download the service key you just created to your local computer by clicking **Download** (see Figure 14.26). You will need the contents of the service key in the next step for configuring the destination.

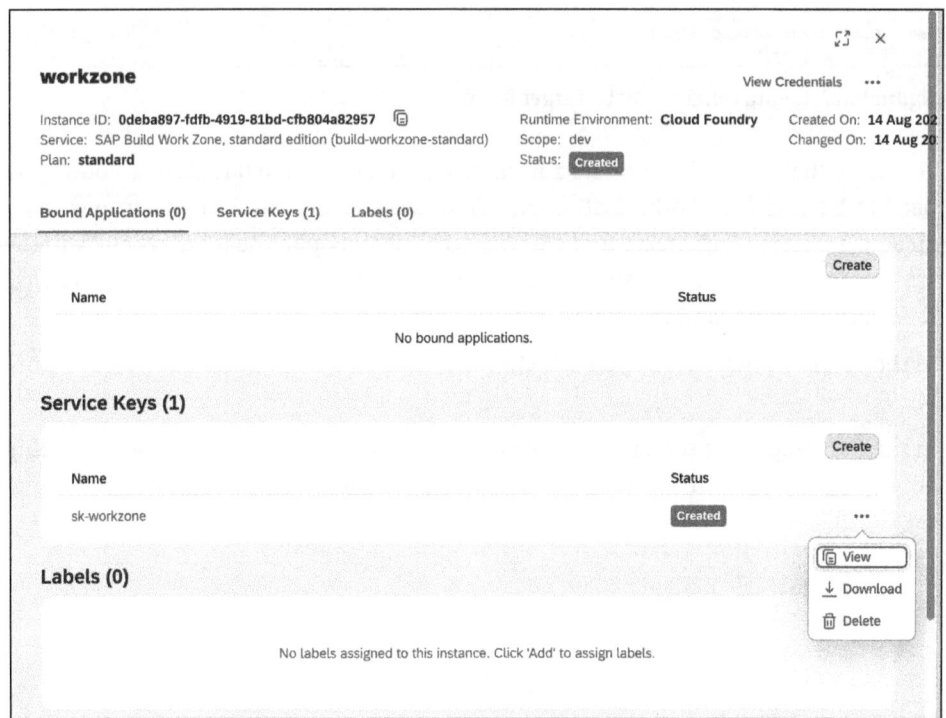

Figure 14.26 Service Key Download

14 Content Transport

Create a new destination in the source system (see Figure 14.27). **Name** can be freely selected. Select **HTTP** for **Type**. In the **URL** field, enter the portal service URL from the service key. Append the string "/cdm_import_service" at the end. Select **Internet** for **Proxy Type** and **OAuth2ClientCredentials** for **Authentication**. Transfer the values for **Client ID** and **Client Secret** from the service key into the corresponding fields. In the **Token Service URL** field, enter the URL from the service key.

Destination Configuration

Name:	SWZ
Type:	HTTP
Description:	
URL:	https://portal-service.cfapps.eu10.hana.ondeman...
Proxy Type:	Internet
Authentication:	OAuth2ClientCredentials
Use mTLS for token retrieval	☐
Client ID:	sb-0deba897-fdfb-4919-81bd-cfb804a82957!b25...
Client Secret:	●●●●●●●●
Token Service URL Type:	Dedicated / Common
Token Service URL:	https://sap-build-workzone-qas-jgz8zo5b.authenti...
Token Service User:	
Token Service Password:	

Additional Properties — ☑ Use default JDK truststore

Edit Clone Export Delete Check Connection

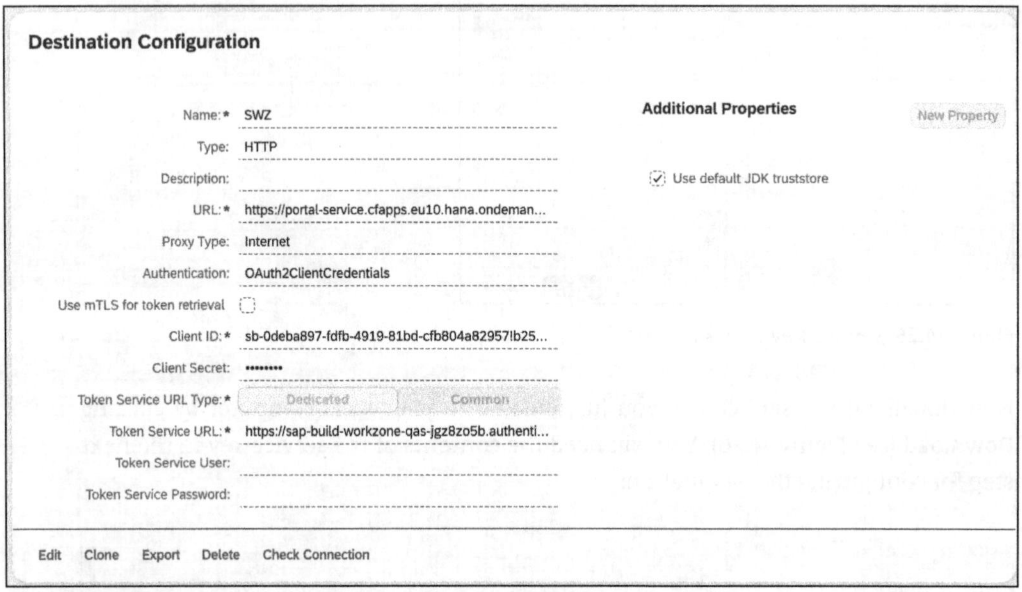

Figure 14.27 Create Destination to Target System

To ensure that role collections are automatically created in the target subaccount, you need to have an SAP Authorization and Trust Management service instance with an **apiaccess** plan type. Navigate to the global account and open the **Boosters** entry in the side menu. Look for the booster named **Configure Access to the Authorization Service**, then click the corresponding tile (see Figure 14.28).

In the details of the service, you can start the installation by clicking **Start** (see Figure 14.29).

In the first step, the prerequisites are checked, including whether the corresponding authorizations are available and whether the global entitlements are available (see Figure 14.30).

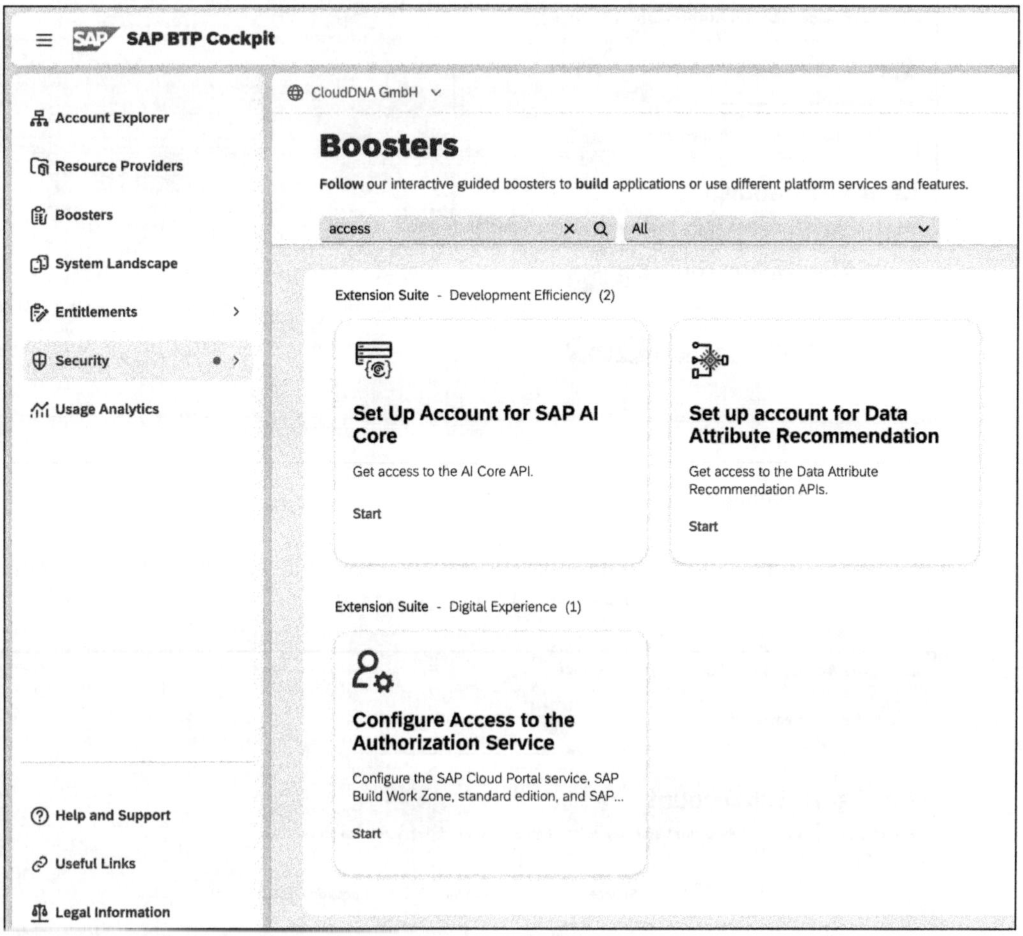

14.2 SAP Cloud Transport Management

Figure 14.28 Authorization Service Booster

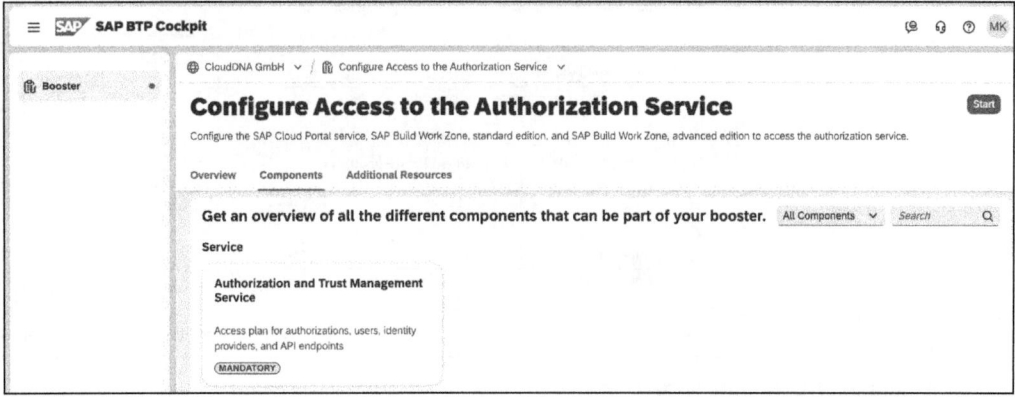

Figure 14.29 Start Booster

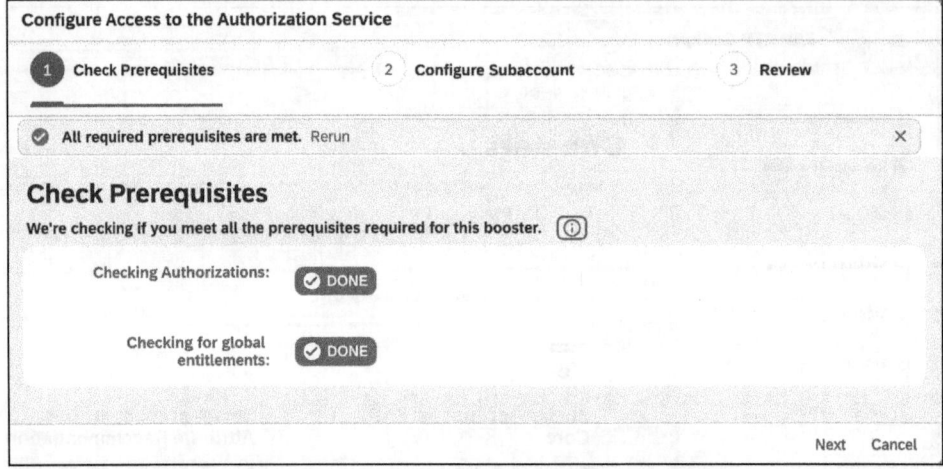

Figure 14.30 Prerequisite Check

The next step is to configure the subaccounts. You must select an existing subaccount. An existing space also must be selected (see Figure 14.31).

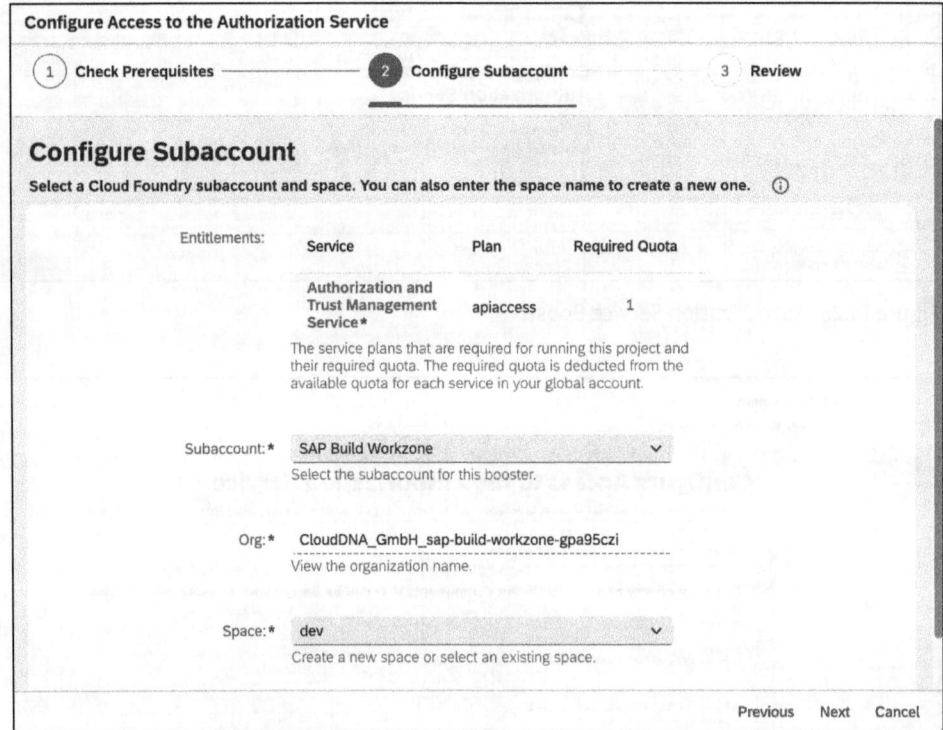

Figure 14.31 Configure Subaccount

14.2 SAP Cloud Transport Management

Finally, you will see a summary. Click **Finish** to complete the process. You will now see a popup with the progress of the setup (see Figure 14.32). Setting up the service may take some time.

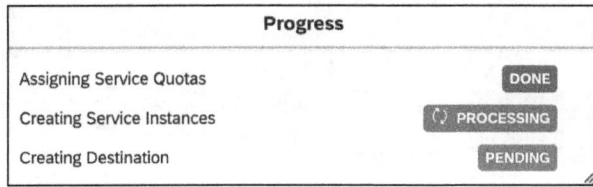

Figure 14.32 Creation Progress

After the setup is successful, you will see a corresponding success message.

Now repeat the setup for the second subaccount in which the next system level is located (see Figure 14.33).

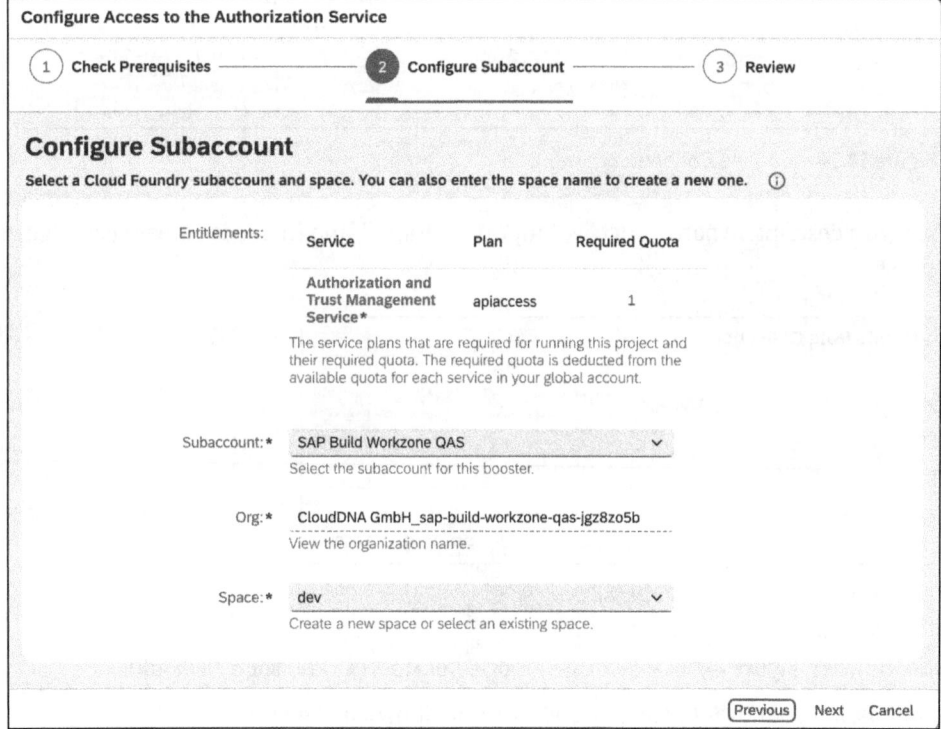

Figure 14.33 Configuration for Target System Subaccount

You must now create a new role collection. To do this, open the subaccount in which the SAP Cloud Transport Management service is located and navigate to the **Security • Role Collections** area in the side menu, as shown in Figure 14.34. Click **+** to create a new role collection.

14   Content Transport

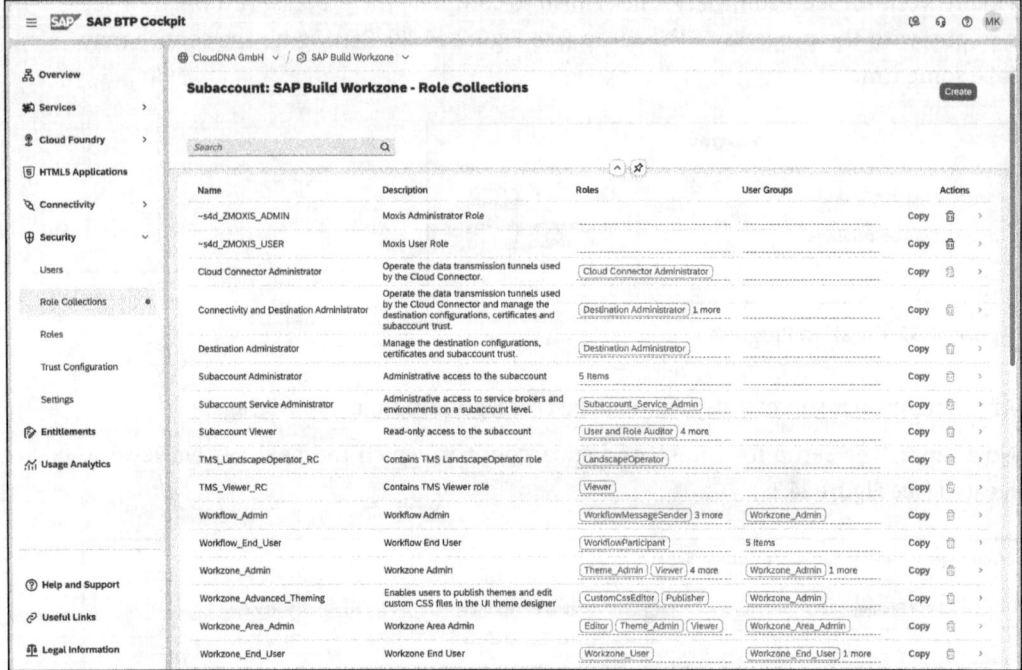

Figure 14.34 Add Role Collection

Assign a descriptive name, such as "Import Operator", and then click **Create** (see Figure 14.35).

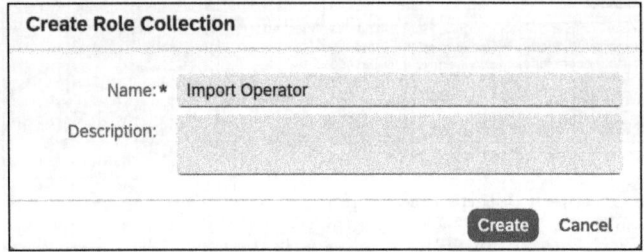

Figure 14.35 Role Collection Details

As shown in Figure 14.36, select the **ImportOperator** role and then click **Add**.

Now assign your user to the role collection, as shown in Figure 14.37.

14.2 SAP Cloud Transport Management

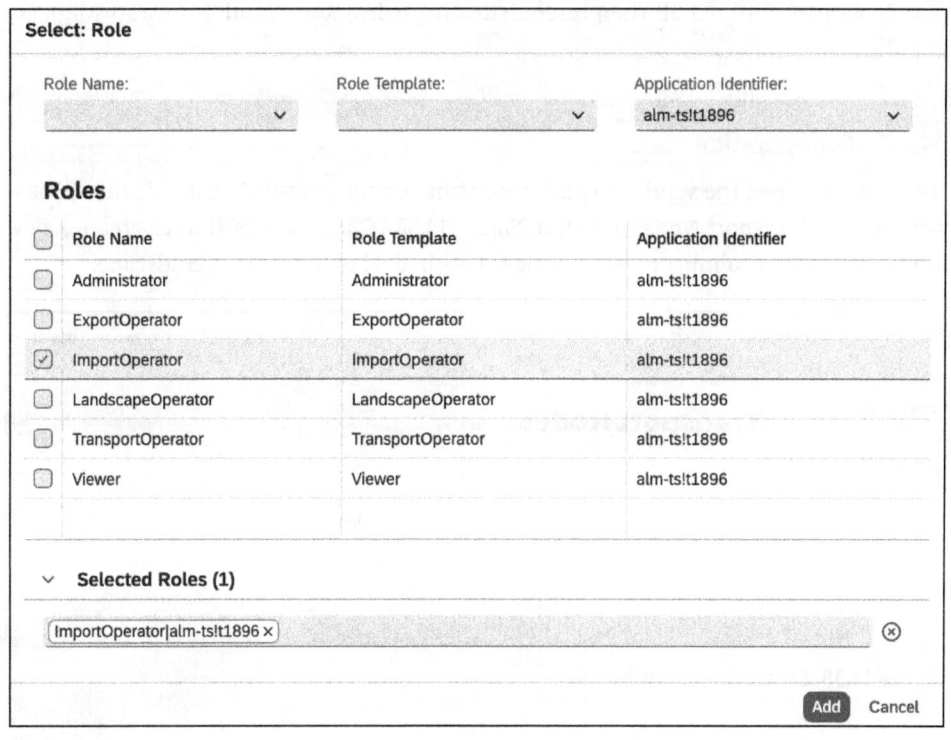

Figure 14.36 Add Role

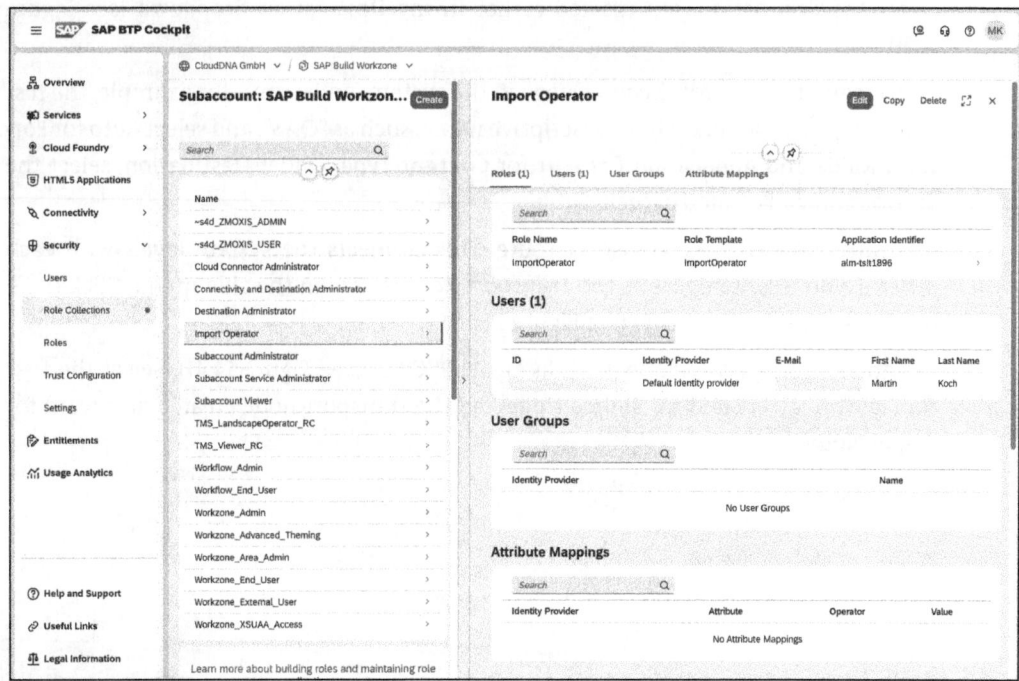

Figure 14.37 Assign User

447

You have now fulfilled all the prerequisites for using SAP Cloud Transport Management.

14.2.2 Configuration

You can now open the service to perform further configuration. In the side menu, navigate to the **Transport Nodes** area (see Figure 14.38). Click the **+** button to create a new transport node. You must create a node for each level of the system landscape.

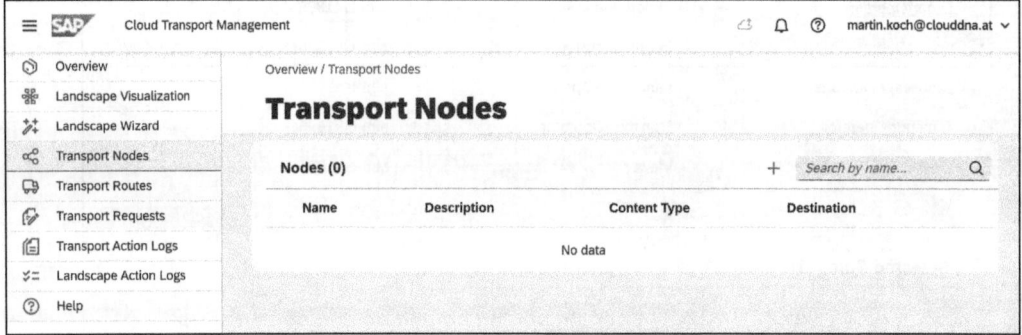

Figure 14.38 Create Transport Node

Now create the node for the development system. Assign a descriptive name, such as "DEV", and select **Auto** for **Forward Mode** and **Application Content** for **Content Type**. Select the destination you created earlier in the **Destination** dropdown (see Figure 14.39).

Now create the node for the next stage in the system landscape—for example, the test or productive system. Assign a descriptive name, such as "QAS", and select **Auto** for **Forward Mode** and **Application Content** for **Content Type**. Under **Destination**, select the destination you created earlier.

You must now create a transport route. This connects the system levels with each other. To do this, navigate to the **Transport Routes** area in the side menu. Click the **+** button to create a new route (see Figure 14.40).

Assign an appropriate name, such as "DEV_2_QAS" (see Figure 14.41). Select the first transport node created for **Source Node** and the transport node created afterward for **Target Node**.

14.2 SAP Cloud Transport Management

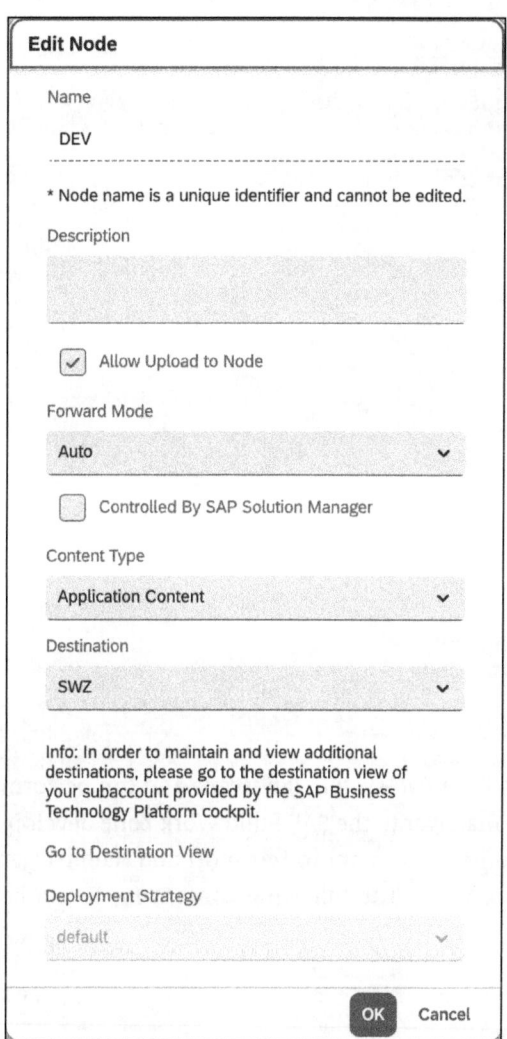

Figure 14.39 DEV Transport Node Configuration

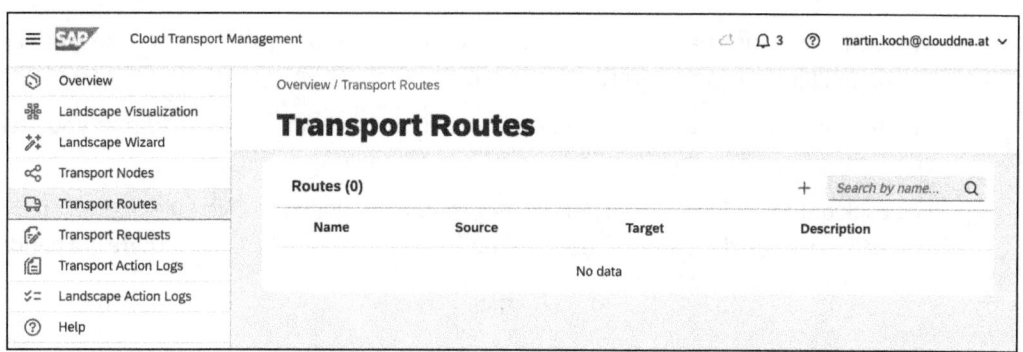

Figure 14.40 Create Transport Route

14 Content Transport

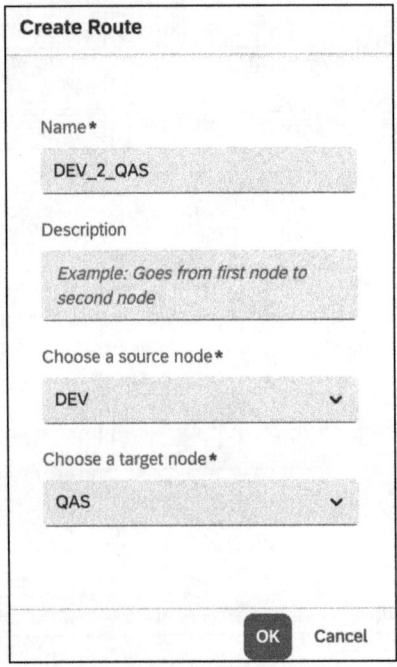

Figure 14.41 Transport Route Details

You are now ready with a configuration for a two-system landscape and can perform transports. To do this, open the content manager in the SAP Build Work Zone development system. You can now select the content you want to transport and then trigger the transport by clicking ⊲ (see Figure 14.42). The following content types can be transported in this way:

- Business content
- Home page
- Workspaces
- Workspace templates

You will see a summary of the selected content. Review it, then click **Transport** as shown in Figure 14.43 to start the transport.

You will now receive a message that the transport has been started and that you can check the progress in the SAP Cloud Transport Management instance.

Log back onto SAP Cloud Transport Management and go to the **Overview** menu. There you will see in the right area of the screen, as shown in Figure 14.44, that there is a pending transport in the QAS system. Click this entry.

14.2 SAP Cloud Transport Management

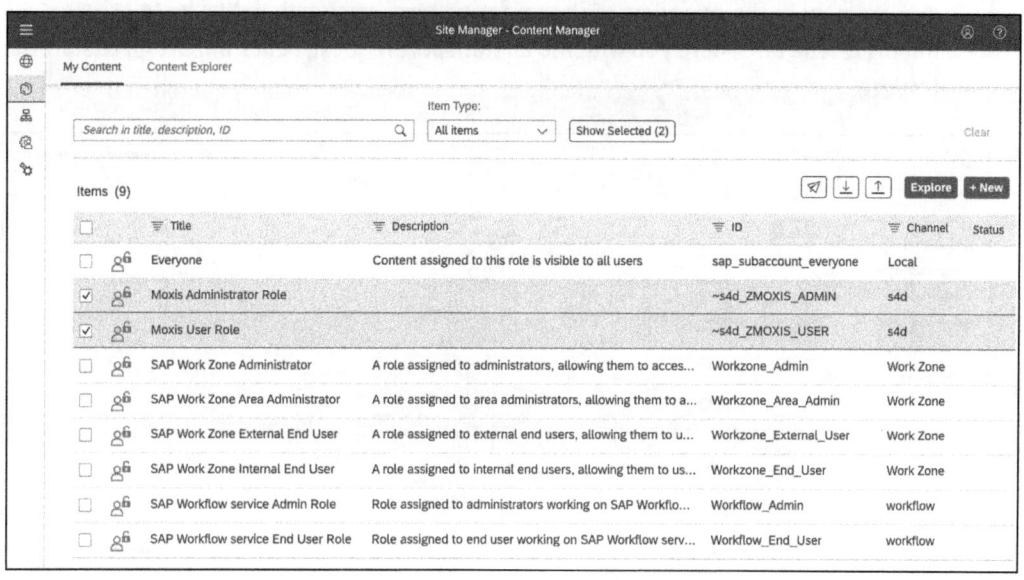

Figure 14.42 Transport Content

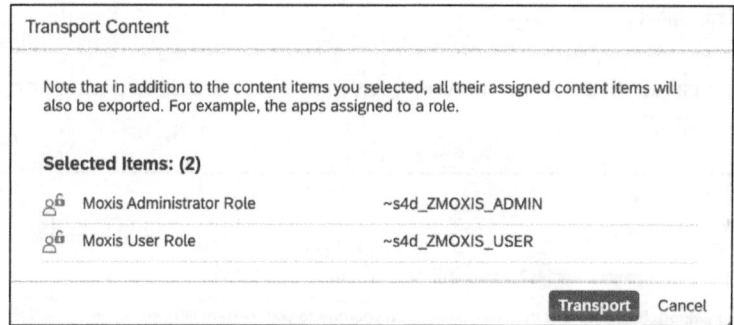

Figure 14.43 Confirm Transport

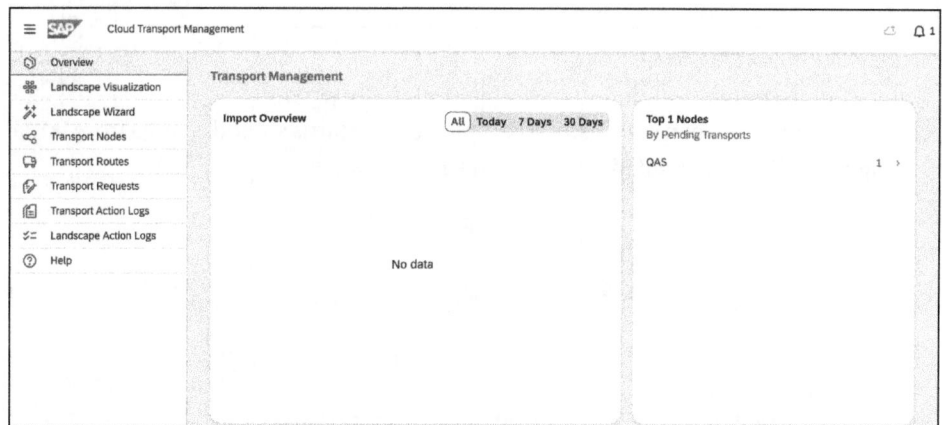

Figure 14.44 Import Overview

451

You are now in the overview of the target system. As shown in Figure 14.45, open the **Import Queue** tab. There you will find all transport requests. Click **Import All** to start the import.

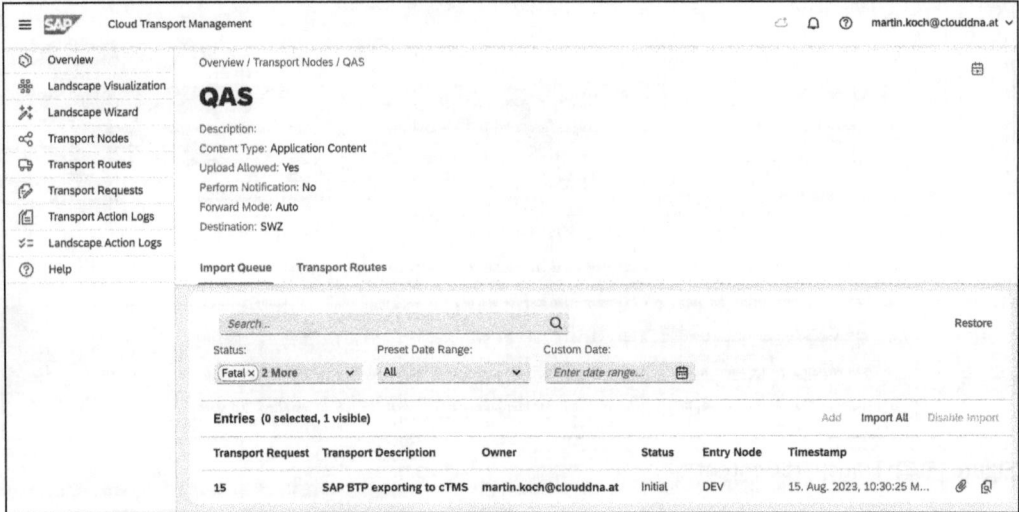

Figure 14.45 Import Transport

You must now confirm again in a popup that you want to import the selected transport requests (see Figure 14.46).

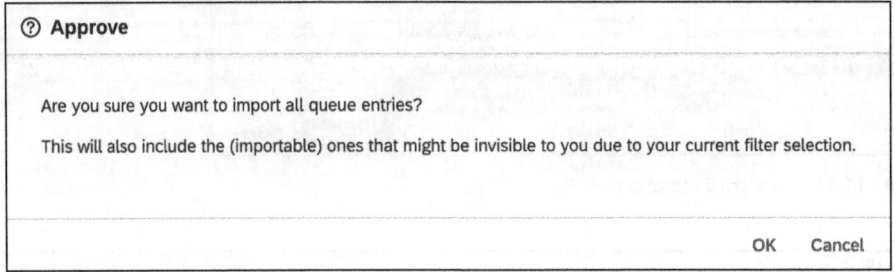

Figure 14.46 Import Approval

After the import has been successfully completed, the status should change from **Initial** (see Figure 14.45) to **Succeeded**, as shown in Figure 14.47.

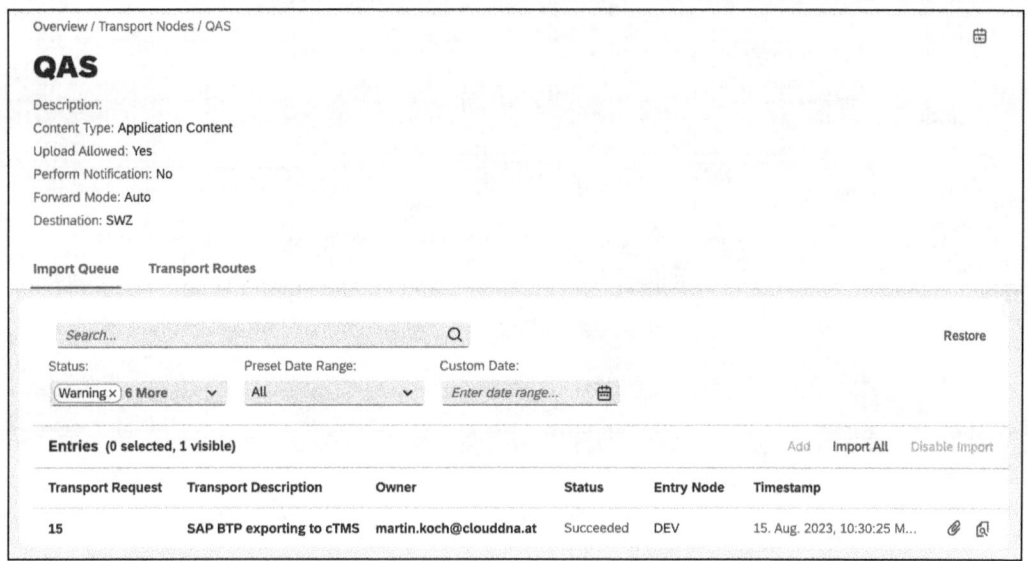

Figure 14.47 Check Import Status

14.3 Summary

SAP Build Work Zone's content transport feature emerges as a pivotal tool, ensuring that your digital workspace remains adaptable and responsive to evolving business needs. Whether it's about optimizing your development processes, streamlining cross-service collaboration, or maintaining content uniformity across subaccounts, this functionality empowers you to harness the full potential of your digital workspace ecosystem.

Chapter 15
Advanced Topics

This chapter explores advanced topics related to SAP Build Work Zone and software development, including notifications, enterprise search, SAP Task Center integration, and SAP Mobile Start. The chapter examines unique features and capabilities of each of these areas, including their ability to enhance the functionality and usability of SAP Build Work Zone.

In this chapter, we will look at advanced topics related to SAP Build Work Zone. For example, in Section 15.1, we'll look at how notifications can be used in the SAP Build Work Zone and what configuration options are available. Following this, in Section 15.2, we'll look at how SAP HANA enterprise search can be used, such as to search for applications or content across all of SAP Build Work Zone.

In Section 15.3, we'll then look at how integration with SAP Task Center can be established. Finally, in Section 15.4, we'll look at how to set up SAP Mobile Start on a smartphone and what additional steps are required in SAP Build Work Zone from administrator and user perspectives.

15.1 Notifications

Nowadays, notifications are an established concept for sending information to users. Usually, business users are familiar with notifications from a variety of social and business networks, which are delivered to mobile devices or via email. In business applications, notifications are an essential part of alerting users to important events or conditions. Often, such notifications involve an action that must be performed by the user. Therefore, it's important that these messages arrive reliably at the user.

A typical example of a notification is of a workflow item, such as an approval, exceeding or falling below various relevant KPI thresholds, or pending invitations in interaction platform groups. SAP Build Work Zone also provides the ability to respond to various notifications depending on the user's business role. These messages are consumed by the notification channel.

15 Advanced Topics

> **[+] Notification Channel**
>
> The *notification channel* is a backend component that aggregates notifications from different notification providers that are configured in each respective environment. For some types of notifications, it's also possible to specify whether they should be delivered via email or on mobile devices. The notification channel is part of the SAP_GWFND software component.

To access notifications in SAP Build Work Zone, a button is available in the header bar next to the user icon (see Figure 15.1). In addition, this button uses an indicator to show how many new notifications are available for the user. In our example, there are no notifications available, so no indicator is visible. If the notification page is opened and new notifications are received, this indicator is updated automatically. It should also be noted that when high-priority notifications arrive, the end user is notified by means of a popup that opens automatically.

Figure 15.1 Icon to Open Notifications Overview in SAP Build Work Zone

Clicking this button will open a new page where notifications, if any, will appear (see Figure 15.2). On this page, you will see all notifications, both new and already read.

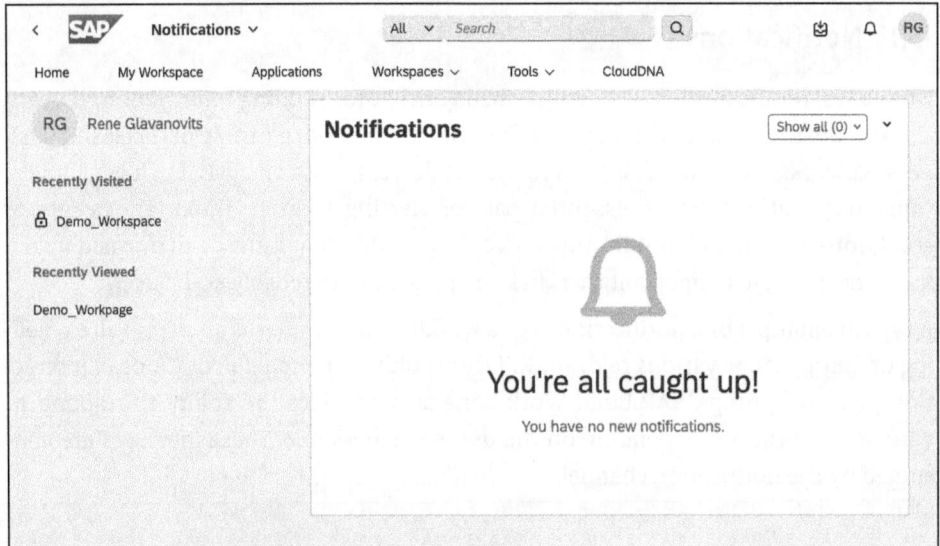

Figure 15.2 Notification Overview in SAP Build Work Zone

In the following sections, we'll look at the types of notifications and how push notifications work.

15.1.1 Types of Notifications

Notifications can basically be divided into two different types: bell notifications and email notifications. The purpose of *bell notifications* is to inform the user about important updates. Furthermore, this type of notification is also used to send invitations or approval requests. A special feature of these notifications is that they do not require any setup steps. As a rule, bell notifications are more urgent than email notifications.

In contrast to bell notifications, *email notifications* are used to notify the user of important and periodic updates via email. And unlike bell notifications, email notifications have to be configured. On the one hand, it's possible to define which notifications are to be delivered, and on the other hand, the frequency with which email notifications are sent for a specific workspace can also be defined. To make this configuration, the **Settings** area must be opened via the user icon (see Figure 15.3).

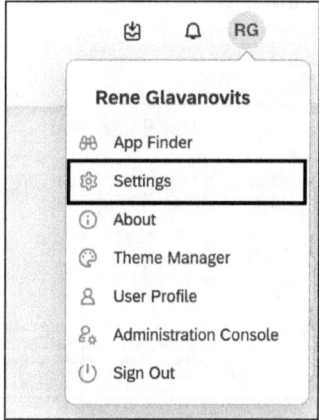

Figure 15.3 Open Settings via User Menu

After clicking **Settings**, a dialog opens. In this dialog, select the **Advanced Settings** tab and then the **Email** link (see Figure 15.4).

After clicking **Email**, you'll see the settings for email notifications. Here, as shown in Figure 15.5, two tabs are available: **Notifications** and **Group Notifications**.

Various configurations can be carried out in the two tabs. In **Notifications**, for example, you can define for which activities notifications are sent. Table 15.1 describes the activities that can be selected here. Notifications for the individual activities also can be activated or deactivated here.

15 Advanced Topics

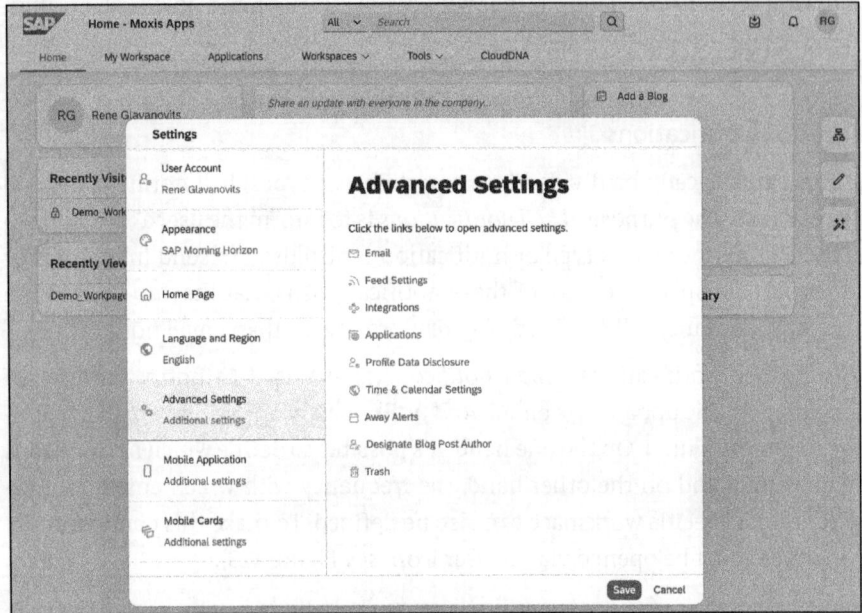

Figure 15.4 SAP Build Work Zone Settings

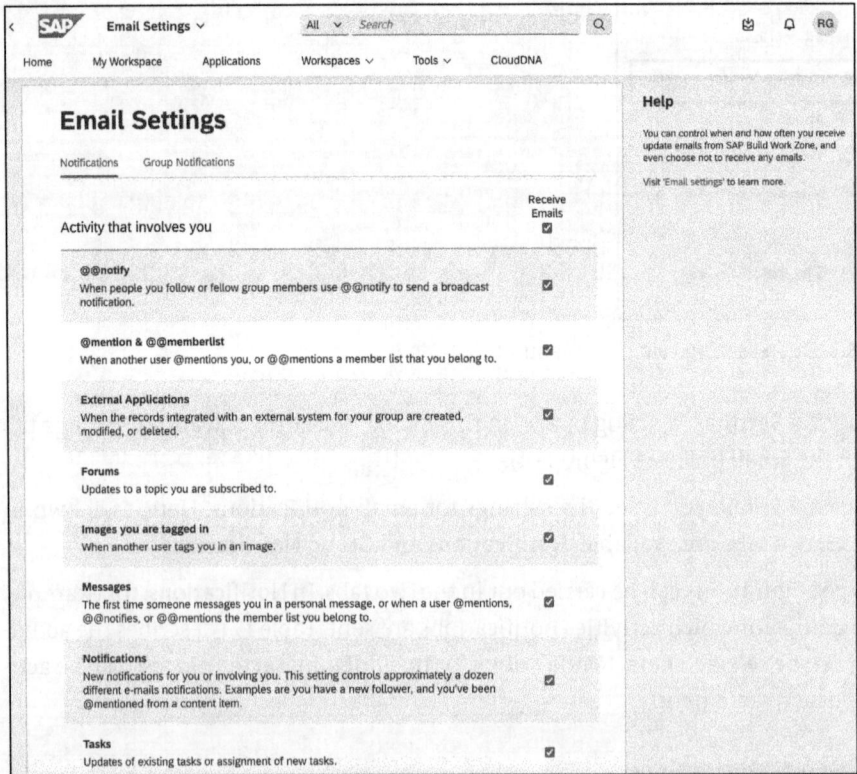

Figure 15.5 Email Notification Settings

15.1 Notifications

Activity	Purpose
@@notify	Applies when people you follow or fellow group members use @@notify to send a broadcast notification.
@mention & @@memberlist	Applies when another user @mentions you or @@mentions a member list that you belong to.
External Applications	Applies when records integrated with an external system for your group are created, modified, or deleted.
Forums	Used for updates to a topic you are subscribed to.
Images You Are Tagged In	Applies when another user tags you in an image.
Messages	Used the first time someone messages you in a personal message, or when a user @mentions, @@notifies, or @@mentions the member list you belong to.
Notifications	Used for new notifications for you or involving you. This setting controls approximately a dozen different email notifications. Examples include notifications that you have a new follower and notifications that you've been @mentioned from a content item.
Tasks	Used for updates of existing tasks or assignments of new tasks.
Videos You Are Tagged In	Applies when another user tags you in a video.
Wall Posts	Applies when another user comments, likes a post, or replies to a feed item on your wall.

Table 15.1 Different Activities for Email Notifications

If you now switch to the second tab, **Group Notifications**, you can define the frequency with which email notifications are sent. This can either be configured generally across all available workspaces or be workspace-specific. Regardless of the level at which this is done, the following options are available:

- **None**
  Means that no email notifications will be sent.
- **Immediate**
  Email notifications are sent immediately.
- **Daily**
  Email notifications are sent once a day.
- **Weekly**
  Email notifications are sent once a week.

15 Advanced Topics

You can do this configuration in general for all workspaces in the **Select Email Frequency for All Workspaces** dropdown. Figure 15.6 shows how to set this configuration for a specific workspace.

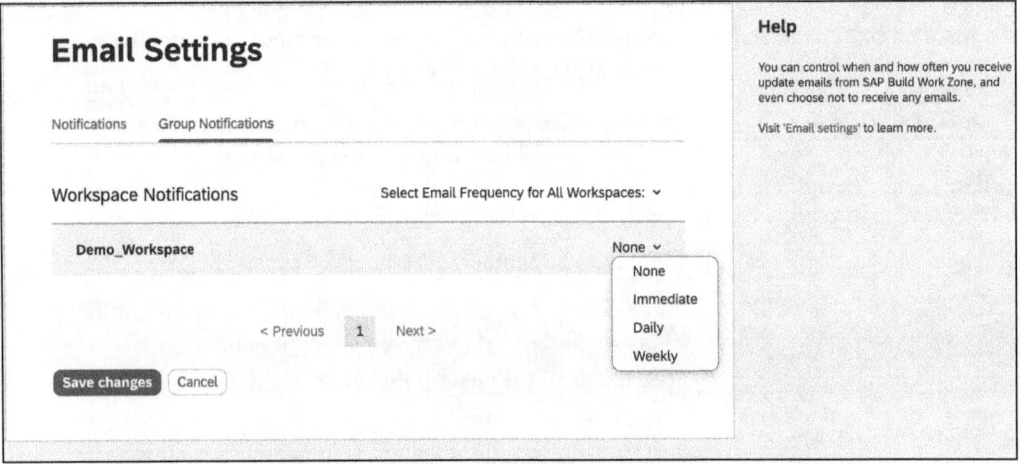

Figure 15.6 Configure Frequency for Email Notifications: Workspace-Specific

15.1.2 Push Notifications with Webhooks

In SAP Build Work Zone, there is also the possibility to set up push notifications, which work via webhooks. Webhooks offer the ability to track certain events that occur in SAP Build Work Zone. Subsequently, the metadata of these events is sent to third-party applications, which process the data and the push notification accordingly.

If you have an enterprise portal with official information, procedures, and policies, and SAP Build Work Zone as a solution to collaborate and work with business applications, this would be a potential use case for webhooks. In this case, webhooks could be used to respond outside of SAP Build Work Zone to updates that occur within it. For example, when blog posts are republished in a dedicated workspace, this change could also be pushed to other applications or to the enterprise portal.

The only thing to keep in mind here is that endpoints and corresponding functions must be developed on the receiving side for this scenario, and these process the incoming data. The creation and configuration of such webhooks can be performed in the administration console. To do this, first go to the **External Solutions • Notifications** menu item (see Figure 15.7).

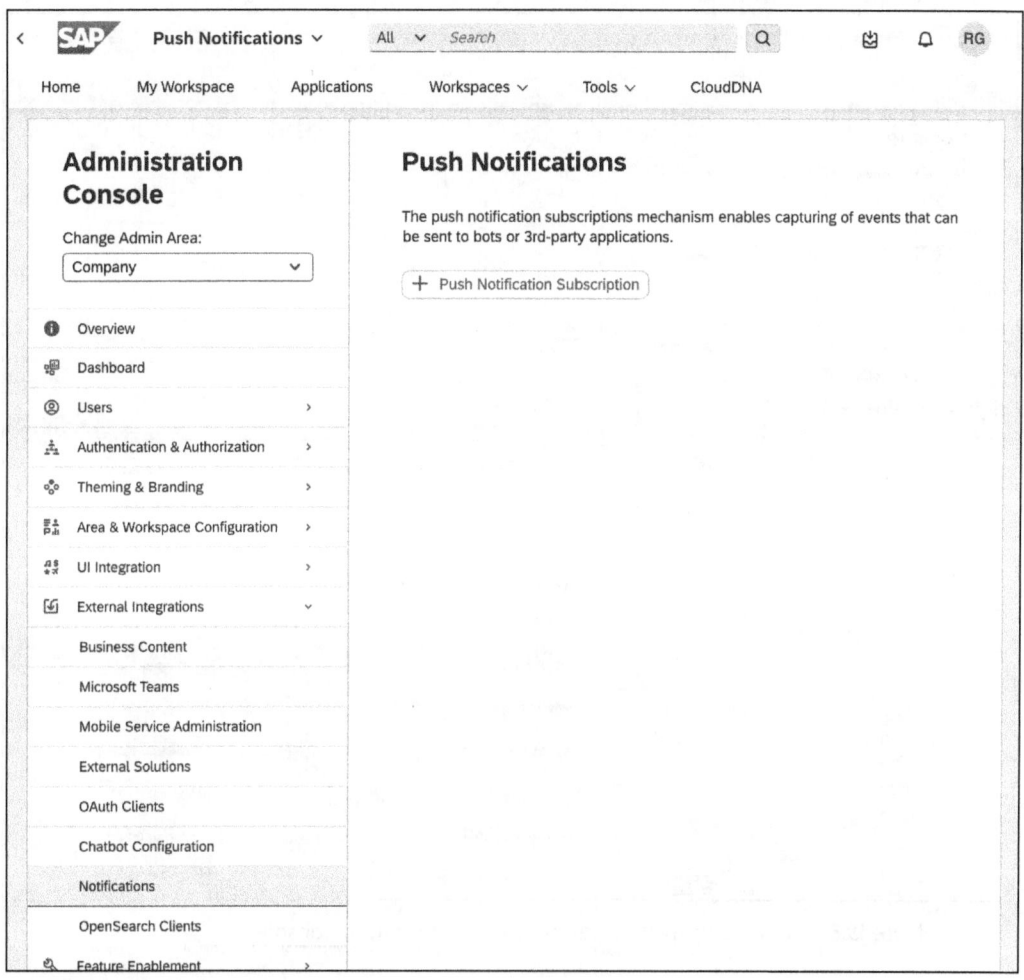

Figure 15.7 Push Notifications in Administration Console

A new configuration can be created by clicking **Push Notification Subscription**. In this configuration, a name, a callback URL, and the scope to which the configuration applies must be specified. In addition, the events for which push notifications are to be sent must be selected. These are divided into the following categories, as shown in Figure 15.8:

- **Alias Account Notification**
- **Content Notification**
- **Workspace Notification**
- **Forum Notification**
- **Company Notification**

15   Advanced Topics

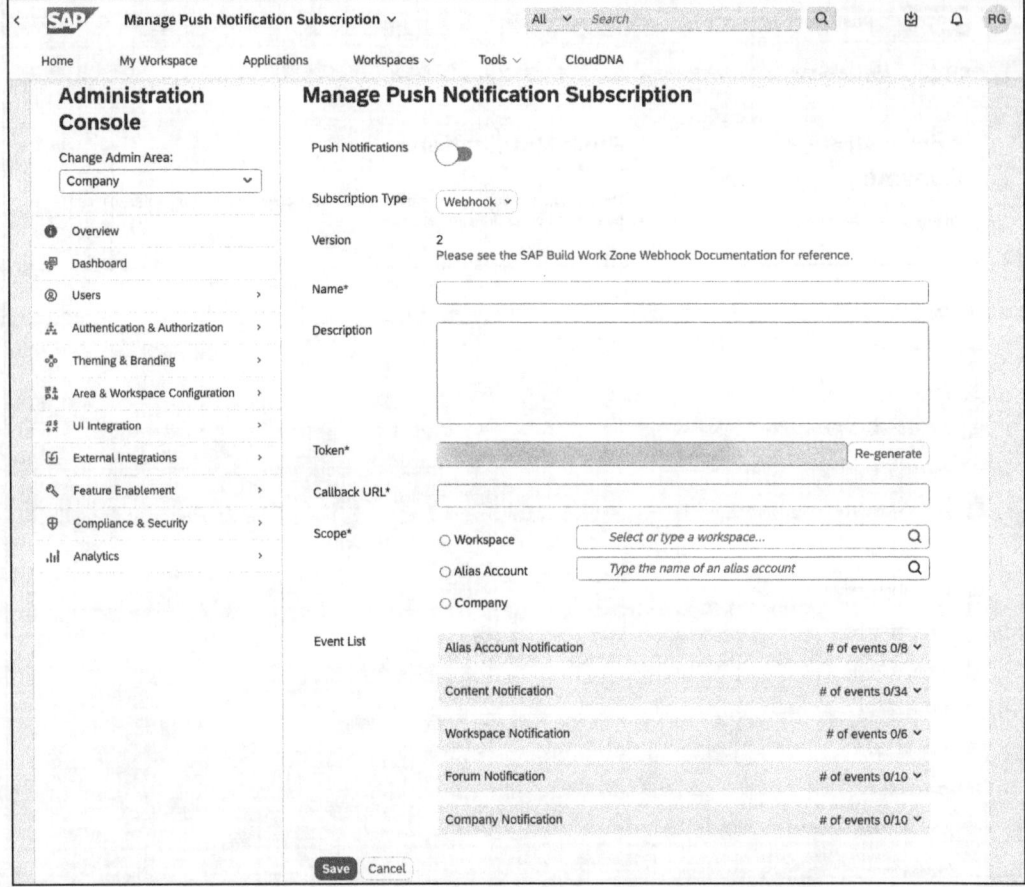

Figure 15.8 Configure Push Notifications in Administration Console

15.2   SAP HANA Enterprise Search

SAP HANA enterprise search is a search solution that provides unified, comprehensive, and secure real-time access to various enterprise data. This provides users with a way to search for structured data and directly access underlying applications or actions. For example, if a user wants to search for a specific object of an entity, they could enter the designated application and search for it there. But with enterprise search, the user could save himself the trouble of entering the designated application. He could search for the respective object directly via enterprise search instead.

Of course, the search has to be configured accordingly so that it works properly. In addition, the searcher only gets back those objects for which he is also authorized. If enterprise search is deactivated, only SAP Build Work Zone content can be searched for. This includes, for example, content, people, and workspaces. A search for applications is not possible in this case. If, on the other hand, you activate enterprise search, you can

15.2 SAP HANA Enterprise Search

search for all applications for which you have the necessary permissions. This includes SAP S/4HANA apps as well as also local applications. A search for SAP Build Work Zone content is not possible in this case.

Enterprise search is located in the header area of SAP Build Work Zone (see Figure 15.9).

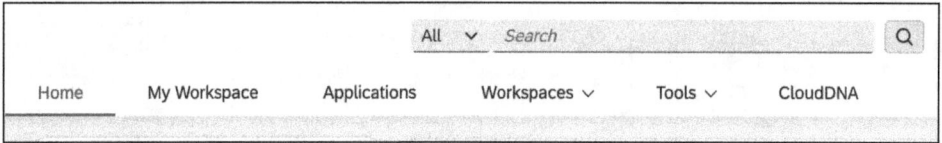

Figure 15.9 Enterprise Search in SAP Build Work Zone

As you can see in Figure 15.9, you can search via a search term and additionally limit the search by category using the dropdown menu to the left of the **Search** field. This allows you to specify that the search should only be performed in certain areas. The dropdown contains the following options:

- **All** (default)
- **Applications**
- **Content**
- **Events**
- **Forums**
- **Workspaces**
- **Knowledge Base**
- **People**
- **Tasks**
- **Tags**

For example, if you search for "Demo" in this case, you'll find hits in workspaces and content (see Figure 15.10).

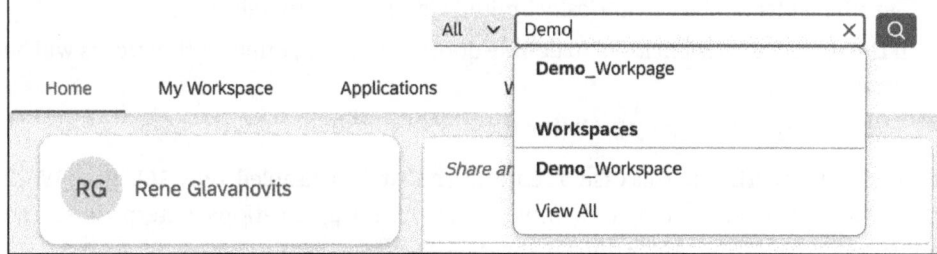

Figure 15.10 Suggestions as Search Results

If the entry is confirmed by clicking the **Search** button or by pressing [Enter], the results list opens as a separate page (see Figure 15.11).

463

15 Advanced Topics

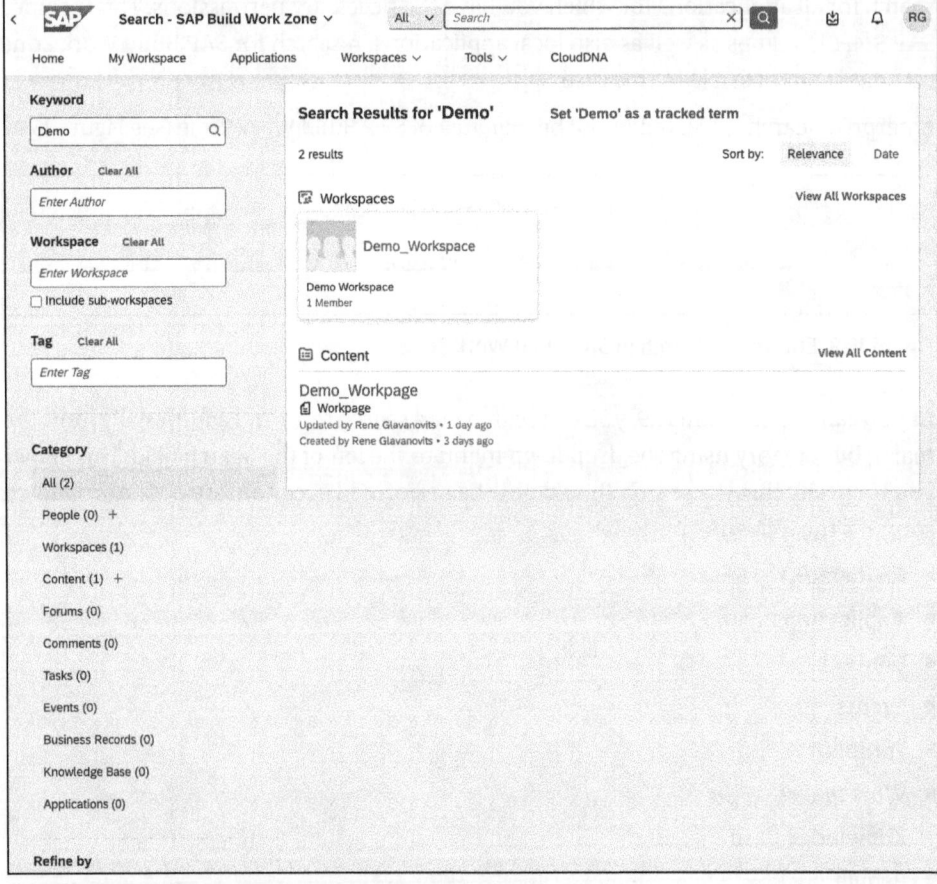

Figure 15.11 Results for Search in SAP Build Work Zone

You can enter letters, numbers, and special characters in the search field. By default, the search mechanism uses wildcards, but when multiple words are entered, only those results are returned that contain all of the words. The following options are available to limit the search:

- Search results can be categorically limited via the category selection.
- If the search term is enclosed in double quotation marks, only exact matches will be found.
- Search for specific tags by prefixing them with a hashtag character (#).

As mentioned earlier, in this case, as enterprise search is disabled, only SAP Build Work Zone content is searched. For example, to search for applications, enterprise search must be enabled. This can be done using the site manager, as follows:

1. Open the administration console of SAP Build Work Zone via the corresponding menu item in the context menu of the user icon.
2. In the administration console, select the **External Integrations • Business Content** menu item. A page like the one in Figure 15.12 should open.

15.2 SAP HANA Enterprise Search

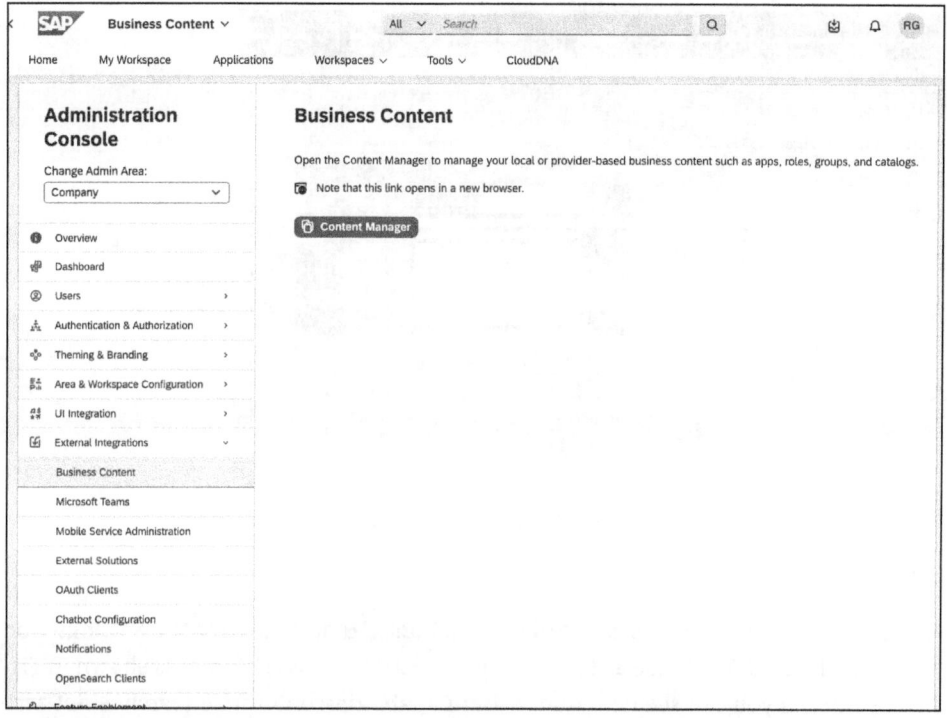

Figure 15.12 External Integrations: Business Content in Administration Console

3. Here, click the **Content Manager** button, which opens the site manager (see Figure 15.13).

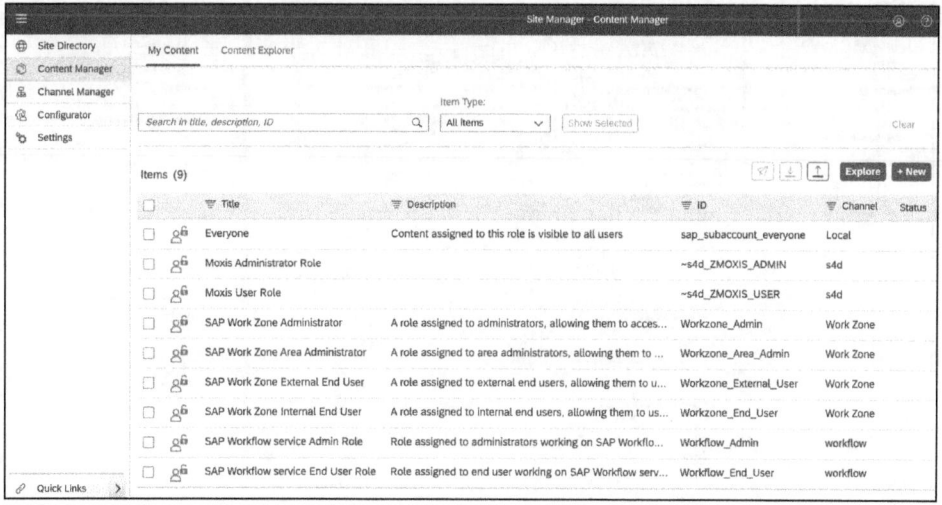

Figure 15.13 Site Manager

4. Once in the site manager, navigate to the **Site Directory** using the entry in the left bar. The page shown in Figure 15.14 should then open, displaying the SAP Build Work Zone card.

15   Advanced Topics

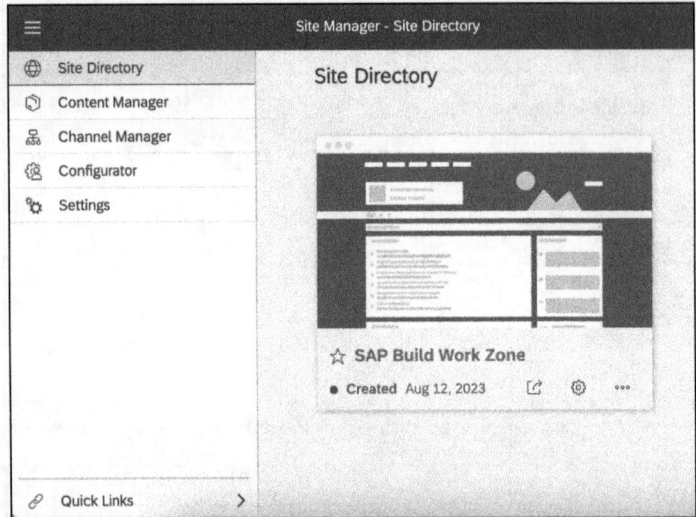

Figure 15.14 Site Directory

5. Open the site settings for SAP Build Work Zone. To do this, click the **Settings** icon inside the card. You'll see a settings page for SAP Build Work Zone, as shown in Figure 15.15. As you can see here under **User Capabilities**, enterprise search is not activated (**Enterprise Search: No**).

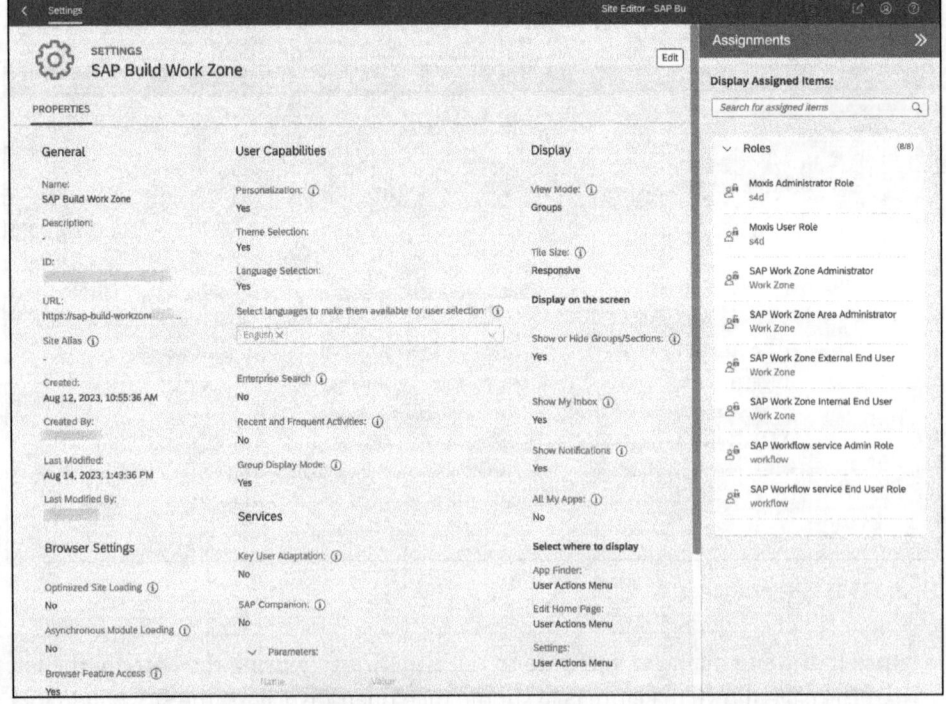

Figure 15.15 Settings of SAP Build Work Zone

15.2 SAP HANA Enterprise Search

6. To activate enterprise search, switch to edit mode using the **Edit** button and activate the **Enterprise Search** toggle, as shown in Figure 15.16.

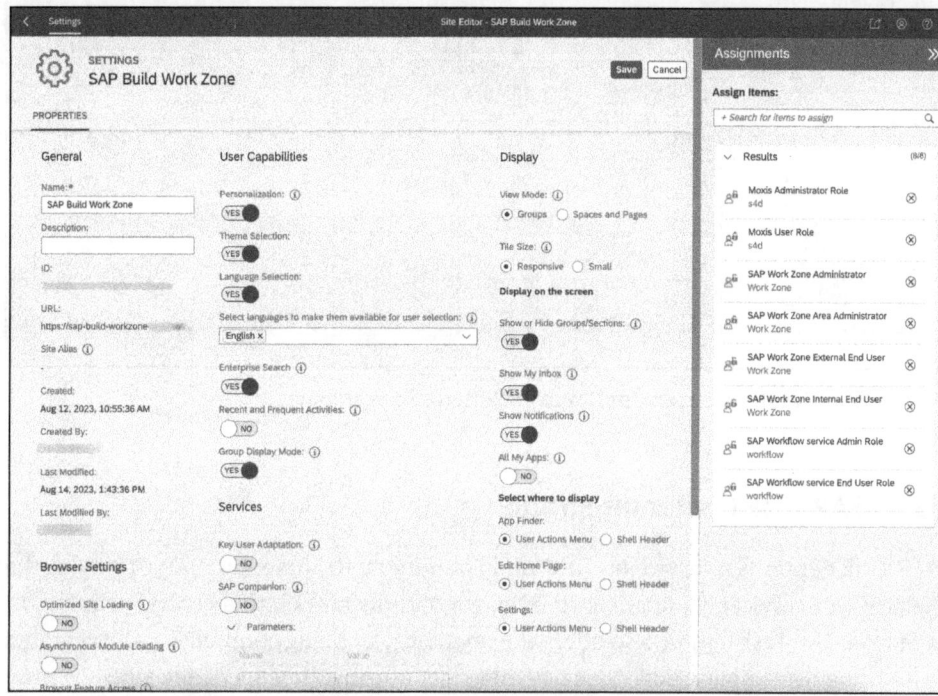

Figure 15.16 Enable Enterprise Search for SAP Build Work Zone

7. Save this change by clicking **Save**. Once this has been done, enterprise search is activated for SAP Build Work Zone. Now the site manager can be closed again and you can navigate back to SAP Build Work Zone.

After you're back in SAP Build Work Zone, you'll notice at a glance that the appearance of the search bar in the header has changed (see Figure 15.17). Also, category selection is no longer available.

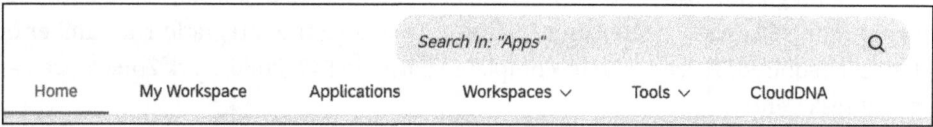

Figure 15.17 Search Bar after Activating Enterprise Search

If you now search for "Demo" again, you won't see any hits. This is because SAP Build Work Zone content is no longer searched. If, on the other hand, you search for a specific application, such as "Inbox", you'll see the My Inbox application returned as a hit, as shown in Figure 15.18. The My Inbox application can also be accessed directly from the hit list. If you want to continue searching for content in SAP Build Work Zone after activating enterprise search, you have to insert a search widget in a work page.

467

Figure 15.18 Result for Search for "Inbox" in Enterprise Search

15.3 SAP Task Center Integration

SAP Task Center is a service in the Cloud Foundry environment of SAP BTP. With the help of this service, it's possible to integrate various tasks into a central solution. In addition, SAP Task Center also offers the possibility of being combined and integrated with other SAP applications. This has the effect of providing end users with a central entry point for accessing tasks assigned to them. End users can access their tasks via the SAP Task Center web application.

SAP Task Center thus can be used as a unified inbox for tasks of all kinds from different applications. Tasks from different SAP solutions are displayed together in a list. After selecting a task, it can be processed with just one click. This reduces the processing time for all tasks, which ultimately affects the efficiency of users. Business users have all their tasks in one system and thus do not have to jump from system to system to process all tasks.

Through integration with SAP Task Center, any task created in a workspace can be displayed in the SAP Task Center user interface. To enable the integration, a number of steps are required. These are to be completed partly in SAP Build Work Zone as well as in SAP Task Center.

15.3.1 Steps in SAP Build Work Zone

First, the transformation code in Identity Provisioning must be updated, as follows:

1. Create an application in the Identity Authentication service. To do this, open the admin console of the Identity Authentication service.

15.3 SAP Task Center Integration

2. Navigate to **Applications & Resources** • **Applications** and click the **Create** button to create a new application of type **SAP BTP Solution** (see Figure 15.19 and Figure 15.20). We recommend assigning a meaningful name, such as "SAP Build Work Zone".

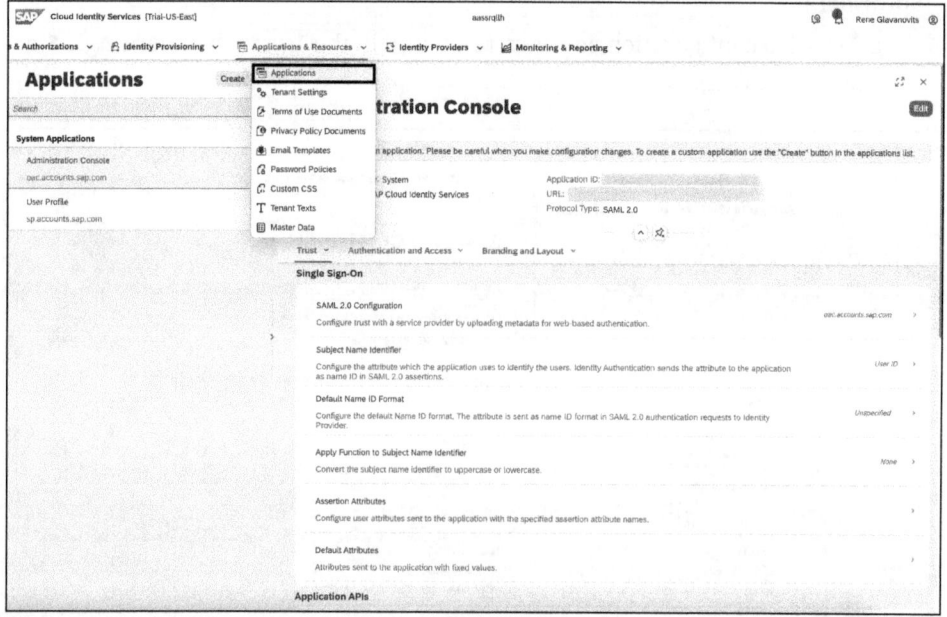

Figure 15.19 Identity Authentication

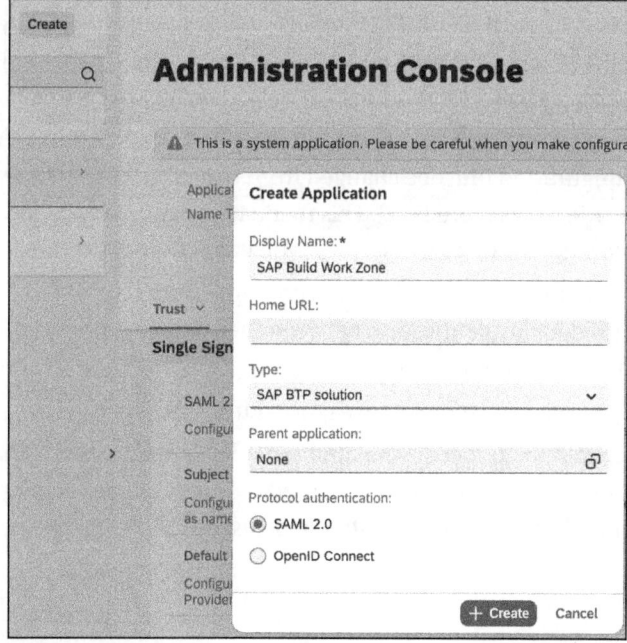

Figure 15.20 Create New Application in Identity Authentication

469

3. After entering your data, confirm it by clicking **Create**.
4. The newly created application appears in the list. Select it, and open the SAML 2.0 configuration editor. In this editor, you have to upload the metadata of the service provider.
5. The SAML 2.0 configuration editor can be found in the **Trust** tab under **Single Sign-On** (see Figure 15.21).

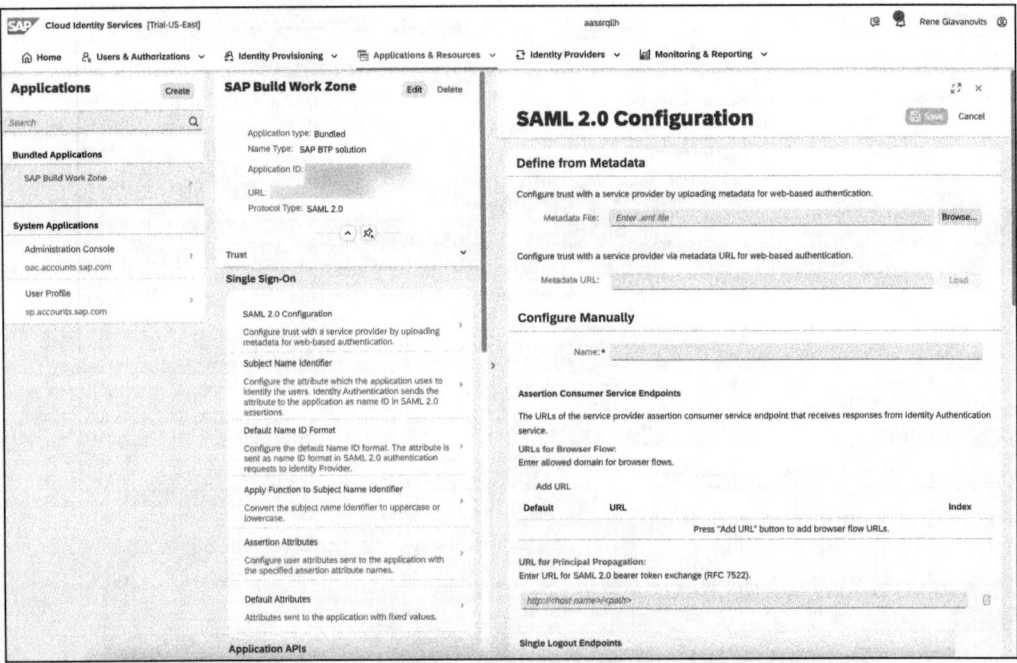

Figure 15.21 SAML 2.0 Configuration Editor

6. Subsequently, the basic configuration must be changed from **User ID** to **Global User ID** in the **Subject Name Identifier** (see Figure 15.22). This menu item can also be found in the **Trust** tab under **Single Sign-On**. Once this is done, you can proceed with **Save**.
7. Now the source and target systems of the Identity Provisioning service must be configured. A technical user is used for authentication between the Identity Authentication service and the Identity Provisioning service.
8. The Identity Authentication service must be set up as the source system.
9. SAP Build Work Zone represents the target system and must be defined as such. In the course of this setup, some values must also be stored. To facilitate debugging afterward, the key-value pairs shown in Table 15.2 should be used.

15.3 SAP Task Center Integration

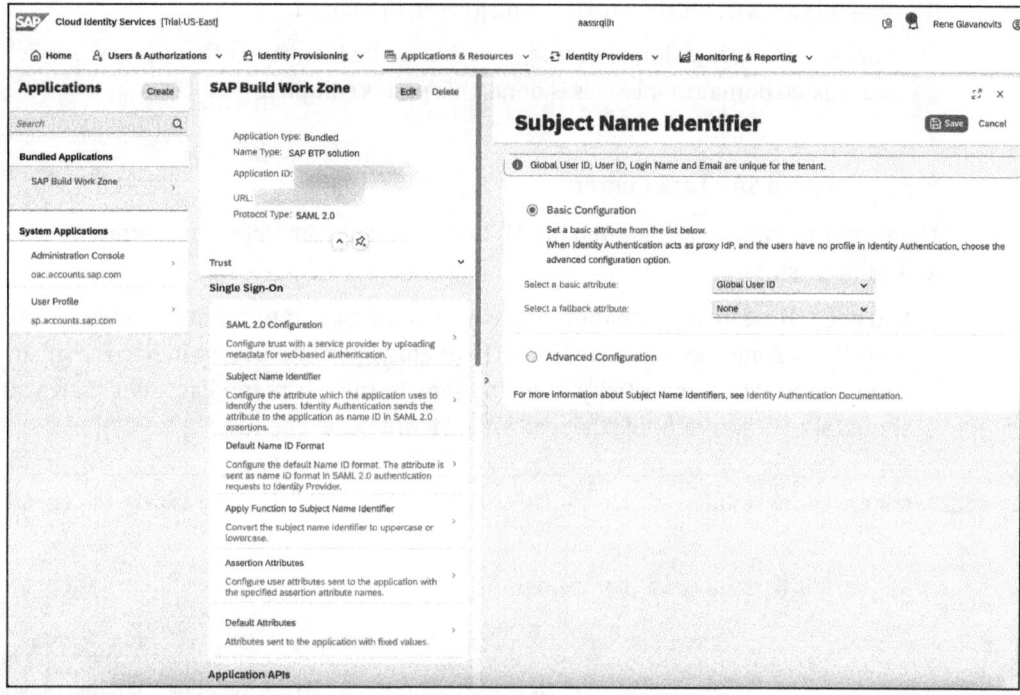

Figure 15.22 Subject Name Identifier Editor in Identity Authentication

Identity Provisioning Property Name	Value
Type	HTTP
ProxyType	Internet
Authentication	BasicAuthentication
ips.failed.request.retry.attempts	3
ips.failed.request.retry.attempts.interval	60
ips.delete.existedbefore.entities	true
ips.trace.failed.entity.content	true

Table 15.2 Properties for Setup of SAP Build Work Zone as Target System

10. Transformations must also be configured. These are used to map the user attributes from the data model of the source system to the data model of the target system and vice versa.

15 Advanced Topics

11. After all this is done, the provisioning job can be started.
12. Finally, the runtime domain must be added to the Identity Authentication service as a trusted domain. Once this is done, this part is complete.

15.3.2 Steps in SAP Task Center

There are also a few steps required in SAP Task Center to complete the integration. Let's look at these steps:

1. You need to create a destination that will be used for communication between SAP Build Work Zone and SAP Task Center. To do this, open SAP BTP, switch to the subaccount, and select **Connectivity • Destinations** in the left menu bar. After that, you should see an overview like that shown in Figure 15.23.

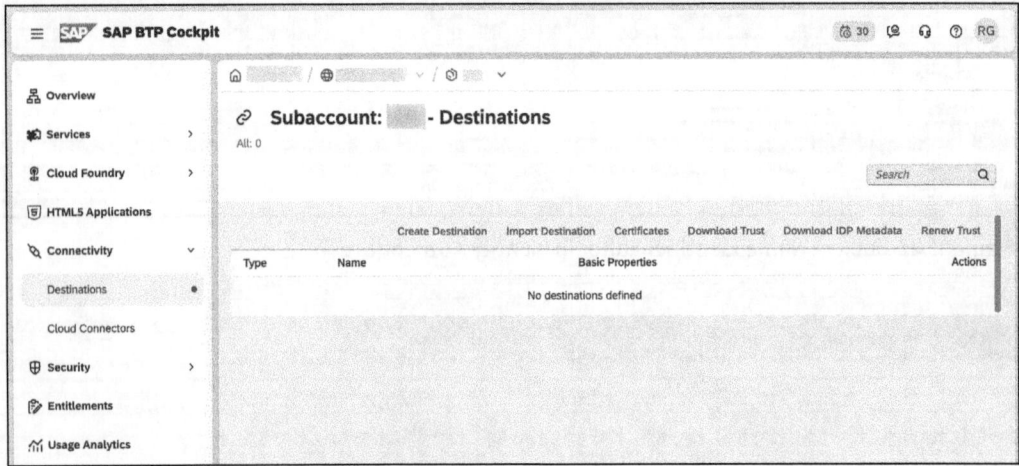

Figure 15.23 Destinations in SAP BTP

2. Click **New Destination** to enter the screen for creating a destination.
3. The values in Table 15.3 must now be entered here.
4. Finally, the destination can be saved with **Save**.

Entry	Value
Name	"WorkZone"
Type	"http"
Description (optional)	"Destination WorkZone Task Center"

Table 15.3 Properties for Creation of Destination in SAP BTP

15.3 SAP Task Center Integration

Entry	Value
URL	API endpoint (can be found in the admin console of SAP Build Work Zone under **Overview • Custom Domain**)
Proxy Type	Internet
Authentication	OAuth2SAMLBearerAssertion
Audience	Audience value (can be found in the admin console of SAP Build Work Zone under **Authentication and Authorization • SAML Local Service Provider**)
AuthnContextClassRef	"urn\:oasis\:names\:tc\:SAML\:2.0\:ac\:classes\:PreviousSession"
Client Key	OAuth client key from the OAuth client that was created
Token Service URL Type	Dedicated
Token Service URL	"<DWS URL>/api/v2/auth/token"
Token Service User	OAuth client key from the OAuth client that you've created
Token Service Password	OAuth client secret from the OAuth client that you've created

Table 15.3 Properties for Creation of Destination in SAP BTP (Cont.)

5. After these properties have been maintained in the destination, further additional properties must be defined under **Additional Properties**. These can be added by clicking **New Property**. These properties are listed in Table 15.4.

Key	Value
"nameIdFormat"	"urn:oasis:names:tc:SAML:1.1:nameid-format:emailAddress"
"tc.enabled"	"true"
"tc.provider_type"	"WorkZone"
"tokenServiceURL.headers.authAttribute"	"user_uuid"
"userIdSource"	"user_uuid"

Table 15.4 Additional Properties for Destination

6. After these values have been maintained, the destination should look as shown in Figure 15.24.

15 Advanced Topics

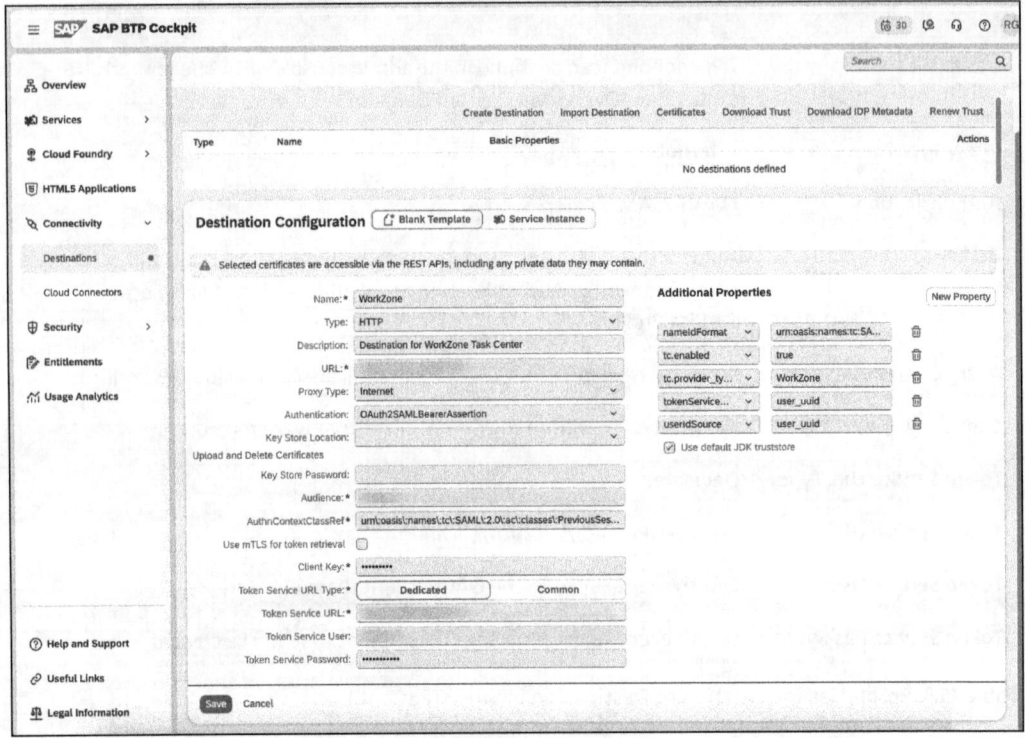

Figure 15.24 Destination in SAP BTP

7. Now the destination can be saved by clicking **Save**.

After performing these steps, everything necessary for integration with SAP Task Center is done.

15.4 SAP Mobile Start

SAP Mobile Start is a native mobile application for all SAP users. It allows access to native or web-based business applications (e.g., SAPUI5 applications), data, and information from SAP Build Work Zone, standard edition. This also allows users to easily access applications and tasks and gives them the ability to complete tasks anytime, anywhere.

> **Availability of SAP Mobile Start**
>
> SAP Mobile Start is available in all public data centers where SAP Build Work Zone, standard edition is available. In addition, SAP Mobile Start does not require a separate subscription as it is already included in the SAP Build Work Zone, standard edition subscription.

15.4 SAP Mobile Start

To set up SAP Mobile Start with SAP Build Work Zone, standard edition, a number of steps are required, as follows:

1. Check whether all system requirements are achieved. These include the following:
   - **General**
     - SAP BTP tenant (provided by SAP)
     - SAP BTP subscription (provided by SAP)
     - SAP Build Work Zone, standard edition subscription
   - **SAP S/4HANA Cloud Public Edition**
     - Cloud connector with principal propagation
     - Embedded deployment for the frontend server on SAP NetWeaver Application Server for ABAP, version 7.54 SP 06
     - Application content for SAP S/4HANA on-premise 1809 or above
   - **SAP S/4HANA and SAP S/4HANA Cloud Private Edition**
     - Application content for SAP S/4HANA Cloud 2108 or above
2. Now activate the SAP Mobile Start setting in SAP Build Work Zone, standard edition. To do this, you must first open SAP Build Work Zone, standard edition.
3. Open the site settings of a specific page. To do this, click the **Settings** icon inside the correct tile (see Figure 15.25).

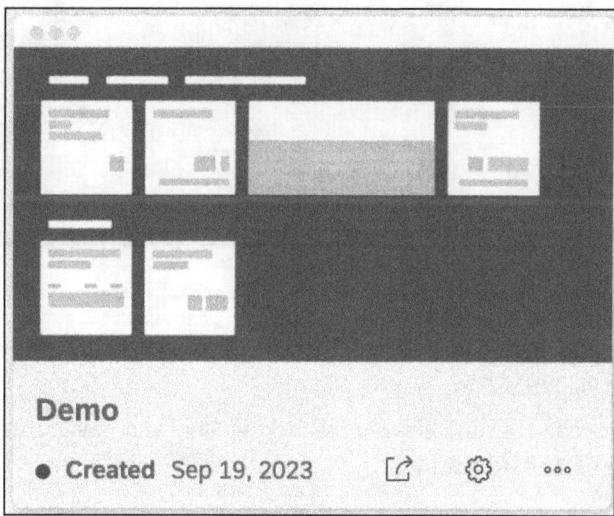

Figure 15.25 Settings Icon for Site

4. This opens a page like the one shown in Figure 15.26.

15   Advanced Topics

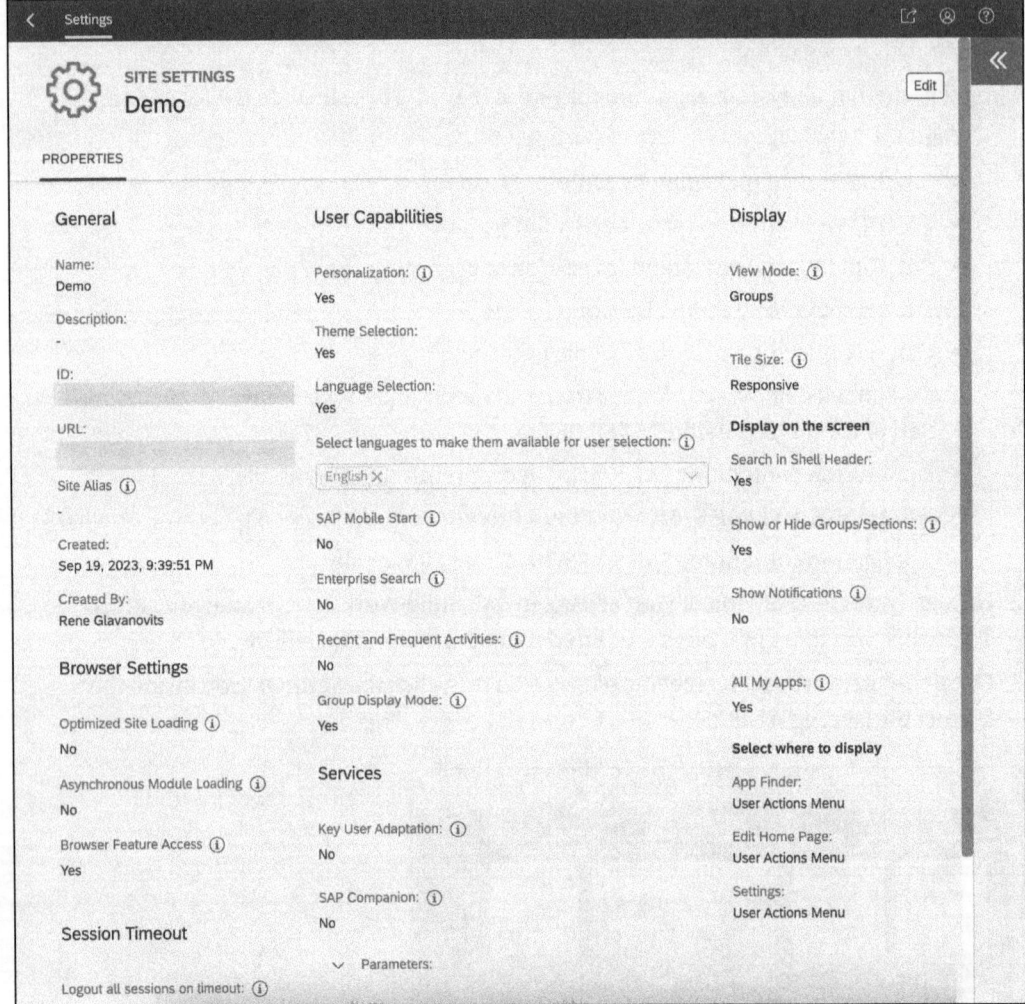

Figure 15.26 Site Settings in SAP Build Work Zone, Standard Edition

5. Now you have to switch to edit mode and activate the switch for **SAP Mobile Start** under **User Capabilities** (see Figure 15.27).

6. After this has been done, the change can be saved by clicking **Save** and navigating back to SAP Build Work Zone, standard edition.

15.4 SAP Mobile Start

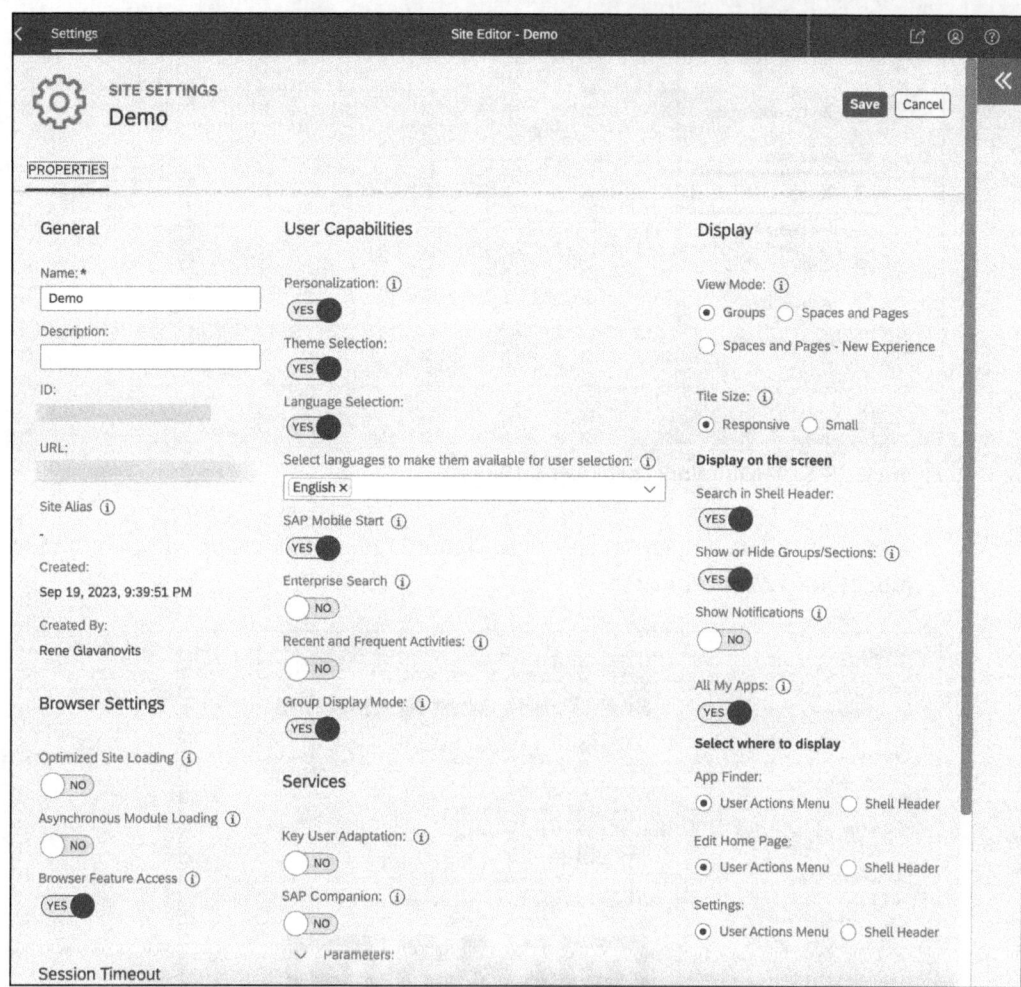

Figure 15.27 Site Settings in SAP Build Work Zone, Standard Edition after Activating SAP Mobile Start

From the admin's point of view, these are basically the items required to enable SAP Mobile Start. But there are still steps required from the end user's point of view, which we will look at next:

1. The end user must now install and register the SAP Mobile Start application on his smartphone. To do this, they simply have to enter SAP Build Work Zone, standard edition and navigate to **Settings • SAP Mobile Start** in the user menu (see Figure 15.28).

477

15   Advanced Topics

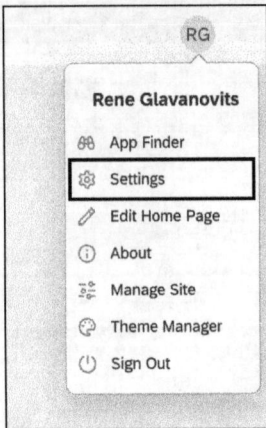

Figure 15.28 SAP Build Work Zone User Settings

2. After that, a dialog opens as shown in Figure 15.29. In this dialog, jump to the **SAP Mobile Start Application** tab.

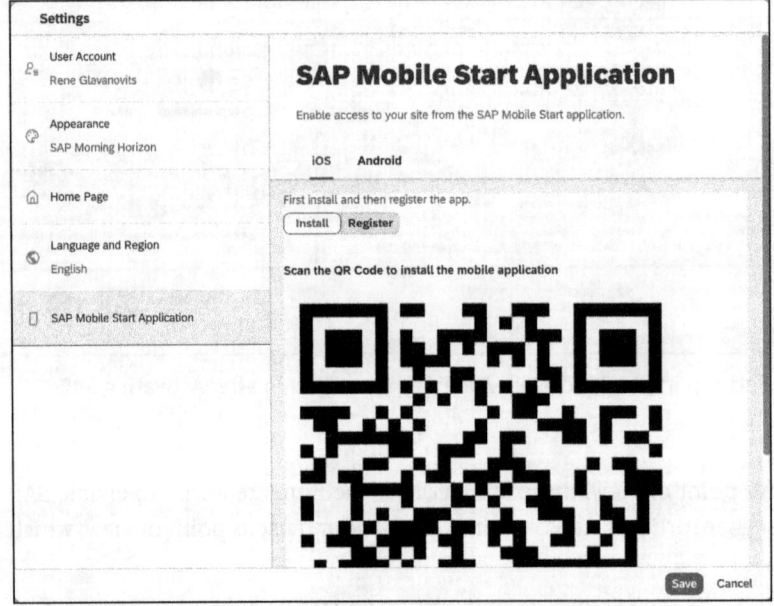

Figure 15.29 SAP Mobile Start Application under Settings

3. To install the app on your smartphone, select either the **iOS** or the **Android** tab, depending on your smartphone's operating system, and click **Install**.

4. The displayed QR code must be scanned with the smartphone on which the application is to be installed. Behind the QR code is a download link through which the app is loaded. Alternatively, you could also manually search for **SAP Mobile Start** in your phone's app store.

15.4 SAP Mobile Start

5. After the download and installation is complete, the app can be opened on your smartphone. At the beginning, the terms of use must be accepted. Once this is done, the app should look as shown in Figure 15.30.

Figure 15.30 Initial View of SAP Mobile Start App

6. Switch back to SAP Build Work Zone. Here, now select the **Register** item (see Figure 15.31). Afterward, you can also scan this QR code with your smartphone. To do this, click the **Scan** button on your smartphone within the SAP Mobile Start application; this opens the camera, which can then be used to scan the code.

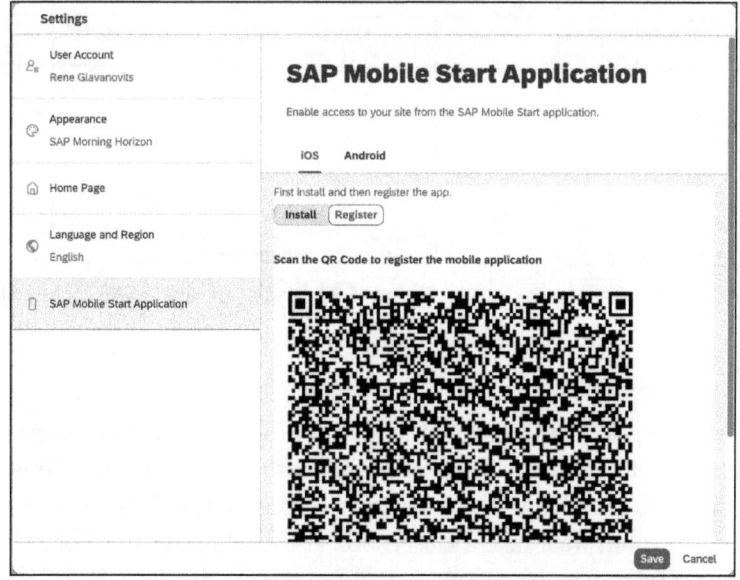

Figure 15.31 SAP Mobile Start Application QR Code for Registration

7. If the QR code has been scanned, this is confirmed with a success message. The user is then prompted on the smartphone to log in via the SAP ID service.

8. After this has also been done, the SAP Mobile Start app can be used on user's smartphone. As shown in Figure 15.32, you now have access to the app, but the administrator still has to integrate tiles, which can then be called up.

The addition of tiles must be handled by the administrator via the site directory of SAP Build Work Zone. Once this has been done, the SAP Mobile Start application can be restarted, and the corresponding tiles should be available in the **Apps** tab. However, it should be noted here that the visibility of each depends on the role assignment behind it. Figure 15.33 shows an example of how such a tile is displayed.

By clicking the tile, the app behind it can be called up. This can be any URL, but also SAPUI5 applications or other web applications. Native apps can also be integrated into SAP Build Work Zone.

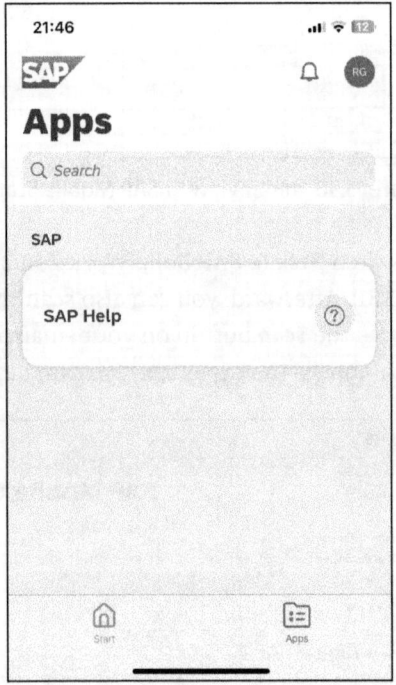

Figure 15.32 SAP Mobile Start after Setup   **Figure 15.33** Display of Tile in SAP Mobile Start

15.5 Summary

In this chapter, you've seen how further topics, such as notifications, enterprise search, and SAP Mobile Start, can be integrated into and used with SAP Build Work Zone. You've also seen what functionalities and options the individual areas of functionality offer. In the following chapter, we'll take a closer look at the administration of SAP Build Work Zone, such as feature management and user management.

Chapter 16
Administration

In this chapter, you'll embark on a comprehensive journey through the intricacies of administering SAP Build Work Zone. The focal points of our discussion encompass feature management, user administration, compliance, and the intricate web of error logs that plays a pivotal role in maintaining system integrity.

In this chapter, you'll learn about administration of SAP Build Work Zone, including feature management, user management, compliance, and error logs. In Section 16.1 and Section 16.2, we look at feature management and user management to see what options are available for restricting, activating, or deactivating certain functionalities. Then, in Section 16.3, we look at the extent to which SAP Build Work Zone supports adherence to guidelines in order to remain compliant. This primarily relates to the publication of content with undesirable vocabulary.

Finally, in Section 16.4, we look at how error logs can be exported from SAP Build Work Zone, what they contain, and what types of reports are available in this context.

16.1 Feature Management

Various configurations can be made in the administration console of SAP Build Work Zone. Among other things, certain features and options can be activated or deactivated if desired. To do this, first open the administration console of SAP Build Work Zone, via the **Administration Console** menu item accessed by clicking the user icon in the top-right-hand corner. The administration console will open as shown in Figure 16.1.

Here, select the **Feature Enablement • Features** entry in the left scroll bar. A page then opens on which various options from different areas can be activated and deactivated. In this case, let's take a closer look at the options relating to **Feature Management** (see Figure 16.2).

16 Administration

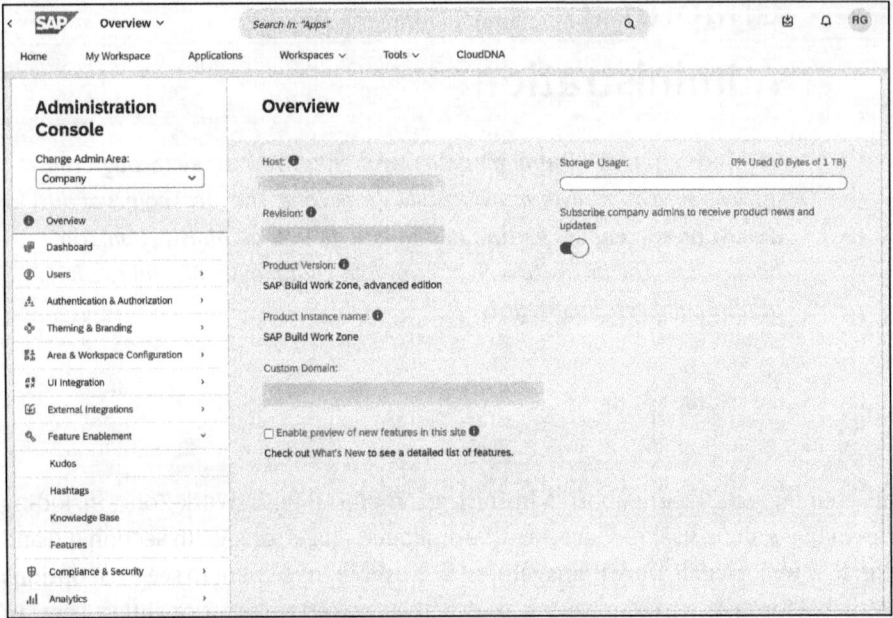

Figure 16.1 Administration Console

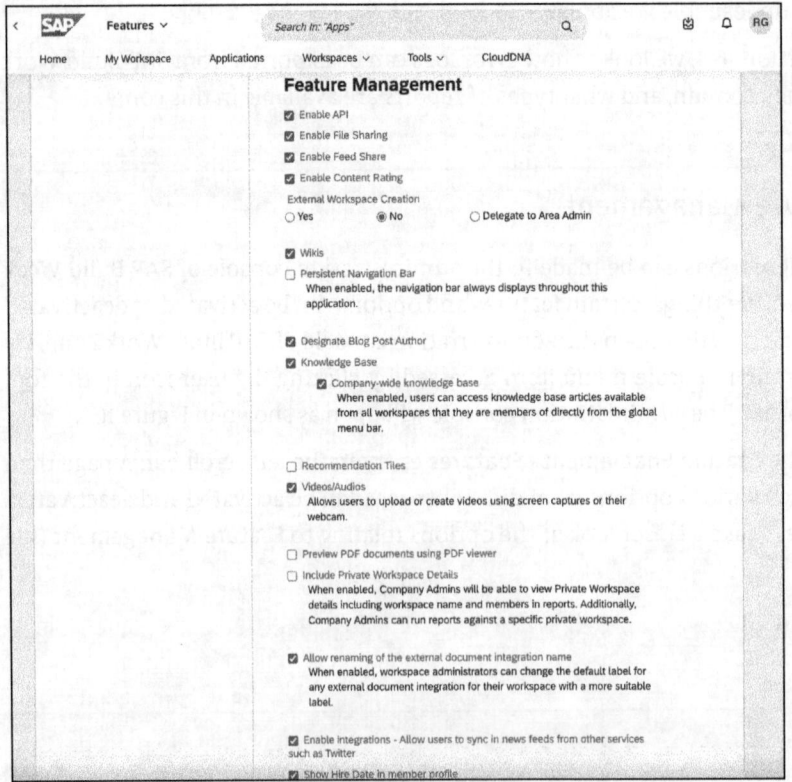

Figure 16.2 Feature Management in Administration Console

In principle, these various options can only be activated or deactivated. No other type of configuration is possible at this point. The following is a list of all possible options and explanations of their use:

- **Enable API**
  This option can be used to activate the SAP Build Work Zone, advanced Edition API for read and write access—for example, for the SAP Build Work Zone mobile app.
- **Enable File Sharing**
  If this option is activated, users are permitted to upload files to pages.
- **Enable Feed Share**
  By activating this option, a **Share** button is displayed for feed entries. This button can be used to share updates with a specific workspace or the entire company.
- **Enable Content Rating**
  This option allows users to rate content that has been uploaded to the site.
- **External Workspace Creation**
  This option enables the creation of private workspaces that are intended for and accessible to users who are not located within the company. The following options are available:
  - Yes
  - No
  - Delegate to Area Admin
- **Allow External Users to Be Created through Workspace Invitation Email**
  This option is only available if **External Workspace Creation** is activated. By activating this option, permission can be granted for external users to be created via an invitation email.
- **Technical Username, Password**
  This option is only available if **External Workspace Creation** is activated. For SAP Build Work Zone to trigger API calls for authentication for user invitations, it's necessary to specify access data—in this case, a technical user and a password.
- **Wikis**
  This option defines whether users are allowed to create wiki pages.
- **Persistent Navigation Bar**
  If this option is activated, the navigation bar is displayed everywhere.
- **Designate Blog Post Author**
  This option can be used to specify whether it should be possible for users to define that other people can also write blog entries in their name. Even if the blog entry is written and published by another person, the user who designated the other author is displayed as the creator.
- **Knowledge Base**
  This option allows knowledge base authors to create knowledge base articles as part

of the employee support for their workspaces. The knowledge base articles can be found on a **Knowledge Base** tab in a workspace. This feature is enabled by default at the company and workspace levels.

- **Company-Wide Knowledge Base**
  With this option, users can be enabled to see all knowledgebase articles of workspaces in which they are members. By default, the knowledge base is activated for a workspace if the company setting for the knowledge base is activated.

- **Recommendation Tiles**
  Controls the display of recommendation tiles in workspaces.

- **Videos/Audios**
  This option controls whether users are allowed to upload video files and audio files made using screen recordings or a webcam.

- **Preview PDF Documents Using PDF Viewer**
  By activating this option, the inline view of PDF files can be enabled within the browser. If certain functions are not available in the inline view, the PDF can be downloaded and opened with a PDF viewer.

- **Include Private Workspace Details**
  By activating this option, administrators have the option of displaying the names of private workspaces in reporting and compliance functions. They also have the option of creating reports on private workspaces.

- **Allow Renaming of the External Document Integration Name**
  Allows workspace administrators to change the default name for all external document integrations in their workspace. If this option is not activated, the default name remains consistent throughout the application and cannot be changed by workspace administrators.

- **Enable Integrations**
  With this option, users can be allowed to synchronize news feeds from other services.

- **Show Hire Date in Member Profile**
  Controls whether the setting date is displayed on profile pages.

- **Show Business Records Feeds Member Profile**
  This option defines whether a feed about user actions is displayed on the users' profile pages. If you do not want the feed to be displayed on the users' profile pages, this option must be deactivated.

- **Send Daily Alert Emails to All Members**
  This option activates the sending of content and updates by email.

- **Send Active Task Reminder to All Members**
  This option activates the sending of reminders to users who are assigned to an active task.

- **Allow User-Level Reporting/Dashboard**
  If you want owners of content elements to see a list of viewers and downloaders, this option can be activated. If this option is deactivated, you can still see the number of viewers and downloaders, but no report can be loaded.

- **Restrict Access to Company Admins Only**
  If this option is activated, only administrators can execute and display workspaces and create a list of viewers of workspace content.

- **Hide Contact Information for External Users**
  By activating this option, user information from external users is hidden if the user is logged in as an external user.

- **Enable Search Appliance Integration**
  Search appliance integration can be activated via this option.

- **Enable SCIM API Support**
  The SCIM API is based on the *cross-domain identity management* system, which provides an open standard for automating the exchange of user data. By activating this option, it is specified that the SCIM API can be used for the provisioning of users.

- **Hide the Change Password Link**
  This option hides the **Change Password** link for users who have been provisioned via the SCIM API.

- **Enable Document Download via CDN**
  If this option is activated, the content delivery network (CDN) temporarily stores documents outside the SAP data center to ensure increased download speed. Customers who have distributed their data centers globally in particular experience a significant improvement in performance as a result.

- **Enable Webcam/Screen Recorder Using WebRTC Technology (Google Chrome and Mozilla Firefox Only)**
  Enabling this option allows users to select an information icon when sharing a video via a feed update, which indicates whether the compatibility prerequisites for screen recording or webcam recording are met. However, webcam recording is only supported by the Google Chrome and Mozilla Firefox browsers. Screen recording is also only supported by these two browsers.

- **Enhanced Widget Styling and Page Layout**
  This option makes it possible to display widgets to make the best use of the available screen space. Furthermore, this feature ensures improved rotating banner and gallery layouts, a larger and bolder font size, and the removal of some profile pictures.

- **Enable Repository Configuration at the Workspace Level for Microsoft Office 365**
  Enabling this option allows workspace administrators to create custom labels for document integration to be used by their workspaces.

- **My Workspace**
  Activating this option means that users can create their own workspaces.

16.2 User Management

Similar to the previous chapter, where you saw which options are available to activate or deactivate features globally in the administration console of SAP Build Work Zone, there is also the option to define certain configurations in the user context. This ranges from the display of certain content, such as the profile picture, to the option of not enabling content creation on certain pages.

As already mentioned, the configuration options can be found in the administration console. Once this has been opened, the **Feature Enablement • Features** entry must be selected in the menu bar. Within this page, there is a **User Management** section in which the corresponding configurations can be made (see Figure 16.3).

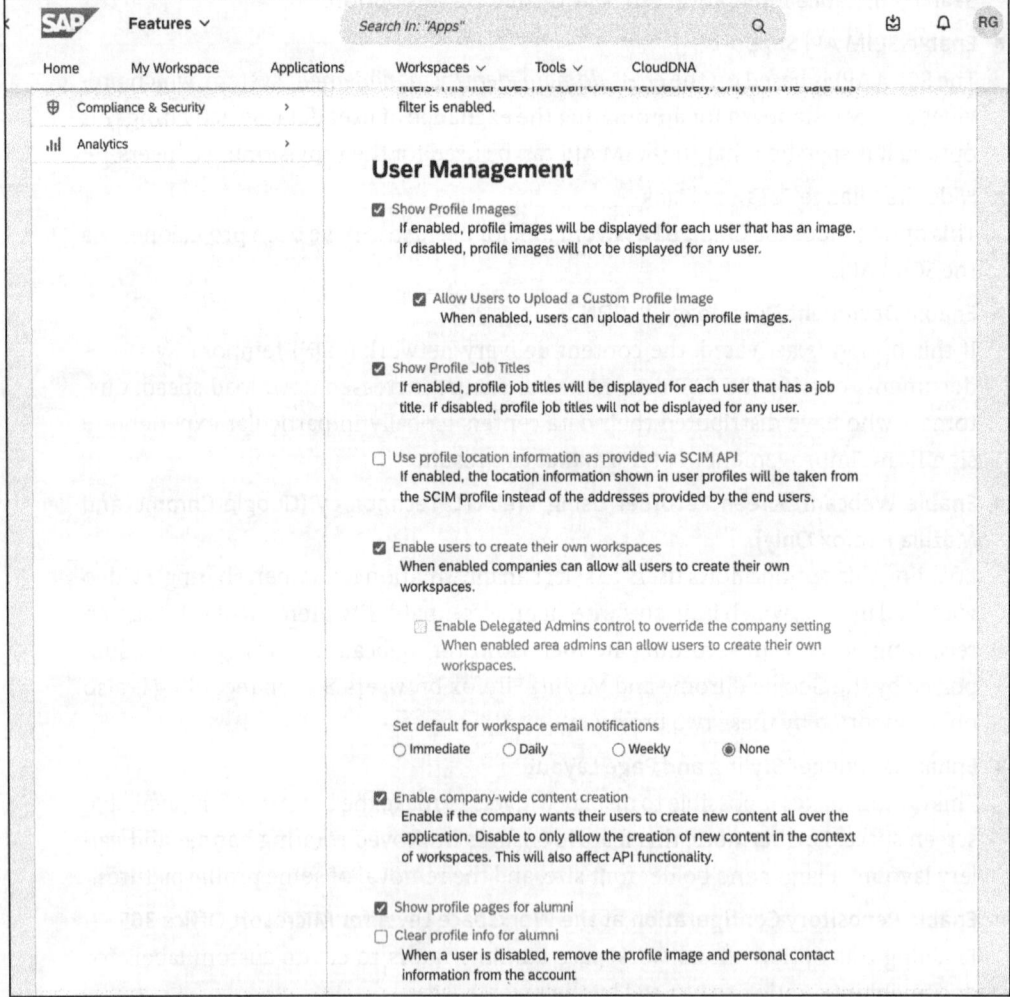

Figure 16.3 User Management in Administration Console

Here too, it's only possible to activate or deactivate the individual options. Further configurations are not possible at this point. The possible configurations are listed and explained in more detail in the following list:

- **Show Profile Images**
  The profile pictures of users are displayed next to the posts of users in forums and in the feed, as well as in other places in SAP Build Work Zone. By activating this option, an image is displayed if one is available. If the option is deactivated, a profile photo will not be displayed under any circumstances.

- **Allow Users to Upload a Custom Profile Image**
  If this option is deactivated, it prevents users from uploading a profile picture. This may be desired if profile images have already been imported by the organization and should not be changed, for example. If the **Show Profile Images** option is deactivated, this option is also deactivated automatically.

- **Show Profile Job Titles**
  If this option is activated, job titles are displayed for all users who have a job title. If deactivated, no job title is displayed for anyone.

- **Use Profile Location Information as Provided via SCIM API**
  If this option is activated, the location information displayed in the user profile is used by the SCIM profile instead of the information specified by the user.

- **Enable Users to Create Their Own Workspaces**
  This option can be used to control whether users can create their own workspaces. If this option is deactivated, workspaces can only be created by administrators.

- **Enable Delegated Admins Control to Override the Company Setting**
  This option controls whether area administrators can allow their users to create new workspaces, even if the previous option is disabled.

- **Set Default for Workspace Email Notifications**
  This option can be used to define the frequency at which email notifications are sent. The following options are available:
  - Immediate
  - Daily
  - Weekly
  - None

- **Enable Company-Wide Content Creation**
  This option can be used to control whether users should be able to create posts only in the workspace or across the entire application. If this option is deactivated, content can only be created within the workspace. This option also has an effect on API functionalities.

16 Administration

- **Show Profile Pages for Alumni**
  Alumni users are users who are no longer in the organization. This option can be used to define whether the profiles of such users should still be findable or not.
- **Clear Profile Info for Alumni**
  This option can be used to specify that personal information of alumni users should be removed. The office location is removed, but the manager and job title are retained.

16.3 Compliance

Compliance means adhering to rules, laws, guidelines, and standards of all kinds. These can be specified at the company level, at the industry level, or by society in general. The aim is to adhere to as many rules as possible that are specified by an organization or a person. In a business context, this usually involves the implementation of guidelines and external regulations that are necessary in order to work in the best possible way.

As SAP Build Work Zone offers the opportunity to exchange information via posts or comments, compliance also plays a not-insignificant role in this context. For example, posts could contain undesirable vocabulary. By defining certain rules, such posts can be identified and users can be made aware of them. With regard to compliance, SAP Build Work Zone offers various configuration options. To access these, the administration console of the SAP Build Work Zone must first be opened, as follows:

1. Click the user icon in the top-right-hand corner of SAP Build Work Zone. A popover then opens as shown in Figure 16.4.

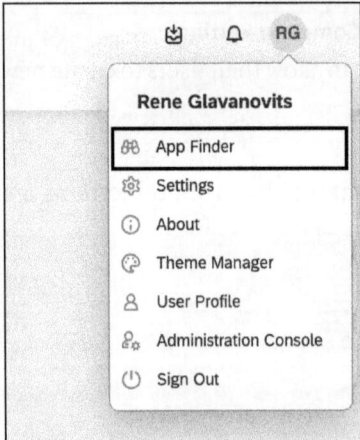

Figure 16.4 User Menu in SAP Build Work Zone

2. Select the **Administration Console** entry to access the admin console.
3. The administration console then opens. Here, various settings can be made. In this case, select the **Feature Enablement • Features** entry.

488

4. After selecting this entry, the view shown in Figure 16.5 opens.

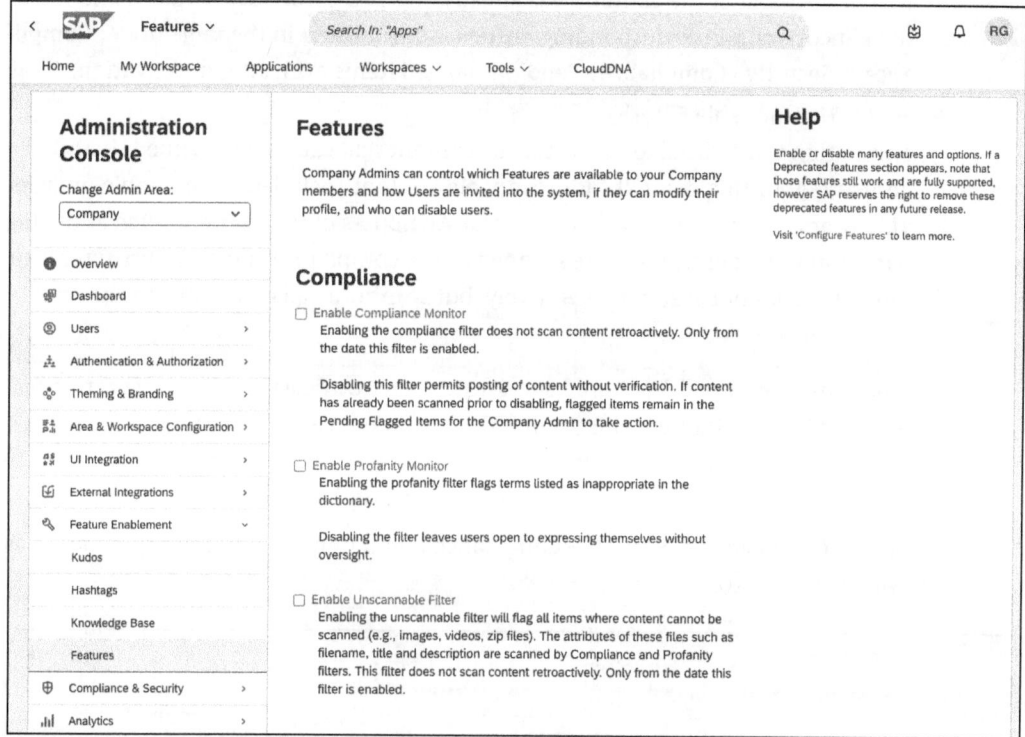

Figure 16.5 Compliance Settings in SAP Build Work Zone Administration Console

Various features can now be configured here. At this stage, however, we are only interested in compliance. The following options are available:

- **Enable Compliance Monitor**
  By activating this option, content published in the SAP Build Work Zone is analyzed and highlights items that contain terms that are not compliant according to the compliance dictionary. However, it should be noted that this function is only effective for entries that are created after activation. This means that obsolete entries that already exist will not be analyzed if this option is activated. In addition, once this function is deactivated, content uploaded by users is no longer tested against the compliance dictionary. This means that as soon as this function is deactivated, content can be published without being scanned. If content was already scanned before deactivation, the flagged items will continue to be displayed under **Compliance & Security** • **Compliance** • **Pending Flagged Items**.

- **Enable Profanity Monitor**
  If this option is activated, the content of the website is monitored and checked against a swear dictionary. If unwanted terms are detected, the corresponding entries are flagged. Here too, if the option is activated, content is not checked retrospectively; only new content from the time the option is activated is checked. If the

option is deactivated, content can also be published without it being checked in any way beforehand. If the content was already checked before deactivation and found to be incorrect, the corresponding entries are still listed in the table under **Compliance & Security • Compliance • Pending Flagged Items** even after deactivation.

- **Enable Unscannable Filter**
  This option can be used to define that all content that cannot be scanned should also be marked. In this context, nonscannable content includes, for example, images, videos, and archive files (such as ZIP files). Attributes of this content, such as the file name, title, and description, are scanned by the compliance and profanity filter. This filter also does not scan retrospectively, but scans only all content from the time of activation.

The options mentioned here only serve to activate or deactivate the respective features. What is still missing at this point is the maintenance of the terms that are considered noncompliant and are therefore also used to check the content. There is a separate interface within the administration console for maintaining these.

Select the **Compliance & Security • Compliance** menu item. An interface then opens as shown in Figure 16.6.

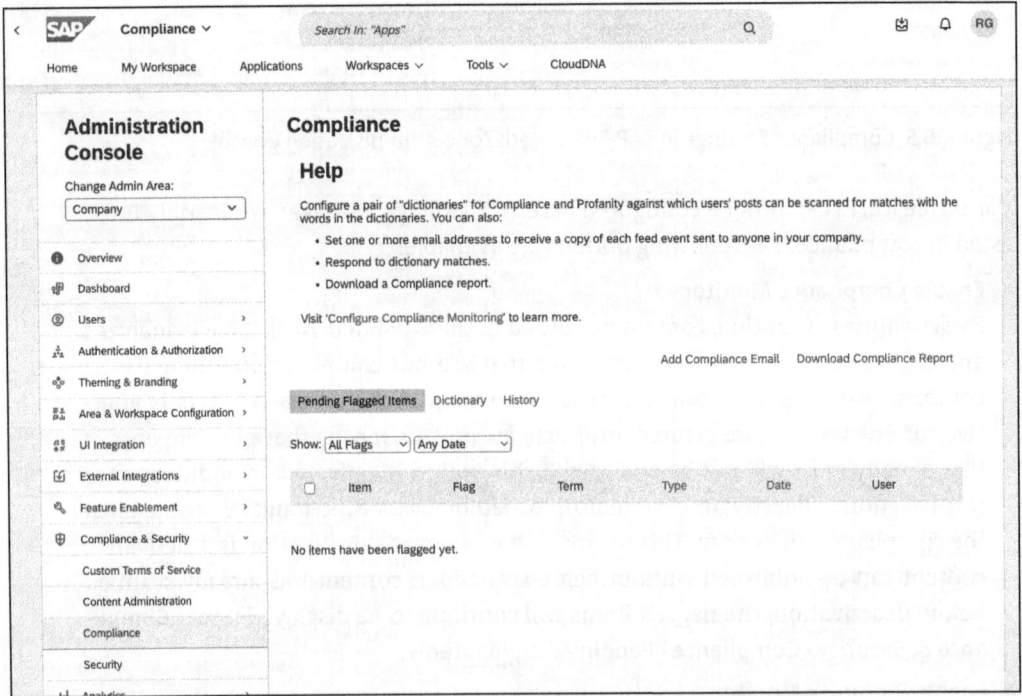

Figure 16.6 Compliance in Administration Console of SAP Build Work Zone

This interface now offers the option of configuring a type of dictionary in which the noncompliant terms are entered. This dictionary is then used to identify possible unwanted content. Furthermore, content that has already been analyzed and recorded can also be viewed on this page in order to obtain an overview of noncompliant posts. A distinction is made between two types when maintaining the dictionary:

- **Compliance**
  This dictionary is used to enter words that could indicate inappropriate disclosure of sensitive company information. This helps to avoid data leaks.
- **Profanity**
  Offensive terms are entered in this dictionary. This helps to ensure professional and respectful communication.

Although a wide variety of undesirable terms can be maintained in the two dictionaries, this does not prevent users from actually using the undesirable terms. Furthermore, no email notifications are sent if compliance is breached. The dictionaries merely serve to provide a way of alerting users to noncompliance. Whether users ultimately take this into account is up to them. To maintain the corresponding terms, proceed as follows:

1. Open the **Compliance & Security • Compliance** menu, if not already done.
2. In this menu, switch to the **Dictionary** tab at the bottom. You'll see the screen shown in Figure 16.7.

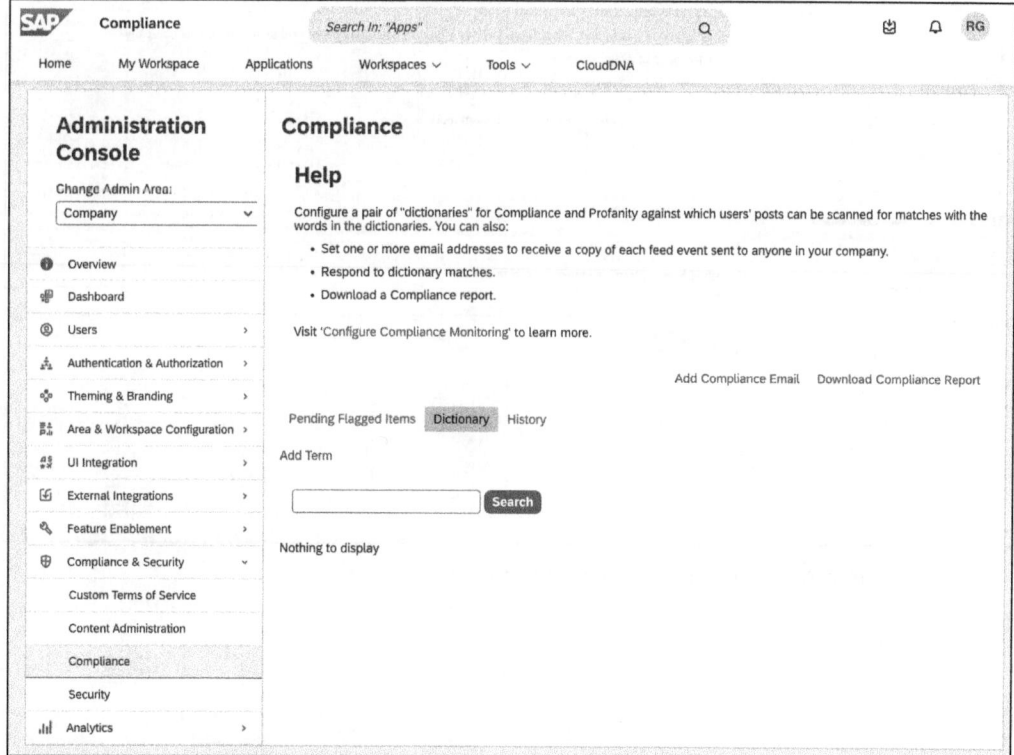

Figure 16.7 Compliance Dictionary in SAP Build Work Zone

16 Administration

3. As you can see here, no terms have yet been entered. To change this, click the **Add Term** button.

4. A kind of form with two input options then appears. First, the type of dictionary you want to maintain must be defined using the radio buttons. **Compliance** and **Profanity** are available for selection. The second input option is an input field in which the unwanted terms are to be entered, separated by commas.

5. Figure 16.8 and Figure 16.9 show how terms can be maintained for the two categories. The terms can then be added to the dictionary by clicking **Add**.

6. Once the terms for the two categories have been maintained and saved, you can see the maintained terms in the overview (see Figure 16.10). In addition, individual terms can be deactivated using the **Disable** button.

Now that you've maintained terms for the two dictionary types and have activated the comparison functionality, content that is added from this point on will be analyzed. For test purposes, add a comment containing the two previously defined terms in SAP Build Work Zone (see Figure 16.11).

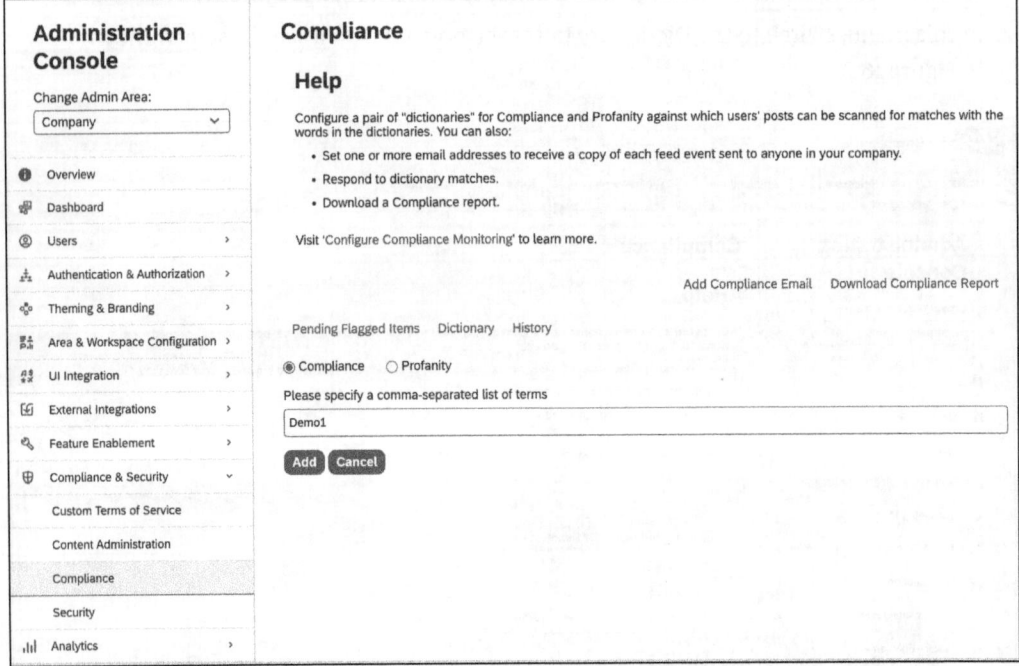

Figure 16.8 Add Compliance Term to Dictionary

16.3 Compliance

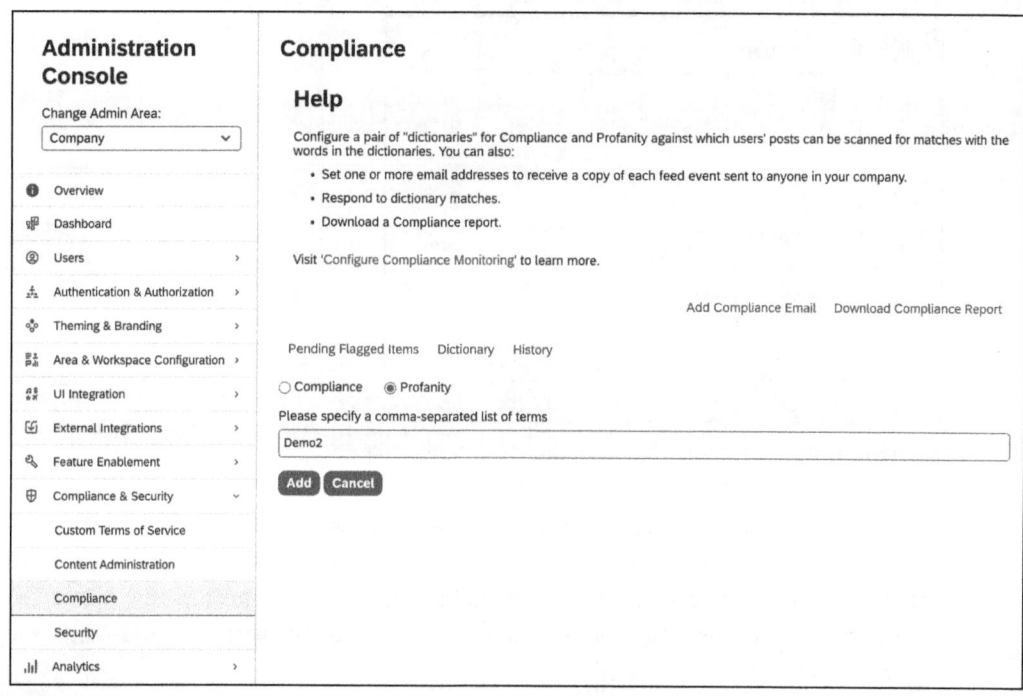

Figure 16.9 Add Profanity Term to Dictionary

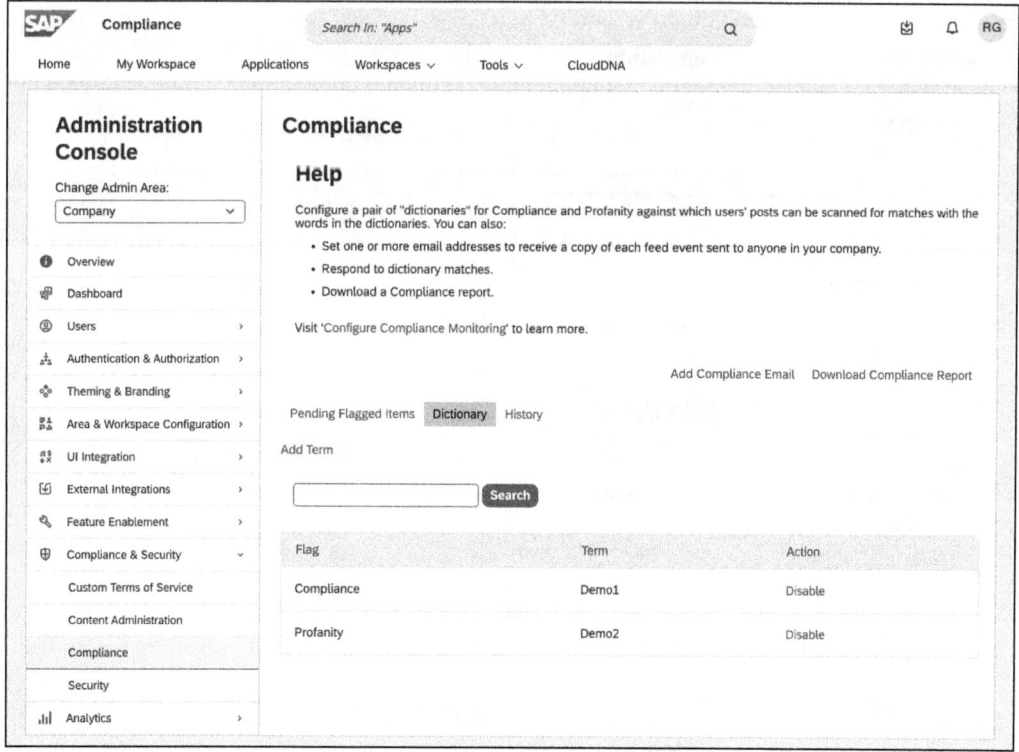

Figure 16.10 Terms Added to Dictionary

16 Administration

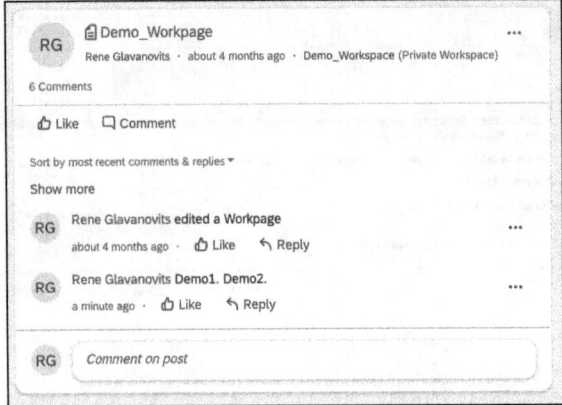

Figure 16.11 Comment with Noncompliant Content

After the comment has been added, you won't initially see that it isn't compliant. Switch back to the administration console in the **Compliance & Security • Compliance** menu item. Here, you can see at a glance that two unwanted terms have been detected in the table in the **Pending Flagged Items** tab in accordance with the guidelines (see Figure 16.12).

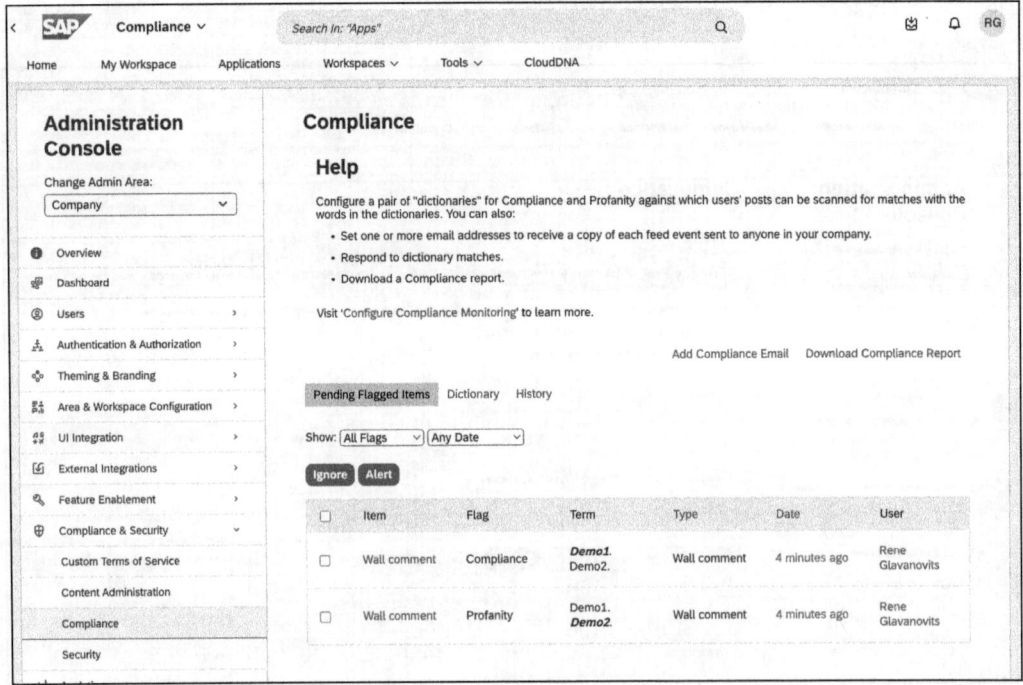

Figure 16.12 Pending Flagged Items in SAP Build Work Zone

As you can see here, the following information is displayed in this tab:

- The item in question
- The category (**Compliance** or **Profanity**)
- **Term**
- **Type**
- **Date**
- **User**

In addition, you can see two options: **Ignore** and **Alert**. First, an entry must be selected from the table, and then the corresponding button must be clicked. If you now select an entry and click **Ignore**, the entry disappears from this table, as shown in Figure 16.13.

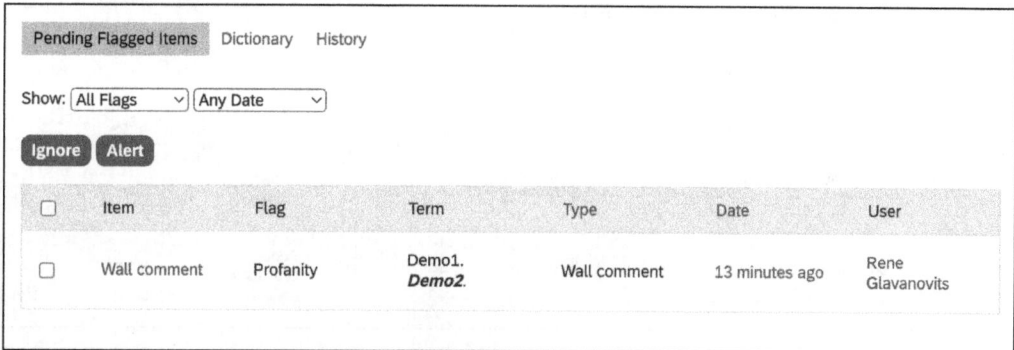

Figure 16.13 Pending Flagged Items after Ignoring Entry

Instead, you'll now find the entry in the **History** tab (see Figure 16.14).

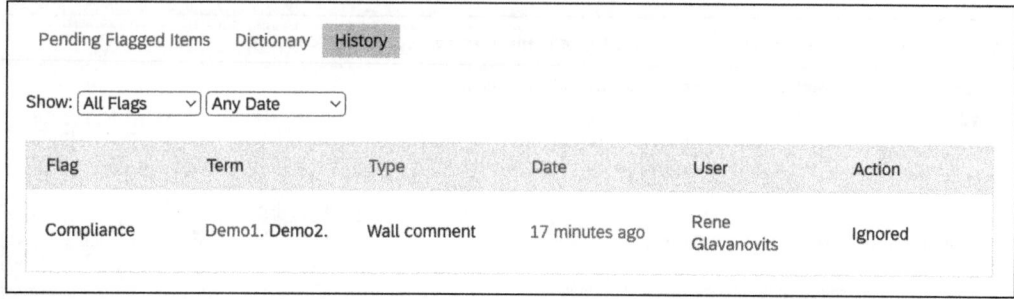

Figure 16.14 History Tab after Ignoring Entry

It's also possible to trigger a notification. To do this, proceed in the same way as before. First select an entry, then click the **Alert** button. Now this entry also disappears from the **Pending Flagged Items** table, as shown in Figure 16.15.

16 Administration

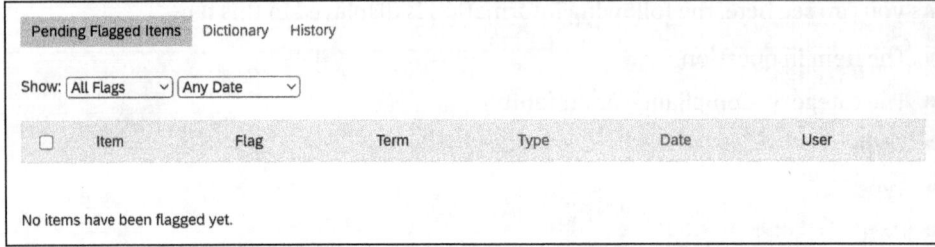

Figure 16.15 Pending Flagged Items after Altering an Entry

Instead, you'll now find this entry in the **History** tab as well (see Figure 16.16). In the history list, you can also see whether the entries have been ignored or alerted.

Figure 16.16 History Tab after Alerting Entry

In addition, after a notification is triggered, an email is sent to the author of the content to inform them that their post violates guidelines (see Figure 16.17).

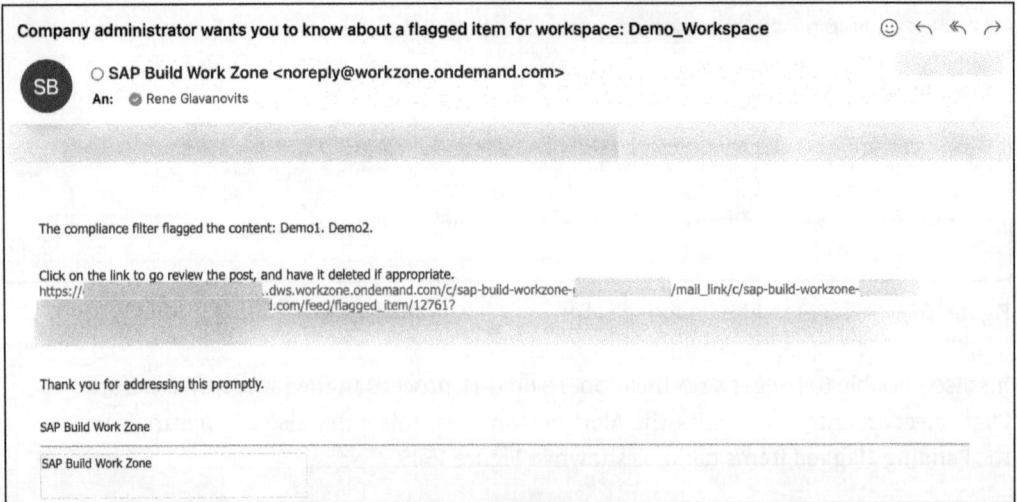

Figure 16.17 Email Sent for Noncompliant Content

16.3 Compliance

As a rule, these emails are only sent to the author of the content. If desired, an email address can also be stored that will receive a copy of these emails. To do this, click the **Add Compliance Email** button. You're then taken to the page shown in Figure 16.18. Any email address can be entered here.

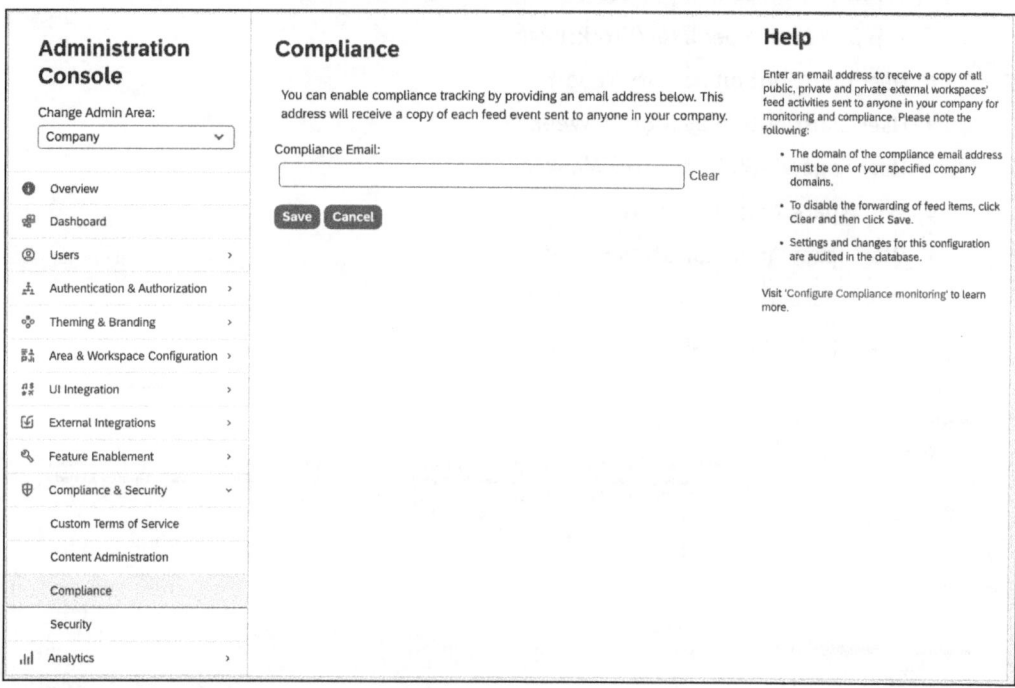

Figure 16.18 Add Compliance Email

Furthermore, a report of noncompliant content can be generated using the **Download Compliance Report** button. Clicking this button opens an interface, as shown in Figure 16.19.

Before a report is created with **Request Report**, various configurations can be made. For example, you can define whether a header should be added to the report, which file type (CSV or XLSX) should be created, and which users (company or external users) should be included. You can also select the type of report to be created. The following options are available:

- Activity Summary by Month/Week
- Company Settings Changes
- Company User Detail Report
- Compliance Report
- Content Views by Month/Week
- Contribution Report by Object by Month/Week
- Engagement Report

497

16   Administration

- Expertise Report
- Kudo Detail
- Search Summary by Month/Week
- Terms of Service Compliance Report
- Top Disk Usage per User/Workspace
- User Contribution Activity Report
- User Contribution by Month/Week
- User Page Views by Month/Week
- Workspace Activity Report
- Workspace Administrators Report
- Workspace Member Activity Report
- Workspace Template Activity Report

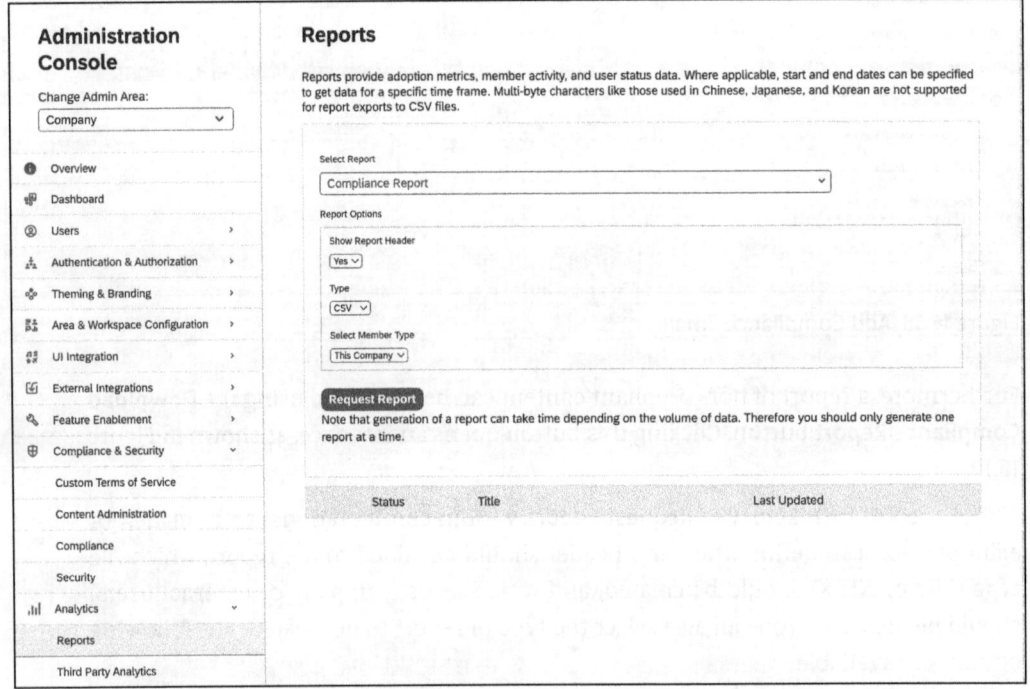

Figure 16.19 Generate Compliance Report

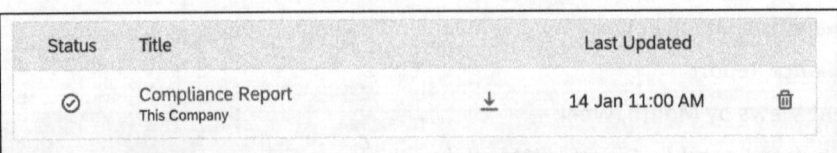

Figure 16.20 Generated Reports

For now, select **Compliance Report** as the report and leave the other settings set to their defaults. Then click **Request Report** to initiate the creation. The generation takes different amounts of time depending on the data volume. As soon as the generation is complete, the creator receives an info email. In the administration console, you'll also see the generated report, which can be downloaded here via the **Download** icon (see Figure 16.20).

In Figure 16.21, you can see what the report looks like. Here you can see the corresponding entries analogous to the history list within SAP Build Work Zone.

	A	B	C	D	E	F	G	H
1	sep=,							
2	Compliance Report							
3	Date	January 14, 2024 11:00 AM PST						
4								
5	Flag	Term	Type	Date	Email	Name	Action	
6	compliance	Demo1. Demo2.	Wall comment	01/14/2024 18:14:59		Rene Glavanovits	Ignored	
7	profanity	Demo1. Demo2.	Wall comment	01/14/2024 18:14:59		Rene Glavanovits	Alerted	
8								
9								
10								
11								
12								

Figure 16.21 Export of Compliance Report

16.4 Error Logs

Error logs are log files in which information on any errors that occur is logged. These enable or simplify the subsequent search for the cause of the error. In most cases, such log files can be viewed by developers and administrators. However, various types of error messages are usually recorded in such logs. These can range from critical errors and warnings to simple informative outputs. An error log entry usually consists of at least the following properties:

- Timestamp
- Type of the log entry (error, info, etc.)
- Error details (e.g., an exception message)
- Correlation identifier

Such error logs are usually also regularly monitored by administrators to identify abnormal behavior. This allows potential problems to be identified, analyzed, and rectified at an early stage. Error logs are an essential tool in the software development and IT management process that ensures the reliability and performance of applications.

Within SAP Build Work Zone, standard edition, it's possible to download an error log file, which can then be used for further analysis. However, it's important to note that this file is not updated indefinitely but only contains entries from the last 24 hours. If

16 Administration

you want to have the log files over a longer period of time, you have to download and save them day by day. To access the log files, proceed as follows:

1. Open SAP Build Work Zone, standard edition. To do this, we navigate to your subaccount in SAP BTP, open the Service Marketplace, and search for SAP Build Work Zone. Select this and jump into it with **Go to Application** (see Figure 16.22).

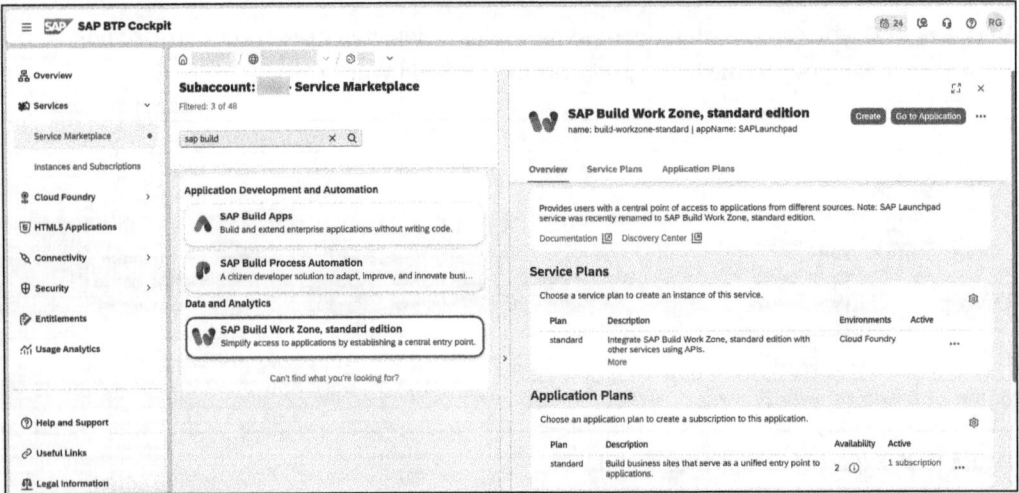

Figure 16.22 Open SAP Build Work Zone, Standard Edition

2. After opening SAP Build Work Zone, you should see a screen similar to Figure 16.23.

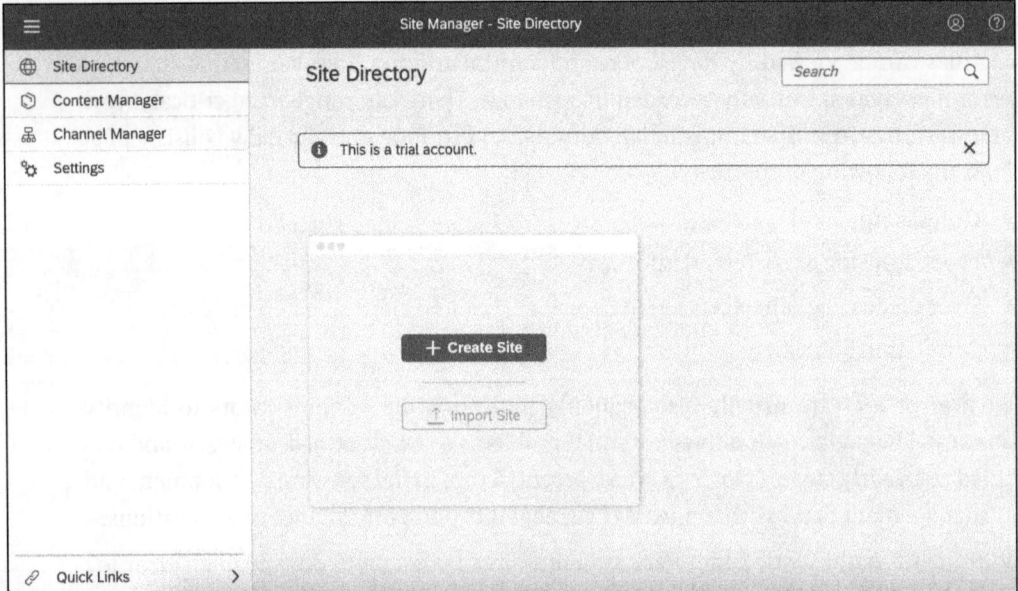

Figure 16.23 Start Page of SAP Build Work Zone, Standard Edition

3. Here, select the **Settings** entry in the menu bar on the left. This takes you to a page where various settings can be made in addition to the error logs (see Figure 16.24).

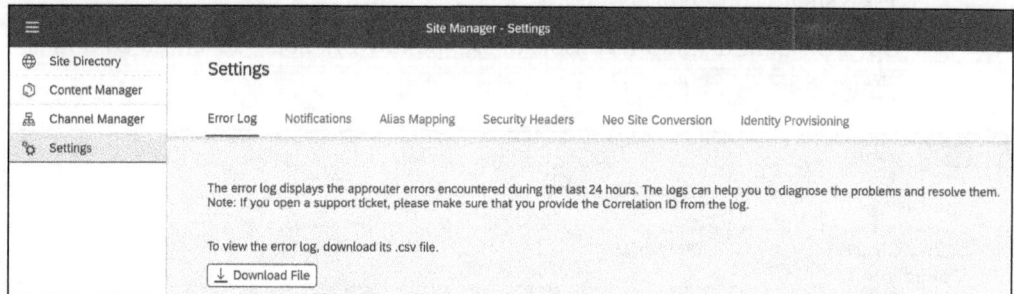

Figure 16.24 Error Log Tab inside Settings of SAP Build Work Zone

4. As you can see in this tab, all the Approuter errors that have occurred in the last 24 hours are recorded in the error log. A CSV file of the log can be downloaded by clicking **Download File**.
5. As you can see in Figure 16.25, in this case, no errors have occurred in the last 24 hours, which means that the error log does not contain any entries.

Figure 16.25 Error Log File of SAP Build Work Zone

The error log always provides assistance when it comes to analyzing, identifying, and resolving errors. However, if it is not possible to rectify or analyze the error, SAP still offers the option of opening a support ticket. In this ticket, it is important that the correlation ID of the error message is also specified.

> **Correlation ID**
> The *correlation ID* is a unique identifier that identifies each request from a microservice. The value is generated by the client and stored as part of the application logs. In turn, the same value is also logged and stored on the server side. The two logs can then be correlated, allowing them to be traced throughout.

Earlier, you saw how the log files can be retrieved in SAP Build Work Zone, standard edition. In SAP Build Work, advanced edition, this basically looks identical, except that the access is slightly different, as follows:

1. Open the administration console of SAP Build Work Zone, advanced edition (see Figure 16.26).

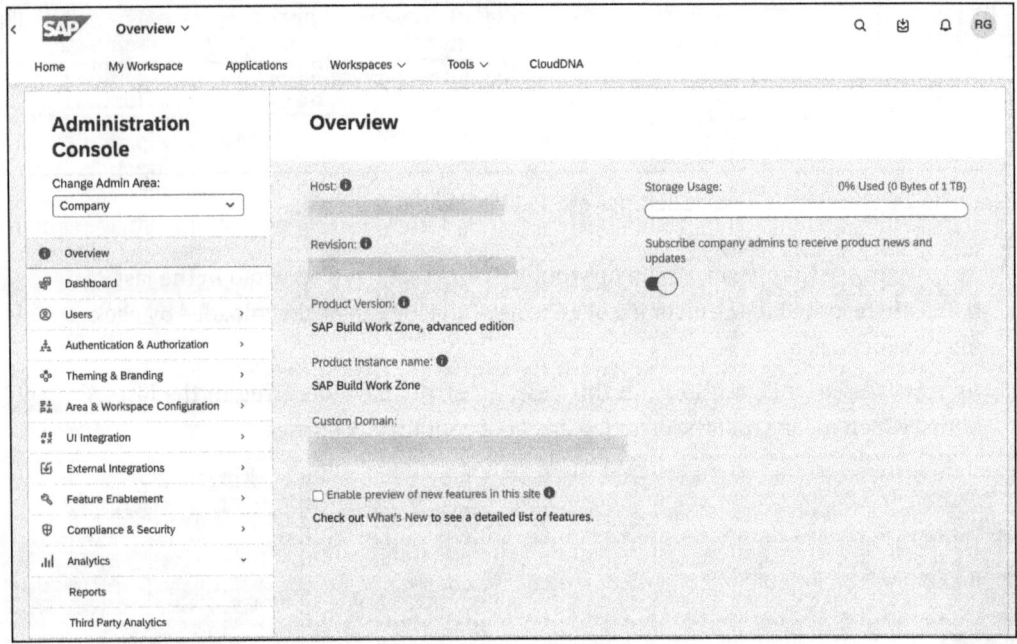

Figure 16.26 Administration Console of SAP Build Work Zone, Advanced Edition

2. Select the **External Integrations • Business Content** entry. A page like the one in Figure 16.27 then opens.
3. Click the **Content Manager** button. This takes you to the site manager for SAP Build Work Zone. From now on, the procedure is the same as in SAP Build Work Zone, standard edition. Clicking **Settings** takes you to the settings and the **Error Log** tab, where the file can be loaded (see Figure 16.28).
4. Click **Download File** to download the log file. As shown in Figure 16.29, the log file is not empty in this case.

As you can see in this log file, in addition to the error message, which also contains information on the call trace of the methods, the correlation ID, and the timestamp of the entry, a log level is written to the log.

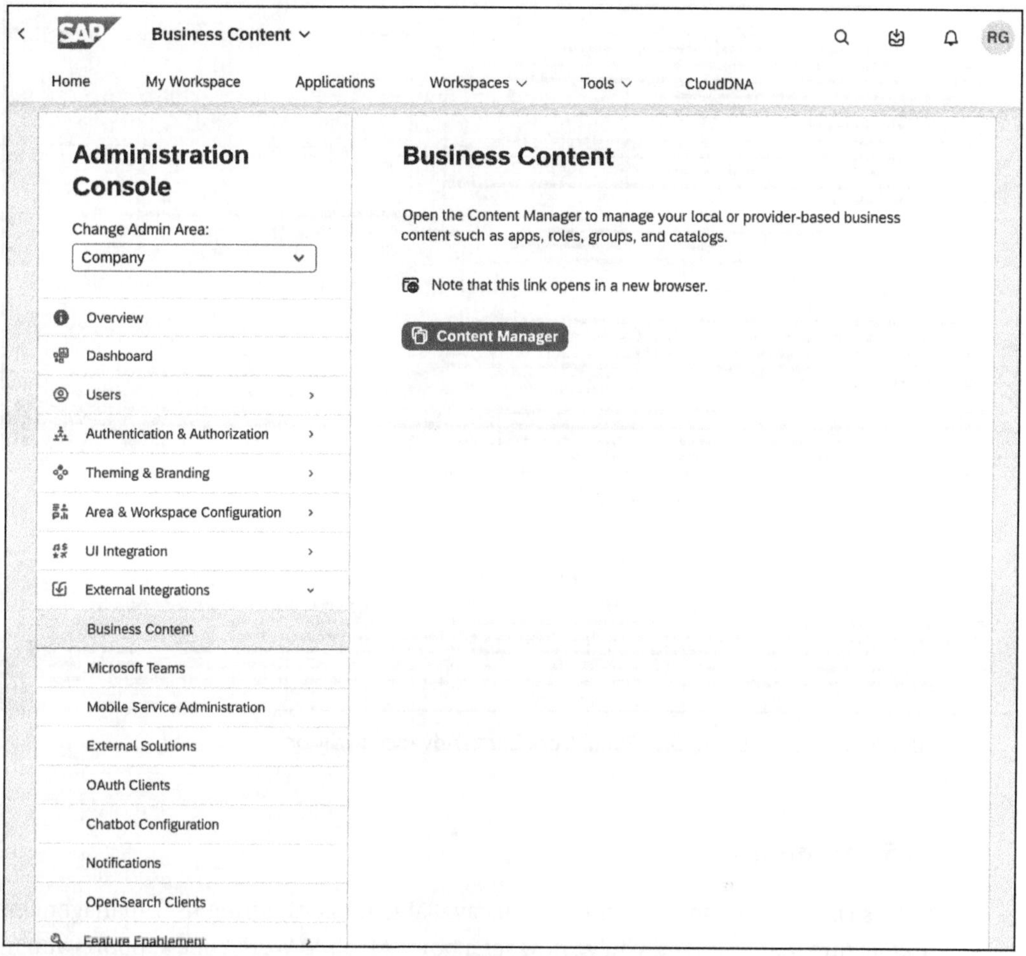

16.4 Error Logs

Figure 16.27 SAP Build Work Zone, Advanced Edition: Business Content

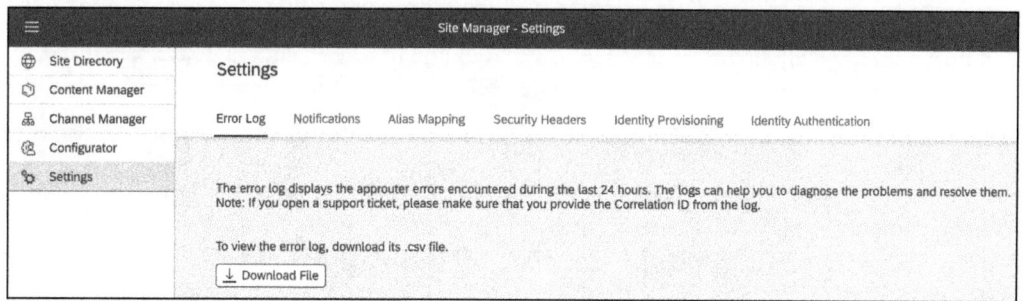

Figure 16.28 Site Manager in SAP Build Work Zone, Advanced Edition

Figure 16.29 Error Log of SAP Build Work Zone, Advanced Edition

16.5 Summary

In this chapter, we looked at the options available for configuring user management and feature management. You've now seen how SAP Build Work Zone supports you in maintaining compliance guidelines and what options SAP Build Work Zone offers for error analysis. Starting with an introduction to the offering in the next chapter, the next couple of chapters will look at SAP Build Process Automation.

PART IV
SAP Build Process Automation

PART IV

SAP Build Process Automation

Chapter 17
Introduction to SAP Build Process Automation

In this chapter, we introduce the SAP Build Process Automation service and give you a brief insight into its features. You'll get to know the components that can be created and what they're used for. We'll show you how lifecycle management and collaboration can be used for the projects to share project content and provide different versions and artifacts. At the end of this chapter, you'll learn where to find the roadmap and how it can be used to gather information about planned improvements to the service.

SAP Build Process Automation is a service in the SAP BTP, Cloud Foundry environment that provides tools and features for designing, executing, and monitoring business processes and process automation. The service can be discovered via SAP Discovery Center. There you can find an overview description of the service and information about pricing. Furthermore, SAP Discovery Center (*https://discovery-center.cloud.sap/serviceCatalog/sap-build-process-automation/?region=all*) provides further information on how to use the service.

In the following sections, we'll walk through subscribing to SAP Build Process Automation, its features and components, and the lobby. We'll also discuss lifecycle management and collaboration for SAP Build Process Automation projects.

17.1 Subscribing to SAP Build Process Automation

Before you can use SAP Build Process Automation, an SAP BTP account administrator must subscribe one of your SAP BTP subaccounts to the relevant application of SAP Build Process Automation. You can choose one of three available subscription options for SAP Build Process Automation:

- **Free plan**
  This plan provides a production environment where all the features of the application can be tested.
- **Standard plan**
  With this plan, you get full production access to the application to create, deploy, run, and monitor your business processes.

- **Starter pack for GROW with SAP S/4HANA Cloud, public edition customers**
  This starter pack provides productive access to SAP Build Process Automation.

After subscribing to one of the plans, additional optional product extensions can be added and configured, as listed in Table 17.1.

Extension	Description
Attachments	If you have an active SAP Document Management service subscription, you can allow participants of a business process to upload attachments to forms within the process. After configuring an attachment destination, the attachment option is visible when editing a form.
Automations	A desktop agent must be installed to run any automation that you build.
Data sources and forms	Using this extension, you enable participants of a business process select data from an external source when interacting with a form by configuring the data sources. Without this configuration, data can only be entered manually.
Mail notifications	You can configure an SMTP mail destination to enable mail notifications inside a business process. The mail notification option is not available while editing a process if this destination is not configured.
SAP Build Work Zone	SAP Build Work Zone provides a central point of entry for your processes. You can use SAP Build Work Zone to start a process, or enter your inbox to approve or decline running processes.
SAP Task Center	SAP Task Center supports integration of your SAP Build Process Automation approval tasks into a central solution alongside tasks from other SAP products.
Store: Live process projects	SAP BTP destinations must be configured to configure and run live process projects that were added from the store.

Table 17.1 Extensions for SAP Build Process Automation

17.2 Features and Components

SAP Build Process Automation provides the following features:

- **Digitize processes**
  Build new or adapt existing end-to-end processes with a modern, intuitive user interface that can be used for low-ode and no-code development.

- **Create interactive forms**
  Using drag and drop, form-based processes can be designed using connected data sources.

- **Build and run automations**
  By creating bots, you can record activities that you can repeat to save time and resources.

- **Manage decisions**
  Develop and manage decision logic using decision tables. Define business rules and use them inside the process.

- **Achieve end-to-end process visibility**
  Process visibility dashboards give you the opportunity to get the real-time statuses of the processes being executed to gain more transparency for the processes in your company.

- **Discover and manage predelivered content**
  Use SAP Business Accelerator Hub to work with predelivered content.

The service of SAP Build Process Automation is available in SAP BTP, Cloud Foundry environment and can be run there. To use these features, SAP Build Process Automation provides a set of components that can be created and managed inside a process automation project. The following list gives an overview of these components and describes their functions:

- **Automation**
  With the automation functionality, you can automate time-consuming, recurring work such as data extraction, data capture, and data preparation. Using a recorder function, you can create bots that do this work for you. Figure 17.1 shows a screenshot of an automation created with SAP Build.

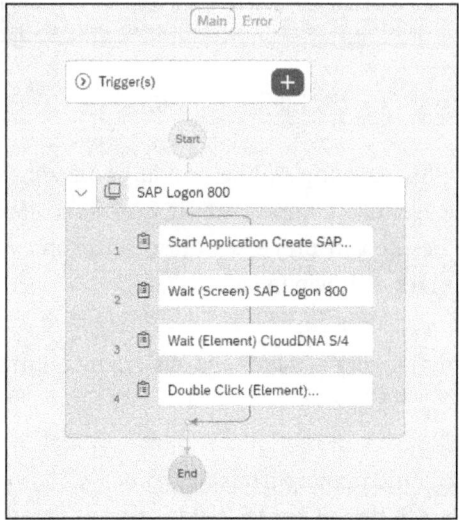

Figure 17.1 Automation Example

- **Alert**
  An alert allows you to raise business events in an automation. By raising an alert, you can pass additional information to the receiver. After a project that includes an alert is deployed, possible receivers can subscribe to this alert by using alert handlers to get the corresponding notifications. As these notifications are sent by email, an SMTP server needs to be configured.

- **Application**
  This component can be used to identify applications during a live automation. This can be a website or a running application on your computer.

- **Project launcher**
  The project launcher helps you to run automations. You can configure the automations to be started and can define if they are started manually or automatically. Figure 17.2 shows the design console of the project launcher.

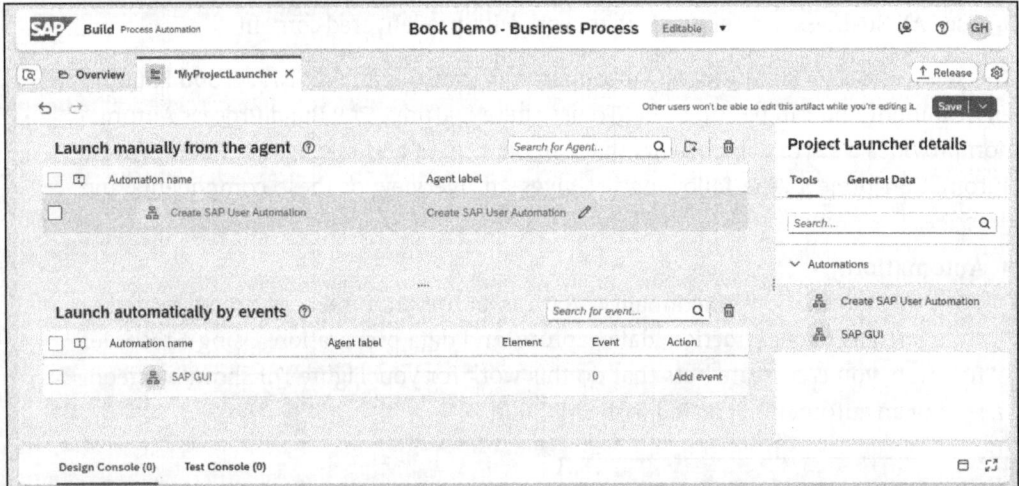

Figure 17.2 The Project Launcher

- **User task**
  Whenever you need human interactivity for data input, like providing personal information or uploading a document, the user task component can be used to create a suitable form. This form can then be used in processes or automations. A screenshot of the design console for a user task is shown in Figure 17.3.

- **Action group**
  An action group enables you to use the functionality of a backend system by calling BAPIs.

- **Process**
  With the process component, you can design and manage a business process. Figure 17.4 shows a screenshot of a business process designed in SAP Build.

17.2 Features and Components

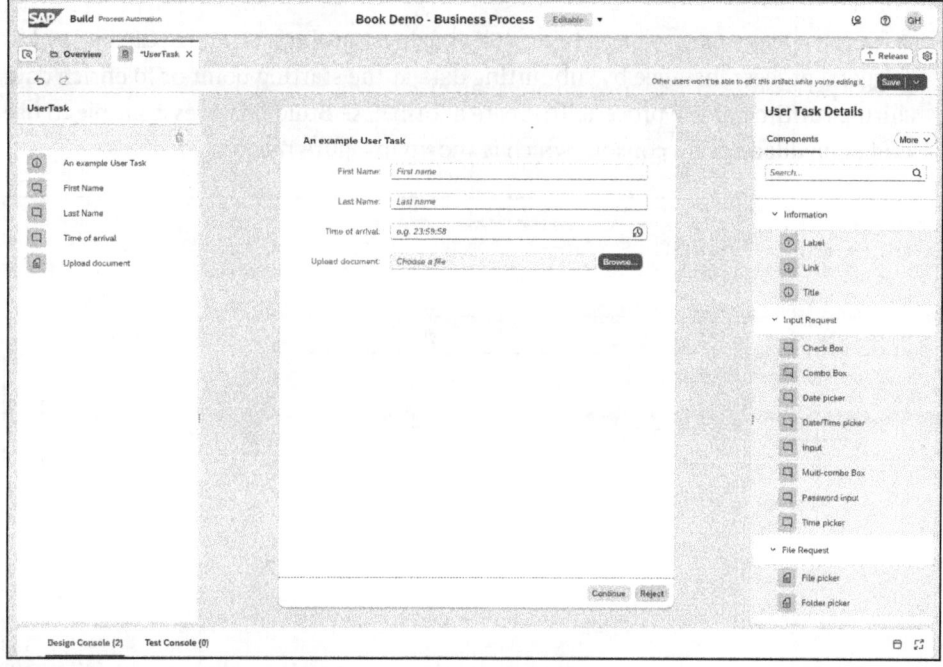

Figure 17.3 Creating User Task

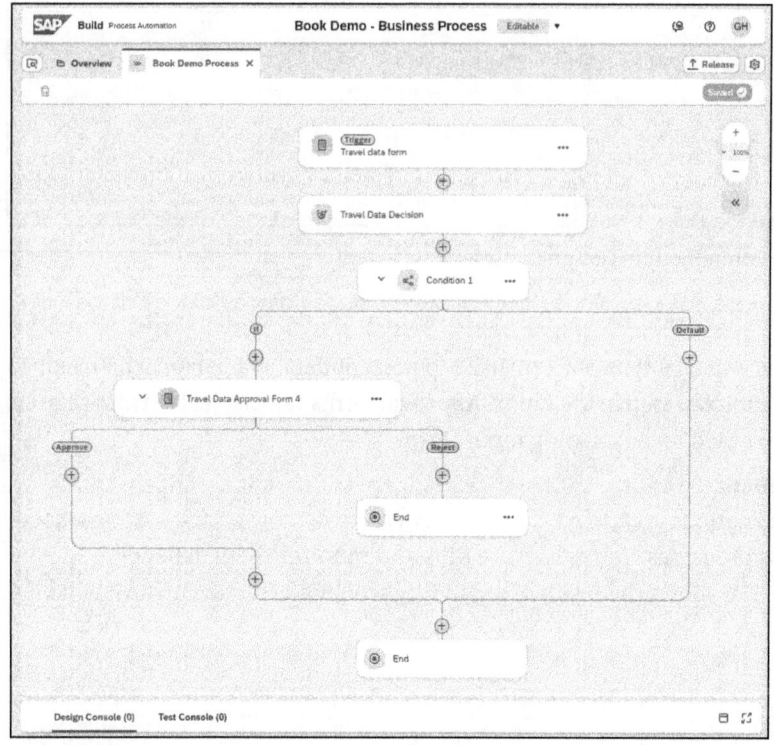

Figure 17.4 Business Process

17 Introduction to SAP Build Process Automation

- **Form**
  A form enables a user to provide data to a business process. The form can be used to trigger a process instance by submitting data at the starting point or to enrich data during runtime of the process. To create a form, SAP Build provides a simple to use and embedded design console, which is shown in Figure 17.5.

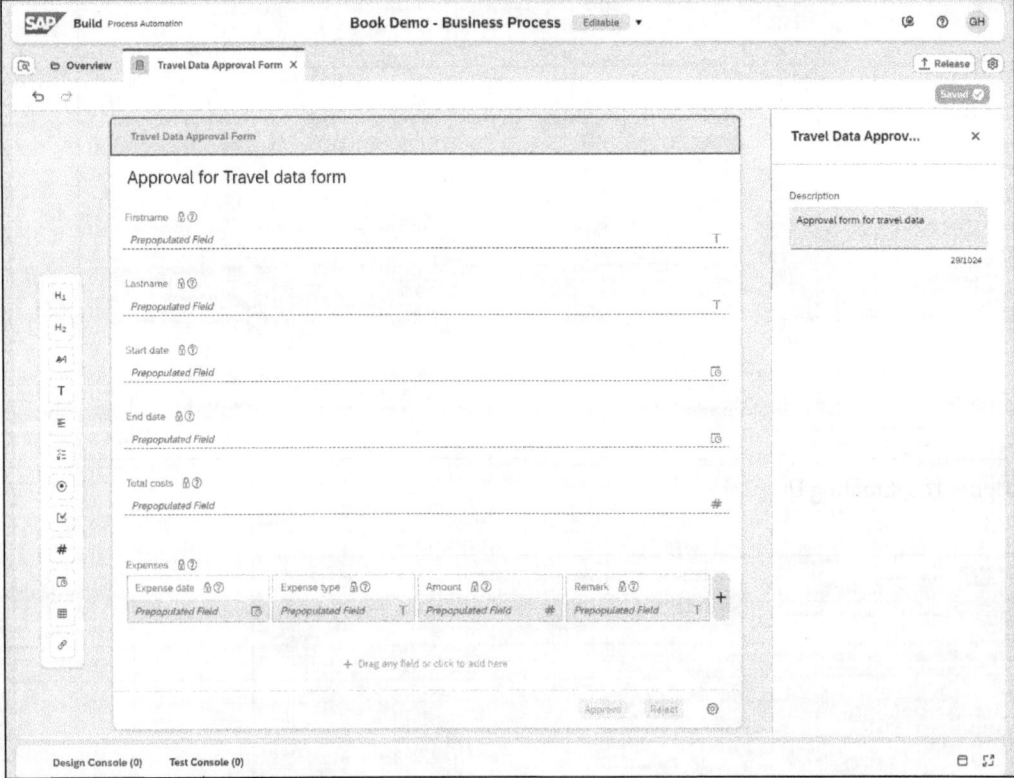

Figure 17.5 Form Component in Design Console

- **Approval**
  An approval is a special type of form, used to present data to a person who needs to complete an approval step in the inbox. Approval forms can be created from scratch, or they can be based on an existing input form.

- **Visibility scenario**
  A visibility scenario is an artifact that enables you to create specific dashboards for your business processes. You can create KPIs and monitor them depending to your needs. Figure 17.6 shows the result of a visibility scenario created with SAP Build.

17.2 Features and Components

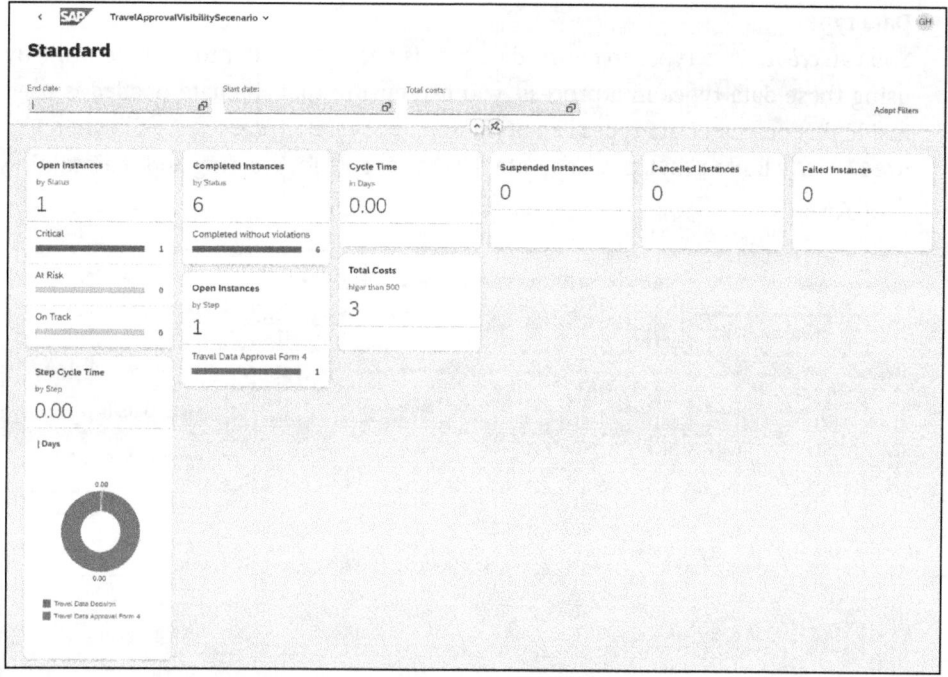

Figure 17.6 Visibility Scenario

- **Decision**

  Decisions allow you to design and manage your business rules in order to reuse them in one or more business processes. Figure 17.7 shows a screenshot of the design console being used to configure a decision.

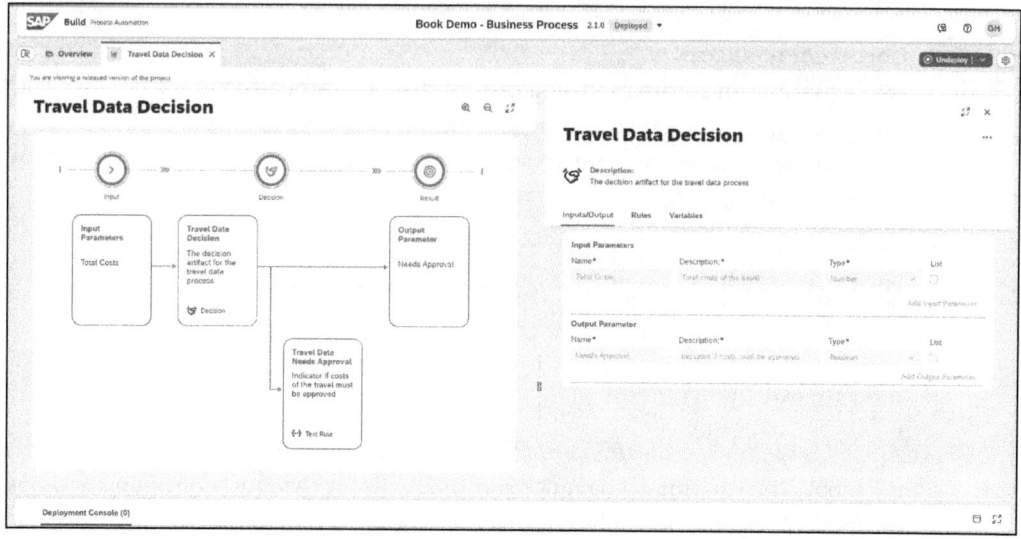

Figure 17.7 Decision Component

- **Data type**
  You can create data types to define data that is needed in order to execute work. By using these data types in a process, you can ensure that the data needed is used inside this process. Data types can be shared by several processes. As shown in Figure 17.8, SAP Build provides a separate design console for creating and maintaining datatypes.

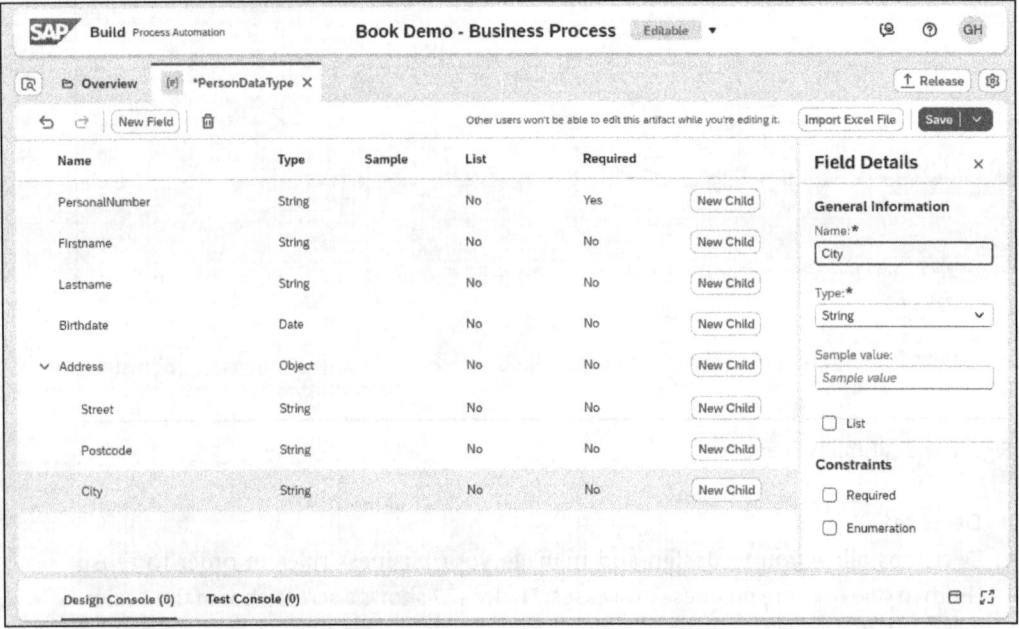

Figure 17.8 Data Type Component

- **Document templates**
  While running an automation, you can extract data from a document. To help with this process, you can create a template and upload it to your project. With such a template, you can help the automation learn how to extract the needed data from the arriving documents.

- **File**
  You can upload files to be used in automations and processes.

17.3 The Lobby

The lobby is the central starting point for development using SAP Build. In the header of the lobby, you can find some quick start manuals for the most common scenarios. These manuals can be hidden by clicking the **Hide** button in the upper-right corner of the header, which will enlarge the space available for showing the list of projects (see

Figure 17.9). Whenever you need the quick start information, you can show it again by clicking the **Show** button (see Figure 17.10).

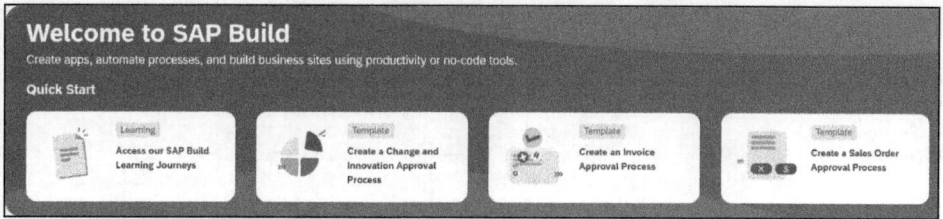

Figure 17.9 Header Showing Quick Start Tiles

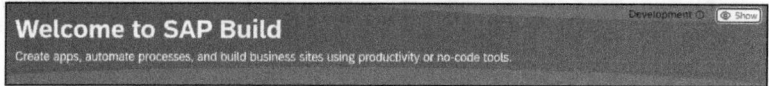

Figure 17.10 Header without Quick Start Tiles

The main element of the start page in the lobby is your list of projects. This list provides information about all projects that are in use. Figure 17.11 shows a list of projects provided in the lobby.

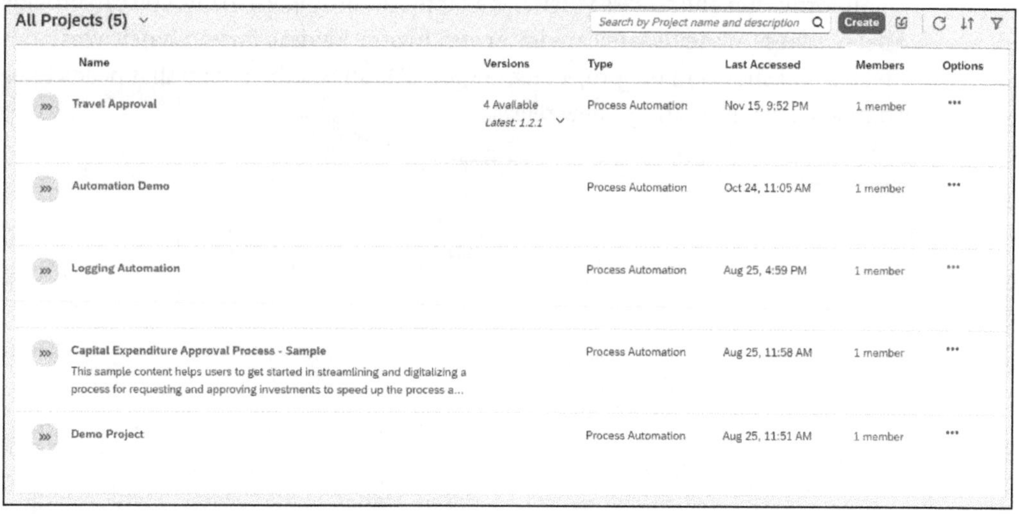

Figure 17.11 Project List

The list provides search functionality as well as functionalities for sorting and filtering. These tools can be found on the right side above the list. Each item in the list represents a single project. You can see information about the type of the project, its version, and when it was last accessed. You can also see the members of the project and their roles within it. In the last column of the list, each item provides a menu with options that can be used for the process. Figure 17.12 shows the list of projects with the options menu opened for the top project.

17   Introduction to SAP Build Process Automation

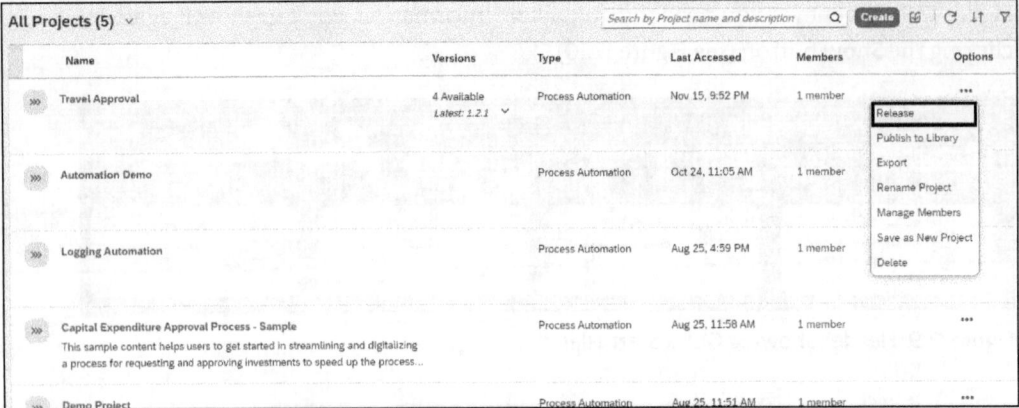

Figure 17.12 Options Menu

These options include the following functionalities:

- **Release**
  Create a new release of the project. When you click this option, a popup form is opened where you can provide information for the new release. You can define if the release version contains a bug fix, a minor change, or a major change. Depending on this information, the release version is adapted accordingly. Finally, you can provide release notes for the new release. After clicking the **Release** button, a new version is created and shown in the project list. Figure 17.13 shows the popup that needs to be completed when creating a new release.

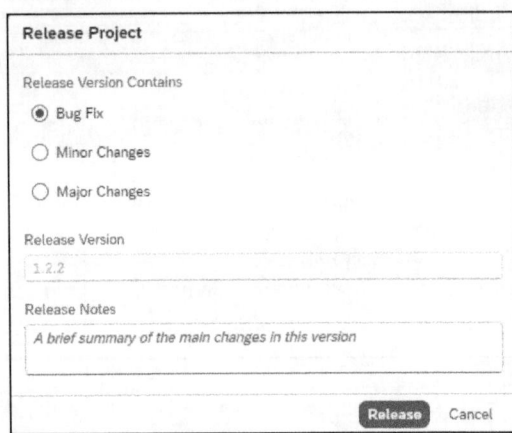

Figure 17.13 Popup to Release Project

- **Publish to Library**
  Publish the content of the project to a library with this option. Once a project is published, it can be shared with all members of the same tenant. If you choose this option, a popup opens for you to provide information about the published content.

516

You can decide which version of your project you want to share. Furthermore, you can categorize the published content by the line of business and product parameters. Finally, you can share the content by pressing the **Publish** button. Figure 17.14 shows an example of the popup to publish a specific version of a project.

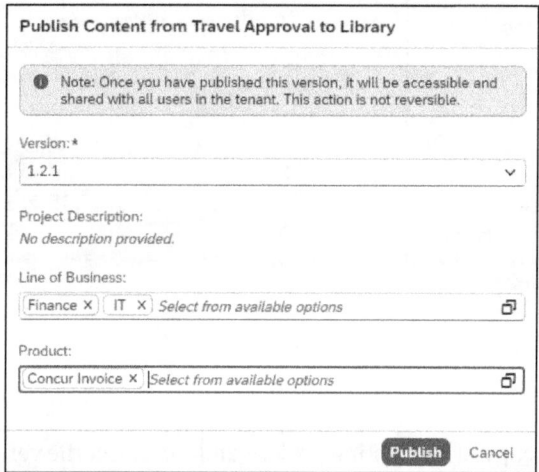

Figure 17.14 Dialogue to Publish Version of Project

- **Export**
  You can export a specific version of the project to download the content for further use. An MTAR archive file is created and downloaded via the browser.
- **Rename Project**
  You can change the name of your project with this option. When choosing this option, a popup form is opened where you can provide the new name and the description for the project.
- **Manage Members**
  Using this option, you can manage the members of the project. By using an email address to identify a member, you can add or delete members and change their types of access. The types you can choose include the following:
  - **Viewer**
    Members of this access type can view and deploy projects.
  - **Developer**
    Developer members can edit, deploy, and release projects. They can manage the dependencies in a project and can publish a project to a library.
  - **Administrator**
    These types of members have full access, including editing, sharing, and deleting a project.

  Figure 17.15 shows an example of the members management dialogue.

17 Introduction to SAP Build Process Automation

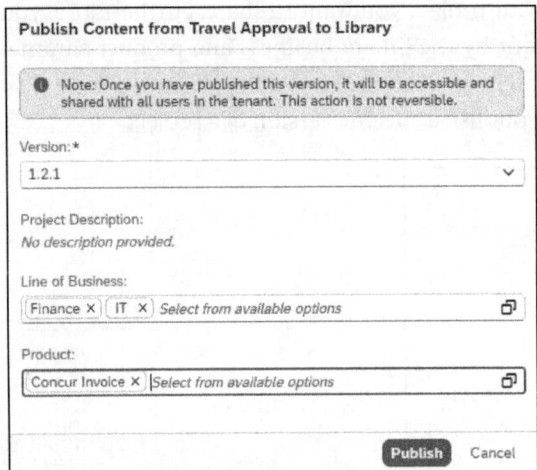

Figure 17.15 Members Management Dialogue

- **Save as New Project**
  You can use this option to create a copy of an existing project. and can select the version of the current project the copy should be made for. For the new project, you can define the name and the description. Figure 17.16 shows the dialogue used to save a new project.

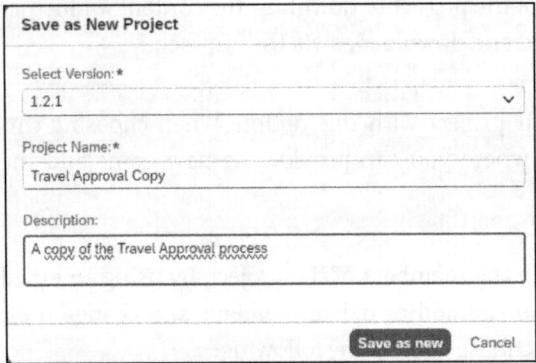

Figure 17.16 Dialogue to Create Copy of Specific Version of Existing Project

- **Delete**
  Whenever you want to delete an existing project, you can use this option. The project and all its contents are deleted and cannot be recovered.

In addition, the lobby provides the functionality to import projects from existing archive files. These files have to be of the type MTAR or ZIP. You can choose an archive file from your local file system and import the content as a new project. Clicking the **Input** button in the menu bar above the project list opens the input dialogue. In this dialogue, you can open your local file browser, or you can drag and drop an archive file to edit to the input list. Once this is done, the project appears in the list of items to be imported. After choosing an item and clicking the **Input** button, the selected project is imported into your SAP Build lobby. Projects loaded into your input dialog exist until you either delete or import them. Figure 17.17 shows the input dialogue with a project available.

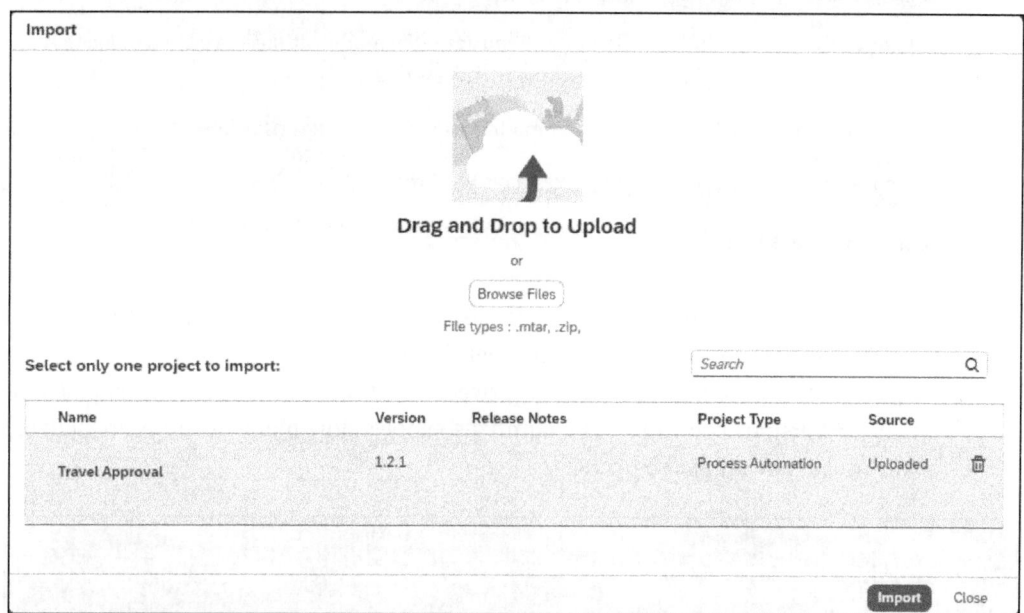

Figure 17.17 Input Dialogue

17.4 Lifecycle Management

Projects that are worked on in SAP Build Process Automation can exist in different versions. These versions are completely independent of each other and can have different statuses. In project lifecycle management, a distinction is made between three different statuses, which are described in Table 17.2.

Status	Description
Editable	Default status when you create a project. An editable project is considered a draft and doesn't appear in the list of projects in the lobby.
Released	Status when you generate or import a project. For an already released project, release it again to create a new project version. You can delete a released project.
Deployed	Change the status of a released project to deployed to use a project in a productive way. For instance, you deploy an automation project to add triggers and execute its automation. You cannot delete a deployed project.

Table 17.2 Possible Project Statuses

17 Introduction to SAP Build Process Automation

Every time a project is released, a new version number is assigned. By specifying this version number, you can declare what has changed compared to the previous version. The first version of a project always has version number 1.0.0. Table 17.3 explains the different versions that can be assigned at release after an update of the initial release.

Release Version	Example	Description
Major version	2.0.0	Important and essential modifications have been made that can lead to inconsistencies.
Minor version	1.1.0	Smaller modifications or bug fixes have been made.
Patch version	1.0.1	Bug fixes have been made.

Table 17.3 Release Versions

So long as a project has not yet been released, it doesn't have a version number and thus the corresponding column in the overview of projects in the lobby is empty. If a project has already been released or deployed, the number of the latest version is visible in the **Versions** column. This column also shows how many versions are actively available, as shown in Figure 17.18.

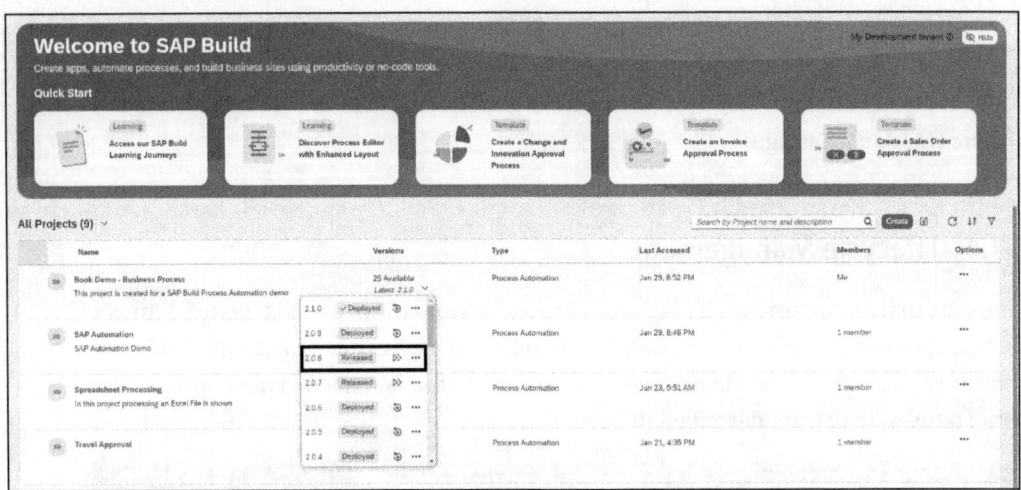

Figure 17.18 Available Project Versions

By clicking the ∨ icon next to the version number, the list of all available versions and their statuses is displayed. Depending on the status, you have the option to carry out further actions for this version. This allows you to deploy or delete versions in **Released** status. Versions that have already been deployed can be redeployed or undeployed. If it is undeployed, its status is changed to **Released**. Both versions in the **Deployed** status and versions in the **Released** status can be transported here, provided you are using the *promote* feature. This feature allows you to transport projects between two tenants. You can access the promote feature by setting up the SAP Content Agent service and the SAP Cloud Transport Management service. Figure 17.18 shows a screenshot of the version list provided in the lobby.

17.5 Collaboration

Projects developed with SAP Build Process Automation enable multiple people to collaborate. At the project level, you can define which people should have access to your project and how this access should be structured. If you use SAP Cloud Identity Services, you can connect to this service and use the users and groups managed there to manage authorizations in SAP Build Process Automation. To establish a connection to the identity directory, you must create a destination in SAP BTP with the settings listed in Table 17.4.

Field Name	Value
Name	"Identity_Authentication_Connectivity_IDS"
Type	HTTP
Description	Add optional description here
URL	"https://<tenant_ID_of_Identity_Authentication>.accounts.ondemand.com/scim"
Proxy Type	Internet
Authentication	Basic Authentication or ClientCertificateAuthentication
User	User ID of the system as administrator (only for basic authentication)
Password	Password of the system as administrator (only for basic authentication)
Use Client Provided Certificate	Not selected (only for client certificate authentication)
Key Store Location	Certificate uploaded (only for client certificate authentication)
Key Store Password	The password for the keystore (only for client certificate authentication)

Table 17.4 Properties of Destination Pointing to Identity Directory

Table 17.5 shows the types of authorizations that can be set up for projects.

17 Introduction to SAP Build Process Automation

Authorization	Description
Viewer	Authorization for read-only actions, such as viewing but not modifying the existing content of the project. Viewers can also export and deploy the project.
Developer	Includes all viewer permissions, plus permission to modify the project, such as modifying existing configurations and releasing and publishing the project.
Administrator	Authorization to carry out various important administrative tasks. Includes all viewer and developer permissions. For example, administrators can assign authorizations and add or remove members. The project creator has administrator authorization by default.

Table 17.5 Types of Authorization

A general authorization setting can be used to define for each project whether only defined project members can work on the project or whether every person with access is authorized to do so. To manage the permissions for a project in the lobby, click the icon in the **Options** column of the affected project. From the menu, select **Manage Members**, as shown in Figure 17.19.

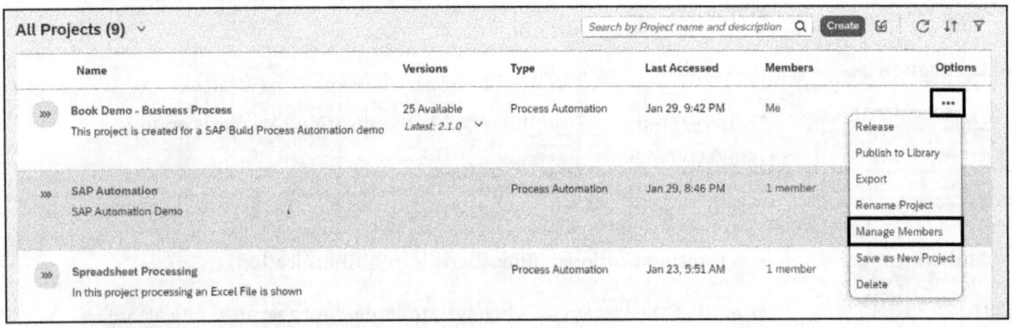

Figure 17.19 Manage Members of Project

A dialog will then open in which you can manage the members of the project. To correct a new member or group for the project, enter the values listed in Table 17.6 in the associated fields.

Field Name	Description
Type	The type of object to be added. You can choose between a single user (**Member**) or a group of users (**User Group**).
Share With	When adding a single user, provide the user name in this field. For a group, provide the technical group name.
Authorizations	This is the type of authorization according to the values in Table 17.5.

Table 17.6 Fields to Provide New Member or Group of Members

17.5 Collaboration

After providing the required data, you can add the entry by clicking the **Add** button, as shown in Figure 17.20.

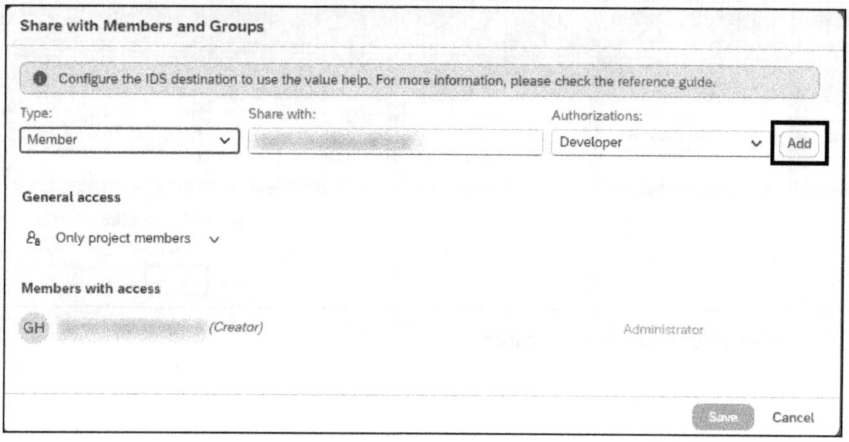

Figure 17.20 Adding New Members

The added user will then appear in the list of members with access. You can change the type of access at any time by adjusting the authorization or removing the member from the list completely, as shown in Figure 17.21.

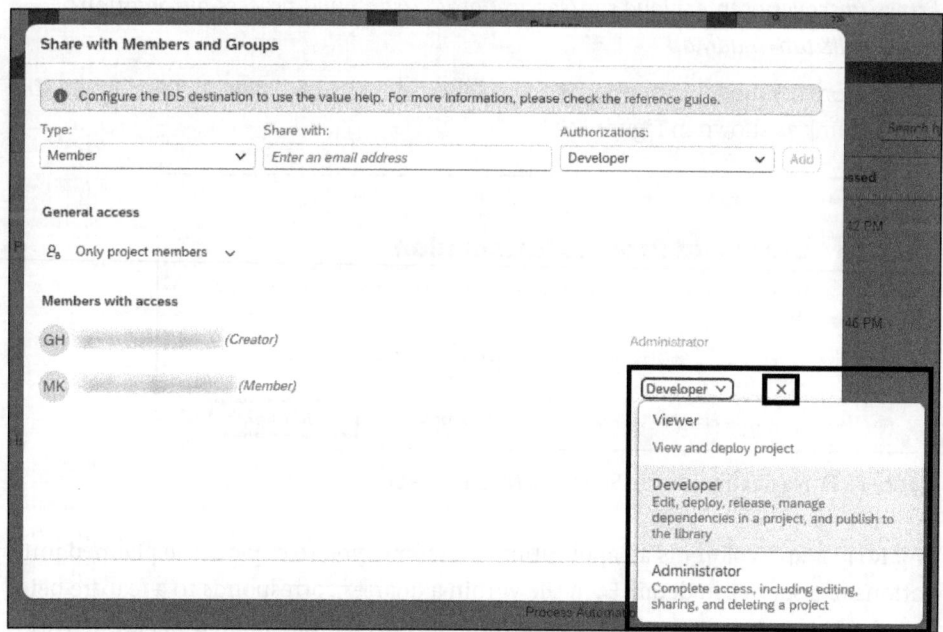

Figure 17.21 Editing or Deleting Existing Members

After clicking the **Save** button in this dialog, the adjustments are saved and the dialog is closed. In the lobby, you can now see that the assignment of members for the

523

affected project has changed. Figure 17.22 shows a screenshot of the lobby after the members have changed.

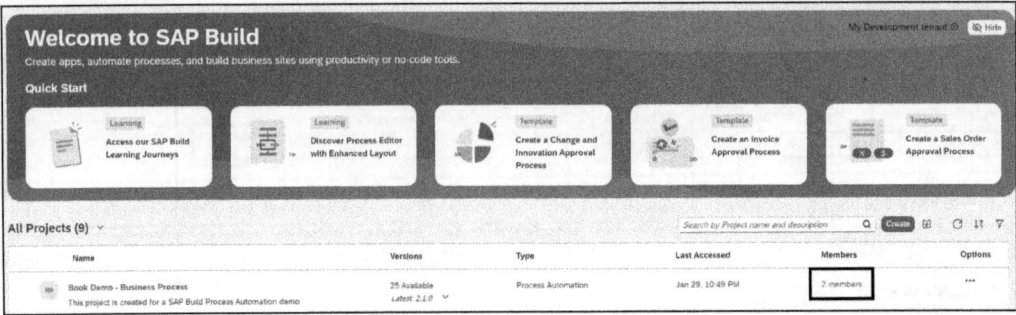

Figure 17.22 Project after Managing Members

17.6 Roadmap

Via SAP Discovery Center, you can find information about the features and pricing of services as well as service roadmaps. The roadmaps show the further development of the services planned for by SAP. To view the SAP Build Process Automation roadmap, which lists the most important innovations and adjustments for each quarter, visit: *https://discovery-center.cloud.sap/serviceCatalog/sap-build-process-automation?region=all&tab=roadmap*

To get more detailed information, you can also use SAP Road Map Explorer. To do this, click the link as shown in Figure 17.23.

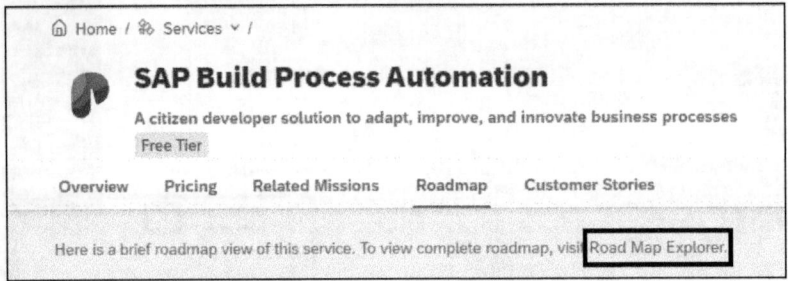

Figure 17.23 Navigation Link to SAP Road Map Explorer

SAP Road Map Explorer is an application that allows you to explore the planned innovations of a service in detail. Each tile within a quarter corresponds to a feature being rolled out that quarter. Features that have already been implemented and are active are marked with a blue bar on the left edge. Tiles for features that will be activated in the future have an orange bar. Figure 17.24 shows a screenshot of two features, one that has already been activated and one that will be implemented this quarter.

17.7 Summary

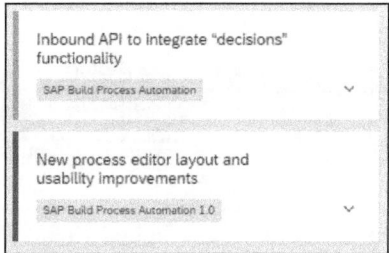

Figure 17.24 Two Features Shown in Roadmap

Selecting a feature opens a dialog in which you can explore the details of this feature. You will receive information about the additional functionalities that the respective feature brings with it. Figure 17.25 shows an example of the detailed view of a selected features.

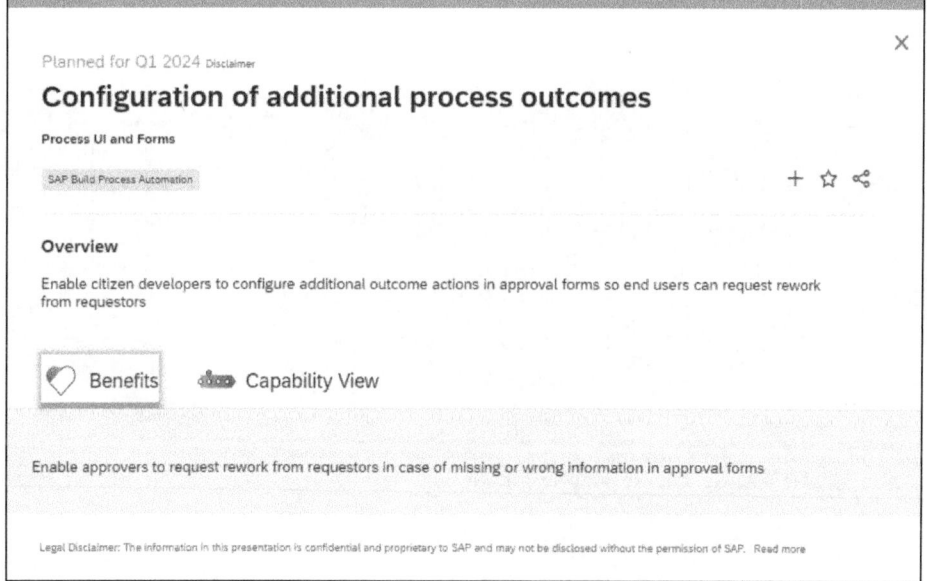

Figure 17.25 Details of Feature in Roadmap

17.7 Summary

This chapter gave a brief overview of the SAP Build Process Automation service. You have gotten to know the features that the service provides. We provided insight into the structure of projects and showed how the lifecycle of developments can be managed. Finally, you learned how to use the roadmap to get important information about service enhancements and changes. In the coming chapters, you'll get to know individual features in detail and see how you can use them to automate your business processes.

Chapter 18
Installing and Configuring SAP Build Process Automation

This chapter covers the installation and configuration of SAP Build Process Automation, focusing on the key steps needed to efficiently set up and customize the tool for your business needs. We aim to provide straightforward guidance in order to facilitate smooth integration.

As mentioned in the previous chapter, SAP Build Process Automation is a solution for citizen developers that enables them to improve and innovate business processes with workflow management and robotic process automation capabilities without code customization. It combines workflow management, RPA functionality, decision management, process visibility, and embedded AI capabilities in an intuitive, low-code experience. With SAP Build Process Automation, business users and technologists can become citizen developers.

To use SAP Build Process Automation effectively, it's first necessary to activate this service on SAP BTP in a subaccount. Detailed instructions on installation using the SAP BTP booster can be found in Section 18.1. This booster not only simplifies installation, but also takes care of most of the necessary configuration steps. Nevertheless, it's essential for troubleshooting and efficient use of the service to have a sound understanding of the configuration processes. For this reason, Section 18.2 is dedicated to explaining this configuration in detail.

Following the successful installation and configuration of SAP Build Process Automation, a further focus is on the security aspects of the service. Section 18.3 therefore presents security-relevant features and best practices. These include securing processes and data as well as adhering to compliance guidelines and data protection regulations. Observing these aspects ensures that SAP Build Process Automation can be used not only efficiently, but also securely and in compliance with applicable regulations.

18.1 Installation

In this guide, we will take you through the seamless installation and setup of SAP Build Process Automation. We start with the creation of a subaccount, which has been greatly simplified in this process. Instead of creating a subaccount manually, we use the integrated function of the booster to do it automatically. If you still want to create a subaccount yourself, there is nothing to stop you. Compared to other services, it isn't necessary to manually assign authorizations (entitlements) to the subaccount—a process that is carried out in the background in both cases, whether manually or automatically created by the booster.

This process ensures that all required authorizations are assigned correctly and completely without you having to worry about the individual technical details. The booster therefore makes the installation process easier for you by automatically making the correct settings. Simply follow these steps to ensure a smooth startup of SAP Build Process Automation (see Figure 18.1):

1. Open the side menu in your SAP BTP cockpit global account.
2. Navigate to the **Boosters** section.
3. Use the search function to find the available SAP Build Process Automation boosters, as shown in the instructions.
4. You will find two different booster options:
    - The first is for the free quota and offers you a free way to test the service. This variant is ideal for familiarizing yourself with the functions and possibilities of SAP Build Process Automation.
    - The second booster is a paid, fully licensed variant designed for permanent productive use.

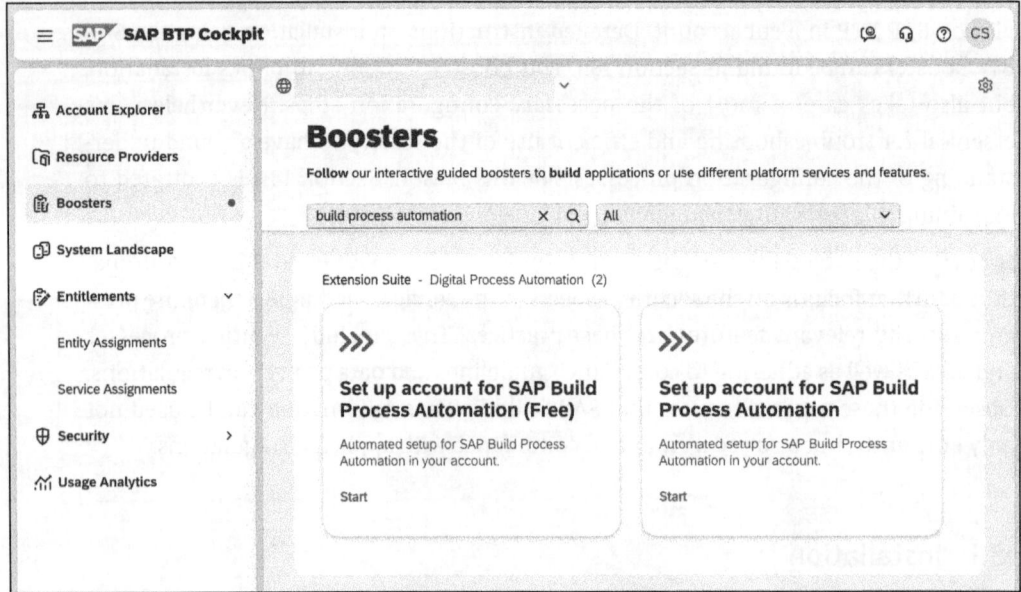

Figure 18.1 SAP BTP Booster for SAP Build Process Automation

5. Select the booster that suits your needs by clicking the corresponding tile. For this demonstration, we're going to choose the paid option. Note that the paid booster offers the full scope of SAP Build Process Automation Services and is therefore recommended for companies planning comprehensive and long-term use.

6. Open the **Components** tab in the booster as shown in Figure 18.2 to see the required components. As you can see, SAP Build Process Automation requires several services and a subscription. Click **Start** to start the booster.

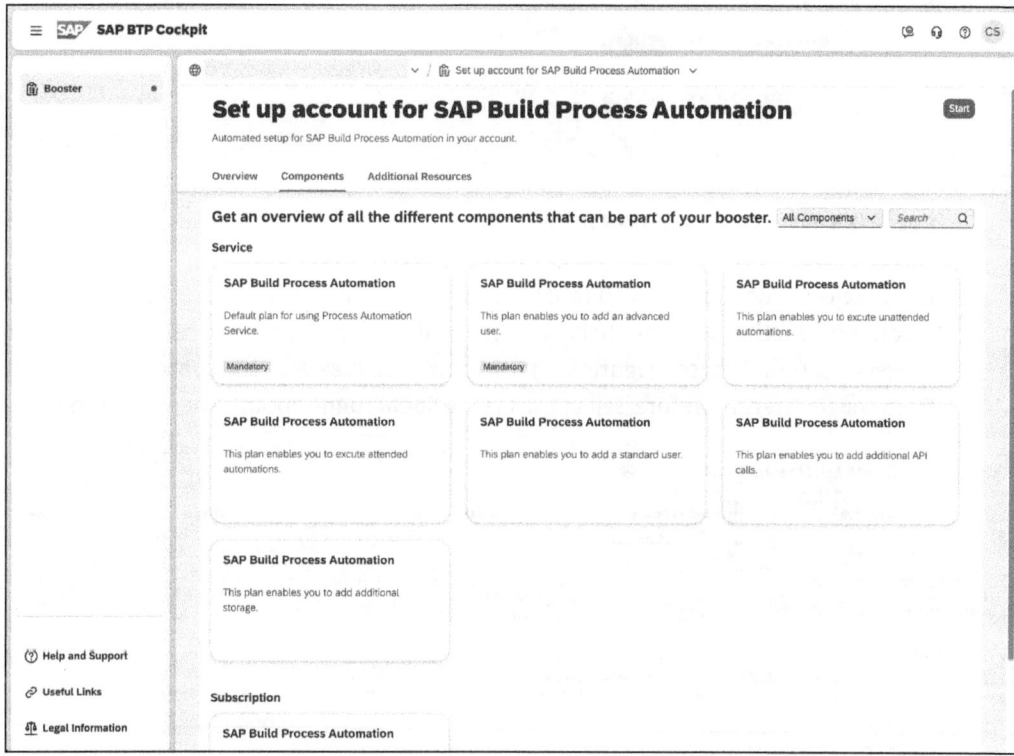

Figure 18.2 Booster Components

When setting up SAP Build Process Automation using the booster, the process begins with a check of the prerequisites, as illustrated in Figure 18.3. In this initial step, the booster will automatically check whether you as a user have the necessary authorizations and whether the required entitlements are available in the global account. These checks cannot be directly influenced by you as a user; they serve to ensure that all basic conditions for a smooth installation are met.

If problems are detected, the system will display a meaningful error message to help you understand which requirements have not been met. This allows you to specifically identify the cause and take appropriate action. Once all checks have been successfully completed, the system will provide the option to click the **Next** button to move to the

18 Installing and Configuring SAP Build Process Automation

next step and continue the configuration. This ensures a smooth and user-friendly process throughout the setup phase.

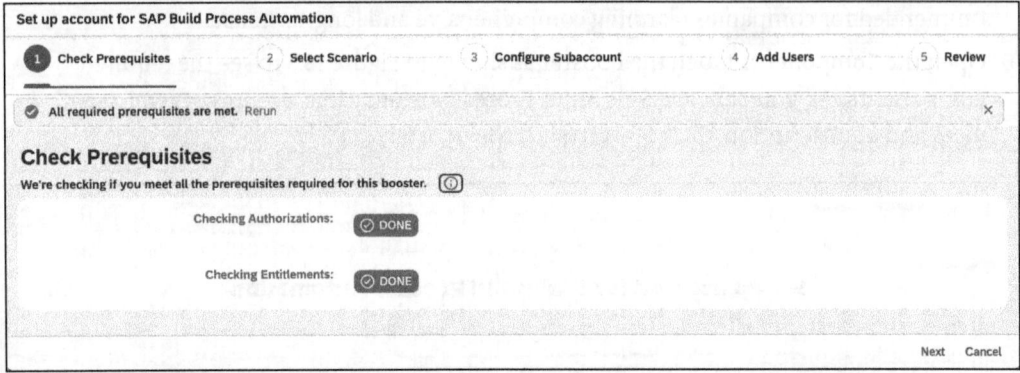

Figure 18.3 Booster Prerequisite Check

In the **Select Scenario** step, you must decide whether you want to use an existing subaccount or whether you want to create a new subaccount for the provision of SAP Build Process Automation (see Figure 18.4). For now, let's create a new subaccount directly from the booster. Therefore, select the **Create Subaccount** option. Then click **Next**.

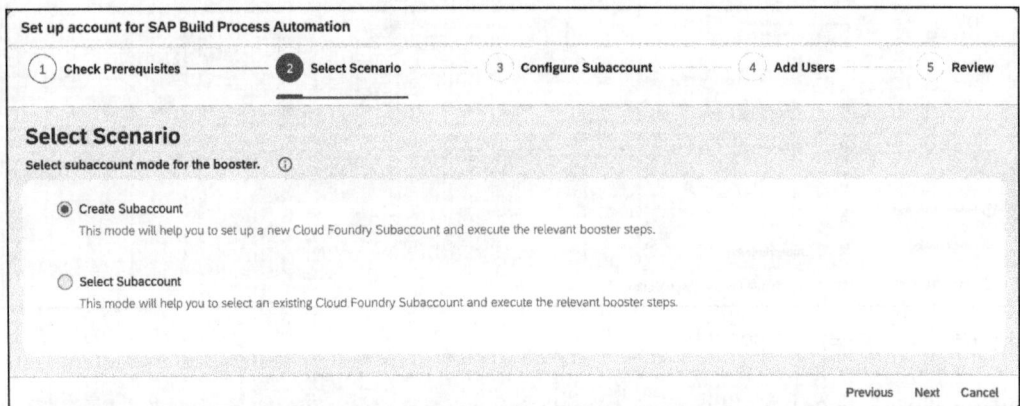

Figure 18.4 Booster Subaccount Selection

The **Configure Subaccount** step takes place as part of the configuration of your subaccount for SAP Build Process Automation. The system will automatically assign all necessary entitlements, both mandatory and optional, as shown in Figure 18.5.

During this process, you have the flexibility to check and adjust the assignment of entitlements. This includes the ability to remove certain entitlements that are not required or that you do not consider relevant in your specific context. This customizability is particularly useful to ensure that your subaccount contains exactly the resources and permissions required for your individual business processes and applications. The ability to modify entitlements gives you, the user, additional control and enables fine-grained customization of the subaccount to your business needs. After completing the configuration and customization of the entitlements, you can continue the process and scroll down to the subaccount details.

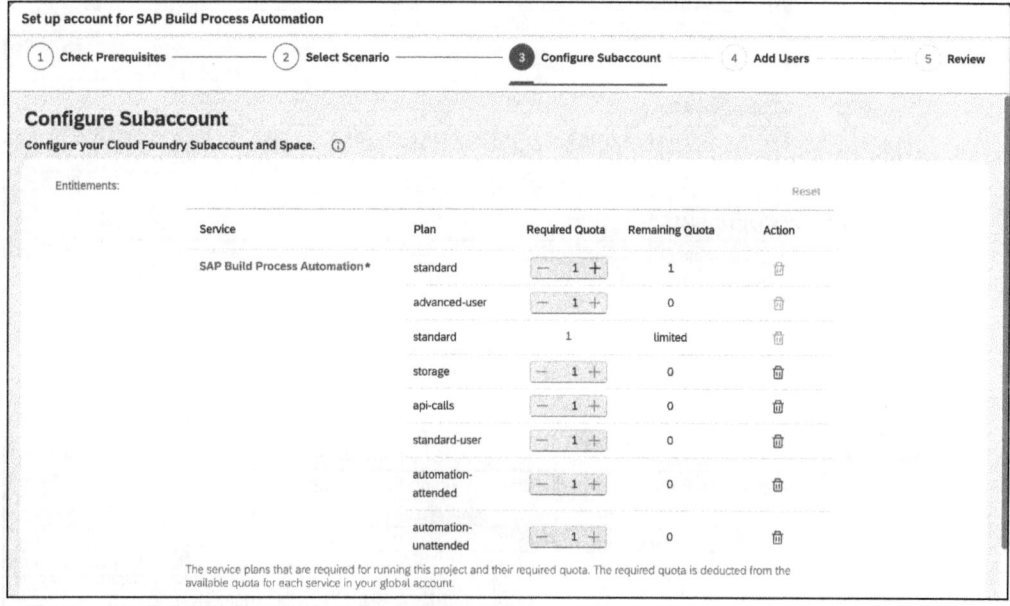

Figure 18.5 Booster Subaccount Configuration: Part 1

When continuing with the setup of SAP Build Process Automation via the booster, you must assign a name for your subaccount in the **Subaccount Name** field. This name is used for identification purposes and should be concise and descriptive to facilitate administration within your organization (see Figure 18.6).

It's also necessary to select an infrastructure **Provider** and an associated **Region**. It's advisable to check the SAP Discovery Center in advance to determine which providers offer the SAP Build Process Automation service and in which regions it is available. Choosing the right provider and region is critical as it can affect latency and data sovereignty. Once you have gathered this information, adjust your **Subdomain** if necessary.

Furthermore, you need to configure the Cloud Foundry organization name in the **Org Name** field and the name of the Cloud Foundry space in the **Space Name** field. These settings structure how your cloud resources are organized and managed. During the installation process, the booster will activate Cloud Foundry in your subaccount and automatically create a space in it. This space is a logical separation within your org where you can deploy and manage your applications and services.

Once you have made all the settings, click **Next** to move on to the next step and continue the setup.

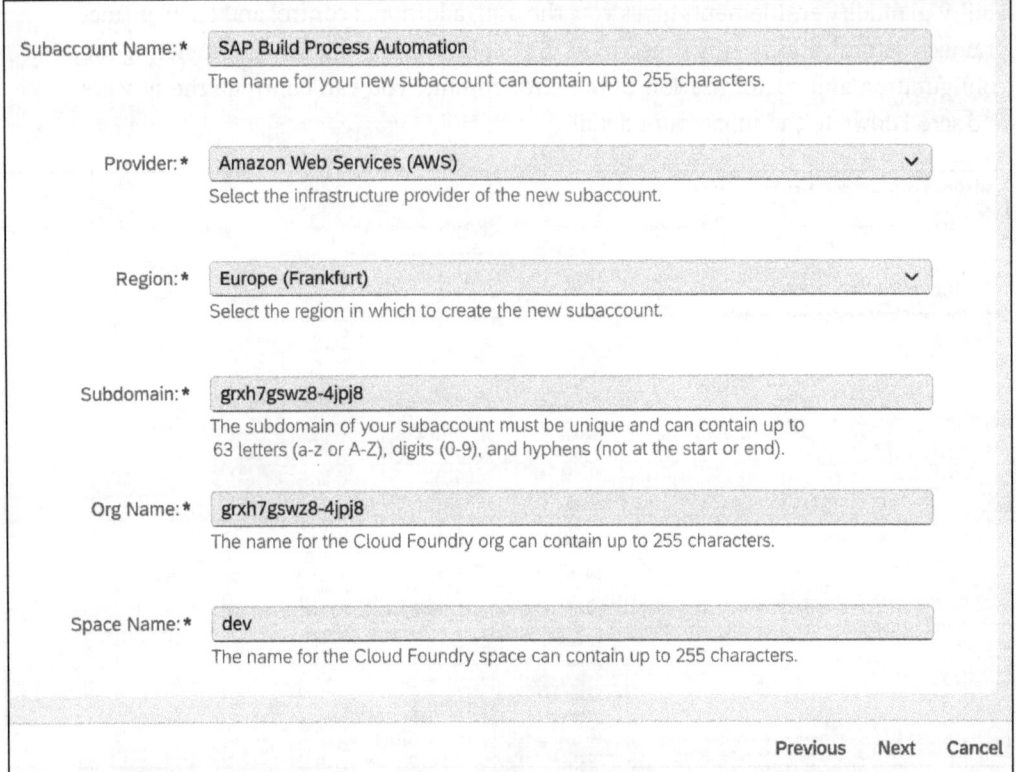

Figure 18.6 Booster Subaccount Configuration: Part 2

In the next step of the SAP Build Process Automation booster, the initial user assignment takes place (see Figure 18.7). You select the appropriate identity provider for the platform users and the application users. This is an essential step as the identity provider controls the authentication of your users and ensures that only authorized persons are granted access. For those users who are to receive administrative authorizations, enter them in the **Administrators** field. The persons listed here are granted extensive authorizations that enable them to carry out important administrative tasks. In particular, they receive the org manager and space manager roles in Cloud Foundry, as well as the subaccount administrator role collection at the subaccount level.

In the **Developers** field, list the users who are to receive developer rights. These users are given access to the tools and resources required to develop and maintain applications. The following role collections are assigned to them:

- ProcessAutomationDeveloper
- ProcessAutomationAdmin
- ProcessAutomationParticipant
- Subaccount Viewer

18.1 Installation

These role collections ensure that developers have the necessary rights to create and manage automation processes and can view the subaccount. Once you have configured all users and their roles, confirm your entries by clicking **Next**. This initiates the process to move on to the next step of the configuration.

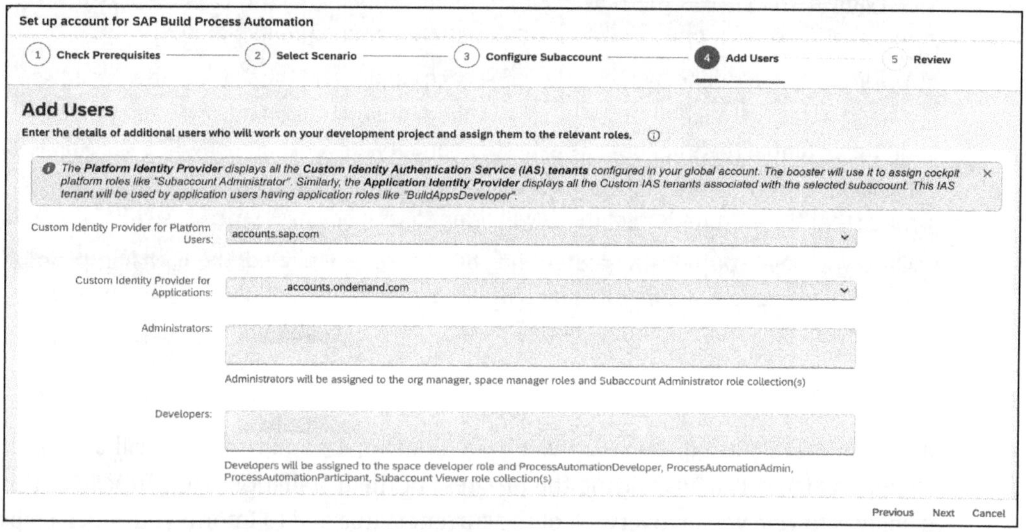

Figure 18.7 Booster User Management Configuration

In the final **Review** step of the booster-based SAP Build Process Automation setup process, you will receive a summary of all the configurations you have previously made. This overview allows you to check all settings again and ensure that everything is set up correctly. The summary typically includes details of the configured entitlements, the assigned user roles, the selected identity providers, and the specific settings for the subaccount, organization, and space.

Once you have checked all the details and ensured that they meet your requirements, you can complete the setup process. Clicking **Finish** will activate the booster, which will then carry out the setup automatically. This includes creating and assigning resources, activating services, and setting up user permissions according to the configurations you have chosen.

> **Subsequent Changes**
> It's important that you are aware that after this step, the setup begins, and no further changes can possibly be made without starting the process from the beginning or making the changes manually in the corresponding subaccount. So, make sure that everything is configured according to your wishes before you click **Finish**.

18 Installing and Configuring SAP Build Process Automation

[!] **License Costs**

It's crucial to be aware that once you have completed these setup steps and successfully run the process with the booster, you will incur license costs. This is particularly relevant if you choose the fully licensed, fee-based variant of the SAP Build Process Automation service, which is intended for permanent and productive use.

The license costs depend on the selected service variant and the scope of use. It's advisable to obtain detailed information about the cost structure in advance and to weigh up the potential costs against the expected benefits for your company. The decision to purchase a paid license should be based on a well-founded assessment of the added value that the automation solutions will bring to your business processes.

Once you have started the booster and the service is activated, the licensing period begins and the fees are calculated according to the terms of your contract or service agreement. Be sure to obtain all necessary internal approvals and plan budgets accordingly before finalizing this step.

After you have started the setup using the booster, a popup window will appear as shown in Figure 18.8, visualizing the progress of the installation. The purpose of this display is to give you an overview of the current status and to inform you which steps have already been completed and which are still being carried out. The installation itself can take a few minutes. During this process, the configured services are set up, necessary components are installed, and authorizations are assigned. It's normal for such processes to take time, as several complex operations are running in the background. As soon as the installation process is complete, you will be notified and can start using SAP Build Process Automation services.

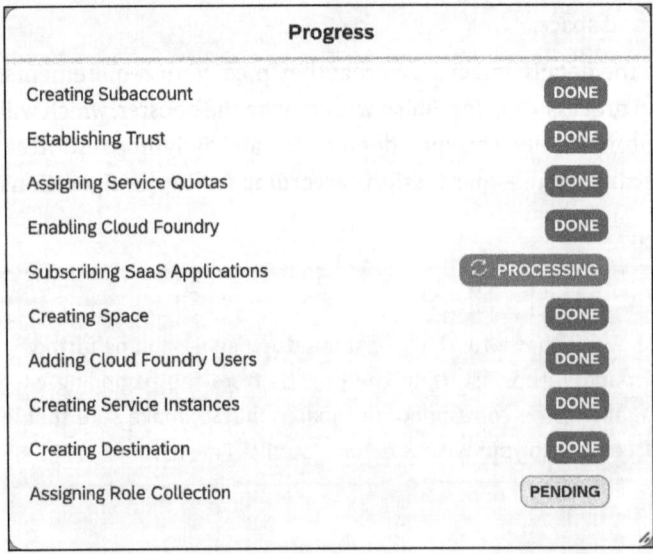

Figure 18.8 Booster Progress

534

18.1 Installation

After completing the installation and configuration of SAP Build Process Automation via the booster, you will receive a confirmation message (see Figure 18.9). This success message is a visual indication that all steps have been completed successfully and the service is now ready for use. Navigate to the subaccount by clicking **Navigate to Subaccount**.

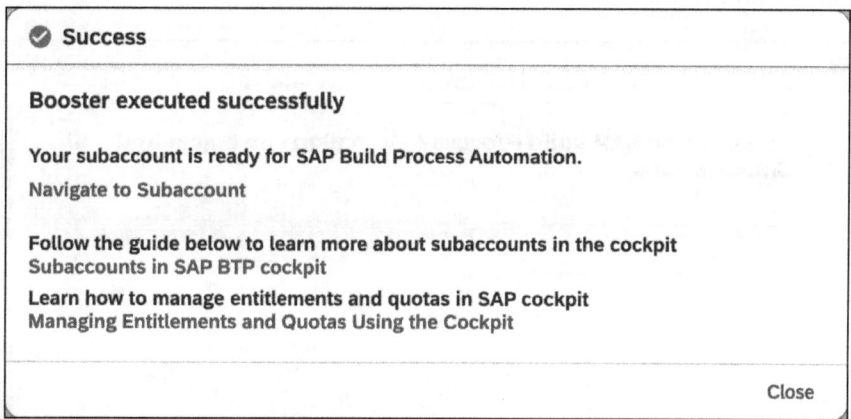

Figure 18.9 Booster Success Message

In the **Overview** section of your subaccount, you can see all the details of the subaccount and the API endpoint, as shown in Figure 18.10.

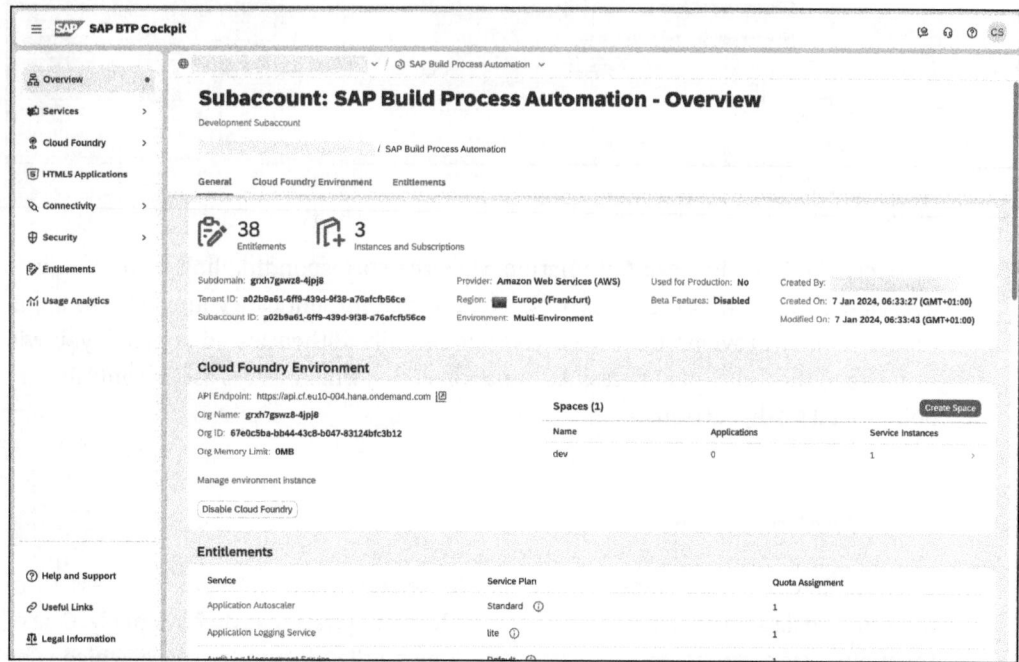

Figure 18.10 Subaccount Overview

535

18   Installing and Configuring SAP Build Process Automation

Next, open the instance and service overview to get an overview of the activated services. To do this, navigate to the **Services • Instances and Subscriptions** area in the side menu. There you will find an application called **SAP Build Process Automation** in the **Subscriptions** area. You can open the application by clicking its name. In the **Instances** area, you will find a service instance called `sap_process_automation` that has been installed in the dev space (see Figure 18.11).

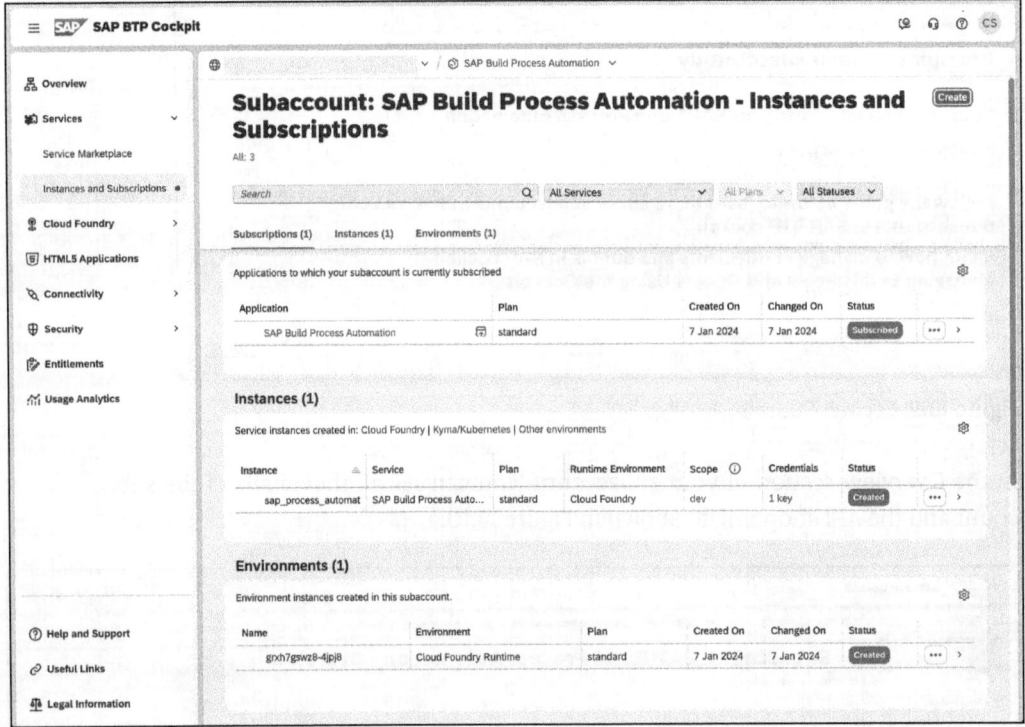

Figure 18.11 Instances and Subscriptions

To open SAP Build Process Automation, click the corresponding link in your service overview. After clicking the link, you will be asked to authenticate yourself to the application identity provider. Once you have successfully authenticated yourself, you will be taken to the SAP Build Process Automation lobby, which serves as the central entry point for all further actions.

18.2 Configuration

Once you have completed the basic setup with the booster for SAP Build Process Automation, it makes sense to get an overview of the configurations you have made. One of the main actions of the booster is the creation of a Cloud Foundry space, which is by default called *dev*. To inspect this space, navigate to the **Cloud Foundry** area in the SAP

BTP cockpit of your subaccount. Within this area, select **Spaces**. Here you will find a list of available spaces, including the **dev** space mentioned, as shown in Figure 18.12. By clicking the name of this space, you will be forwarded directly to it. In the dev space, you can get a precise overview of the service instances running there.

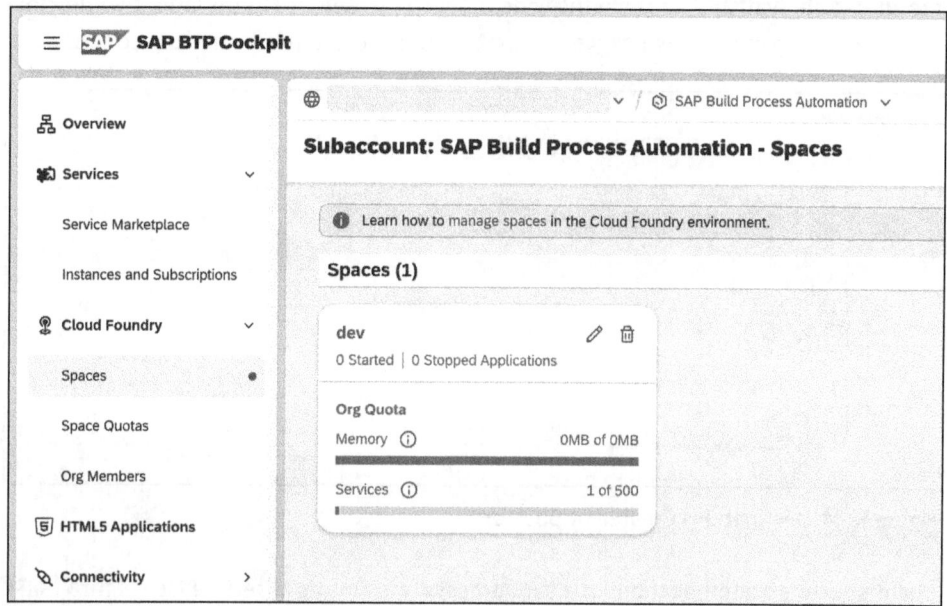

Figure 18.12 Cloud Foundry Spaces

In the next step, you can navigate to the instances in the space. There you will see all instances of services that you have either created yourself or that were created by the booster. This means that you will also see the services that you see in the overview at the subaccount level. As you can see in Figure 18.13, a service key has already been created for this service. This contains a client ID and a client secret and can be used in the OAuth context to authenticate and request a token.

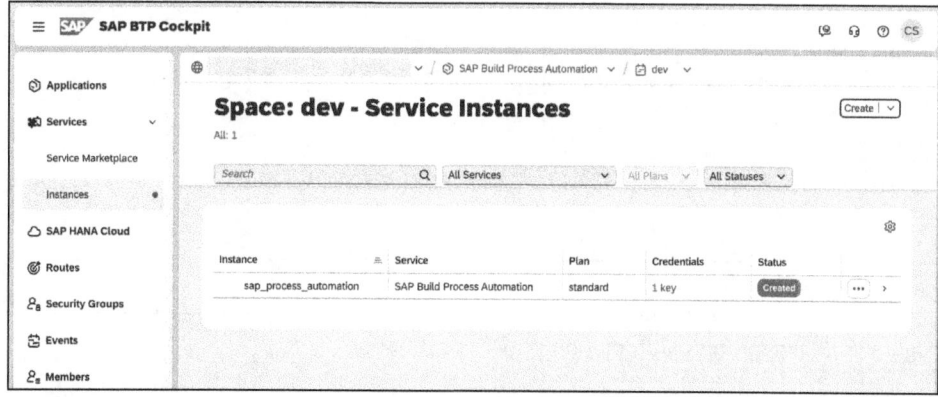

Figure 18.13 Service Instances

18 Installing and Configuring SAP Build Process Automation

The booster process for SAP Build Process Automation also includes setting up destinations, which are essential to establish a connection between your SAP BTP instance and internal and external services or systems. However, these destinations are not specific to an individual Cloud Foundry space but are configured at the subaccount level and are therefore available subaccount-wide. As you can see in Figure 18.14, two destinations have been created, which we will look at in more detail later in this section.

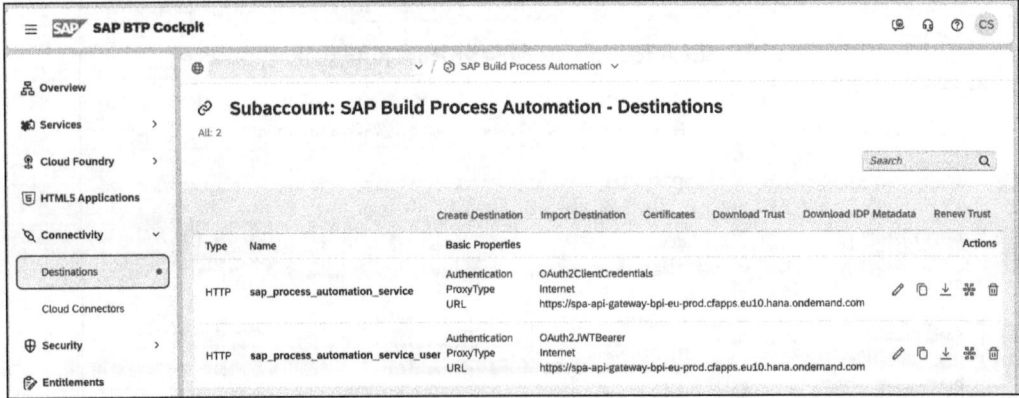

Figure 18.14 Destinations Created by Booster

The first destination has the name **sap_process_automation_service** (see Figure 18.15). This destination is used to call the SAP Build Process Automation service programmatically via APIs.

Figure 18.15 sap_process_automation_service Destination

As shown in Figure 18.16, a destination called **sap_process_automation_service_user_access** also was created. Unfortunately, the official SAP Build Process Automation documentation does not contain any information about the purpose of this destination.

Figure 18.16 sap_process_automation_service_user_access Destination

As previously discussed, the license costs for a service usually start to accrue when it is activated or when a subscription is created. To keep control of these costs and ensure that they remain within your budget, it is important to regularly view usage analytics. Usage analytics provide you with a detailed insight into the usage and associated costs for the services in your SAP BTP. This data is provided at the individual subaccount level and is also aggregated at the global account level for an overall picture. To access usage analytics, open the side menu in your SAP BTP cockpit and navigate to the **Usage Analytics** section as shown in Figure 18.17.

In this section, you will find valuable information about the current usage of services, which will help you to monitor and control spending. Here you can view reports on service usage, analyze how different services are being used, and understand how usage affects overall costs. This is an essential cost management tool that helps you make decisions about scaling, optimizing, or possibly reducing services to optimize your spend.

Within SAP Build Process Automation, it's sometimes necessary to send messages or notifications to affected users by email. To do this, you need to create a corresponding destination. This step is not carried out by the booster; that is, you need to manually create the destination. To do this, navigate to the **Connectivity • Destinations** area in the SAP BTP cockpit of the subaccount and click **Create Destination** in the destination overview (see Figure 18.18).

18 Installing and Configuring SAP Build Process Automation

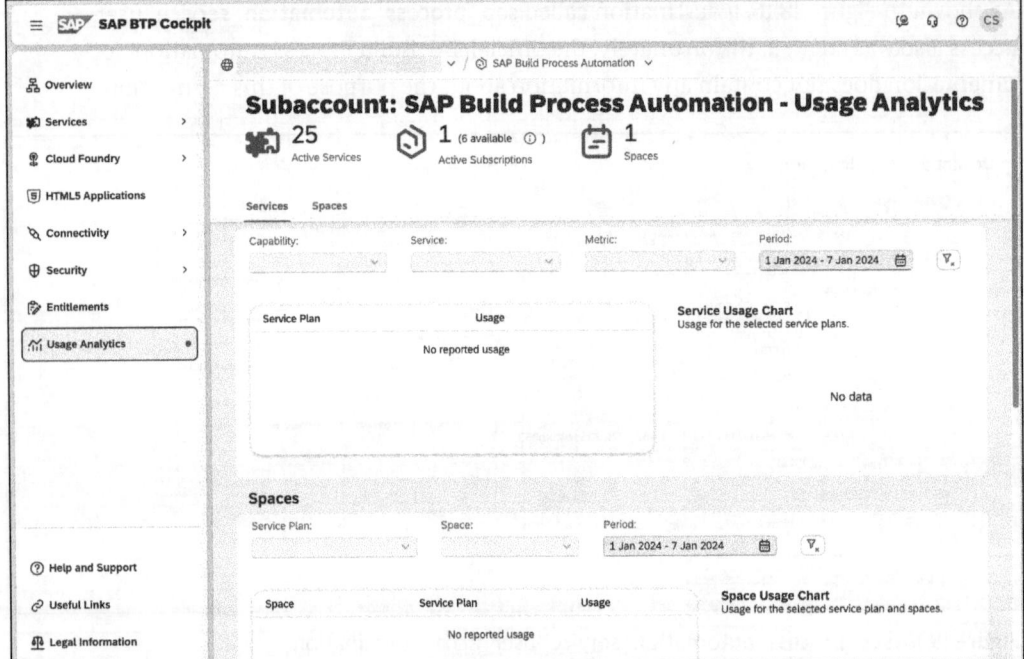

Figure 18.17 Subaccount Usage Analytics

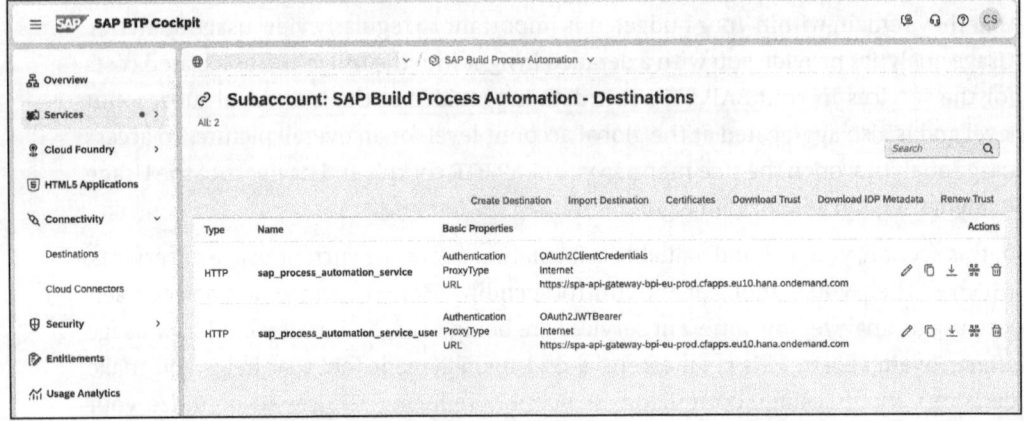

Figure 18.18 Destination Overview

Create a destination with the basic properties listed in Table 18.1. Be aware that the name of the destination is defined by SAP and must not be adjusted.

Field	Value
Name	sap_process_automation_mail
Proxy Type	Internet

Table 18.1 Mail Destination Properties

Field	Value
Type	MAIL
Description	An optional short description
User	The user for logging into the mail server
Password	The password for logging into the mail server
Authentication	**BasicAuthentication** or **OAuth2Password** method is supported

Table 18.1 Mail Destination Properties (Cont.)

After the basic properties are assigned, the additional properties listed in Table 18.2 must be added to this destination.

Name	Value
mail.transport.protocol	smtp.
mail.smtp.auth	true.
mail.smtp.starttls.required	true.
mail.smtp.host	Host name of your mail server.
mail.smtp.port	Port on which your mail server listens for connections (usually 587).
mail.smtp.from	Email address to use as the "From" address of emails sent by SAP Build Process Automation—for example, user@example.com. *Hint*: This address must belong to an existing mailbox because it receives the replies to emails that the workflow capability sends.
mail.smtp.ssl.trust	Either an asterisk (*) or a space-separated list of acceptable host names. If you don't provide a value, trust is based on the certificate provided by the server. This must be part of the SAP JVM default trust store in the corresponding SAP BTP subaccount.
mail.smtp.ssl.enable	false.
mail.smtp.starttls.enable	true.
mail.bpm.send.disabled	false.

Table 18.2 Mail Destination Additional Properties

Figure 18.19 shows an example of a working destination. Whenever you choose **MAIL** as the **Type** of the destination, it is not possible to perform a connection check for this destination directly from the SAP BTP cockpit.

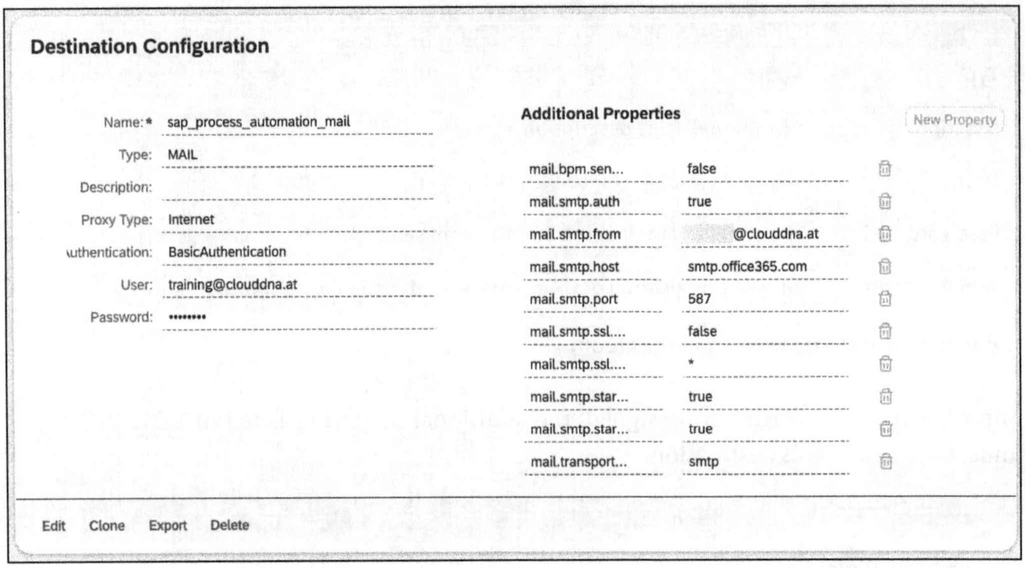

18 Installing and Configuring SAP Build Process Automation

Figure 18.19 Create Mail Destination

SAP Build Process Automation allows you to perform a test for mail dispatch. To do this, open the SAP Build Process Automation application with a user with administrator authorizations, as shown in Figure 18.20. Then navigate to **Control Tower** in the side menu. In the **Backend Configuration** area, you will find a tile with the name **Mail Server**. Click this tile to open the configuration.

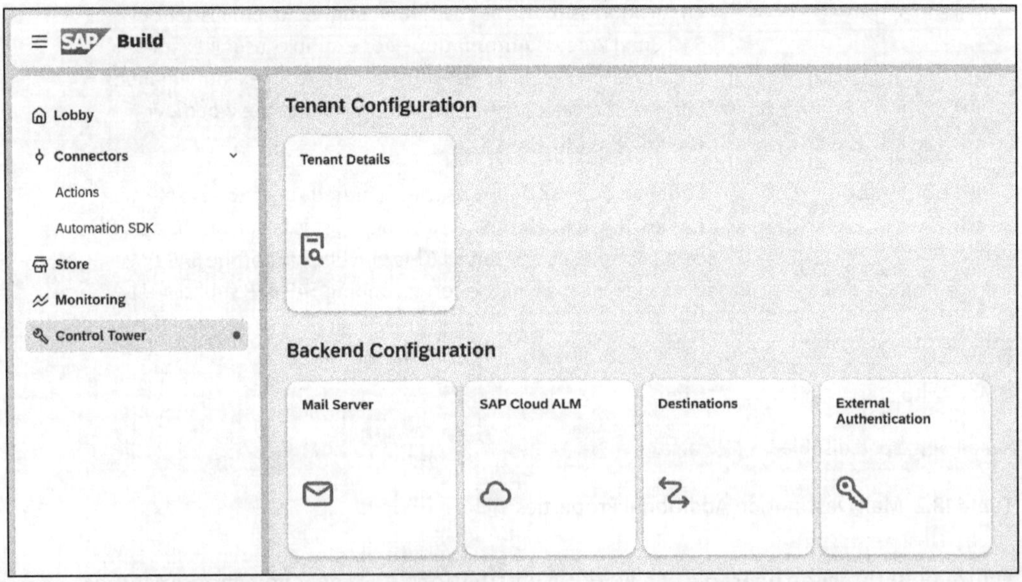

Figure 18.20 SAP Build Control Tower

In the mail server configuration, you will find the basic attributes like the **Name** of the destination being used, the **Host Address** and **Port** of the mail server, and the **Sender Address** (see Figure 18.21). To open the test mail dialog, click the **Send Test Mail** link.

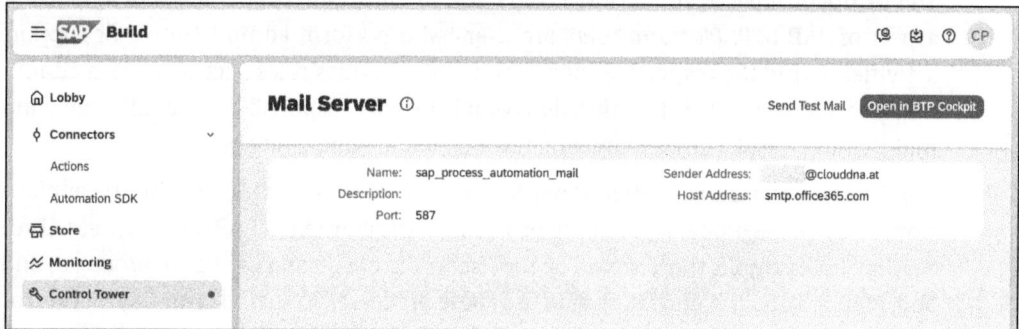

Figure 18.21 Mail Server Configuration: Test Mail

As shown in Figure 18.22, you must provide a recipient **Email address** and a **Context Text**. Clicking the **Send Test Mail** button triggers the submission of the test mail. If there is a technical error, the error message will be shown. Otherwise, if you get a success message, check the recipient's mailbox for the incoming mail.

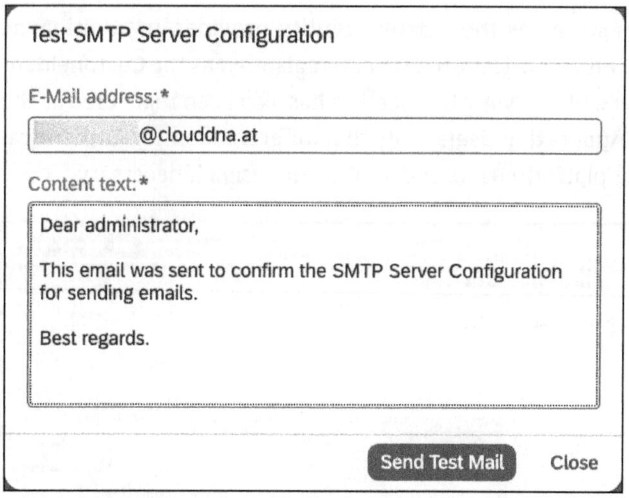

Figure 18.22 Send Test Message

18.3 Security

SAP provides several authentication options on SAP BTP to meet the security and access requirements of companies. These authentication methods include the SAP ID service, Identity Authentication (part of SAP Cloud Identity Services), and SAML-based third-party identity providers. These different services enable flexible and secure authentication of users at the subaccount level while providing the ability to customize identity services to meet specific business needs.

The distinction between platform users and application users is another important aspect of SAP BTP. *Platform users* are users who perform administrative or support activities within the respective subaccount. This includes roles such as system administrators and support staff. This user group is usually responsible for configuring and maintaining the platform, managing resources, and supporting other users.

Application users, on the other hand, are those who interact directly with the applications and use the processes running on them. In the context of SAP Build Process Automation, for example, these would be the users who create and configure processes and automations and the end users who use these processes in their day-to-day work—for example, to carry out approval processes or to interact with automated workflows.

This division makes it possible to control authorizations and access in fine detail and ensure that each user only receives the access rights they need for their specific tasks. This guarantees a high level of security and compliance, while at the same time ensuring the user-friendliness and efficiency of the platform. In your SAP BTP subaccount, you should therefore carefully plan and assign roles and access rights according to the tasks and responsibilities of the users.

The SAP ID service is always stored as the default identity provider. In addition, as shown in Figure 18.23, Identity Authentication has been registered as the **Custom Identity Provider for Platform Users**. Identity Authentication has also been registered as the **Custom Identity Provider for Application Users**. Note that different Identity Authentication instances can be used for platform users and application users if necessary.

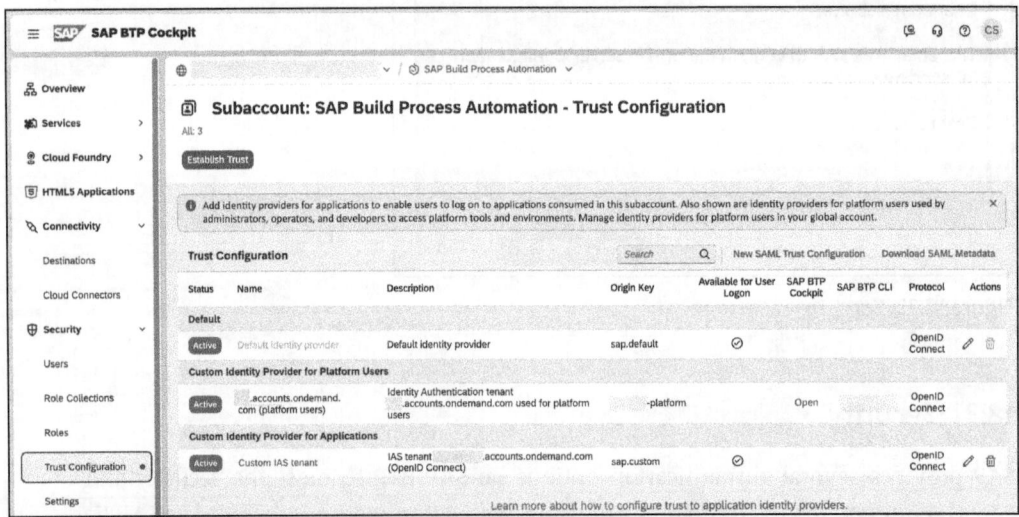

Figure 18.23 Trust Configuration

18.3 Security

With the activation of SAP Build Process Automation, the following role collections, shown in Figure 18.24, have been created in the subaccount:

- ProcessAutomationAdmin
- ProcessAutomationDeveloper
- ProcessAutomationParticipant

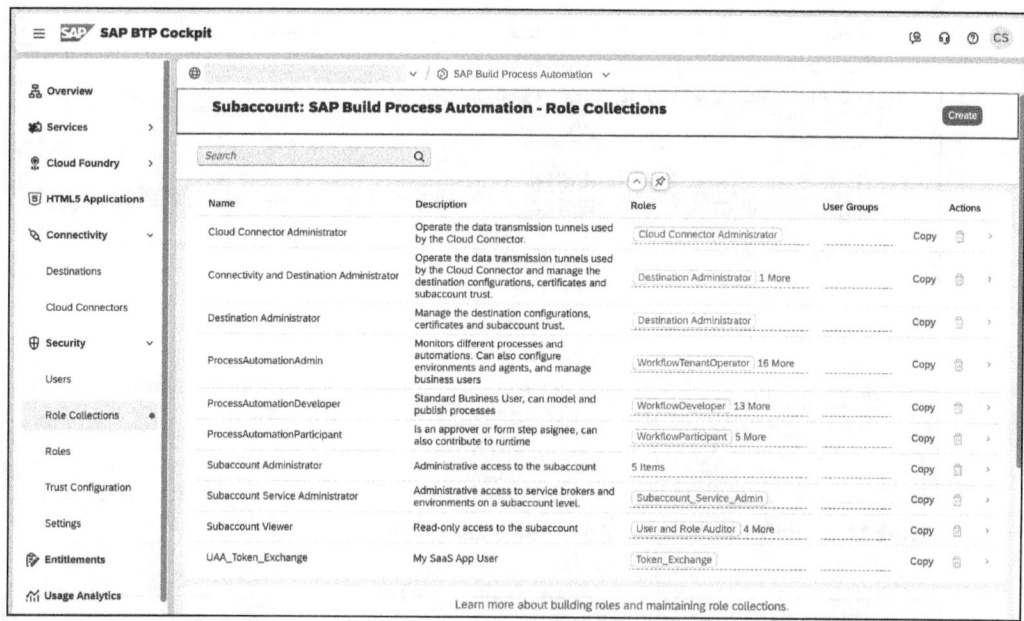

Figure 18.24 SAP Build Process Automation Role Collections

The ProcessAutomationAdmin role collection allows users to manage the process configuration, permissions, and authorizations within SAP Build Process Automation. As shown in Figure 18.25, the role collection combines among others important roles from the SAP Intelligent RPA functionality and from process visibility.

The ProcessAutomationDeveloper role collection allows users to manage the creation, editing, and publishing of individual processes and automations within SAP Build Process Automation. As shown in Figure 18.26, the role collection combines among others important roles from the SAP Intelligent RPA functionality, process visibility, SAP Workflow Management, and SAP Business Rules Management.

The ProcessAutomationParticipant role collection allows users to manage the creation, editing, and publishing of individual processes and automations within SAP Build Process Automation. As shown in Figure 18.27, the role collection combines among others important roles from the SAP Intelligent RPA functionality and SAP Workflow Management.

18   Installing and Configuring SAP Build Process Automation

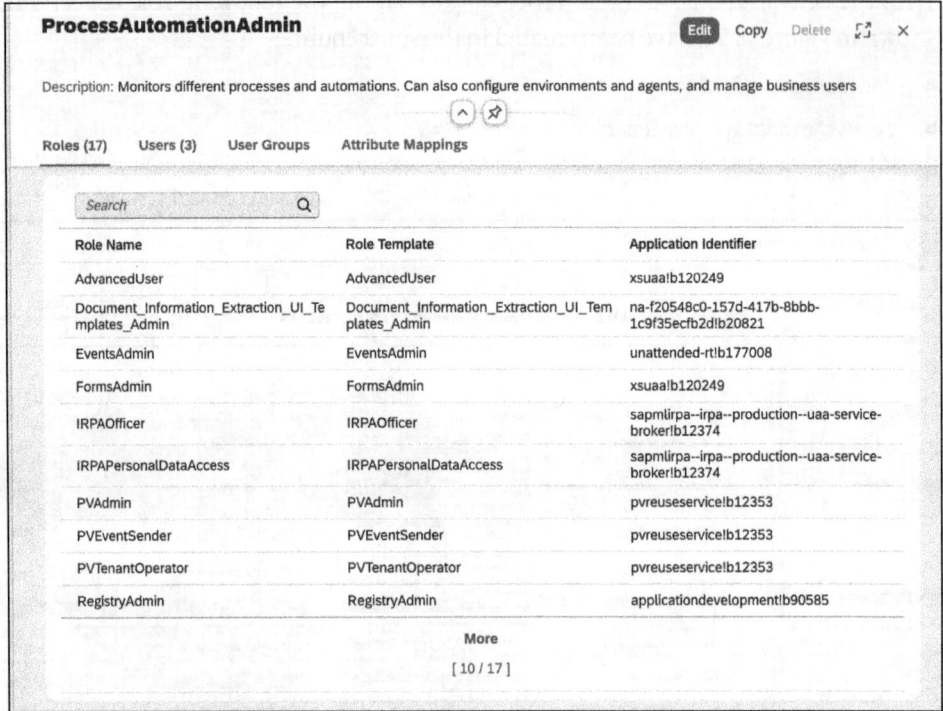

Figure 18.25 ProcessAutomationAdmin Role Collection

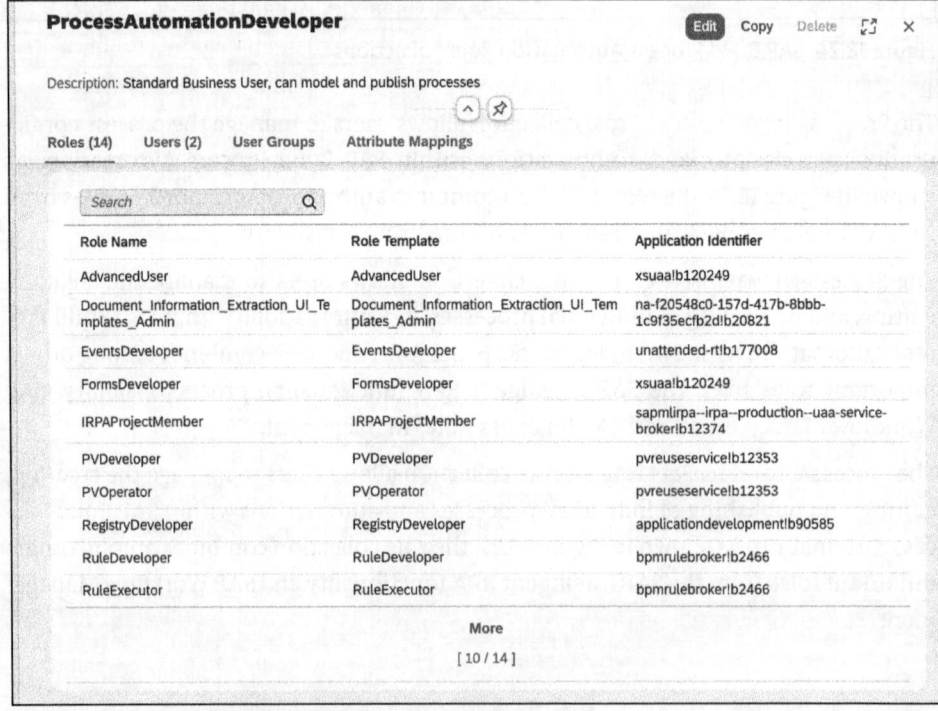

Figure 18.26 ProcessAutomationDeveloper Role Collection

546

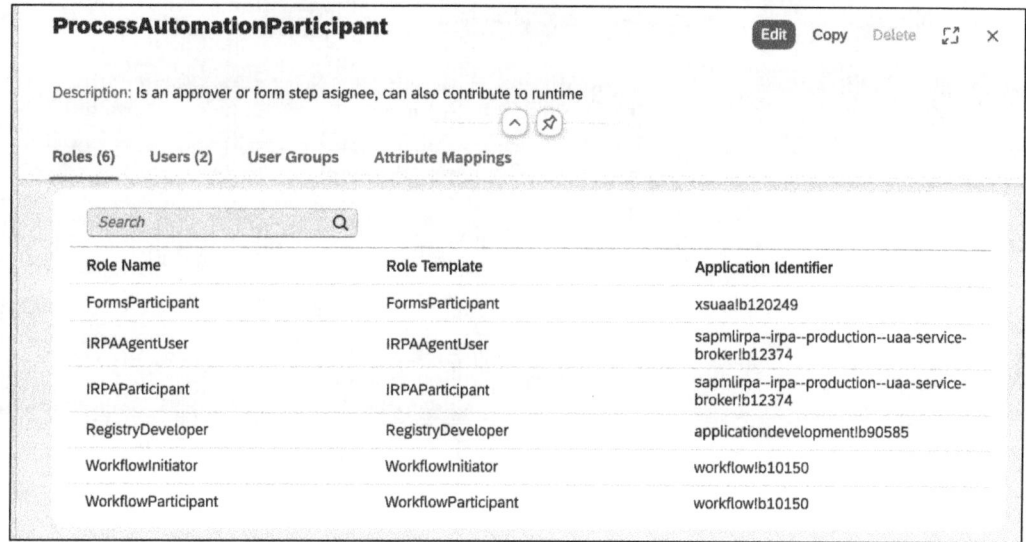

Figure 18.27 ProcessAutomationParticipant Role Collection

18.4 Summary

In this chapter, we focused on the installation of SAP Build Process Automation on SAP BTP. It begins, as explained in Section 18.1, with the provision of the service in a subaccount using the SAP BTP booster. This process simplifies the installation process by automatically creating a subaccount and assigning the necessary entitlements. The booster functionality guides users through the steps of the installation and takes care of configuring the basic settings to ensure a quick and efficient setup. Once installed, users can access the application directly via the SAP BTP console and start using it.

After installation, the configuration takes place, which is largely automated in the booster process. This was shown in Section 18.2. However, users can check and adjust the entitlements assignments and the configuration of Cloud Foundry spaces. In addition, monitoring license costs by incorporating usage analytics is an essential security and control mechanism to keep usage transparent and within budget.

The setup differentiates between platform and application users, with the corresponding roles and authorizations assigned according to responsibilities. Security aspects are an integral part of the installation and configuration of SAP Build Process Automation. These elements were discussed in detail in Section 18.3. The platform supports authentication using the SAP ID service, Identity Authentication, and third-party SAML identity providers. This enables secure and controlled access control. The definition of user roles and the use of different instances of Identity Authentication for platform and application users enables finely tuned access control.

Chapter 19
Processes

One main part of SAP Build Process Automation is the digitalization of business processes. Requesting a leave, confirming an invoice, and similar activities are part of every company's daily business. This chapter will give an insight into the tools that SAP Build Process Automation provides to digitize such processes, and you will learn how to use them in order to digitize your own business processes.

This chapter is all about creating processes. You will get familiar with the SAP Build lobby and learn about its functionalities and how to use them. You will create a new project and learn how to use the tools provided to design your business process and define the rules to manage the data. You will see in detail how the designing editor is used and what elements are provided to design your process. Furthermore, you'll use the elements that are necessary for designing your business needs. The elements used in the process will include the following:

- Triggers
- Forms
- Conditions and branches
- Actions

To show you how these elements are used in context, we'll create a process scenario that will follow us throughout the remaining chapters. So now, let's dive into the topic.

19.1 Creating a Project with a Business Process

The first step will be to navigate to the lobby, where you will see the welcome screen, including all the projects that are currently in use (see Figure 19.1).

In the left panel, a menu is provided that enables you to navigate to other important sections. In the **Connectors** section, you can create actions to be used in your processes or you can make use of the Automation SDK. The **Store** section offers many predefined processes that might already provide the functionality you need. In this section, you can search for solutions by defining filters for special parameters, like the project type, the line of business, the content type, or the industry, for example. Once you have found a project that might fulfill your needs, you can import it into your lobby and use it out of the box or even enhance its functionalities depending on your needs.

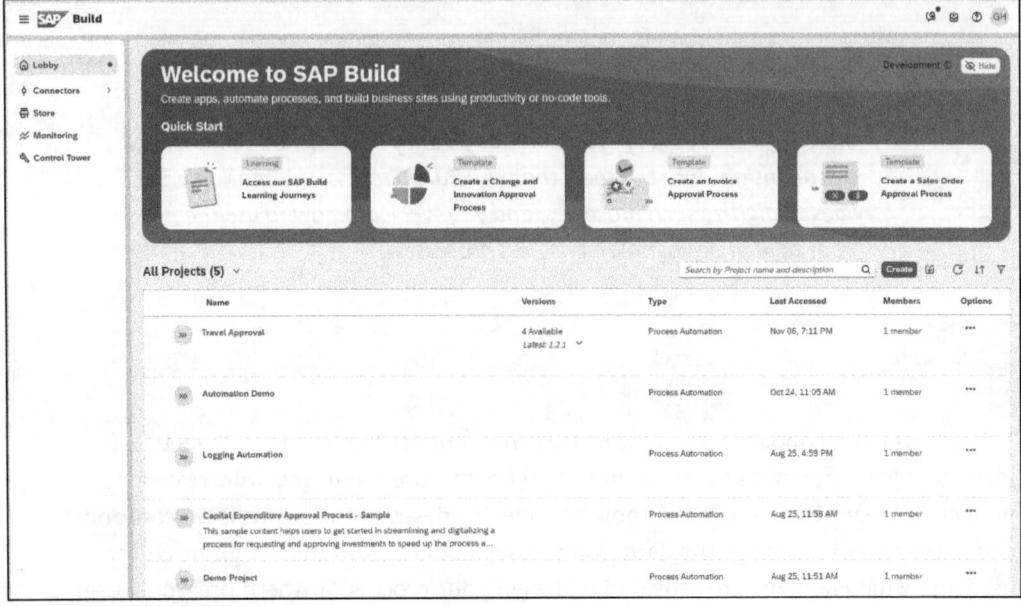

Figure 19.1 Lobby of SAP Build

In the **Monitoring** section, you're able to see the status of your processes and get information about the currently running or already run process instances. You see the processes that are deployed and ready to use and you can create new process instances by using triggers.

The **Control Tower** section provides a collection of functionalities to configure services that might be useful for your processes, like mail servers, destinations, and many more.

The next step is to create a new project. Therefore, press the **Create** button in the header of the project list. Figure 19.2 shows the position of the **Create** button.

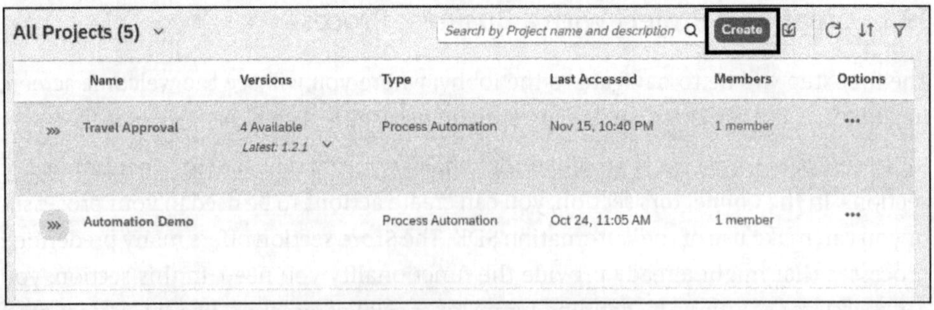

Figure 19.2 Create Button

After pressing the **Create** button, a new dialogue comes up where you define what kind of project to create. In this case, you want to create an automated process. By selecting the **Build an Automated Process** tile in the middle of the screen, you can start creating an automated process project. Figure 19.3 shows the dialogue that comes up after pressing the **Create** button.

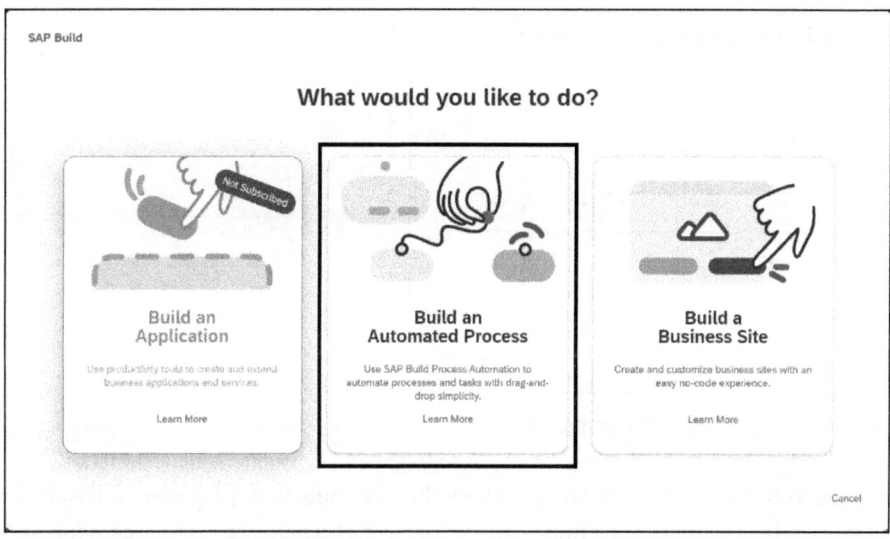

Figure 19.3 Dialogue to Create New Project

The next step is to define which kind of process you want to create. For now, you want to create a business process, so select the **Business Process** tile on the left side of the next dialogue, as shown in Figure 19.4.

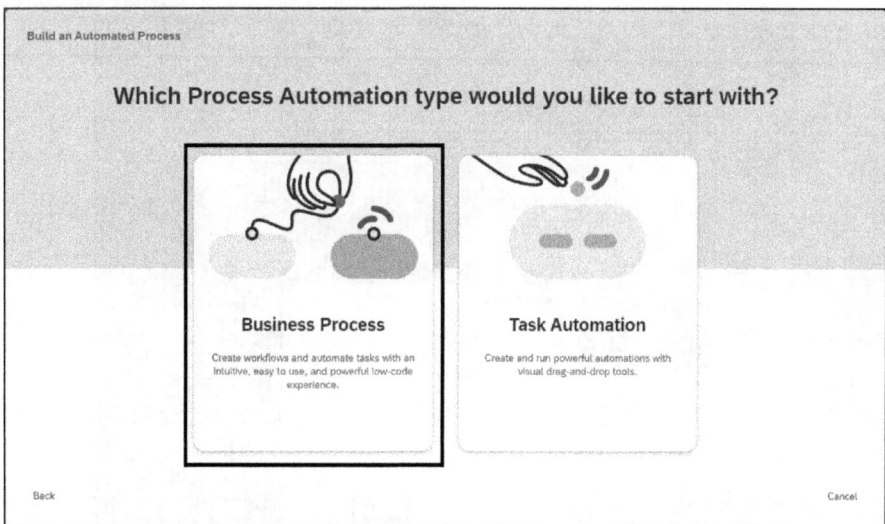

Figure 19.4 Selecting Type of Process Automation to Be Created

After selecting the tile for creating a business process, you can define the **Project Name** of your new business process project and enter a short **Description**. Confirm your entries by pressing the **Create** button. Figure 19.5 shows a screenshot of this dialogue.

Figure 19.5 Dialogue to Enter Name and Description for Your Project

In your browser, a new tab is opened to show the overview of your project artifacts. As you have decided to create a business process project, the next step is to create the process itself. Therefore, the first thing to do is to give the process a **Name** and an optional **Description**. The application shows a dialogue to provide this information. As you enter the name for the new process, an **Identifier** is automatically created. You can rename this identifier, but be aware that it has to be unique inside your project. Figure 19.6 shows the new tab with the dialogue to create the process.

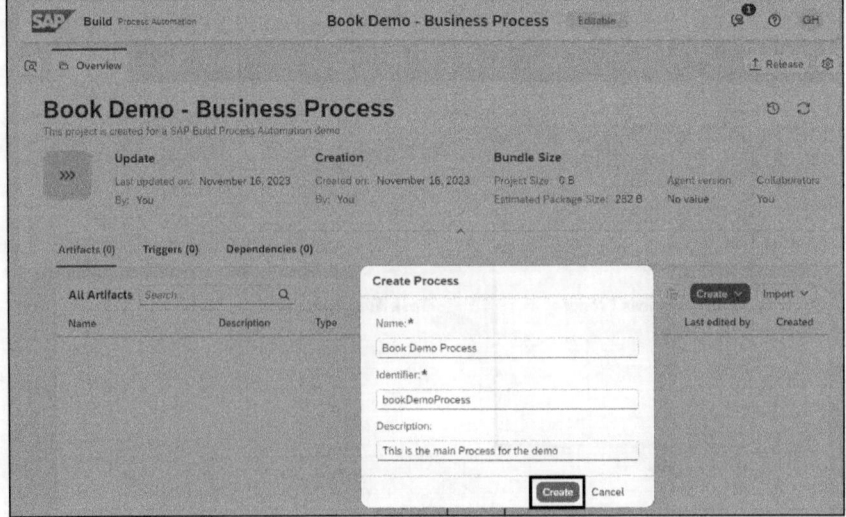

Figure 19.6 Dialogue to Create New Business Process

19.1 Creating a Project with a Business Process

After providing the name and the description for your new business process, you confirm your entries by pressing the **Create** button. As a result, the business process is created, and a new panel is opened where you can start to design your process. Whenever you create a new process, the end event is automatically created and added to the designing view. Figure 19.7 shows a screenshot of the opened designing panel.

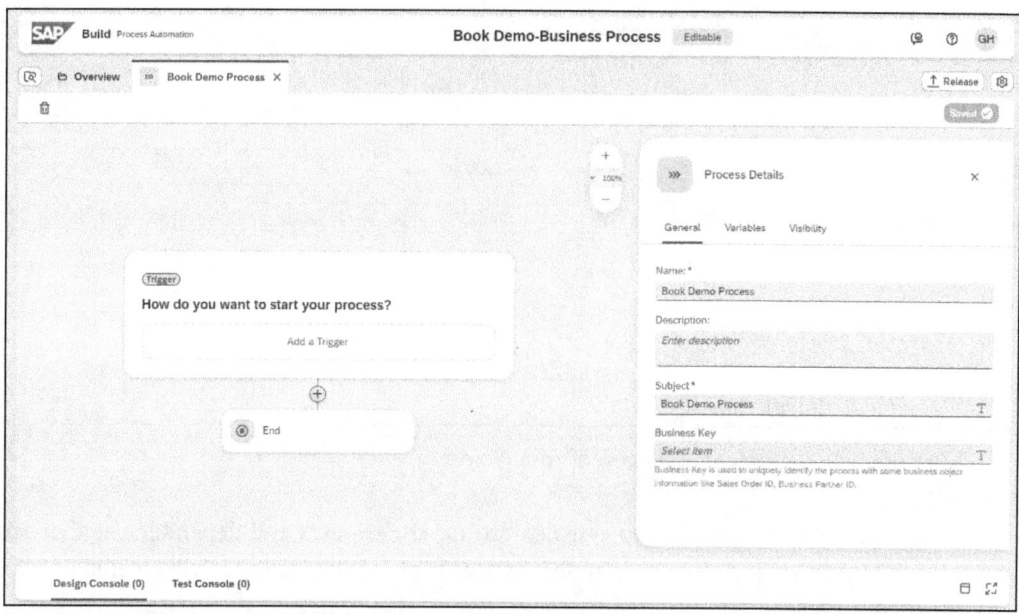

Figure 19.7 Design Console for Business Process

As this panel is opened in a new tab in the project window, you can switch back to the project overview at any time. Therefore, you only need to select the **Overview** tab that is situated on the left of the currently opened tab. Press the header of the **Overview** tab to navigate to your project artifacts, as shown in Figure 19.8.

Figure 19.9 shows a screenshot of the overview page of your project. On this page, you see the metadata of your project in the upper part of the view. Underneath, the list of artifacts inside the project is presented. In this list you always see the objects that are part of your project. These can be objects of various types, like processes, alerts, user tasks, and many more. You will get to know many of these throughout the next chapters. For the moment, you can only see one artifact in this list—the process you just created. You see the name of your newly created artifact and the description you entered. You also see the username of the last editor of the artifact and the date when the artifact was created. By clicking the menu icon on the right side of the artifact, you have the option to delete it.

19 Processes

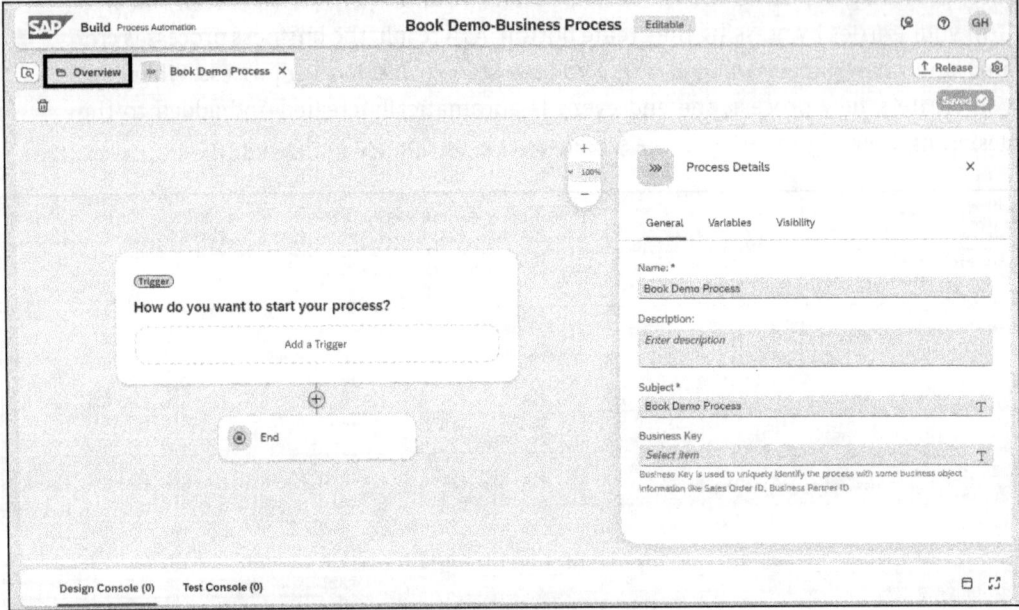

Figure 19.8 Navigate to Overview of Your Project

Apart from the artifacts, you can also browse the triggers and dependencies that are part of your project. For now, these lists are empty as you have not yet created objects of these types.

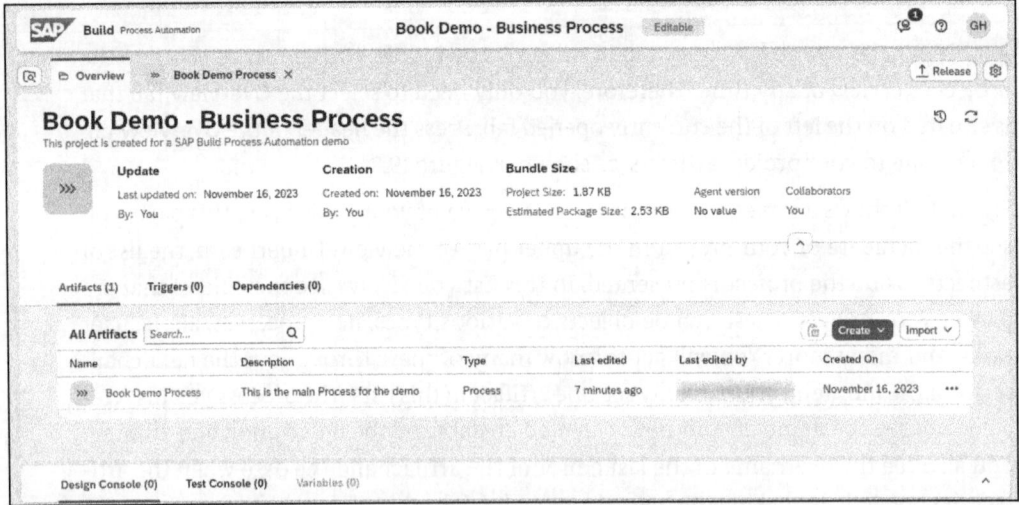

Figure 19.9 Overview Page of Your Project

By selecting the tab of your new process, you navigate back to the designing panel. In this panel, you see the process with its tasks, events, and other elements used. On the right side of this view, you see the properties panel that always shows the properties of the currently selected element of the process. The properties panel automatically adapts to the selected element in the process and shows the properties that can be set. The selected element in the designing editor can be identified by a thick border surrounding it. As you have just opened the designing editor and no element is selected yet, the process details are shown in the properties panel. So, whenever you need to view or edit the process details, you just unselect the selected element in your process by clicking on the design panel somewhere outside the process. Now, it's time to create a trigger.

19.2 Triggers

To start the business process, you need to define a trigger. A *trigger* is an event that indicates the start of the process instance. A business process can always have only one trigger defined. There are three types of triggers that can be used: form triggers, API triggers, and event triggers.

19.2.1 Form Triggers

Whenever a form is needed to enable the user to provide the request data for a business process, a form trigger can be used. In the following sections, we'll walk you through first how to define a form and then how to use it as a form trigger in the example process.

Adding a Form Trigger

To define a form as the trigger, click the **Add a Trigger** link on the start element of the process, as shown in Figure 19.10.

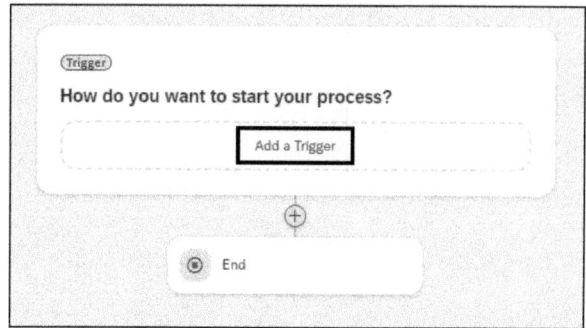

Figure 19.10 Adding Trigger

In the displayed menu, select the **Submit a Form** item, as shown in Figure 19.11.

19 Processes

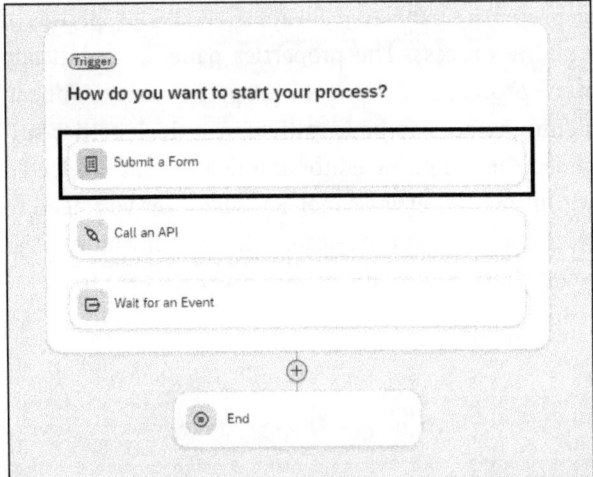

Figure 19.11 Selecting Trigger Type

Next, you can either select **Blank Form** to add a new trigger form or select one of the existing forms listed in the displayed panel. Figure 19.12 shows how to add an existing form as a trigger.

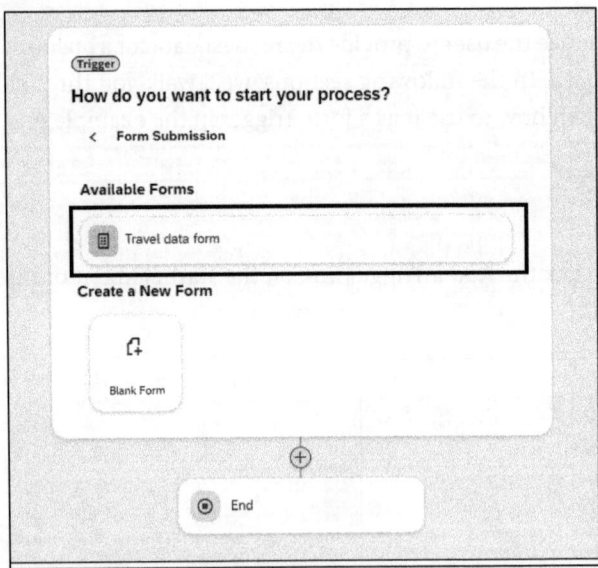

Figure 19.12 Creating Form Trigger

If you select an existing form, it's directly added to the trigger. By selecting the new form option, a new tab is opened, and you can create a form that will then be used as the trigger form. On deployment of the process, a form link will be generated that can be used to provide data for a new process instance. A launchpad configuration parameter also is generated that can be used to integrate your request form into your launchpad.

If you use a form as the trigger, the fields that are part of the form are automatically used as the variables of the process input. You can click anywhere on the canvas outside the process elements to switch to the process details and see the **Variables** created, as shown in Figure 19.13.

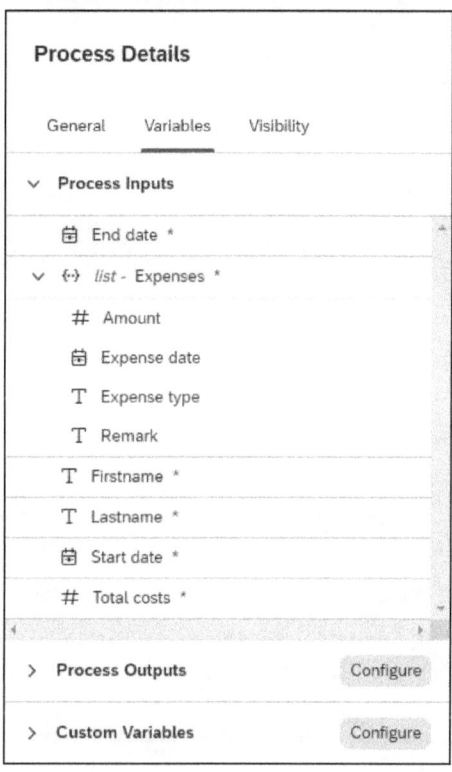

Figure 19.13 Process Input Variables Created on Adding Form Trigger

Using Form Triggers in the Example Process

In this section, you'll learn how to use a form trigger to start an instance of the example travel expenses process and enable a user to provide information about his travel. The form used in this example is created in the chapter describing the form element of a business process. To see how a form is built and configured, refer to Section 19.3. The form used in the example is described in detail in Section 19.3.2.

To add a form trigger for a business process, click the **Add a Trigger** link of its start element. In the displayed menu, select the **Submit a Form** item and select the available **Travel Data Form** as you want to use this form as the input trigger. Refer back to Figure 19.12 to see how this is done. As a result, a placeholder for the travel data form is put into the trigger element, as shown in Figure 19.14. Here you can also see that the process inputs are automatically added to the process details.

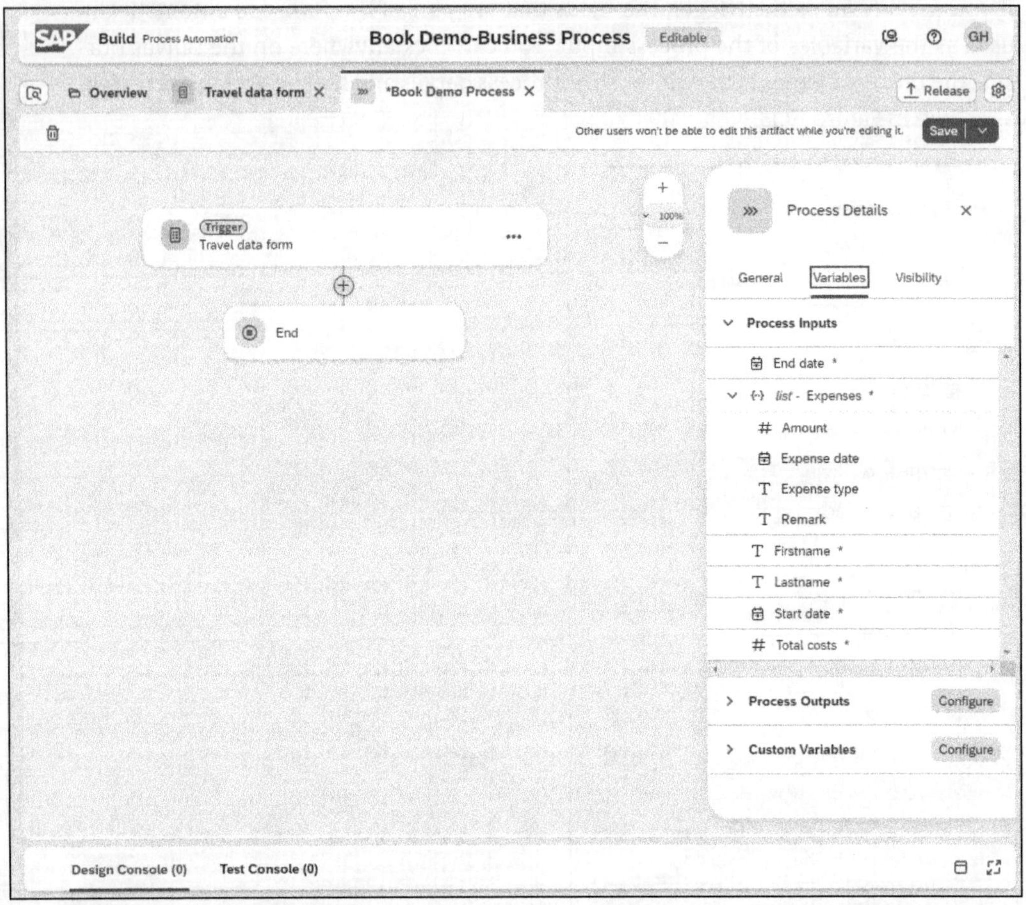

19 Processes

Figure 19.14 Form Trigger Added to Process

Save the process by clicking the **Save** button in the upper-right corner. The caption of the button is changed to **Saved**. In the next step, you need to create a release that can be deployed. Therefore, click the **Release** button in the upper-right corner above the **Save** button.

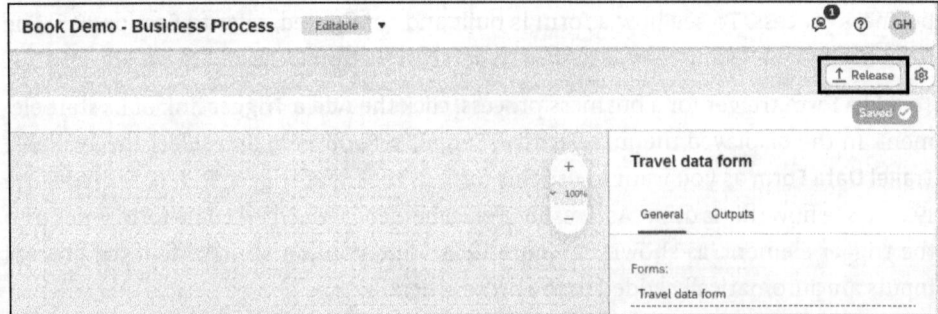

Figure 19.15 Release Button

19.2 Triggers

In the display dialog, you can enter a **Version Comment** to describe the content of the version to be released. Click the **Release** button to confirm your entry and create a release version of your project. Figure 19.16 shows this release dialogue.

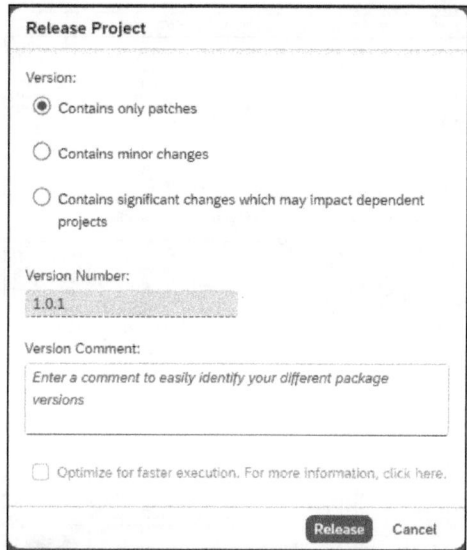

Figure 19.16 Release Dialog

The tab where you developed your process is automatically closed and the overview of your project is shown. In the header of your project, you see that you have released a version. In this view, click the **Deploy** button in the upper-right corner to deploy your project to the server.

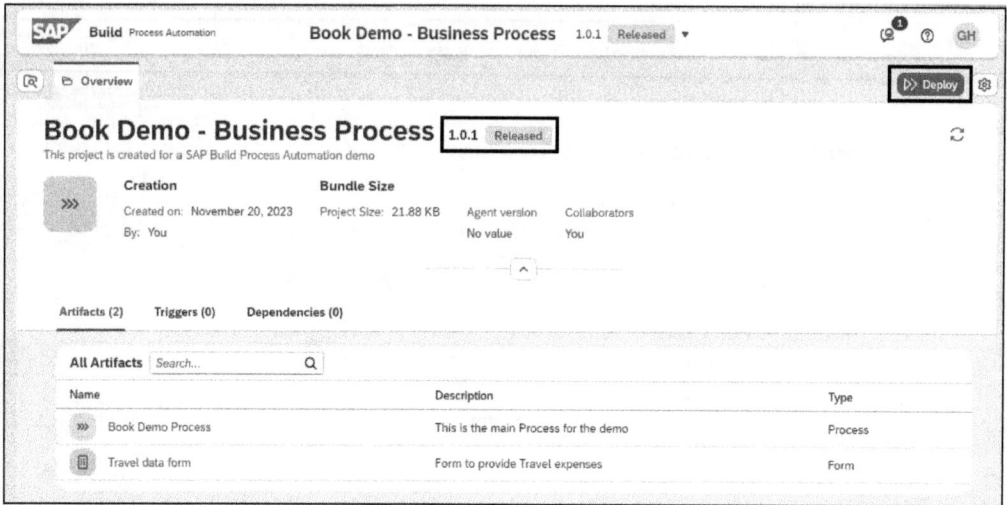

Figure 19.17 Overview of Released Project

559

On clicking the **Deploy** button, a new dialogue is opened that supports the deployment. As you deploy your process for the first time, no changes need to be done and the wizard can be followed without any action. The first page of the wizard shows the artifacts to be deployed, as shown in Figure 19.18.

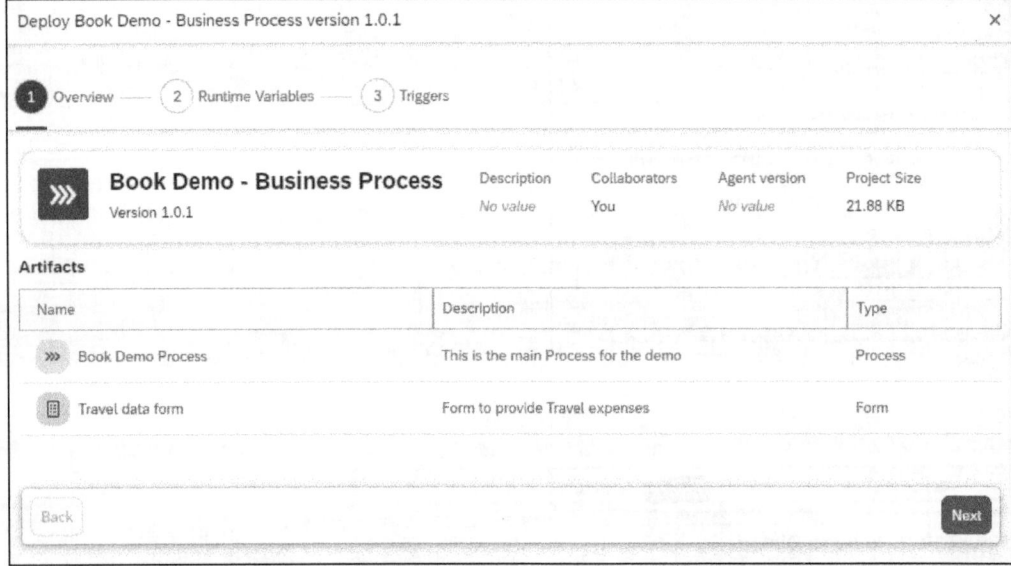

Figure 19.18 First Page of Deployment Wizard

Click **Next** to navigate to the second page. This page lists the runtime variables used in the process (see Figure 19.19). As currently no runtime variables are used, no action is needed on this page.

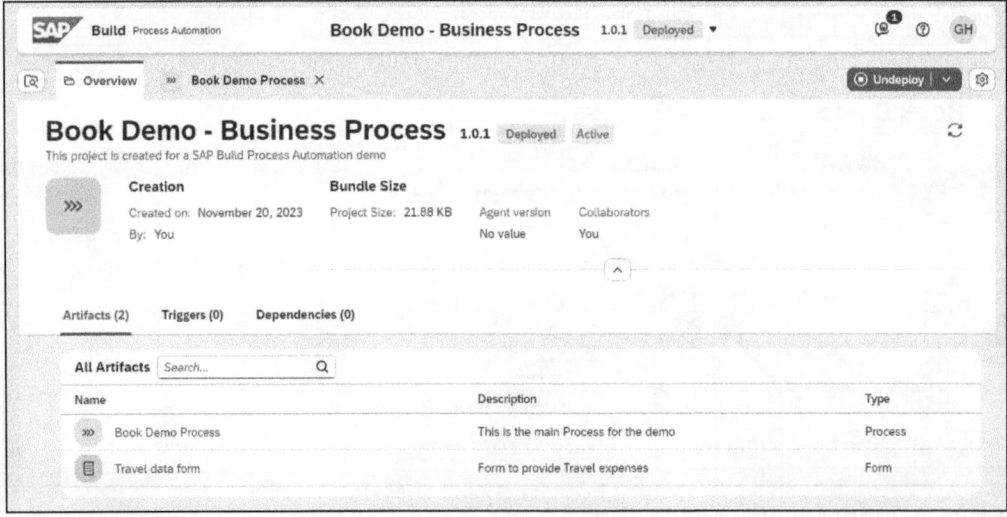

Figure 19.19 Overview of Deployed Project

19.2 Triggers

Click **Next** to proceed to the last page of the wizard. Here, the changes to existing triggers are listed. As you've created a new trigger, no changes are shown. Click the **Deploy** button to start the deployment of the project.

After the project has been deployed, you can see in the **Overview** of your project that the status has changed. The project is now deployed and active.

Navigate to the process by clicking the line representing your process in the artifacts list. In the opened process view, ensure that the deployed version is the version shown. Select the trigger form of the process to show the properties of the trigger. You can see that the **Form Link** and **Launchpad Configuration Parameter** were created, as shown in Figure 19.20.

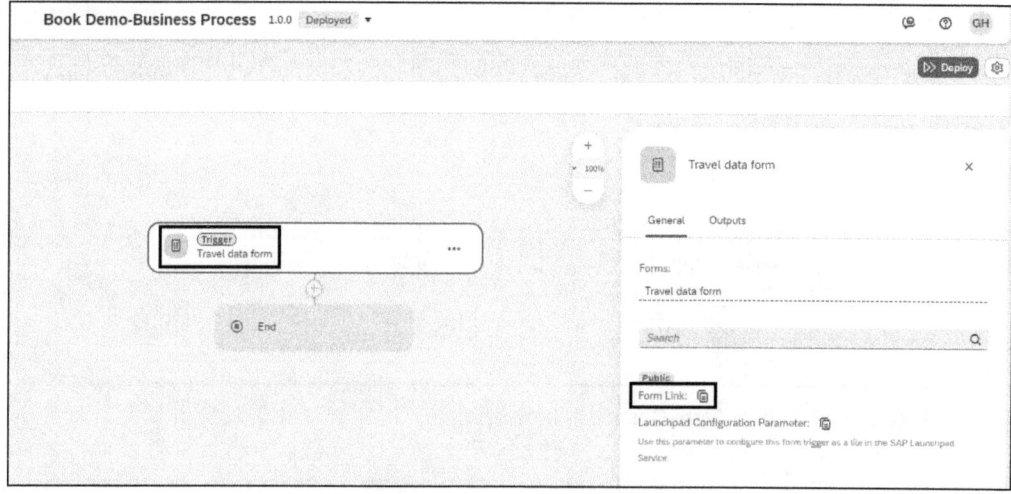

Figure 19.20 Form Link for Trigger Form

Copy the form link by clicking the 📋 icon next to the form link and paste it into a new browser tab. The form is displayed and can be used to enter the travel data and start a new process instance by clicking the submit button in the bottom-right corner. Figure 19.21 shows the generated form that can be used to trigger a process instance of your created business process.

After clicking the **Submit** button, a new process instance is started. As you currently do not have any logic implemented in the process, the data submitted by the form is fetched by the new instance and the process is immediately finished. To check if the process instance was started and worked correctly, you can look at the monitoring of the process automation. Therefore, navigate to the lobby and click the **Monitoring** menu item in the left menu.

In the **Monitoring** view, click the **Processes and Workflow Instances** tile, as shown in Figure 19.22.

19 Processes

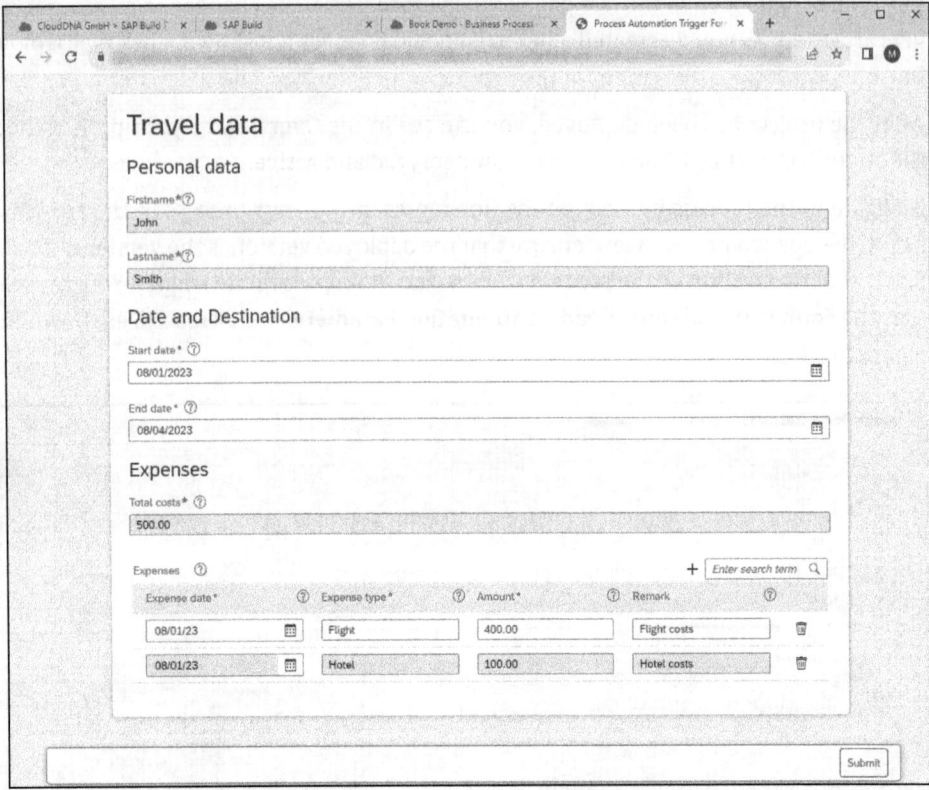

Figure 19.21 Trigger Form Opened in Browser

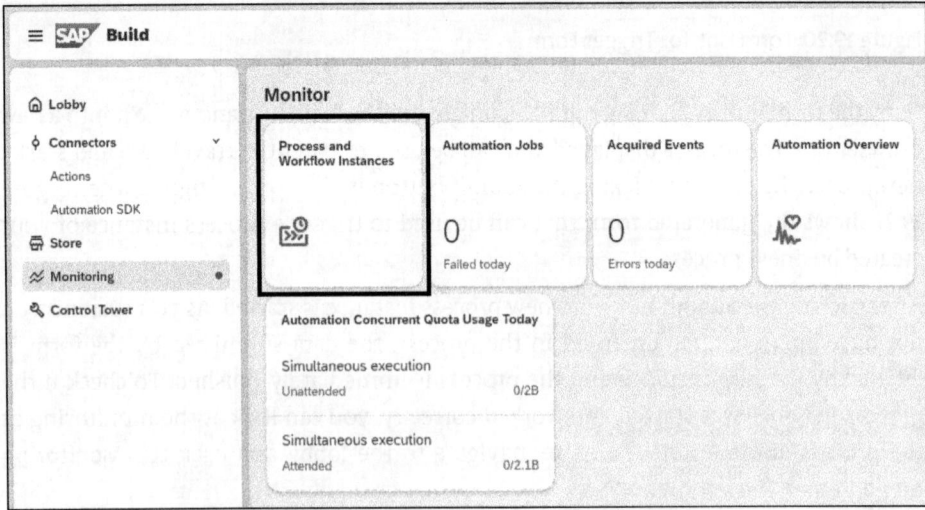

Figure 19.22 Workflow Instance Monitoring Tile

In the **Monitoring** view, use the filter for the status to show only completed workflow instances. You can also use the **Search** field to find workflow instances by the name of

19.2 Triggers

the business process. In the result list, you should find an item with the name of your workflow. Figure 19.23 shows a list of completed workflow instances in the monitoring overview.

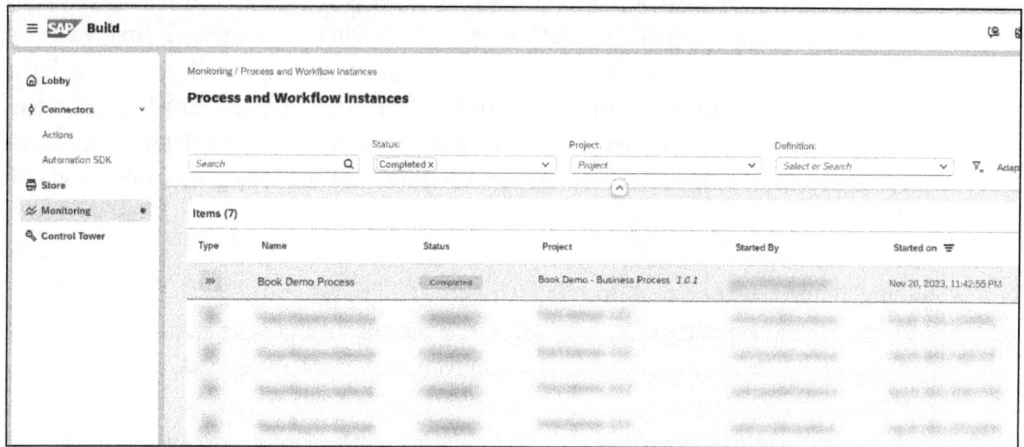

Figure 19.23 List of Completed Workflow Instances in Monitoring

Click the item that represents the instance of your process workflow. In the detail view that is opened, you can see the data that was sent from the trigger form and used to start a new workflow instance. Figure 19.24 shows an example of a workflow instance triggered by the form.

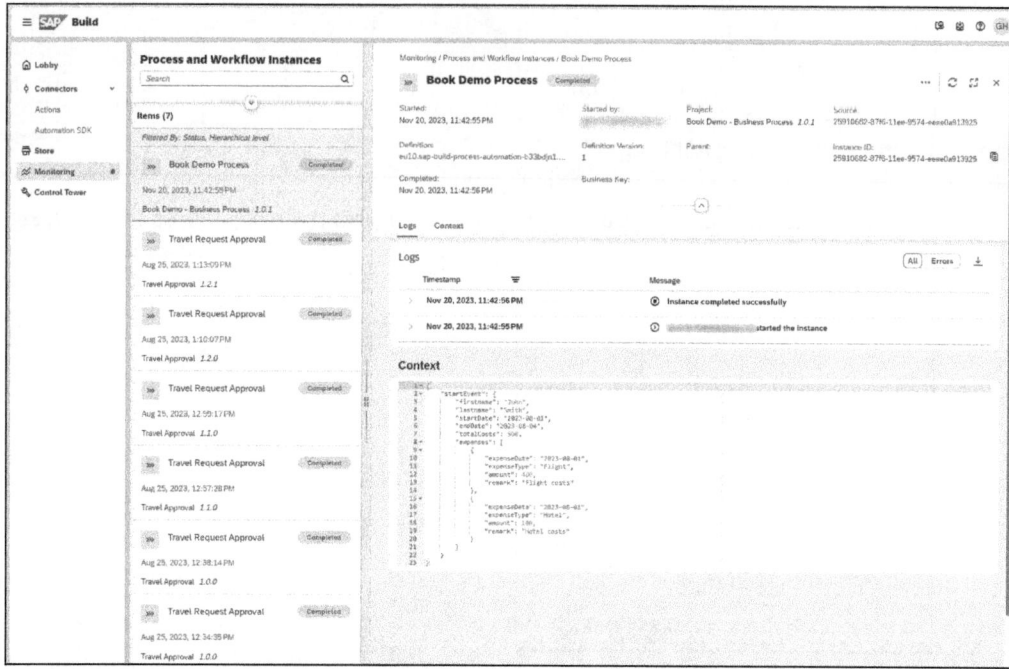

Figure 19.24 Monitoring Details of Process Instance

563

19.2.2 API Triggers

An application programming interface (API) allows two applications to share information within and across organizations. SAP Build Process Automation provides API triggers for business processes. This means that process instances can be triggered by sending data to the process automation. In the following sections, we'll first walk you through adding an API trigger to your project in general and then walk you through the process in practice for our running example. We'll focus on the following key steps: adding properties to an API trigger, creating and deploying a new release, creating an API key, creating an instance and service binding, and triggering the process via an HTTP request.

Adding an API Trigger

To add a new API trigger, select the **Add a Trigger** link on the start element of the process, as shown in Figure 19.10. In the displayed menu, select the **Call an API** item, as shown in Figure 19.25.

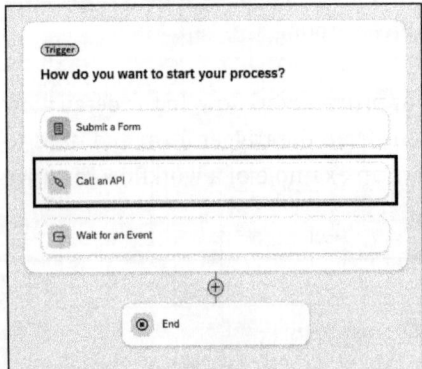

Figure 19.25 Selecting API Trigger

Finally, provide a name and a description for the trigger to be created, as shown in Figure 19.26.

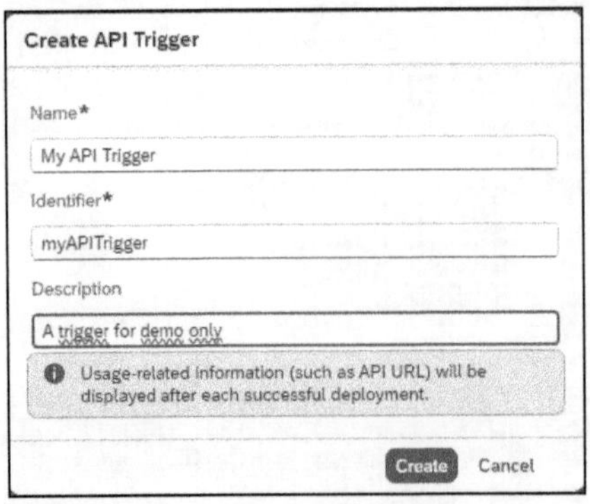

Figure 19.26 Creating API Trigger

Whereas a form trigger automatically creates the process input variables, you must create these variables manually whenever you use an API trigger. This can be done in the **Variables** section of the **Process Details** area. Click anywhere in the canvas to show the process details and select the **Variables** tab. In the **Process Inputs** section, click the **Configure** button, as shown in Figure 19.27.

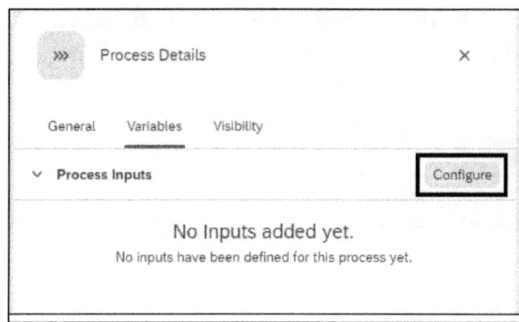

Figure 19.27 Configure Process Input

In the displayed dialog, you can add the variables needed as an input for your business process, as shown in Figure 19.28.

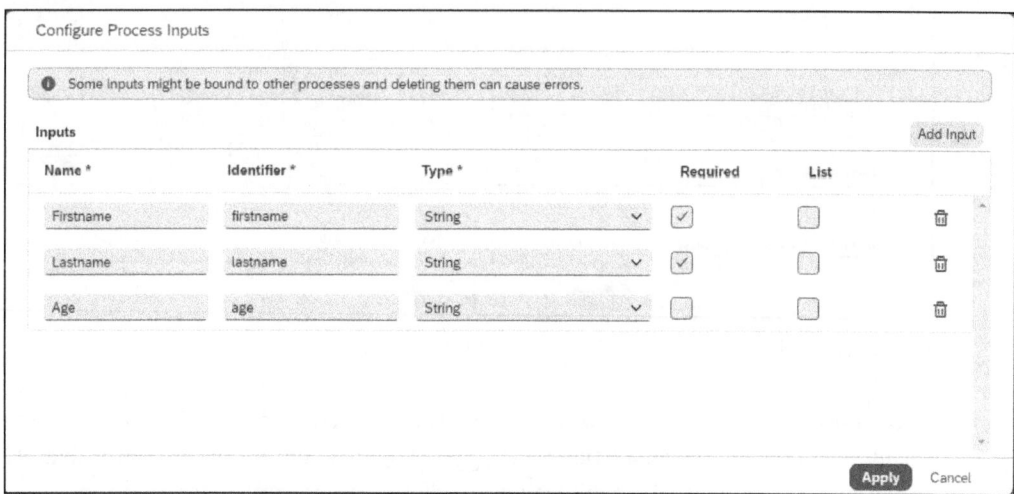

Figure 19.28 Process Inputs for API Trigger

After the process has been released and deployed, the API trigger can be used to start a new instance of the process by sending a JSON-formatted message to the endpoint that was generated during deployment. In the walkthrough for the example process in the following sections, you will see in detail how this is done.

Adding Properties to an API Trigger

In Section 19.2.1, you learned how to trigger a process instance using a form. In this section, we want to show you how to start the process instance by using an API trigger to enable an external system to start the process instance by sending a message.

Navigate to the lobby and create a new process by clicking the **Create** button on the right side above the artifacts list and selecting the **Process** menu item. In the open dialogue, enter the values given in Table 19.1 and click the **Create** button to confirm your entries.

Field	Value
Name	"Book Demo Process API"
Identifier	"bookDemoProcessAPI"
Description	"Example process using an API trigger"

Table 19.1 Properties of New Process

Figure 19.29 shows a screenshot of this dialogue.

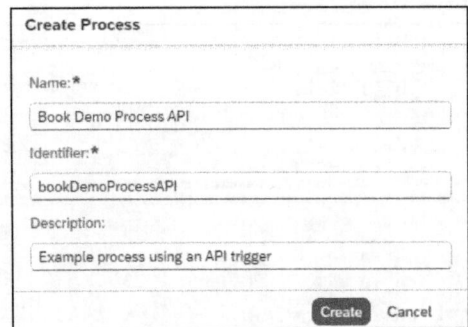

Figure 19.29 Dialogue for Properties of New Process

In the new process, create a new API trigger as explained in the previous section. As a result, the dialogue to create a new API trigger is opened. Enter the values in Table 19.2 in the corresponding fields of the dialog and click create to confirm your entries.

Field	Value
Name	"Travel Data API Trigger"
Identifier	"travelDataAPITrigger"
Description	"API trigger to send travel data"

Table 19.2 Properties of API Trigger

19.2 Triggers

Figure 19.30 shows a screenshot of the dialogue to create the API trigger.

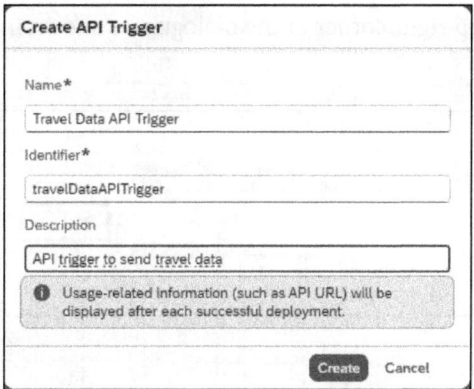

Figure 19.30 Properties of API Trigger

Whenever you use an API trigger, the process input variables need to be set manually. To do so, click anywhere in the canvas of your process development view and select the **Variables** tab in the **Process Details** panel. Click the **Configure** button in the **Variables** section, as shown in Figure 19.31.

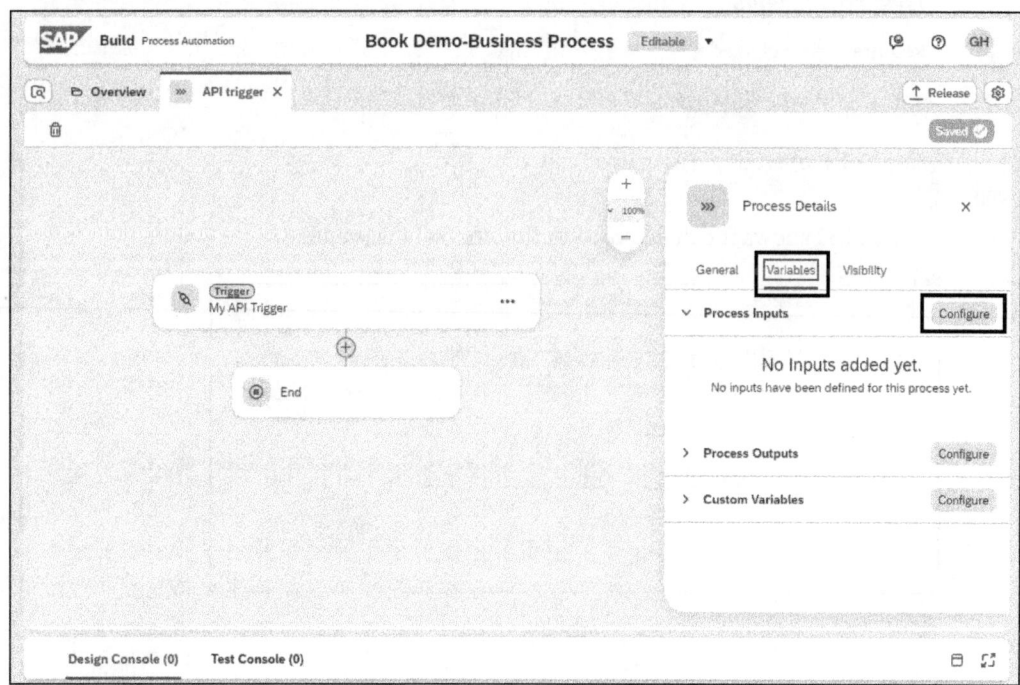

Figure 19.31 Configuration of Process Inputs

19 Processes

In the displayed dialog, you can now create the structure of the data needed in the process. The fields added to your process input will define the interface of the API. To add a field, click the **Add Input** button in the top-right corner of the dialogue, as shown in Figure 19.32.

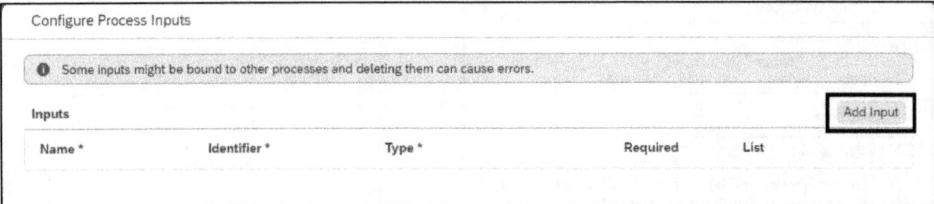

Figure 19.32 Adding new Input Field

In the line that is added to the list of the inputs, enter the values according to Table 19.3.

Field	Value
Name	"Firstname"
Identifier	"firstname"
Type	String
Required	Selected
List	Not selected

Table 19.3 Properties of Firstname Field

Figure 19.33 shows the dialogue after the first field was added.

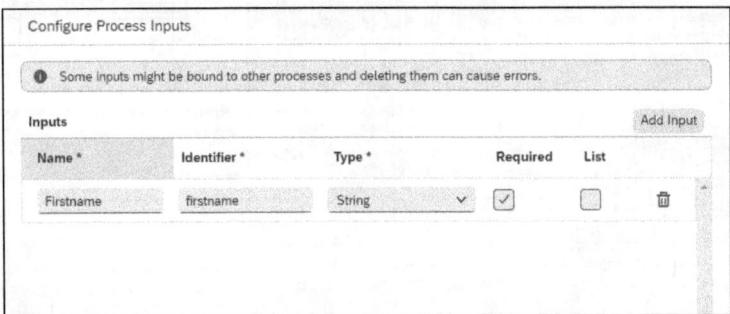

Figure 19.33 Dialog with First Field Added

Repeating this action, enter the fields listed in Table 19.4.

19.2 Triggers

Field Name	Field 2	Field 3	Field 4	Field 5
Name	"Lastname"	"Start date"	"End date"	"Total costs"
Identifier	"lastname"	"startDate"	"endDate"	"totalCosts"
Type	String	Date	Date	Number
Required	Selected	Selected	Selected	Selected
List	Not selected	Not selected	Not selected	Not selected

Table 19.4 Fields of Process Input

Your process input dialogue should look like the screenshot shown in Figure 19.34. Press the **Apply** button to confirm your entries.

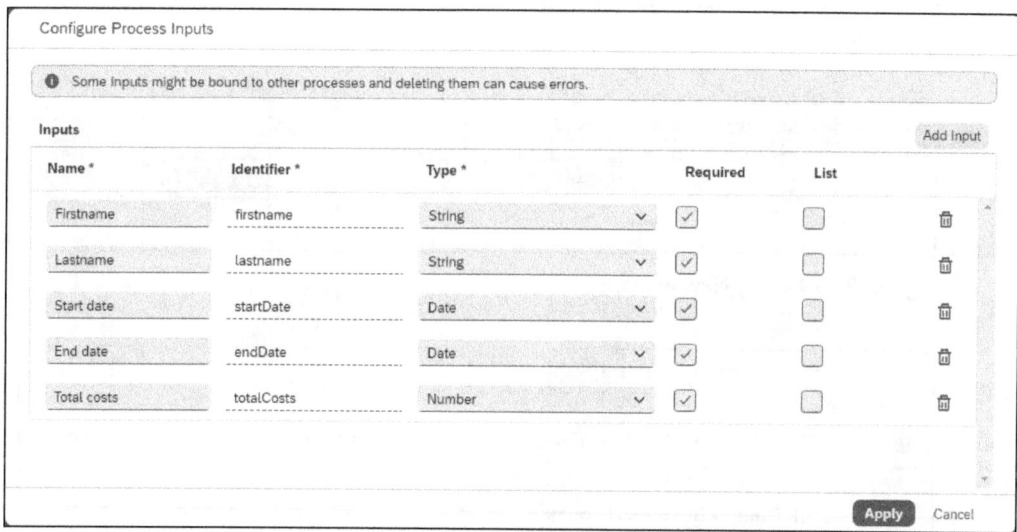

Figure 19.34 Process Inputs Dialogue after Adding Fields

Creating and Deploying a New Release

As the API trigger is now created, it must be released and deployed to call it via an HTTP POST request. To release a new version of your project, click the **Release** button in the top-right corner. In the displayed dialog, add a **Version Comment** and click the **Release** button to create the new version, as shown in Figure 19.35.

The new version of your project can now be deployed to the server. Therefore, click the **Deploy** button in the top-right corner of the view, as shown in Figure 19.36.

As a result, the deployment wizard is displayed in a dialog. In the first step, the artifacts that are part of the project are displayed. Click the **Next** button to proceed to the next step of the wizard, as shown in Figure 19.37.

19 Processes

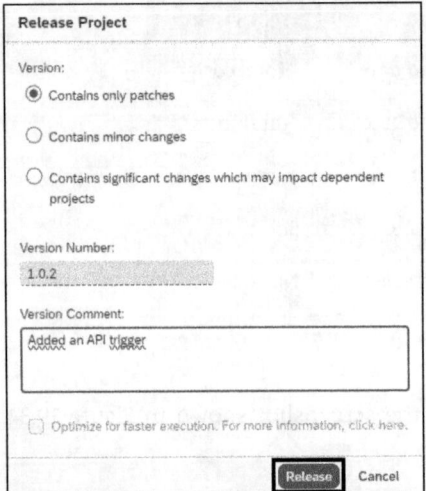

Figure 19.35 Release New Version

Figure 19.36 Deploy New Release

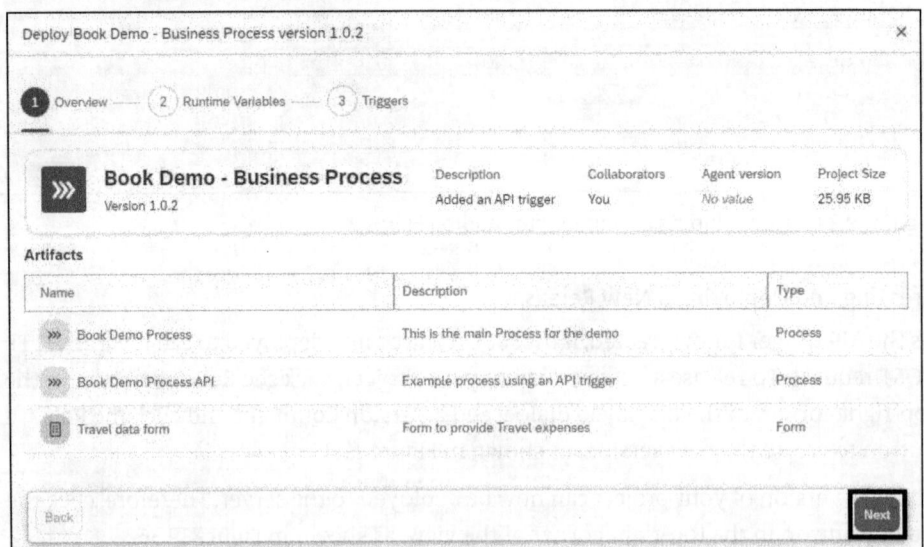

Figure 19.37 Artifacts to Be Deployed in New Release

The next step of the wizard displays the runtime variables of the project. As you aren't using any variables yet, press **Next** to proceed to the final step. In this step, the triggers

19.2 Triggers

are displayed. You can see that your new API trigger will now be added due to the deployment. Click the **Deploy** button to start the deployment of the new version.

During the deployment, a new endpoint was created for the API trigger. This endpoint can be found in the monitoring section of SAP Build. Navigate to the landing page of the lobby and select the **Monitoring** item in the left menu. In the monitoring view, you can find a tile with the caption **Triggers** in the **Manage** section, as shown in Figure 19.38. You can see that a trigger is available. Click the tile to navigate to the overview of triggers deployed on the server.

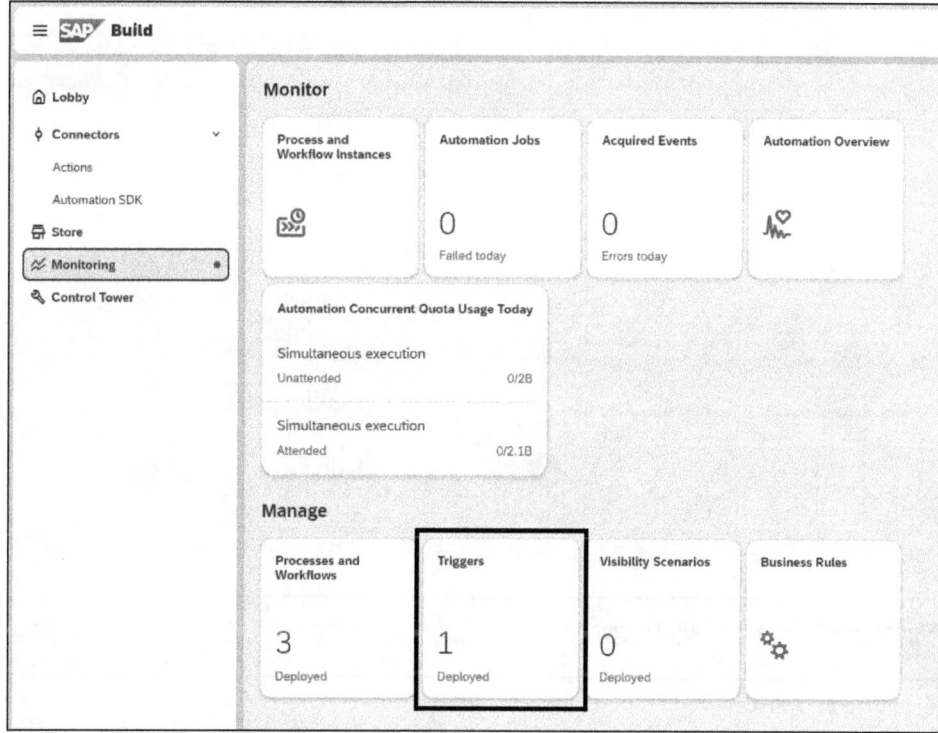

Figure 19.38 Triggers Available on Server

You can see that your trigger was added to the list. Click the three-point menu in the action column of the line item that represents your trigger, then select **View** to display the details of your trigger, as shown in Figure 19.39.

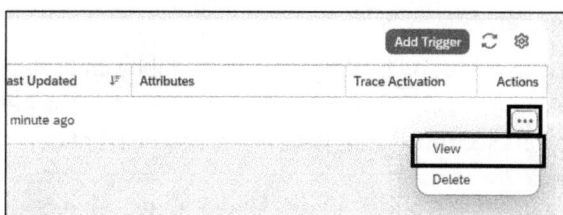

Figure 19.39 Display Details of API Trigger

571

19 Processes

As a result, a dialogue is opened that shows details of the API trigger. In this dialog, you can find the endpoint **URL** that can be used to call the trigger via HTTP POST request, and an example for a request used to call the trigger is displayed, as shown in Figure 19.40.

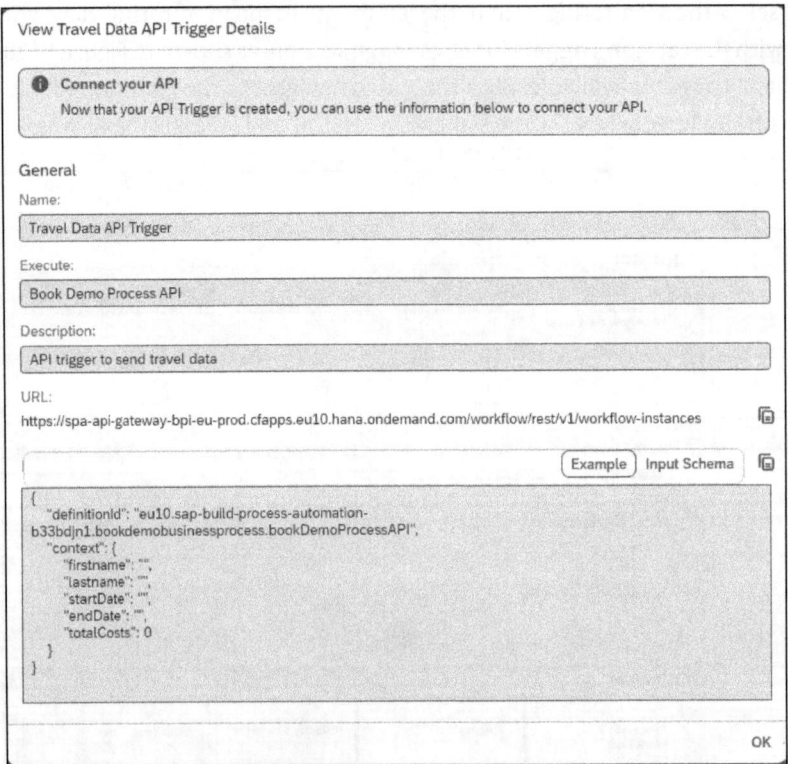

Figure 19.40 Details of API Trigger

Creating an API Key

To call the API trigger of a business process, the user or the program executing the call needs to be identified and authorized. Therefore, API keys are used. An *API key* is a unique key that is used to identify, authenticate, and authorize the calling service. When an API key is passed to an API server, this server checks the validity of the API key and identifies the authorizations of the calling service. SAP Build Process Automation also uses API keys for authentication and authorization. Whenever a business process should be started using an API trigger, an API key is needed.

To create an API key in SAP Build, navigate to the landing page of the lobby and select the **Control Tower** menu item in the left menu. In the **Control Tower** view, select the **API Keys** tile that can be found in the **Others** section. Figure 19.41 shows a screenshot of the **Control Tower** view.

19.2 Triggers

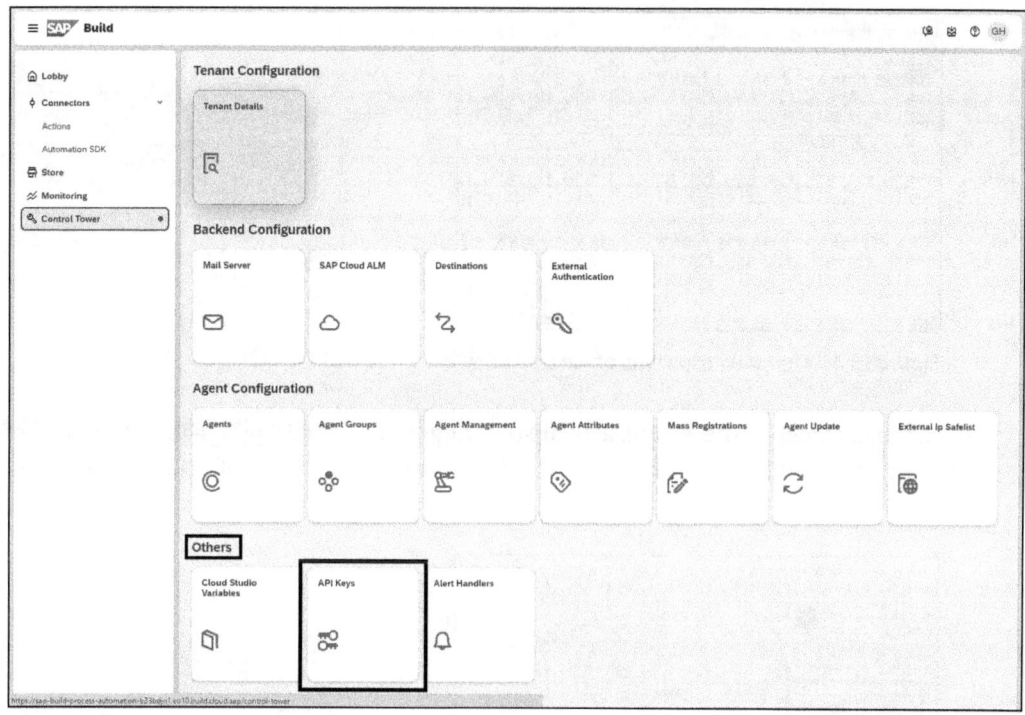

Figure 19.41 Tile for API Keys

The list of created API keys is displayed. In this case, there are no API keys created yet, so the list is empty. To create a new API key, click the **Add API Key** button at the bottom right of the list, as shown in Figure 19.42.

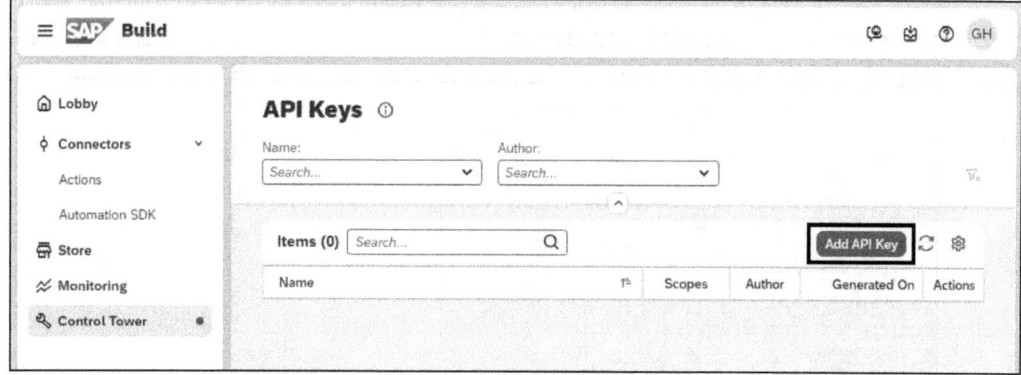

Figure 19.42 Creating New API Key

In the displayed dialog, enter a **Name** and a **Description** for your new key, then click the **Next** button to confirm your entries, as shown in Figure 19.43.

573

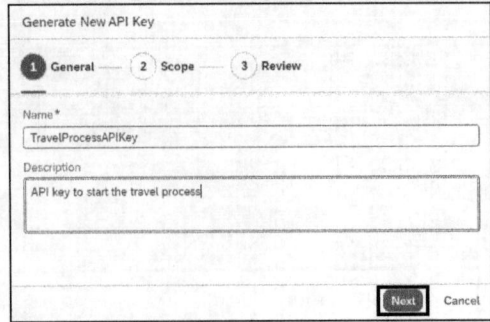

Figure 19.43 General Properties of New API Key

In the next step of the wizard, activate the **trigger_read** and **trigger_execute** options of the scope runtime. Again, click the **Next** button to confirm your selections, as shown in Figure 19.44.

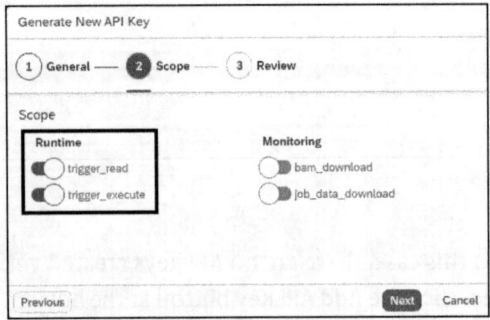

Figure 19.44 Setting Scope of API Key

In the third step of the wizard, the summary of the properties for the new API key is displayed and can be reviewed. If the data you entered is correct, click the **Add** button to finally create a new API key, as shown in Figure 19.45.

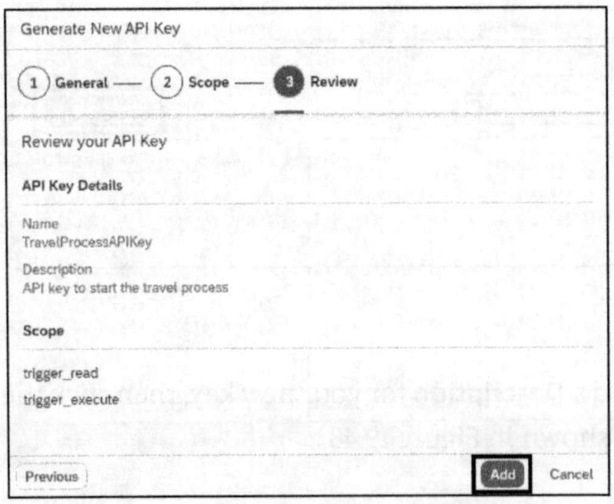

Figure 19.45 Summary of API Key Details

19.2 Triggers

As a result, a new API key is created and displayed in the next dialog. Be aware that this API key is only shown once. Copy the key by clicking the **Copy** button, and paste it to a safe place as you will need it later to call the API trigger, as shown in Figure 19.46.

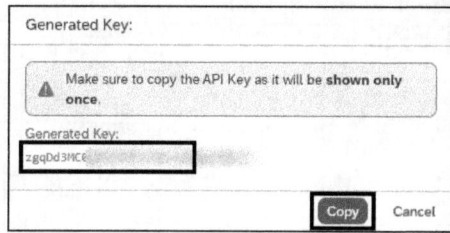

Figure 19.46 New API Key

After you copy the API key, you'll be returned to the list of API keys, where the new key is displayed, as shown in Figure 19.47.

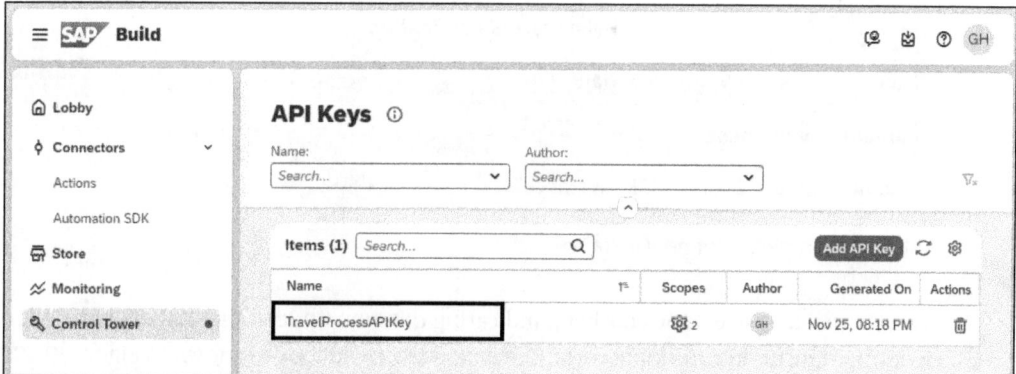

Figure 19.47 List with Your New API Key

Creating Instance and Service Bindings

To access the SAP Build Process Automation service from an external application, you need to create the service binding where credentials and the links for access are created.

This needs to be done in the subaccount of SAP BTP itself. Navigate to the subaccount where the SAP Build Process Automation service is subscribed and open the **Instances and Subscriptions** view. In the top-right corner of this view, click the **Create** button to add a new instance, as shown in Figure 19.48.

As a result, a dialog is displayed to specify the instance to be created. Fill in the fields according to Table 19.5.

575

19 Processes

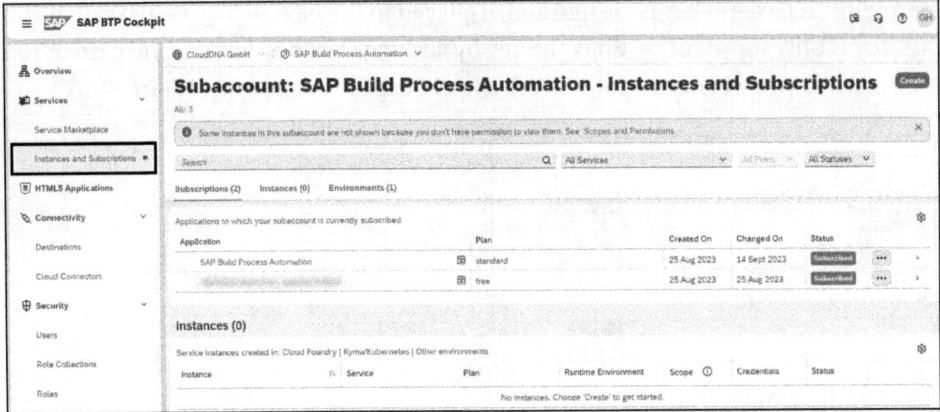

Figure 19.48 Create Instance

Field Name	Value
Service	SAP Build Process Automation
Plan	Standard
Runtime Environment	Other
Instance Name	"SPA-instance"

Table 19.5 Properties of new Instance

You also need to select the checkbox indicating that you understand that enabling a service might result in additional costs. The dialogue should look as shown in Figure 19.49.

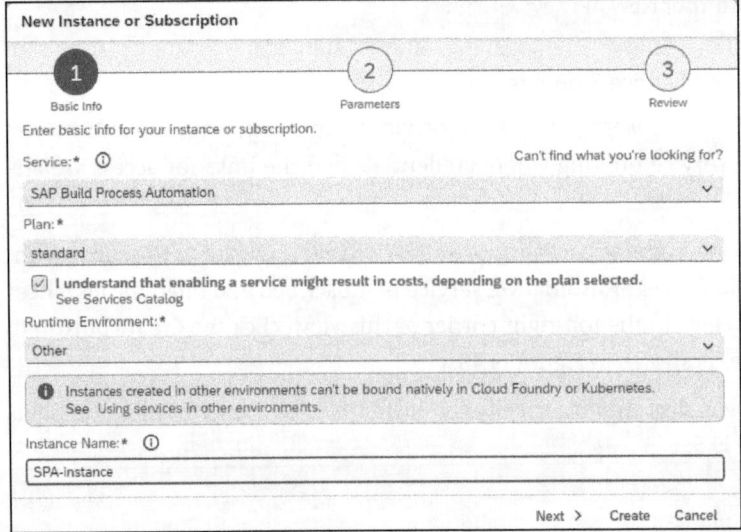

Figure 19.49 Dialogue to Create New Instance

19.2 Triggers

Click the **Create** button at the bottom of the dialog to confirm your entries. As a result, the new instance is now displayed in the overview of the instances and subscriptions in your subaccount, as shown in Figure 19.50.

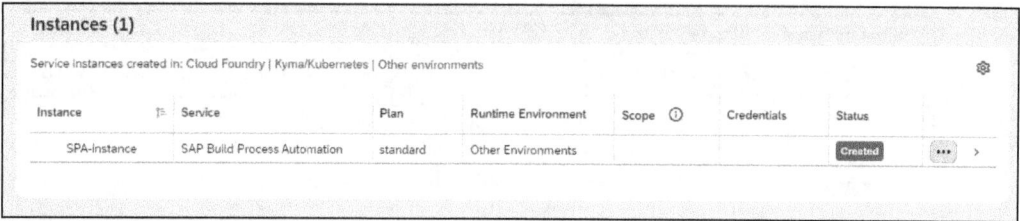

Figure 19.50 New Instance

The list entry for the new instance provides a three-point menu button. Click this button and select **Create Service Binding** to create the service binding needed for the access. Figure 19.51 shows the menu items provided.

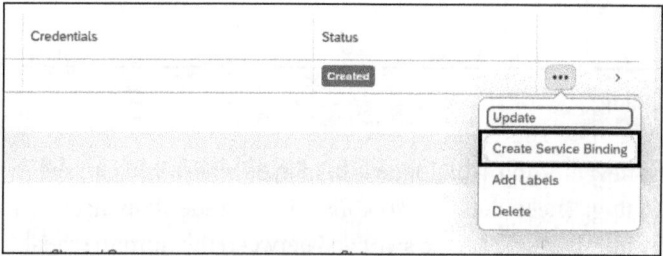

Figure 19.51 Function Menu of New Instance

In the display dialog, enter a **Binding Name** and click the **Create** button to confirm your entry, as shown in Figure 19.52.

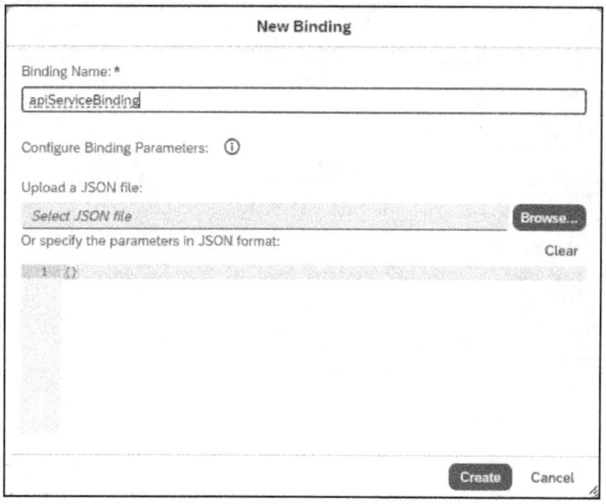

Figure 19.52 Creating New Service Binding

577

19 Processes

As a result, the service binding is created, and the view is split with the details of your instance shown on the right side (see Figure 19.53). Here you can see the entry of the new service binding you just created.

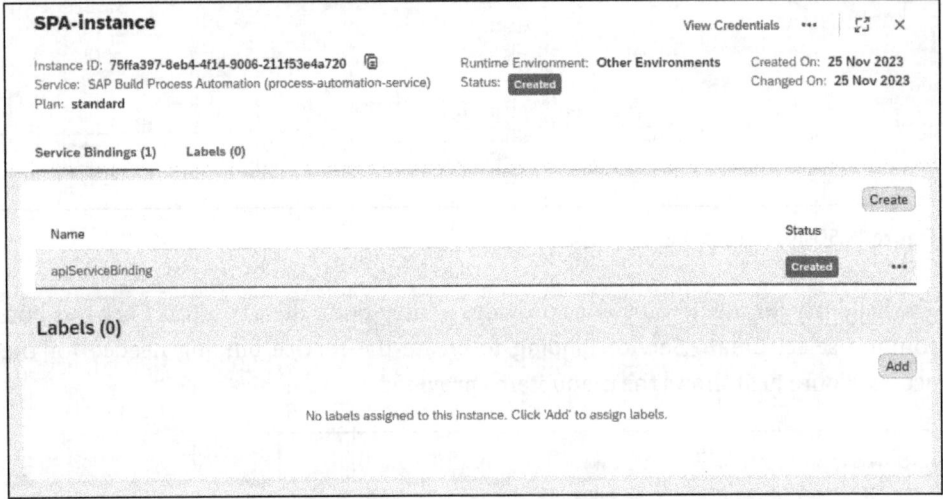

Figure 19.53 New Service Binding

By clicking the name of the new binding, a dialogue is displayed where you can see the details created for this binding. These details also contain the credentials needed to access the service. The view in this dialog can be switched between the human readable form or a representation in JSON format. Figure 19.54 and Figure 19.55 show the details dialogue in both forms.

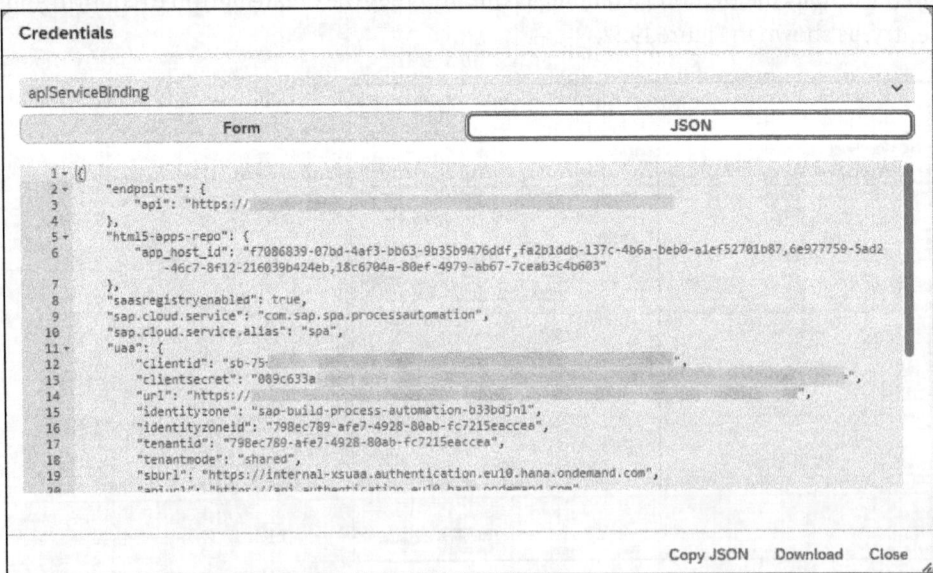

Figure 19.54 Binding Details in JSON Format

19.2 Triggers

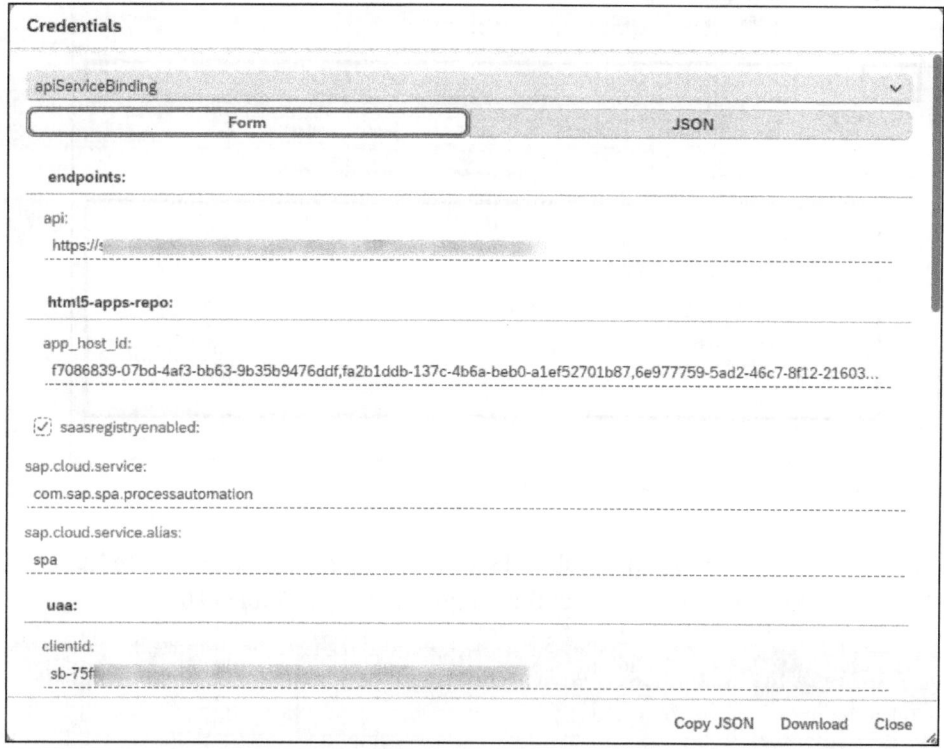

Figure 19.55 Binding Details Displayed in Form

Now, all prerequisites are done, and you can proceed to create an HTTP request to trigger your workflow from an external application. As long as you have access to your SAP BTP subaccount, you can access this information any time you need it.

Triggering the Process via an HTTP Request

Postman is an application created for developers to design, build, and test APIs. In this example, we'll use Postman to send an HTTP request to SAP Build Process Automation to call the API trigger and start a new instance of the workflow.

At the beginning of the example in this chapter, you deployed a business process including the API trigger. After deployment, a URL representing this trigger and an example message to be sent was created. You can see a screenshot of these details in Figure 19.40. Now you need these details to create a valid request in Postman. Starting from the lobby, navigate to **Monitoring** and click the **API Triggers** tile. In the list of the triggers, view the details. Copy the URL and the example message to your new Postman request. Set the type of the request to **POST**. Finally, enter data for the fields to be provided. Your final request should look like the screenshot shown in Figure 19.56.

19 Processes

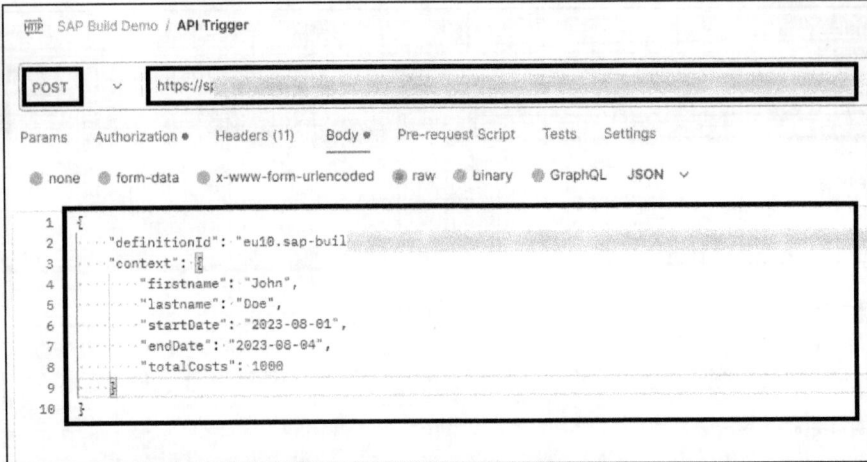

Figure 19.56 Request in Postman

Now switch to the **Authorization** tab and set the **Authorization Type** to **OAuth 2.0**. In the **Configure New Token** section, enter the values according to Table 19.6.

Field Name	Value
Token Name	"API token".
Grant Type	Client Credentials.
Access Token URL	From your service binding, enter the URL that can be found in the uaa/url JSON path; see Figure 19.54. Add the /oauth/token path to the URL.
Client ID	From your service binding, enter the URL that can be found in the uaa/clientid JSON path; see Figure 19.54.
Client Secret	From your service binding, enter the URL that can be found in the uaa/clientsecret JSON path; see Figure 19.54.
Scope	Leave this field blank.
Client Authentication	Send as Basic Auth Header.

Table 19.6 Properties for OAuth 2.0 Authorization

> **Note**
>
> Ensure that you add the /oauth/token path to the access token URL as this is the correct endpoint for the token service that will respond with an access token once it is called.

Figure 19.57 shows a screenshot of the configuration for a new token.

19.2 Triggers

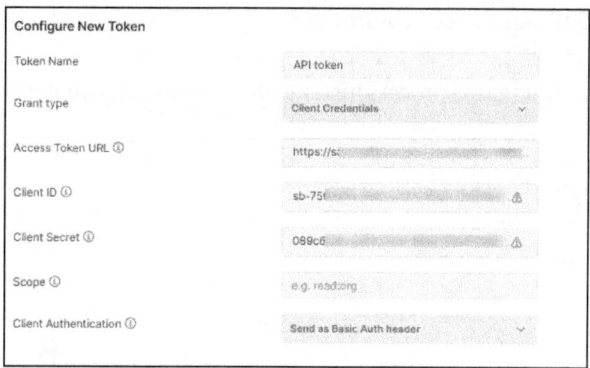

Figure 19.57 Configuration for New Token

Scroll to the bottom of the configuration view and click the **Get New Access Token** button. Be sure to add the token to your request by clicking the **Use Token** button, as shown in Figure 19.58.

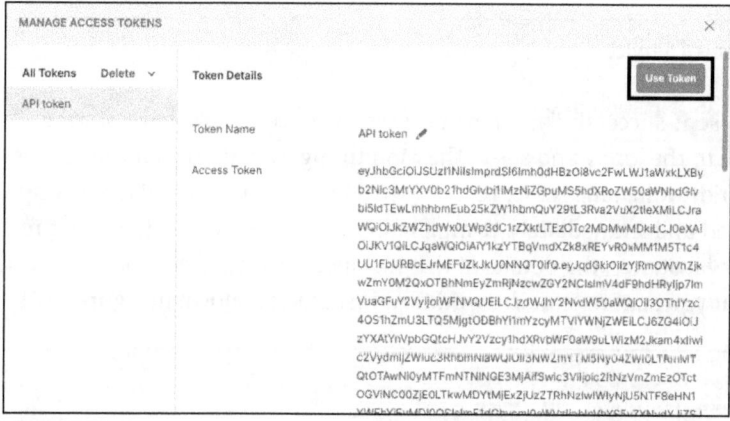

Figure 19.58 Adding Token to Your Request

Now select the **Headers** section of your request and enter a new header item using the API key that was generated in the Section section. Therefore, enter a random name as the key for the header item, and as the value use the API key that was created in the service instance inside SAP BTP.

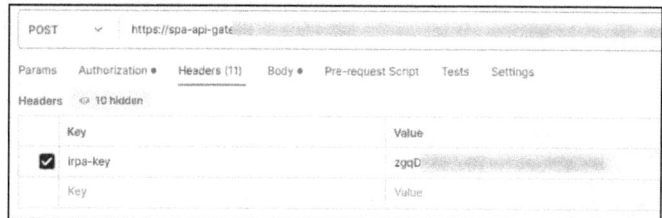

Figure 19.59 Sending API Key via Request Header

581

19 Processes

Now you can send the request to SAP Build Process Automation by clicking the **Send** button next to the request URL. As a result, the service that was called responds with a JSON message that includes metadata of the created process instance, as shown in Figure 19.60.

```
Body  Cookies  Headers (20)  Test Results

Pretty  Raw  Preview  Visualize  JSON

1  {
2      "id": "bfe0b216-8f9f-11ee-8084-eeee0a914aa1",
3      "definitionId": "eu10.sap
4      "definitionVersion": "1",
5      "subject": null,
6      "status": "RUNNING",
7      "businessKey": null,
8      "parentInstanceId": null,
9      "rootInstanceId": "bfe0b216-8f9f-11ee-8084-eeee0a914aa1",
10     "applicationScope": "own",
11     "projectId": "eu10.sap-buil
12     "projectVersion": "1.0.2",
13     "startedAt": "2023-11-30T16:44:37.681Z",
14     "startedBy": "sb-clone-0b9866cb-1ce3-44f9-9e24-ab7aaa38041f!b120249|workflow!b10150",
15     "completedAt": null
16  }
```

Figure 19.60 Response of Service

As the request was sent successfully, a new process instance should have started. To prove that, navigate to the lobby and select the **Monitoring** item in the left menu. Click the **Process and Workflow Instances** tile. You can see that a new instance of the example workflow was started and immediately completed. By clicking the line item that represents the workflow instance, the details of this instance are displayed. There you can see the message that you sent to start the workflow instance, as shown in Figure 19.61.

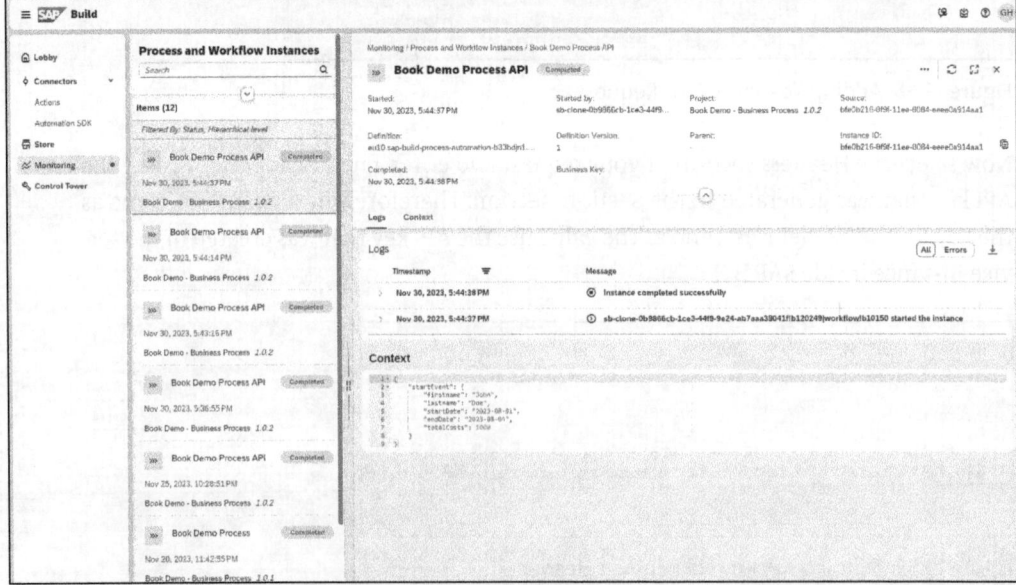

Figure 19.61 Monitoring for Started Workflow Instance

19.2.3 Event Triggers

If you use an SAP S/4HANA Cloud system and you have subscribed to and set up the SAP Event Mesh service in the SAP BTP cockpit, then you can use events raised by the SAP S/4HANA Cloud system to trigger processes in SAP Build Process Automation. Once an event trigger is added to a process, the process automation listens to the subscribed queue of the event mesh and triggers a new process instance whenever an appropriate event is raised. As part of the event, a message based on a predefined structure is sent to the process as the incoming content.

To use an event trigger to start a process instance, create a new process artifact in the overview of the example project. In the design console, you can see the start event and the end event as the only parts that build the process at the beginning. In the start event, click the **Add a Trigger** link to select the type of trigger to be used. Select **Wait for an Event**, as shown in Figure 19.62.

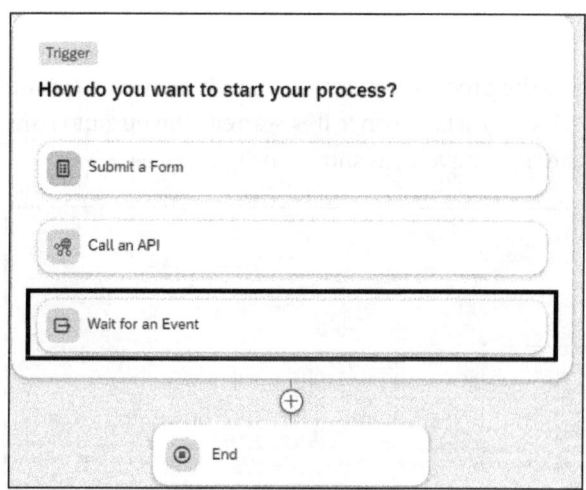

Figure 19.62 Adding Event Trigger

In the displayed dialog, you can define the event that is used to trigger new instances of the process. Enter your data according to Table 19.7.

Field Name	Description
Name	The name of the event trigger.
Executes	Set automatically. The name of the current process. This field is read-only.
Add Description	You can add an optional description for the trigger.
Event Object	A predefined list of event objects provided.
Event	Available events for the selected event object.

Table 19.7 Fields to Be Provided for Creating Event Trigger

19 Processes

Figure 19.63 shows a screenshot of the dialog to create an event trigger. Once you have provided the data, click the **Create** button and the event trigger is added to your process.

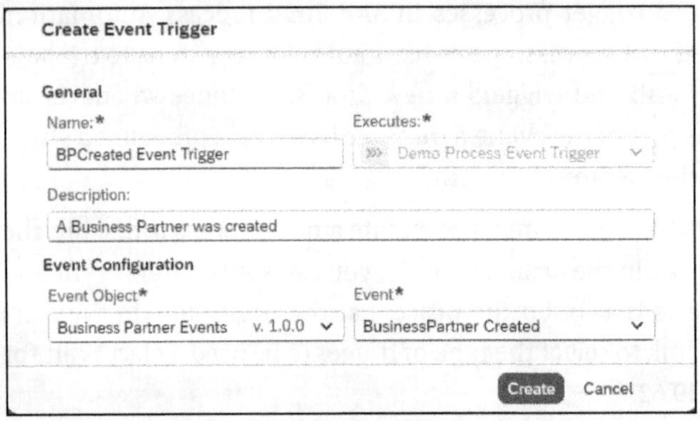

Figure 19.63 Create Event Trigger Dialog

Once the event trigger was added to the process, you can see that it delivers a certain output, which is available for the process instance once it is started. The output is displayed in the properties panel of the trigger event, as shown in Figure 19.64.

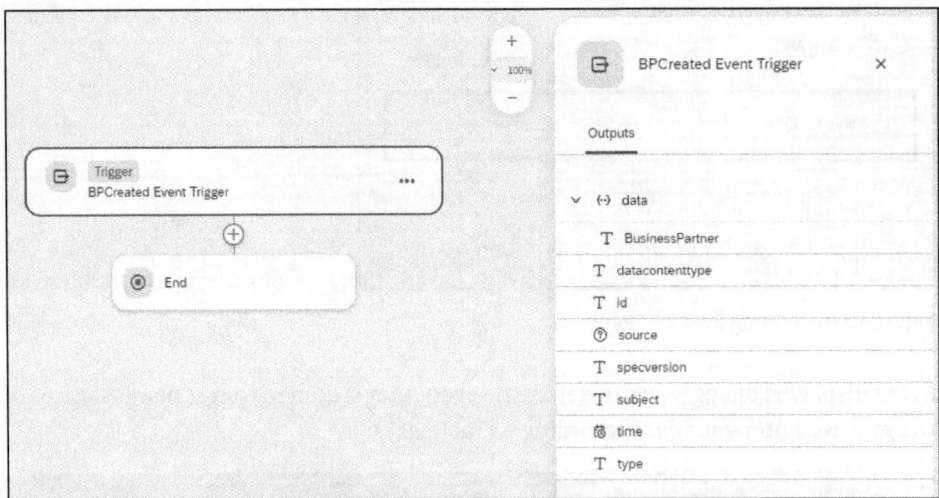

Figure 19.64 Output of Event Trigger

Now you've learned how to configure and use the options that are available to start a workflow instance. In the next section, we'll show you how to create a form and how to use the different field types that are provided.

19.3 Forms

In SAP Build Process Automation, you can create interactive forms to collect and share information throughout the process. These forms can be used as triggers or during the process as additional steps. Forms can include simple data like textual information or numeric values, but you can also use tables to capture lists. Finally, forms can be used to provide information for approval. In this section, you'll learn how to create a form element and then use a form in the running example process. We'll also cover a special type of forms: approval forms.

19.3.1 Creating a New Form

There are two options to create a new form. In the **Overview** page of your project, you can use the **Create** menu to add a new form. To do so, click the **Create** button at the top of your artifacts list and select the **Form** option from the elements list as shown in Figure 19.65.

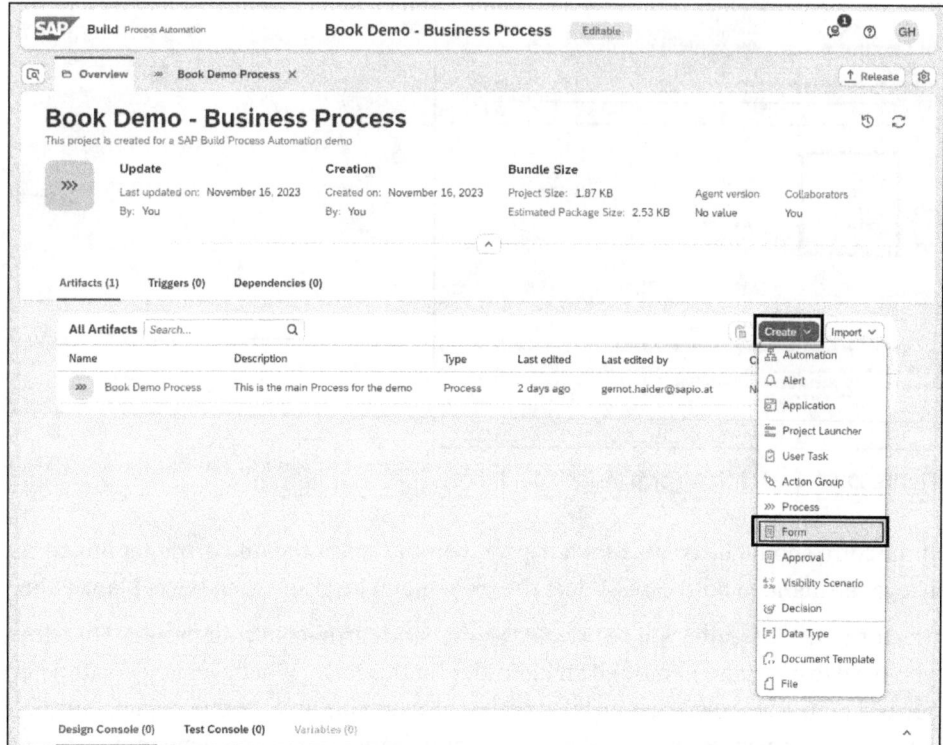

Figure 19.65 Adding Form in Overview of Project

The second option is to create the form directly inside the business process design console. To do so, click the ⊕ icon on the connector between two existing elements, as shown in Figure 19.66.

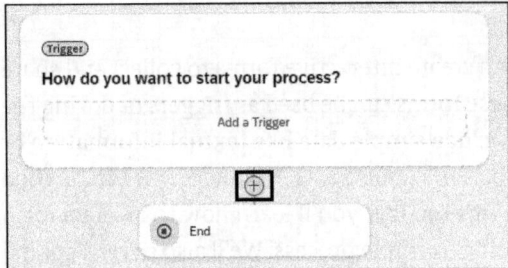

Figure 19.66 Adding New Step to Process

In the panel displayed, select the **Forms** menu item and then select a new **Form**, as shown in Figure 19.67.

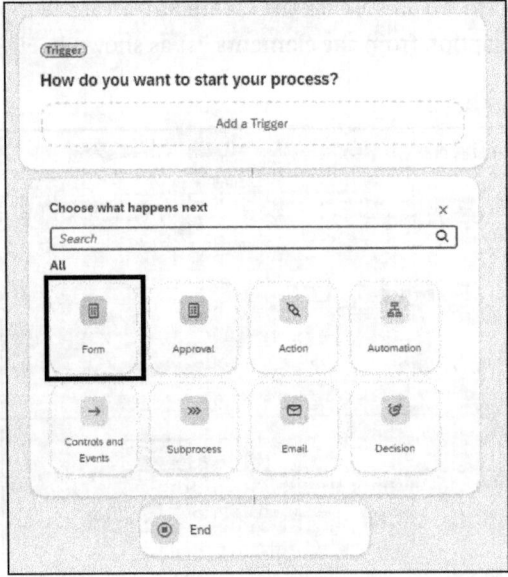

Figure 19.67 Create New Form inside Your Process

If the form should be created for a trigger, you can select the **Add a Trigger** link of the trigger element. In both cases, select the **Form** menu item and then select **Blank Form**.

The form that is created will be handled as a separate artifact in your project. Therefore, you need to provide a name and an identifier for this form. Whenever you create a new form, a popup is shown that enables you to provide this data. In this popup, enter the name of your new form. The identifier is automatically created depending on your name, but it can be modified for your needs. Optionally, you can add a description to provide additional information about this form. Figure 19.68 shows the popup to provide information for creating a new form.

For the example process, you need a form to provide data about recently completed business travel. Therefore, create a new form and enter the information in Table 19.8.

19.3 Forms

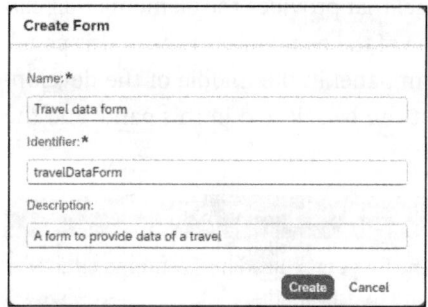

Figure 19.68 Providing Data for New Form

Input Field	Value Provided
Name	"Travel data form"
Identifier	"travelDataForm"
Description	"A form to provide data of a travel"

Table 19.8 Data Provided for New Form

Press the **Create** button to confirm your input. As a result, a new tab is opened in your project view that is titled with the name of your newly created form. Inside this tab, the form development canvas is shown. This editor enables you to design your form by adding formatting elements like headings or elements like text fields, checkboxes, tables, and many more that can be bound to the data used in your business process. Figure 19.69 shows the empty canvas of your created form.

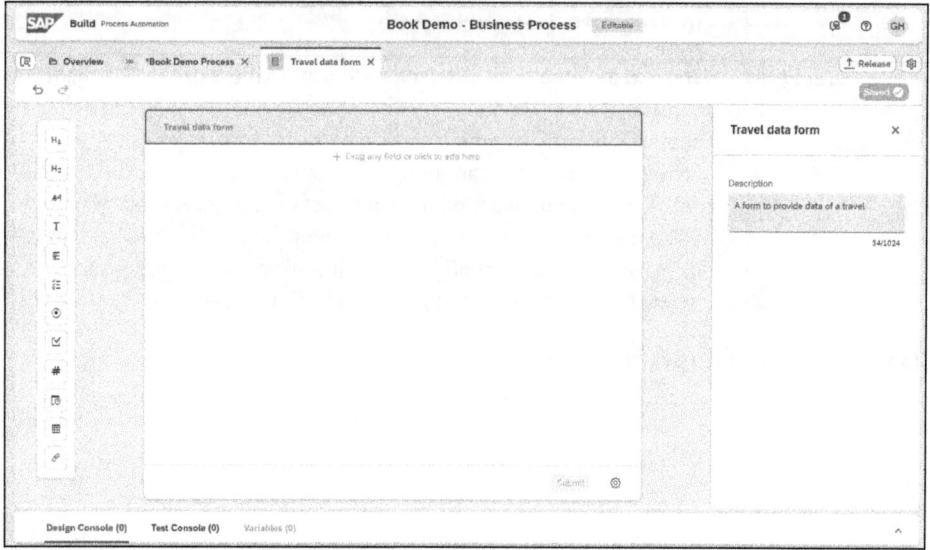

Figure 19.69 Canvas for Editing Forms

On the left side of the canvas, you can see a panel that provides the elements that can be added to your form. Whenever you select one of the element buttons in this panel, the corresponding element is added to the form panel in the middle of the development canvas. Table 19.9 describes the tools that can be selected in this panel and the coding elements to be used in the form.

Icon	Name	Description
H₁	Headline 1	Adds a headline to the form. The text of the headline is entered directly in the form and cannot be bound two a variable.
H₂	Headline 2	Clicking this button adds a subheadline to the form. The subheadline behaves the same as the headline. The content is fixed and cannot be bound to data from outside.
A	Paragraph	The paragraph element can be used to add textual content to the form using standard formatted text. It can be used to add descriptions or additional information for the user. The content of the paragraph cannot be bound to data from outside the form.
T	Text	The text tool adds a text field to the form. This element consists of a label and the field that can be used to enter information. A description optionally can be added. If so, a question mark icon is shown next to the label of the text field, and your description is shown on moving the mouse cursor over this icon. In the properties of the text field, you can define a minimum and maximum character limit. Text fields should only be used for text up to 256 characters. You can also configure input validation. Therefore, you can choose from a set of predefined rules, or you can create your own validation rule using regular expressions. You can also define an error message that is shown in case of an invalid input. Furthermore, the text field can be set as a required field, or it can be set to read-only to avoid data input.
E	Text Area	The text area tool adds a text area element to the form. This element consists of a label and a text area itself to provide data. You can define a character limit by configuring a minimum and maximum value for the content provided. You can configure input validation by selecting from a set of predefined validation rules or by defining your own validation rule using regular expressions. In case of a wrong input, an error message is displayed that can be configured for this text area. Finally, you can define if the text area is a required field or if it is read-only.

Table 19.9 Elements to Be Used in Forms

Icon	Name	Description
	Dropdown	The dropdown tool is an element in your form that enables the user to select a value from a predefined list. The element shown on the form consists of a label and the dropdown field itself. In the properties of the dropdown field, you can define if the options to be selected are configured value by value or if the options refer to a data source. Furthermore, you can define the dropdown field as a required field or you can set it to read-only. The dropdown field can also be set to multiple selection, another property that can be defined. As for the other input fields, you can add a description to the element that is shown on moving the mouse over the question mark icon displayed next to the label.
	Choice	The choice tool adds a set of radio buttons to your form according to the configuration. In the selection options of the choice element, you can enter the options that should be shown in the form. You can define the alignment of the choices and switch between vertical alignment and horizontal alignment. This element can be set to read-only, and you can choose if multiple selection should be allowed or not. Again, you can add a description for the user. The description will be shown by moving the mouse over the question mark icon that is shown next to the label of the element.
	Checkbox	The checkbox tool adds a checkbox to the form. In the properties, you can define the element as required. You can also set the checkbox to read-only. By adding a description in the properties of the checkbox, you can provide additional information about this field to the user. Providing a description adds the question mark icon to the label of the checkbox.
	Number	This tool adds a number field to your form. A number field is a text field that can be used to view and edit numerical values in special numeric format. In the properties of this field, you can define decimal places for the numeric format and you can enter a minimum value and a maximum value, used for input validation. Beyond that you can set the field as a required field, and you can define if the field is editable or read-only. As for all other input fields, a description can be added to give additional information to the user.
	Date	The date tool adds a date field to the form. The field consists of a label and a date selector control. In the properties of this field, you can define the dates that can be chosen. Therefore, you can define if the user should be able to select from all dates or only from past dates or future dates. For presentation, you can select from a set of format options that are used to show the date in the form. Furthermore, you can define the field as a required field, and you can set the field to read only if no data input should be allowed. Adding a description here is an option.

Table 19.9 Elements to Be Used in Forms (Cont.)

Icon	Name	Description
▦	Table	This tool is used to add a table to the form. In the properties, you can only edit the description and define if the table should be read-only or not. Defining the columns of the table must be done in the table control itself. Therefore, the table control has its own toolbox with elements that can be added as columns. Using this toolbox, you can add columns of specific types to your table and configure their properties the same way you configure fields on the form itself.
🔗	Link	The last element provided by the toolbox is the link. Using this tool, you can add a link to your form. In the properties of this element, you can enter the link and you can enter a link text that is displayed in place of the link itself. If no link text is provided, the link itself is shown in the form. For the link element, you can define if a label should be shown or not. Finally, you can edit the description of the link element.

Table 19.9 Elements to Be Used in Forms (Cont.)

Once an element is added to the form, it's shown in the preview section in the middle of the development panel. By clicking a button in the toolbox panel, the new element is always added at the end of the elements list. Each element provides a three-point menu that can be used to delete an element from the form or to rearrange the order of the elements. Figure 19.70 shows this menu for the **Sub header** element that needs to be moved beyond the **Main header** element.

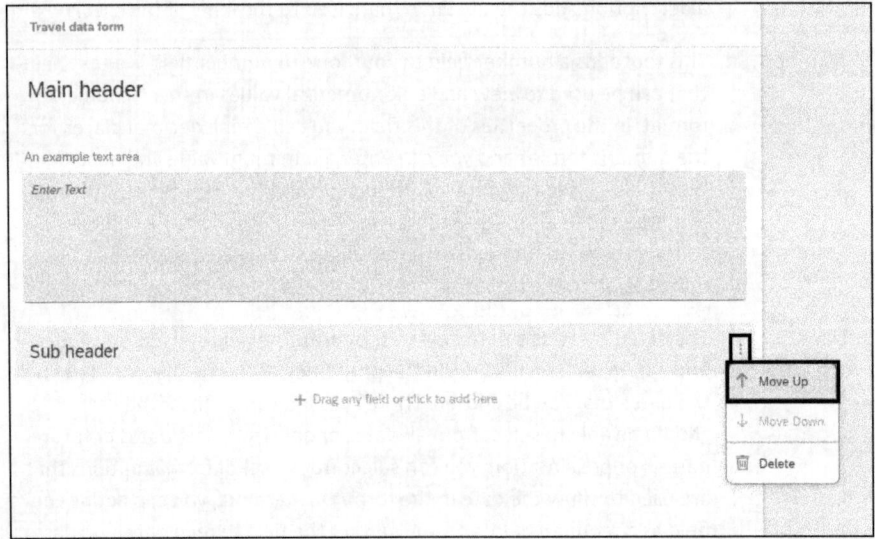

Figure 19.70 Changing Order of Elements

After clicking the **Move Up** menu item, the **Sub header** element is arranged between the **Main header** element and the text area, as shown in Figure 19.71.

19.3 Forms

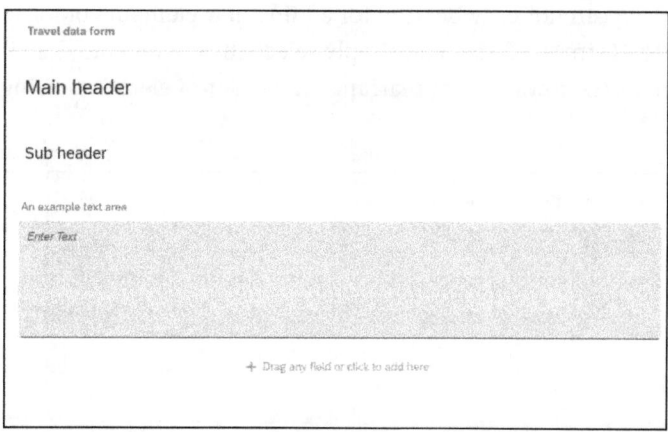

Figure 19.71 Form after Changing Order of Elements

Another option to add and arrange the elements in the form is to use drag and drop. This option enables you to directly set a new element to its final position and change the order of already edited elements without using the elements menu functionality. Figure 19.72 shows how a text field is added by using drag-and-drop functionality. You simply press an element in your toolbox panel and drag it to the position in the form where the element should be created. After dropping, the element is added at the correct position inside your form.

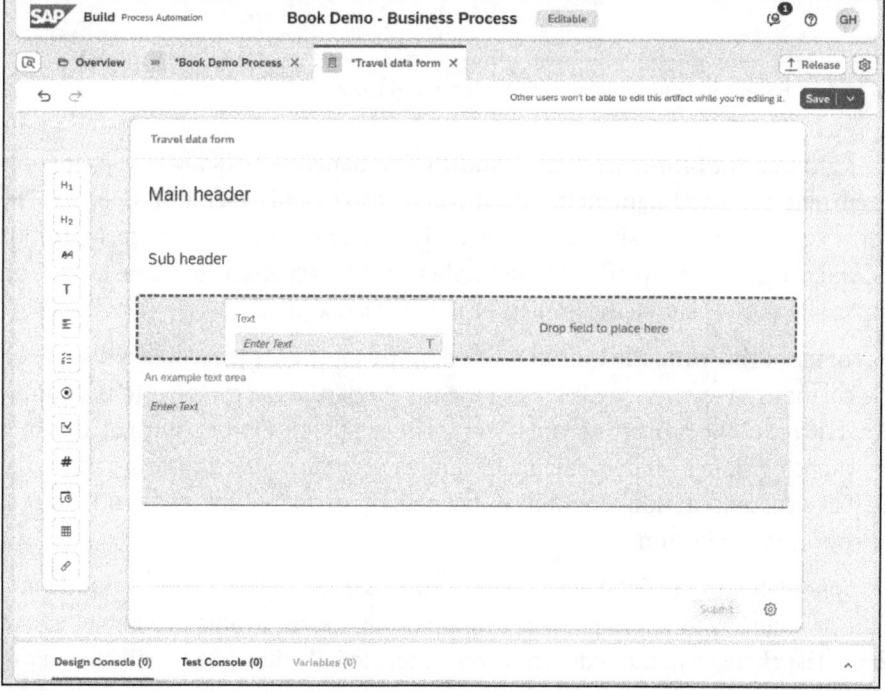

Figure 19.72 Adding Element by Drag and Drop

As mentioned, drag and drop can not only be used for adding new elements but also to rearrange them inside the form. To do so, you simply select an element and drag it to its new position. Figure 19.73 shows how to rearrange the order of elements using drag and drop.

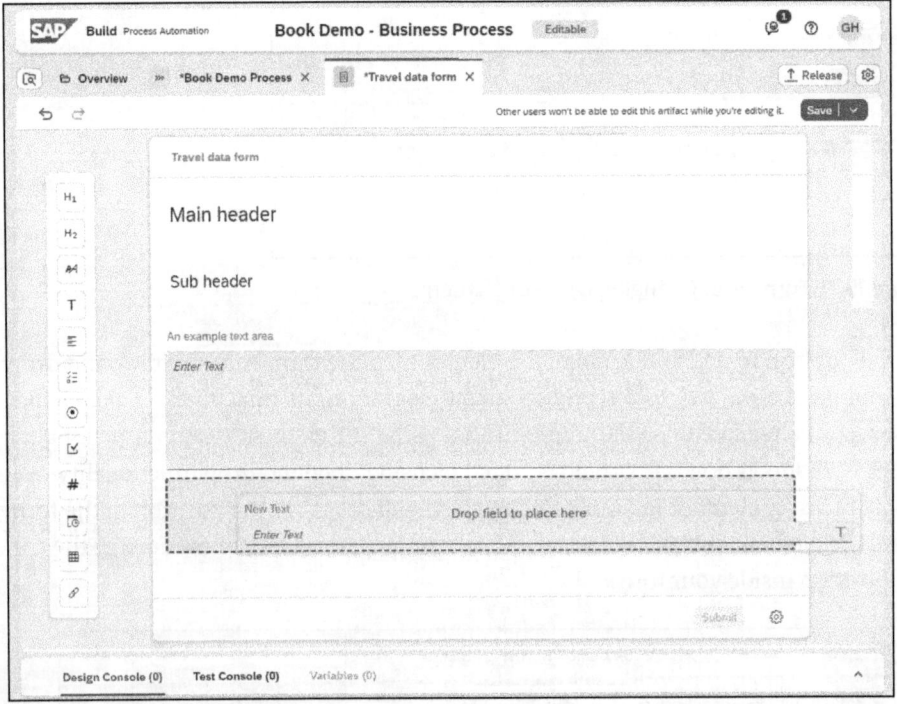

Figure 19.73 Changing Order of Elements by Drag and Drop

On the right side, the properties panel is shown. This panel always shows the properties of the currently selected element in the form and enables you to set the properties. The properties panel always adapts to the currently selected element in the form. In this panel, you can set the properties of the single elements according to Table 19.9. As an example, Figure 19.74 shows the properties panel of a text field.

In the footer of the form, the button to confirm the entries is presented. This button can be customized to your needs by clicking the **Properties** button next to the confirmation button. In the properties panel, you can select the title to be shown on this button by choosing from a set of predefined values or you can define a custom caption. Be aware that a custom caption cannot be translated. Figure 19.75 shows the configuration of the confirmation button.

In the upper-left corner of the development canvas, the functionalities for undo and redo are provided by the icons shown in Figure 19.76. The undo function enables you to undo the last changes in the order they were executed. If you undo a change, you can use the redo function to repeat the step that was just revoked.

19.3 Forms

After your changes to the form are completed, you can save the form by clicking the **Save** button in the upper-right corner of the development canvas as shown in Figure 19.77. You can choose if you want to keep the artifact locked after saving the changes or if you want to reset the lock so that other members of the project can work on the artifact.

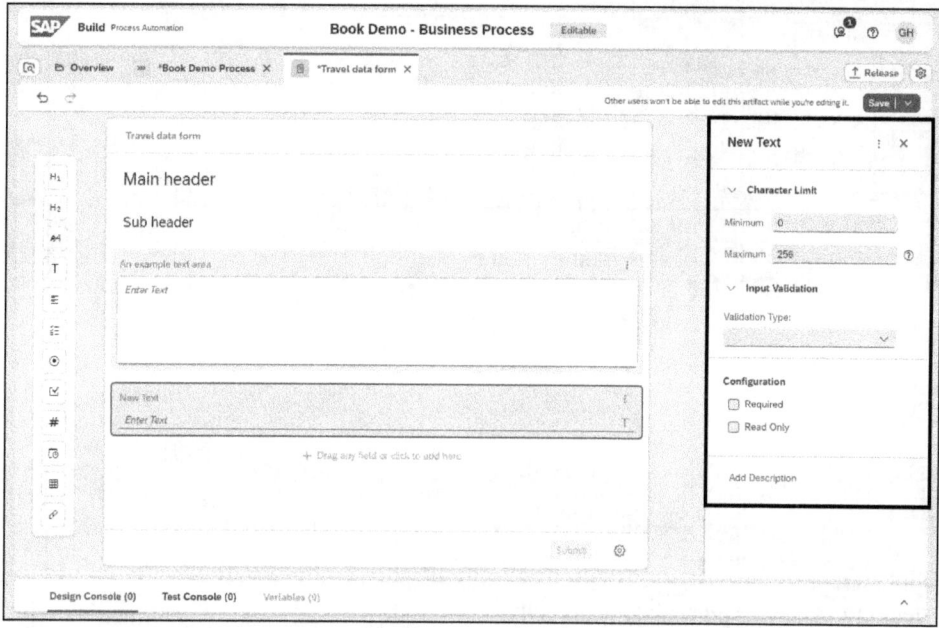

Figure 19.74 Properties Panel of Text Field

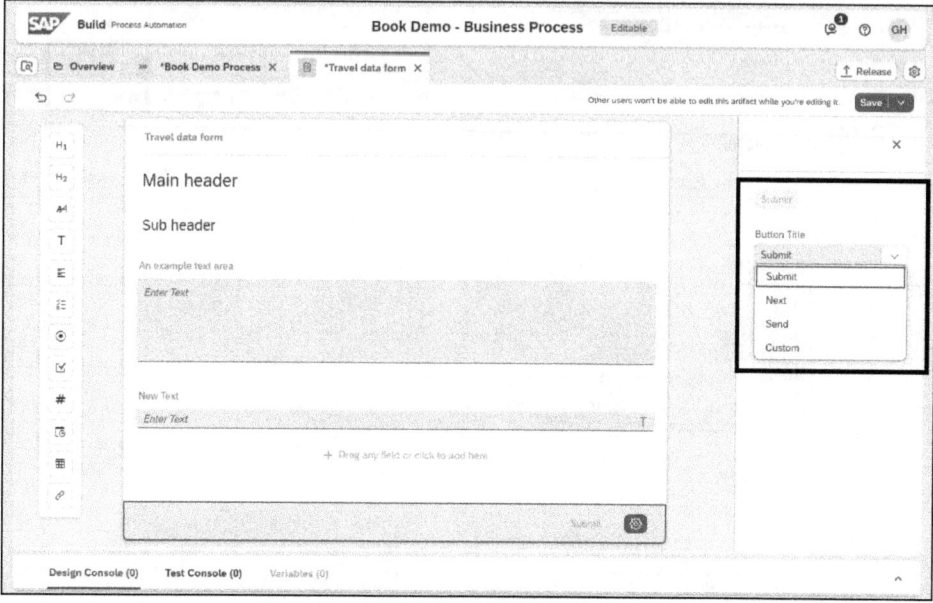

Figure 19.75 Changing Caption of Confirmation Button

19 Processes

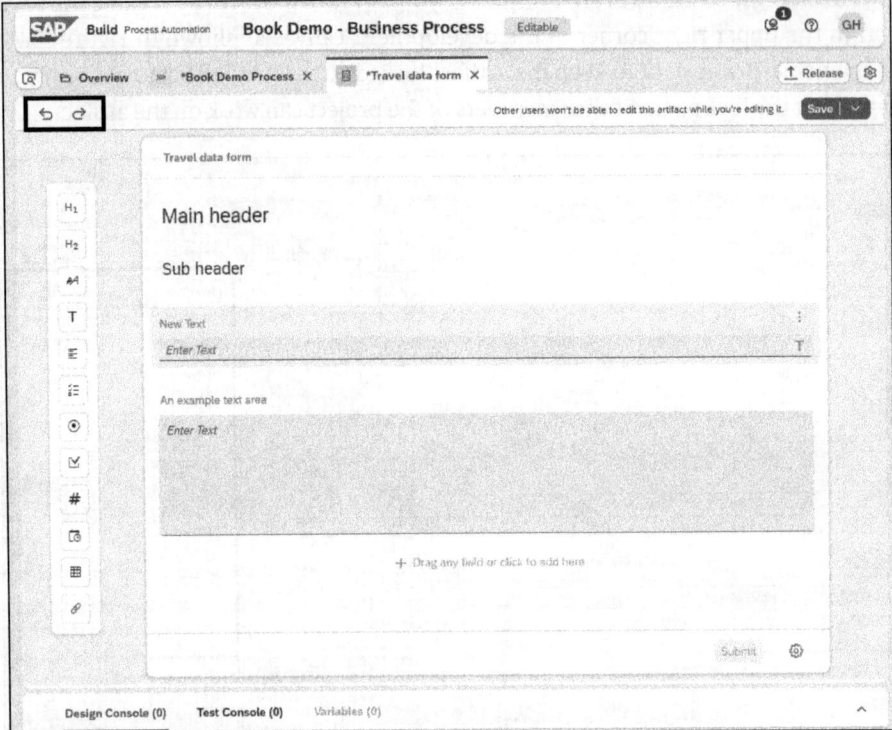

Figure 19.76 Undo and Redo Functionality

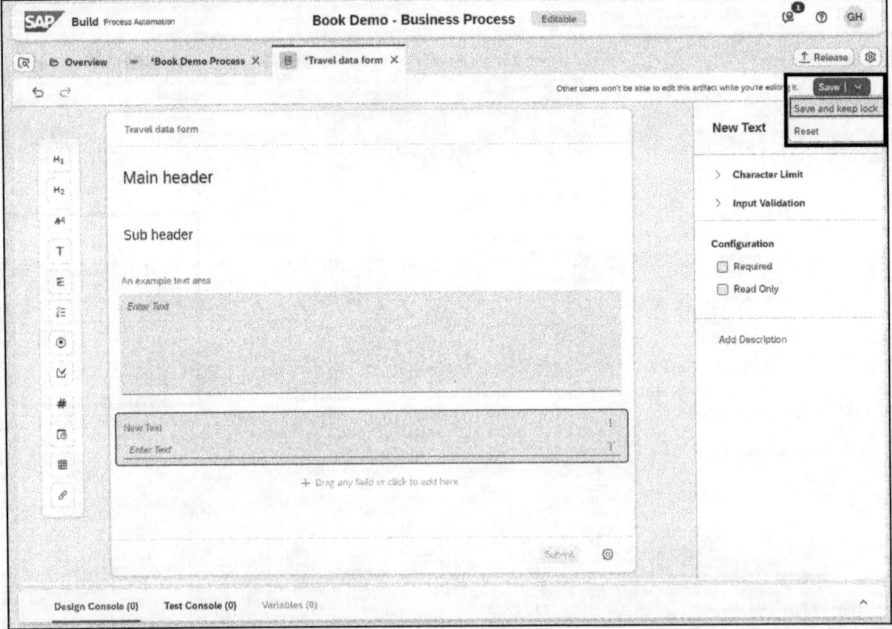

Figure 19.77 Saving Your Form

19.3 Forms

For as long as you work on the form, the artifact is locked and cannot be edited by other members. Therefore, you have the option to either keep the lock in place after saving—if you want to save a temporary status, for example—or to reset the lock once your work is finished, or to give others the opportunity to work on the artifact. By directly clicking the **Save** button without using the menu option of the button, you reset the lock of the artifact. By creating a change in the form you just saved, the lock is immediately set again.

19.3.2 Using Forms in the Example Process

Now that you have learned about the theory of creating forms for your business processes, you can create a form that will be used in our example process. As the example business process is about confirming travel expenses, you need to create a form that enables the user to provide data for the expenses of his business travel.

The first step is to create an artifact of type form. Therefore, navigate to the overview of your project and click the **Create** button above the artifact list. From the list that is shown, select the **Form** item, as shown in Figure 19.78.

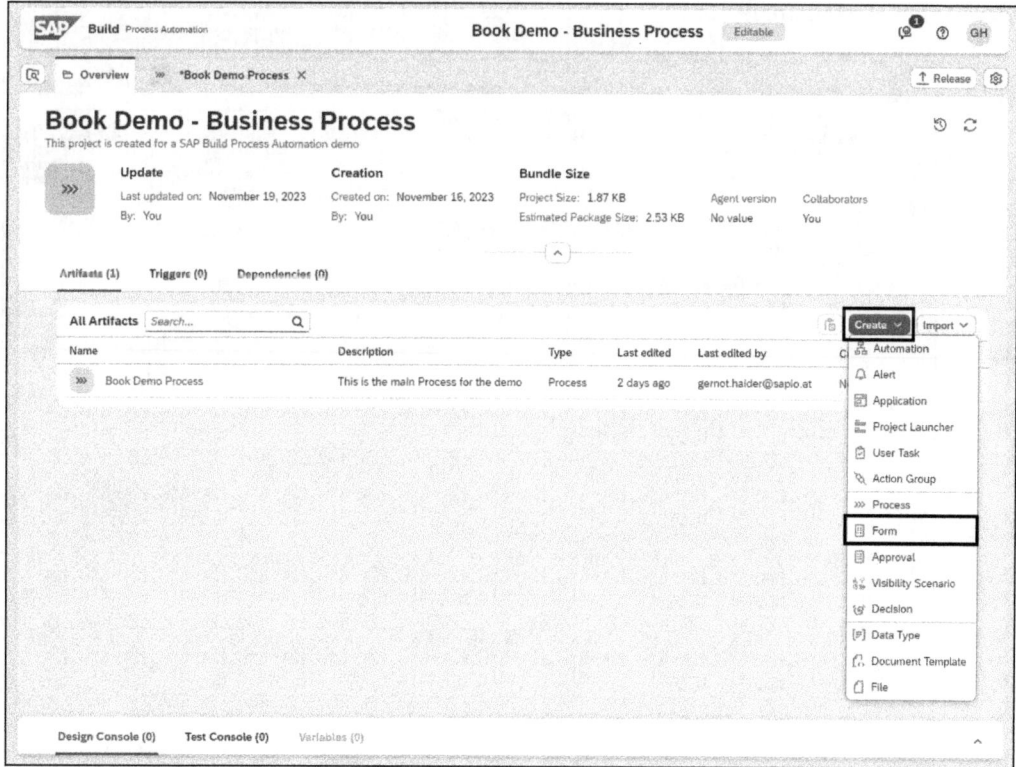

Figure 19.78 Create New Form

19   Processes

The dialogue to create a new form is shown. Enter the **Name**, **Identifier**, and **Description** in the corresponding input fields as shown in Figure 19.79, and press the **Create** button to confirm your entries.

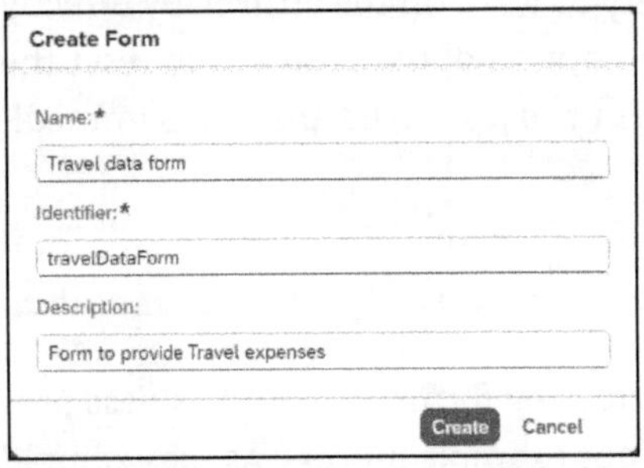

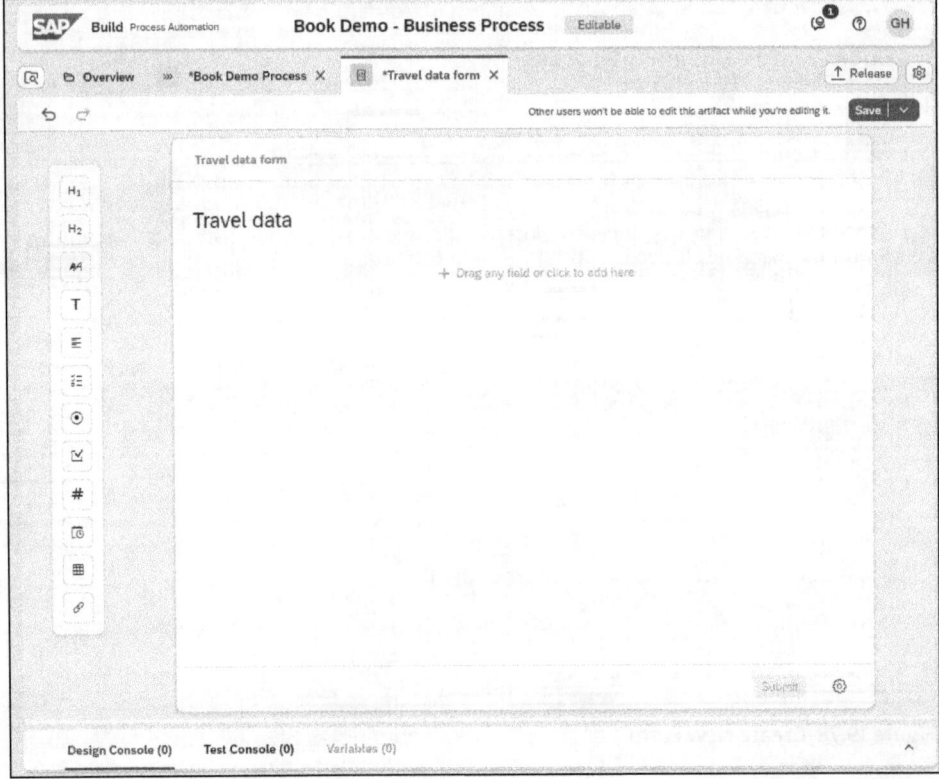

Figure 19.79 Data for New Form

After you confirm your input, a new panel is opened, and you'll see the development canvas for your new form. From the toolbox on the left side of the development canvas, select the **Headline 1** element and drag it to your form. Click the created headline element in your form and enter the text "Travel data". Your form now has a header, as shown in Figure 19.80.

Figure 19.80 Header Line of Form

19.3 Forms

In the next step, you create sections for the personal data, the destination data, and the expense of the travel. Therefore, select the **Headline 2** tool from the tools panel and drag it to your form. Select the created subheader element and enter the text "Personal data". Repeat the step to add two more elements of type **Headline 2**. For these elements, set the titles to "Date and Destination" and "Expenses". Your form should now look like the screenshot shown in Figure 19.81.

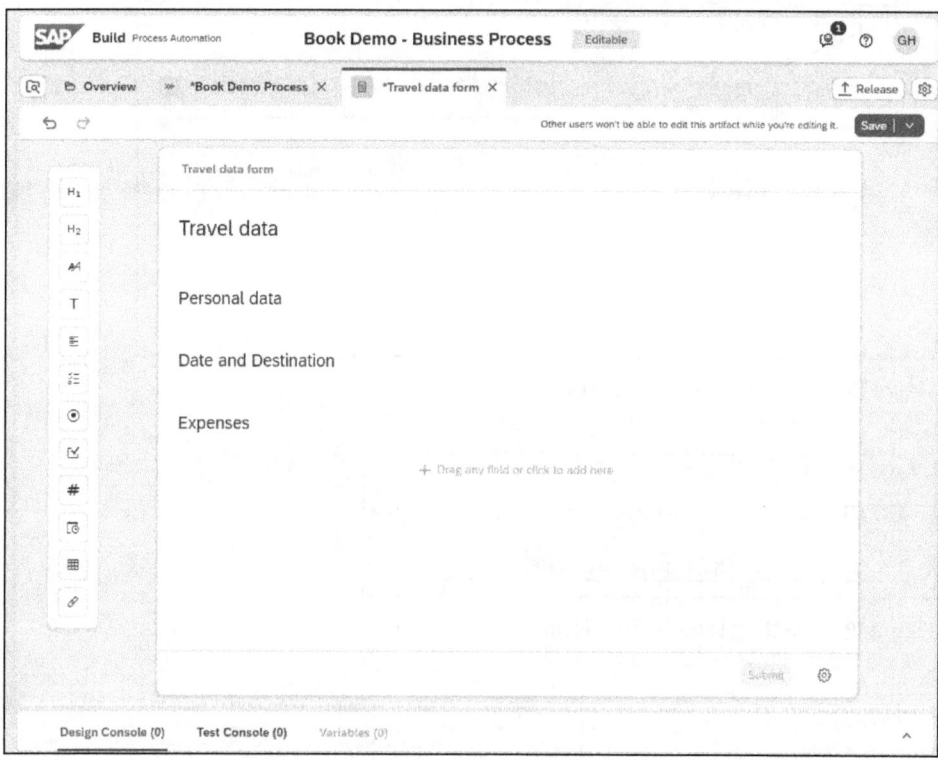

Figure 19.81 Structure of Form

Now you can start to add the input fields to your form. From the left panel, select the **Text** item and drag it to your form to place it between the **Personal data** and **Date and Destination** headers, as shown in Figure 19.82.

Select the label of the added text field element and enter "Firstname" as the label text, as shown in Figure 19.83.

In the properties of the text field, select the **Required** checkbox to indicate that the field is mandatory and add a description. Figure 19.84 shows the properties panel of the text field.

597

19   Processes

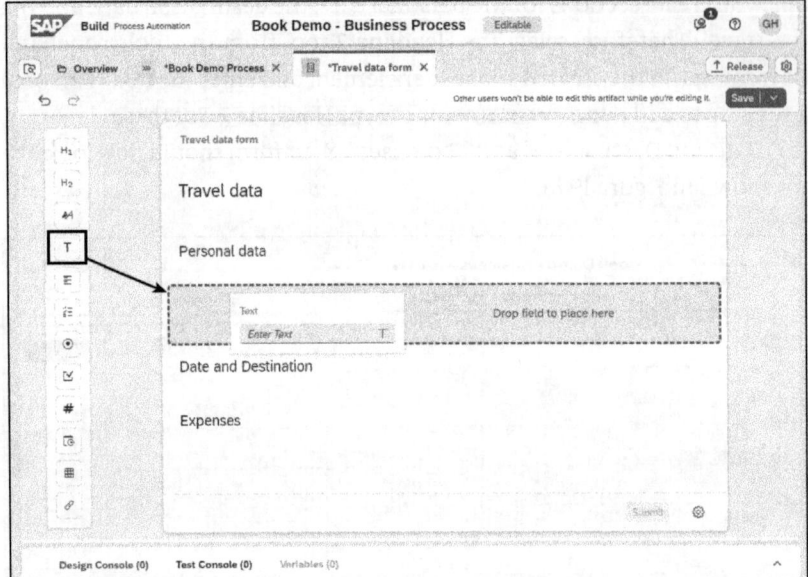

Figure 19.82 Adding Text Field to Form

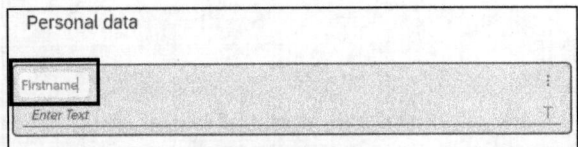

Figure 19.83 Setting Label of Text Field

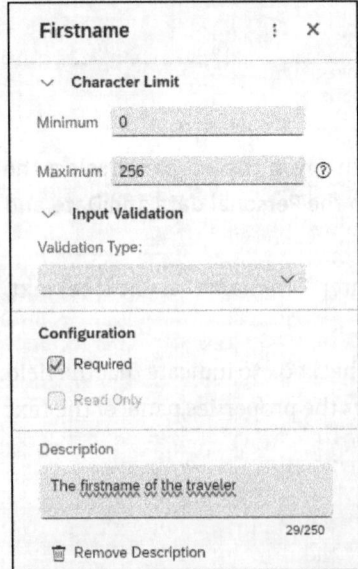

Figure 19.84 Properties of Firstname Field

19.3 Forms

Add another text field underneath the **Firstname** field, and set the properties listed in Table 19.10.

Property	Value
Label text	"Lastname"
Required	Select
Description	"The last name of the traveler"

Table 19.10 Properties of Lastname Field

Save the form by clicking the **Save** button in the upper-right corner. After adding the two text fields, your form should look like Figure 19.85.

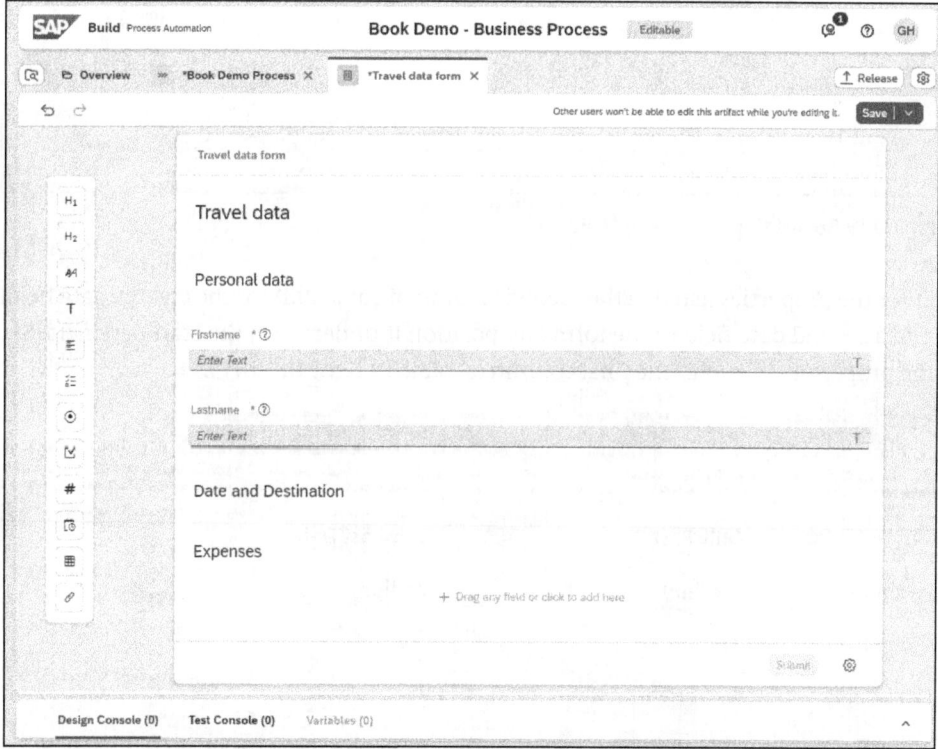

Figure 19.85 Fields for Personal Data

In the next step, you add the fields for providing information about the start date and end date of the travel and about the destination. From the toolbox panel on the left side, select the **Date** tool and drag it underneath the **Date and Destination** header, as shown in Figure 19.86.

599

19 Processes

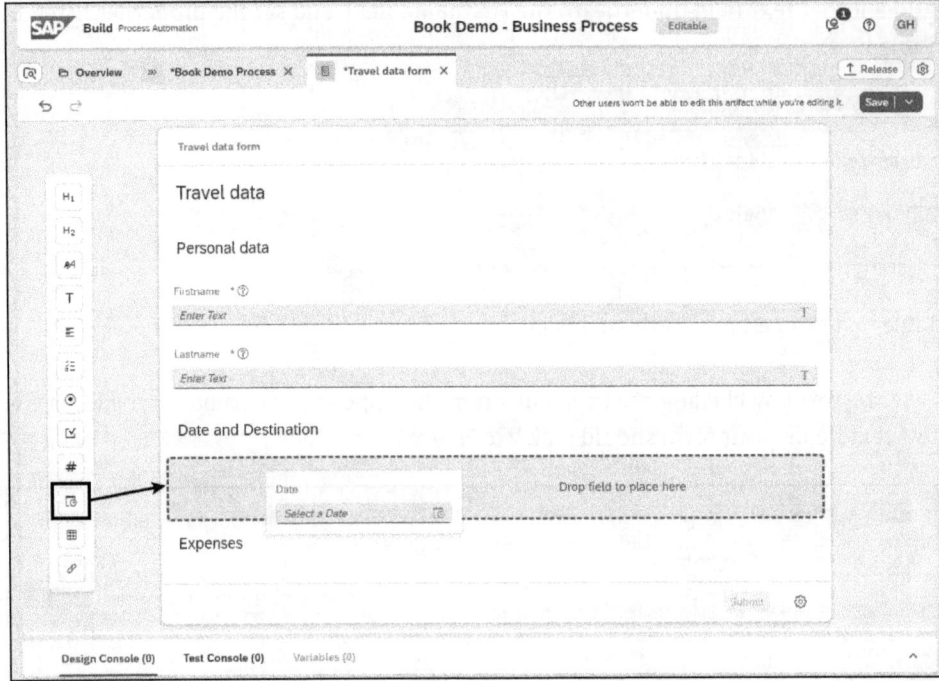

Figure 19.86 Adding Date Field to Form

Enter the properties listed in the second column of Table 19.11 for the created date field. Add a second date field to the form and position it underneath the start date field. Set the properties shown in the final column of Table 19.11 for this field.

Property	Date Field 1	Date Field 2
Label text	"Start date"	"End date"
Date Options	Only Past Dates	Only Past Dates
Format	Medium	Medium
Required	Yes	Yes
Read only	No	No
Description	"The start date of the travel"	"The end date of the travel"

Table 19.11 Properties for Start Date Field

In the expenses section, you need a numeric field and a table to enable to the user to provide his data. Add a **Number** field to the expenses section by selecting the corresponding icon in the left panel and dragging it underneath the header of the expenses section. Figure 19.87 shows how the number field is added.

19.3 Forms

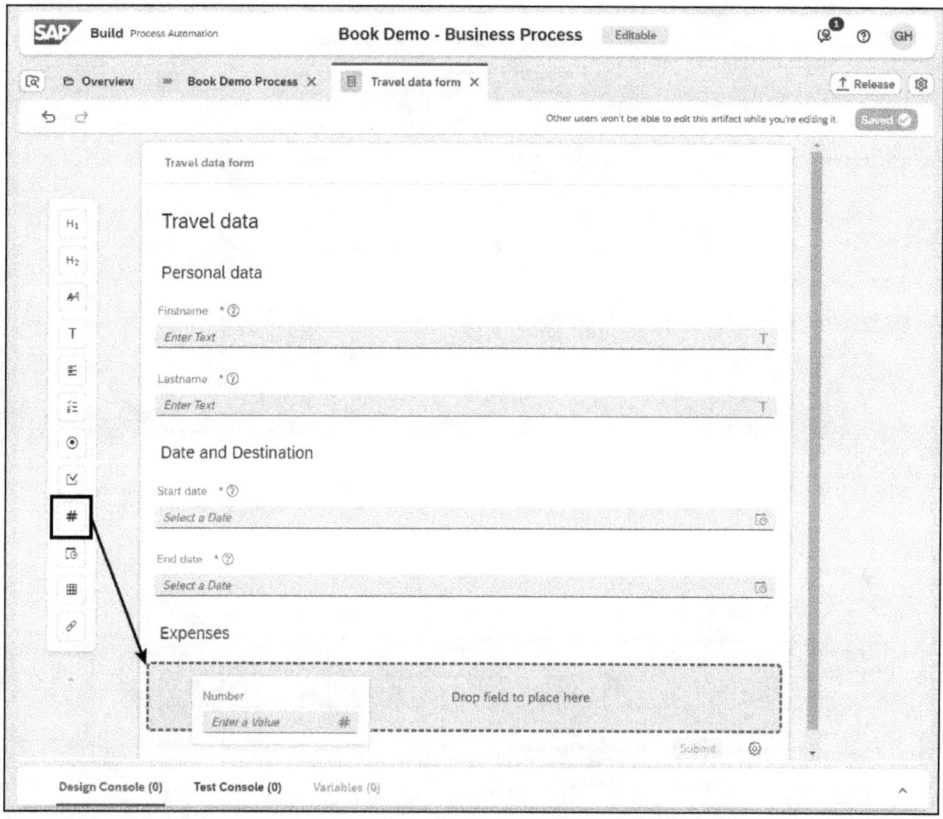

Figure 19.87 Adding Number Field to Form

Set the properties shown in Table 19.12 for the created field.

Property	Value
Label text	"Total costs"
Fixed Decimal Places	Selected
Decimal Places	2
Required	Yes
Description	"The total costs of the travel"

Table 19.12 Properties of Number Field

To create the table, select the **Table** icon in the left panel and drag it underneath the expenses header, as shown in Figure 19.88.

19 Processes

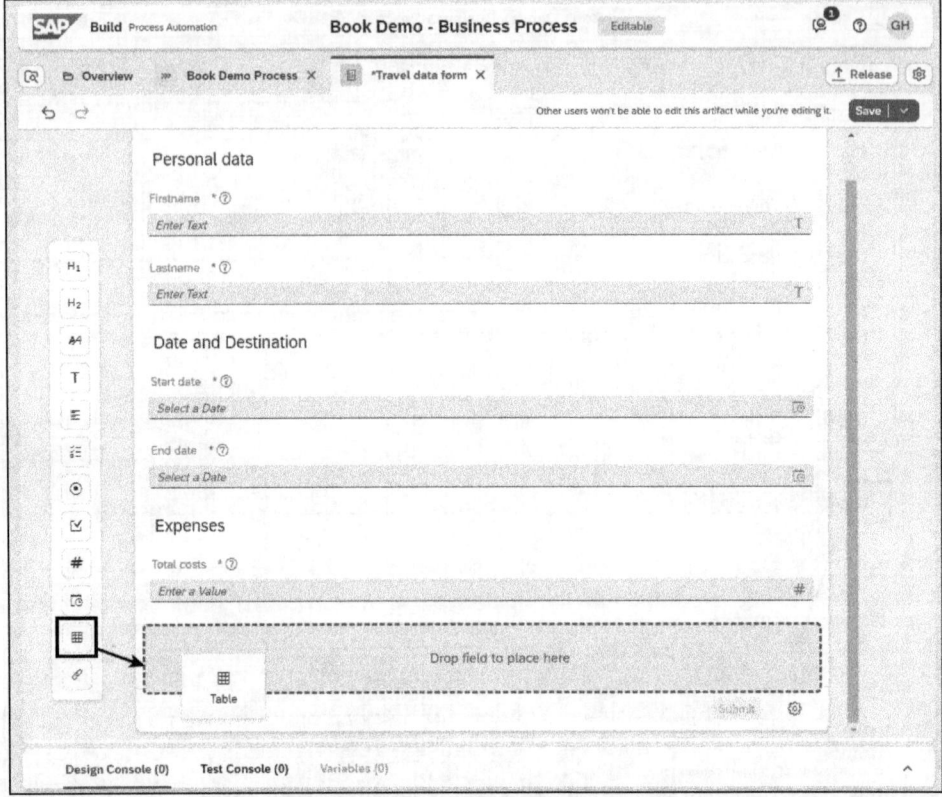

Figure 19.88 Add Table to Form

Add the properties shown in Table 19.13 to the created table.

Property	Value
Label Text	"Expenses"
Description	"The expenses of the travel"

Table 19.13 Properties of Expenses Table

To add a column to the expenses table, click the plus icon (**+**) inside the table element. In the menu that is shown, select the **Date** item, as shown in Figure 19.89.

19.3 Forms

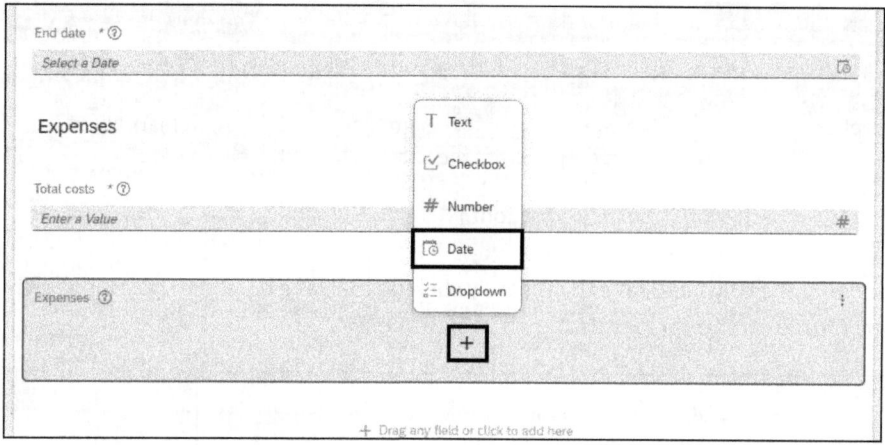

Figure 19.89 Adding Column to Table

The column is added to the table, and you can set the properties. Adjust the settings in Table 19.14 for the date column.

Property	Value
Label Text	"Expense date"
Date Options	Only Past Dates
Format	Short
Required	Yes
Description	"The date of the expense"

Table 19.14 Properties of Expense Date Column

As the date field is the only column added to the table at the moment, it requires the full width. The icon to add a new column has now moved to the right. To add a new column, click the item and repeat the actions done for the date column. Add three more columns according to the properties shown in Table 19.15.

Property	Column 1	Column 2	Column 3
Field Type	Text	Number	Text
Label Text	"Expense type"	"Amount"	"Remark"
Fixed Decimal Places	-	Select	-
Decimal Places	-	2	-

Table 19.15 Properties of Table Columns

19 Processes

Property	Column 1	Column 2	Column 3
Required	Yes	Yes	No
Description	"The type of the expense"	"The amount of the expense"	"Your remark about the expense"

Table 19.15 Properties of Table Columns (Cont.)

After adding the columns, save the form by clicking the **Save** button in the upper-right corner. Your form should look as shown in Figure 19.90.

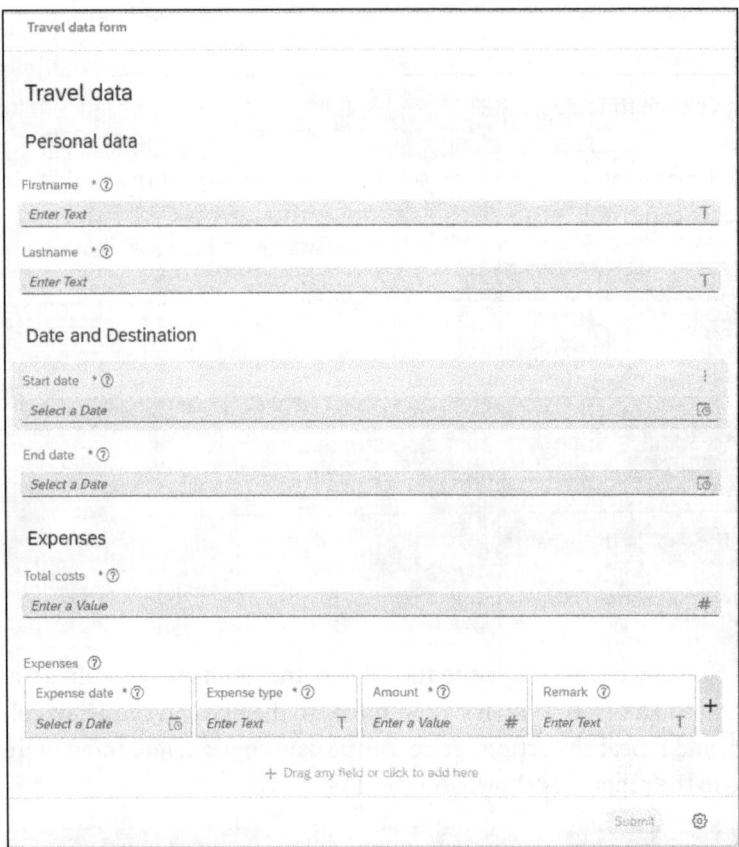

Figure 19.90 Form after Adding Required Elements

19.3.3 Approval Forms

One special type of form is the approval form. Whenever an approval is needed during the business process, this type of form can be used to display the data the approval relies on. Approval forms can be created from scratch by adding new fields and configuring the binding, or they can rely on created forms that are used somewhere else

19.3 Forms

within the business process, like in the trigger. When a user opens an approval task in his inbox, he will see the data presented by the approval form and he can approve or reject the task with the buttons automatically created in the form. Let's now add an approval form to the example process to show how the form is added and how the binding for the fields is done.

To add an approval form to the example process, navigate to the lobby and open the project, including the example process that is started by the form trigger. Click the artifact in the list that represents the process to open it in the development canvas. Right now, the process only consists of the trigger and the **End** event, as shown in Figure 19.91.

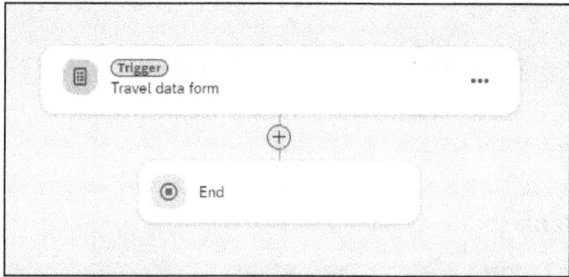

Figure 19.91 Example Process

To add a new approval form to your process, click the ⊕ icon between the trigger and the end event. In the displayed menu, select the **Approval** item and then **Blank Approval**, as shown in Figure 19.92 and Figure 19.93.

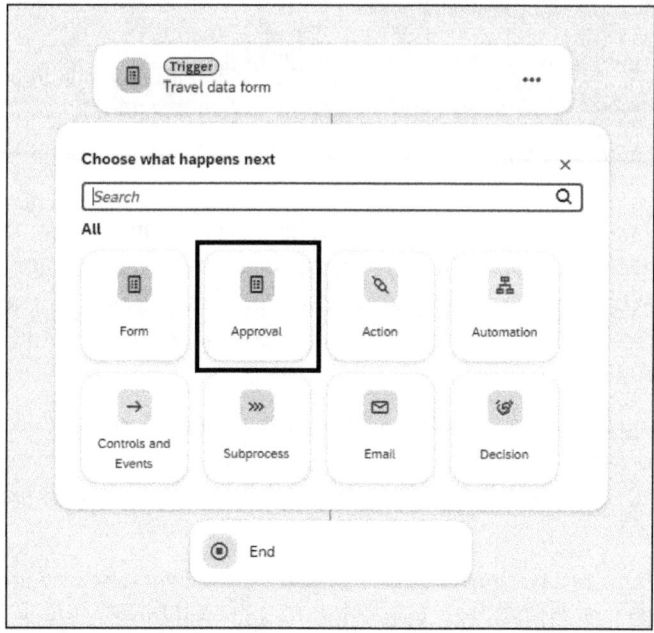

Figure 19.92 Selecting Approval Element

19 Processes

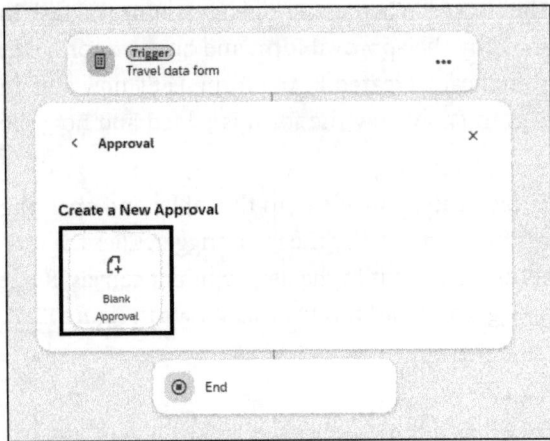

Figure 19.93 Adding Blank Approval

As a result, a popup dialog is displayed where you can provide the data of your new approval form. Fill in the fields according to Table 19.16.

Field Name	Value
Name	"Travel Data Approval Form".
Identifier	**travelDataApprovalForm** (the value of this form is automatically created depending on the value you provide for the name of the form, but can be changed to fit your needs if necessary).
Description	"Approval form for travel data".
Based on the form	Select this checkbox as the approval form will be based on the input form you created.

Table 19.16 Values for New Approval Form

When you select the checkbox to define that the approval form is based on an existing form, a combo box is displayed below, providing a list of forms available. In our project, currently only one form exists, so this form is selected automatically. The dialogue should now look like the screenshot in Figure 19.94. Click the **Create** button to confirm your entries.

In the development canvas, a new element that represents the approval form is added between the trigger and the end event. This element is currently marked with an error flag as the configuration still needs to be finished. Figure 19.95 shows the current status of the business process.

Save the changes by clicking the **Save** button, and navigate to the overview of your project. You can see that the new artifact was created for the approval form, as shown in Figure 19.96.

19.3 Forms

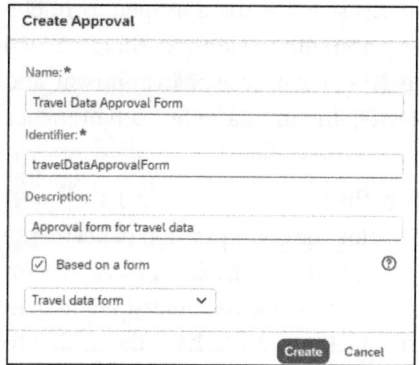

Figure 19.94 Dialogue to Create New Approval Form

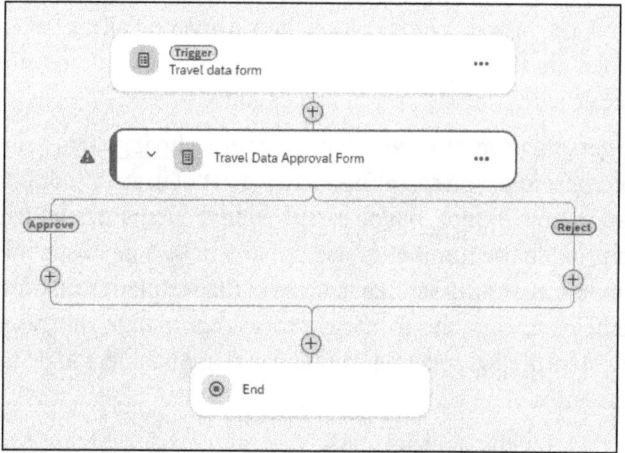

Figure 19.95 Process with Approval Form Added

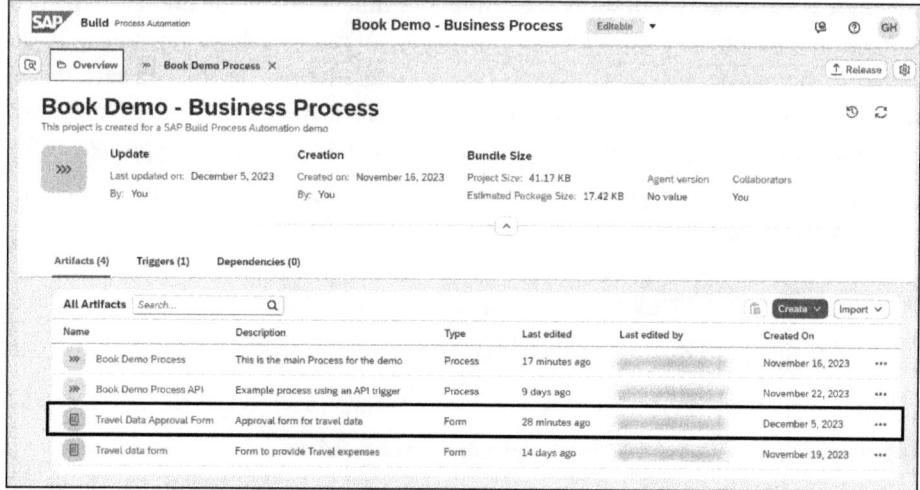

Figure 19.96 Approval Form in Artifacts List

Navigating back to the development canvas, you can see that the approval form element has one incoming branch at the top. At the bottom of the element, there are two outgoing branches. The left branch represents the data flow in case of an approval, and the right branch is executed if the approval is rejected. In our example, both branches will immediately lead to the end of the process.

The next step is to create the binding for the fields of the approval form. Click the travel data approval form element to load the properties inside the configuration panel on the right side of the development canvas. The configuration panel consists of three tabs for setting the general data of the form and to define the input and the output binding. In the **General** tab, you can set the name for the current step, and you have the option to edit the form that was automatically created based on our input form. Furthermore, you can define the subject of the form and you can enter a description. Finally, you can set the priority of the task that will be shown in the inbox. Whenever you enter a field for which the value could be bound, the process context is opened, providing all variables that are available. These variables are shown in a structured manner, and you can only choose values that are compatible with the field you currently edit.

In the recipient section of the general properties, you can configure who is going to see the approval task in his inbox. Therefore, you can either define a list of users by adding their user names in the corresponding textbox, or you can provide group names, which are then used to identify and provide the number of users that will be able to see the task in their inboxes. For the moment, we will set fixed values in the recipient's section, but it's possible to bind these fields to variables from the process content. In this case, the recipients' user names are identified at runtime, and you can use business logic to determine the concerned usernames.

Finally, you can define a due date for the approval task. You can choose whether the duration should be static or dynamic. By choosing a static duration, you can enter a fixed number and the time unit. The smallest duration that can be set is one minute. If you configure a dynamic duration, the number of time units can be bound to the process context. This can be used whenever you need different due dates depending on the context of the process. The last option is to use a reference date as the due date. In this case, you can calculate the due date inside your process and bind this date to the due date property of the approval form. By default, no due date is set for the approval task.

Configure the general properties according to Table 19.17 and save your changes by clicking the **Save** button.

Property Name	Value
Step Name	"Approval of travel data"
Subject	"Approval of travel data"
Description	"The approval task of the provided travel data"

Table 19.17 General Properties of Approval Form

19.3 Forms

Property Name	Value
Priority	Medium
Users	Enter your SAP BTP user name

Table 19.17 General Properties of Approval Form (Cont.)

The next step is to configure the inputs for the approval form. In the properties panel, select the **Inputs** tab. You can see the list of fields that are used in the approval form, and you can now bind values from the process context to these fields. The fields displayed in the **Inputs** tab are sorted alphabetically, and you can see their data types. By clicking on a field, the process content is automatically opened, providing all variables available for binding. As for the general properties, only values with a compatible data type can be bound to the corresponding fields of the approval form. Figure 19.97 shows the configuration panel of the input data for the approval form and the process content, displaying the available values for binding.

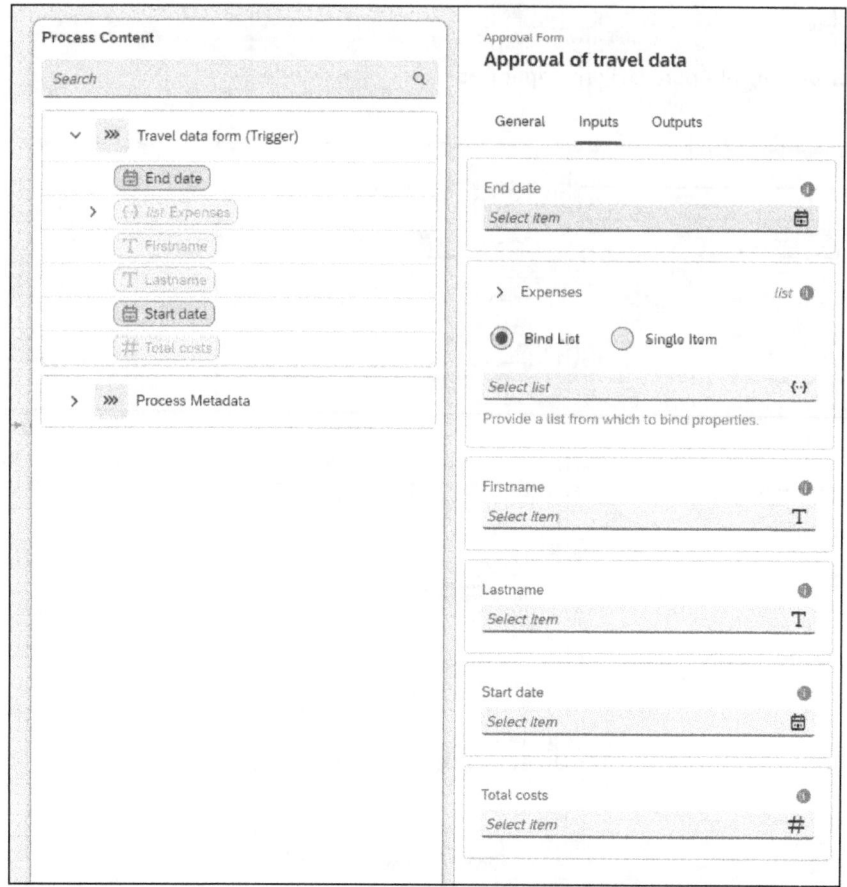

Figure 19.97 Using Process Content for Binding Input Values of Approval Form

To set a binding on the field of the approval form, simply click the needed field from the process content. The field is automatically set to the selected field of the form input to represent the binding. By selecting the fields of the process content for each field of the approval form as shown in Table 19.18, you can now define the binding for the data to be shown in the approval task of the inbox. After completing the binding, the input properties should look as shown in Figure 19.98.

Input Field Name	Name of Process Content Field
End Date	End date
Expenses	Expenses (select the option bind list to show all expenses in the approval form)
Firstname	Firstname
Lastname	Lastname
Start Date	Start date
Total Costs	Total costs

Table 19.18 Binding of Approval Form Input Data

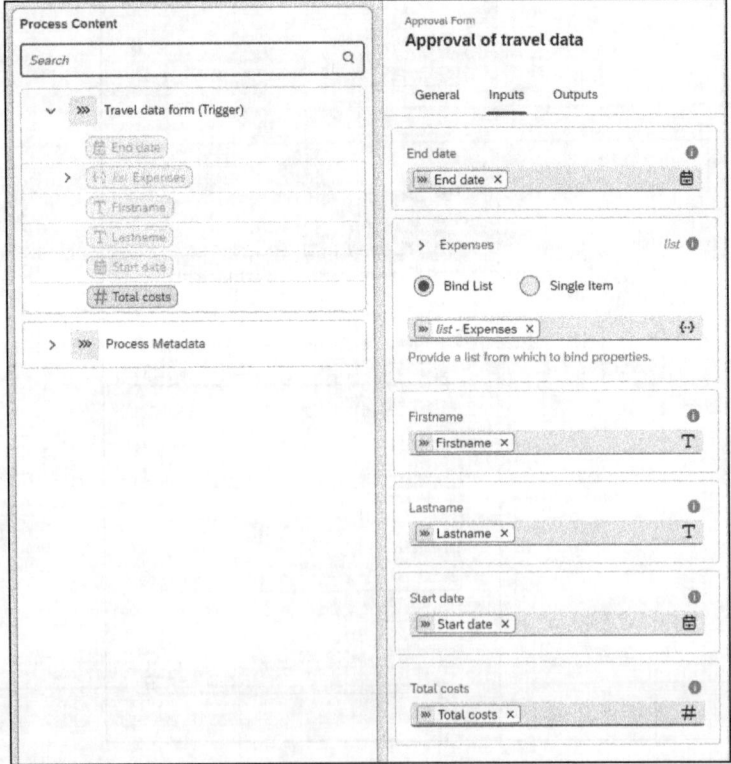

Figure 19.98 Binding for Approval Form

Save your changes by clicking the **Save** button and create a new release by clicking the **Release** button. In the overview of your project click the **Deploy** button in the top-right corner to deploy this new version and make it available for runtime.

After deployment, you can now start a new process instance by loading the link of the trigger form and submitting data to the process automation. Navigate back to your process and click the trigger form element to load its properties in the properties panel on the right side of the development canvas. Copy the form link and paste it to a new browser tab. In the displayed input form, enter some data representing travel and submit the data by clicking the **Submit** button in the bottom right corner.

From the lobby, navigate to **Monitoring** and click the tile for **Process and Workflow Instances**. You can see that an instance is currently running. Figure 19.99 shows a screenshot of the monitoring displaying the running instance.

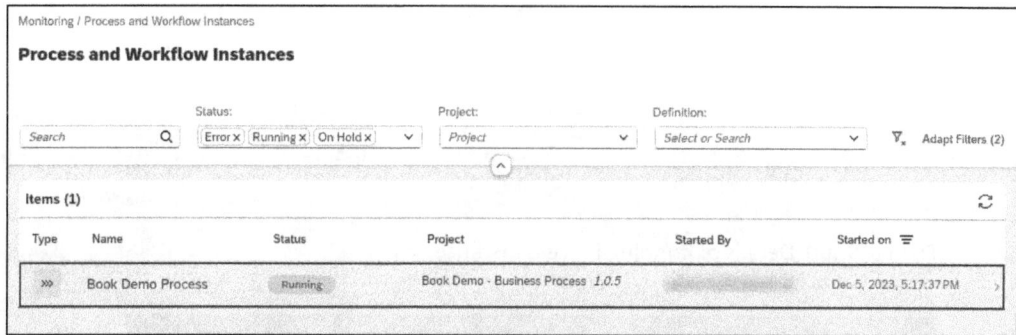

Figure 19.99 Running Process Instance

Click the item representing the running instance to load the details. You can see that an approval task is available and that you are the recipient for this approval task, as shown in Figure 19.100.

In the top-right corner of the current view, click the 📧 icon that represents the inbox. As a result, **My Inbox** is opened in a new browser tab. In the task list, you can see the approval task. As it is the only task in the inbox, the details are automatically displayed, and you can see the data you entered in the input form, but the data is read-only as you're now viewing the approval form, as shown in Figure 19.101.

In the bottom-right corner, the buttons for **Approve** and **Reject** are displayed. Click the **Approve** button to approve the travel data. As a result, the approval is executed, and the task is deleted from the inbox. Navigate back to the monitoring view and click the item that represents your running process instance. The data for the instance is reloaded, and you can see that the status of the instance changes to **Completed**. In the details, you can see that the approval task was completed and that subsequently the process instance has completed successfully as well. Figure 19.102 shows a screenshot of the completed process instance.

19   Processes

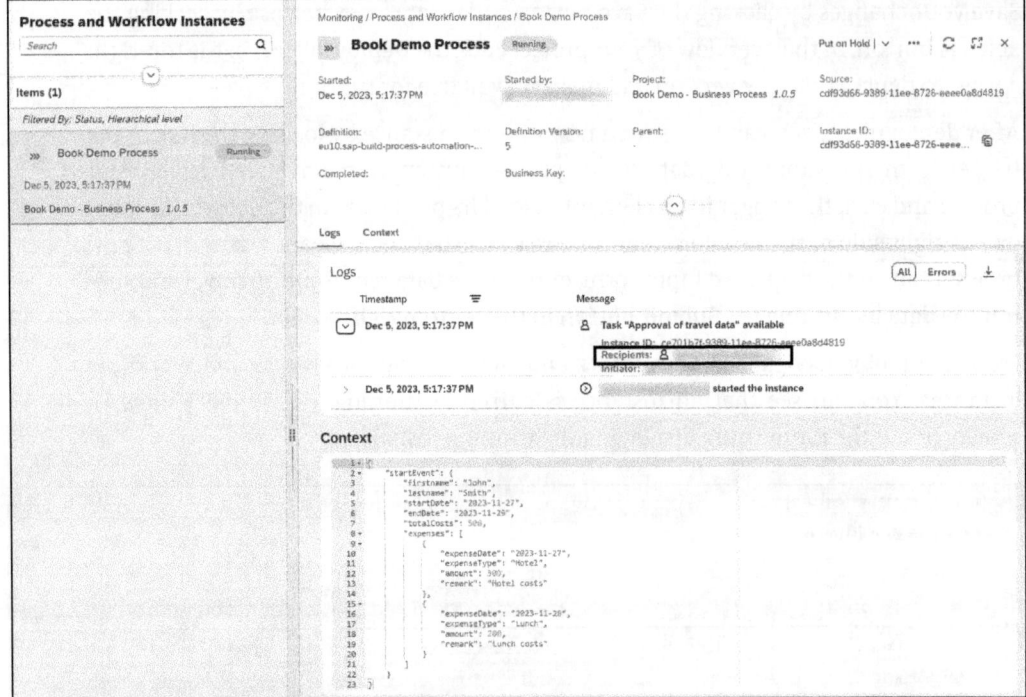

Figure 19.100 Details of Running Process Instance

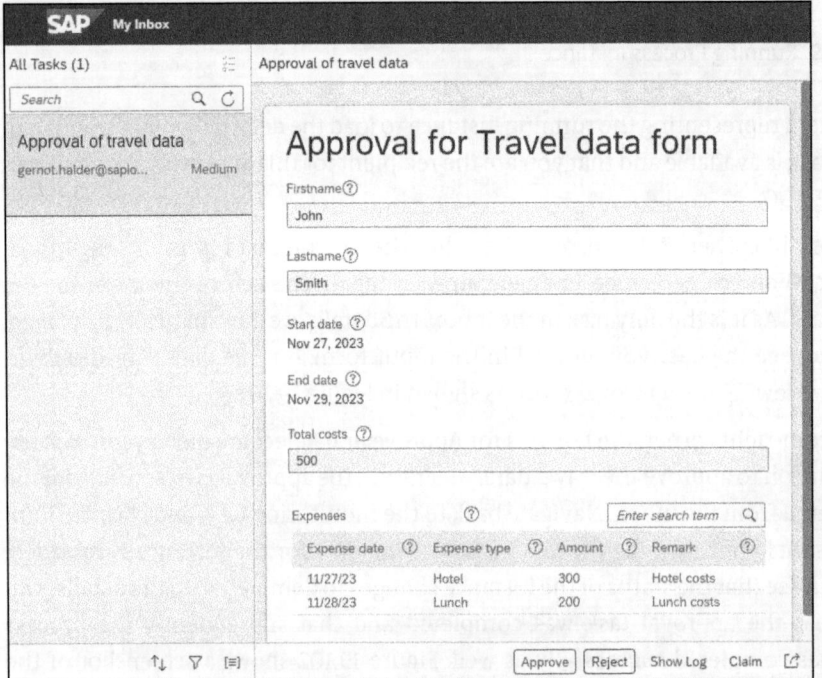

Figure 19.101 Approval Form Displayed in Inbox

19.4   Conditions and Branches

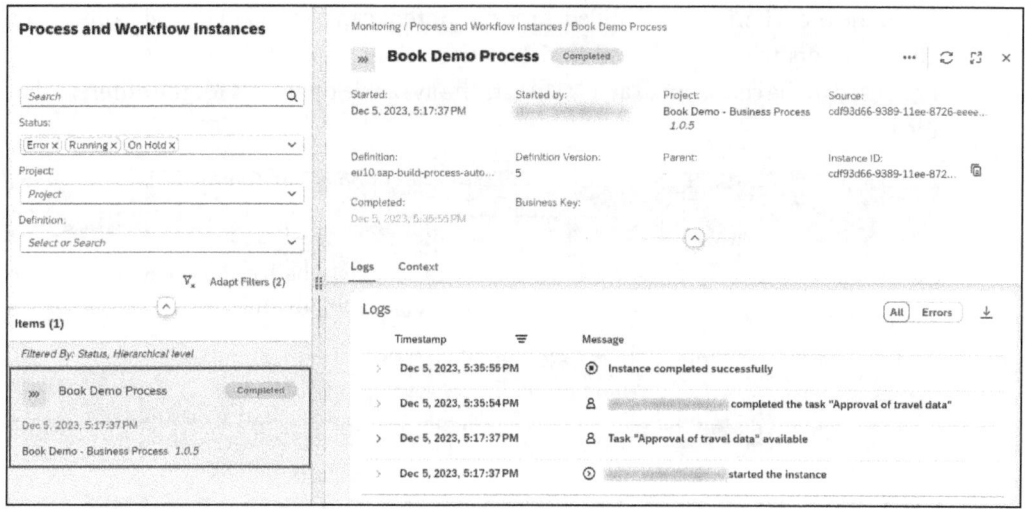

Figure 19.102 Completed Process Instance

You have now learned how input forms can be created and used inside your process. You have seen how approval forms can be added to the process and how a running process instance behaves in such a case. In the next section, you'll learn how to add conditions and branches to a process in order to design different branches and tasks depending on values of the process content.

19.4   Conditions and Branches

Conditions and branches are tools that help you control the data flow in certain sections of your process. Whereas *conditions* are used to decide which single way the data is to be sent by defining a simple decision based on an if-then-else clause, *branches* are executed in parallel, with data sent over all routes. The process waits for all parallel branches to be ended and then proceeds with the next step. The examples in the next sections will show how to use these tools.

19.4.1   Conditions

Using a condition, a business process can be routed based on defined criteria. SAP Build Process Automation provides a design element that can be used to accomplish data-based routing. The condition element has one incoming branch that routes the business process to the condition. During runtime, the condition element acts like an if-then-else condition that is familiar from standard programming languages. If the defined criteria is met, the element routes to the *if* branch. If the criteria is not met, the business process proceeds to the *else* branch, which is also called the *default* branch.

To define the criteria to be fulfilled, process content can be used in logical expressions. The operators that can be used in these expressions depend on the data type of the item used in the condition. Table 19.19 lists the available operators for the different data types.

Operator	Available for Data Type	Description
Is equal to	**Text**, **Date**, **Number**	Equals to true, if both compared items have the same value, otherwise false
Is not equal to	**Text**, **Date**, **Number**	Equals to false, if the compared items have the same value, otherwise true
Is empty	**Text**	Equals to true, if the item to be checked has no value, otherwise false
Is not empty	**Text**	Equals to false, if the item to reject has a value, otherwise true
Contains	**Text**	Equals to true, if the item to check contains the given text, otherwise false
Does not contain	**Text**	Equals to false, if the item to check does not contain the given text, otherwise true
Is less than	**Number**	Equals to true if the item to be checked has a smaller value than the item compared with, otherwise false
Is greater than	**Number**	Equals to true if the item to be checked has a greater value than the item compared with, otherwise false
Is less than or equal to	**Number**	Equals to true, if the item to be checked has a smaller or the same value than the item compared with, otherwise false
Is greater than or equal to	**Number**	Equals to true, if the item to be checked has greater or the same value than the item compared with, otherwise false
Is earlier than	**Date**	Equals to true, if the item to be checked has an earlier date than the item compared with, otherwise false
Is later than	**Date**	Equals to true, if the item to be checked has a later date than the item compared with, otherwise false

Table 19.19 Operators Available for Defining Condition

19.4 Conditions and Branches

Operator	Available for Data Type	Description
Is earlier than or equal to	Date	Equals to true, if the item to be checked has an earlier or the same date as the item compared with, otherwise false
Is later than or equal to	Date	Equals to true, if the item to be checked has a later or at the same date as the item compared with, otherwise false
Number of items is greater than	List	Equals to true, if a list has more entries than defined in the item compared with, otherwise false

Table 19.19 Operators Available for Defining Condition (Cont.)

In the condition editor that is used to define a condition, multiple simple conditions can be joined by either a logical AND or a logical OR. If your condition needs a mix of AND *and* OR, you have the option to create groups. Inside a group, you can define if all conditions or any condition must be met to meet the group. The same logic can then be used to join the group's results.

We will now add a condition to our business process in order to only create an approval step if the total cost exceeds an amount of 500. Therefore, open the example process and switch to the editable version if necessary. Currently, the process consists of a trigger form and then an approval form that is activated immediately after the process instance was started. As you only want to show the approval form if the criteria for the total costs are met, you need to add the condition between the trigger and the approval form. In the process, click the ⊕ icon between the trigger and the approval form and select the **Controls and Events** tile, as shown in Figure 19.103.

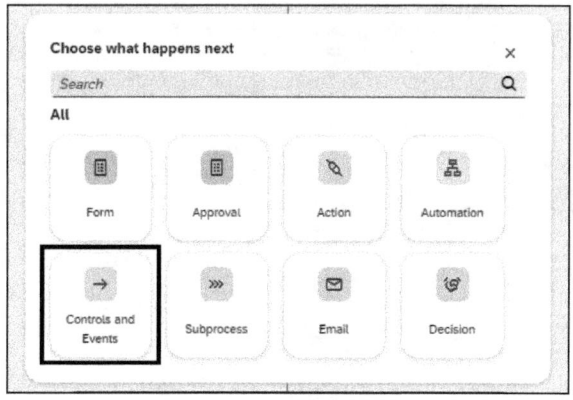

Figure 19.103 Selecting Element Type

Select the **Condition** item in the displayed menu, as shown in Figure 19.104.

19   Processes

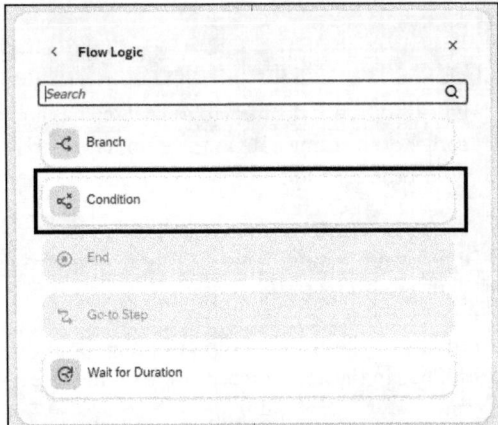

Figure 19.104 Adding Condition Element

As a result, a condition element is added between the trigger element and the approval form. The condition element is marked with the error flag as the condition still needs to be defined. In addition, only the if branch—that is, the branch that is chosen if the condition is fulfilled—is connected with the process element. The default branch currently has no connection. To configure the condition element properly, a logical condition must be defined, and the default branch must be connected with an element of the process. Figure 19.105 shows the process after the condition element was added.

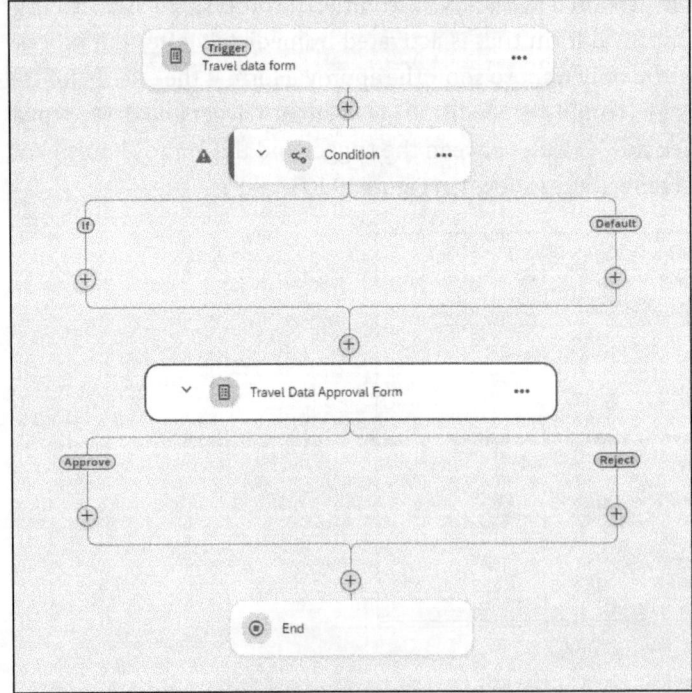

Figure 19.105 Condition Element Added

19.4 Conditions and Branches

If the condition is not met, the process should immediately come to an end. Therefore, click the plus icon (+) of the default branch and drag it to the end event. The default branch is automatically connected with the end event.

The next step is to define the condition that must be met to create an approval form. In this example, the approval must be done for travels that have a total cost greater than 500. To create this condition, click the condition element to display the properties in the configuration panel on the right side. In the properties, you can define a name for the condition element and for the branch of the fulfilled condition.

To create the condition itself, open the condition editor by clicking the **Open Condition Editor** button, as shown in Figure 19.106.

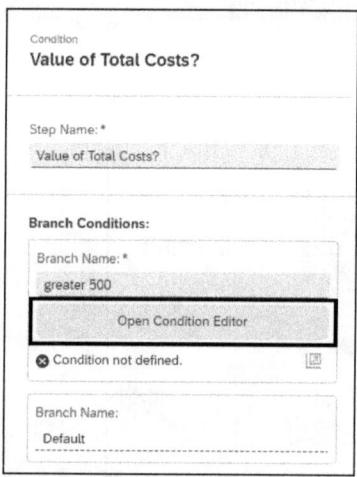

Figure 19.106 Button to Open Condition Editor

In the condition editor, click the left field to select the item that needs to be checked by condition. In the opened process content, select the **Total costs** field. In the field for the operator, select the **is greater than** value. In the third field, enter the fixed value "500", as shown in Figure 19.107.

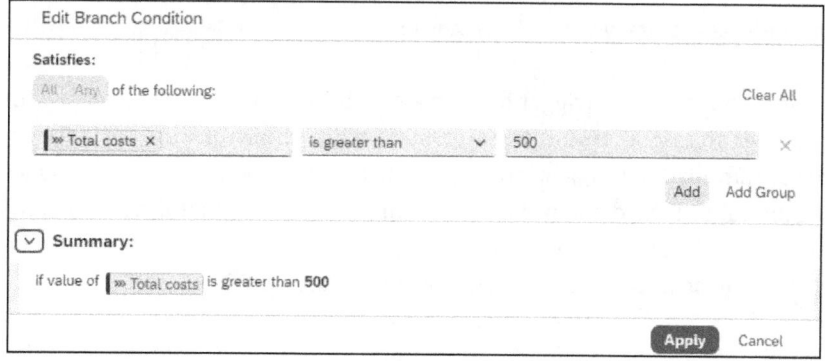

Figure 19.107 Created Condition

19 Processes

At the bottom of the dialog, a summary section is shown, where a human-readable version of your conditions is displayed. If the condition meets your requirements, click the **Apply** button to confirm your entries. The error flag is not displayed anymore, and, in the properties, you can see that the condition is properly defined. The process can now be saved, released, and deployed again. Figure 19.108 shows the business process with the configured condition.

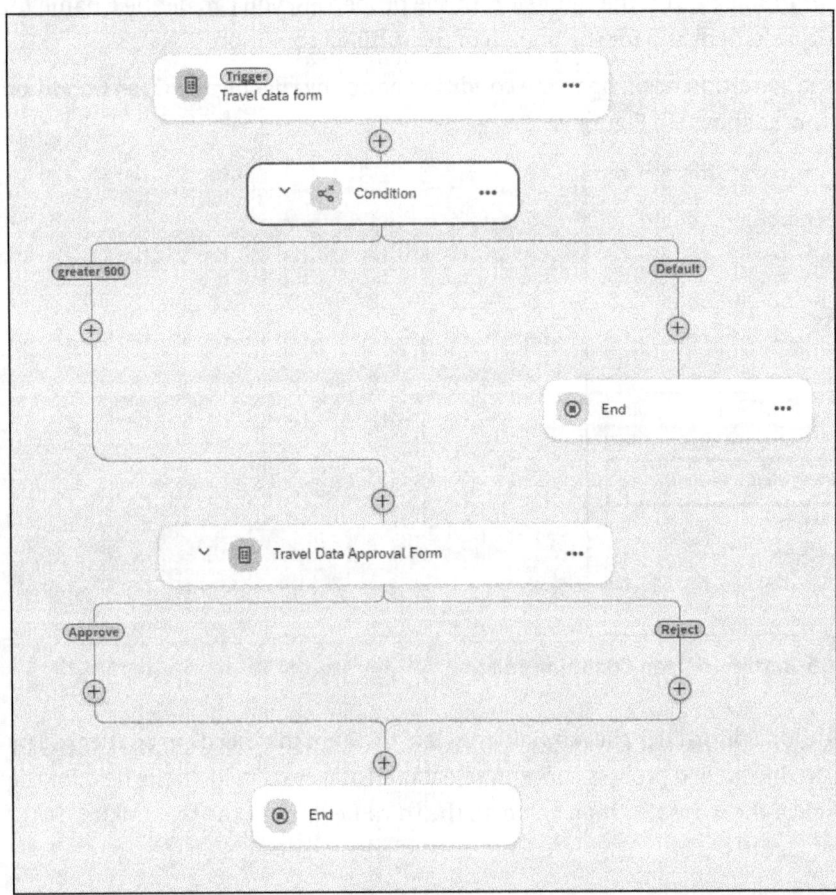

Figure 19.108 Business Process with Condition Added

To test the condition, open the trigger form in a new browser tab and enter data with total costs greater than 500. Submit the form to start a new process instance. In another browser tab, load the trigger form again and enter data with total costs smaller than 500. Again, submit the data. In the monitoring, you can see that the second process instance was immediately completed, whereas for the first process instance the approval task is waiting to be processed, as shown in Figure 19.109.

19.4 Conditions and Branches

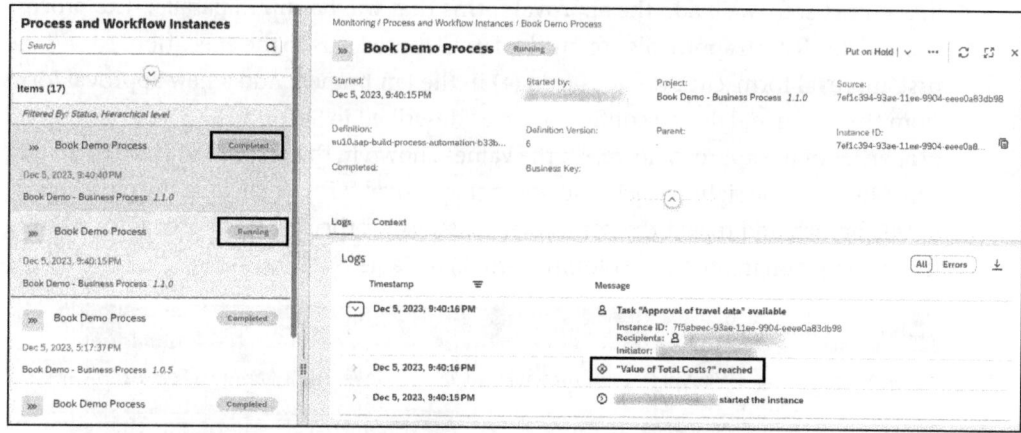

Figure 19.109 Monitoring Showing Condition Was Fulfilled

19.4.2 Branches

In a business process, branches can be used to design tasks that need to be processed in parallel. The process will not continue processing until all parallel tasks have been completed. In an SAP Build Process Automation process, up to 10 branches can be run in parallel. To see how branches behave when a process is running, we will again use our example process. Navigate to the development canvas and delete the approval form and the decision control so that only the trigger and the end event are left.

Next, click the ⊕ icon between the trigger element and the end event, and select the **Controls and Events** item, as shown in Figure 19.103. Finally select the **Branch** item from the panel displayed, as shown in Figure 19.104. Click the branch element to load the properties panel of the control. In the properties, you can set the names of the element itself and of the branches leading from the element. You also can add branches if needed. Set the values of the properties according to Figure 19.110. You can see that the labels of the element in the development canvas are changed accordingly.

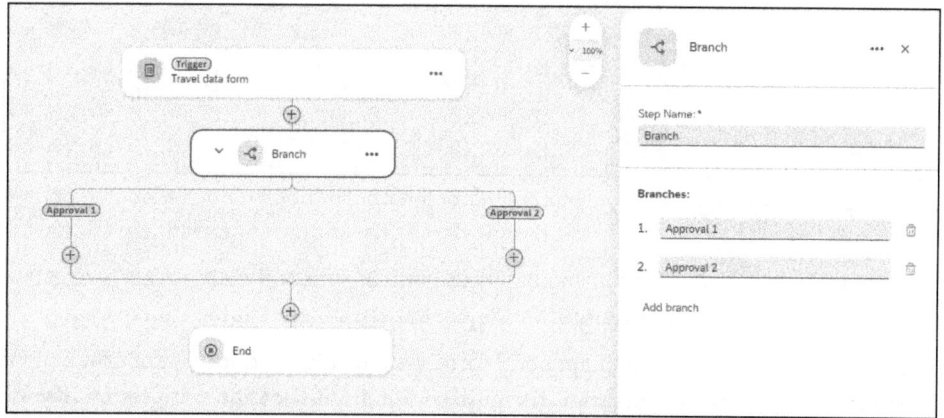

Figure 19.110 New Branch Element

19 Processes

In the next step, we'll add the approval forms that will be run in parallel. The process will wait until both approvals are finished until it proceeds to the execution. To add the first approval form, click the plus icon (+) in the left branch. Add a new approval form from the menu and do the configuration as described in Section 19.3.3. For the general properties of the approval form, set the values shown in the second column (Form 1) of Table 19.20. In the right branch, add another approval form by clicking the plus icon (+) of this branch and repeat the configuration steps for the second approval form using the values given in the third column (Form 2) of Table 19.20.

Property Name	Form 1	Form 2
Step Name	"Travel Data Approval Form 1"	"Travel Data Approval Form 2"
Subject	"Approval 1"	"Approval 2"
Priority	Medium	High
Users	Your SAP BTP user name	Your SAP BTP user name

Table 19.20 Properties for Approval Form in Upper Branch

The process will now look as shown in Figure 19.111.

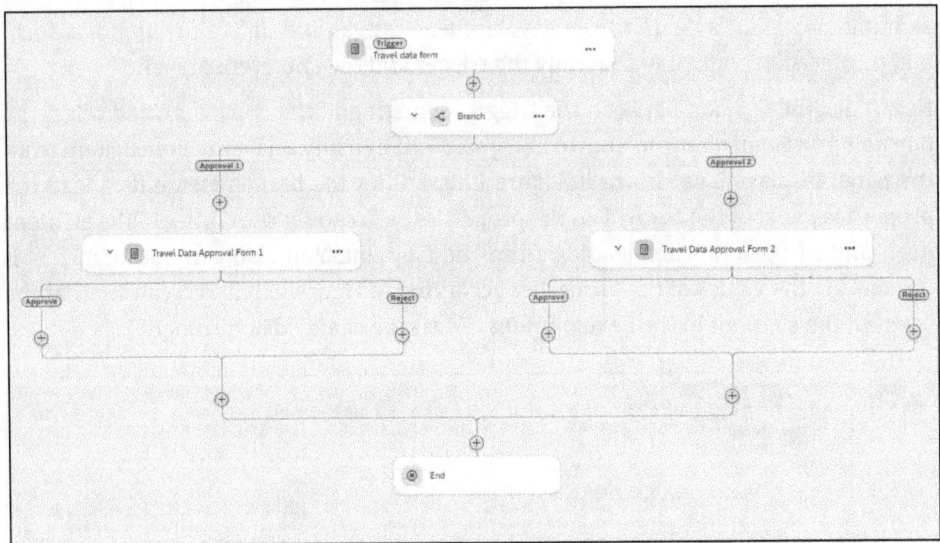

Figure 19.111 Parallel Approvals

You want the process to stop immediately as soon as one of the approvals is rejected. To ensure this, add an end event for the **Reject** branch of **Approval 1**.

Repeat this step for the second approval form. As a result, both forms are now free of errors and the process can be saved and released. Finally, deploy the new version of your process. Figure 19.112 shows the final state of the process with the parallel approvals.

19.4 Conditions and Branches

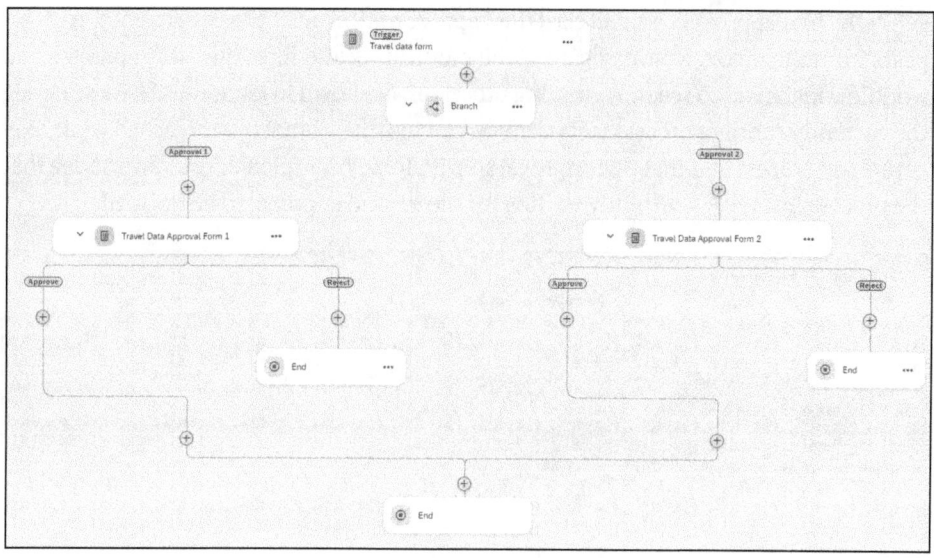

Figure 19.112 Process with Parallel Approvals

To test a new process, open the trigger form in a browser window and submit the filled-in form to start the new process instance. Once the new instance is running, open the inbox by clicking the **Inbox** icon in the upper-right corner of the lobby. You can see that two approval tasks were created and wait for approval or rejection, as shown in Figure 19.113. According to the configuration that was set for the different approval steps in the process, the priority of the two tasks is set.

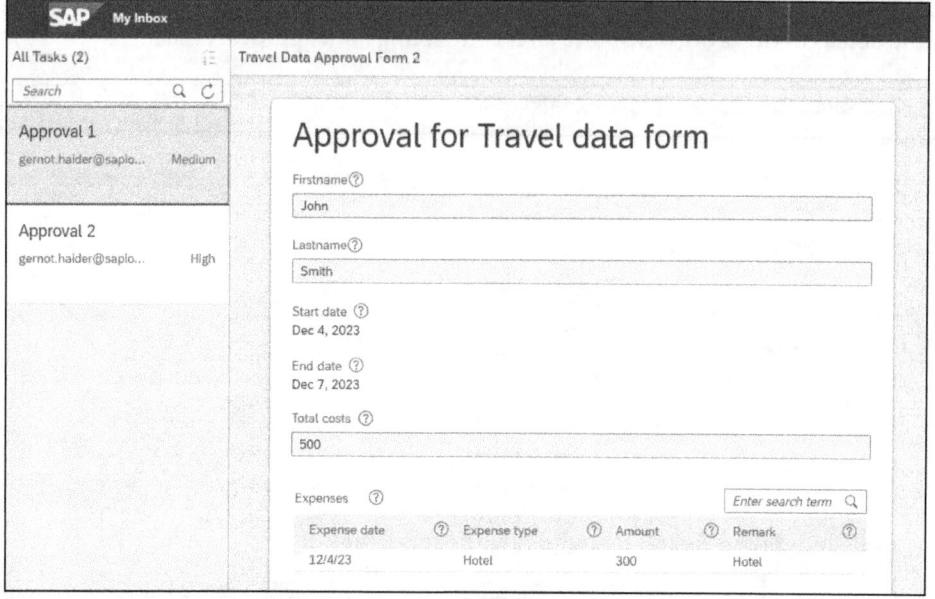

Figure 19.113 Inbox Showing Approval Tasks for Both Branches

19 Processes

Select the first approval task and click the **Approve** button. As the task is finished, it disappears from the inbox. Now navigate to the monitoring and filter the list for process and workflow instances to see only running processes, as shown in Figure 19.114. By selecting the currently running process instance, you can see that a branch was reached while executing the process and that one approval step is already completed. You can also see that the process instance is currently waiting for the second approval to be finished.

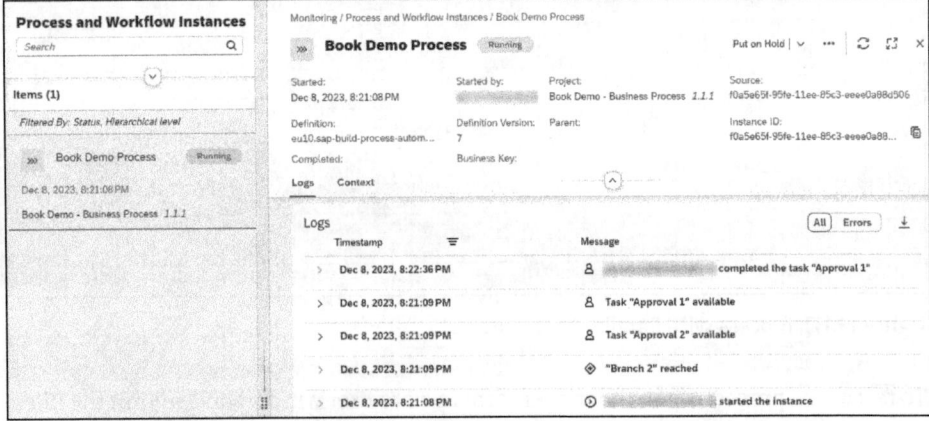

Figure 19.114 Monitoring of Process Instance

Now navigate back to your inbox and complete the second task by clicking the **Approve** button. Refresh the details view of the process instance in the monitoring. You can see that the second approval was completed, and as a result the merge of the branches was reached, and the process instance could be completed successfully. Figure 19.115 shows the details of the process instance after completing both approval tasks.

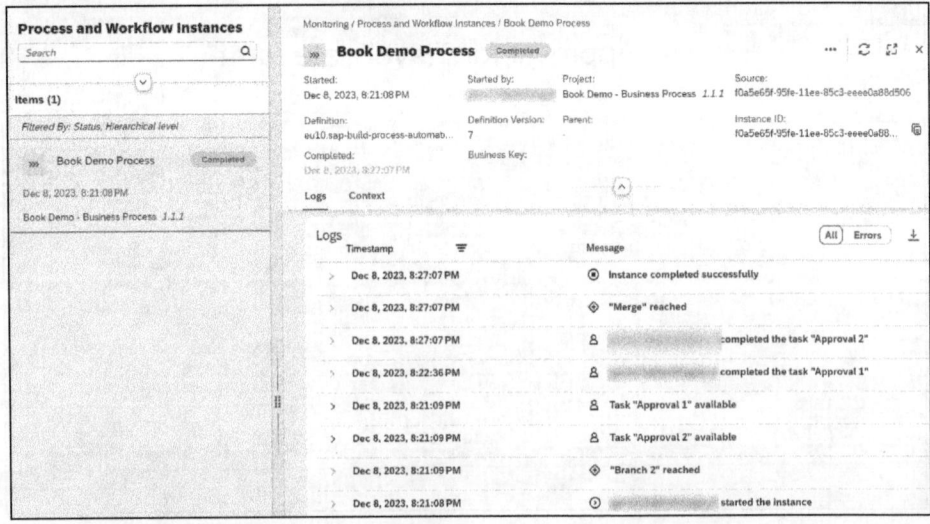

Figure 19.115 Monitoring of Completed Process Instance

You have seen now that the process instance waits for both branches to be completed before the execution proceeds.

19.5 Actions

With the use of actions, you can connect your business process with the backend system and call functionality to be executed there. In Chapter 21, you will learn how an action can be created to be used in a process automation. For now, we just want to show how an existing action can be added to a business process and how this action needs to be configured.

To add an action to a business process, click the ⊕ icon between two process steps and select the **Action** element from the displayed panel. As a result, a dialog is displayed where you can browse a library of existing actions, each action represented by a tile. To add an action, click the **Add** button of the corresponding tile. For your example process, click the **Add** button of the tile with the name **Add New Entity to Log Entry** to add this action as a new step to your business process, as shown in Figure 19.116.

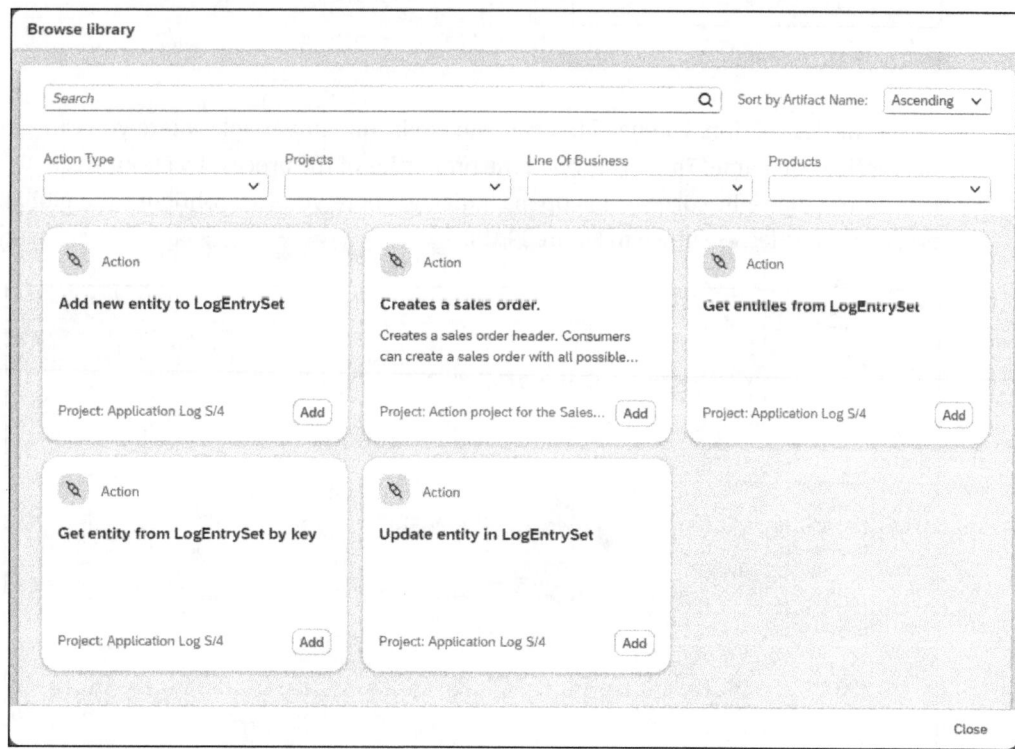

Figure 19.116 Browse Dialog for Existing Actions

The action is added to the business process as a new step and can be configured. By clicking the action step in the business process, the properties are displayed in the right-side panel of the development canvas. In the **General** section of the properties, you can see that a destination variable has to be set in order to establish a connection to the backend system, as shown in Figure 19.117.

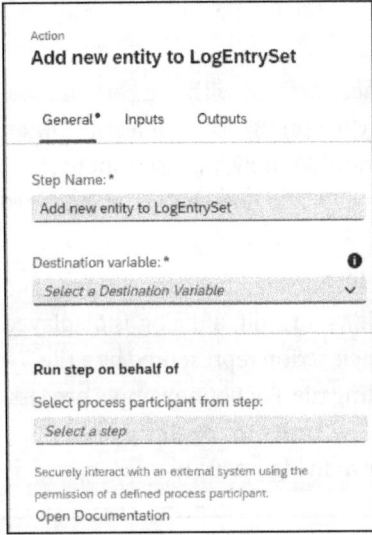

Figure 19.117 General Properties of Action

To set the destination, you need to create an environment variable where you can set the destination name. Therefore, open the properties of the process by clicking the ⚙ icon in the top-right corner. This opens a dialog where you can configure the global project properties, as shown in Figure 19.118.

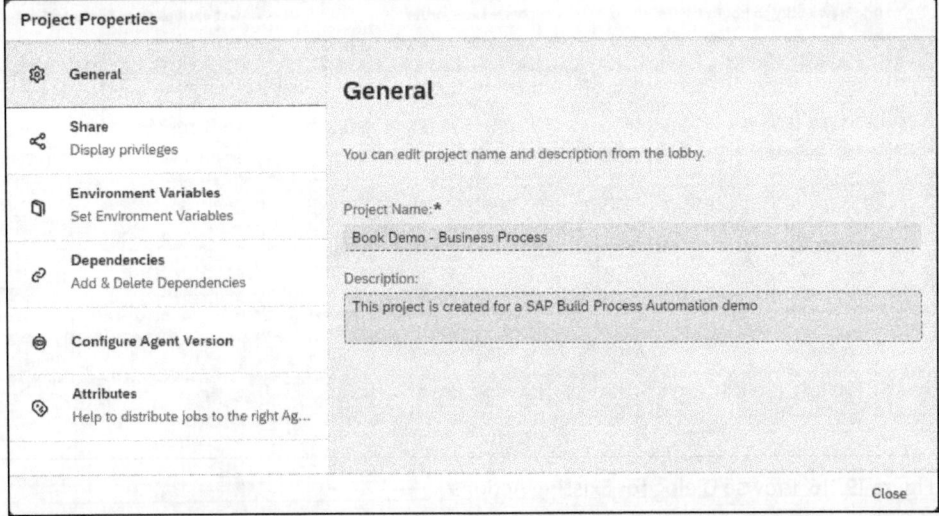

Figure 19.118 Project Properties

Click the **Environment Variables** item in the left panel of the dialog to show the environment variables of the project. In the view of the environment variables, click the **Create** button to create a new variable, as shown in Figure 19.119.

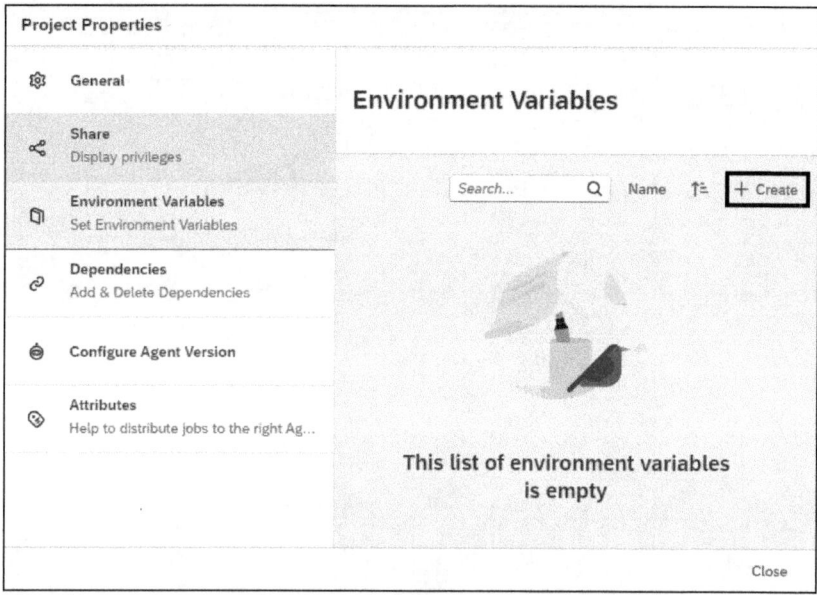

Figure 19.119 Creating New Environment Variable

Next, enter an identifier for the environment variable, representing the destination for the action, and write a short description. As the type of this variable, choose the value destination from the list. Click the **Create** button to confirm your entries and create a new environment variable. Finally, click the **Close** button in the bottom-right corner of the project properties dialog.

This new environment variable can now be used to set the destination of the action in your business process. Therefore, click the field for the destination variable in the **General** settings of the action, and select the variable you've just created. If this variable is not present yet, save the project by clicking the **Save** button, then select the field for the destination variable again. As the destination is configured, the action is now ready to go.

19.6 Summary

In this chapter, you learned how to create a business process and how to use the development canvas. You now know how to add the most common elements and how to configure their properties. You have also seen how a process instance behaves depending on the elements used and how the monitoring represents the data. For a deeper dive into monitoring, refer to Chapter 25.

Chapter 20
Rules and Decisions

Decisions accompany us through life. Especially in the context of business processes, decisions are made as to which steps should be taken and when. In this chapter, we focus on decision-making in automated processes.

Rules and decisions can be used to let a process decide what to do next based on the data that is processed. Whereas *rules*, also known as *text rules*, are collections of expressions in a simple if-then-else format, *decision tables* are collections of input and output rule expressions that are represented as a table. This gives the designer the option to configure multiple conditions in one table that are combined to calculate an outcome of the decision. In this chapter, we will show you how the different types of decisions are created and how they can be used to control the process and the method of data processing.

20.1 Decision Editor

With the decision editor, you can create and configure your decisions according to your needs. Figure 20.1 shows a screenshot of the decision editor. Starting from the project's overview page, you can access the editor by creating a new decision artifact using the **Create** button above the artifact's list. Another option that will be presented throughout the following pages is to add a new decision directly to your process. The decision element also provides the option to open the decision editor in order to design the rules.

The editor is divided into two parts. On the left side, the decision diagram is displayed. This diagram is a flow chart that describes the execution flow of the decision logic between input and output. Using this diagram, you can view the input data, the output data, the policies configured, and the rules within these policies. By clicking an element of the diagram, you can directly access the data that is part of the selected component. Figure 20.2 shows a screenshot of a completely configured text rule and how the components are displayed.

20   Rules and Decisions

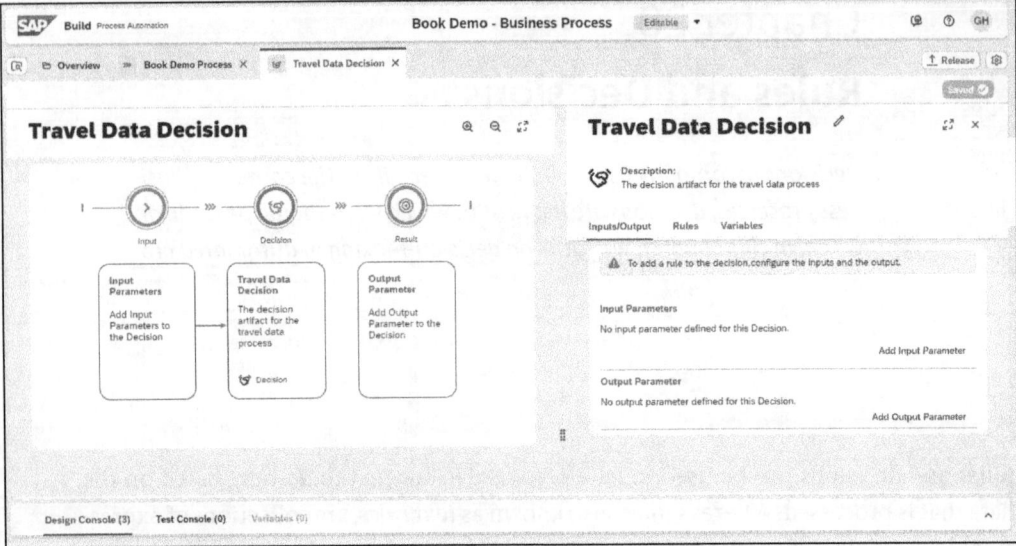

Figure 20.1 Decision Editor

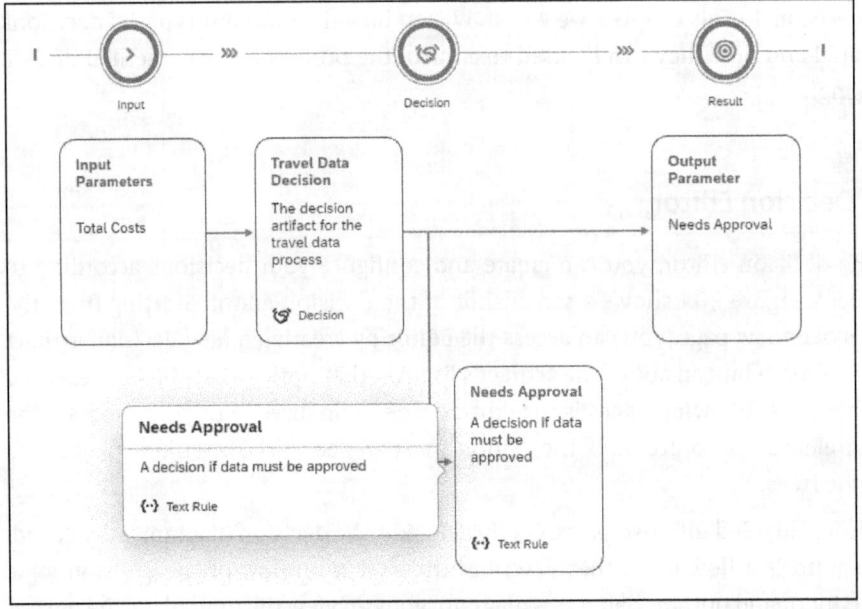

Figure 20.2 Decision Diagram of Configured Text Rule

Note that as soon as all sections of the decision are configured correctly, the steps representing the input to the decision and the result are colored green. If a configuration is missing or has an error, the color of the corresponding step changes to orange or even to red. Details of missing data or an erroneous configuration are shown in the design console at the bottom of the decision editor, as shown in Figure 20.3.

20.2 Creating a Decision Artifact

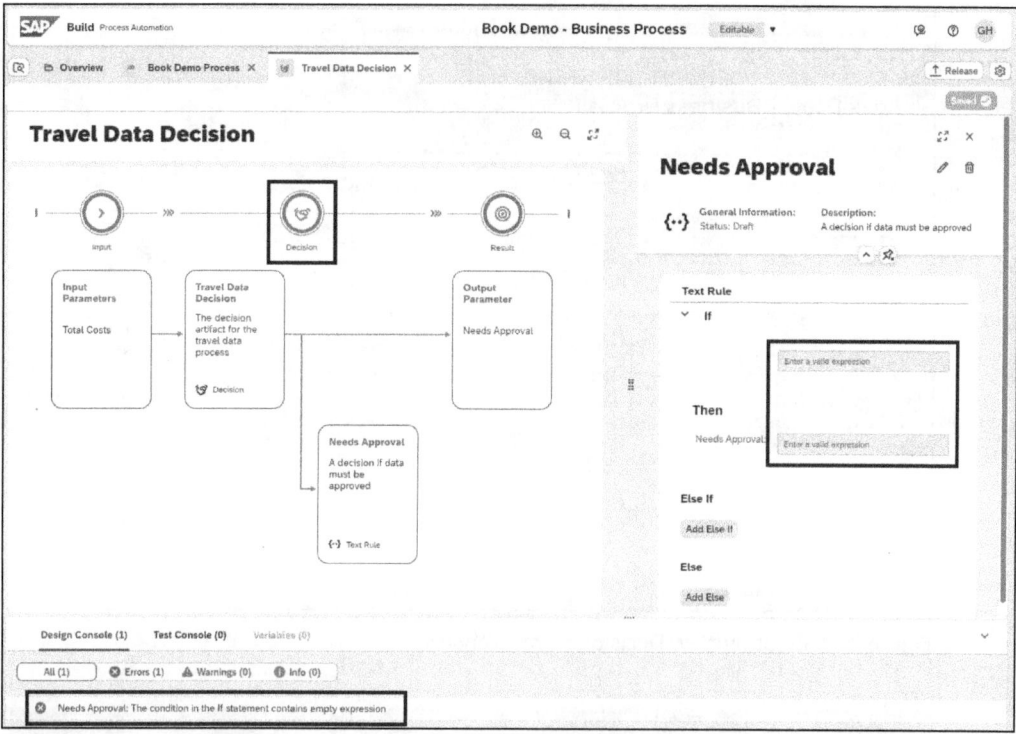

Figure 20.3 Error Details in Decision Editor

The lifecycle of decision follows the same rules as a business process. You can create releases to save different versions of a decision, and you can revert to an older decision if necessary.

20.2 Creating a Decision Artifact

In SAP Build Process Automation, you have two options to create a new decision. The first option is to create a new decision artifact directly from the overview. To do this, select your business project in the lobby to load the artifacts list. In this list, click the **Create** button on the right side above the artifacts list and select **Decision** from the displayed items menu, as shown in Figure 20.4.

As a result, a dialog is displayed where you can provide a name, a unique identifier, and a description for the new decision. Fill in the fields according to Table 20.1.

20 Rules and Decisions

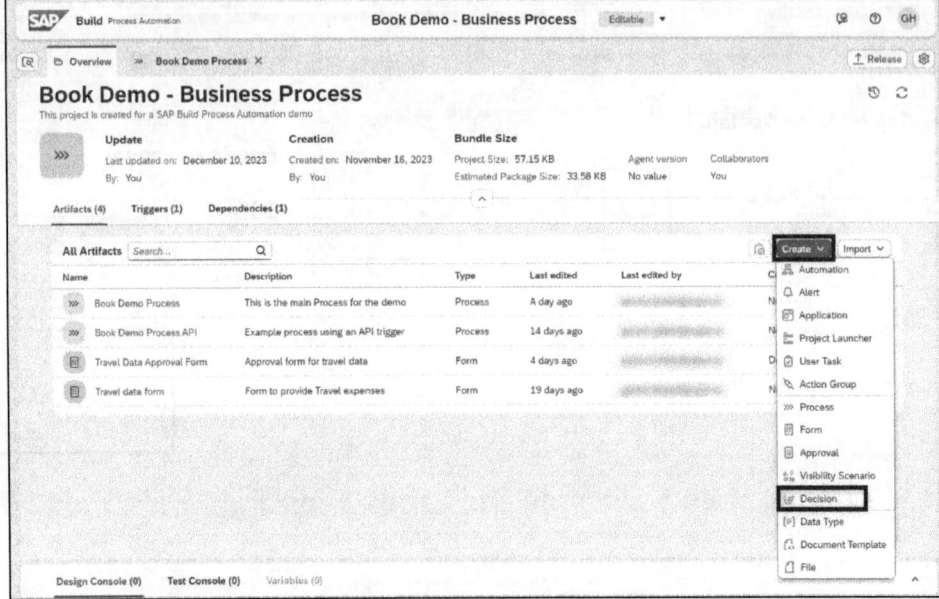

Figure 20.4 Creating New Decision from Overview

Field Name	Value
Name	"Travel Data Decision"
Identifier	"travelDataDecision" (created automatically as you type the name of the decision, but can be changed if necessary)
Description	"The decision artifact for the travel data process"

Table 20.1 Values for New Decision

A screenshot of this dialog is shown in Figure 20.5. Click the **Create** button to confirm your entries and to create a new decision.

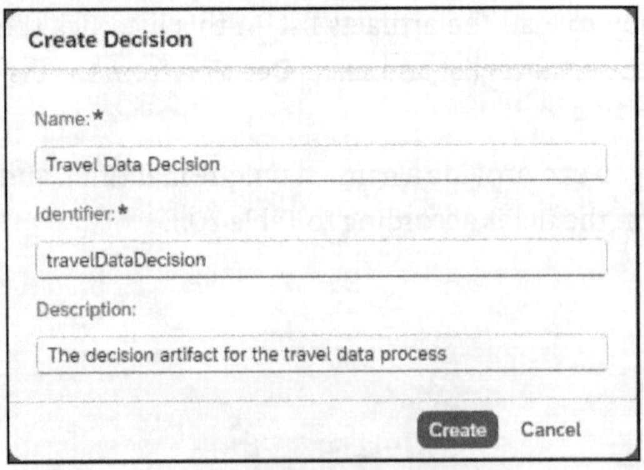

Figure 20.5 Create Decision Dialog

630

20.2 Creating a Decision Artifact

After the **Create** button is clicked, a new tab is opened showing the development tool for decisions. Figure 20.6 shows a screenshot of this development tool.

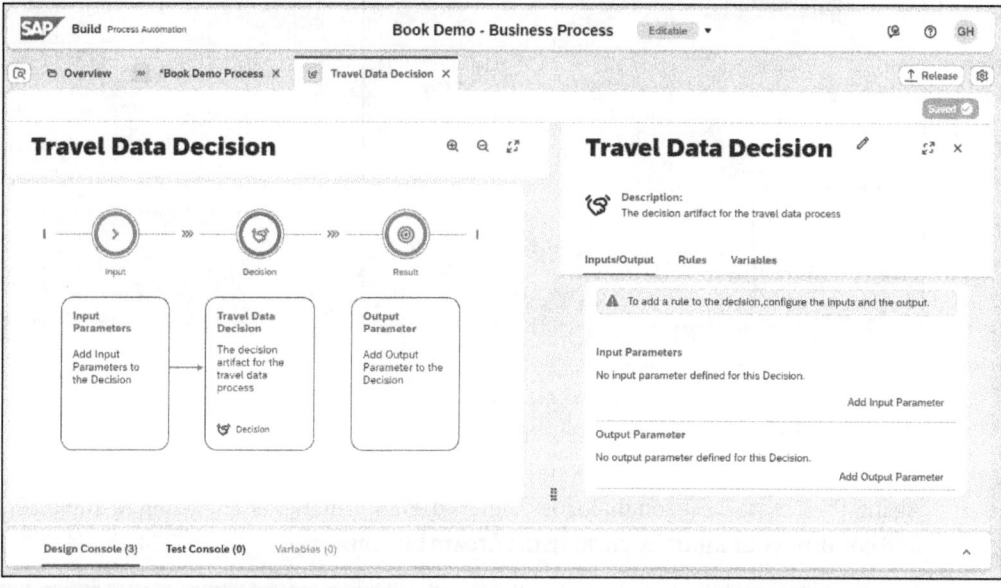

Figure 20.6 Development Tool for Decisions

The second option is to directly create the decision inside your business process. To do this, first open the process in a new tab if it is not opened yet, then click the ⊕ icon right after the process trigger. In the displayed panel, select the **Decision** item, as shown in Figure 20.7.

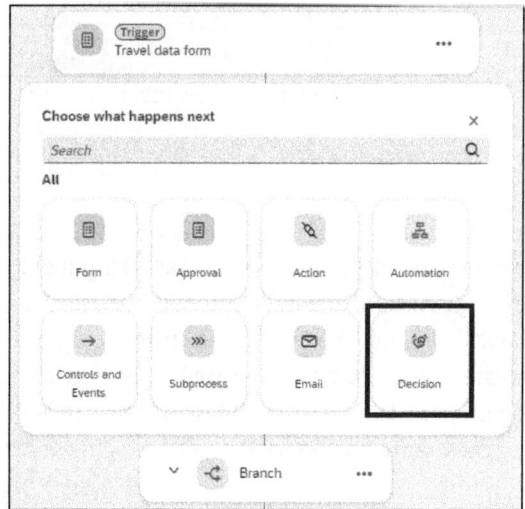

Figure 20.7 Selecting Decision Element

20 Rules and Decisions

Finally, select **Blank Decision**, as shown in Figure 20.8.

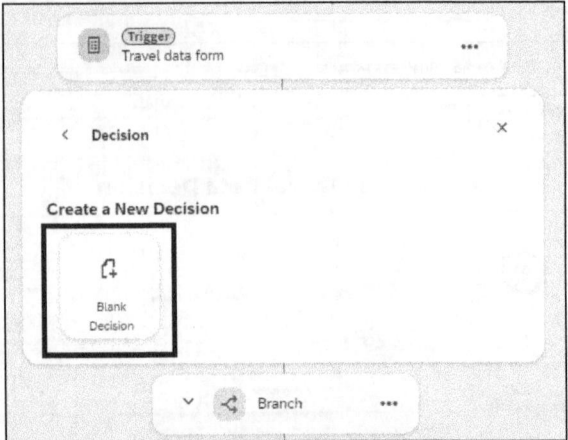

Figure 20.8 Creating Decisions inside Processes

Again, the **Create Decision** dialog is displayed. Fill in the fields according to Table 20.2 and confirm your input by clicking the **Create** button.

Field Name	Value
Name	"Determine Approvers"
Identifier	"determineApprovers" (value created automatically as you type the name of the decision, but can be changed if necessary)
Description	"The decision to determine possible approvers for the travel data"

Table 20.2 Values for New Decision

As a result, the design console for the decision will be opened in a new tab.

20.3 Creating a Data Type

A *data type* is an object that describes a data structure that can be used as an input or as an output parameter in an automation, in a decision, or even in processes. To create a data type, navigate to your project overview and click the **Create** button above the artifacts list. Select the **Data Type** item to create a new data type to be used in the decisions. As a result, the **Create Data Type** dialog is opened. Fill in the fields according to Table 20.3 and confirm your input by clicking the **Create** button in the bottom-right corner of the dialog (see Figure 20.9).

20.3 Creating a Data Type

Field Name	Value
Name	"Approver"
Identifier	"approver" (value is created automatically as you type the name of the decision, but can be changed if necessary)
Description	"The approver for the given travel data"

Table 20.3 Values Describing New Data Type

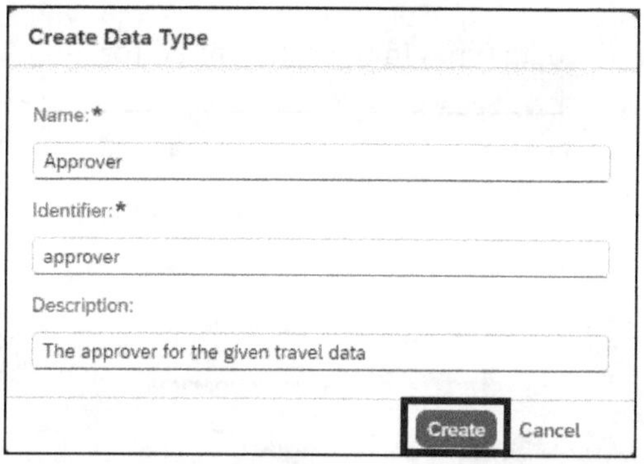

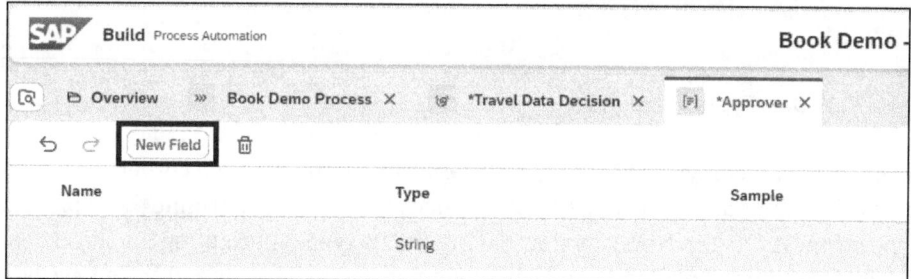

Figure 20.9 Create Data Type Dialogue

As a result, the data type editor is opened in a new tab. This editor enables you to define the structure of your data type by adding fields describing their types and setting additional properties. To add a new item to the list of fields in a part of the data type, click the **New Field** button on the left side above the list, as shown in Figure 20.10.

Figure 20.10 Adding New Field to Data Type

In the properties panel of the new field, enter the values according to the second column (Field 1) of Table 20.4. Add another field by clicking the **New Field** button and entering the properties in the third column (Field 2).

633

20 Rules and Decisions

Field Name	Field 1	Field 2
Name	"Username"	"UserGroup"
Type	String	String
Sample Value	"somebody@domain.com"	"Approver_Asia"
Required	Yes	Yes

Table 20.4 Properties of Username and UserGroup Fields

Figure 20.11 shows a screenshot of the created data types in the data type editor.

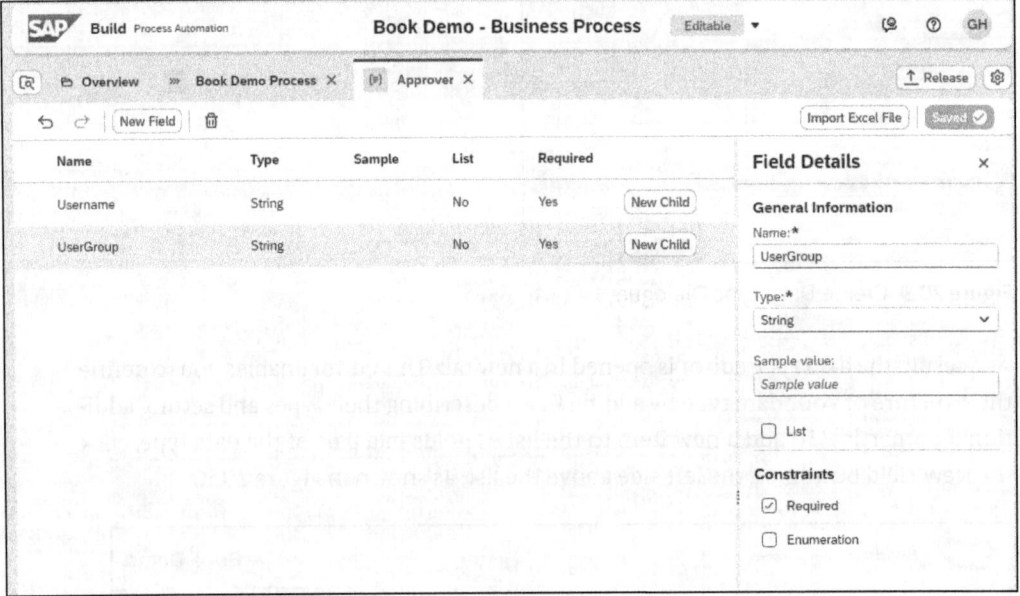

Figure 20.11 Data Types for Approver

Finally, save the data type by clicking the **Save** button in the top-right corner. In Section 20.6, we'll use this data type in the result of a decision table to determine the approvals in the business process. Now let's create another data type representing the travel data that you need for the decision. Navigate to your artifacts overview and add a new data type with the properties shown in Table 20.5.

Field Name	Value
Name	"Travel Data"
Identifier	"travelData"
Description	"Travel data needed for the approver determination"

Table 20.5 Values Describing Travel Data Data Type

Add two fields to the data type according to Table 20.6.

Field Name	Field 1	Field 2
Name	"Location"	"TotalCosts"
Type	String	Number
Required	Yes	Yes
Enumeration	Yes	-
Enumeration Values	USA,Europe,Asia (use the comma as the delimiting character)	-

Table 20.6 Properties of Fields Representing Travel Data Data Type

Finally, save the data type by clicking the **Save** button in the upper-right corner. Figure 20.12 shows a screenshot of the data type created for the travel data that that will be used in the decision table.

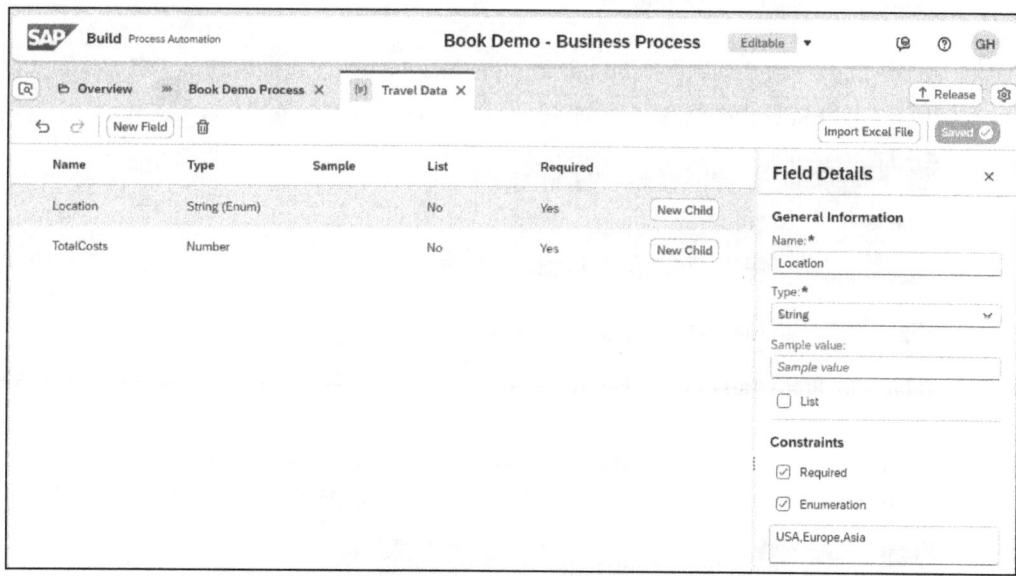

Figure 20.12 Fields Representing Travel Data Data Type

20.4 Creating a Text Rule

In Section 20.2, you created a decision artifact named *travel data decision*. Navigate back to the project overview and open this decision by clicking the corresponding item in the artifacts list. In the decision diagram, click the component that represents the input parameters. In the right panel, the tab for input and output parameters is displayed. Add a new input parameter by clicking the **Add Input Parameter** button as shown in Figure 20.13.

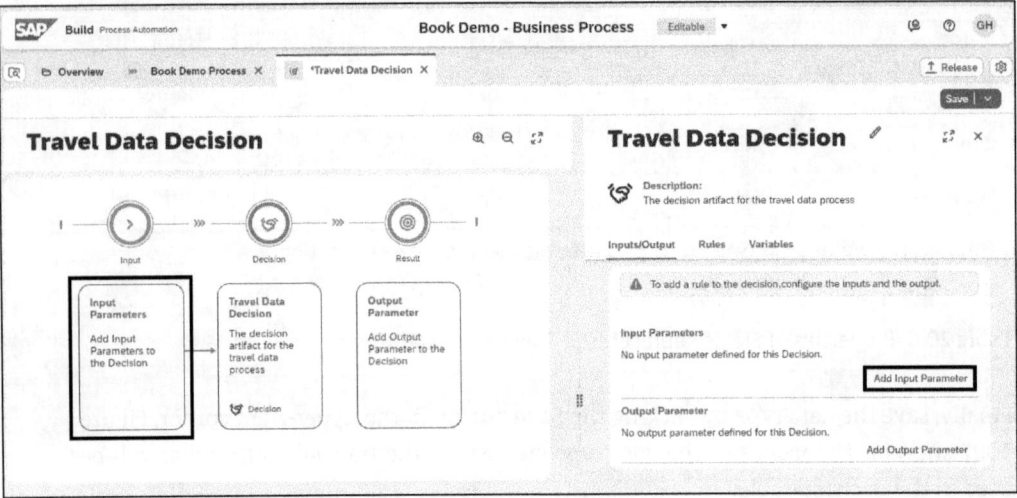

Figure 20.13 Add Input Parameter

In the line created for the input parameters, add a name, a description, and a data type for the input parameter to be created, according to Table 20.7.

Field Name	Value
Name	"Total Costs"
Description	"Total costs of the travel"
Type	Number

Table 20.7 Properties of Input Parameter

In the next step, create an output parameter for the decision. Click the **Add Output Parameter** button and in the line created, fill in the properties as shown in Table 20.8.

Field name	Value
Name	"Needs Approval"
Description	"Indicator if costs must be approved"
Type	Boolean

Table 20.8 Properties of Output Parameter

Figure 20.14 shows the decision editor after the input parameter and the output parameter have been created. You can see that the components representing the parameters

20.4 Creating a Text Rule

are now colored green as they are already configured, whereas the decision itself is still colored red.

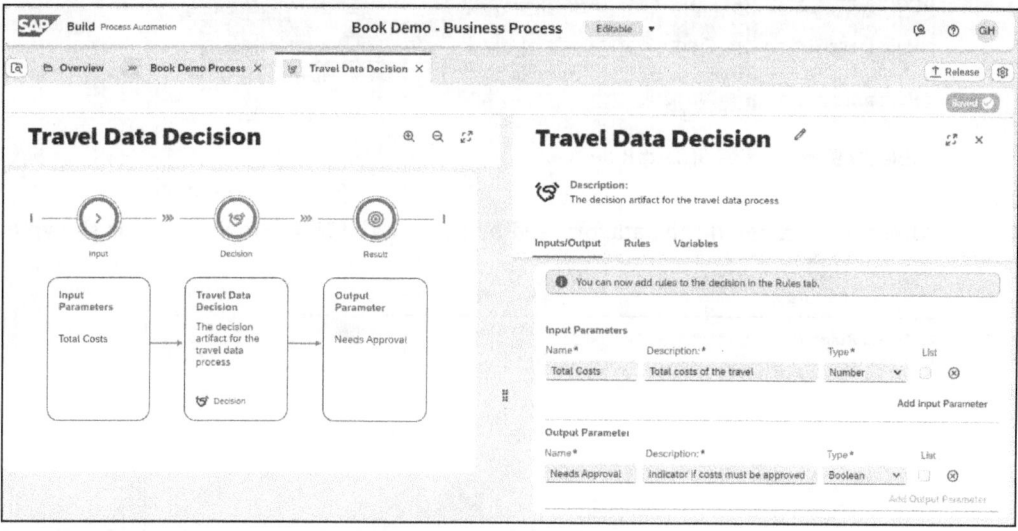

Figure 20.14 Decision after Input and Output Were Configured

Next, let's create the rule that will be used to determine if the travel data needs approval or not. Click the component representing the decision in the decision diagram, or select the **Rules** tab in the configuration panel on the right side of the editor to switch to the rule's configuration. Click the **Add Rule** button as shown in Figure 20.15 to create the rule that will be used in this decision.

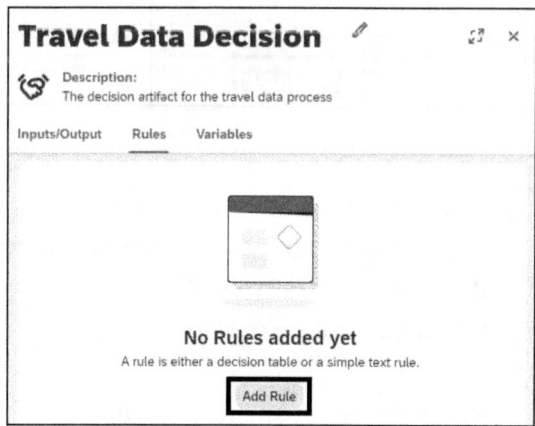

Figure 20.15 Add New Rule to Decision

As a result, the wizard to create a rule is displayed in a dialog. Select **Text Rule** as the rule type and enter a name and a description for the rule as shown Table 20.9.

20 Rules and Decisions

Field Name	Value
Rule Name	"Travel Data Needs Approval"
Rule Description	"Indicator if costs of the travel must be approved"
Reusable Rules	No

Table 20.9 Properties of Text Rule

After you've entered the data, proceed by clicking the **Next Step** button, as shown in Figure 20.16.

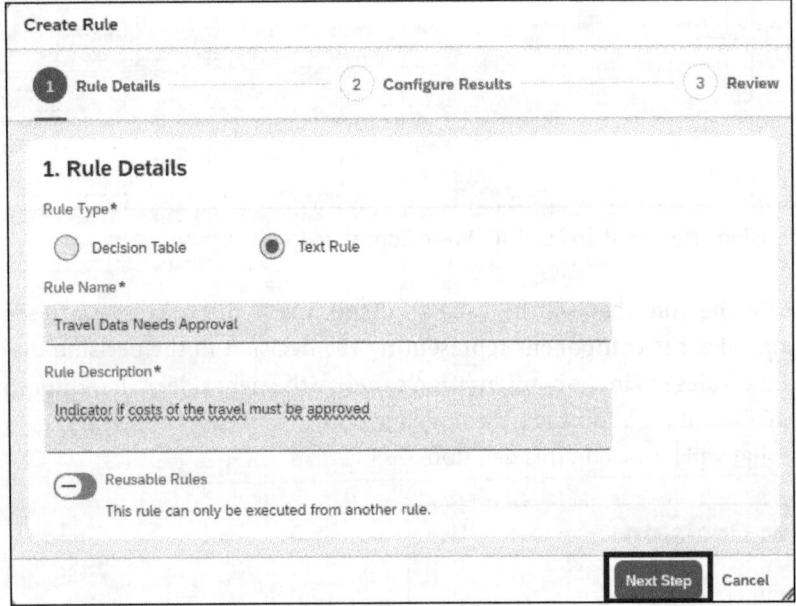

Figure 20.16 Rule Details

The next step is to configure the results of the decision. From the vocabulary, click the **Needs Approval** parameter (see Figure 20.17). The parameter is automatically added to the result details. Click the **Next Step** button to proceed to the creation of the text rule.

The last step of the wizard is the review of the configured parameters. Click the **Create** button to create the rule and add it to the decision.

As a result, the text rule is created and added to the decision component in the decision diagram. In the right panel, the text rule is displayed and can now be configured by adding a logical expression and the result value. Figure 20.18 shows the created but still empty text rule.

20.4 Creating a Text Rule

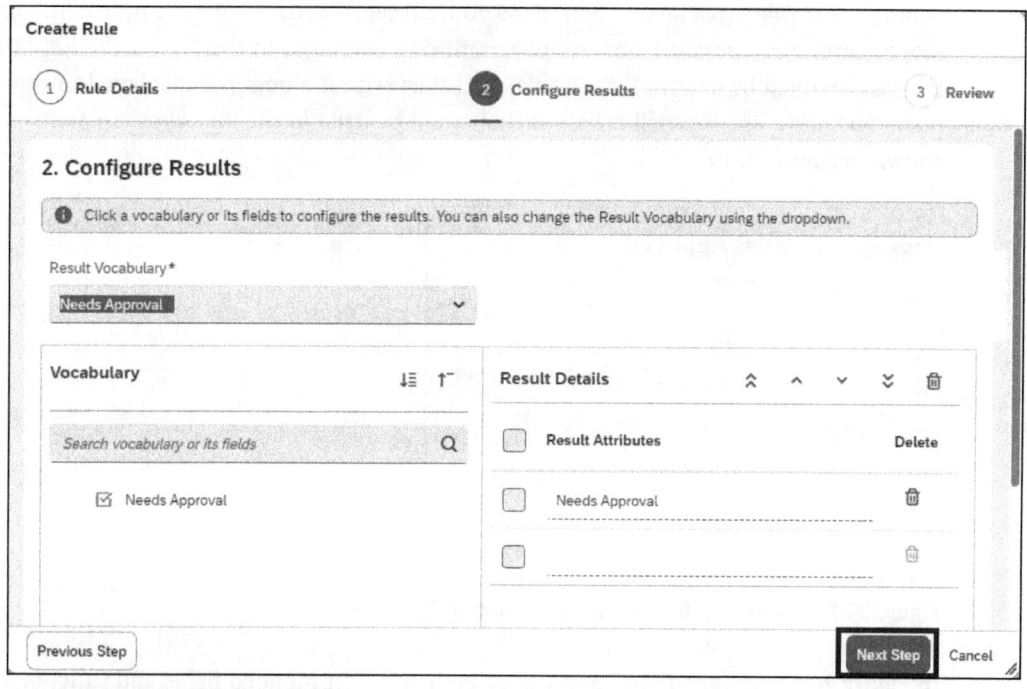

Figure 20.17 Result Details of Rule

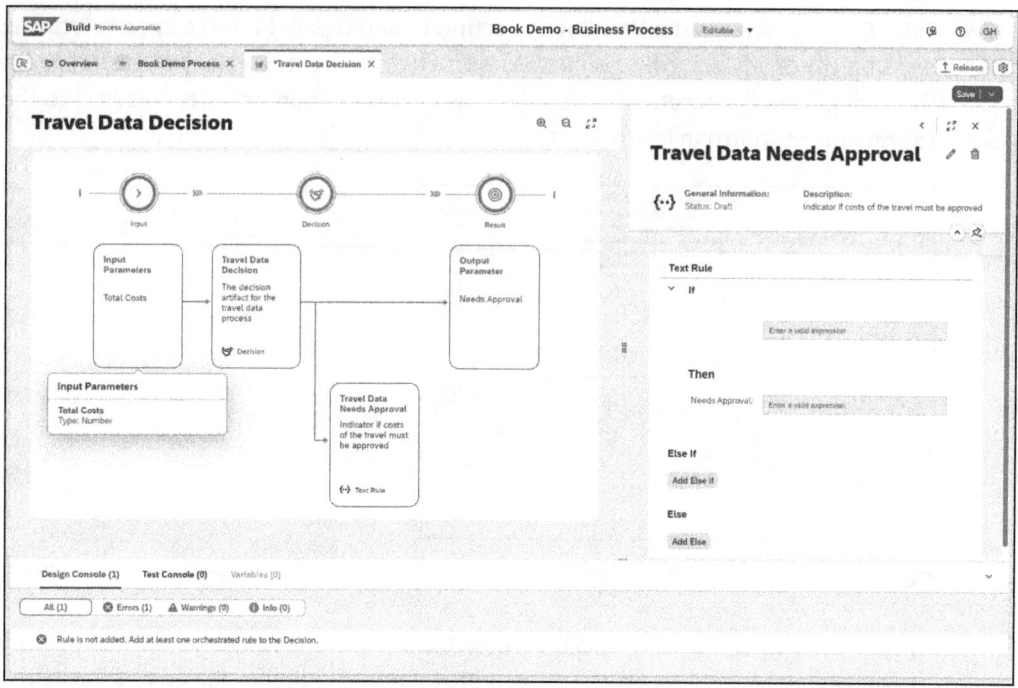

Figure 20.18 New Text Rule in Decision Editor

20 Rules and Decisions

Now the text rule must be configured. As you can see, it behaves like a simple if/then/else assignment, familiar from any programming language. In the *if* part, the logical expression must be declared. As soon as you start typing, a popup is displayed to support you with a list of possible values that could be used in the logical expression, as shown in Figure 20.19.

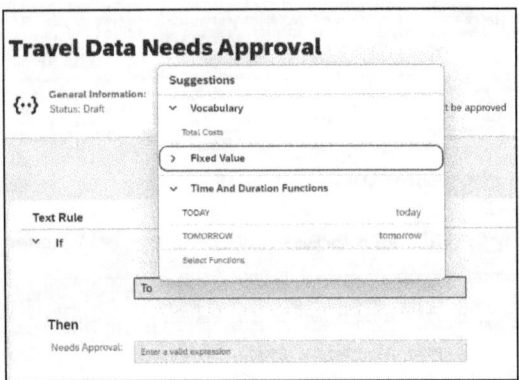

Figure 20.19 Value Help for Logical Expression of Text Rule

In Figure 20.19, we typed "To", and the value help recommended fields and functions that were found according to the input. You can either continue typing, or select a value from the given suggestions by clicking the item in the list. In this case, select the **Total Costs** item from the **Vocabulary** section to add this field to the expression. The decision editor adjusts the possible values to help you create a valid expression. From the **Comparison Operators** section, select the **is greater than** operator to add it to the expression, as shown in Figure 20.20.

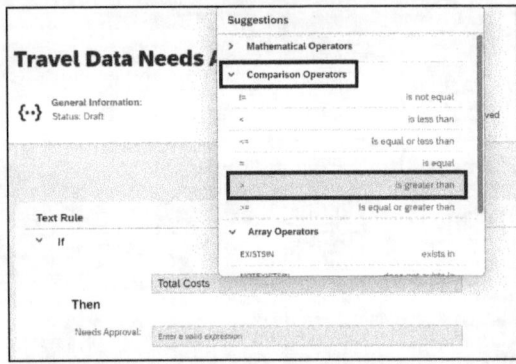

Figure 20.20 Add Operator to Logical Expression

Again, the decision editor adjusts the suggestions according to the current entry in the logical expression. In this example, an approval task should be created as soon as the total costs of travel exceed $500. Therefore, we now want to enter a fixed value to be compared with the total costs entered in the input form of a process instance. As

20.4 Creating a Text Rule

shown in Figure 20.21, expand the **Fixed Value** section by clicking it and enter "500" in the text field provided.

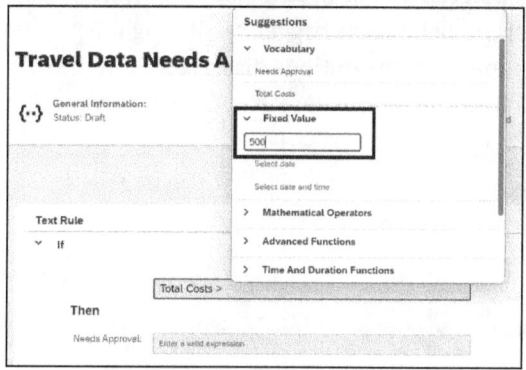

Figure 20.21 Fixed Value: Part of Expression

If the logical expression is fulfilled, the value of the **Then** branch is taken as the result of the decision; otherwise, the value of the **Else** branch is taken as the result. As the result value here is of a Boolean data type, you can use fixed values for the **Then** branch and the **Else** branch. Click in the field of the **Then** branch, and enter the fixed value "True". Now add an **Else** branch by clicking the **Add Else** button, and enter the fixed value "False". The text rule is now defined properly. Save the decision by clicking the **Save** button in the top-right corner of the decision editor. After the save, all components in the decision diagram are colored green. Figure 20.22 shows the final state of the decision.

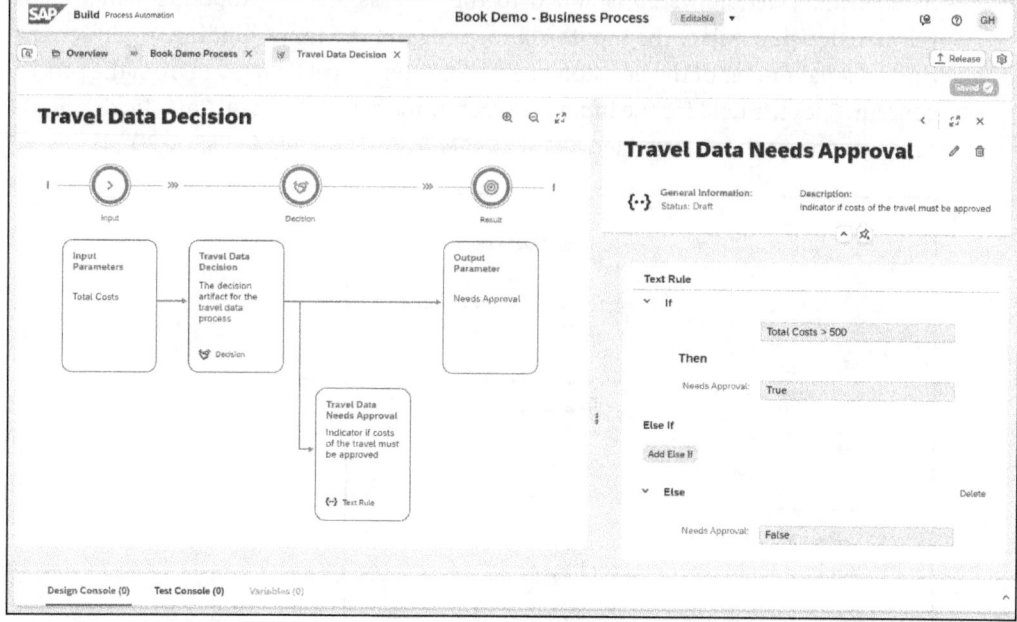

Figure 20.22 Final Decision State

641

20.5 Using a Text Rule in a Process

In the last section, we created a text rule that can now be used inside a process. Navigate to your example process and add a decision behind the trigger by clicking the plus (+) icon and selecting the **Decision** menu item from the options displayed. From the displayed submenu, select the **Travel Data Decision** item. Figure 20.23 shows a screenshot of the **Decision** menu.

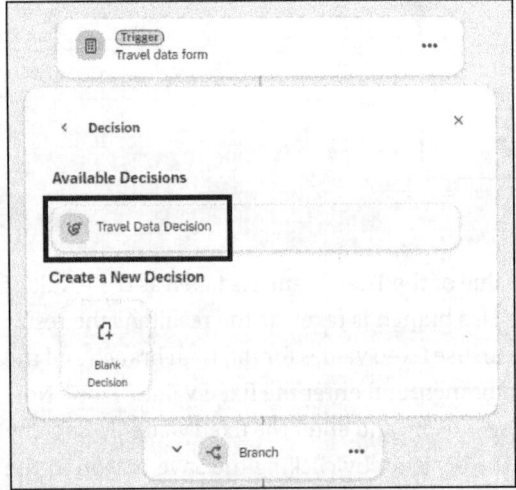

Figure 20.23 Menu to Add Decision

As a result, a decision step is added to the process. In the properties panel of the selected decision, select the **Inputs** tab to configure the input binding. You'll see that the input parameter of the decision needs to be bound to a field given from the process content. Click the field for the input parameter, and select the **Total Costs** field from the process content that is provided by the trigger form, as shown in Figure 20.24.

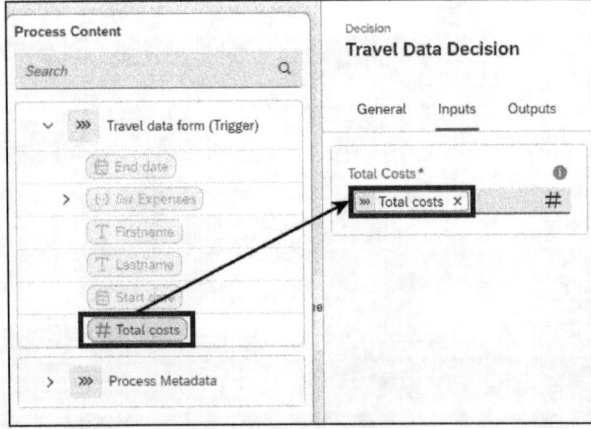

Figure 20.24 Input Binding of Decision

20.5 Using a Text Rule in a Process

Now you need to bind the outcome of the decision to a variable in the process in order to use it subsequently. Select the **Outputs** tab in the properties panel to show the binding configuration for the output of the decision. You can see that the **Needs Approval** variable that represents the result of the decision can be bound. Open the combo box of the custom variables and select **Add Variable**, as shown in Figure 20.25.

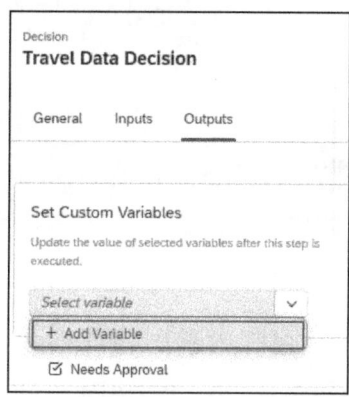

Figure 20.25 Add Custom Variable

In the displayed dialog, shown in Figure 20.26, provide the values according to Table 20.10, then click the **Add** button to create the new variable.

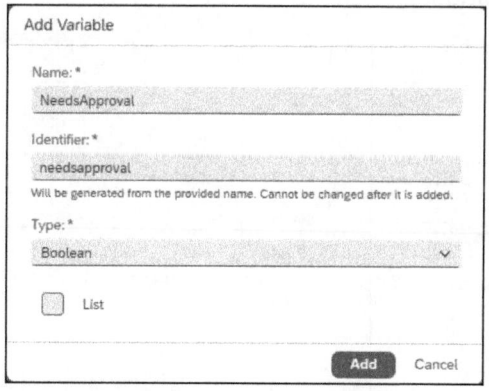

Figure 20.26 Dialogue to Add Variable

Field Name	Value
Name	"NeedsApproval"
Identifier	"Needsapproval" (value of this field is generated from the provided name and can only be changed so long as the variable was not added)
Type	Boolean

Table 20.10 Values for New Variable

After the variable is created, it appears in the output section of the decision and can be bound to the outcome of the decision. Click the field that represents the new variable and select the **Needs Approval** field from the travel data decision as the binding source. As a result, the field is written to the new custom variable, and the binding is accomplished. Figure 20.27 shows a screenshot of this binding.

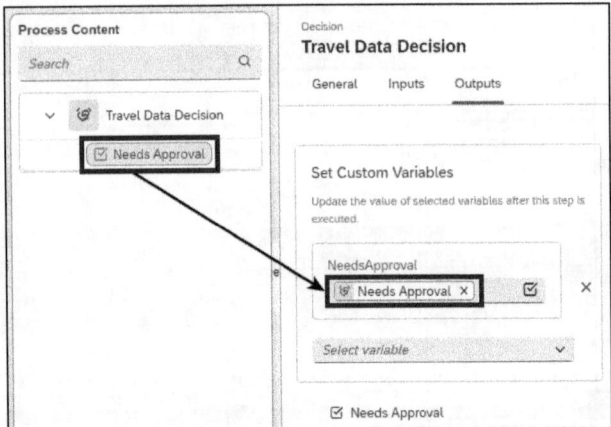

Figure 20.27 Output Binding of Decision

Now create the condition control right after the decision and use to create it the **Needs-Approval** custom variable inside this condition, as shown in Figure 20.28.

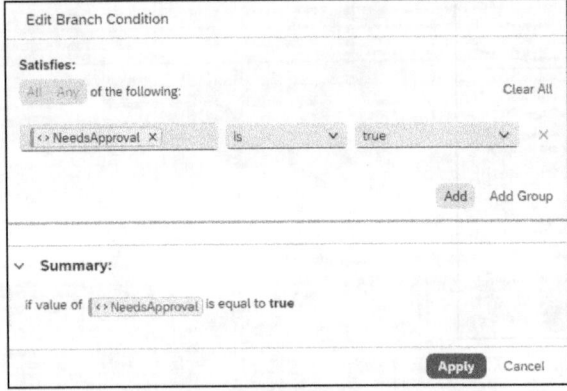

Figure 20.28 Condition Using Decision's Output

In the if branch of the condition, add an approval based on the travel data form and create the binding as shown in Chapter 19, Section 19.3.3. The process now looks like shown in Figure 20.29. Save the process and create a new release. Finally, deploy the release.

To test the process, open the trigger form in the browser and enter travel data with total costs greater than 500. Submit the form to start a new process instance. Navigate to the inbox, and you will see that a new approval task was created with the data that was

20.5 Using a Text Rule in a Process

submitted in the trigger form. As the total costs exceeded the value of 500, the decision in the process resulted in a *true* outcome, and the condition leads into the branch that created the approval task. Figure 20.30 shows the monitoring details of the process instance. In the logs of the process instance, you can see that the travel data decision was completed successfully and that the condition leads to the creation of an approval task.

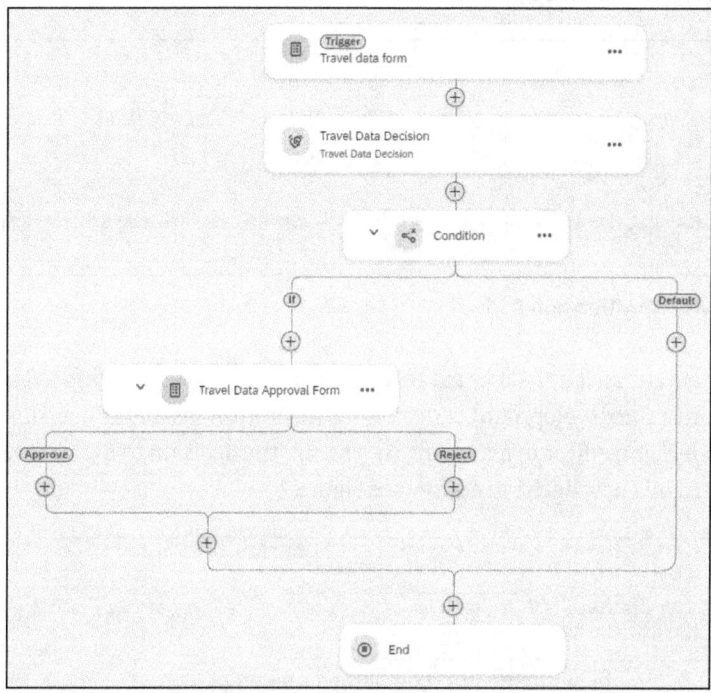

Figure 20.29 Process Using Text Rule

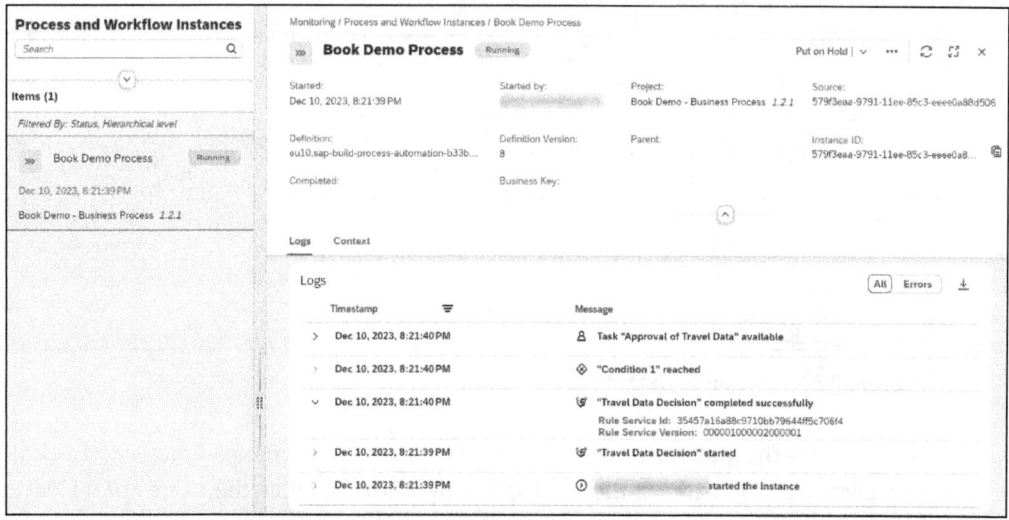

Figure 20.30 Monitoring Details of Process Instance

645

20 Rules and Decisions

20.6 Creating a Decision Table

Let's now create another decision using a decision table to determine the approver of the travel data. Therefore, open the development canvas of your process and add a new decision inside the positive condition branch, as shown in Section 20.2. With the new decision, the process looks like the screenshot in Figure 20.31.

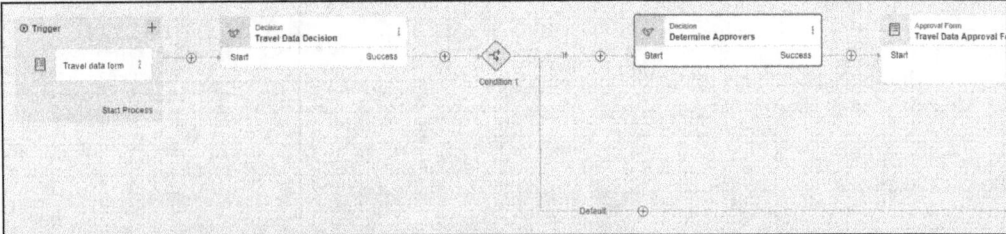

Figure 20.31 Process with New Decision Added

Select the decision element inside the process to show its properties in the properties panel on the right side of the development canvas. Click the button inside the decision element to open the decision editor in a new tab. As a result, the decision to determine the approvers is shown in a new decision editor (see Figure 20.32).

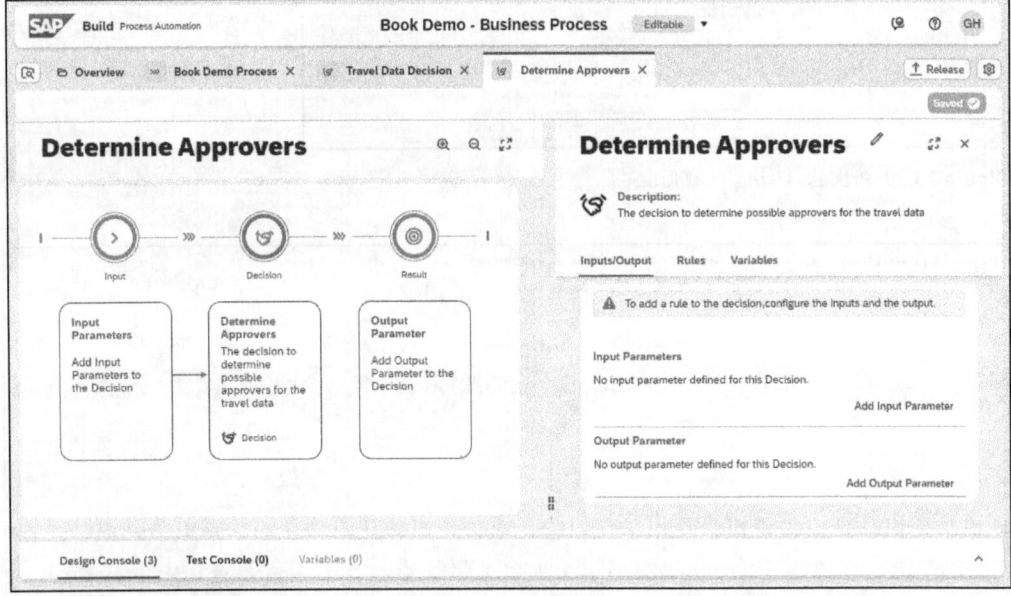

Figure 20.32 Decision to Determine Approvers

In the tab for the input and the output parameters, add an input parameter and an output parameter by clicking the corresponding buttons. Enter the values provided in Table 20.11 to define the parameters.

20.6 Creating a Decision Table

Field Name	Value
Input Parameter	
Name	"Travel Data"
Description	"The data of the travel to determine the approvers"
Type	Travel Data
Output Parameter	
Name	"Approvers"
Description	"The approvers for the travel data"
Type	Approver

Table 20.11 Input and Output Parameters

Figure 20.33 shows a screenshot of the configured parameters. You can see that we used the data types created in Section 20.3 to facilitate the creation of the binding parameters for the decision.

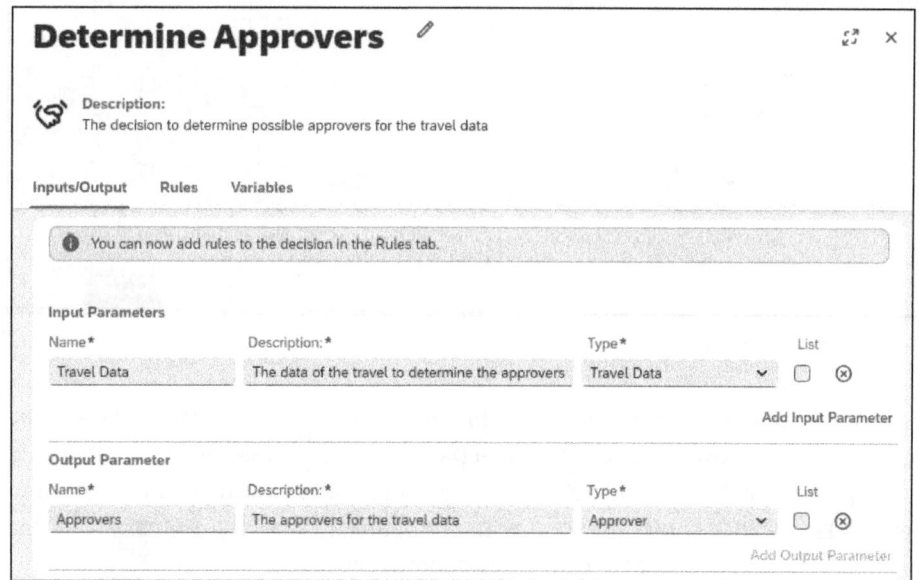

Figure 20.33 Input and Output Parameters of Decision

Now, switch to the **Rules** tab and add a new rule by clicking the **Add Rule** button. Again, the wizard to create a new rule is displayed in a dialog. For **Rule Type**, choose **Decision Table**, then enter the values provided in Table 20.12 for the corresponding fields. Click **Next Step** to proceed to the next step of the wizard.

647

20 Rules and Decisions

Field Name	Value
Rule Name	"Approver determination"
Description	"A rule to determine the approvals for the travel data"
Reusable Rules	No
Hit Policy	First Match

Table 20.12 Values for Decision Table

Figure 20.34 shows a screenshot of the rule details defined in the wizard.

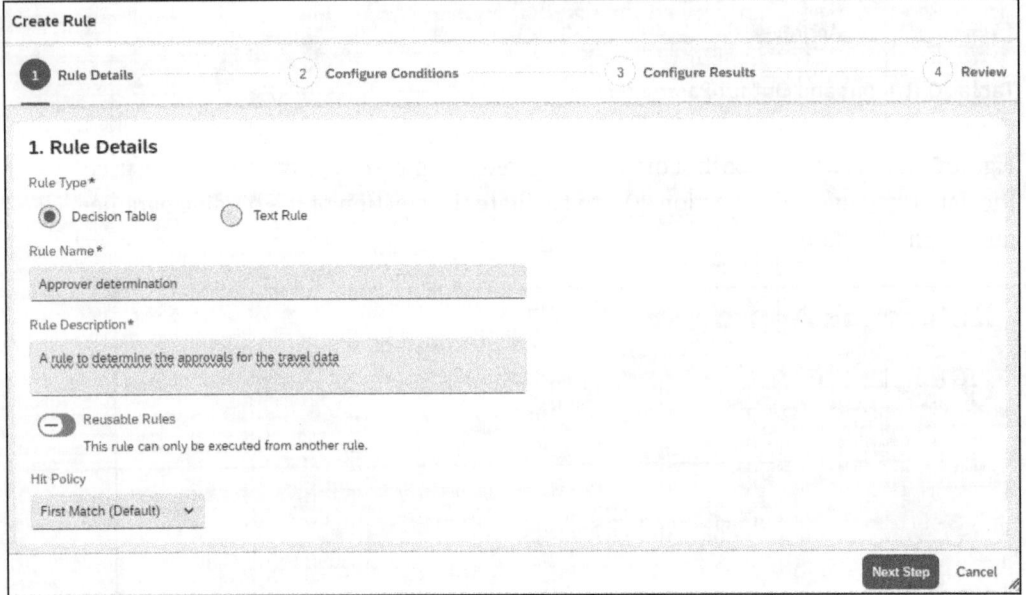

Figure 20.34 Rule Details of Decision

In the next step of the wizard, the condition details are defined. In the table, add the **Location** and **TotalCosts** fields of the **Travel Data** data type to the condition attributes. As the operator for the **Location** field, select the = sign. For **TotalCosts**, do not provide an operator. Figure 20.35 shows the conditions configured in the wizard. Finally, click the **Next Step** button to proceed to the configuration of the results.

As the result of the decision, you want to add the user name and the user group of the **Approvers** data type. Click the data type node in the **Vocabulary** area on the left side to add all its elements to the result details table. Finally, click the **Next Step** button to proceed with the review of the created rule. Figure 20.36 shows a screenshot of the result's configuration.

20.6 Creating a Decision Table

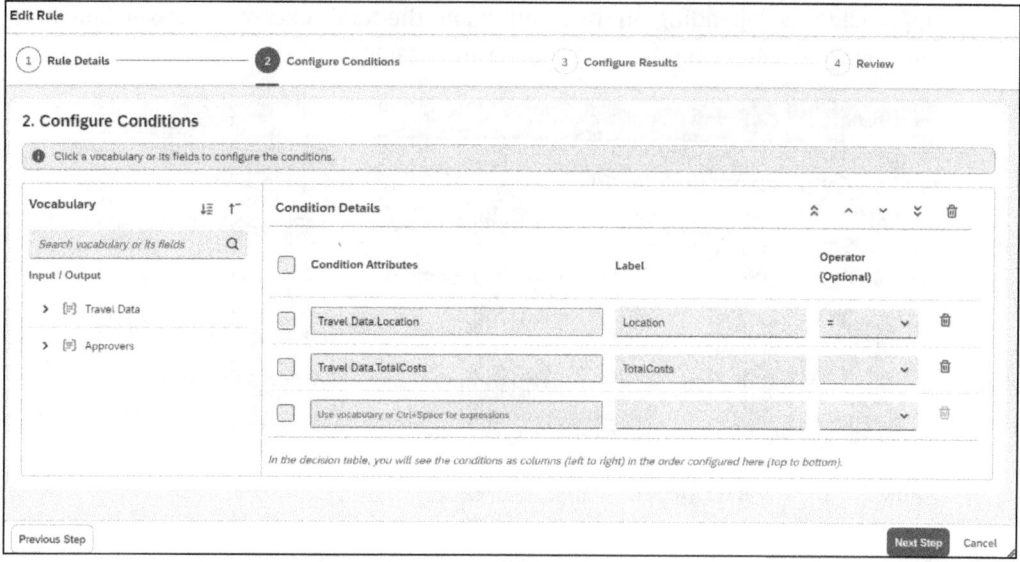

Figure 20.35 Condition Details of Rule

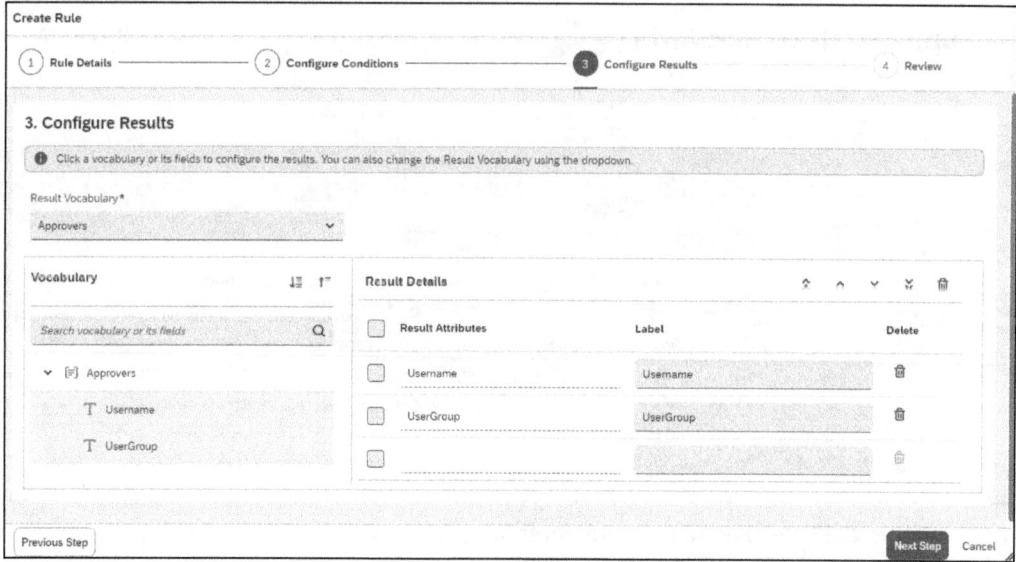

Figure 20.36 Results of Decision

In the last step of the wizard, a summary of the created rule is shown. Click the **Create** button to confirm the configuration and to create the rule inside your decision.

After the rule is created, the decision table can now be used to define the possible combinations of input values and the results these combinations lead to according to your business rules. In this example, we'll define the user names that will be attached to the approval tasks depending on the location and the total costs of the travel data. Enter the lines provided in Table 20.13 to the decision table.

Location	TotalCosts	Username	UserGroup
'USA'	>1000	Some other SAP BTP user name	'appr_USA_large'
'USA'	>500	Your SAP BTP user name	'appr_USA_small'
'Europe'	>1000	Your SAP BTP user name	'appr_Europe_large'
'Europe'	>500	Some other SAP BTP user name	'appr_Europe_small'
'Asia'	>1000	Your SAP BTP user name	'Appr_Asia_large'
'Asia'	>500	Your SAP BTP user name	'appr_Asia_small'

Table 20.13 Criteria for Decision Table

Figure 20.37 shows the decision table for the approver determination.

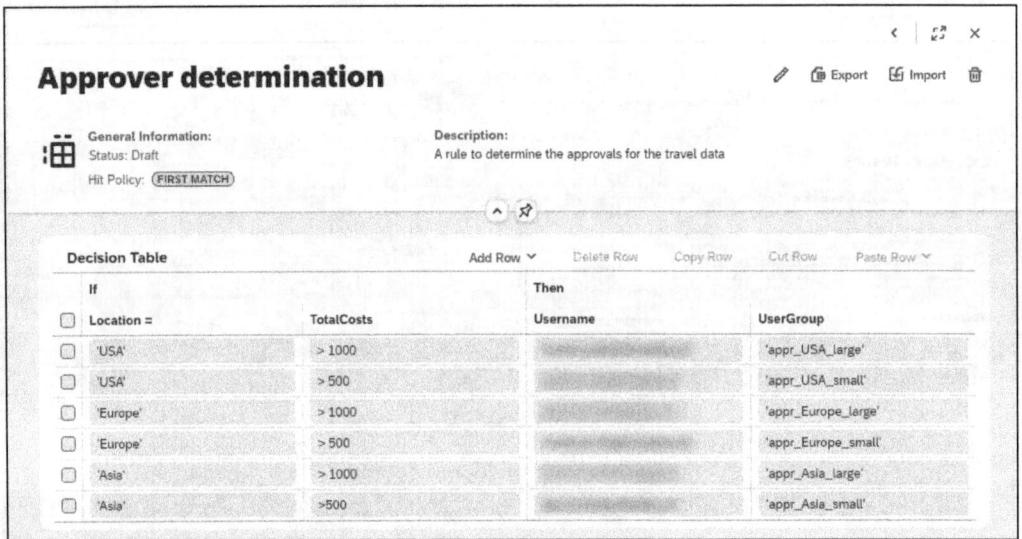

Figure 20.37 Decision Table for Approver Determination

> **String Variable**
>
> Make sure that all values that correspond to a string variable are placed between apostrophes ('') in order for them to be valid.

To test the decision table, you need to add a new field to the trigger form. Open the **Travel Data Form** form and add a dropdown field underneath the **End Date** field. Set the

20.6 Creating a Decision Table

label as **Location** and add **USA**, **Europe**, and **Asia** as the field's options, as shown in Figure 20.38. Save the form.

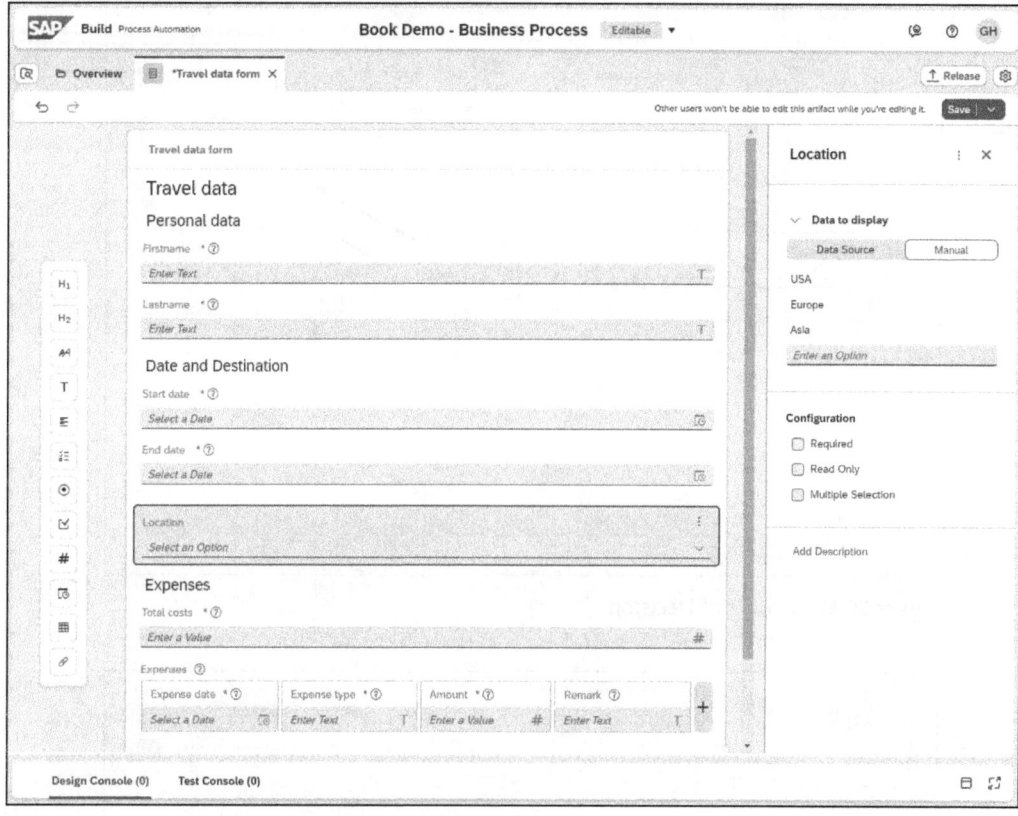

Figure 20.38 Dropdown Added to Trigger Form

Open the process and set the binding for the decision table. Select the decision step and open the properties. In the **Inputs** section, select the **Single Properties** option, then enter the binding of the variables to be used in the decision, as shown in Figure 20.39.

The decision returns the user name for the approval task. This binding must be done as well. Select the travel data approval form and open the properties panel. In the **General** section, bind the **Users** field to the **Username** defined by the decision table, as shown in Figure 20.40.

Save the process, create a new release, and deploy it. Open the trigger form and enter a new travel considering the rules defined in the decision table. Move to the **Monitoring** area to see that the approval tasks were assigned to the users according to the rules defined in the decision table, as shown in Figure 20.41.

651

20 Rules and Decisions

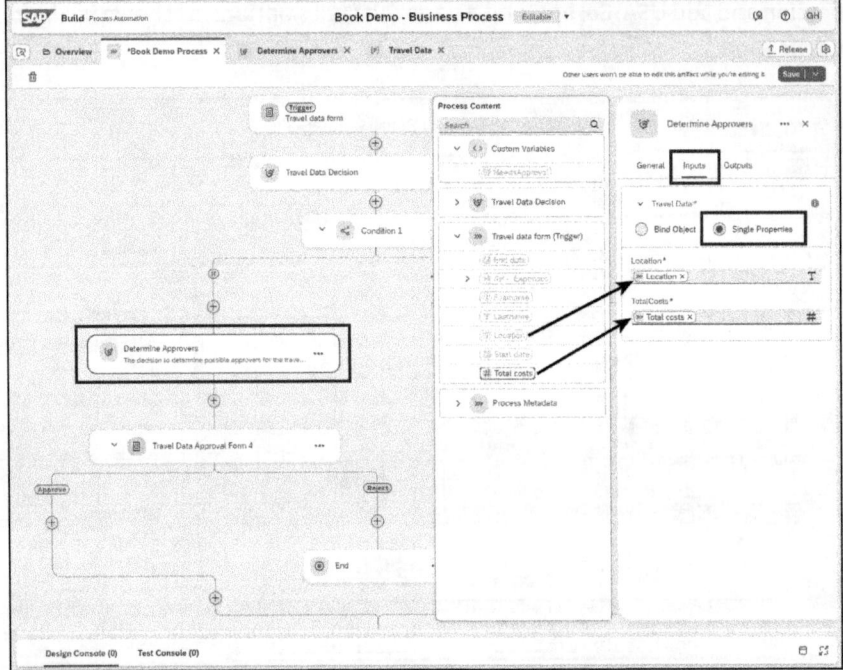

Figure 20.39 Binding of Decision

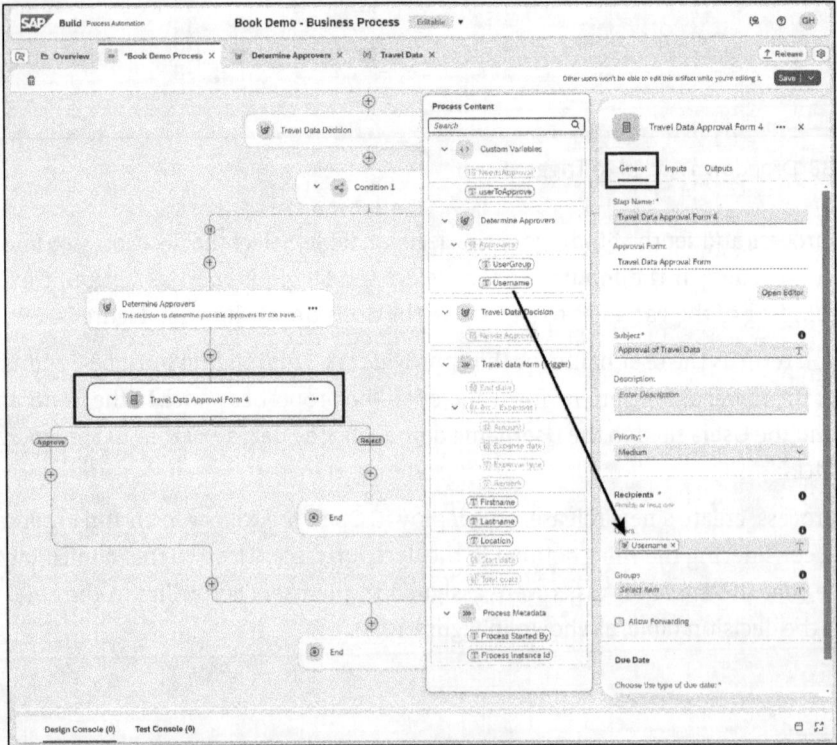

Figure 20.40 Binding of User to Approve Request

20.7 Summary

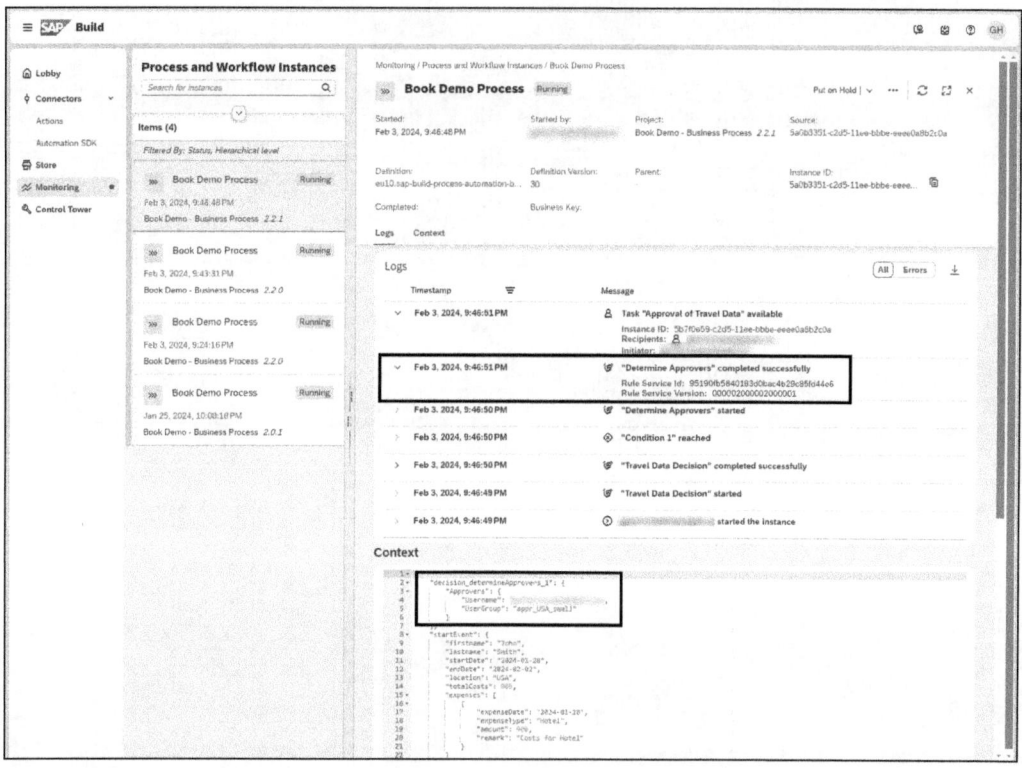

Figure 20.41 Monitoring Including Decision Table Result

20.7 Summary

In this chapter, you learned how to create decisions and how to use them in a business process in order to execute process steps depending on the data processed. We showed the usage of text-based decisions, so-called text rules that can be used to create simple decisions. When it comes to complex decisions based on multiple factors, the usage of decision tables is preferable. You learned how to create such tables and how to manage the bindings inside a process in order to provide the decision table with the data needed to calculate a result. Finally, we used these tools in a sample process to show how the process runtime uses them to process the data and how the decisions are presented in the monitoring area.

Chapter 21
Action Projects

In modern times, data is spread among a huge number of systems. Business processes need to access these systems in order to retrieve data needed at runtime or to execute functionalities. With action projects, SAP Build Process Automation provides the tools to access foreign services and read or provide data during the runtime of a business process.

With action projects, you can create artifacts to communicate with external systems that are connected with SAP BTP. Actions created in such a project can be used to query data or to execute functions depending on your needs. An action project provides a design console to define the data structure of the messages sent and received by external services and to test them before you use them in your processes. As actions are separate artifacts in SAP Build Process Automation, they can be reused in multiple processes. In this chapter, we will create an action project in order to query data from an external service. We will use this data to fill a combo box inside a request form and to query detail information depending on the selected item of the combo box.

21.1 Create an Action Project

In the following sections, we will show you how to create an action project, how to define the services you want to consume, and how to use these services in your processes. We'll also describe the input parameters, the output of the action, and how to test the action.

21.1.1 Project Creation

To begin the process of action project creation, let's use the sandbox system of SAP Business Accelerator Hub. SAP Business Accelerator Hub offers an API that can be used for testing purposes. At *https://api.sap.com/api/API_BUSINESS_PARTNER/overview*, you'll find the description of this API and can download the specification. To do this, click the **API Specification** tile in the **API Resources** area. In the table that is now displayed, click the ⤓ icon to the right of **JSON** to download the specification in JSON format. Figure 21.1 shows a screenshot of SAP Business Accelerator Hub.

21   Action Projects

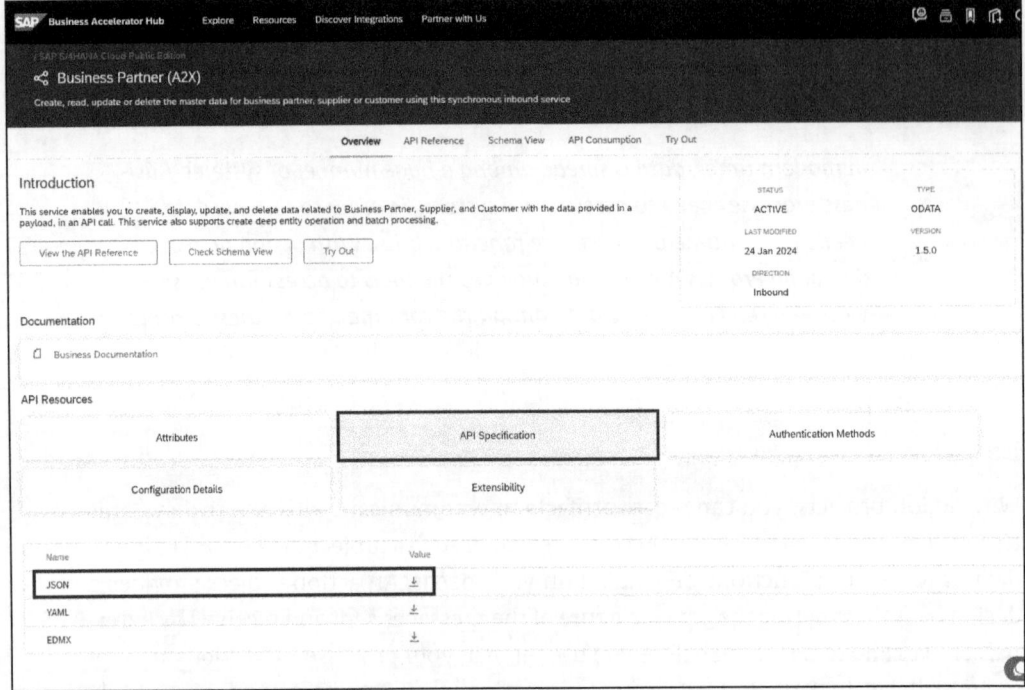

Figure 21.1 SAP Business Accelerator Hub

The starting point for creating an action project is the SAP Build Process Automation lobby. In the menu panel on the left, expand the **Connectors** menu item and select the **Actions** menu item. You will be taken to the action projects overview. All available projects are displayed here and can be used in your developments. To create a new project, click the **Create** button above the project list, as shown in Figure 21.2.

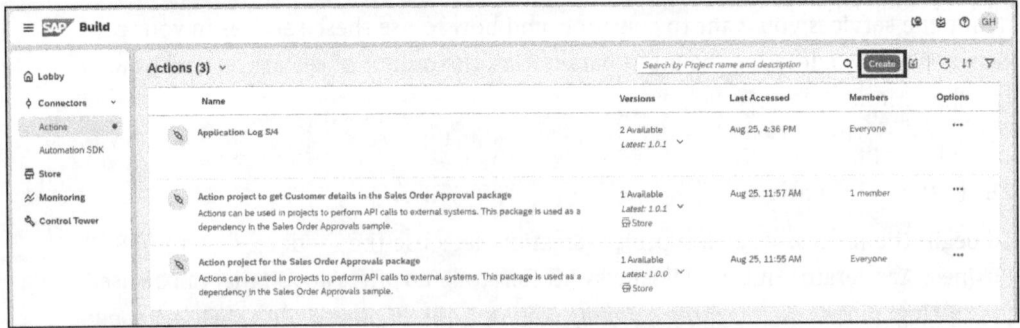

Figure 21.2 Action Projects Overview

A dialogue will then be displayed in which you can specify which API source you would like to use for your action project. Select the **Upload API Specification** tile, as shown in Figure 21.3.

656

21.1 Create an Action Project

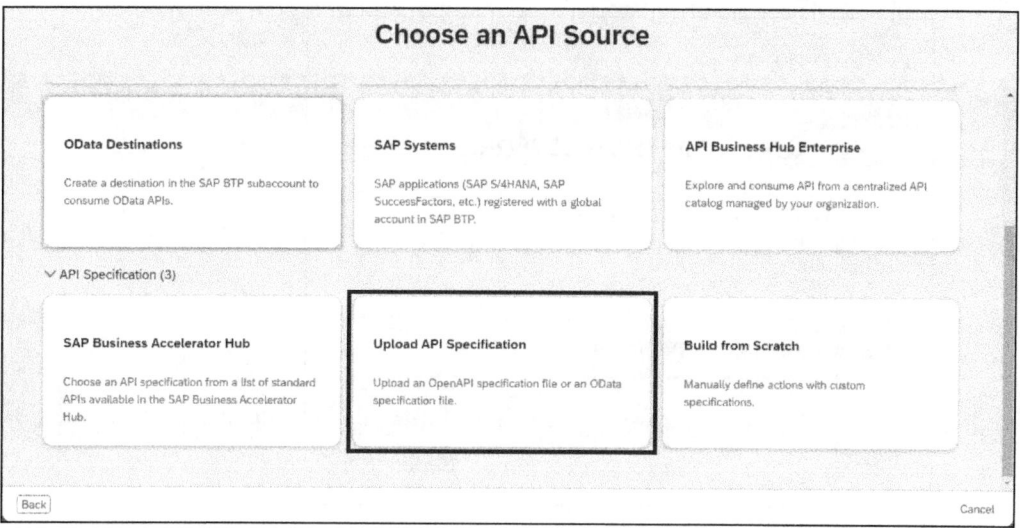

Figure 21.3 Selecting API Source

In the following dialogue, enter the API specification file you have just downloaded and confirm your entry by clicking the **Next** button, as shown in Figure 21.4.

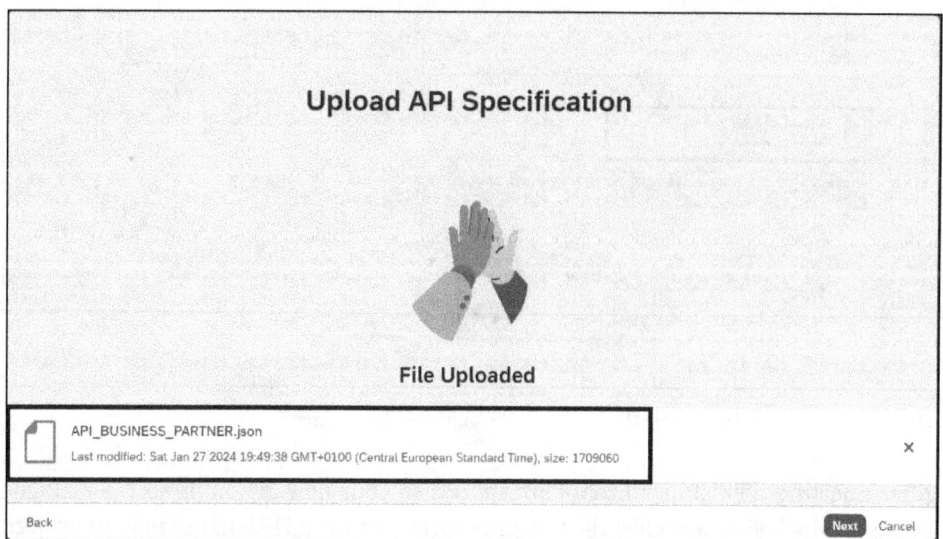

Figure 21.4 Upload API Specification

After the specification has been successfully uploaded, another dialogue appears in which you have to define a name for your project. Optionally, you can also provide a description for the project. Confirm your entry by clicking the **Create** button. Figure 21.5 shows a screenshot of this dialogue.

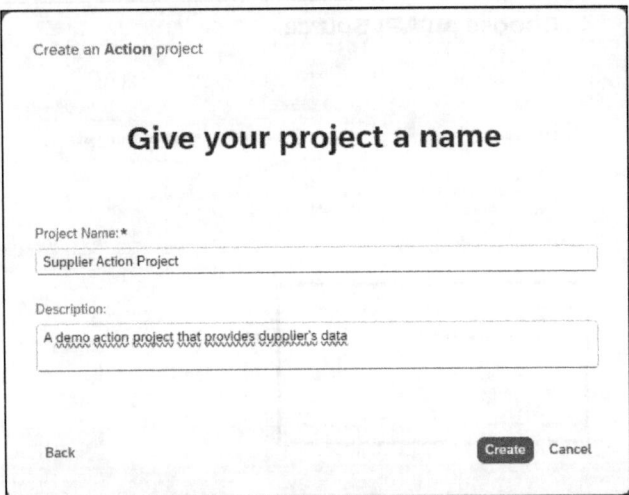

Figure 21.5 Dialog to Provide Action Project Name

The action project is now created and appears in the list of the overview of all available actions, as shown in Figure 21.6.

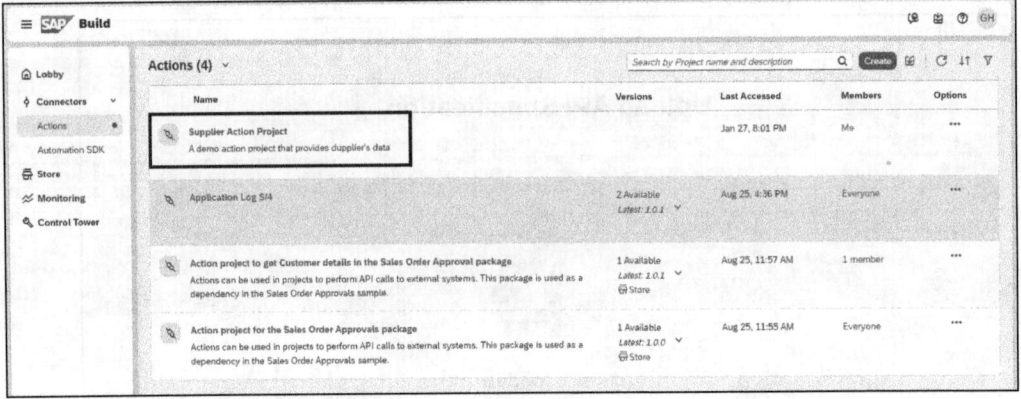

Figure 21.6 New Action Project

After selecting the list entry that corresponds to your new action project, a dialogue opens in which you can select the OData services you would like to use in your project. Expand the **Supplier** entry and select **GET** the **/A_Supplier** and **GET /A_Supplier('{Supplier}')** services, as shown in Figure 21.7. The first service method provides a list of supplier data. In the second method, you can specifically determine a supplier's data by specifying its supplier number. These will be the two services that we will use to create actions that can be added to a process in order to query data from external systems.

Confirm your selection by clicking the **Add** button. You will now see the overview of the available actions within your action project. On the left side of this view, the actions that were previously selected are shown in the form of a list. The actions are described by their name, and you can see what form of action each is. The selected action is displayed on the right. Here you can see the context path of the action at the top. Below there are three tabs for describing the input parameters, describing the output of the action, and testing the action.

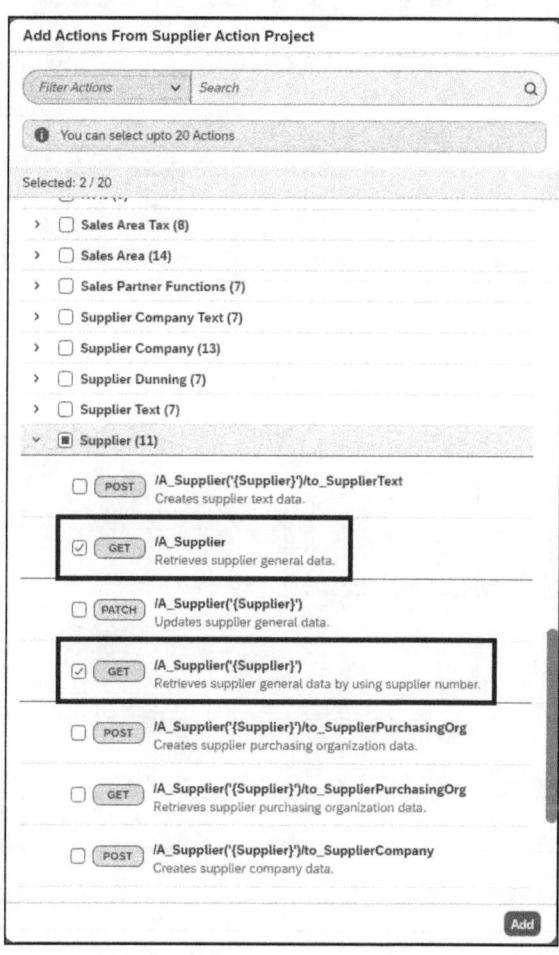

Figure 21.7 Selecting Service Methods to Be Used

21.1.2 Input

Select the **Retrieves Supplier General Data** action in the list of the available actions. Figure 21.8 shows the action to receive the list of the suppliers.

The input describes the parameters that can or must be specified when calling the respective action. By clicking the > icon in the rightmost column of the respective parameter, its properties can be viewed and edited, as shown in Figure 21.9.

21 Action Projects

Retrieves supplier general data.

Retrieves general data of all the supplier records available in the system.

GET /A_Supplier

Input Output Test

Parameter

Key	Parameter	Type	Label	Static	Value	API Format	Tags	
$top	query	integer	$top	No				>
$skip	query	integer	$skip	No				>
$count	query	boolean	$count	No				>
$filter	query	string	$filter	No				>
$orderby	query	array	$orderby	No				>

Figure 21.8 Input Parameters of Action

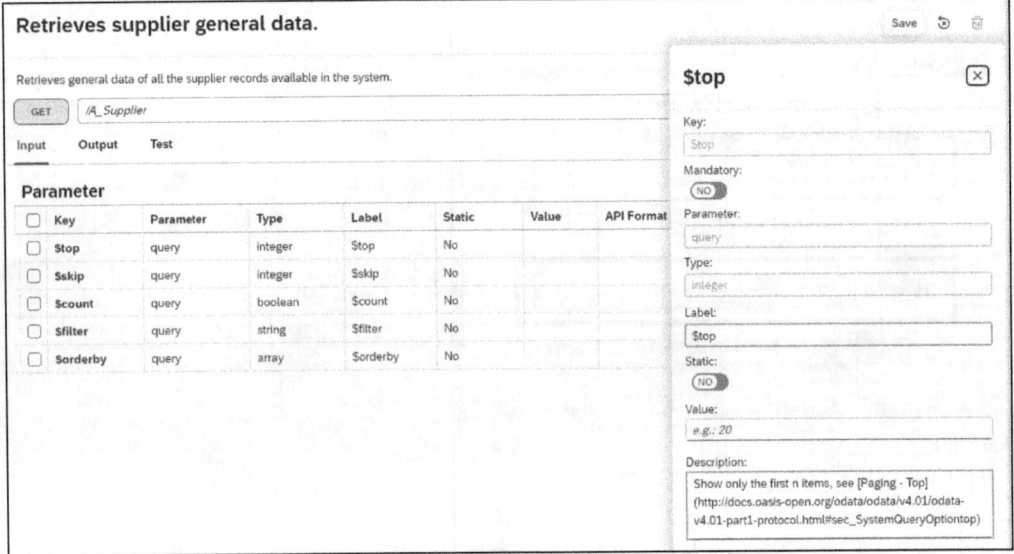

Figure 21.9 Properties of Input Parameter

By clicking the **Add** button above the parameter list, it's possible to add additional input parameters that will be taken into account when calling the action. In the panel displayed, either an existing parameter can be selected, or you can define a new parameter. Depending on the characteristics of this parameter, it's then possible to specify further detailed information in the properties.

Click the **Add** button above the parameter list. In the panel that appears, select the **$select** field from the list of available fields. Then click the **Add** button to add the parameter to the list of input parameters, as shown in Figure 21.10.

21.1 Create an Action Project

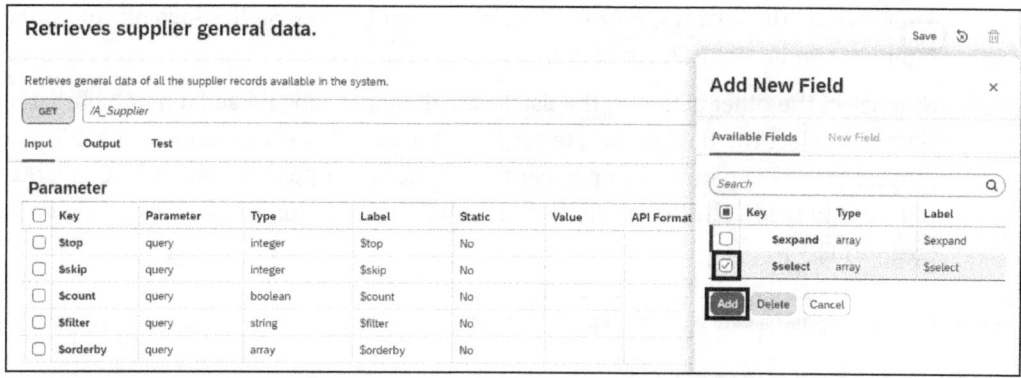

Figure 21.10 Adding Input Parameter

A GET request normally returns all available fields of an entity that are marked as deliverable. The $select parameter can be used to limit the number of fields by specifying the field names. Open the properties of the $select parameter by clicking the corresponding line in the list. In the **Value** field, select the **Supplier** and **SupplierName** values, as shown in Figure 21.11.

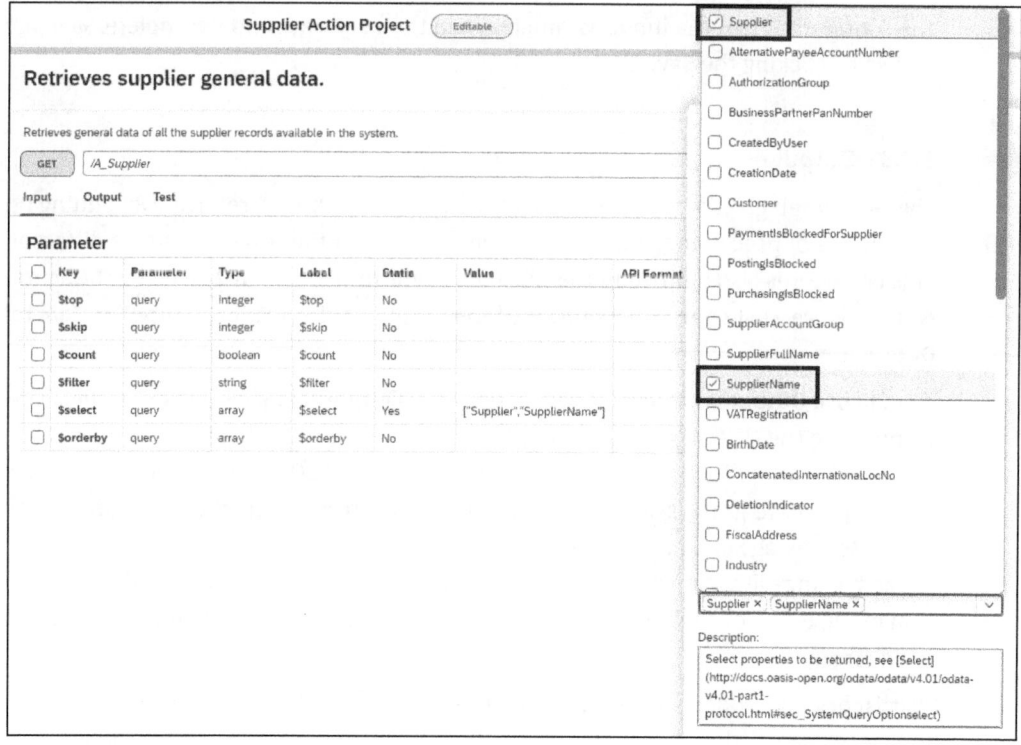

Figure 21.11 Definition of $select Fields

661

21 Action Projects

After making the settings, save your action project by clicking the **Save** button in the top-right corner of the action editor.

Now select the other action in the list. Select the **Input** tab and add a new parameter. Here too, select the field called **$select** from the list of available fields. Then edit the properties of this parameter and select the **Supplier**, **SupplierName**, and **SupplierAccountGroup** fields. The list of input parameters for this action should look as shown in Figure 21.12.

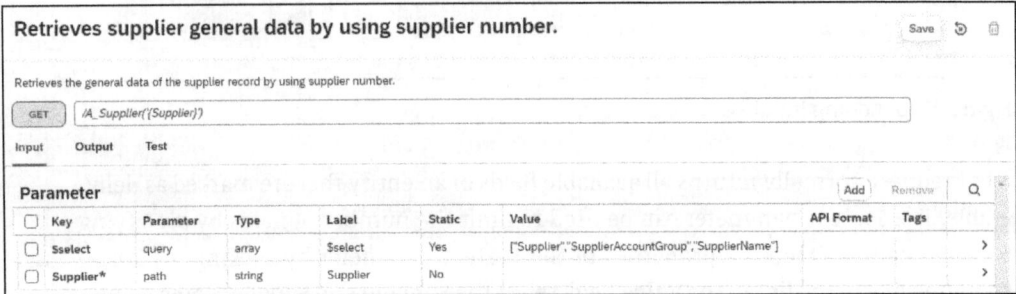

Figure 21.12 Input Parameters of Second Action

The configuration of the input parameters and their actions is now complete. Save the project by clicking the **Save** button.

21.1.3 Output

The **Output** tab gives you an overview of the fields that the API returns. Depending on the success of processing your request, the structure of the response can be different. This behavior is also defined in the API and is determined and displayed by the action editor. Figure 21.13 shows an example of the output of the **Retrieves Supplier General Data** action.

The **Choose Output** field appears at the top of the **Output** tab. The code in this field corresponds to the HTTP code that is sent as a response to processing in the header of the response message. In Figure 21.13, this field has a value of 200. This corresponds to successful processing. By selecting a different value in this list, the structure of the response message can be displayed for each respective case. In our example, all HTTP codes starting with 4 are covered by their own message structure to provide a description of the error for these cases. Figure 21.14 shows the message structure in the event of an error.

The list of status codes can be edited by clicking the **Edit List** link below the output field. In the opened dialogue, existing status codes can be deleted or newly added. The screenshot in Figure 21.15 shows a new status with the value 500. This is returned by the API if an internal server error occurs while processing the request. Clicking the **Save** button saves the new status.

21.1 Create an Action Project

Figure 21.13 Action Output

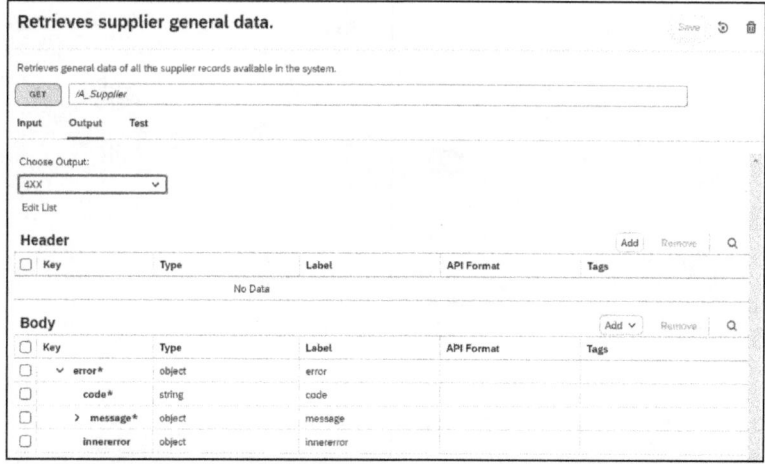

Figure 21.14 Output in Case of Error

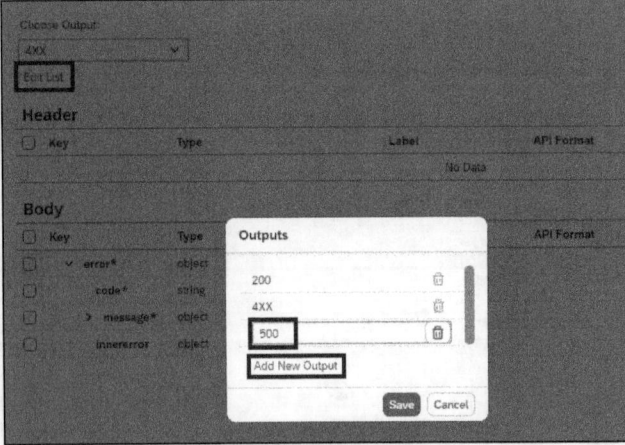

Figure 21.15 Adding New HTTP Status

21.1.4 Test

The **Test** tab offers an interface with which you can test the configuration of the selected action. The application sends a request to other connected external systems based on the input values entered. This system can be specified by setting a defined SAP BTP destination or manually by setting the endpoint and authentication information. Figure 21.16 shows a screenshot of the test environment for the **Retrieves Supplier General Data** action.

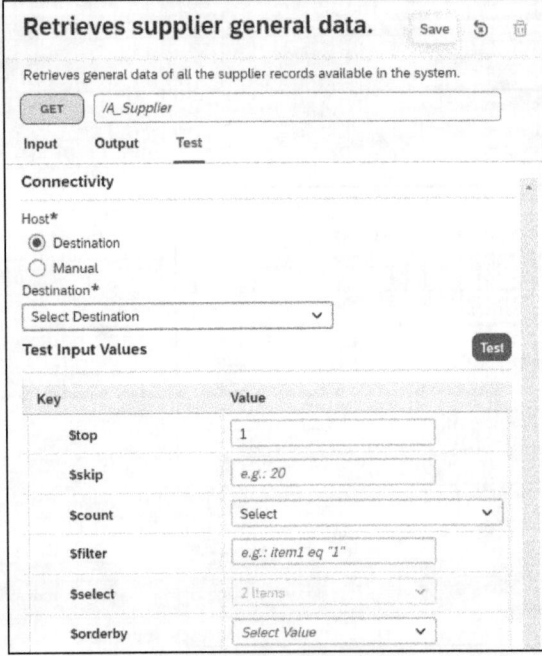

Figure 21.16 Test Tool of Action Editor

21.1 Create an Action Project

> **Authentication**
>
> When manually configuring the external system, it should be noted that only basic authentication is supported as an authentication method. If the external system does not support this method, the external system must be specified via an SAP BTP destination.

In our example, as already mentioned, we've used the sandbox system of SAP Business Accelerator Hub. In order to be able to test the actions, a corresponding destination must be created in the subaccount, through which the connection to the external system on which the API is to be called can be set up. To do this, open the subaccount in which SAP Build Process Automation is subscribed and navigate to **Destinations**, as shown in Figure 21.17.

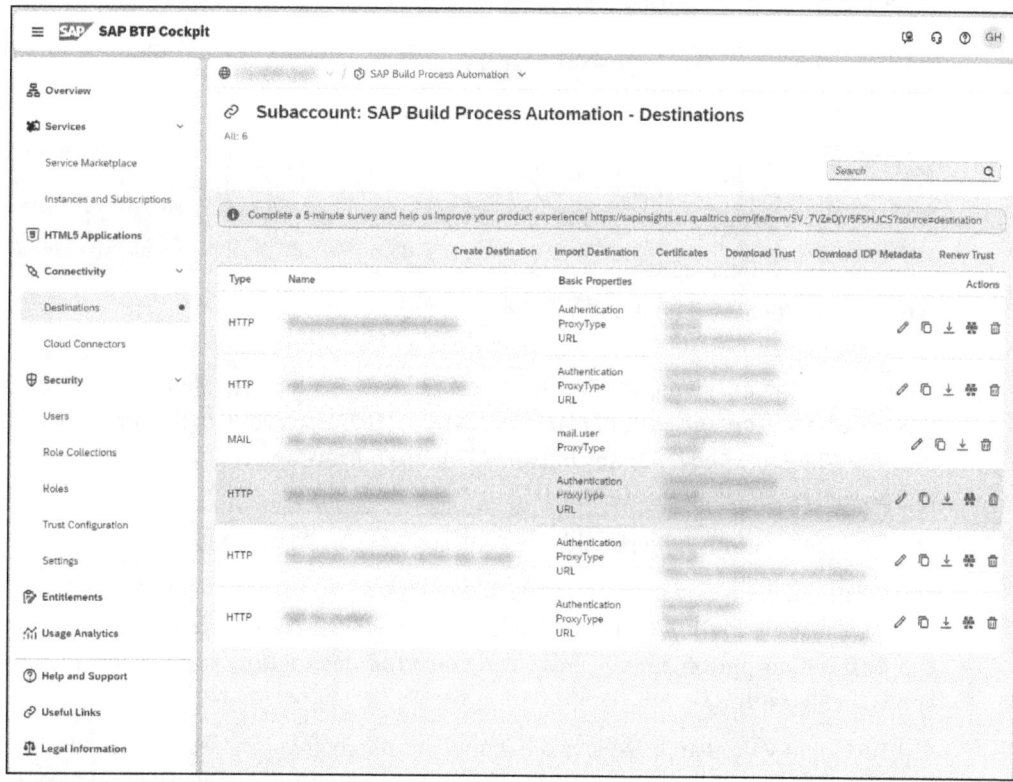

Figure 21.17 Destinations of Subaccount

Create a new destination and specify the values listed in Table 21.1.

Field Name	Value
Name	"SAP_S4_Sandbox"
Type	"http"

Table 21.1 Values of Destination to Sandbox System

Field Name	Value
Description	"S/4 cloud system"
URL	"https://sandbox.api.sap.com/s4hanacloud/sap/opu/odata/sap/API_Business_Partner"
Proxy Type	Internet
Authentication	NoAuthentication

Table 21.1 Values of Destination to Sandbox System (Cont.)

For the destination to be used in your action project, the additional properties listed in Table 21.2 must be configured for the destination.

Property Name	Property Value
"sap.applicationdevelopment.actions.enabled"	Enter the value "true"
"sap.processautomation.enabled"	Enter the value "true"
"URL.headers.APIKey"	Enter the API key for SAP Business Accelerator Hub

Table 21.2 Additional Destination Properties

You can find the API key in the **Try Out** section of SAP Business Accelerator Hub. To do this, visit *https://api.sap.com/api/API_BUSINESS_PARTNER/tryout*. In the top-right corner, you will find a **Show API Key** button. Click this button, and the **API Key** will appear in a popup, as shown in Figure 21.18.

The configured destination should now look as shown in Figure 21.19.

Now return to the action editor and select the **Retrieves Supplier General Data** action in the list of actions. In the details panel, select the **Test** tab. In the **Connectivity** area, select the **Destination** option for the host and enter the destination you just created, as shown in Figure 21.20.

The **Test Input Values** area displays the input parameters that can be transmitted to the external system for this action. You can see that the input parameter named $select already has a value specified. This is because we have already filled this parameter with values when defining the input.

21.1 Create an Action Project

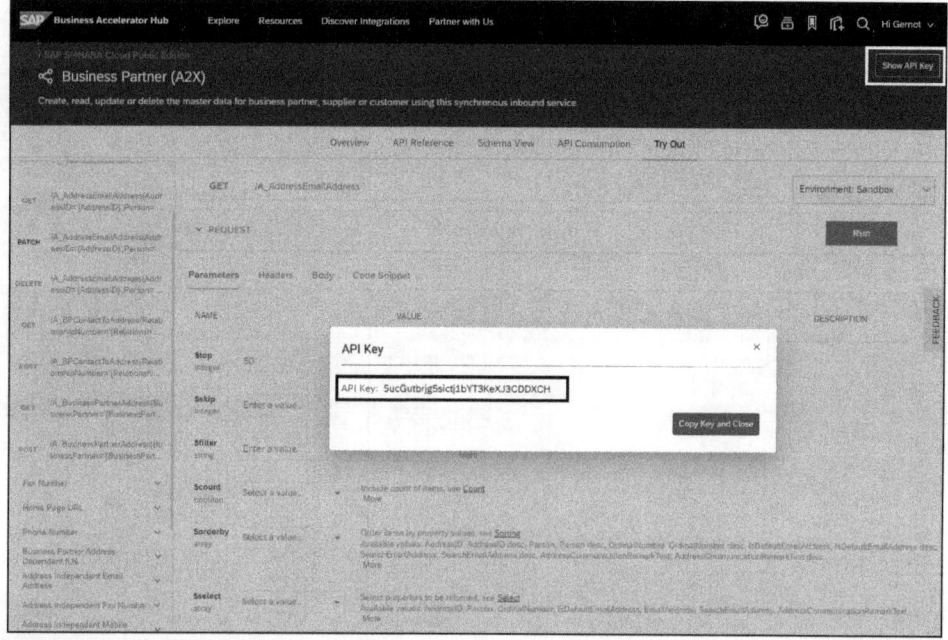

Figure 21.18 Finding API Key Information

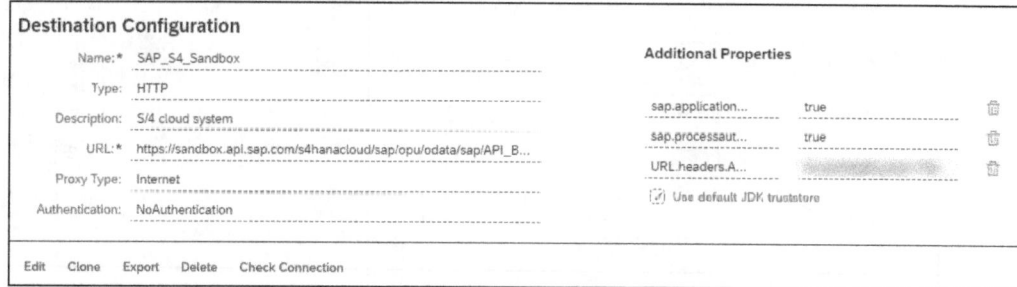

Figure 21.19 New Destination

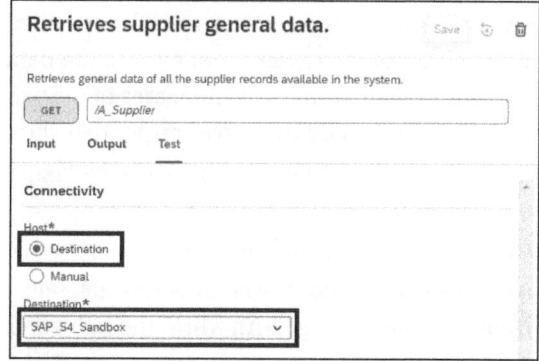

Figure 21.20 Defining External System in Test Tab

667

Now specify the value "5" for the parameter named $top. This parameter is used to provide the API with information about how many data records should be delivered as a result. For a first test, it's enough to get five suppliers. Figure 21.21 shows a screenshot of the test configuration.

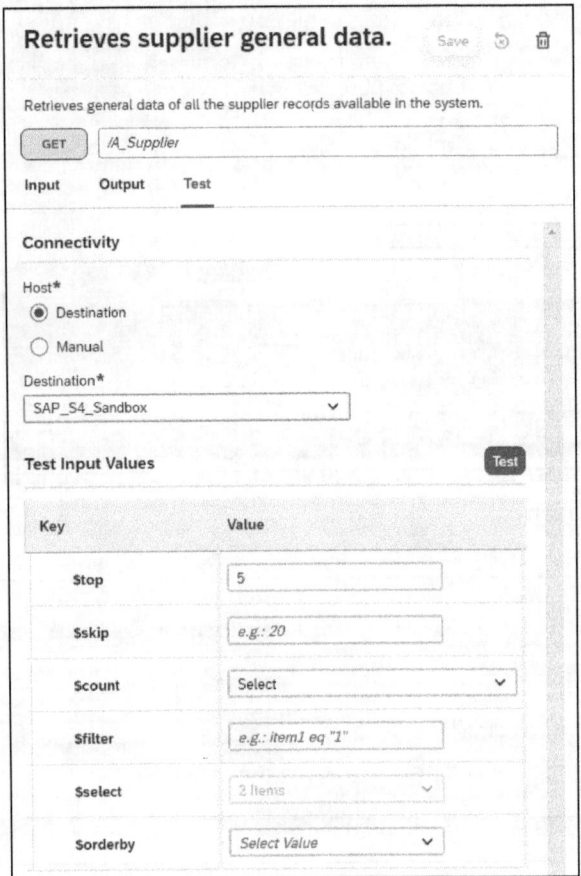

Figure 21.21 Test Configuration

Now click the **Test** button. The test environment sends a request with the specified information to the sandbox system. This processes the request and returns the data based on the request. The result of the call is now displayed in the lower area of the test environment. On the right, you can see which HTTP code the server responded with. On the left side, you can switch between the outgoing and incoming messages. Figure 21.22 shows a screenshot of the result.

By clicking the **View API** link on the right, you can view the response from the API. Click this link and go to the response message. In the **API** window that now appears, select the **Body** tab, as shown in Figure 21.23. You'll see the message with which the API server responded.

21.1 Create an Action Project

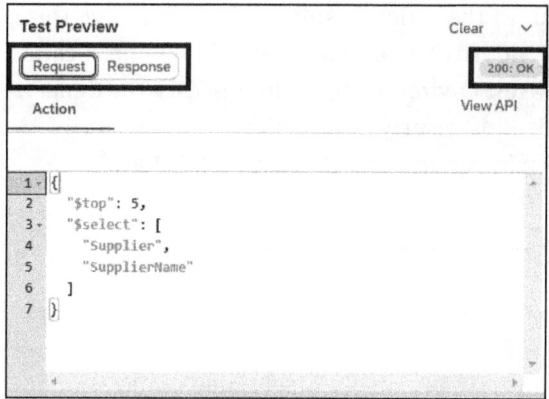

```
1  {
2    "$top": 5,
3    "$select": [
4      "Supplier",
5      "SupplierName"
6    ]
7  }
```

Figure 21.22 Result of Test Request

Figure 21.23 Response Message of API Server

21 Action Projects

As you can see on the left, the result of the action is still blank. In this case, this is because the result of the API call does not structurally correspond to the specification. However, this can be corrected in the test environment. To do this, click the **Generate Output** button. As a result, a dialogue is displayed in which the structure of the response message is shown, as shown in Figure 21.24. Click the **Add Output** button to add the response message structure to the output.

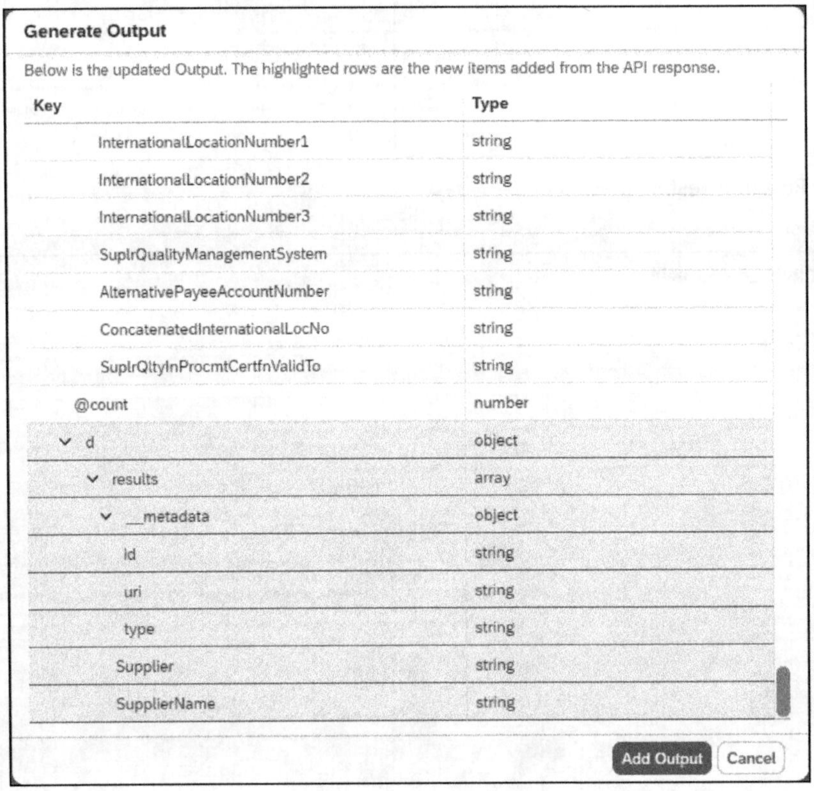

Generate Output	
Below is the updated Output. The highlighted rows are the new items added from the API response.	
Key	**Type**
InternationalLocationNumber1	string
InternationalLocationNumber2	string
InternationalLocationNumber3	string
SuplrQualityManagementSystem	string
AlternativePayeeAccountNumber	string
ConcatenatedInternationalLocNo	string
SuplrQltyInProcmtCertfnValidTo	string
@count	number
⌄ d	object
⌄ results	array
⌄ __metadata	object
id	string
uri	string
type	string
Supplier	string
SupplierName	string

Figure 21.24 Add Response Message Structure to Output

Navigate back to the output to see that the structure has been added to it, as shown in Figure 21.25.

Delete the **value** and **@count** fields as they are not part of the response structure. Select the **results** field and enter the value "Main Output Array" for the **Tags** field in the properties. In this case, this setting must be made so that the results can be used to fill the dropdown element in the trigger form. Figure 21.26 shows a screenshot of this configuration.

Run a new test. Now the action delivers the final result, as shown in Figure 21.27.

21.1 Create an Action Project

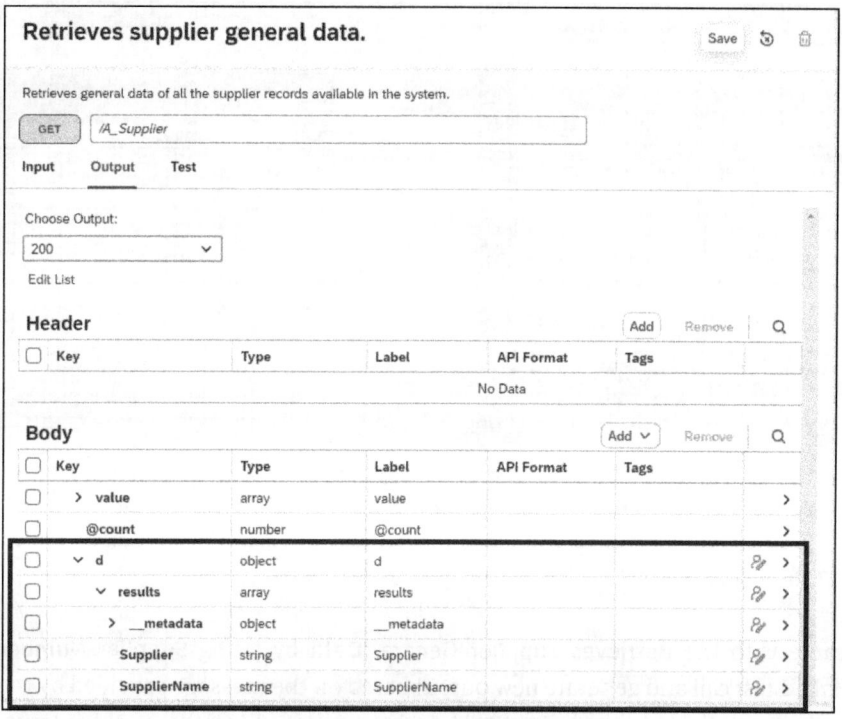

Figure 21.25 Output after Adding Structure of Response Message

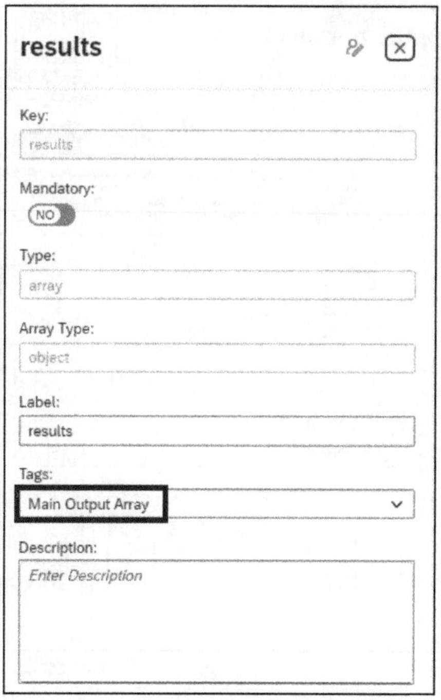

Figure 21.26 Configuration of Output

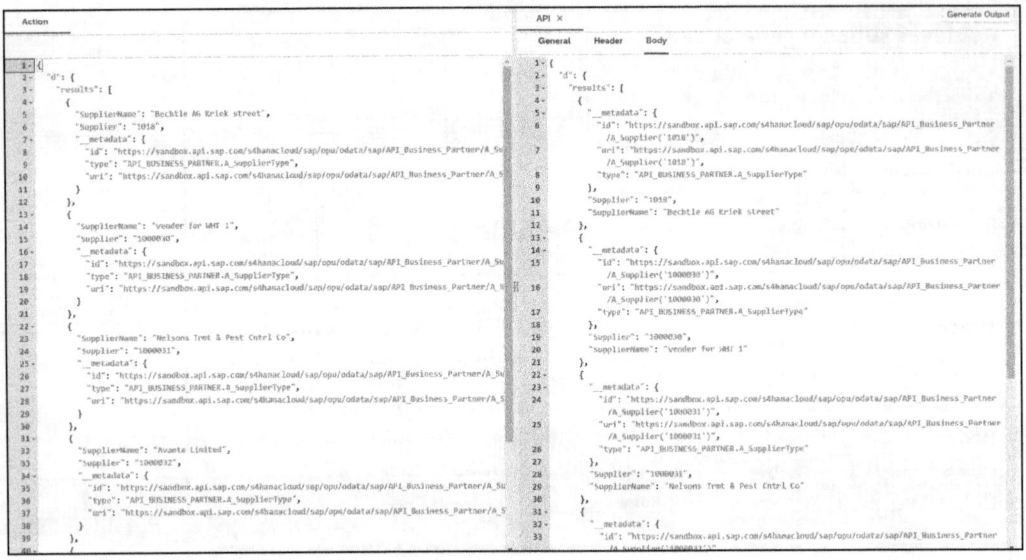

21 Action Projects

Figure 21.27 Result of New Test

Do the same with the **Retrieves Supplier General Data by Using Supplier Number** action. Make a test call and generate new output based on the message provided by the API. Then delete all fields in the output that do not correspond to the message structure. As a result, the new output should look as shown in Figure 21.28.

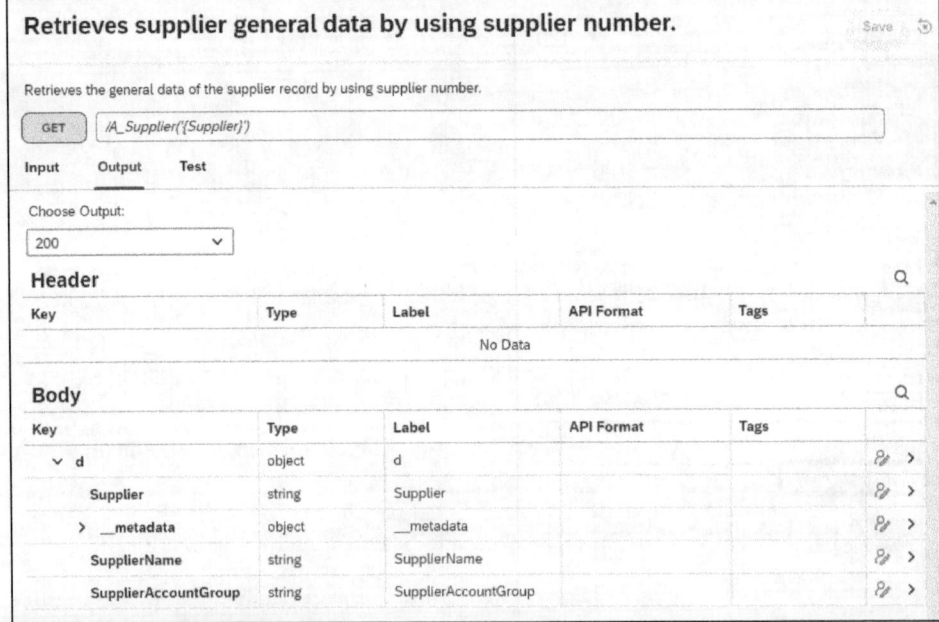

Figure 21.28 New Output Structure

672

For the action to be used in a business process, it must be published. To do this, click the **Release** button in the upper area of the action editor, as shown in Figure 21.29.

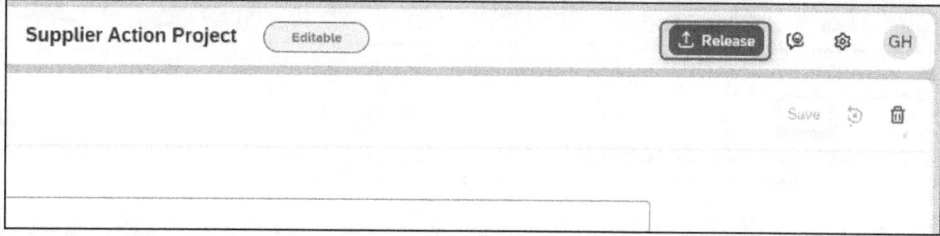

Figure 21.29 Release Action Project

In the following dialogue, enter a short description of your project and confirm your entry by clicking the **Release** button, as shown in Figure 21.30.

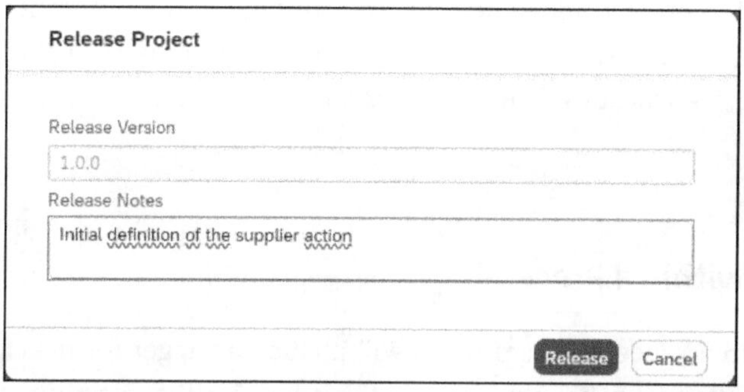

Figure 21.30 Release Dialogue

The project has now been released and can be published later. To do this, click the **Publish to Library** button, which has now replaced the **Release** button. Figure 21.31 shows that your action project has received a version number, and it is in the released state.

Figure 21.31 Released Version of Project

Navigate back to the overview of action projects. You can see that your project is now available in version 1.0.0, as shown in Figure 21.32.

21   Action Projects

Name	Versions	Last Accessed	Members	Options
Supplier Action Project A demo action project that provides dupplier's data	1 Available Latest: 1.0.0	Jan 27, 11:58 PM	Everyone	...
Application Log S/4	2 Available Latest: 1.0.1	Aug 25, 4:36 PM	Everyone	...
Action project to get Customer details in the Sales Order Approval package Actions can be used in projects to perform API calls to external systems. This package is used as a dependency in the Sales Order Approvals sample.	1 Available Latest: 1.0.1 Store	Aug 25, 11:57 AM	1 member	...
Action project for the Sales Order Approvals package Actions can be used in projects to perform API calls to external systems. This package is used as a dependency in the Sales Order Approvals sample.	1 Available Latest: 1.0.0 Store	Aug 25, 11:55 AM	Everyone	...

Figure 21.32 Published Action Project in Overview List

You have now created an action project, added actions, and configured the properties for input and output. You then successfully tested the action to determine the list of suppliers. In the next section, we will show how this action can be used in a business process.

21.2   Using the Action within a Process

We will now use the action in a process. This process will include a trigger form as a starting step by connecting a dropdown element to an action to enable the selection of a supplier number. We will then use a second action with this number to read more data via the API. In the final step of the process, we will send this information in the form of an email.

The first step is to make the destination that you set up to test the actions visible to the new process. To do this, go to the lobby and navigate to the **Control Tower** area. In the control tower, select the **Destinations** tile to go to the configuration of the destinations, as shown in Figure 21.33.

You will be taken to the overview of the registered destinations, which is currently empty. Click the **New Destination** button to enter a new destination, as shown in Figure 21.34.

In the following dialogue, all destinations are displayed in which SAP BTP is configured and the additional `sap.processautomation.enabled = true` parameter has been set. In this case, this is the destination to the SAP S/4HANA sandbox, as shown in Figure 21.35. Select this destination and confirm your selection by clicking the **Add** button.

The destination is added to the list and is now available in the processes.

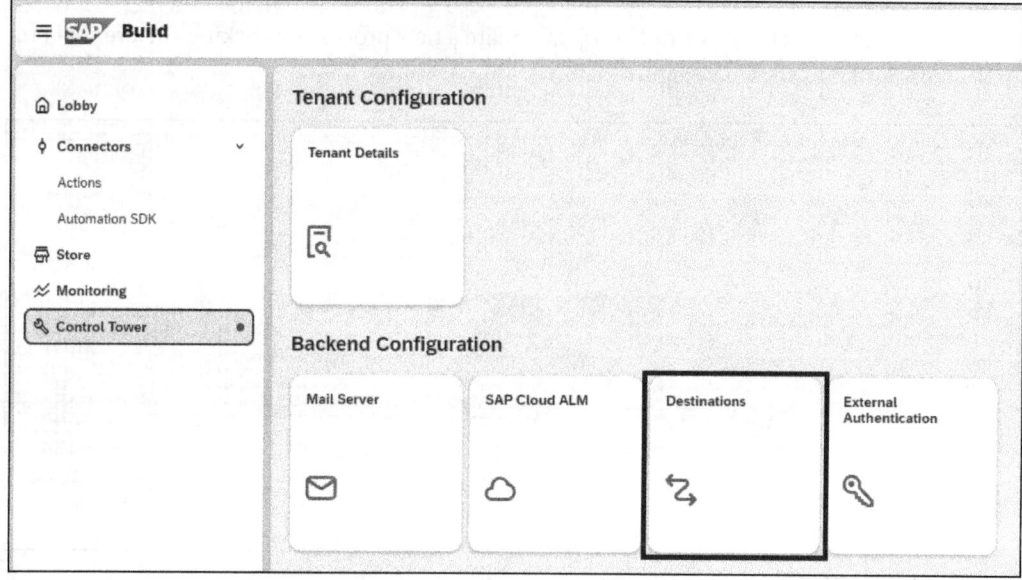

Figure 21.33 Navigate to Destinations Configuration

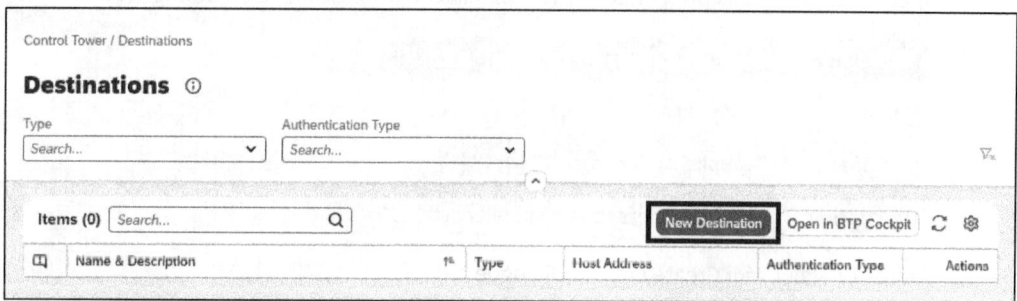

Figure 21.34 Adding New Destination

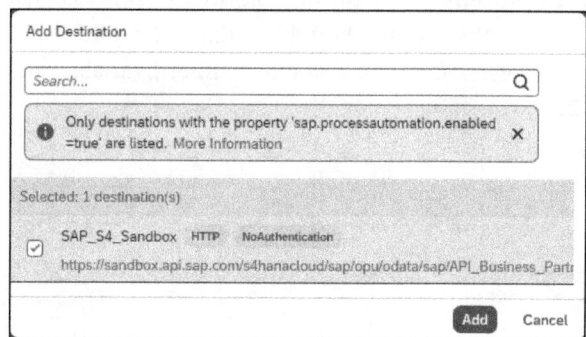

Figure 21.35 Adding Sandbox Destination

In the next step, we will create the process by using the actions. Navigate back to the lobby and select the example project. Create a new process by clicking the **Create • Process** entry, as shown in Figure 21.36.

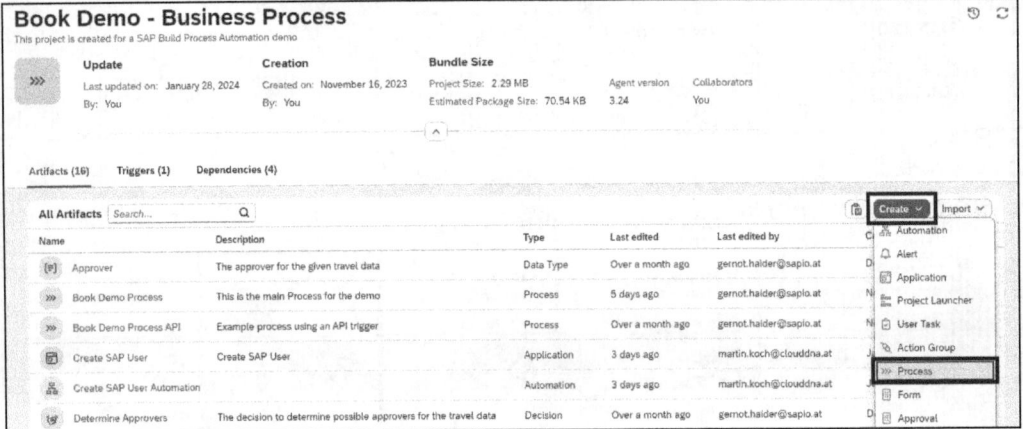

Figure 21.36 Creating New Process

Specify the values as provided in Table 21.3 in the process creation dialogue.

Field Name	Value
Name	"Supplier Action Process"
Identifier	"supplierActionProcess"
Description	"Demo process to test supplier actions"

Table 21.3 Values for Create Process Dialogue

By clicking the **Create** button, the process is created and opened in the design console. In the first step, we have to create an environment variable in the process that describes the destination through which the actions should be called. To do this, click the ... button to go to the project properties. Create a new environment variable, and provide the values listed in Table 21.4.

Field Name	Value
Identifier	"SAP_S4_Sandbox"
Description	"SAP_S4_Sandbox"
Type	"Destination"

Table 21.4 Properties of New Environment Variable

Figure 21.37 shows a screenshot of the new environment variable.

21.2 Using the Action within a Process

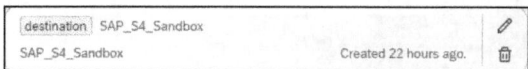

Figure 21.37 Environment Variable for Destination

Let's now create a trigger form, with which the process can be initiated. To do this, click the **Add a Trigger** link, as shown in Figure 21.38.

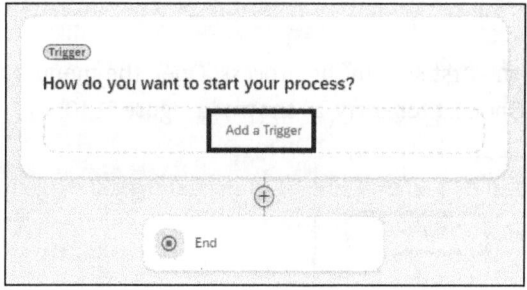

Figure 21.38 Adding Trigger

In the menu that appears, select **Submit a Form** to create a form. After you have decided to use a form as a trigger, the next step is to display the available forms in the project that could be used as a carrier. There is also the option to create a new form. Click the **Blank Form** tile to create a new form for the trigger, as shown in Figure 21.39.

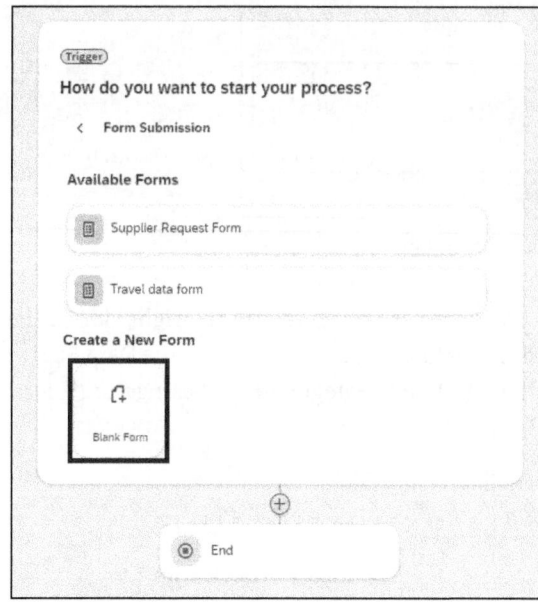

Figure 21.39 Creating New Trigger Form

In the dialogue for creating the new form, enter the values from Table 21.5 and click the **Create** button.

677

Field Name	Value
Name	"Supplier Action Trigger Form"
Identifier	"supplierActionTriggerForm"
Description	"Trigger form to start the supplier demo process"

Table 21.5 Values for Create Form Dialogue

The trigger form is now displayed as the first step in the process. Open the menu by clicking the ••• icon, then select the **Open Editor** entry, as shown in Figure 21.40.

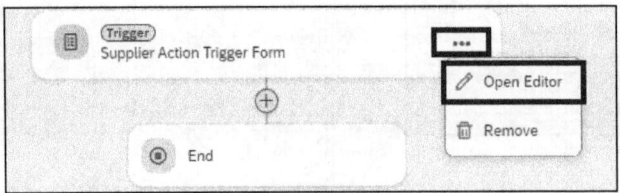

Figure 21.40 Opening Form Editor

In the form editor, add a dropdown element and change the value of the label to "Supplier Number", as shown in Figure 21.41.

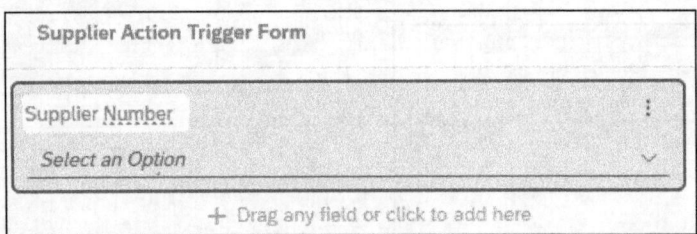

Figure 21.41 Dropdown Element for Supplier Number

By selecting the added element, the properties are displayed on the right side of the design console. In these properties, in the **Data to Display** area, select the **Data Source** option. Then open the value help for the action to be integrated by clicking the ⌕ icon in the **Data Source** field, as shown in Figure 21.42.

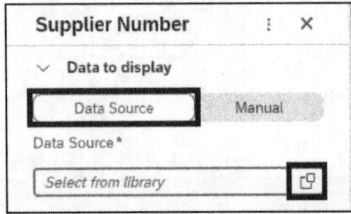

Figure 21.42 Selecting Data Source

21.2 Using the Action within a Process

As a result, a dialogue is displayed in which you can select the appropriate action from the library. In our case, only the action we created will be displayed. Click the **Add** button in the action tile to define it as the data source of the dropdown element, as shown in Figure 21.43.

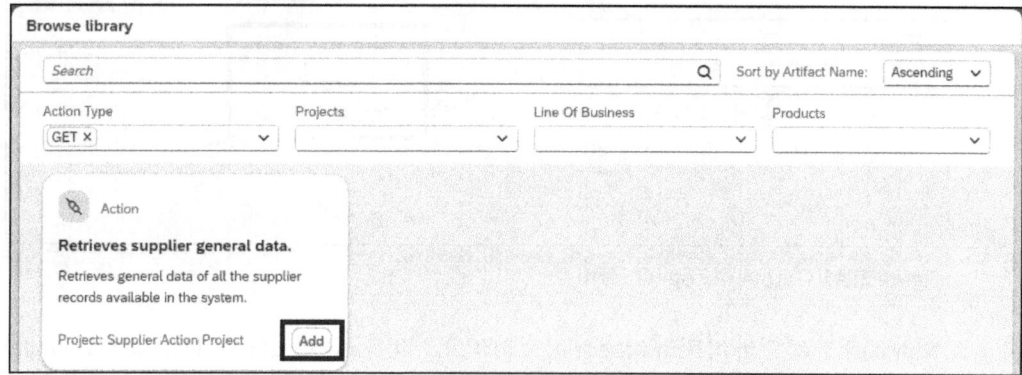

Figure 21.43 Adding Action to Dropdown Element

After selecting the data source, two additional fields are displayed with which the destination and the available fields from the action that should be displayed in the dropdown element can be defined. For the destination, select **SAP_S4_Sandbox**, which you previously maintained in the other variables. Open the list of values for the **Available Data** field, and select both fields that are displayed by selecting their checkboxes, as shown in Figure 21.44.

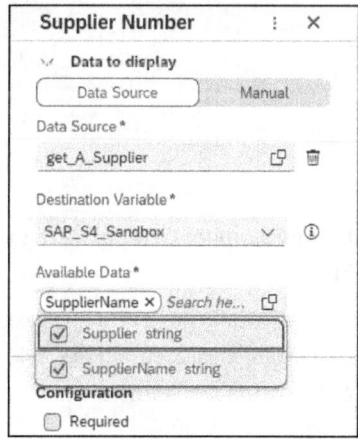

Figure 21.44 Configuration of Destination and Available Fields

Save the form and return to the process. In the process, select the trigger form. In the form's properties, in the **Outputs** section (see Figure 21.45), you can see that the **Supplier** and **SupplierName** fields are passed to the process.

679

21 Action Projects

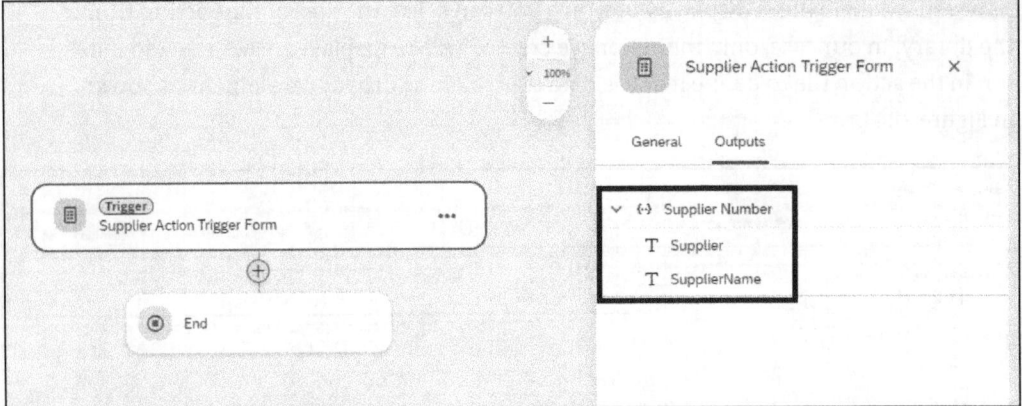

Figure 21.45 Output of Trigger Form

Now click the ⊕ icon below the trigger form. In the opened panel, select the **Action** tile, as shown in Figure 21.46.

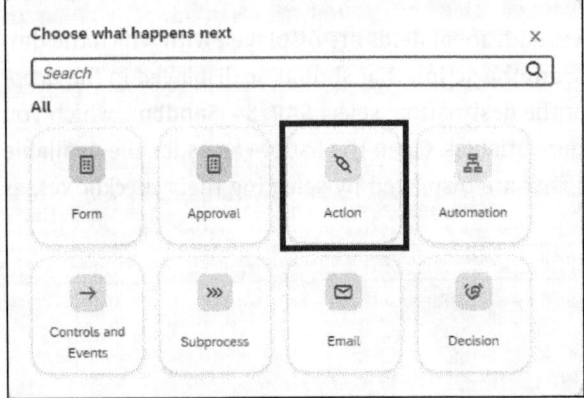

Figure 21.46 Adding Action to Process

In the opened dialogue, click the **Add** button in the **Retrieves Supplier General Data by Using Supplier Number** action tile, as shown in Figure 21.47.

The action is added as the second step of the process, and the properties of the process step are displayed on the right. In the **General** section, specify the destination (**SAP_S4_Sandbox**) that should be used to call the API. Figure 21.48 shows a screenshot of this configuration.

21.2 Using the Action within a Process

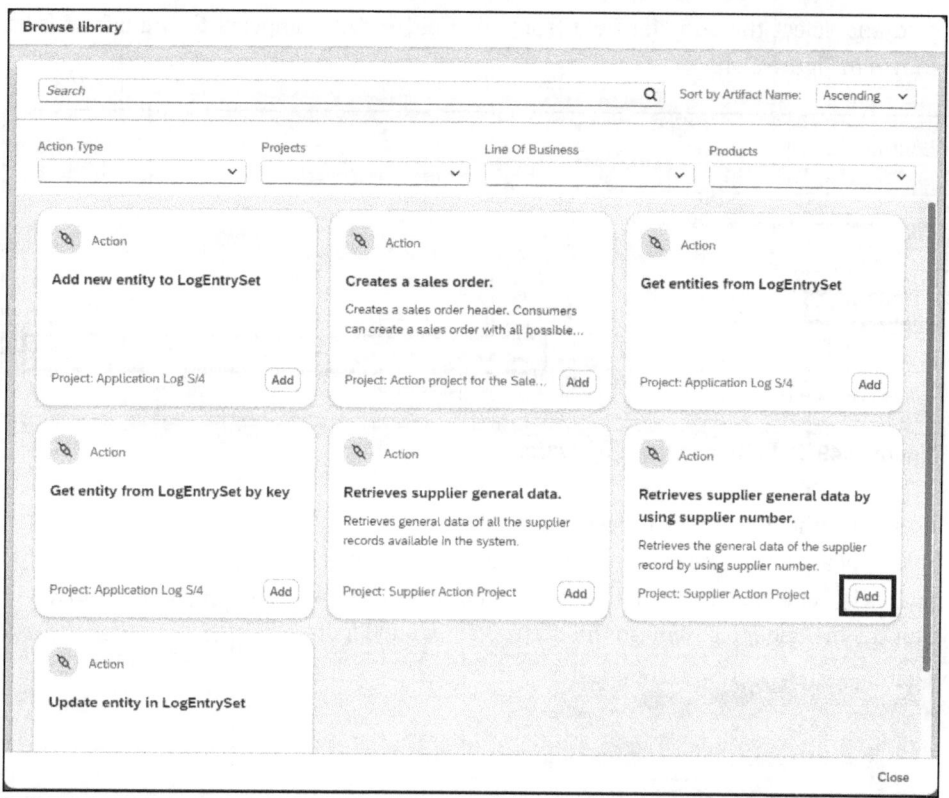

Figure 21.47 Selecting Action

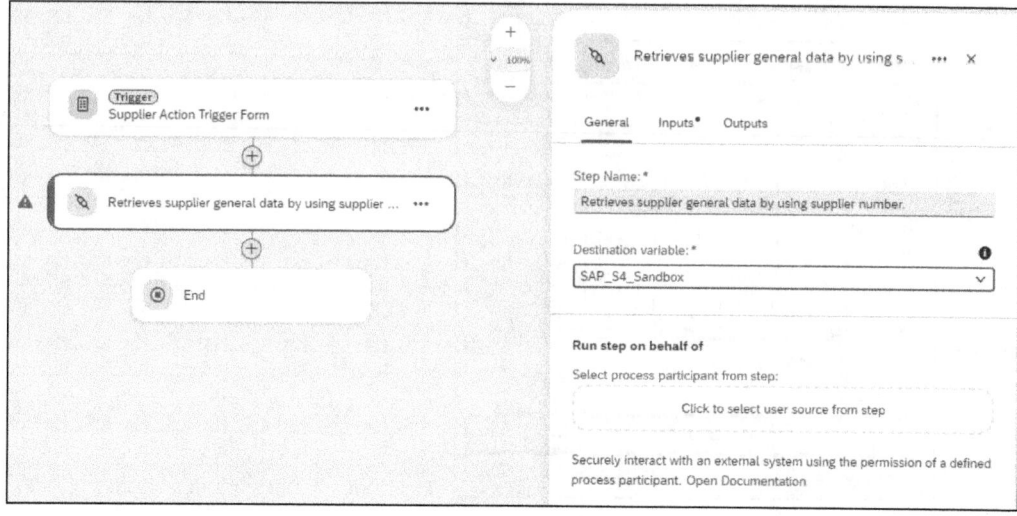

Figure 21.48 Configuration of Destination

In the **Inputs** section, you must now define which attribute from the context of the process should be used as an input parameter for the action. To do this, click the **Supplier** field and select the **Supplier** field from the field list that appears to the left of it, as shown in Figure 21.49.

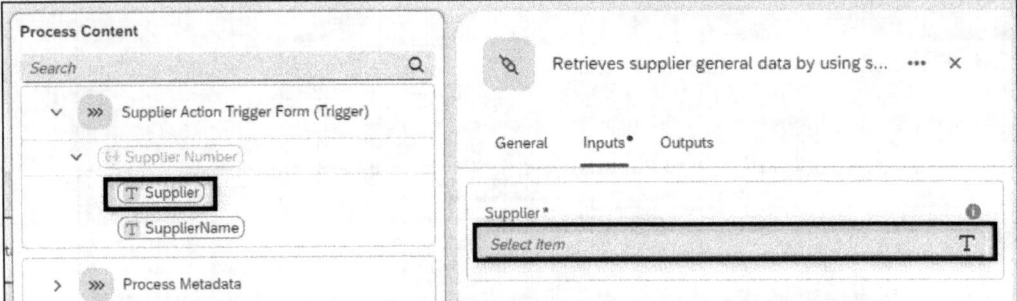

Figure 21.49 Definition of Input for Action

Create another step in the process by clicking the ⊕ icon below the action. In the panel that appears, select the **Email** tile. In the properties panel of this process step, enter the values in Table 21.6 for the fields. After that, click the **Open Mail Body Editor** button to edit the message to be sent in this step, as shown in Figure 21.50.

Field Name	Value
To	Your email address
Subject	"Supplier data"

Table 21.6 General Data of Email Step

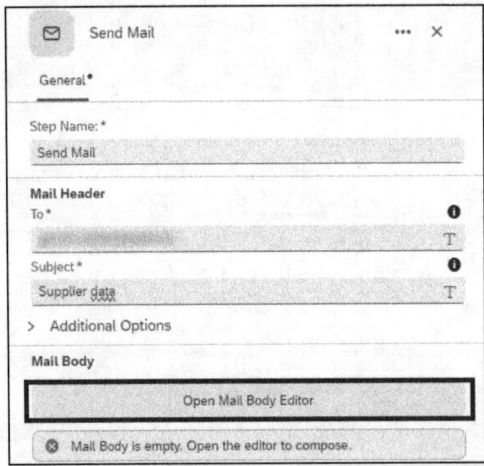

Figure 21.50 Opening Mail Body Editor

Define a message to be sent using the fields from the action result, as shown in Figure 21.51.

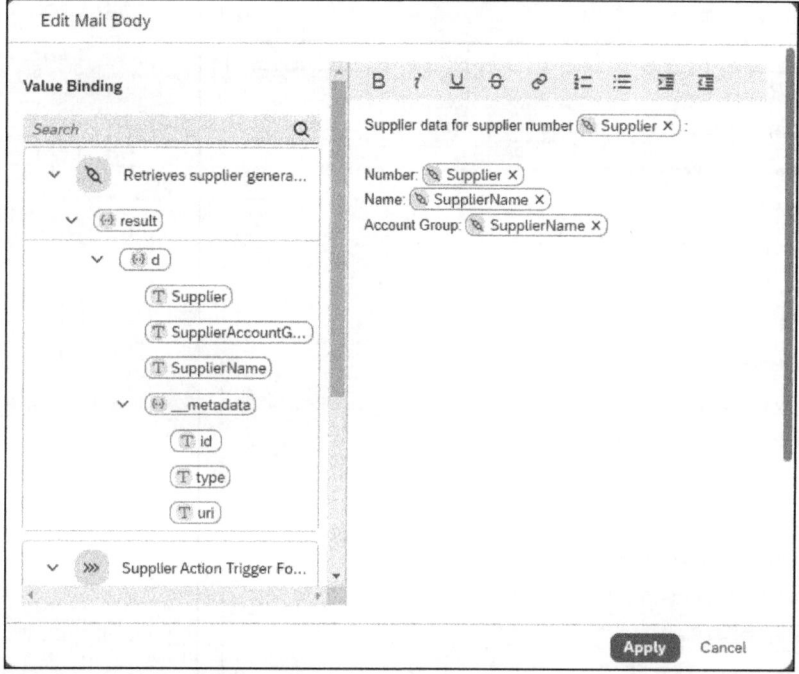

Figure 21.51 Defining Mail Body

Finally, click the **Apply** button. The process for demonstrating the use of actions is now complete. You can save the process and create a release that you can then deploy.

Before deployment, you must specify a value for the destinations environment variable that is configured to call the APIs through the actions. Select the value **SAP_S4_Sandbox** here, as shown in Figure 21.52. Finally, deploy your project.

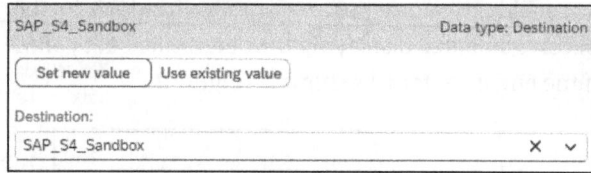

Figure 21.52 Defining Destination before Deployment

After deployment, open the process again and select the trigger form. A web link was generated for this form during deployment, which can be used to access the form. Copy this link and open it in a new browser tab or browser window. Figure 21.53 shows a screenshot of the dropdown element in the trigger form. As you can see, the options action was added to the dropdown element.

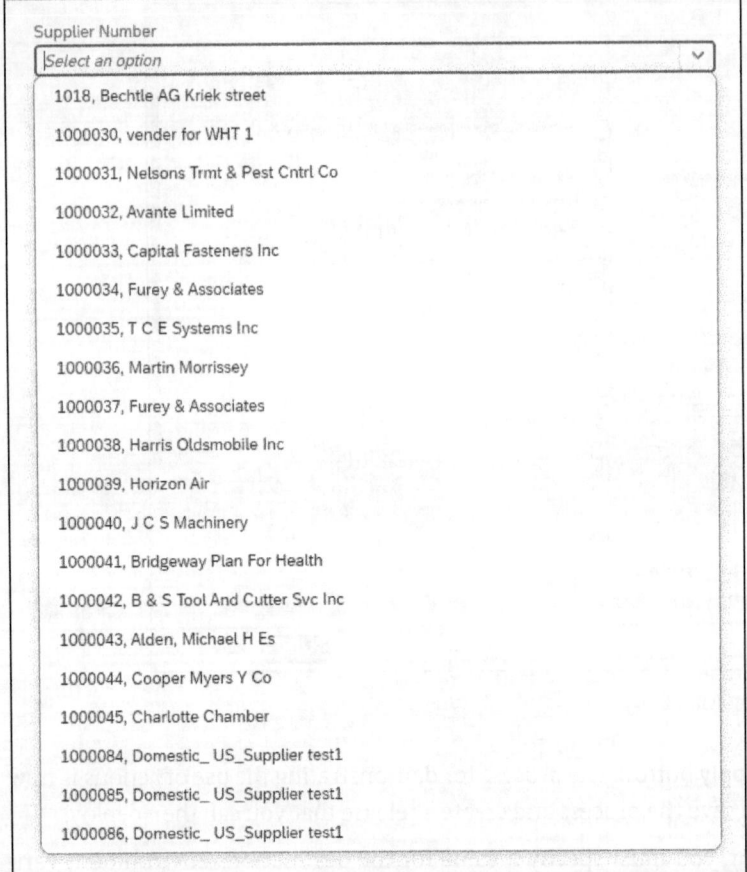

Figure 21.53 Dropdown Element in Trigger Form

Select an item and submit the form. This triggers a process instance in which further processing takes place. You will receive an email displaying the determined data of the action based on the definition of the email content, as shown in Figure 21.54.

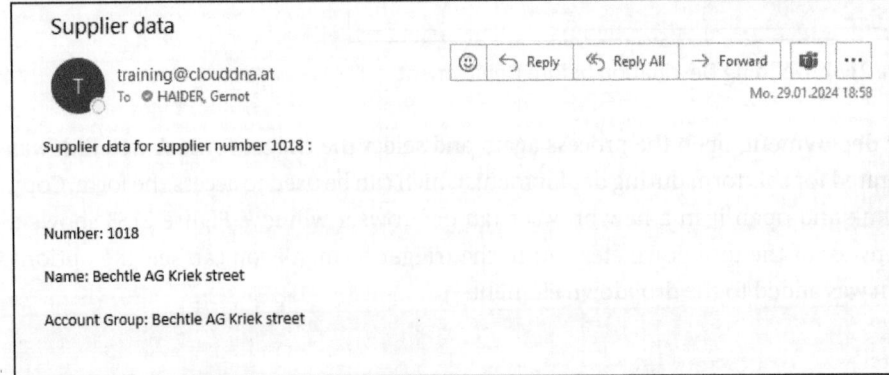

Figure 21.54 Email Sent from Actions Example Process

21.3 Summary

In this chapter, we showed how actions can be used in a process. We used an action to populate a dropdown element with selection options. We then used the selected value in a further action in the process to send a request to an API and thus receive detailed information about the selection.

In the next chapter, we turn to visibility scenarios, a powerful tool used for visualizing key figures.

Chapter 22
Visibility Scenarios

Process visibility refers to the clarity and transparency in understanding the various stages, steps, and activities within a process. It involves ensuring that relevant stakeholders have the necessary insights and information to track progress, identify bottlenecks, and comprehend how each step contributes to the overall outcome. Achieving high process visibility typically involves employing methods such as documentation, visualizations, status updates, reports, and effective communication channels to provide stakeholders with a clear understanding of the process dynamics and performance.

With visibility scenarios, SAP provides a tool that makes it possible to monitor workflows during ongoing operations and visualize their data. This is possible not only with the processes built in SAP Build Process Automation, but also with processes that are active in your on-premise landscape. This makes it possible to monitor all business workflows used in SAP in a central location. The following sections will show how such scenarios can be created, how the processes are bound, and how the data can be visualized. You will see how KPIs can be defined in order to create a dashboard that shows all relevant information at a glance. Finally, we'll run an example process and see how single parts of the visualization can be used to dive deeper into the relevant instances that are the basis of the statistics.

22.1 Creating a Visibility Scenario

In SAP Build Process Automation, *visibility scenarios* are an artifact within a project. To create a scenario, navigate from the lobby to the project in which you want to create the scenario by selecting the relevant project from the list of all projects, as shown in Figure 22.1.

You'll see the overview of the artifacts of the selected project. To add a new visibility scenario, open the menu of available artifact types by clicking the **Create** button. Figure 22.2 shows the opened menu. Create a new visibility scenario by selecting the **Visibility Scenario** menu entry, as shown in Figure 22.2.

22 Visibility Scenarios

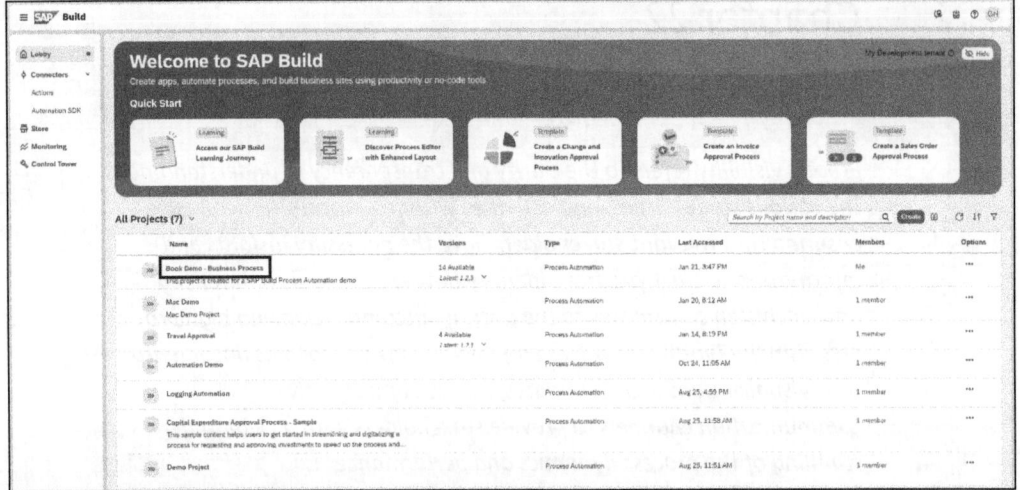

Figure 22.1 Navigate to Project

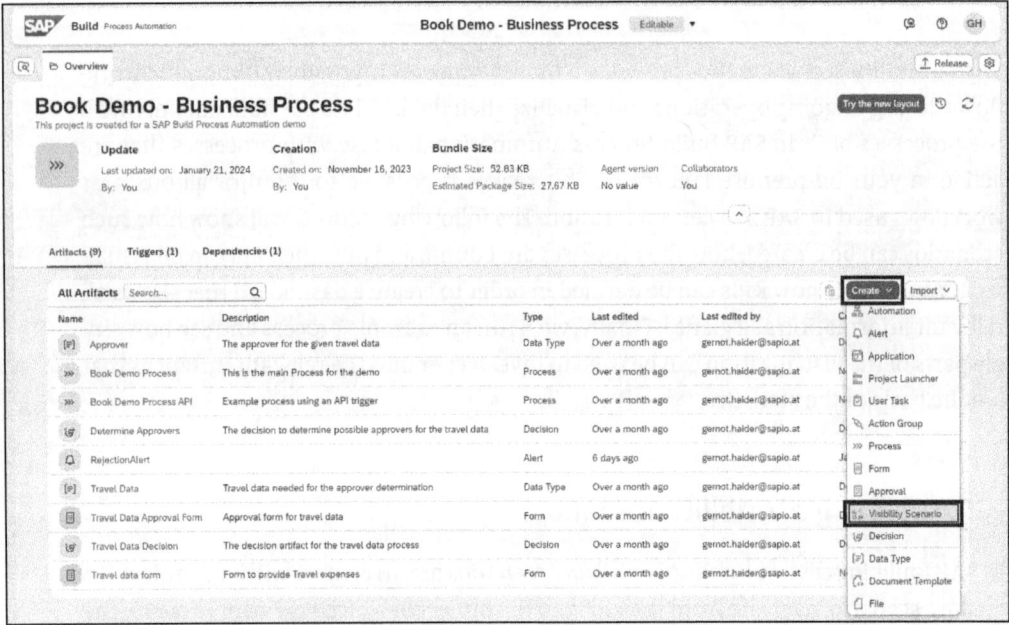

Figure 22.2 Creating New Visibility Scenario

For the scenario to be created, a name and an identifier must be specified. Optionally, a description can also be entered. Fill in the fields in the dialogue shown in Figure 22.3 with the values from Table 22.1, then click the **Create** button to create the new visibility scenario. Initially, the value for the identifier field is derived from the name for the scenario. However, this value can be adjusted if necessary.

22.1 Creating a Visibility Scenario

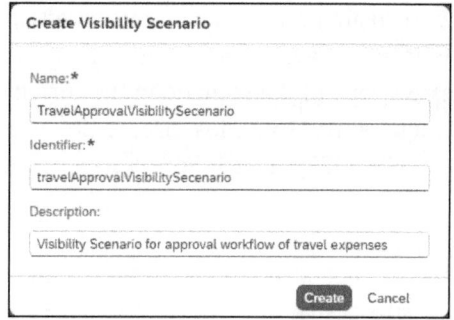

Figure 22.3 Dialogue to Create New Visibility Scenario

Field Name	Value
Name	"TravelApprovalVisibilitySecenario"
Identifier	"travelApprovalVisibilitySecenario"
Description	"Visibility Scenario for approval workflow of travel expenses"

Table 22.1 Values for Dialogue to Create New Visibility Scenario

The created scenario will open in a new tab. If you navigate to the **Overview** area, you can see that a new entry for the visibility scenario you just created has been added to the list of artifacts. Figure 22.4 shows a screenshot of the overview with the new entry for the visibility scenario that was just created.

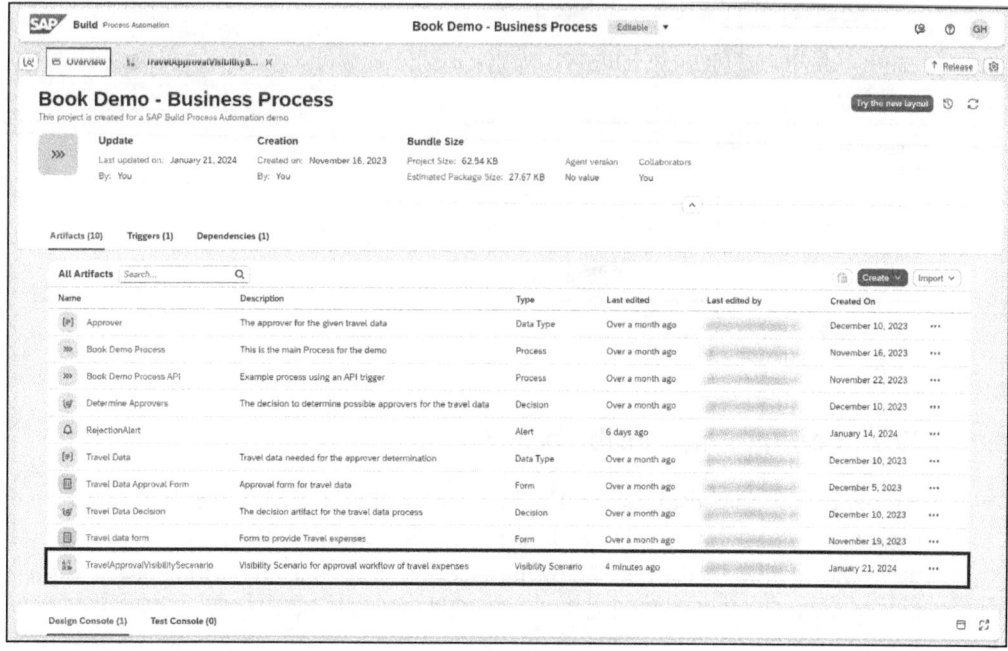

Figure 22.4 New Visibility Scenario in List of Project Artifacts

689

22 Visibility Scenarios

Existing scenarios can be copied in this view to create new configurations based on them in other projects. To do this, click the ⋯ icon in the **Actions** column, then select the **Copy** menu item. If you would like to create a copy of the scenario in the current project, you can do this using the **Duplicate** menu entry. Scenarios that are not activated can be deleted by clicking the **Delete** menu item. Figure 22.5 shows a screenshot of the action menu of a visibility scenario.

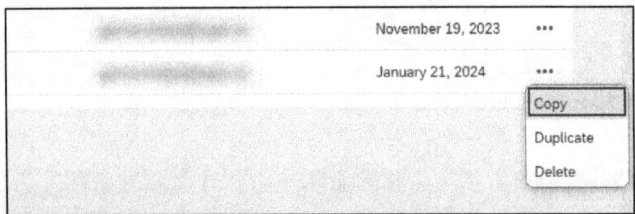

Figure 22.5 Action Menu of Visibility Scenario

If an existing scenario available in the form of an export file is to be imported into the current project, this can be done using the **Import** menu, as shown in Figure 22.6. After specifying the ZIP file that represents the exported scenario, as shown in Figure 22.7, it will be included in the list of available scenarios.

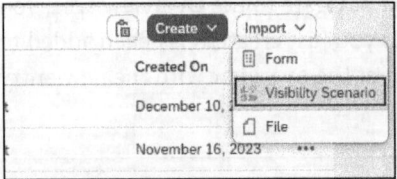

Figure 22.6 Import Menu in Artifacts List

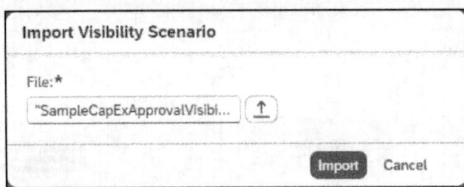

Figure 22.7 Importing Existing Scenario

22.2 Configuring a Visibility Scenario

A separate design console is available for configuring a visibility scenario, as shown in Figure 22.8. This editor is divided into several tabs in which the properties of the scenario can be described. In the following sections, we'll walk you through each of these tabs in turn.

22.2 Configuring a Visibility Scenario

22.2.1 General

In the **General** tab, you can specify how the instance should be displayed in the process workspace by specifying values for the **Instances Label** that is displayed to represent a group of instances and the **Instance Label** that is used to represent a single instance. The name and description of the scenario can also be subsequently edited here. The scenario ID cannot be changed here; it was created when the scenario itself was created and can no longer be changed. In addition to this information, the dashboard link to the scenario is also displayed in this tab. It is generated during deployment and is available from this point on. Figure 22.8 shows a screenshot of the **General** tab.

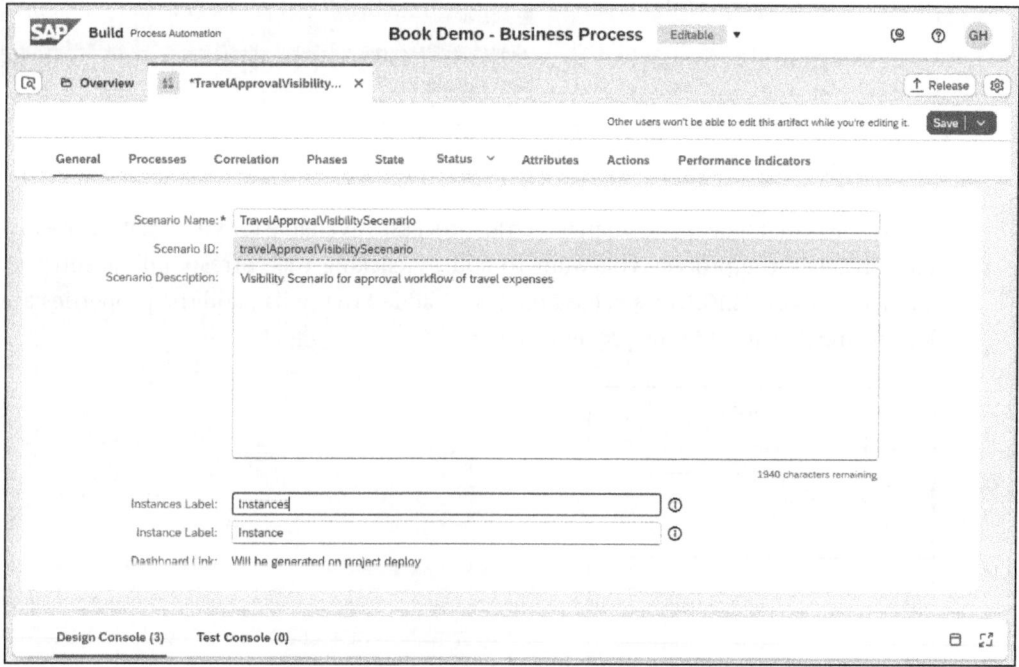

Figure 22.8 General Tab of Visibility Scenario

22.2.2 Processes

The **Processes** tab is divided into two sections. On the left, a list of processes used in the opened scenario is displayed. The right side shows the properties of the selected process from the list. In the process list, one or more processes that are to be visualized or monitored are added to the scenario. To add a process, click the + icon above the list, as shown in Figure 22.9. In the opened menu, select the **Add Process** entry to add a process with SAP Build Process Automation.

22 Visibility Scenarios

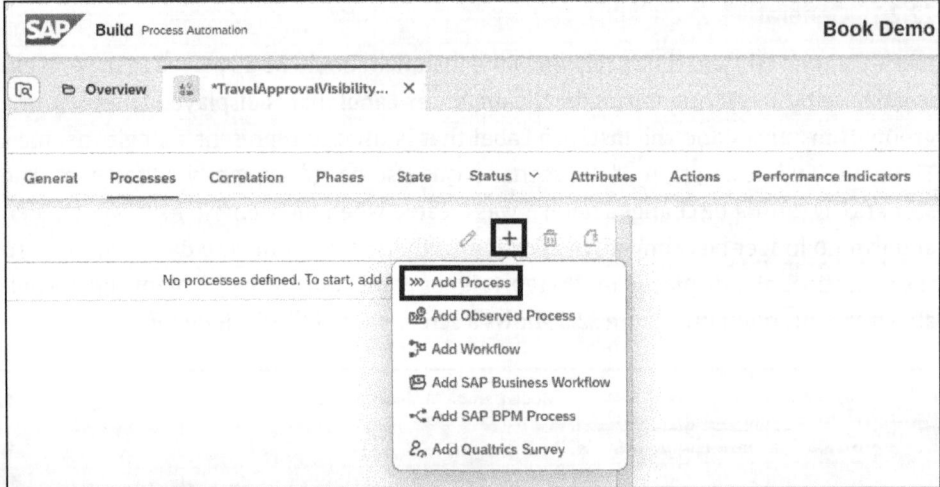

Figure 22.9 Adding Process to Visibility Scenario

As a result, a dialogue is shown in which all processes available for visualization are displayed. Select the process you want to add by clicking the corresponding entry, as shown in Figure 22.10. The selected process is added to the list, and the properties are loaded and displayed in the process data area.

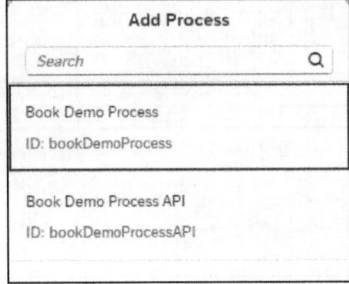

Figure 22.10 Selecting Process to Be Used in Scenario

The contents of the loaded process are now visualized on the right side of the design console. The upper area consists of the events that take place as part of the processing of the process and can be visualized. Each step that is part of the selected process is mapped with two events: one event for the creation and one for the termination of the step. The start event and the end event of the process itself are exceptions. Only the event itself is available for these. In addition, there are higher-level events that occur when the process is suspended, resumed, or aborted. Figure 22.11 shows a screenshot of the events that can occur in the case of the example process.

The context available for visualization is displayed in the lower part of the process data. At the current time, the context data list is empty, as shown in Figure 22.12.

22.2 Configuring a Visibility Scenario

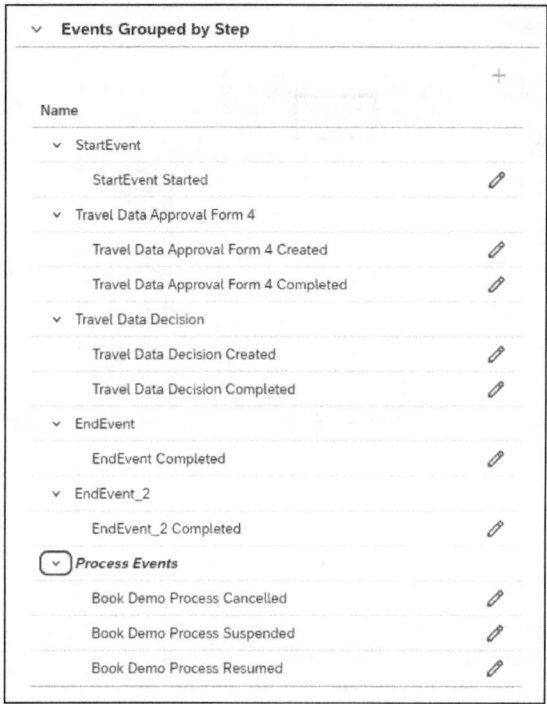

Figure 22.11 Events of Selected Process

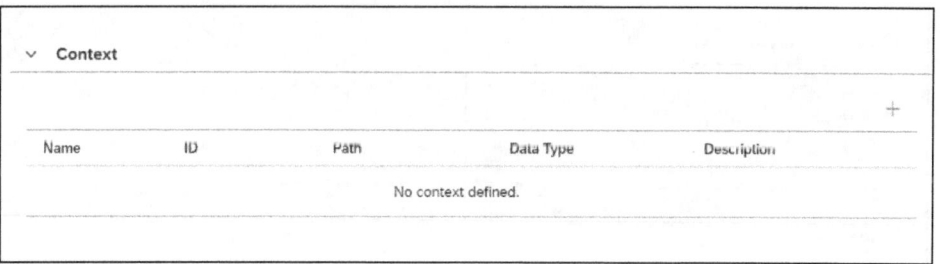

Figure 22.12 Context of Selected Process

For data to be available in this section for use in the visibility scenario, the process must be configured accordingly. To do this, navigate back to the artifacts overview and open the selected process to view it in the design console. Then open the panel on the right to view the properties of the process. Select the **Visibility** tab and click the + icon to make process data visible for a visibility scenario, as shown in Figure 22.13.

All attributes that the process can provide for the visibility scenario are then displayed. Select the **End Date**, **Start Date**, and **Total Costs** properties in turn from the list of attributes. The attributes are added to the visibility and can be removed again, if necessary, by pressing the **Remove** button. Figure 22.14 shows the configuration of the **Visibility** tab after adding these properties.

693

22 Visibility Scenarios

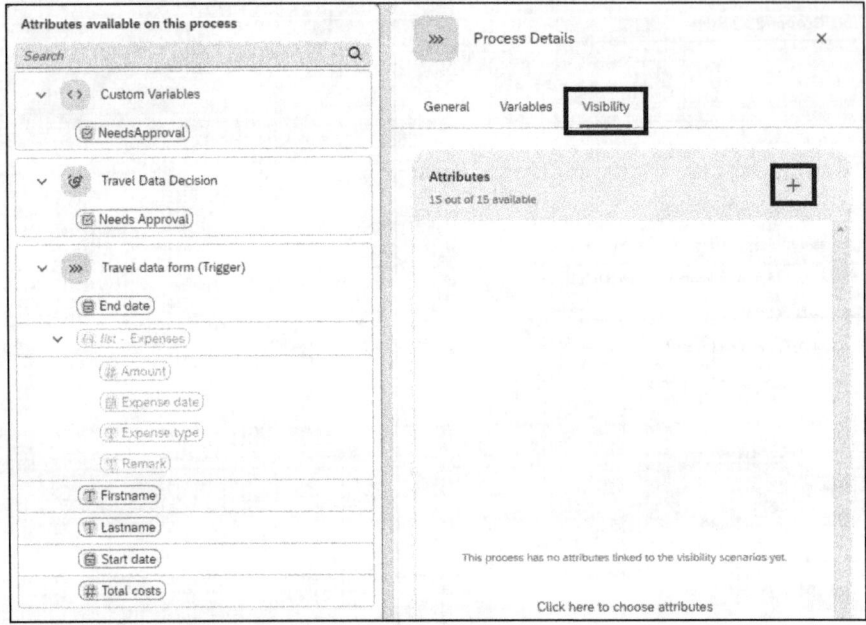

Figure 22.13 Visibility Configuration in Process

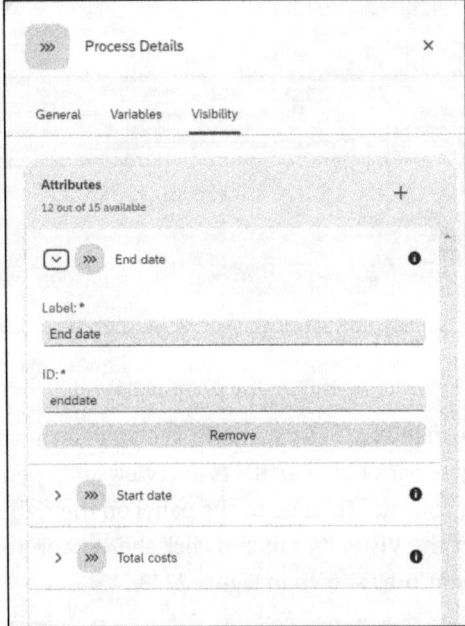

Figure 22.14 Configured Visibility of Process

After you finish editing, save the process. Changing the process causes the visibility scenario to respond with an error message because the inserted process has changed properties that affect the scenario, as shown in Figure 22.15.

22.2 Configuring a Visibility Scenario

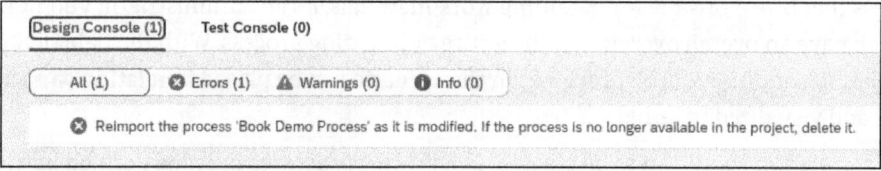

Figure 22.15 Error Caused by Modifying Process

To correct this error, the affected process must be reimported. To do this, click the 🗘 icon above the process list. In the **Reimport** dialogue, select both checkboxes, then click the **Reimport** button, as shown in Figure 22.16. The process will be updated, and the error will be resolved after saving the scenario.

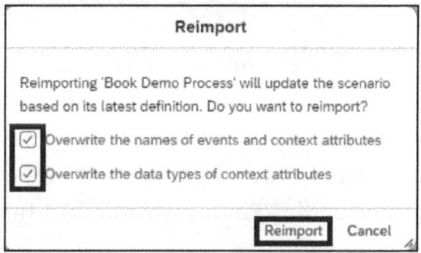

Figure 22.16 Reimport Process

As a result, the attributes that were made visible for the scenario are now displayed in the context area of the process, as shown in Figure 22.17.

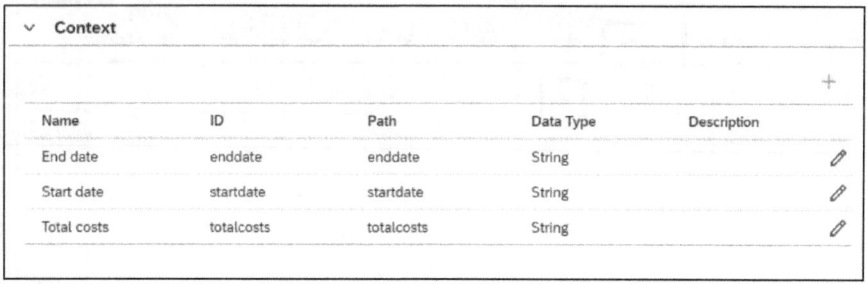

Figure 22.17 Context Variables Available in Visibility Scenario

You have now successfully created a visibility scenario, added a process, and configured the process so that content data can be used in the scenario. In the next sections, you will learn how this information can be used for visualization within the scenario.

22.2.3 Correlation

It often happens that different processes are related to each other. As an example, a process for onboarding a new employee can be used. This process can trigger another

22 Visibility Scenarios

process that is responsible for obtaining work materials. As an administrator, you now want to have an overall overview of the entire onboarding process. With the help of the correlation conditions, both processes in the scenario can be placed in relation to each other and visualized together.

For a correlation condition to be created, at least two processes must be defined in the scenario. These processes must each have a context. The contexts of the correlating processes must also have at least one property that has the same meaning and through which the processes can be related to one another. In our example of the onboarding processes, this could be the personnel number of the new employee. You can use the correlation conditions to map both 1:1 relations and 1:n relations.

22.2.4 Phases

By defining phases, the visualization of the process in the visibility scenario can be divided into sections. The individual phases are limited by events that occur during the process. If a start event defined in a phase is reached in the process, this phase comes into effect. It will end again when the defined end event comes into force. To create phases in the visibility scenario, switch to the corresponding **Phases** tab and click the + icon, as shown in Figure 22.18.

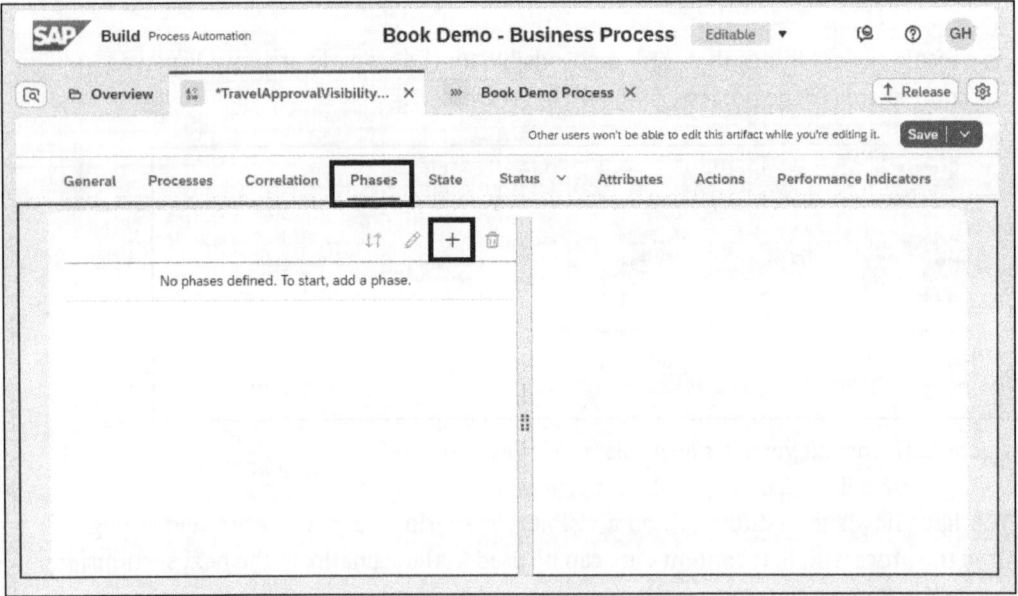

Figure 22.18 Creating Phases

In the dialogue that appears, enter your **Name** and **ID** for the phase to be created. The ID is automatically generated based on the name but can be changed if necessary. Click the **OK** button to create the new phase, as shown in Figure 22.19.

22.2 Configuring a Visibility Scenario

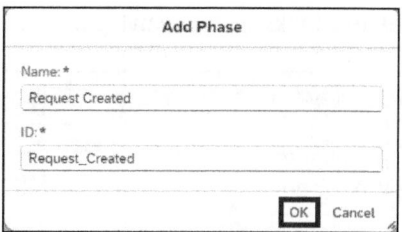

Figure 22.19 Provide Name of New Phase

The new phase is added to the list, and the properties can be edited. You can now specify start events and end events that define the start and end of the phase. To do this, expand the combo box of the start event and select the **Travel Data Approval Form 4 Created** entry, as shown in Figure 22.20.

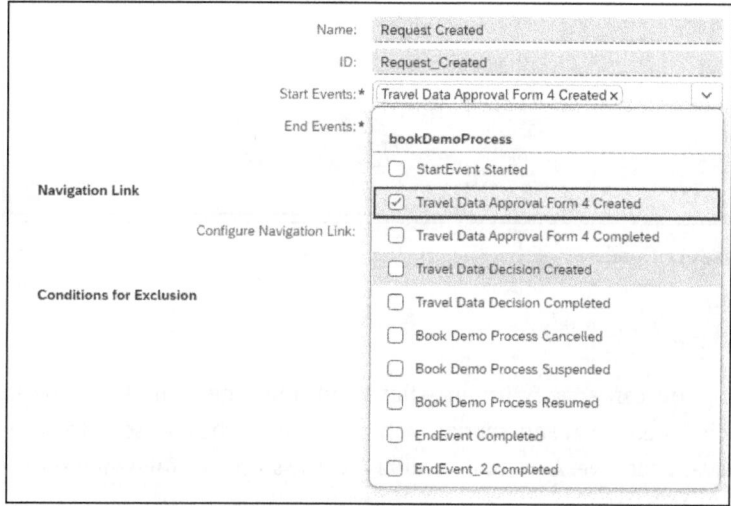

Figure 22.20 Defining Start Event of Phase

To define the end of the phase, proceed in the same way and select the **Travel Data Approval Form 4 Completed** event for the end of the phase.

Add a second phase to the scenario and configure it with the data shown in Table 22.2.

Property	Value
Name	"Request Approval"
ID	"Request_Approval"
Start Events	Travel Data Decision Created
End Events	EndEvent Completed, EndEvent_2 Completed

Table 22.2 Properties of New Phase

697

22 Visibility Scenarios

After specifying the data from Table 22.2, the **Phases** area looks as shown in Figure 22.21.

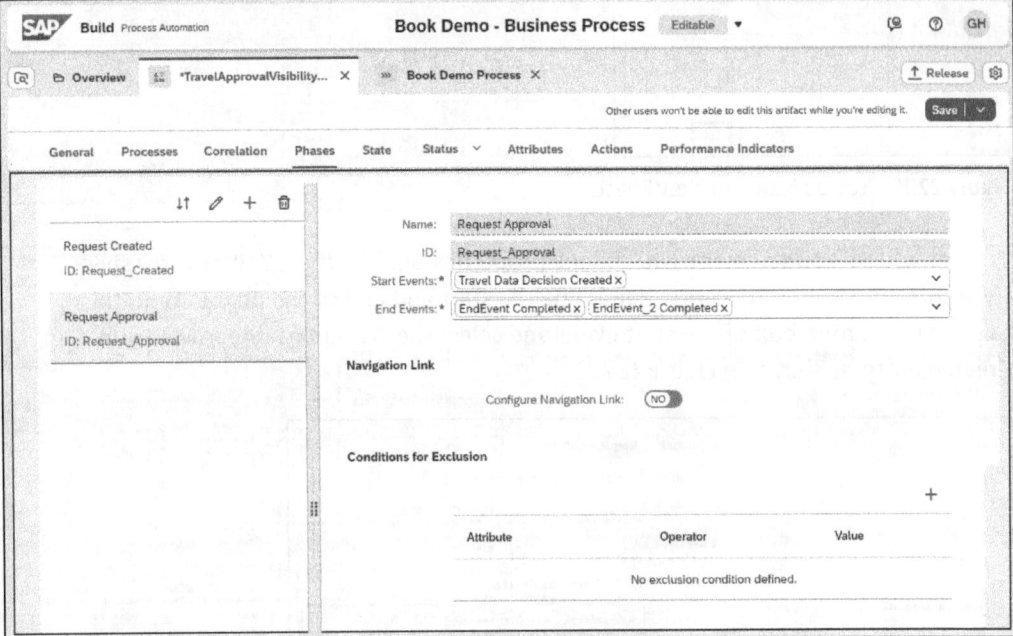

Figure 22.21 Two Phases Defined

22.2.5 State

In the **State** section, you can specify under what conditions the visibility scenario should assume which state. The events that are achieved during the course of the process processing are again decisive. A visibility scenario can assume the following states:

- **Open**
  The events that mark the process as finished or aborted have not yet been reached.
- **Completed**
  An event was reached that marked the process as ended, but the process was not aborted.
- **Abruptly Ended**
  The process was terminated with an event that marked it as aborted.

The names of the individual states can be changed if necessary. For the **Completed** and **Abruptly Ended** states, the events to be considered can be expanded or reduced if necessary. Figure 22.22 shows a screenshot of the configuration of the states.

22.2 Configuring a Visibility Scenario

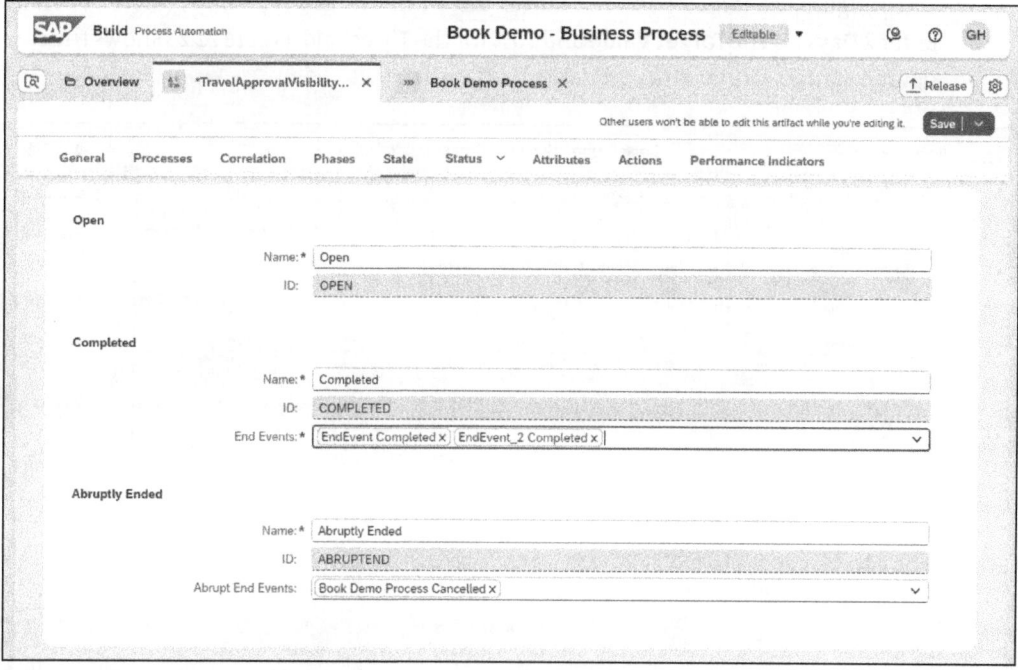

Figure 22.22 State of Visibility Scenario

22.2.6 Status

In contrast to the states defined in Section 22.2.5, which look at the technical aspect of an instance, the **Status** section assesses an instance from a technical point of view. An instance can assume the following technical states:

- **Critical**
  An instance is considered critical when the defined time for the target has expired or when a defined substatus is reached.
- **At Risk**
  An instance is considered at risk when the defined time for the target threshold is close or when a defined substatus is reached.
- **On Track**
  If an instance is neither **Critical** nor **At Risk**, it is in the **On Track** status. This means that the threshold and the target time have been exceeded again.
- **Completed without Violation**
  If an instance has ended within the defined time, it assumes this state.
- **Completed with Violation**
  The instance was terminated, but the permitted duration was exceeded.

22 Visibility Scenarios

To configure the target, go to the **Status** tab. In the **Target Type** field, select **Constant**. Enter **2 Days** for the **Target Value** and **50%** for the **Threshold**. Figure 22.23 shows the configured status section of the scenario.

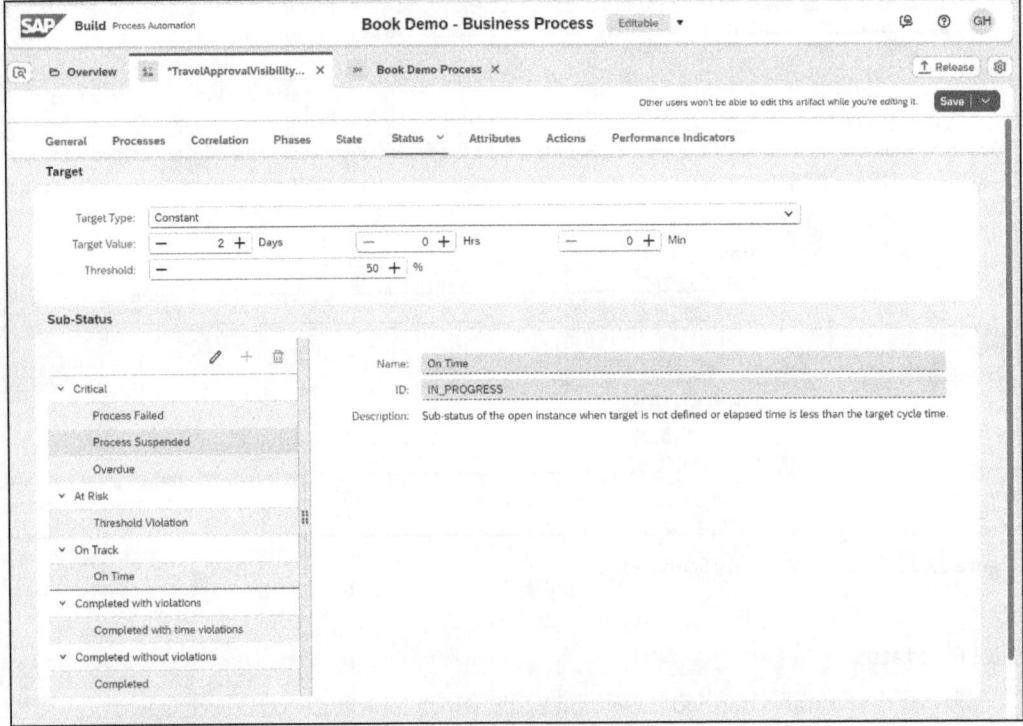

Figure 22.23 Status of Visibility Scenario

22.2.7 Attributes

Attributes help users find a specific instance in the monitoring. Visibility scenarios offer a variety of attributes that a process brings with it. In addition to these standard attributes, calculated attributes or context attributes can be used. The standard or default attributes are values that come with every process out of the box. They offer information about the duration of a process or the current status, for example, and can be used as they are. Context attributes describe values that form the content of the process. They are the data being processed. The calculated attributes can be used to form new information by executing expressions on default or context attributes. Depending on the data types of the attributes, various functions can be used to calculate new attributes to be displayed on a visibility dashboard. Finally, you can use entities to group context attributes and calculated attributes in order to define new performance indicators.

Default Attributes

The visibility scenario comes with a variety of default attributes. Table 22.3 shows the list of these attributes and describes them.

Attribute ID	Attribute Name	Description
SC_State	State	The state of the instance based on start, end, and abrupt end events of a process.
SC_Status	Status	The status of the instance based on how an instance is progressing according to the defined target.
SC_SubStatus	Substatus	Detailed information on why the instance is in a certain status.
SC_Start_Time	Start Time	The start time of a scenario instance. It is defined by the first event observed.
SC_End_Time	End Time	Defines the end timestamp of a scenario instance. If there are multiple end events, it's the timestamp of the first end event observed.
SC_Elapsed_Time	Elapsed Time	The duration between the start time and the current time for running processes. If an end event occurred, this is the duration between the start time and the end time.
SC_Number_Of_Instances	Number of Instances	Determines the total number of instances for a given filter criteria or breakdown. For a single instance this value is 1.
SC_Active_Phases	Active Phases	The active phases for an instance.
SC_Completed_Phases	Completed Phases	The completed phases of an instance.
SC_Active_Steps	Active Steps	The steps currently active or failed.
SC_Completed_Steps	Completed Steps	The completed steps of an instance.
SC_Target_Time	Target Time	The time when the instance should be completed. It is derived from the start time and target defined.
SC_Threshold_Time	Threshold Time	Defines the time after which an open instance is determined to be at risk. This is derived from the start time, target, and defined threshold.

Table 22.3 Default Attributes

For the standard attributes, you can define how they should be displayed in the scenario. The decisive factor is the value specified in the **Importance** field. To define the display, remember that the values in Table 22.4 can be specified.

Name	Display Area
Key	The attribute is displayed in the overview page of the instances view as a column in the instances table.
Significant	The attribute is displayed as a field in the **Information** tab of the instances view details page.
Internal	The attribute is not displayed but can be used within the performance indicators.

Table 22.4 Configuration of Default Attributes

Figure 22.24 shows the configuration of the **Start Time** attribute.

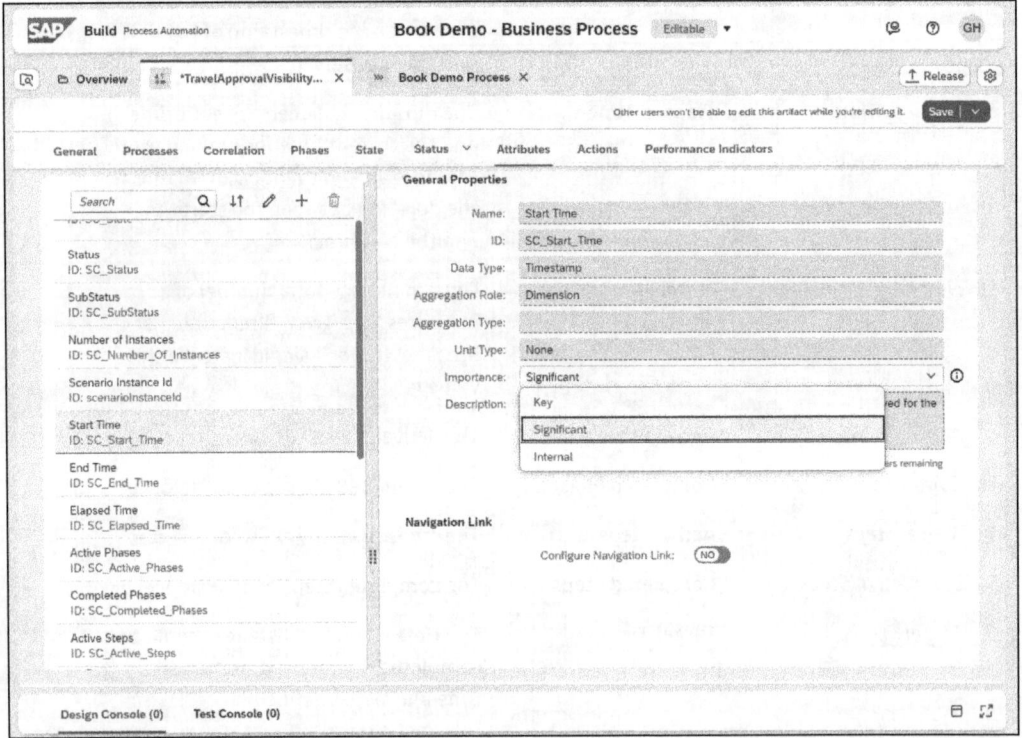

Figure 22.24 Configuration of Start Time Attribute

Context Attributes

Context attributes provide information that form the technical content of the process and are processed by the process. For numeric attributes, an aggregation can be

22.2 Configuring a Visibility Scenario

selected in the configuration, which is carried out when a search or filter returns multiple results in monitoring. As with the default attributes, the form in which these attributes are displayed can also be defined for the context attributes by specifying the importance. Optionally, a navigation link can be specified to navigate to further information about this attribute. Figure 22.25 shows the configuration for the **Total Costs** context attribute.

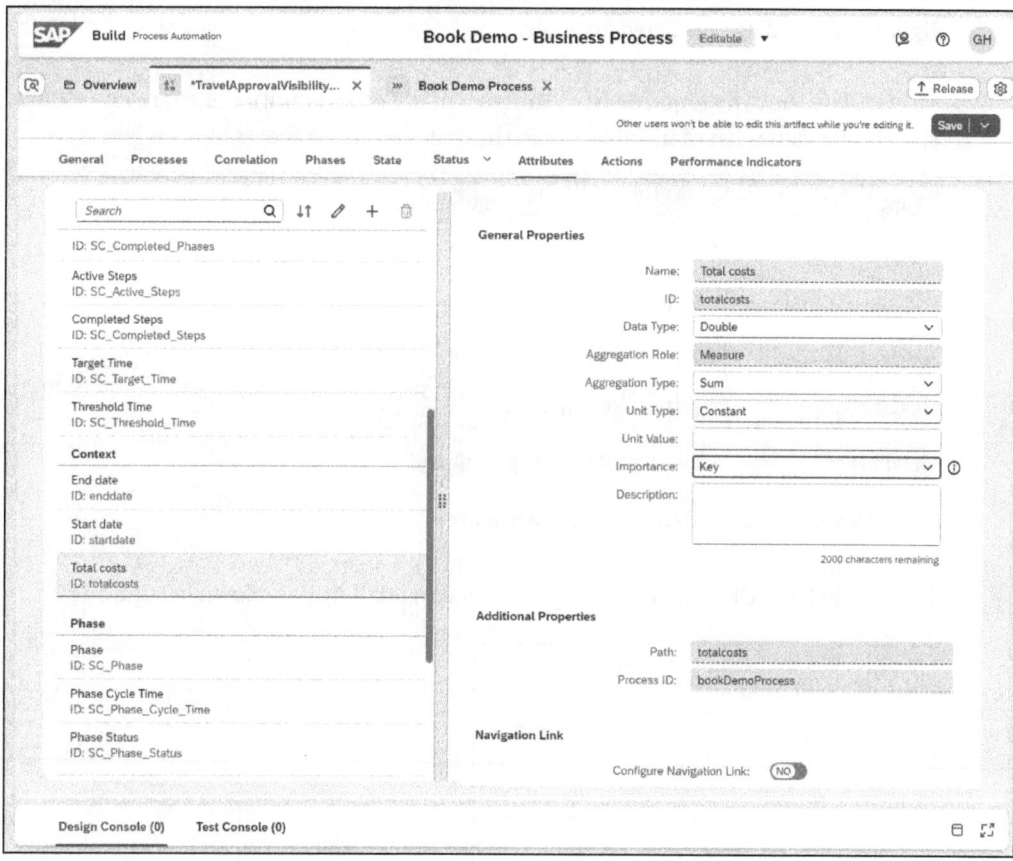

Figure 22.25 Configuration of Context Attribute

Calculated Attribute

In addition to the default attributes and context attributes, calculated attributes can be added to the scenario. These calculated attributes are based on the events that can occur in the process or attributes used in it and are calculated using predefined rules. To create a calculated attribute, click the + icon above the list of attributes and select the **Add Calculated Attribute** entry, as shown in Figure 22.26.

22 Visibility Scenarios

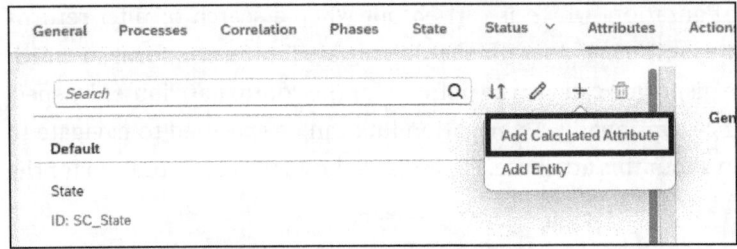

Figure 22.26 Adding Calculated Attribute

To calculate the time required from the start of the process until the approver is found, specify the values listed in Table 22.5 in the dialogue that appears.

Field Name	Value
Name	"Approver Finding"
ID	"Approver_Finding"
Expression Type	Duration between Events
Start Event	StartEvent Started
End Event	Travel Data Decision Completed

Table 22.5 Values to Define Calculated Attribute

Finally, click the **OK** button to add the calculated attribute, as shown in Figure 22.27.

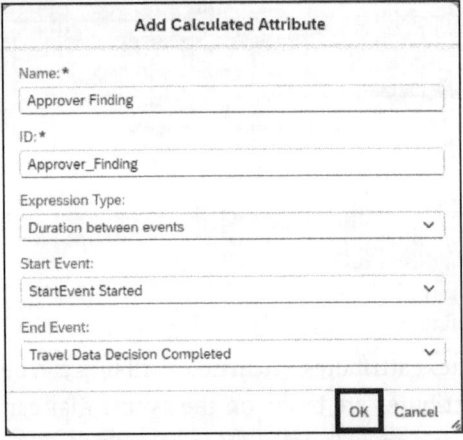

Figure 22.27 Adding Calculated Attribute

The new attribute is added to the end of the attribute list and can be configured as shown in Figure 22.28.

22.2   Configuring a Visibility Scenario

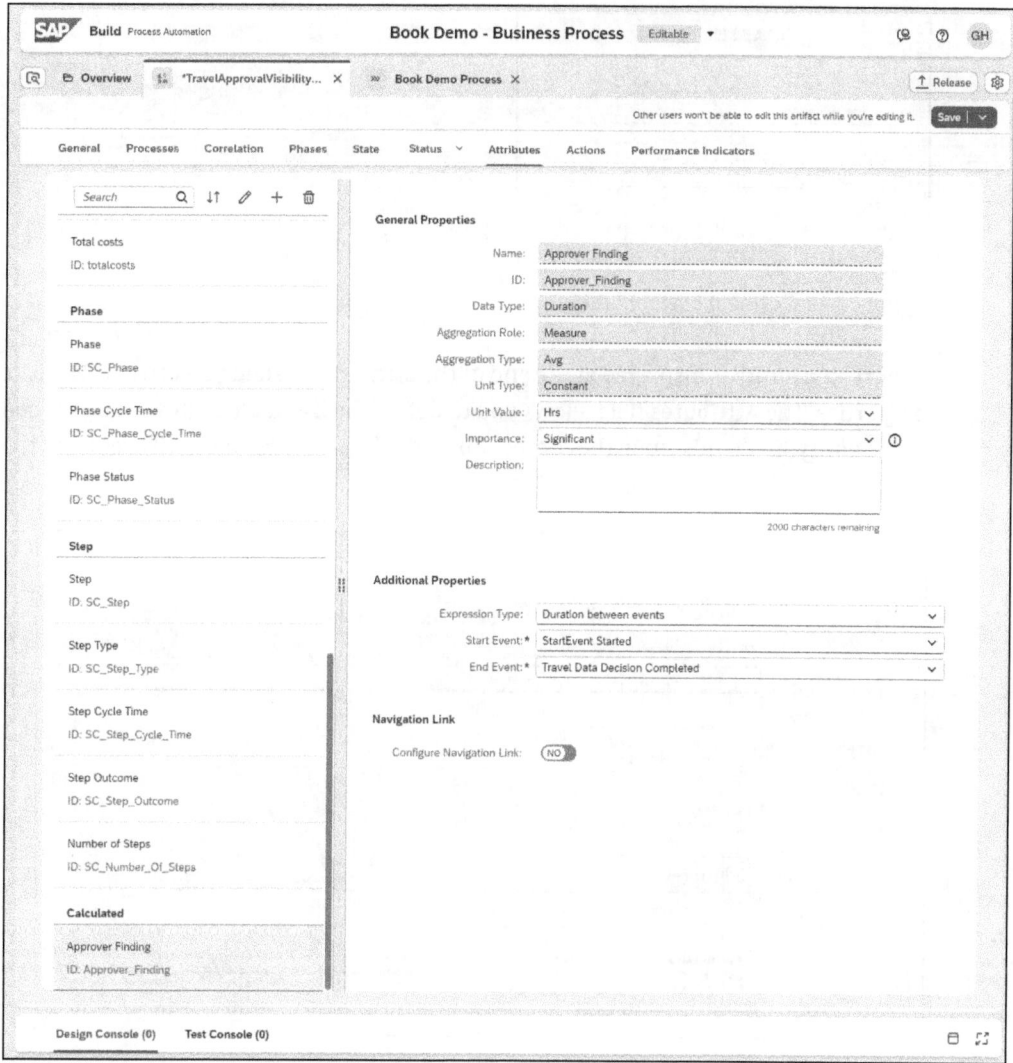

Figure 22.28 Configuration of Calculated Attribute

Entities

In a visibility scenario, context attributes and calculated attributes can be combined into one entity. This allows the business user to create performance indicators based on this entity. To create an entity, click the icon above the attribute list and select the **Entity** entry, as was shown back in Figure 22.26.

In the dialogue that appears, enter a name and an ID for the new entity. The ID is automatically generated based on the name but can be customized if necessary. Figure 22.29 shows an example of this dialogue. Click the **OK** button to confirm your entries and to create the entity.

705

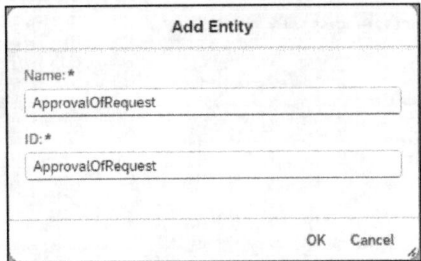

Figure 22.29 Defining Entity Name

The created entity is displayed at the end of the attribute list, and you can configure its properties. The **Attributes** field allows you to define the list of attributes to be grouped in the entity. To do this, open the combo box and select the relevant attributes from the list, as shown in Figure 22.30.

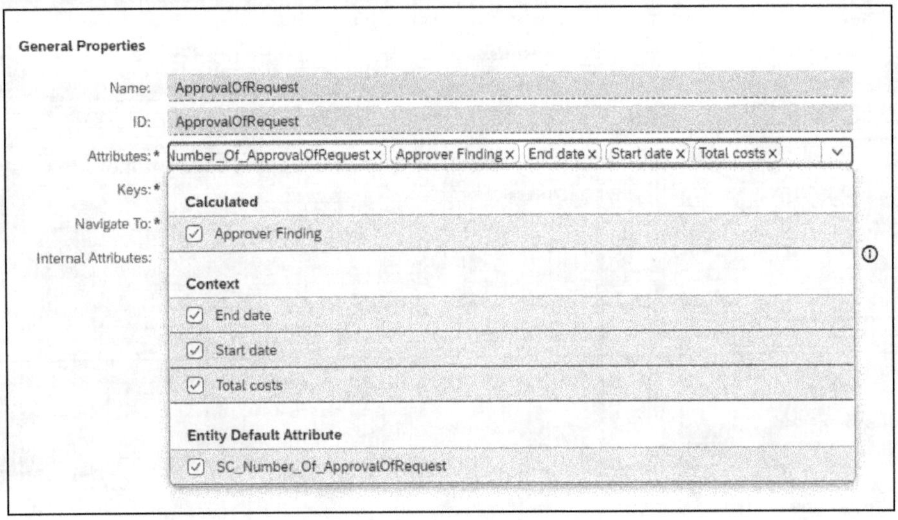

Figure 22.30 Selecting Attributes for Entity

The **Keys** field contains one or more attributes that uniquely identify an entity in the set of all similar entities. As with the attributes of an entity, the attributes that form the key are selected by selecting them from a list.

The **Navigate To** field defines which detailed information should be navigated to. By default, **Instances** is selected here, which means that the associated scenario instance is navigated to. If there are several entities in your scenario that have a parent-child relationship, you can navigate from the parent entity to the child entity by appropriately configuring this field. It should be noted that the key fields of the parent entity are also part of the content of the shared entity.

22.2 Configuring a Visibility Scenario

Optionally, one or more attributes that are defined as internal and are not displayed in the scenario instances view can be specified in the **Internal Attributes** field. Figure 22.31 shows an example configuration of an entity.

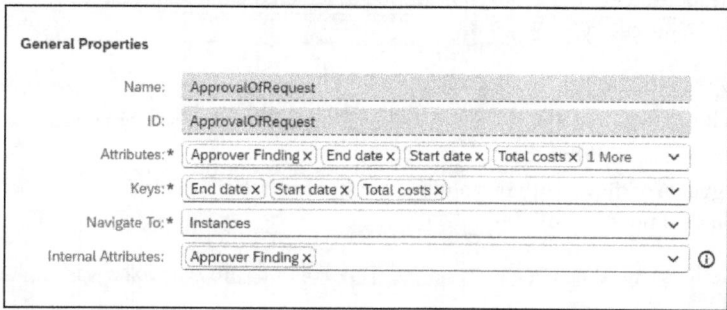

Figure 22.31 Configuration of Entity

22.2.8 Actions

Actions allow the user to react to specific situations in a process. They are created at the visibility scenario level and are therefore available to all instances, provided the nature of the data and the current status of the process allow this. There is a separate tab in the design console for configuring actions, where actions can be edited and deleted. On the left side of this view, the defined actions are displayed in the form of a list. In the right part, the properties of the selected action from the list are displayed and can be edited. Figure 22.32 shows the actions editing interface.

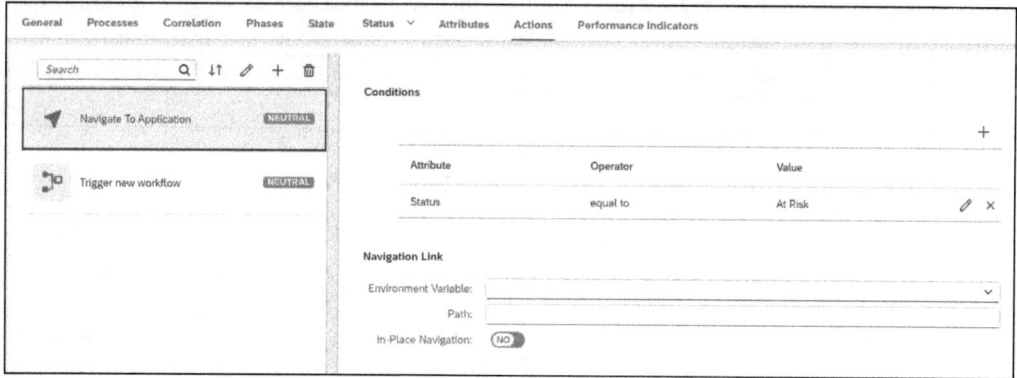

Figure 22.32 Design Console of Actions

To create an action of this type, click the + icon above the action list. The properties of the action are specified in the following dialogue. The meaning of the fields are listed in Table 22.6.

22 Visibility Scenarios

Field Name	Value
Name	Name of the action
ID	Unique identifier of the action. It is derived based on the name but can be adapted if necessary.
Type	The type of the action. Two options are available: ■ **Navigational**: This option enables the user to navigate to another application to take further actions. ■ **Trigger Workflow**: With this option, the user can initiate a workflow in certain situations.
Sentiment	This defines how the action is visualized. Three options are available: ■ **Neutral**: The action is displayed as a neutral action. ■ **Positive**: The action is displayed as a positive action. ■ **Negative**: The action is displayed as a negative action.

Table 22.6 Fields for Creation of Action

There are two types of actions that can be used in the visibility scenario, which we'll discuss in the following sections.

Navigational

The **Navigational** action makes it possible to navigate based on a condition to an application in which further action is taken. To create an action of this type, specify data in the dialogue as shown in Figure 22.33. Then confirm your entry by clicking the **OK** button.

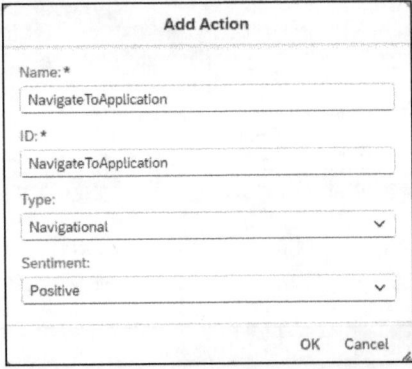

Figure 22.33 Adding New Navigational Action

The new action is added to the list on the left side of the design console, and detailed information about the action's configuration is displayed on the right. In the first step of configuration, conditions can be defined that must be met for the action to be created. In this example, the action should only become visible when there are more than three instances. To begin, click the icon + above the list of conditions to add a new condition, as shown in Figure 22.34.

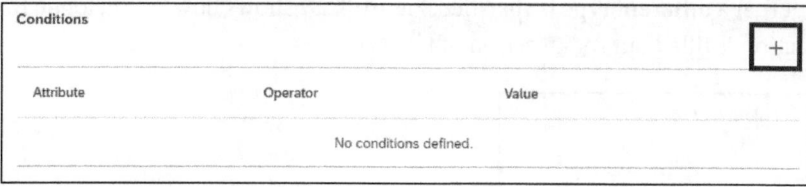

Figure 22.34 Adding Condition to Action

Fill in the fields of the displayed dialogue as shown in Figure 22.35, then confirm your entry by clicking the **OK** button.

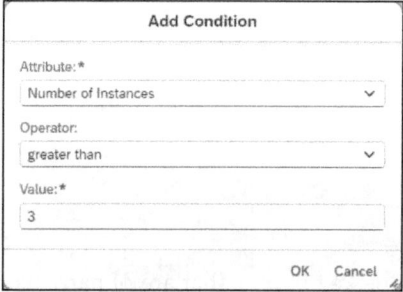

Figure 22.35 Properties of Condition

In the second part of the configuration for this type of action, the goal of the navigation is defined. In the **Environment Variable** field, you can specify a variable that is defined in the business process project of type **Destination**. A link to jump to an external application can be specified in the **Path** field. Finally, by activating or deactivating the **In-Place Navigation** toggle button, you can set whether the application to be navigated to should be opened in the same browser window or in a new browser window. Figure 22.36 shows an example configuration for the navigation link.

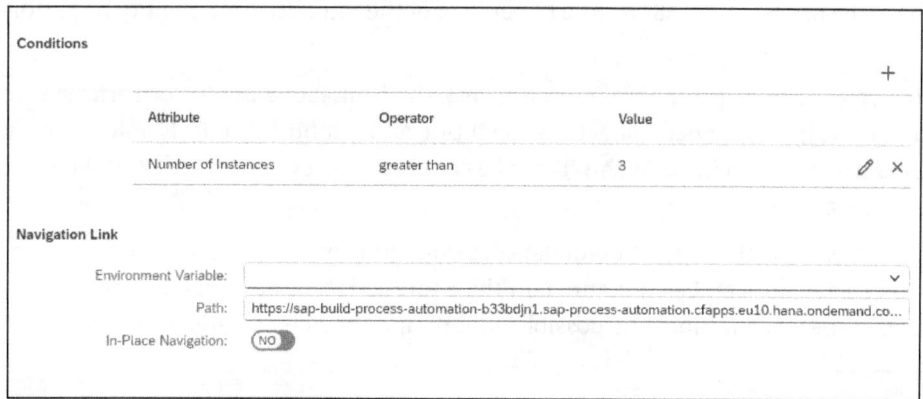

Figure 22.36 Configuration of Navigation Action

Trigger Workflow

The **Trigger Workflow** type of action is created in the same way as the navigation action, except that a different type is specified. Figure 22.37 shows how the dialogue for creating an action is filled out for an action of this type.

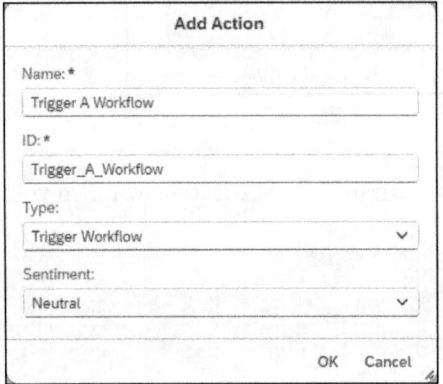

Figure 22.37 Adding Trigger Workflow Action

22.2.9 Performance Indicators

By defining performance indicators, you can define key figures that are of particular importance to the business user. These key figures are then calculated based on the existing process instances and displayed in the form of a map on the overview page of the workspace process. A set of general metrics is automatically delivered in the performance indicators when creating a visibility scenario. However, the list of these key figures can be edited and expanded to include your own key figures. Like many other interfaces for configuring the properties of a visibility scenario, the **Performance Indicator** editing interface is divided into two halves. On the left side, the existing indicators are displayed in the form of a list; on the right, you can view and edit the properties of the currently selected list entry. To create a new performance indicator, click the icon above the list. Figure 22.38 shows a screenshot of the interface to configure the performance indicators.

To create a new performance indicator, click the icon above the list of performance indicators. In the opened dialogue, enter a title and a subtitle. An ID is automatically generated based on the title; this ID can be changed if necessary. Figure 22.39 shows an example of this dialogue.

In the next step, the properties for data determination and visualization of the performance indicator can be configured. In the **Representation** field, you can choose the form of the visualization. The possible options and their characteristics are described in Table 22.7.

22.2 Configuring a Visibility Scenario

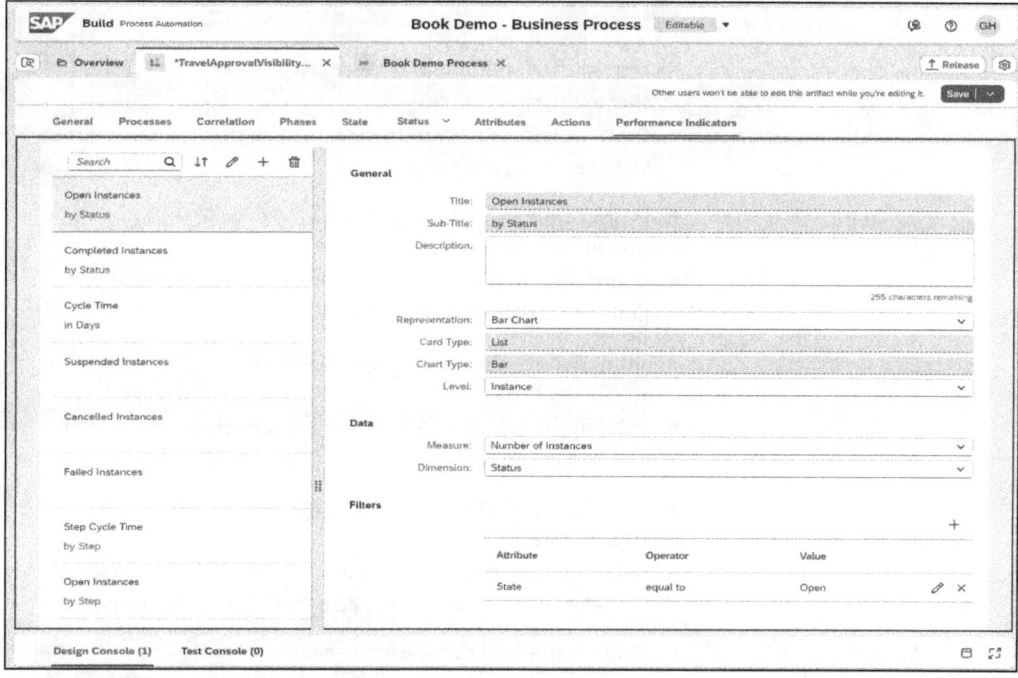

Figure 22.38 Interface of Performance Indicators

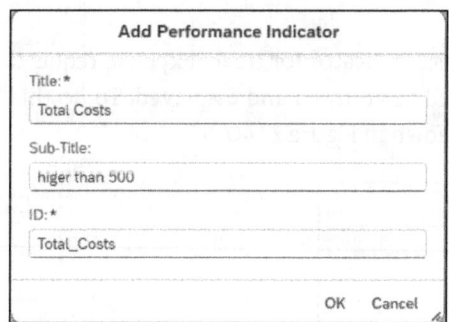

Figure 22.39 Adding Performance Indicator

Representation Name	Displayed Objects
Header	■ Displays title, subtitle, and the aggregated value of the selected measure in the header area
List	■ Displays title, subtitle, and aggregated value of the selected measure in the header area ■ Selected measure grouped by the selected dimension in the chart area

Table 22.7 Representation of Performance Indicator

Representation Name	Displayed Objects
Bar Chart	- Displays title, subtitle, and aggregated value of the selected measure in the header area - Bar chart in the chart area where the selected measure is grouped by the selected dimension
Donut Chart	- Displays title, subtitle, and aggregated value of the selected measure in the header area - Donut chart in the chart area where the selected measure is grouped by the selected dimension
Column Chart	- Displays title, subtitle, and aggregated value of the selected measure in the header area - Column chart in the chart area where the selected measure is grouped by the selected dimension
Line Chart	- Displays title, subtitle, and aggregated value of the selected measure in the header area - Chart represented in the form of a line in the chart area where the selected measure is grouped by the selected dimension
Table	- Displays title and subtitle in the header area - List of records in which the columns are the selected attributes

Table 22.7 Representation of Performance Indicator (Cont.)

In the example shown, we create a performance indicator for travel expense requests for which total costs exceed 500. These should be counted and displayed. To do this, enter the values f or the individual fields as shown in Figure 22.40.

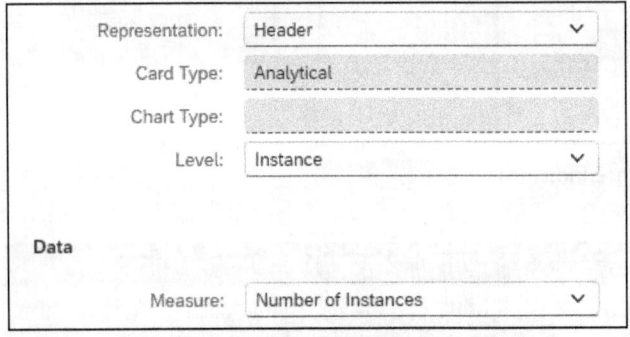

Figure 22.40 Data Determination and Representation

Finally, we define the filter to only take into account the instances that exceeded the limit. To do this, create a filter and enter the values shown in Figure 22.41.

Finally, save your visibility scenario. The configuration is now complete. You can create a new release of the project and deploy it.

22.2 Configuring a Visibility Scenario

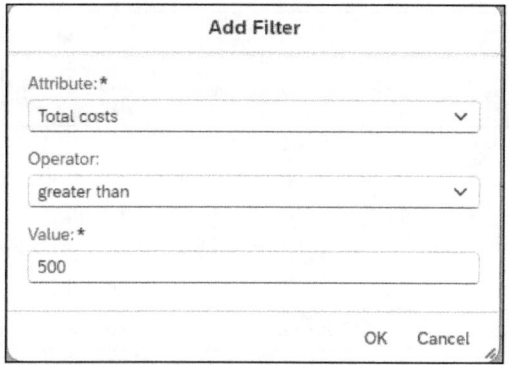

Figure 22.41 Filter for Performance Indicator

After deployment, open the new visibility scenario and navigate to the **General** tab. You will see that a link has been generated in the **Dashboard Link** field. Copy this link and open it in a new browser window. The dashboard of your visibility scenario will now be displayed. Because no workflow instances have been started since these scenarios were first deployed, no values are displayed yet.

Start a few workflow instances like the trigger form. Make sure to specify total costs higher than 500 for some applications. You can also approve or reject some of the launched instances in your inbox. Then return to the visibility dashboard to see how the values have changed, as shown in Figure 22.42.

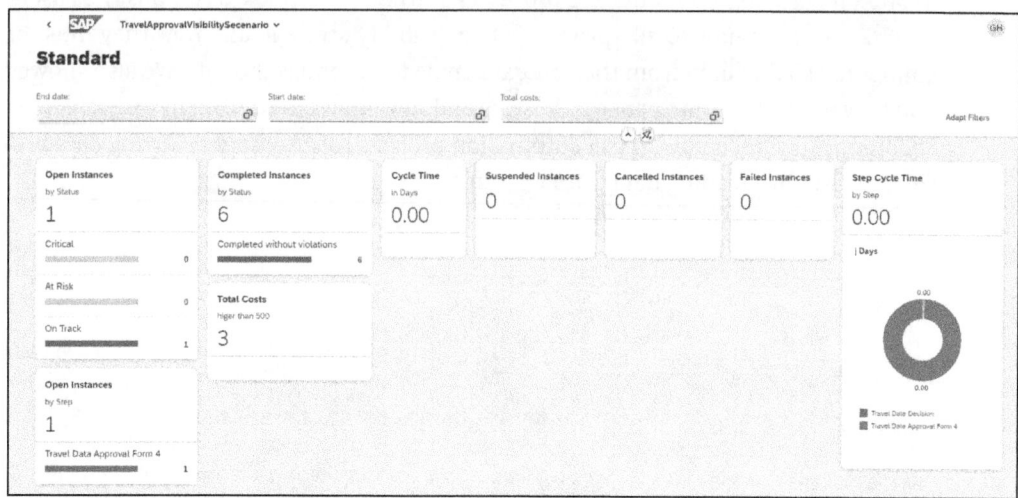

Figure 22.42 Visibility Dashboard

By clicking an individual card, you can open a detailed view in which the instances that make up the information for the respective performance indicator are displayed. For example, click the map of completed instances to see which applications have already been processed, as shown in Figure 22.43.

713

22 Visibility Scenarios

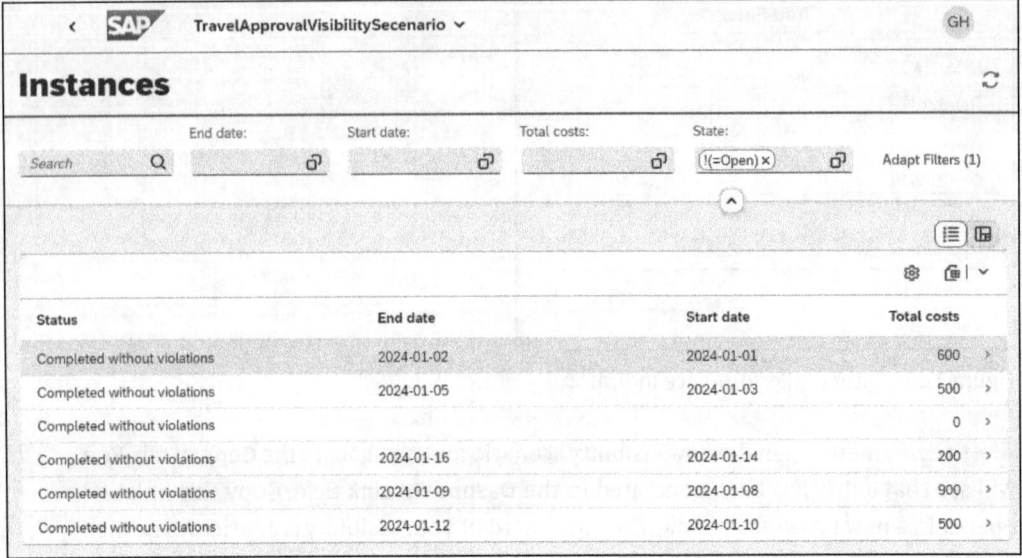

Figure 22.43 Details of Performance Indicator

22.3 Summary

In this chapter, you have seen how you can use visibility scenarios to enable instances in order to visualize their business processes and create dashboards for business users. You have learned how to add processes to a visibility scenario and how they must be configured so that data from the context can be used for dashboards. We also showed you how to create performance indicators and how they can be visualized. In the next chapter, you will see how to use automation projects to automate recurring data processing tasks like reading data from Excel files.

Chapter 23
Automation

Automating repetitive tasks can save a lot of time and work. Therefore, SAP Build Process Automation comes with an automation functionality that allows you to create automations. This can be done either manually or by using a recorder.

Automations in SAP Build Process Automation are meticulously crafted sequences of steps, designed to streamline and orchestrate a variety of tasks across multiple applications and interfaces. Each automation acts as a conductor, seamlessly coordinating activities on different screens and programs, specifically tailored to operate on a designated computer. This intricate orchestration of processes not only enhances efficiency but also ensures that complex workflows are executed with precision and ease.

In our comprehensive guide, we begin by laying the groundwork with an in-depth look at the installation process and an overview of the SAP Build Process Automation desktop agent in Section 23.1. Here, we'll uncover the intricacies of setting up the desktop agent, a vital tool for capturing workflows. Besides recording, it's also needed to execute automations, either attended or unattended. You'll learn not only how to install it but also understand its functionalities and the pivotal role it plays in automating repetitive tasks.

Moving forward, we'll delve into the nuances of capturing application data in Section 23.2, where the focus will shift to how SAP Build Process Automation can seamlessly interface with a multitude of applications to streamline your business processes. This sets the stage for Section 23.3, which tackles the automation of file operations, an essential component of any digital workflow. We'll discuss how to automate the lifecycle of files, from creation to deletion, thereby enhancing efficiency.

As we proceed to Section 23.4, the spotlight turns to environment variables, the silent enablers of flexible and dynamic automation. We will guide you through setting up and utilizing these variables to ensure your automations are both adaptable and scalable, regardless of the environment they are deployed in.

Section 23.5 is dedicated to the crucial stage of deployment, where your automation comes to life. It's here that we will address how to effectively transition your automated processes from the testing ground to live environments, making sure they perform optimally in real-world scenarios.

23 Automation

23.1 Desktop Agent

The *desktop agent*, version 3, is a local software module that is installed on user workstations. Its core function is to execute automation projects with finesse, engaging and operating a variety of applications. But it is also the core component in recording applications. This includes adeptly reading screen content, inputting data, making selections, and processing information efficiently.

These *automation projects* are linked to specific tenants that operate via the desktop agent. Users can effortlessly monitor the activities of the desktop agent through a conveniently accessible menu located in the computer's taskbar, provided that the desktop agent is in a ready state or currently active.

Upon installation, the desktop agent is designed to initiate automatically upon Windows startup. This feature is customizable to suit individual user needs. The versatility of the desktop agent allows for two distinct operational modes: attended and unattended. In *attended mode*, users have the autonomy to activate automations manually, selecting them as required. Conversely, in *unattended mode*, the system is engineered to perform automations autonomously, thereby requiring no manual input from the user, epitomizing seamless operation.

The previous version of the agent was version 2, but here we have opted to make use of the advanced features and improvements of version 3. If you are still working with version 2, we invite you to look at the comprehensive documentation of SAP Build Process Automation. There you will find a detailed guide that explains the migration process from version 2 to the optimized version 3 step by step. This resource has been carefully compiled to ensure a seamless transition and efficient continuation of the automated processes with the latest version of the agent.

The desktop agent is required both to record applications and to run automations on local computers or servers. The agent was originally only available for Windows but was also released for macOS at the end of 2023. We will show you how to install it on macOS in this book. Installation on Windows is basically identical. The agent is downloaded via the control tower. To do this, open SAP Build Process Automation as shown in Figure 23.1 and navigate to the **Control Tower** option in the side menu. Then click the tile named **Agents** in the **Agent Configuration** area.

There you will see an overview of all agents already registered (see Figure 23.2). In the toolbar, you will find a button called **Downloads Agent**. Click it to open the download page.

The download page describes a three-stage procedure, as shown in Figure 23.3. First, the desktop agent must be installed, then the browser extension must be enabled, and finally the desktop agent is registered in the SAP screen. First select the desired operating system in the **Install the Agent** tile, and then click the button to start the download to your local computer or a server.

23.1 Desktop Agent

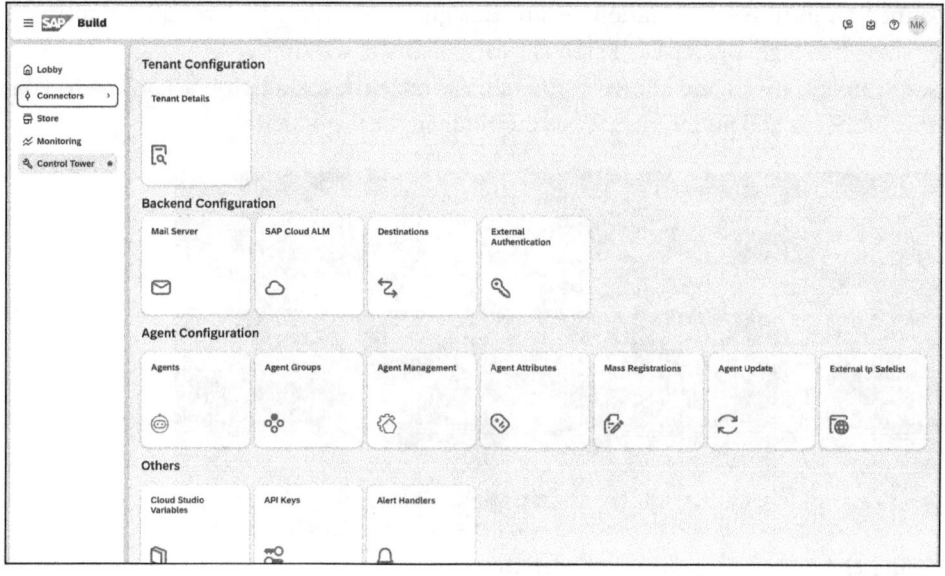

Figure 23.1 Control Tower

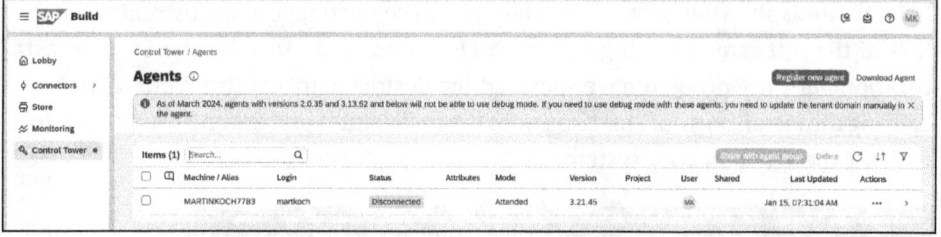

Figure 23.2 Agent Overview

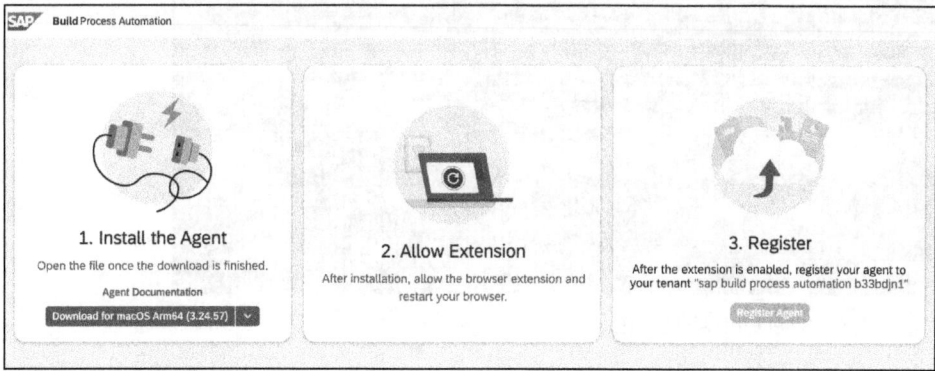

Figure 23.3 Agent Download

Once the desktop agent software is downloaded to your local machine, initiating the installation is as straightforward as double-clicking the installation file. This action will launch an intuitive installation wizard, designed to guide you through a multistage process (see Figure 23.4). Each step is clearly defined, walking you through the necessary configurations and allowing you to customize the installation according to your specific needs and preferences. Click **Continue** in the **Introduction** step.

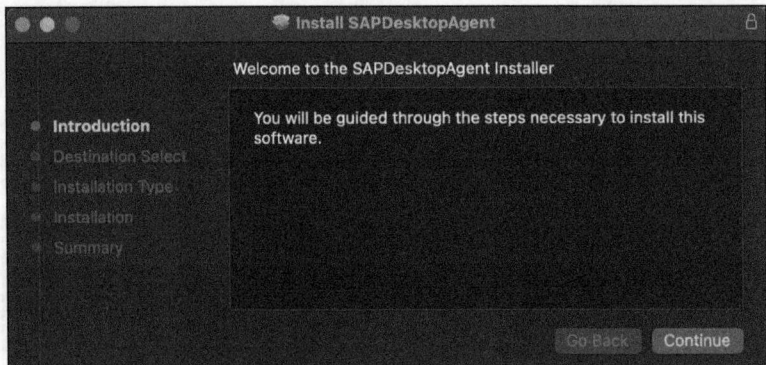

Figure 23.4 SAP Desktop Agent Installer: Overview

In the next step of the installation process, you have the option of selecting the installation path, as shown in Figure 23.5. Here you have the freedom to customize the location of the software according to your preferences. To do so, click the **Change Install Location** button. Once you have specified the desired path, confirm your selection by clicking the **Install** button. This continues the installation process and brings you closer to using the agent on your system.

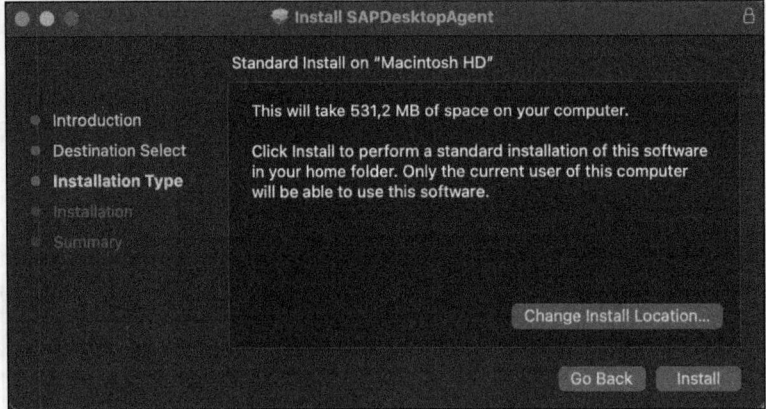

Figure 23.5 Select Install Location

Following the selection of the installation location, you will witness the installation progress. Should there be any hiccups or errors during this process, you will be promptly notified at this juncture. Typically, the installation should proceed smoothly and conclude without any issues, leaving the system primed for the agent's utilization.

Upon completion of the installation process, you will arrive at the **Summary** step, which confirms the successful installation of the software. To finalize the process, simply click **Close** to exit the dialogue, and you will be ready to start using the newly installed desktop agent on your system.

In its standard configuration, desktop agent version 3 is adeptly set up for automatic updates, ensuring that the latest version is seamlessly delivered and applied to your system. This feature guarantees that your agent remains up to date with the newest enhancements and functionalities, without the need for manual intervention.

Once the main installation is complete, the next step is to activate the browser plugin. A popup window will appear in your browser asking for permission to activate the plugin. Once you have confirmed the activation, the last step is to register the agent in SAP Build Process Automation. To do this, first open the desktop agent you have just installed. After opening the desktop agent, you will notice that it has the status **Disconnected**, as illustrated in Figure 23.6.

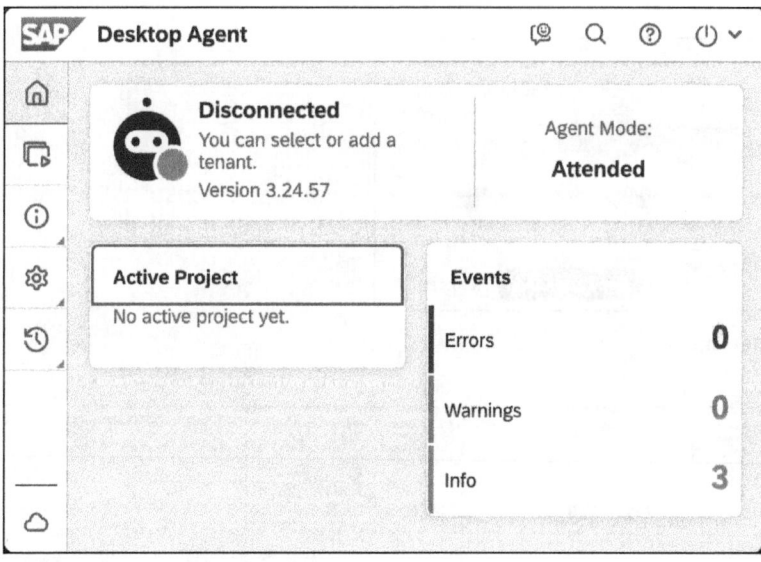

Figure 23.6 Desktop Agent Home

Click the cloud icon in the side menu. In the dialog that appears, you have the option of registering the agent with the SAP Build Process Automation service. To do this, click **Add Tenant** (see Figure 23.7).

The next step is to return to SAP BTP. There you will find the **Register Agent** option (refer back to Figure 23.3). A popup with the link to register the agent will now open, as shown in Figure 23.8. Click **Copy and Close** to copy the link to the clipboard.

Click **Add Tenant** in the desktop agent. Then assign a **Name** and copy the link from the clipboard into the **Domain** field (see Figure 23.9). Then click **Save**.

23 Automation

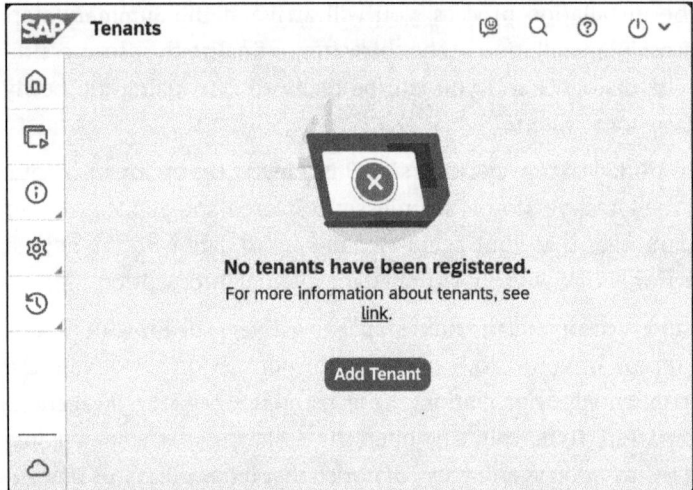

Figure 23.7 Tenant Overview

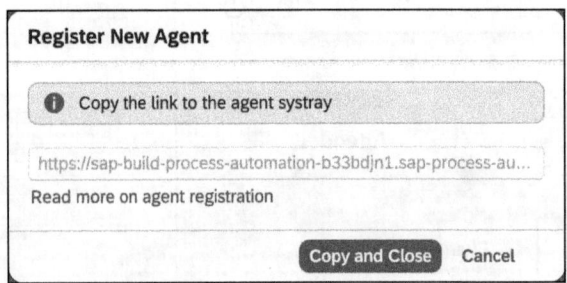

Figure 23.8 Register Agent

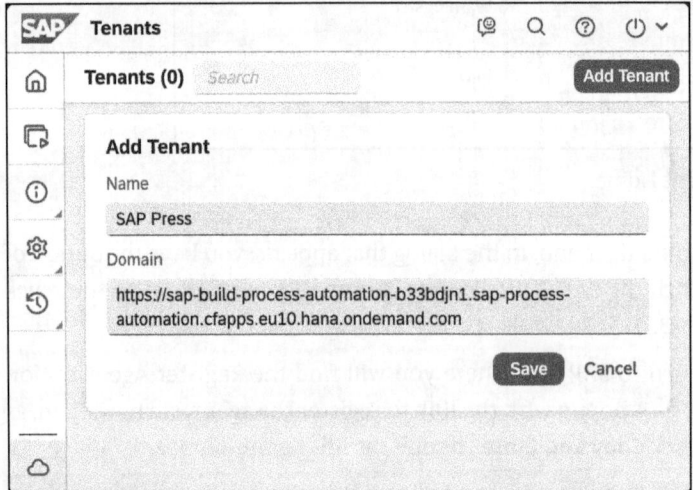

Figure 23.9 Add Tenant

23.1 Desktop Agent

You will now see the tenant you have just added in the tenant overview. Click the ... button, then select **Activate** from the context menu, as shown in Figure 23.10.

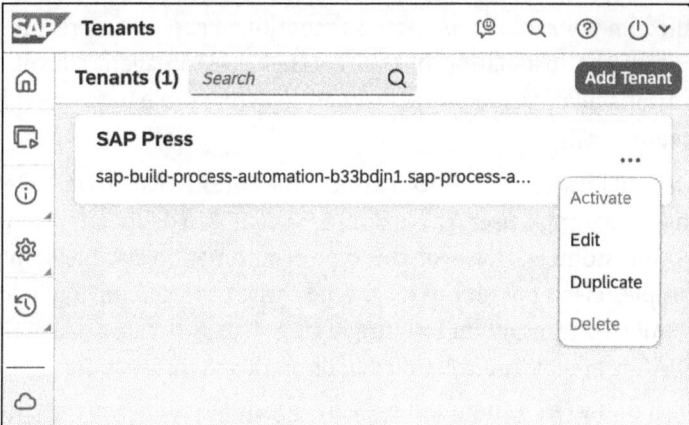

Figure 23.10 Tenant Overview

You must then confirm in a popup that you want to switch to this tenant (see Figure 23.11). To do this, click **OK**.

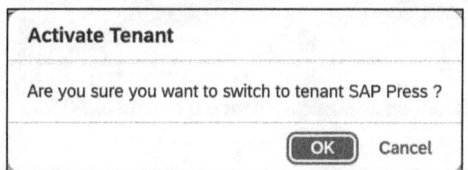

Figure 23.11 Confirm Tenant Activation

Once registration is complete, you should be able to locate your agent in the desktop agent app within the SAP Build Process Automation control tower, as shown in Figure 23.12. This step confirms that your agent has been successfully registered and is now ready for use.

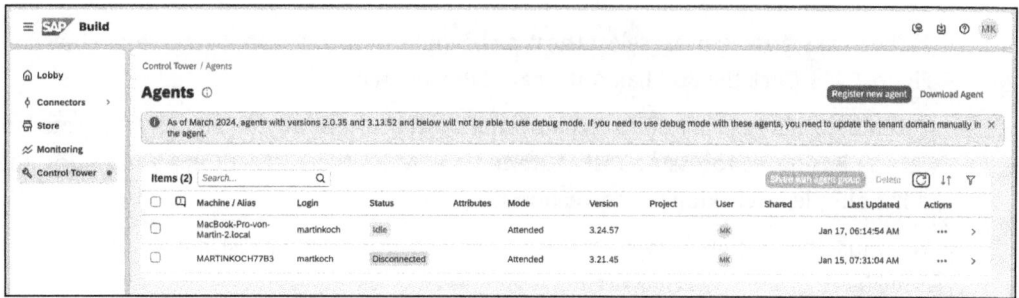

Figure 23.12 Check Agent Connection Status

23.2 Capture Applications

The easiest way to automate applications is to use the desktop agent as a recorder. This is software that is installed on your local computer, as mentioned earlier, and records your activities step by step. The installation of the required desktop agent was discussed in Section 23.1. This installation is a prerequisite for you to be able to follow the example presented ahead.

In this section, we'll show you how you can automate activities in SAP GUI. In this case, we're specifically looking at creating a user. This example was chosen deliberately as it is a typical use case for SAP customers. However, this type of automation also has some special features. For example, it isn't possible to record the entire process from logon to the activity within the SAP GUI transaction in a single step. This is because SAP GUI works with SAP's own DIAG protocol. You will find details on this in this section.

In the first step, you must enter the SAP Build lobby, as shown in Figure 23.13. There, click the **Create** button to create a new automation.

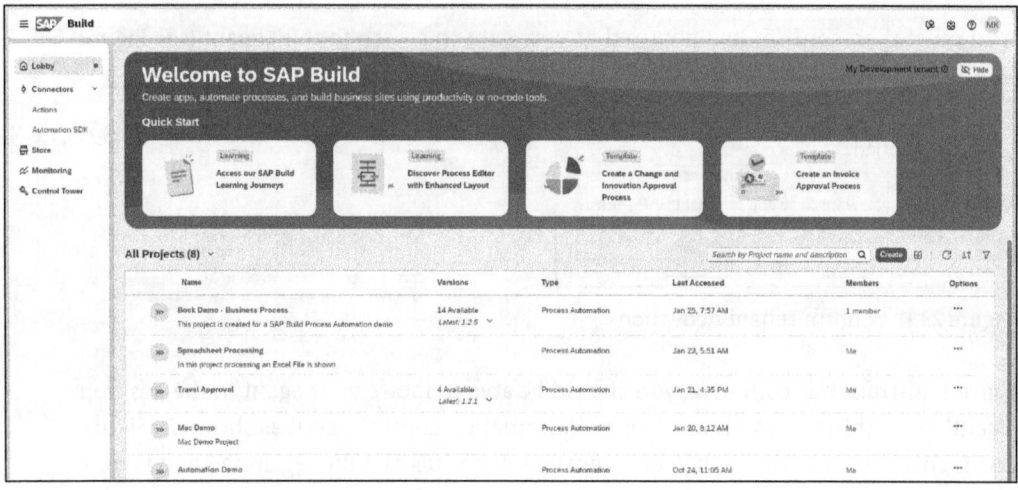

Figure 23.13 SAP Build Lobby

In the next step, you must select the type of application you want to create, as shown in Figure 23.14. Click the **Build an Automated Process** tile.

You must then select the process automation type. As shown in Figure 23.15, you can choose **Business Process** or **Task Automation**. Select the **Task Automation** option for this example by clicking the corresponding tile.

You must then assign a **Project Name** and a **Description** for the task automation project (see Figure 23.16). Then click **Create**.

23.2 Capture Applications

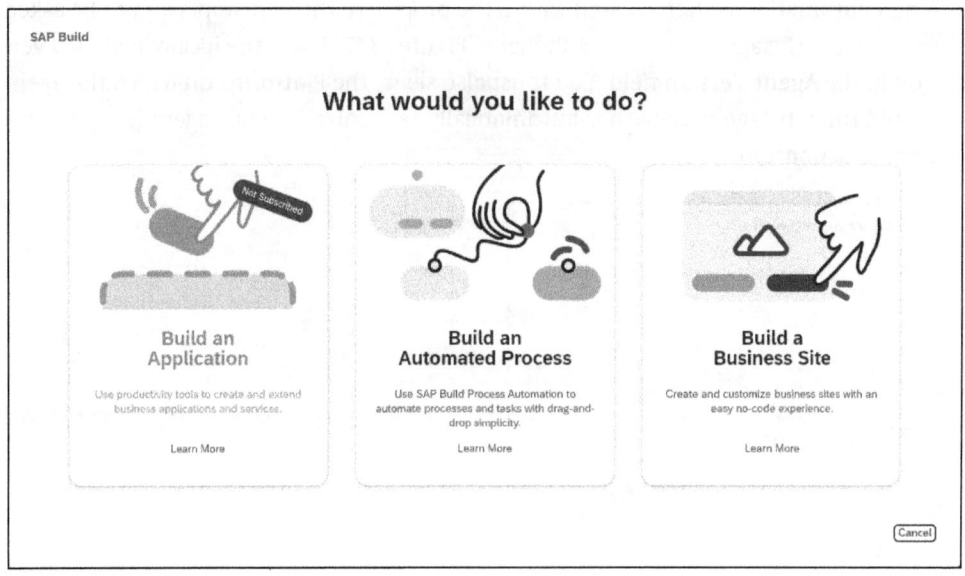

Figure 23.14 Select Application Type

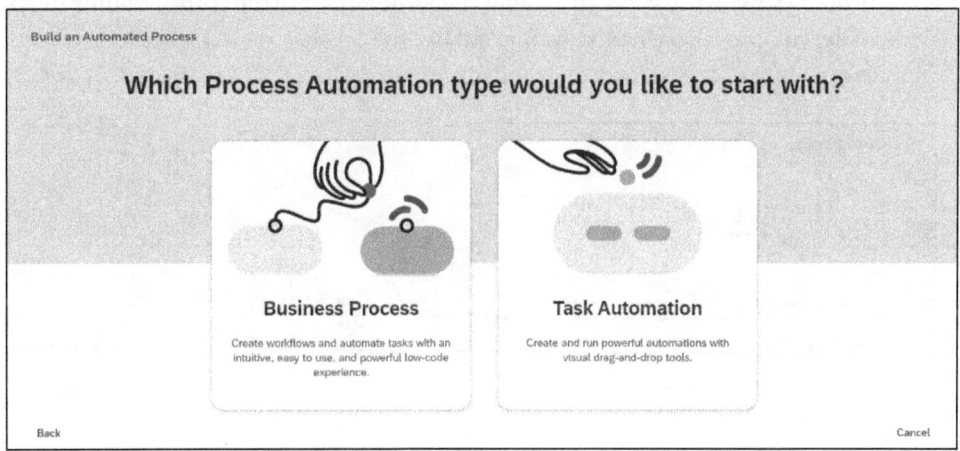

Figure 23.15 Select Process Automation Type

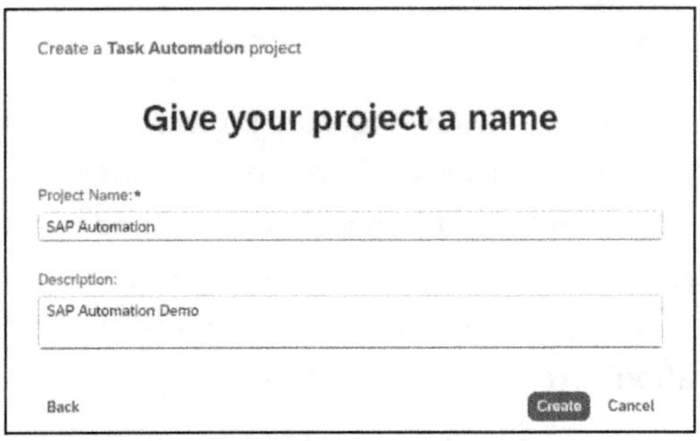

![Provide Project Details form]

Figure 23.16 Provide Project Details

723

A new automation is then created within the project. In the first step, you will be asked to configure the agent version, as shown in Figure 23.17. Select the locally installed version in the **Agent Version** field. You must also select the **Platforms** on which the agent should run. The agent version is automatically recognized if your agent is registered. Then click **Confirm**.

Figure 23.17 Configure Agent Version

You must then maintain the parameters of the automation artifact as shown in Figure 23.18. In doing so, you must assign a **Name** and an **Identifier** in the corresponding fields. Optionally, you can assign a **Description** and use the toggle to select whether the automation can also be started from another automation. Then click **Create**.

Figure 23.18 Provide Automation Details

You will now be taken to the detailed view of the previously created project, as shown in Figure 23.19. There, on the **Artifacts** tab, you will see a list of all the artifacts associated with the project. You will also find the automation artifact previously configured in the dialog. However, this will not help you to record your activities at this point. You must create an artifact of the application type to record the application. To do this, click the **Create** button and then select the **Application** entry in the popover.

23.2 Capture Applications

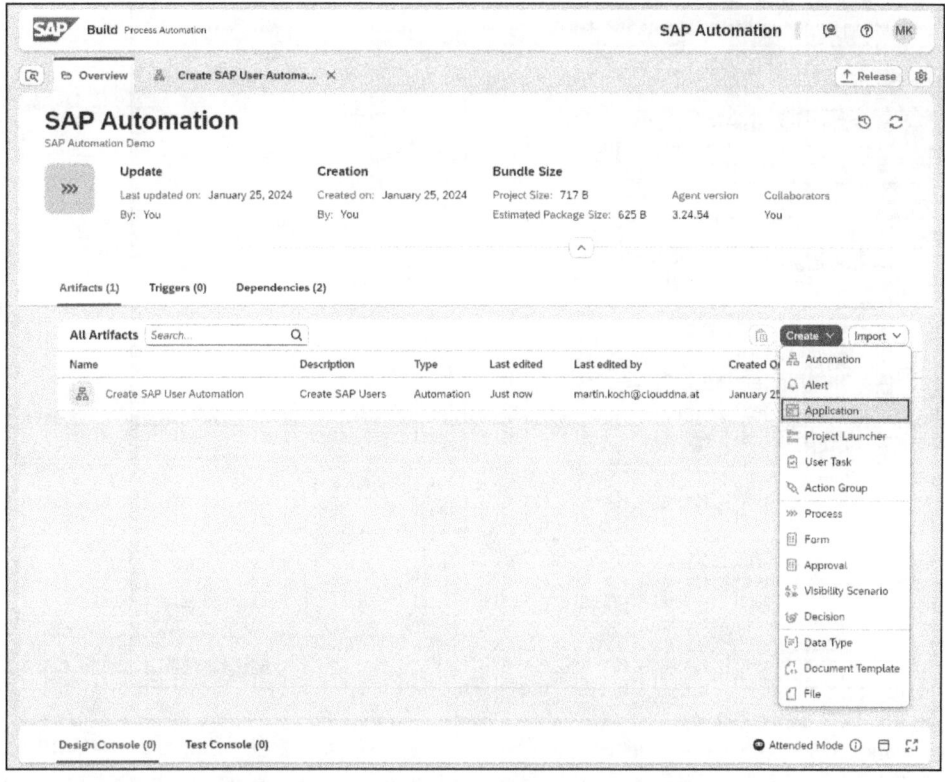

Figure 23.19 Create Application

A dialog then opens in which you must assign an **Application Name** and an **Application Identifier**. Optionally, you can enter a **Description** (see Figure 23.20). Then click **Create** to create the application artifact.

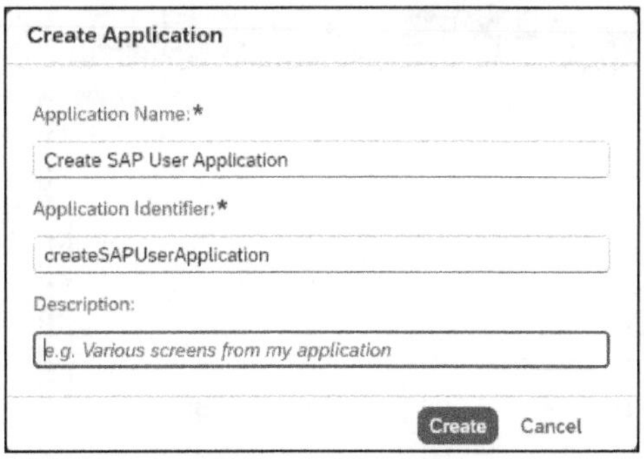

Figure 23.20 Provide Application Details

All apps open on your computer for which an automation can be recorded are then recognized. As shown in Figure 23.21, a distinction is made between **UI Automation** and **Web**. In the **UI Automation** area, click the **SAP Logon** entry.

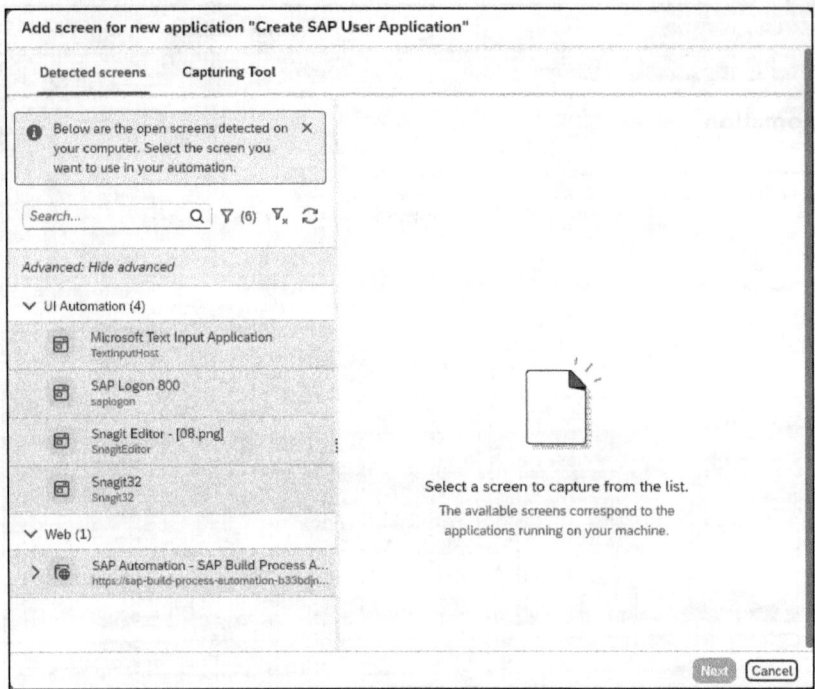

Figure 23.21 Select Screen

A screenshot of SAP GUI is then displayed. Check whether this is the correct screen, and then click the **Next** button (see Figure 23.22).

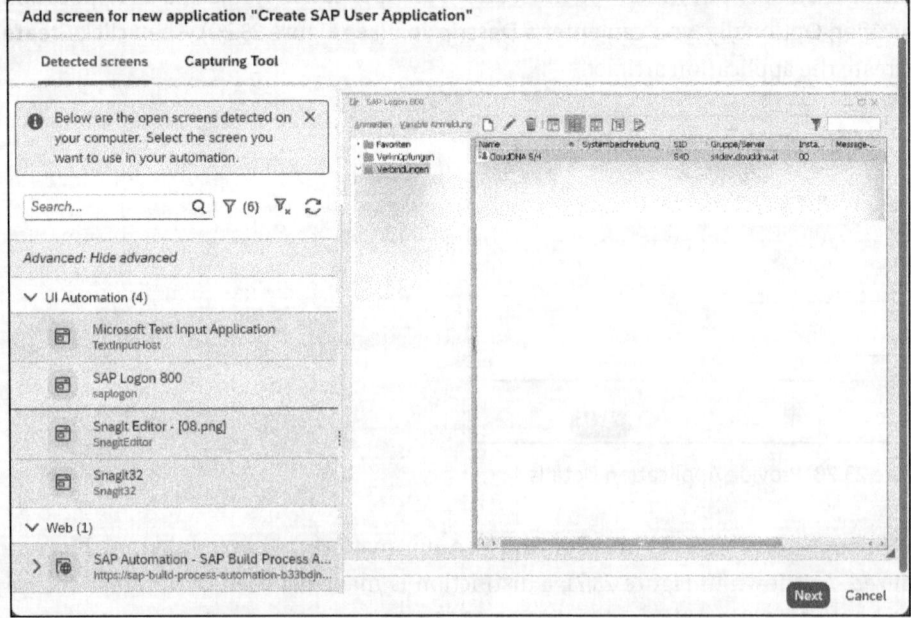

Figure 23.22 Check Recorded Screen

23.2 Capture Applications

You must then select whether the screen should be recorded via the **Recorder** or whether you want to carry out a **Manual Capture** (see Figure 23.23). Select the **Recorder** option for the type, and in the **Technology** dropdown, select **UI Automation**. Then click **Record**.

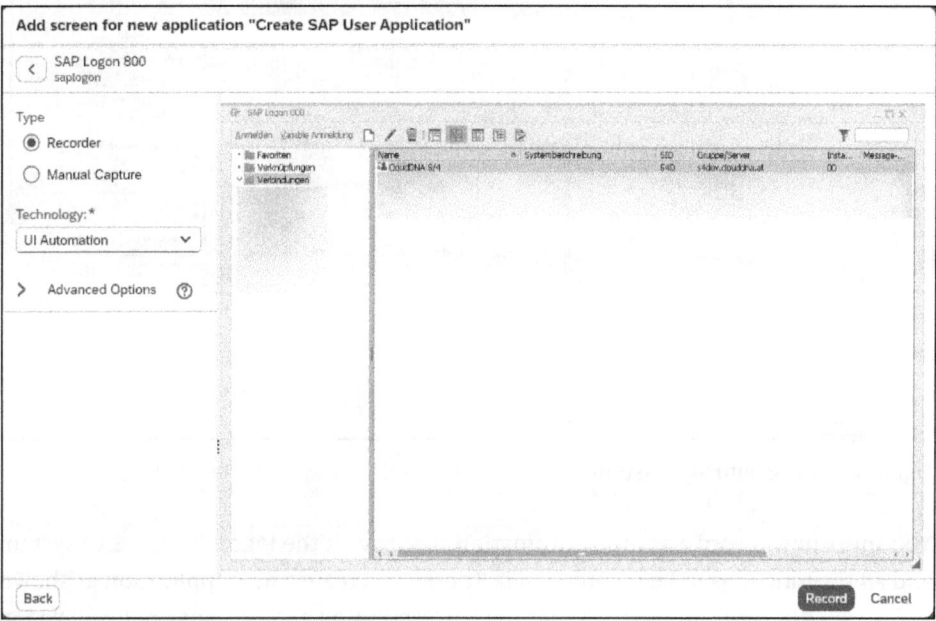

Figure 23.23 Select Capturing Type

The recorder will then open on your local computer, as shown in Figure 23.24. Click the ⊙ button to start the recording. Be sure to verify that the mode is set to **Automatic**.

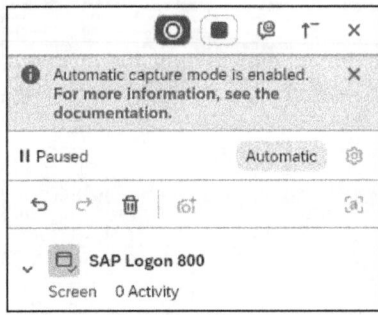

Figure 23.24 Recorder

You must now select the desired SAP system in SAP GUI by double-clicking it. This opens the logon screen for the SAP system. At this point, SAP GUI switches to the DIAG protocol. You can see the result in the automation editor, as shown in Figure 23.25. If you now log on from the SAP system using a user and password, this will result in an error in the recording, as the UI application does not support the DIAG protocol.

23 Automation

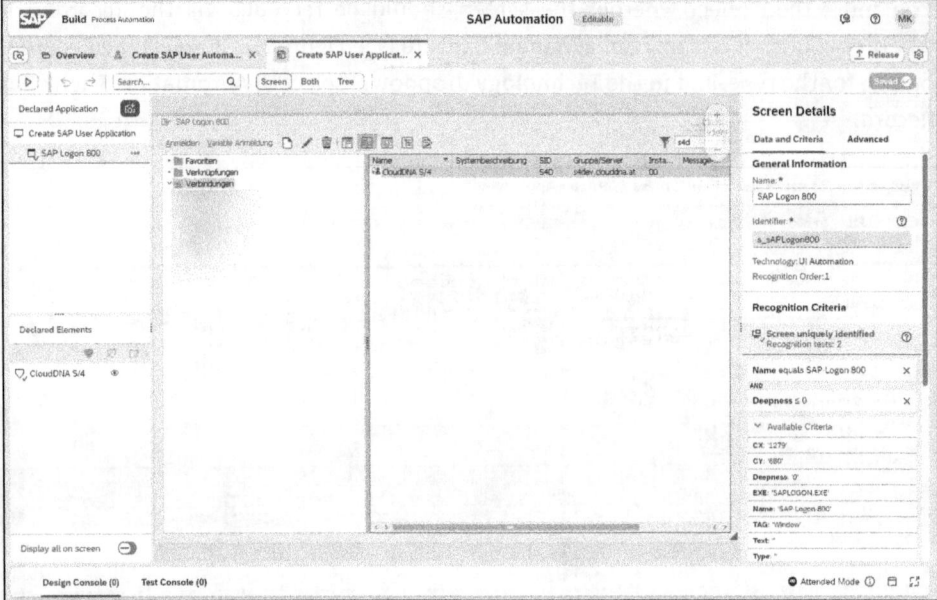

Figure 23.25 Log onto SAP System

You must now record a second automation that covers the logon for the SAP system and all the functions to be recorded in it. Therefore, create a new application as shown in Figure 23.19. You can now see that the corresponding applications are displayed in the list in the **SAP GUI** area. This is because you have previously opened the logon screen. Select the application for which you see the logon screen in the overview, as shown in Figure 23.26, and then click **Next**.

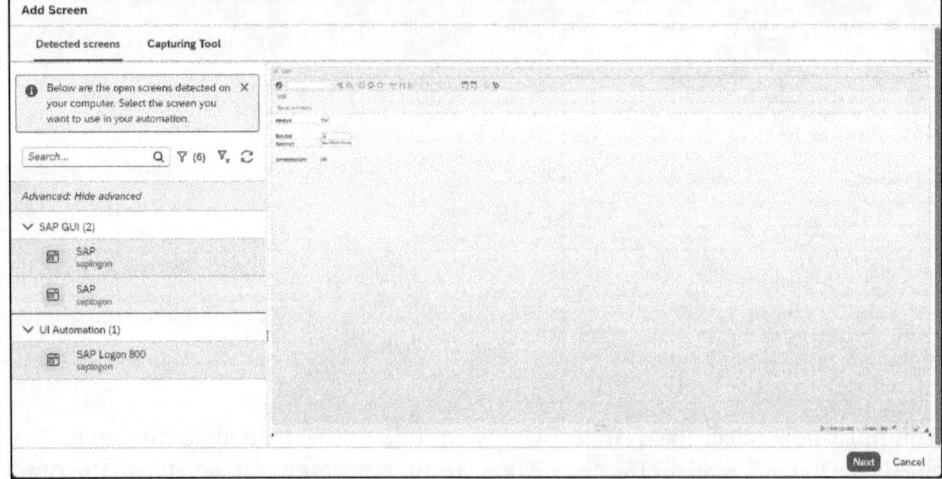

Figure 23.26 Create SAP GUI Automation

728

As shown in Figure 23.27, select the **Recorder** option as the **Type**. The selection for **Technology** should already be preset to **SAP GUI**. You therefore do not need to make any changes. Now click **Capture**.

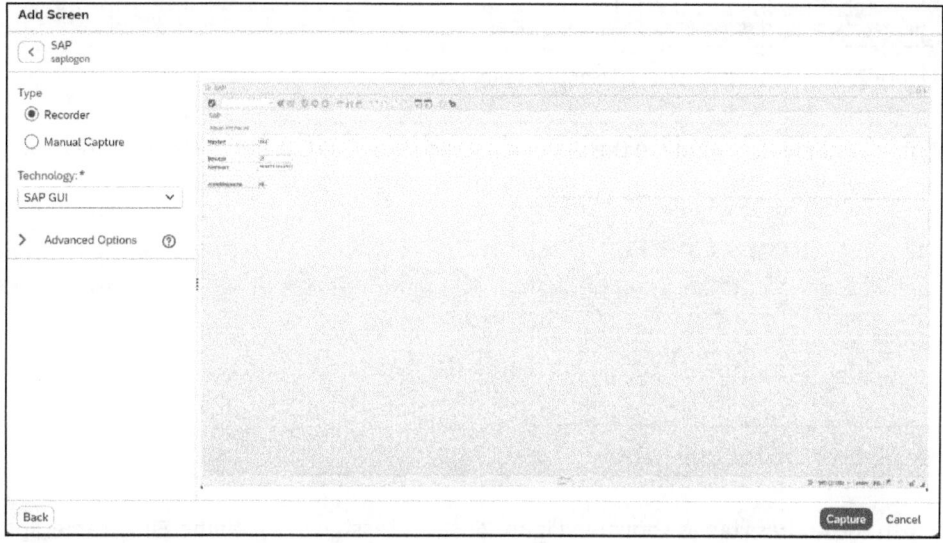

Figure 23.27 Select Type and Technology

Next, assign a **Screen Name**, as shown in Figure 23.28. The **Screen Identifier** is assigned automatically and can be adjusted if necessary. Now click **Go to Application**.

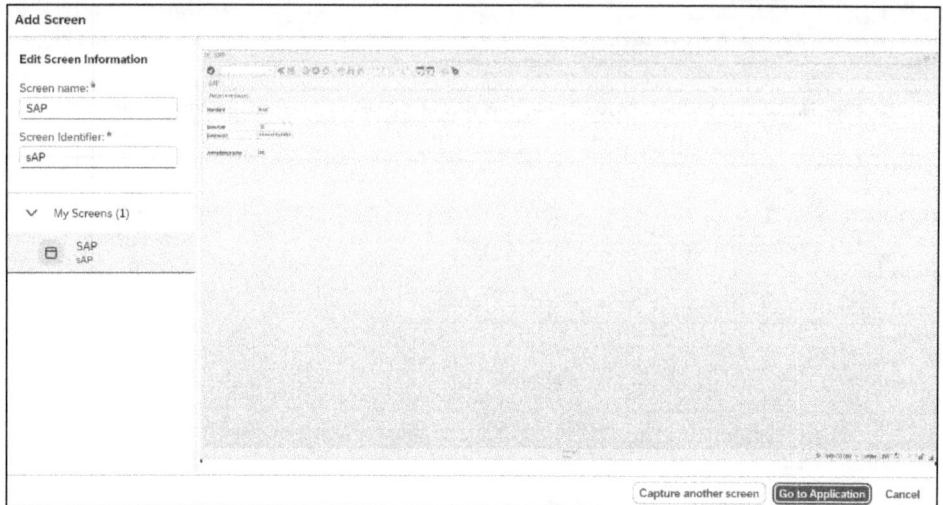

Figure 23.28 Go to SAP GUI Application

The recorder will then open. Click the ⊙ icon as before.

23 Automation

Once you have logged onto the SAP system, you can call up Transaction SU01 to maintain the users (see Figure 23.29).

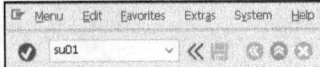

Figure 23.29 Transaction SU01: User Maintenance

Enter a unique user name in the **User** field, then click the **User** button (see Figure 23.30).

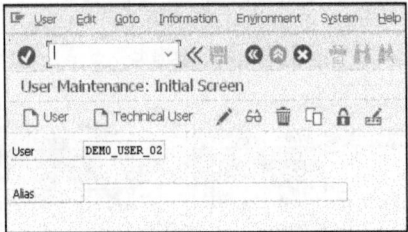

Figure 23.30 Provide User Name

Open the **Address** tab as shown in Figure 23.31, and assign a **Last Name**, **First Name**, and **Email Address**.

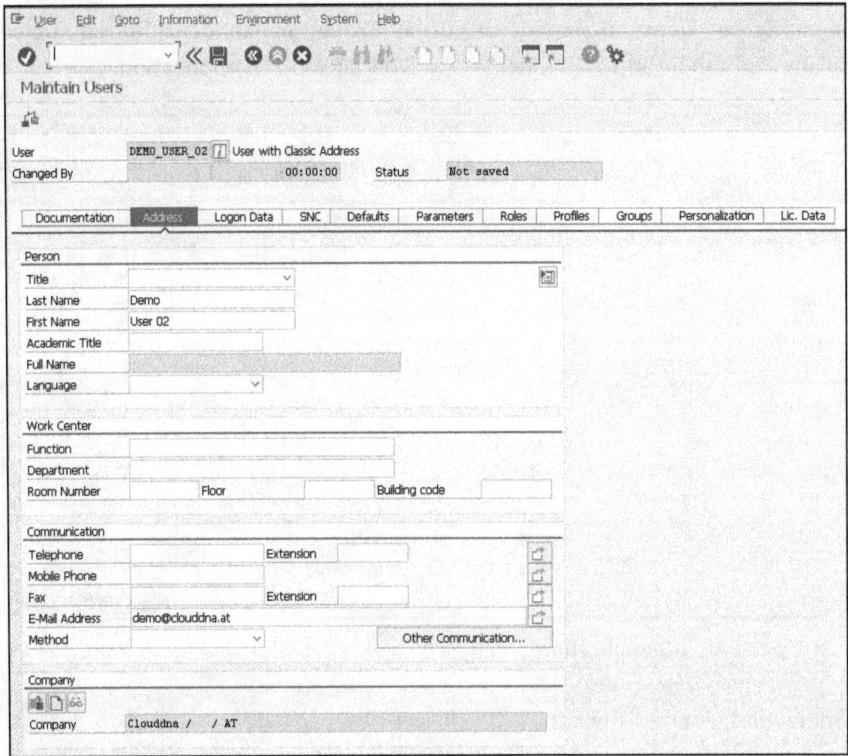

Figure 23.31 User and Address Data

Then open the **Logon Data** tab as shown in Figure 23.32. Enter a **New Password** that corresponds to the password guidelines of your system. Repeat the entry in the **Repeat Password** field. Then click **Save**.

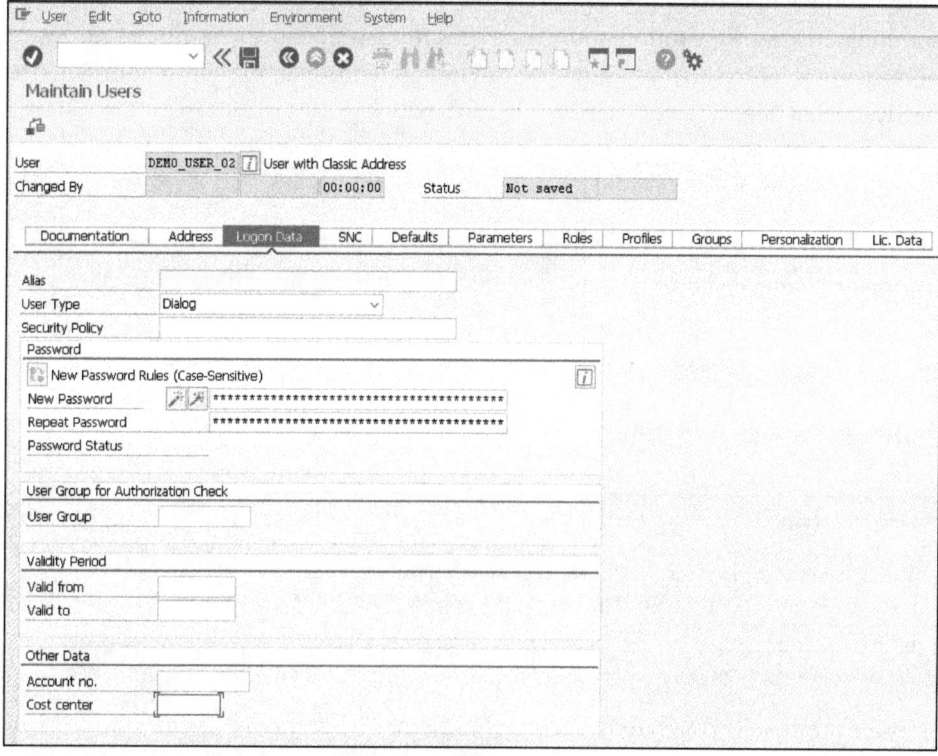

Figure 23.32 Provide Logon Data

Now stop the recorder by clicking the ■ button (see Figure 23.33).

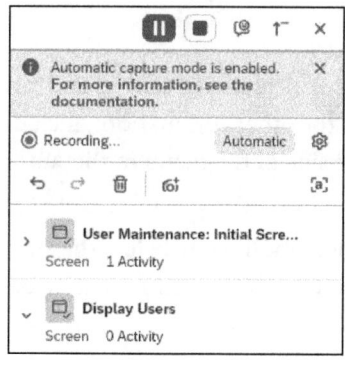

Figure 23.33 Stop Recording

You will be asked whether you would like to export the recording you have just made to the cloud studio (see Figure 23.34). Click **Export**.

You can now export the recording as shown in Figure 23.35. You can either create a new automation or update the previously created automation. Here, pick the first option by selecting **Create a New Automation**, then click **Next**. We deliberately chose this option because we want to show you how you can integrate the existing automation into the newly created one.

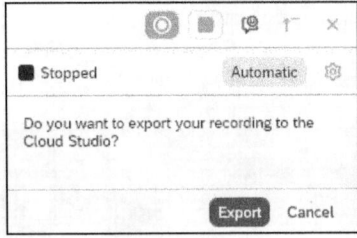

Figure 23.34 Export Recording

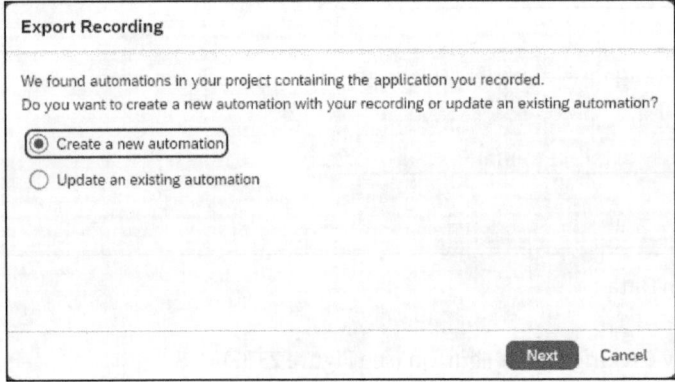

Figure 23.35 Export Recording

You will now be taken to the automatic editor. You will see all recorded actions and can adjust them here. As shown in Figure 23.36, you will find the other automations, including the **Create SAP User Automation**, on the right-hand side of the menu in the **Automations** area. We will reuse these in the next step.

Now drag and drop the **Create SAP User Automation** element directly below the start activity. The result should look like Figure 23.37. You have now learned how to reuse automations in different places.

23.2 Capture Applications

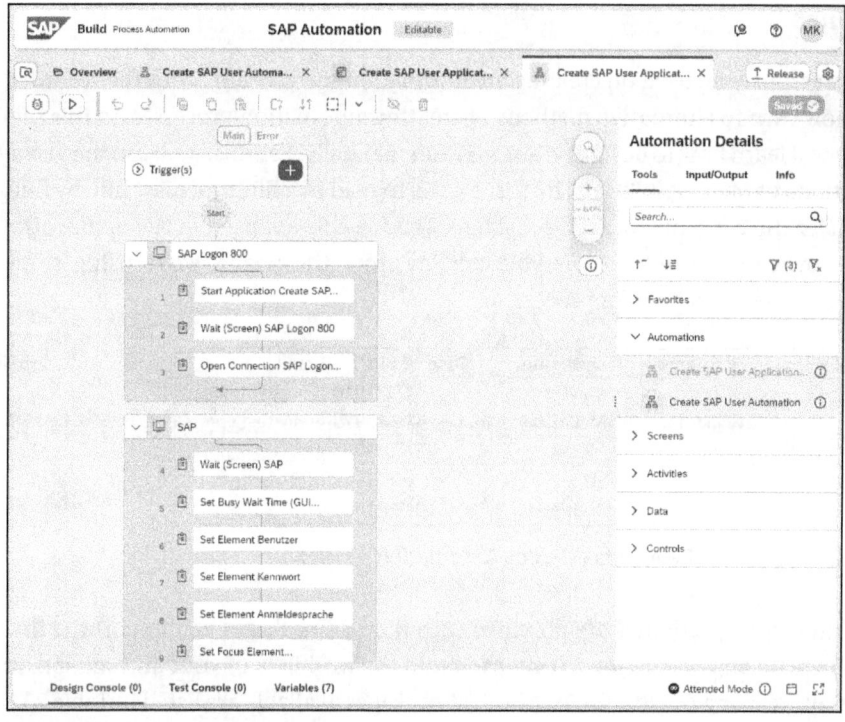

Figure 23.36 Automation Editor

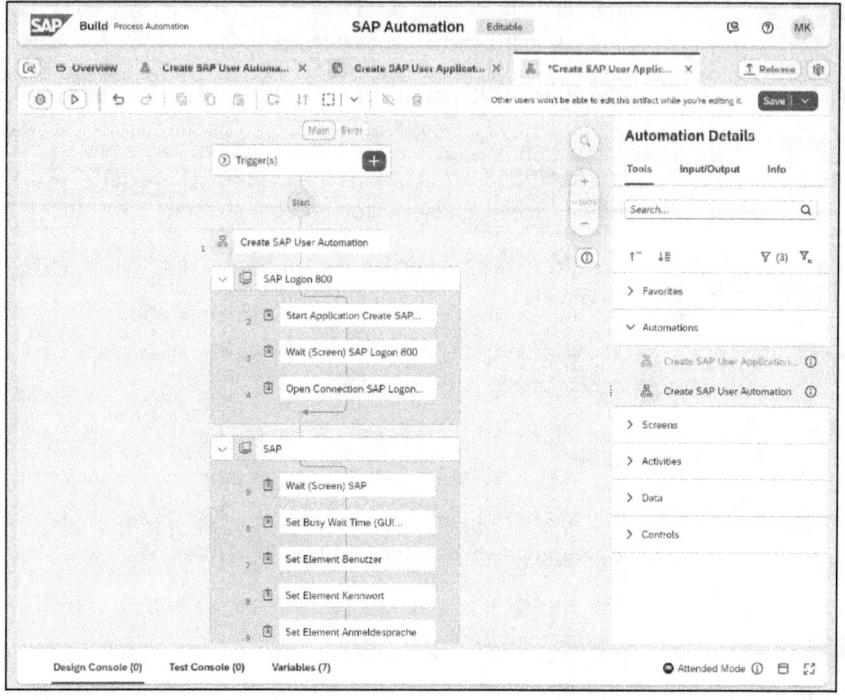

Figure 23.37 Add Existing Automation

23.3 Files

Having previously focused on creating an automation by recording with the recorder, we will now turn to alternative methods of creating automation processes. In this section, you will learn how to design an automation manually. For this, we'll use the versatile functions of Microsoft Excel. The first step is to read in a file, process it line by line, and transfer the contents to a target system. The file content used in the demonstration is shown in Figure 23.38. As you can see from the image, we're using some employee master data here for processing.

Personnel Number	Firstname	Lastname	Date of Birth	Email
001	Mike	Meyers	1/1/1970	mm@test.com
002	William	McCaffrey	4/16/1979	wm@test.com
003	Anita	Moses	8/10/1997	am@test.com
004	Kenneth	Strong	6/8/1989	ks@test.com
005	Virgie	Rodriguez	10/18/1969	vr@test.com

Figure 23.38 Raw Data Being Processed by Automation

In the cloud studio, activities are provided with the SDK packages you import the first time you create an automation in order to build the workflow of your automation. In this case, we're using the Excel SDK. The SDK packages available are listed in Table 23.1.

SDK Name	Description
Core SDK	The Core SDK contains main methods and functions to build and run your automation projects.
SAPUI5 SDK	The SAPUI5 SDK is a collection of activities that allow you to create automations using SAPUI5.
SAP GUI for HTML SDK	The SAP GUI for HTML SDK is composed of activities for accessing and manipulating applications using SAP GUI for HTML.
Excel SDK	The Excel SDK is a collection of activities that allow you to create automations using Microsoft Excel.
Word SDK	The Word SDK is a collection of activities that allow you to create automations using Microsoft Word.
Outlook SDK	The Outlook SDK is a collection of activities that allow you to create automations using Microsoft Outlook.
PowerPoint SDK	The PowerPoint SDK is a collection of activities that allow you to create automations using Microsoft PowerPoint.
PDF SDK	The PDF SDK is a collection of activities that allow you to create automations using PDF.

Table 23.1 Available SDKs

SDK Name	Description
BAPI SDK	The BAPI SDK is a collection of activities that allow you to create automations in the SAP system by calling BAPIs, the standard interfaces to the business object model in SAP products.
SAP Ariba Solution SDK	The SAP Ariba Solution SDK is a collection of activities that allow you to create automations using SAP Ariba.
SAP SuccessFactors SDK	The SAP SuccessFactors SDK is a collection of activities that allow you to create automations using SAP SuccessFactors.
Document Information Extraction SDK	The Document Information Extraction SDK is a collection of activities for extracting and enriching information using the Document Information Extraction service.
Google Workspace SDK	The Google Workspace SDK is a collection of activities that allow you to automate Google Workspace products such as Google Drive, Gmail, Google Sheets, Google Calendar, Google Docs, and Google Slides.
Google Authorization SDK	The Google Authorization SDK is a collection of activities that allow you to acquire authorization with Google to perform activities in the Google Workspace and the Google Document AI SDK.
Google Document AI SDK	The Google Document AI SDK is a collection of activities that allow you to automate document processing using Google Document AI.
Google Vision AI SDK	The Google Vision AI SDK is a collection of activities that allow you to extract data from documents using Google Vision AI capabilities.
Google Cloud Storage SDK	The Google Cloud Storage SDK package is a collection of activities that allow you to automate the Google Cloud Storage product.
Microsoft 365 Cloud SDK	The Microsoft 365 Cloud SDK package is a collection of activities that allow you to create automations on Office objects in Microsoft Cloud.
Web SDK	The Web SDK package is a collection of activities that allow you to create automations for web applications.
Java SDK	The Java SDK package is a collection of activities that allow you to automate Java applications made based on frameworks such as Java Swing and JavaFX.

Table 23.1 Available SDKs (Cont.)

To start manually creating an automation, it's first necessary to create a new project. Start this process by opening the SAP Build Process Automation lobby. In the detailed view there, you will find a **Create** button, which you can use to initiate the new project. Clicking this button starts the creation of your automation project, whereupon you will be taken to the configuration view to define your automation tasks.

In the next step, you must select a suitable template. There are three options available to you; the first option is deactivated (refer back to Figure 23.14) as the SAP Build Apps service is not available in this instance. Click the **Build an Automated Process** tile.

Once you have initiated a new project, the next step is to select the automation type. There are two options available to you: **Business Process** and **Task Automation**. For this project, which is intended to automate a specific task, **Task Automation** is the right choice. Therefore, click the corresponding tile to continue.

It is now necessary to define a meaningful name and a detailed description for your automation project. Enter the relevant information in the **Project Name** and **Description** fields, as shown in Figure 23.39. This information will later help you to identify and understand the purpose of the project. Once you have entered this information, confirm your entries by clicking the **Create** button.

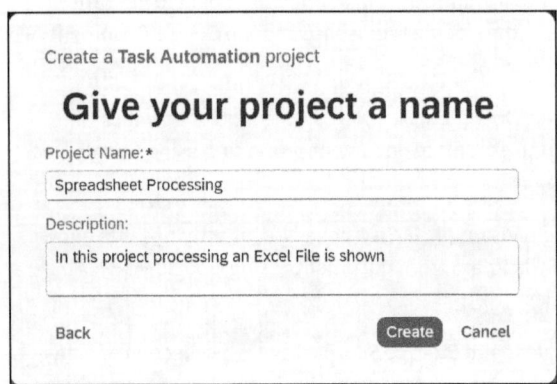

Figure 23.39 Define Project Details

In the further course of the setup process, you must select the version of the agent and the supported platforms, as illustrated in Figure 23.40. Note that not all SDKs are available for both platforms. This selection is crucial to ensure that your automation is compatible with the appropriate technical resources and environments and works optimally. Make your choice based on the requirements of your automation project and the specifications of the target platforms. Finally, click the **Confirm** button.

Once you have created an automation project, the next step is to create an automation. You can create several automations within a project. You must assign a **Name**, **Identifier**, and **Description** to each automation, as shown in Figure 23.41.

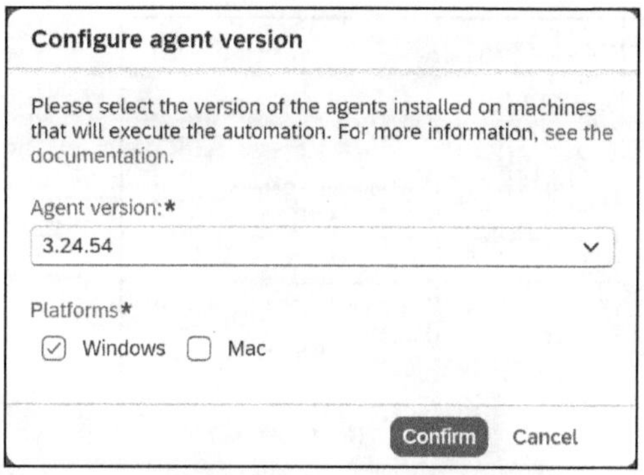

Figure 23.40 Configure Agent Version

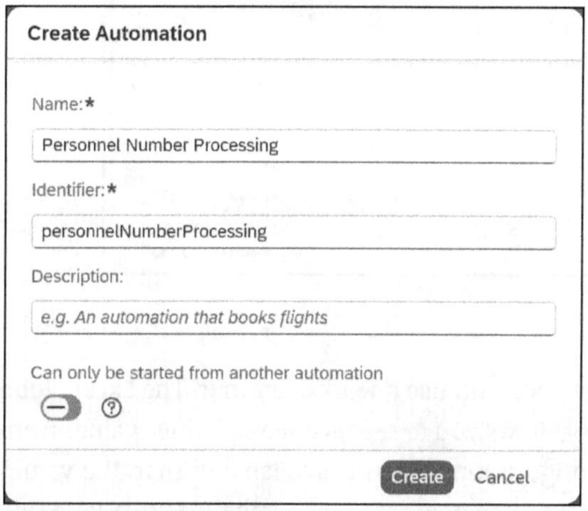

Figure 23.41 Provide Automation Basic Data

Once you have made your selections, you are taken directly to the automation editor, shown in Figure 23.42. In this environment, the drag-and-drop principle allows you to drag elements from the menu on the right-hand side into the workspace to configure them for your automation. Due to the large selection of available elements, we recommend using the search function to find the functions you need quickly and efficiently.

In this example, we will start with the **Open Excel Instance** element. Drag this element into the workspace and place it directly after the trigger of the process. This will be the first step of your automation to start processing within Excel. From there, you can add more actions to build the automation sequence you want. When you use the **Open Excel Instance** activity in your automation, you must always add either the **Release Excel Instance** activity or the **Close Excel Instance** activity to your automation to properly close Excel.

To be able to run this automation at a later time, it's necessary that Microsoft Excel is installed and functional on the computer on which the automation is running.

23 Automation

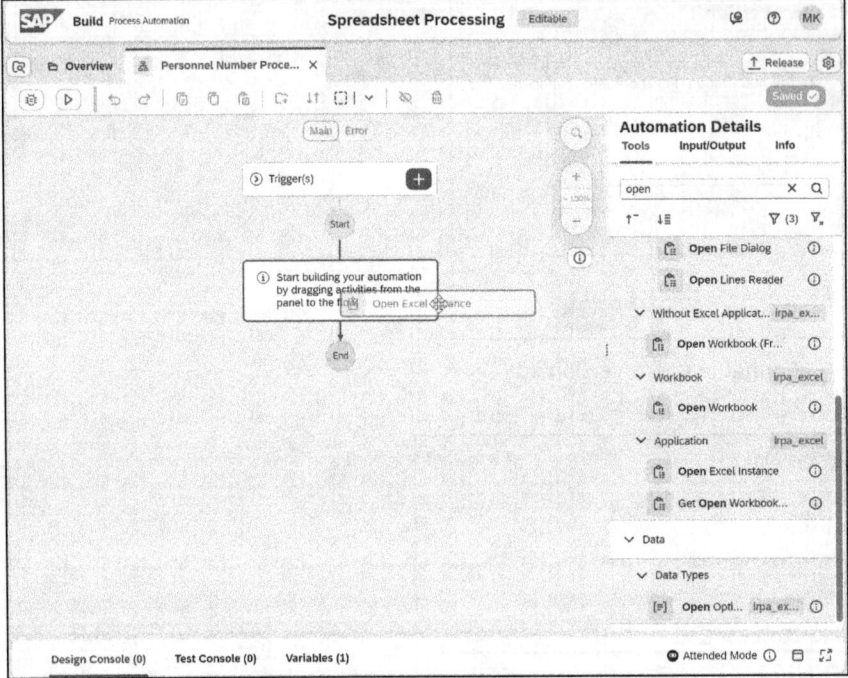

Figure 23.42 Open Excel Instance

In the next step, you must open a workbook and use a worksheet in it. The **Excel Cloud Link** activity is used for this. This activity allows you to retrieve the cell range values from a specified worksheet. A dedicated configuration screen is available to map the values from the worksheet to an existing data type or a new data type. Drag the **Excel Cloud Link** element directly to the previously added **Open Excel Instance** activity (see Figure 23.43).

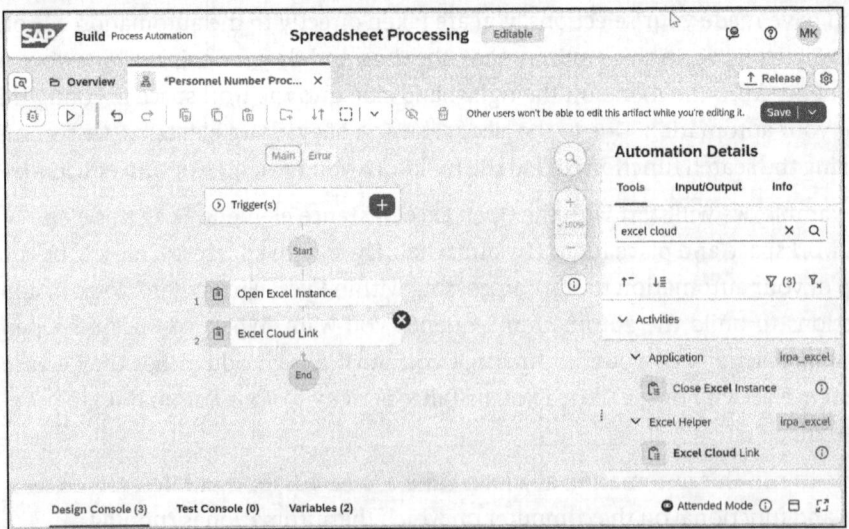

Figure 23.43 Add Excel Cloud Link

23.3 Files

Then click the activity. This opens the configuration for this activity in the right-hand menu (see Figure 23.44). You can click **Edit Activity** to adjust the parameters of the activity.

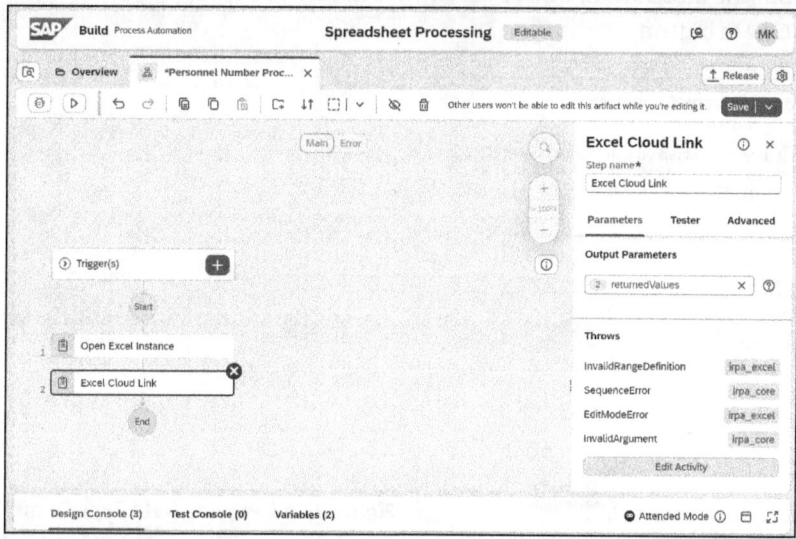

Figure 23.44 Edit Activity

You will now see the **Excel File** and **Data Type** columns, as shown in Figure 23.45. Select the desired Excel file in the **Workbook** field. Also configure the path to the file in the **Workbook Path** field. You can also set the workbook buffer using environment variables, for example. We'll go into this in detail in Section 23.4. Now switch to the **Data Type** column. There you have the option of defining the data types used. In this case, we want to derive the data type directly from the structure of the Excel file. To do this, click **+ From Excel Data**.

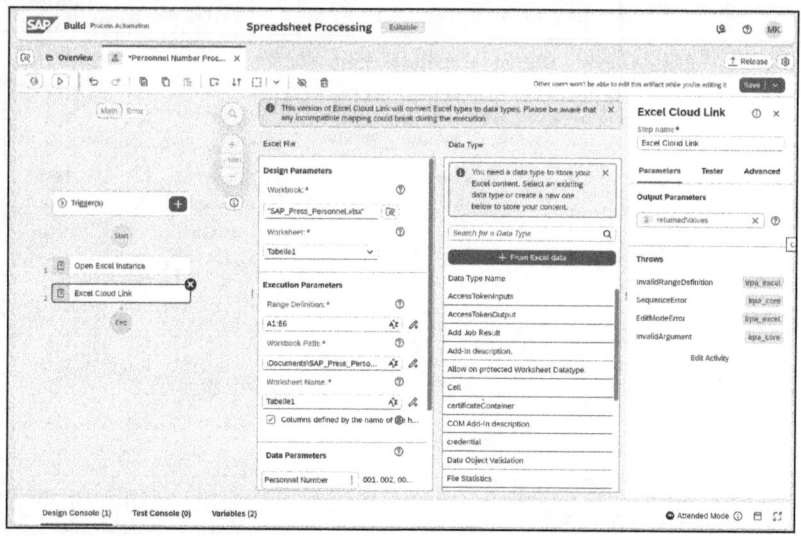

Figure 23.45 Set Workbook Path and Generate Data Type

23 Automation

A dialog with the title **Create Data Type from Your Excel Data** now opens (see Figure 23.46). Maintain the **Name** and **Alias** attributes as well as an optional **Description**. As you can see from the illustration, we use descriptive names in our employee example. Now click the **Create** button.

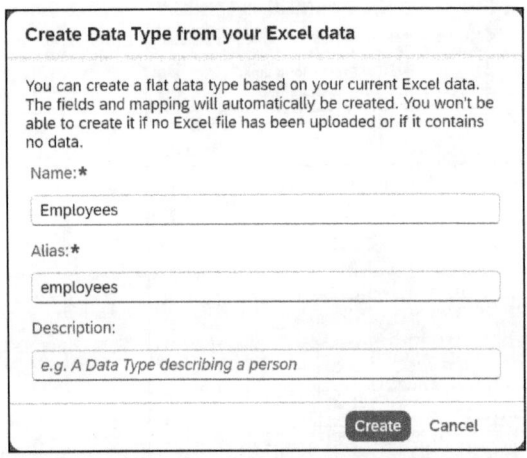

Figure 23.46 Provide Data Type Details

As you can see in Figure 23.47, the data type you have just created is now displayed. At this point, you have the option of making corrections. For example, you can adapt the names to your requirements. For this example, do not make any changes here; accept the generated data type proposed by the system.

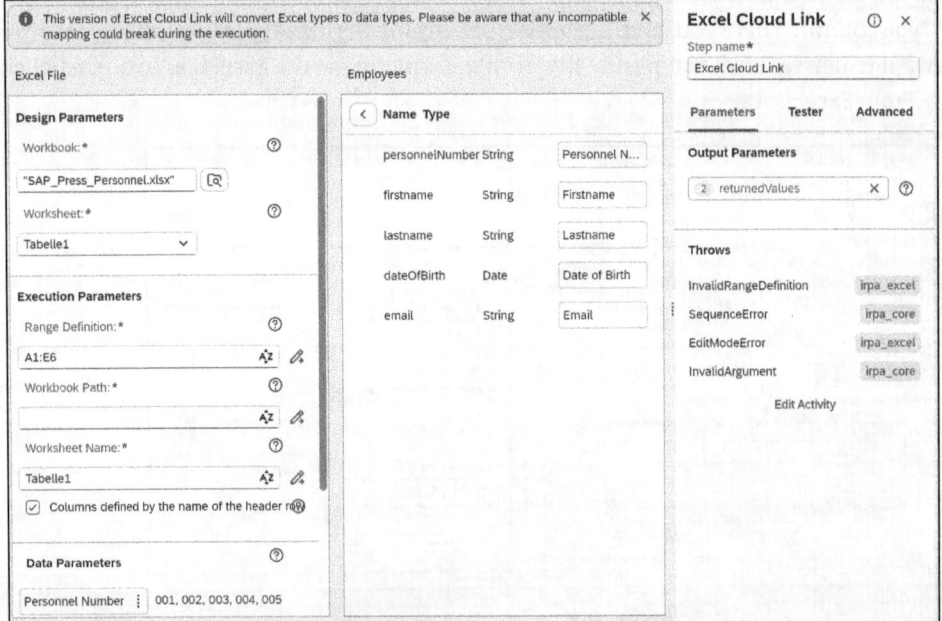

Figure 23.47 Check Generated Data Type

23.3 Files

In the next step, you want to iterate over the individual rows of the Excel worksheet. A **For Each** loop is used for this purpose. To do this, drag the **For Each** control from the right-side menu into the automation, as shown in Figure 23.48.

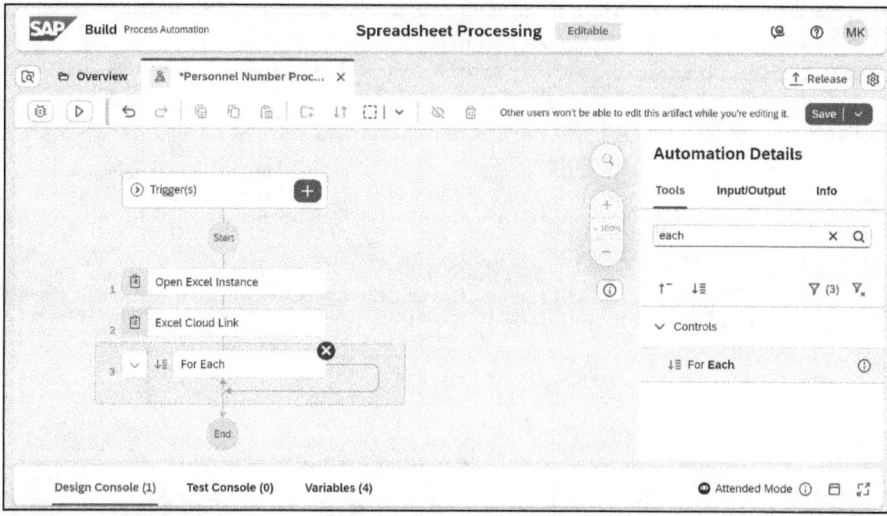

Figure 23.48 Add For Each Loop

In the next step, you need to configure the loop. To do this, click the **For Each** activity. The configuration then opens in the right-hand menu as usual. In the **Set Looping List** field, select the **Returned Values** entry. Select **currentMember** and **index** as **Loop Parameters**, as shown in Figure 23.49.

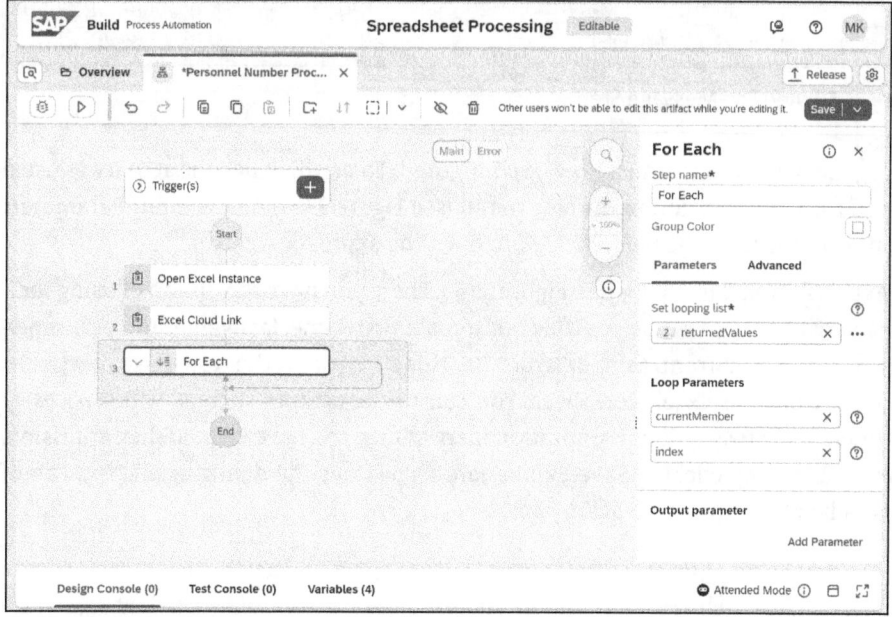

Figure 23.49 Customize For Each Loop

23   Automation

In the next step, we'll write each individual line to the locomotive of the automation. To do this, drag the **Log Message** activity from **Core SDK** into the loop, as shown in Figure 23.50.

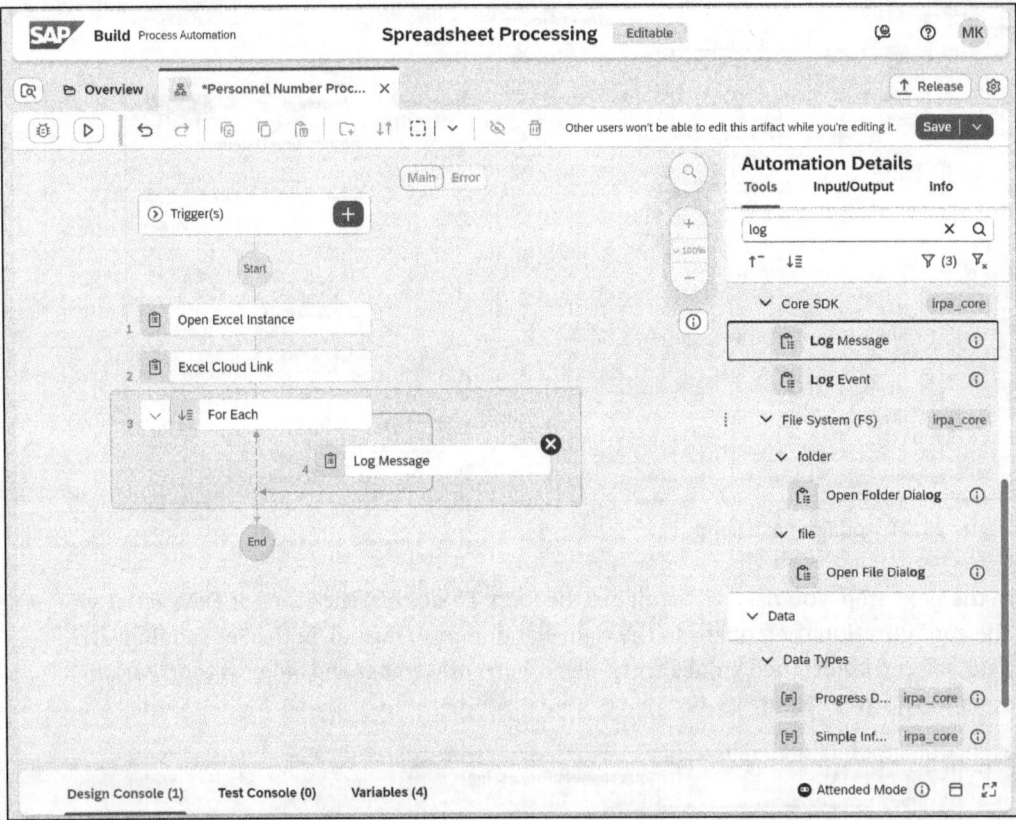

Figure 23.50 Add Log Message Step

The log message activity must now be configured. To do this, click the activity as usual to open the configuration in the side menu (see Figure 23.51). In the **Input Parameters** area next to the message field, click to open the expression editor.

The expression editor is shown in Figure 23.52. There you have the option of using individual variables. In this case, we will concatenate **firstName**, **lastName**, and **personnelNumber** from the **currentMember** structure. This structure is the respective row that is extracted from the Excel worksheet. You can use constants such as whitespaces or identifiers set between double quotation marks. Once you have created the expression, you can save it by clicking **Save Expression**. This closes the dialog again. You can of course adapt the expression again later.

23.3 Files

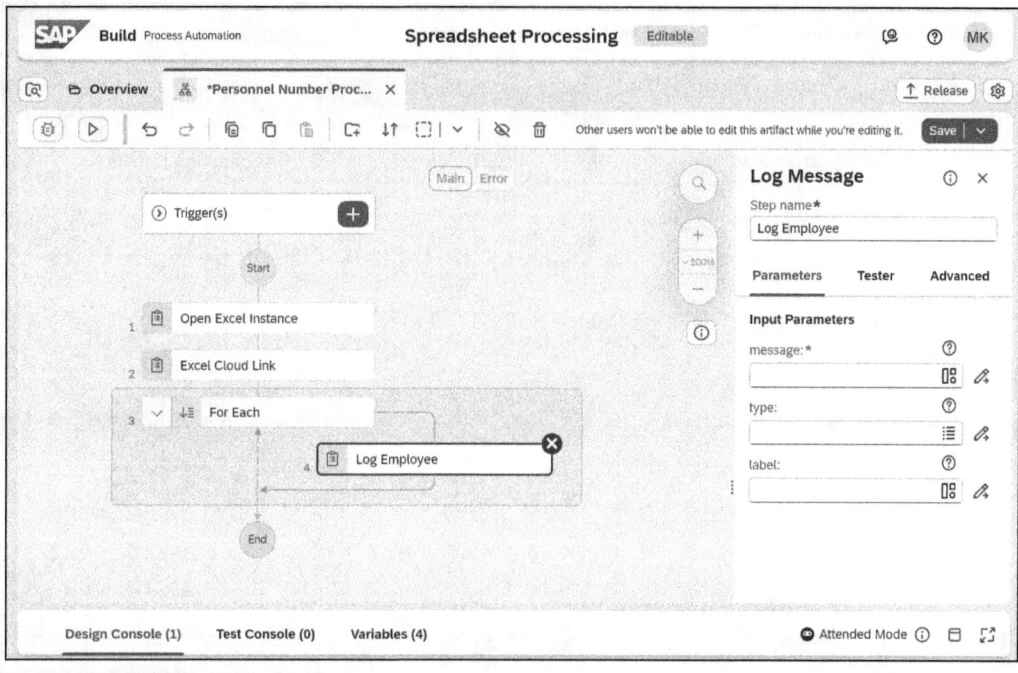

Figure 23.51 Customize Message

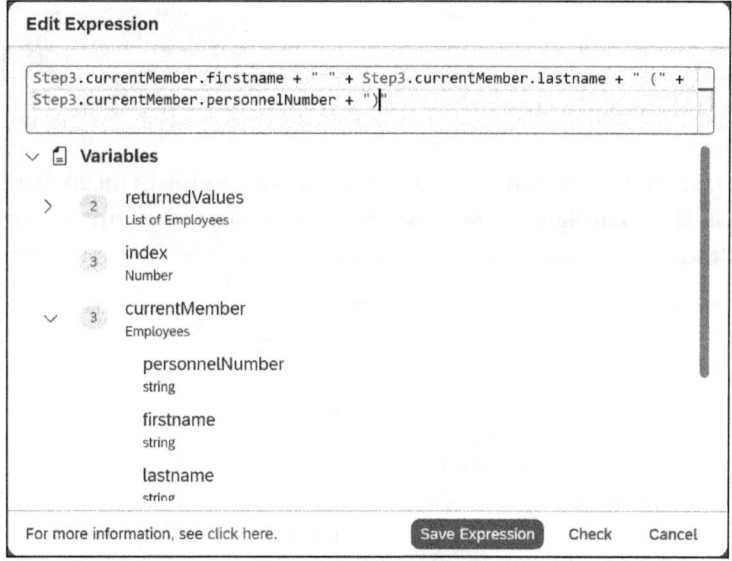

Figure 23.52 Use Expression Editor

Finally, as previously mentioned, you must close the Excel instance so that the automation can be executed successfully. To do this, drag the **Close Excel Instance** activity into the automation as the last step, as shown in Figure 23.53. Then click the **Save** button in the toolbar in the top-right-hand area.

743

23 Automation

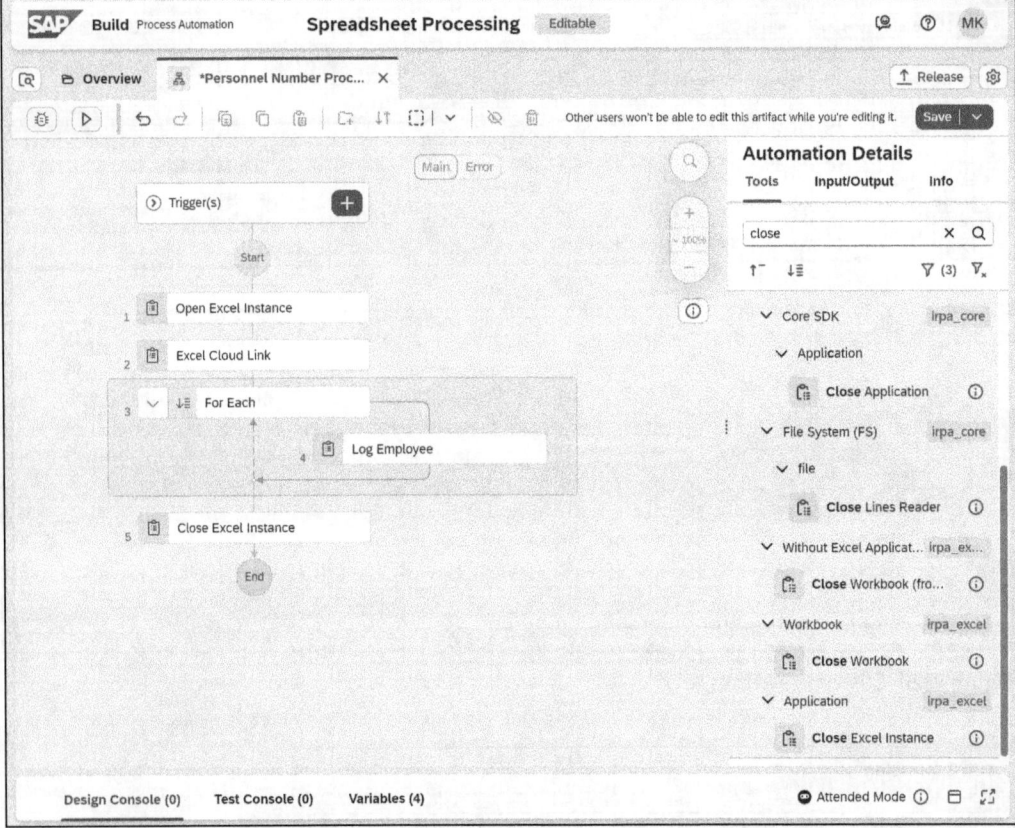

Figure 23.53 Add Close Excel Instance Step

You can now run or test the automation. To do this, click the **Run** button in the toolbar. This opens a dialog, as shown in Figure 23.54, in which you can start the automation by clicking the **Test** button.

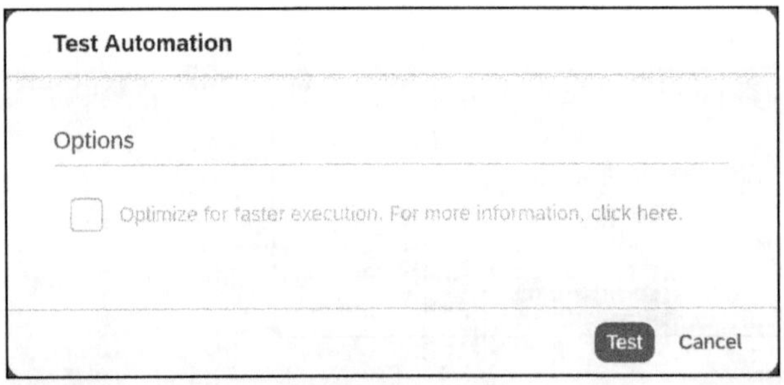

Figure 23.54 Test Automation Popup

The desktop agent is then configured. This step may take some time. The automation is then executed. You will notice that an Excel instance is opened on your computer in the background and is closed again at a later time after the automation has been

23.4 Environment Variables

successfully run. You should have opened Microsoft Excel at least once beforehand to ensure that no dialogs such as a license activation or a login screen appear.

Once the test has run successfully, you should see a screen similar to the one shown in Figure 23.55. On the left-hand side, you will find a list display of the automation log that has just been run.

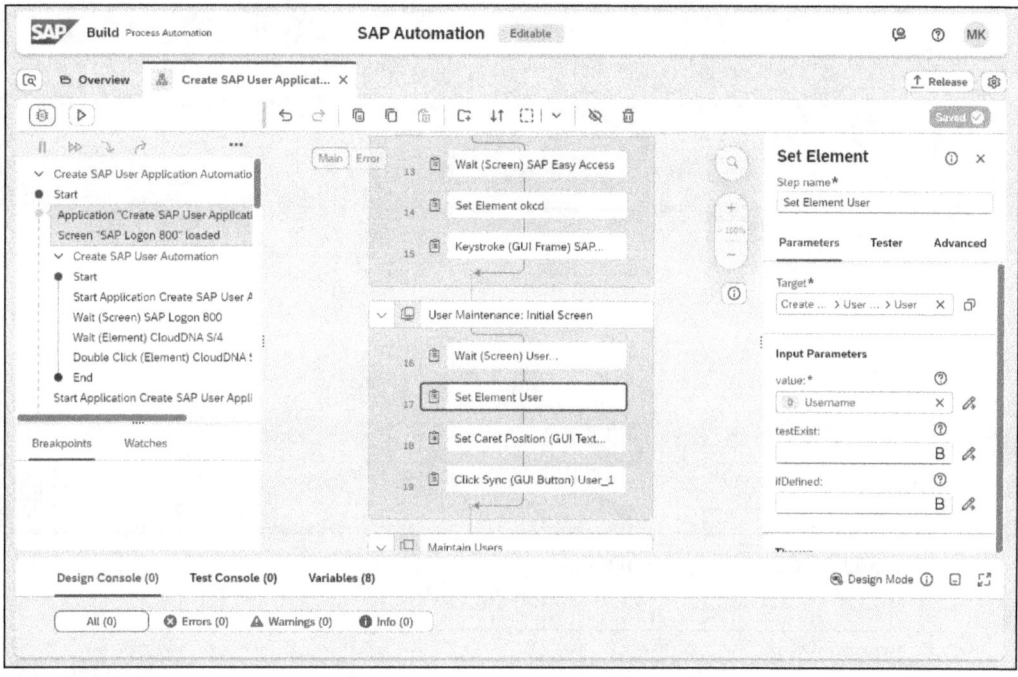

Figure 23.55 Test Run

23.4 Environment Variables

SAP has also considered the possibility of dynamically adjusting or externally controlling certain parameters in automations. This is done via environment variables. In this section, we will show you how you can adapt the previously created automation, in which you create a user in the SAP system, in this way.

In the first step, open the previously created automation in the automation editor (see Figure 23.56). Then click the ⚙ button in the top right-hand area to access the project properties.

In the **Project Properties** dialog, open the **Environment Variables** entry in the side menu as shown in Figure 23.57. As you can see from the illustrations, no variables have currently been created. Click the **+ Create** button to create a new variable.

23   Automation

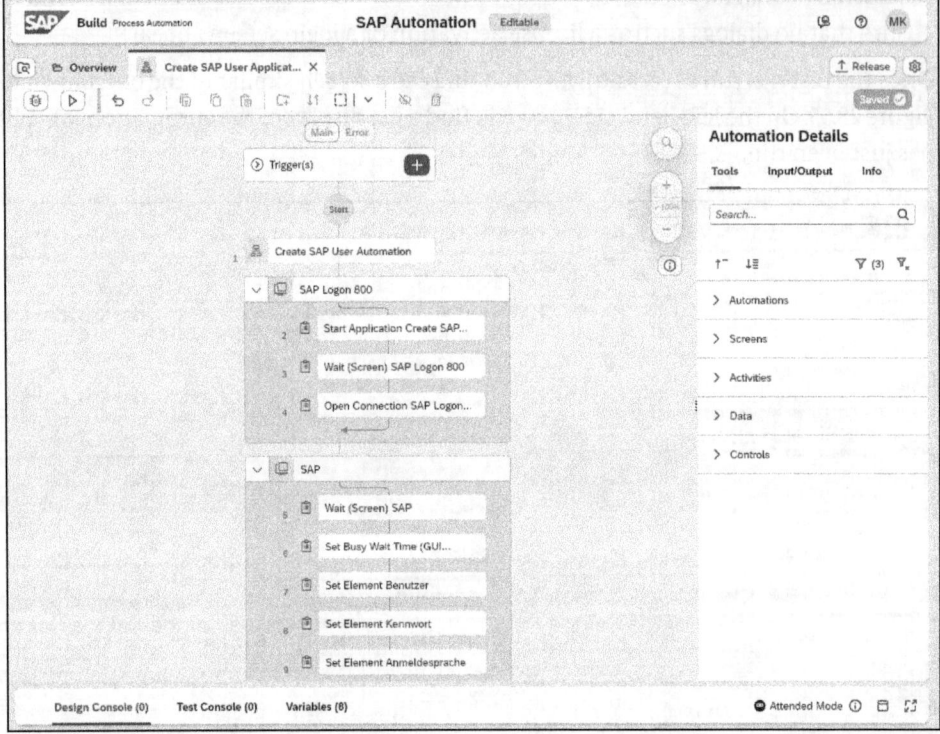

Figure 23.56 Open Automation Editor

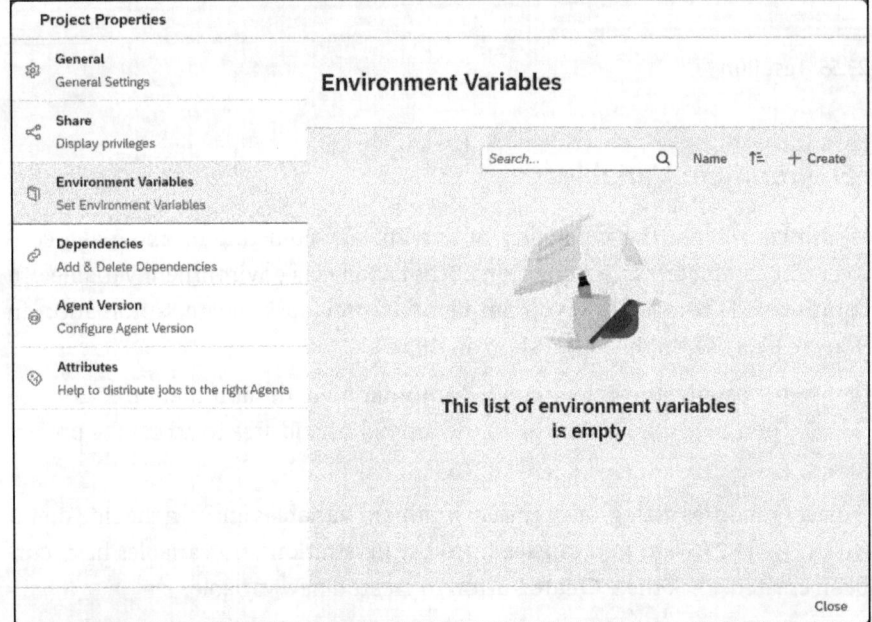

Figure 23.57 Open Environment Variables

Enter the value "DefaultPassword" in the **Identifier** field and a suitable description in the **Description** field. Then select the **Password** option for the **Type** field. Finally, click **Create** (see Figure 23.58).

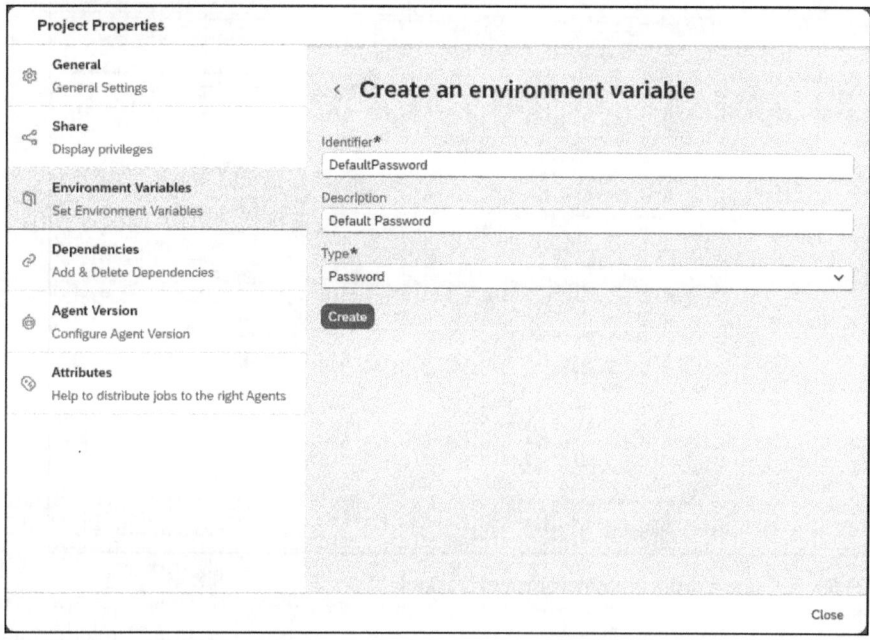

Figure 23.58 Create Environment Variable

You can now use the environment variable in your automation. To do this, search for the **Set Element New Password** activity in the previously created automation, as shown in Figure 23.59. Click it to open the parameters. These are displayed as usual in the right-hand area. You will find an area with the heading **Input Parameters** inside the **Parameters** tab. Select the previously created environment variable for the **value** input parameter. Repeat the steps for the activity named **Set Element Repeat Password**.

In addition to environment variables, you also have the option of defining the input and output parameters. These are values that are passed to the automation and returned to the caller after successful execution. To do this, click in the whitespace in the editor. This makes the **Automation Details** area visible on the right. Open the **Input/Output** tab as shown in Figure 23.60. Create a new parameter with the name **Username**. Select **String** as the type. Then click **Save**.

23   Automation

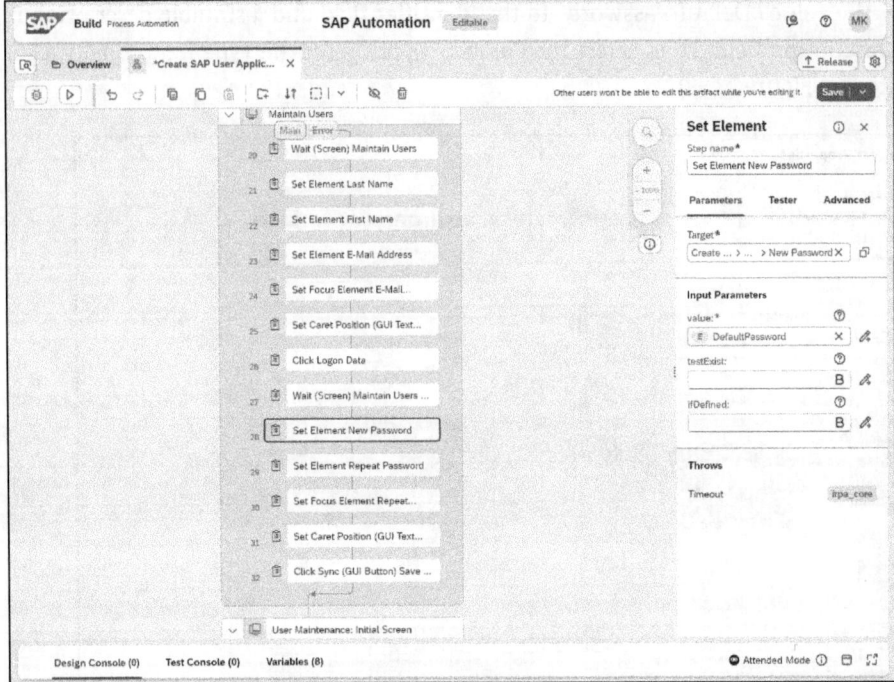

Figure 23.59 Set Password from Environment Variable

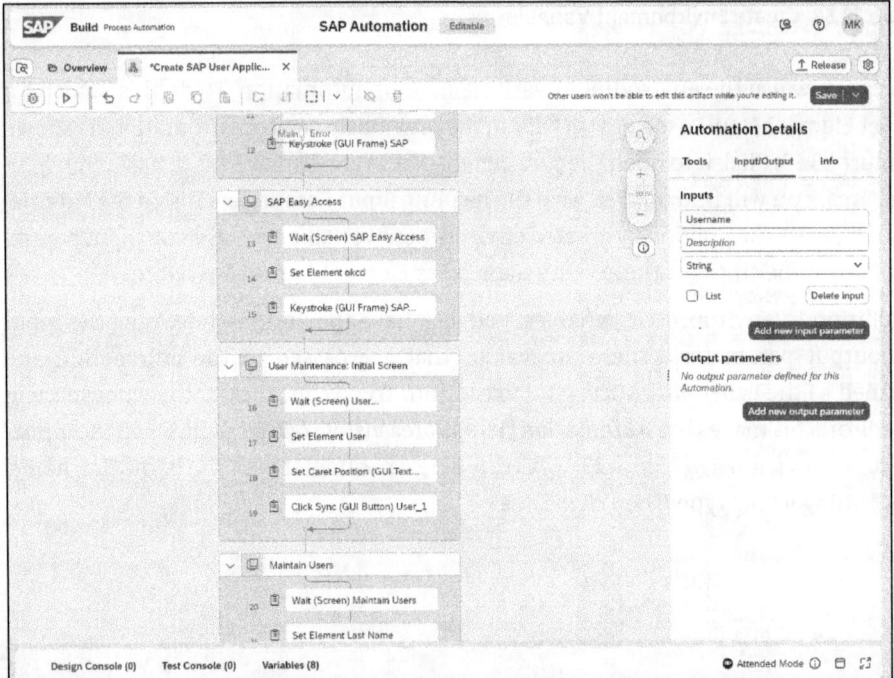

Figure 23.60 Create Input Parameter

23.4 Environment Variables

Now search for the activity called **Set Element User** in the automation. Select it and then assign the previously created input parameter to the **value** parameter in the **Input Parameters** area (see Figure 23.61).

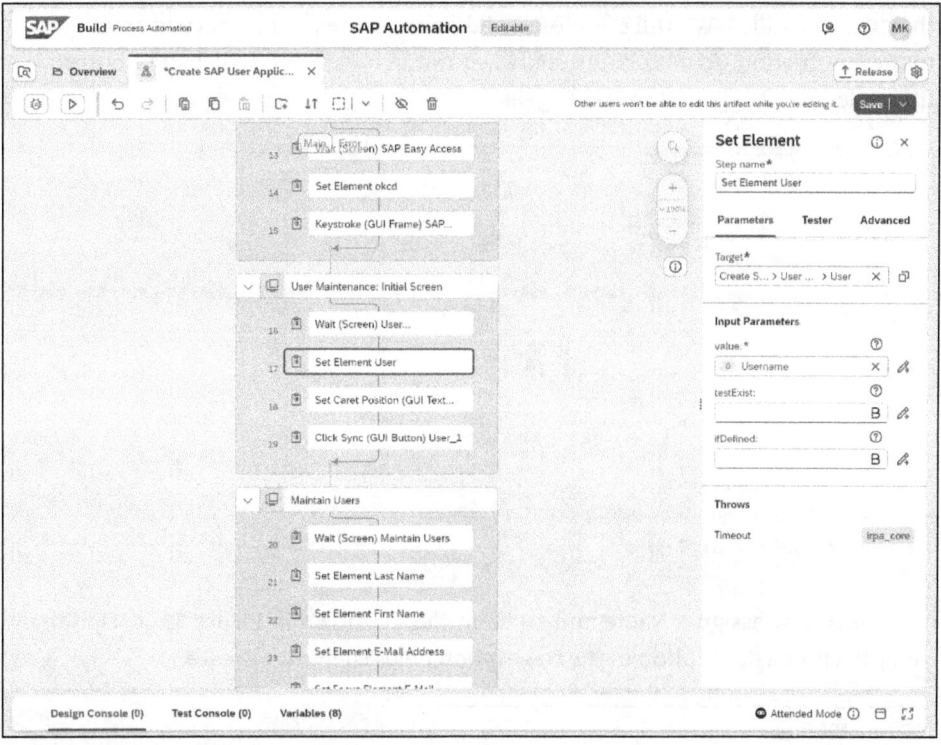

Figure 23.61 Assign Input Parameter

Now save your changes by clicking **Save**. You can then test the automation as usual. To do this, click ▷. A dialog will then appear as shown in Figure 23.62, in which you must maintain both the input parameters and the environment variables. Then click **Test** to execute the automation with the parameters provided.

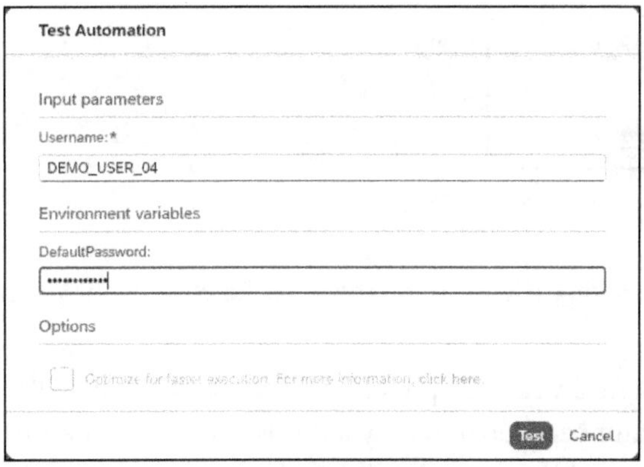

Figure 23.62 Provide Data for Test Automation

749

23.5 Deployment

To be able to use an automation, you must also make it available in SAP BTP within SAP Build Process Automation. Ideally, this requires your automation to have a trigger so that it can be called. We will therefore build on the last example and add an API trigger to the automation. To do this, proceed as shown in Figure 23.63. Click the **+** button, and then select **API Trigger • New API Trigger**.

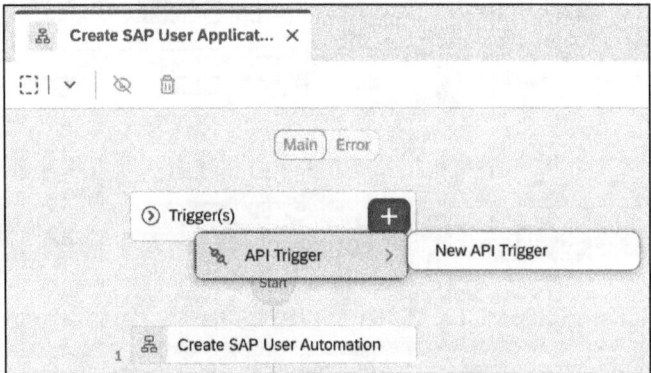

Figure 23.63 Add an API Trigger

You must now assign a **Name** and an **Identifier** as shown in Figure 23.64. Optionally, you can enter a description in the **Description** field. Then click **Create**.

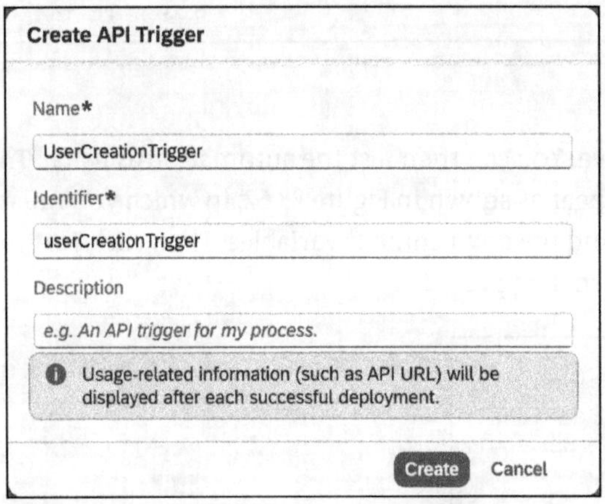

Figure 23.64 Create API Trigger

You must then release your development artifacts. To do this, click **Release**, as shown in Figure 23.65.

A new **Version Number** is always assigned when the project is released, as shown in Figure 23.66. A combination of three numbers separated by a dot is used here. The first number indicates the major version and the second number the minor version. The

23.5 Deployment

third number indicates the patch level. The version numbers are automatically incremented with each release, but you can also adjust them manually. You can also enter a meaningful comment for each version in the **Version Comment** field. Then click **Release**.

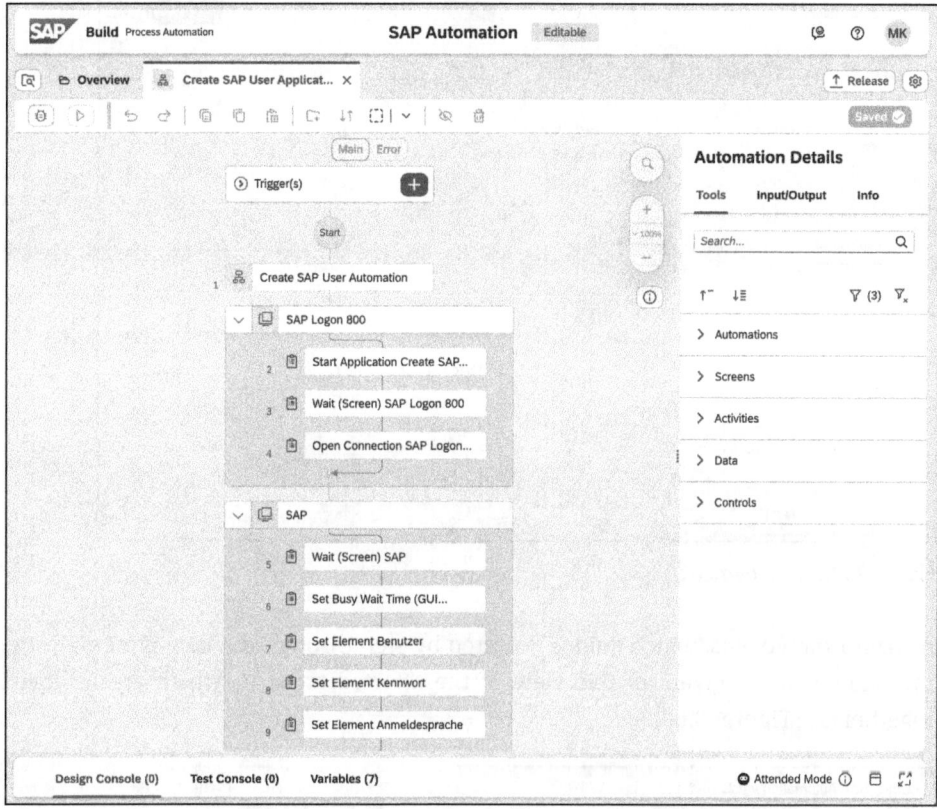

Figure 23.65 Release Project

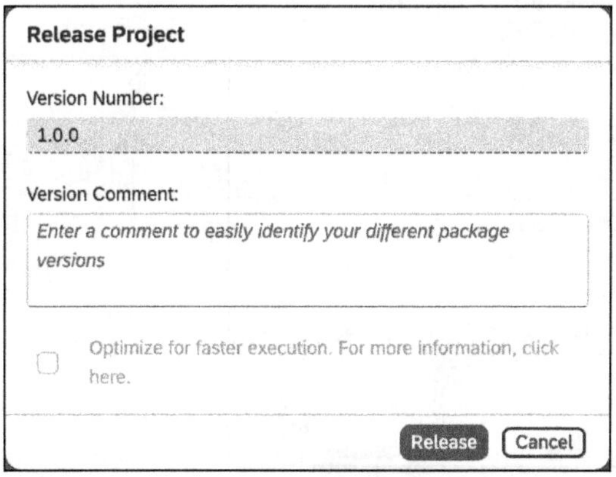

Figure 23.66 Version Information

Once you have released your artifacts, you still need to deploy them so that they can subsequently be used. To do this, click **Deploy**, as shown in Figure 23.67.

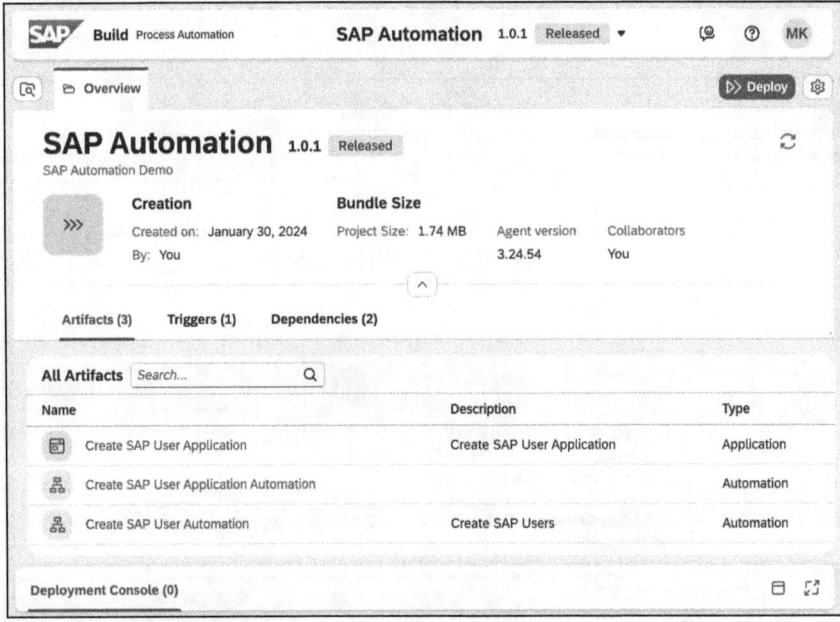

Figure 23.67 Deployment

A wizard then opens, which guides you step by step through the deployment. In the first step, you are given an **Overview** of the affected artifacts, which are deployed together (see Figure 23.68).

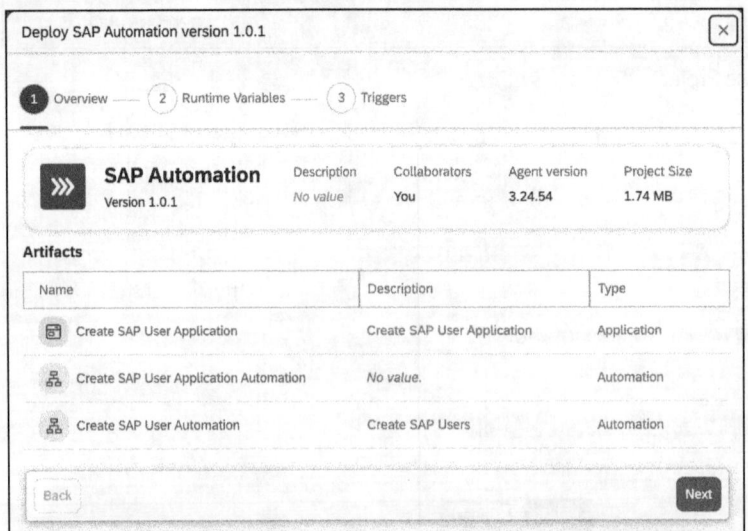

Figure 23.68 Deploy Wizard

23.5 Deployment

As you can already see at this point, the deployment can only ever be for the entire project, not individual artifacts. You should therefore bear in mind not to make the projects too large and pack in too many artifacts. Check the list of artifacts, then click **Next**.

In the second step, you must configure and set the runtime variables as shown in Figure 23.69. Once you have done this, you can jump to the last step of the wizard by clicking **Next**.

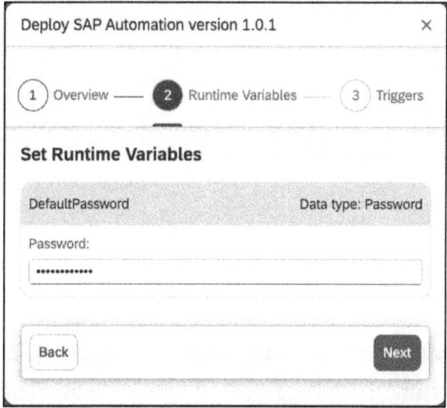

Figure 23.69 Maintain Runtime Variables

Finally, you will see a list of all triggers in the last step. You cannot make any adjustments at this point. Click **Deploy** to start the deployment. Depending on the size of the artifacts, this may take a few seconds.

Once the deployment has been successfully completed, you will see a dialog informing you of the status, as shown in Figure 23.70. You can jump to the SAP Build Process Automation monitoring by clicking on **Go to Monitor • Triggers**.

Figure 23.70 Deployment Success Message

All triggers for all projects are listed in the monitor ring (see Figure 23.71). Here you can add triggers yourself, and you can also execute existing triggers.

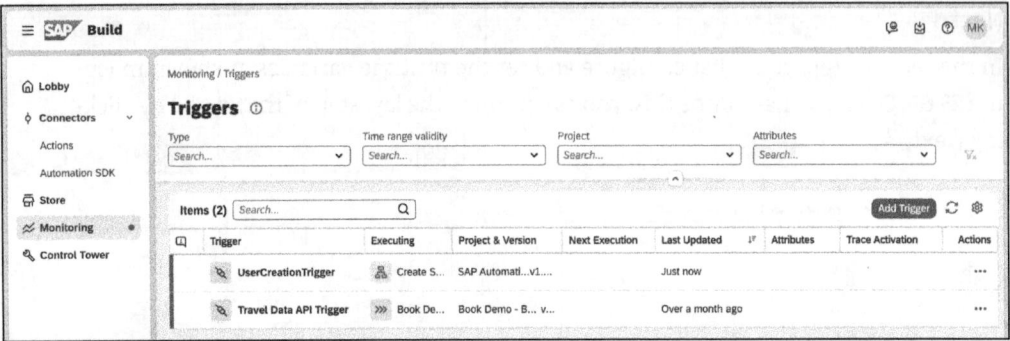

Figure 23.71 Monitoring Trigger

23.6 Summary

In this chapter, we provided a systematic introduction to the key components and processes of SAP Build Process Automation. Section 23.1 focused on the desktop agent, a local application that is installed on user computers or servers and that acts as an execution environment for automation projects. In Section 23.2, you learned how to capture interactions with various automation applications using the recorder.

Section 23.3 showed you how to automate Microsoft Excel, for example. The handling of local files was also covered in this context. There you learned how to create automations without the recorder. In Section 23.4, we looked at environment variables, which are important for adapting automations to different execution contexts. Finally, in Section 23.5, we looked at the deployment of automations in a production environment.

Chapter 24
My Inbox

Task management involves the process of organizing, planning, tracking, and completing tasks efficiently and effectively. It encompasses various steps and strategies to ensure that tasks are executed in a timely manner and aligned with broader goals and objectives. My Inbox is an application that was built by SAP in order to support users in task management. Tasks are brought together from various sources in order to provide a single point of entry for the responsible users.

My Inbox might already be familiar to you if you use the SAP Fiori launchpad on your on-premise system or SAP Task Center in the cloud. With this application, SAP provides a single point of entry for all workflow tasks that need human interaction. The application can be used in a browser as well as on mobile devices. It has an intuitive user interface that enables a user to explore the necessary data for his work. My Inbox uses different connectors to collect data on workflow tasks from a wide variety of systems and combine them in a central task list. These connectors are based on an interface that can also be implemented for customer-specific developments in order to individually add user tasks to the inbox. Non-SAP systems can also be integrated using an OData service.

In the following sections, we'll walk through using My Inbox, implementing substitution functionality, and integration with SAP Build Work Zone.

24.1 Using My Inbox

The inbox has a variety of functionalities that make it possible to view and edit assigned tasks. The metadata underlying a task can be sorted, filtered, or grouped to provide a user with the ability to treat specific tasks separately or restrict large sets of tasks. A search function makes it possible to find certain tasks by specifying a search criterion. In the following sections, we'll walk through these functions.

24.1.1 Task Functionality

The simple but extensively expandable user interface of My Inbox can be used for a variety of activities. For complex and individual activities, the user interface for displaying and editing workflow tasks can be completely replaced and provided with its own logic. The people responsible for tasks can create profiles for representatives in different areas of responsibility and specify their activities. Figure 24.1 shows a screenshot of the My Inbox application.

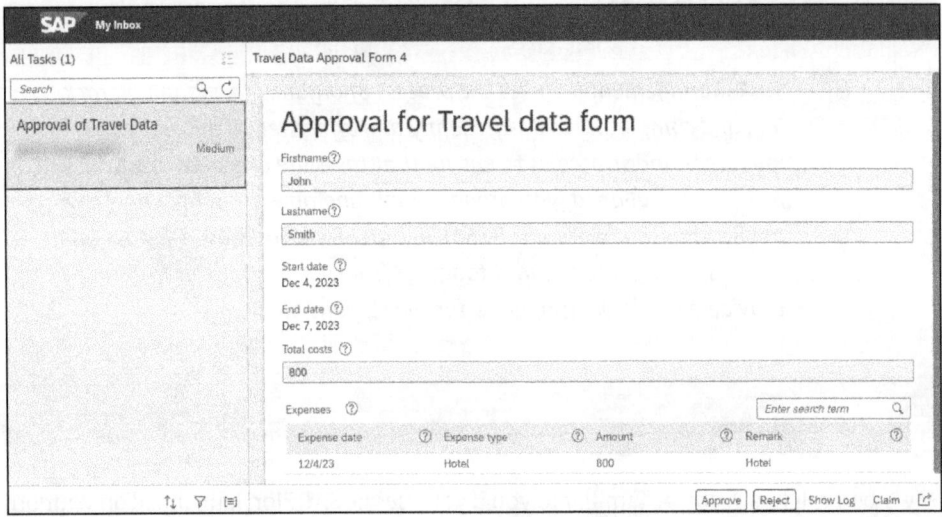

Figure 24.1 My Inbox Application

The application is based on the master-detail view and is divided into three sections. The section on the left contains the list of tasks assigned to the logged in user. The main area of the application displays the data for the selected task. The bottom area shows the actions available to the user depending on the selected task. In most cases, the tasks involve approving applications. In this case, the **Approve** or **Reject** actions are available to make the decision. In addition to these actions, the user has the option of viewing the workflow log by clicking the **Show Log** button, as shown in Figure 24.2.

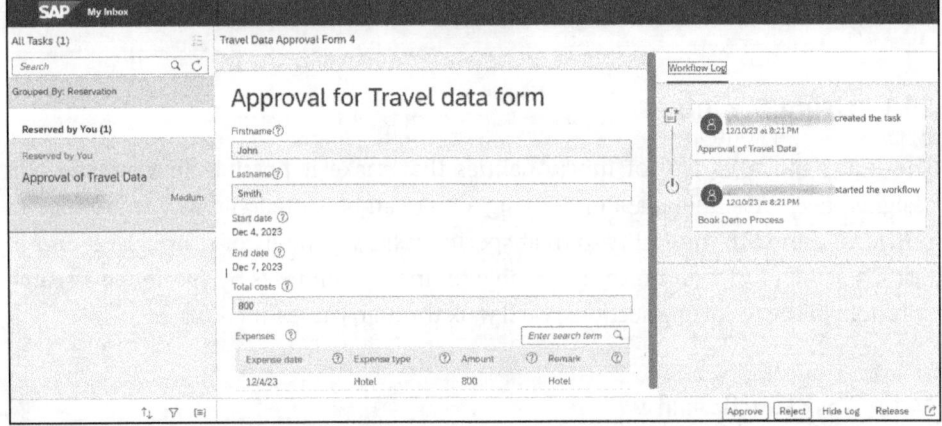

Figure 24.2 Workflow Log for Selected Task

A task can be reserved by clicking the **Claim** button in the bottom area (available or not, based on context). This task is then only available to the reserving user and is removed from the list of all other users to whom this task was assigned. Reserved tasks can then be released again by clicking the **Release** button, which makes them editable again by other responsible users.

24.1.2 Search Functionality

Above the list of tasks in the inbox, there is a text field that represents the search function. By specifying a search literal, the items in the task list are filtered with each key press. The inbox checks whether the specified search literal is part of the displayed data of the listed tasks. In the example given, the search literal "Approval" can be used to find all items that have this word in the task title, as shown in Figure 24.2.

24.1.3 Sorting

Using the sorting function, the list of displayed tasks can be sorted according to different properties. The function is located below the list of tasks and is activated using the ↧ button. Pressing this button will display a popup where the sorting method can be selected, as listed in Table 24.1.

Sort Criteria	Description
Created By (A on Top)	Sorting the list according to the creator of the tasks starting with A
Priority (Highest on Top)	Sorting the list according to the priority
Title (A on Top)	Sorting the list according to the title beginning with A
Created On (Newest om Top)	Sorting the list according to the creation date, with the newest item the first item in the list
Due On (Earliest on Top)	Sorting the list according to the due date, with the item next to be due the first item in the list

Table 24.1 Sorting Criteria of My Inbox

Figure 24.3 shows the popup to select the sorting criteria.

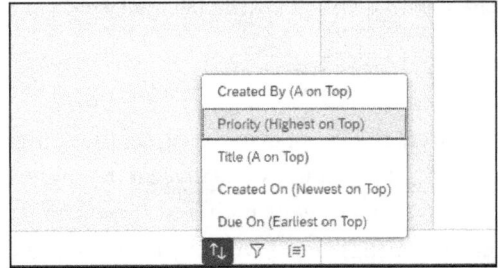

Figure 24.3 Sorting Criteria of My Inbox

24.1.4 Filter

The tasks assigned to the user in the My Inbox can be filtered according to specified criteria. The filter function is structured in two stages, with the first step specifying the property by which the filter is to be filtered. In the second step, a value is specified for the selected property, which determines the filter criterion. The filter function allows multiple properties and criteria to be specified at the same time, allowing the filter combinations to be varied as desired. The filter option is made available by pressing the ▽ button next to the list of tasks assigned to the user. The tasks can be filtered according to the properties listed in Table 24.2.

For most of these properties, predefined value lists are configured in My Inbox, which can be adjusted in Customizing if necessary. The contents of these value lists form the criteria for the filter of the respective property. Table 24.2 lists the values that can be used as criteria for each respective property out of the box.

Filter Property	Available Standard Values
Priority	<ul><li>Very High</li><li>High</li><li>Medium</li><li>Low</li></ul>
Due On	<ul><li>Overdue</li><li>Within a Week</li><li>Within a Month</li><li>All</li></ul>
Status	<ul><li>Ready</li><li>In Progress</li><li>Executed</li><li>Reversed</li></ul>
Created On	<ul><li>Today</li><li>Past 7 Days</li><li>Past 30 Days</li><li>All</li></ul>
Task Type	Task type values are calculated at runtime according to the tasks in the list, allowing you to filter the current list based on the task types available

Table 24.2 Available Values for Filter Criteria

The following example declares a filter that filters the list of tasks based on the status, priority, and due date properties. Only tasks that have a status of **Ready** and **In Progress** should be displayed. In addition, only tasks that have very high and high priority should be displayed. With the third filter criterion, these tasks will only be displayed if they are due within a week.

Figure 24.4 shows the dialog for selecting the properties to filter by. Here you can see that criteria are specified for the **Priority**, **Due On**, and **Status** properties.

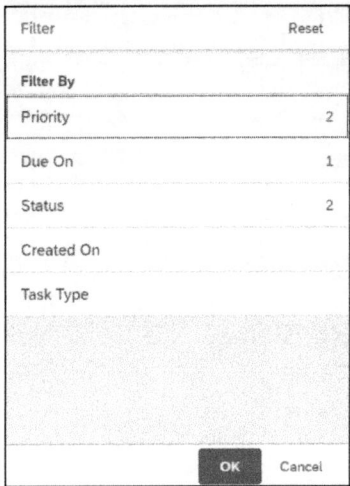

Figure 24.4 Properties to Filter By

By selecting each respective property, the list of values for selecting the filter criteria is displayed. One or more values can then be specified or selected from this list as a filter criterion. Figure 24.5 shows the set of possible values for the priority of the tasks. Here you can see that the values selected are **High** and **Very High**.

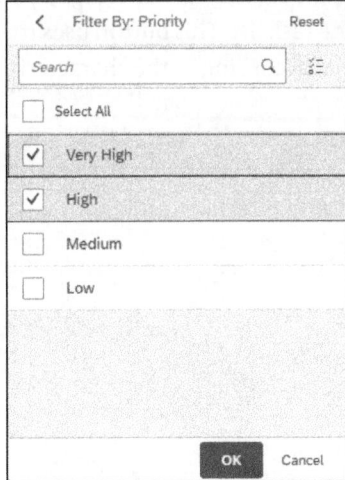

Figure 24.5 Values Chosen for Priority Property

As already mentioned, the values for the task type are calculated at runtime based on the tasks assigned in the inbox. Figure 24.6 shows the list of possible task types for the example project from Chapter 21 described in this book. In this case, only the **Travel Data Approval Form 4** task type is available.

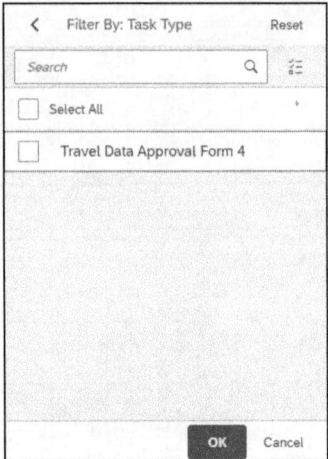

Figure 24.6 Filter Values for Task Type Property

In this way, filter criteria can be combined as desired to limit the tasks that should be displayed in the inbox list to those that are the processor's focus.

24.1.5 Grouping

Another feature that My Inbox offers is grouping tasks. Like sorting and filtering, copying can also be selected using a separate button below the task list. This button uses the [≡] icon. As shown in Figure 24.7, the tasks can be grouped according to the characteristics listed in Table 24.3.

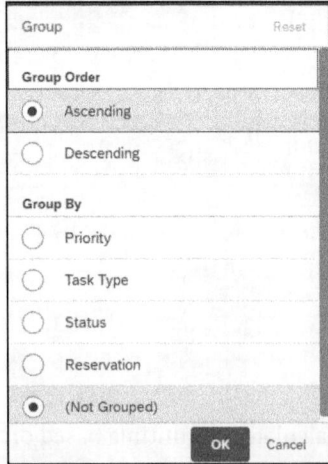

Figure 24.7 Grouping of Tasks

Group By	Description
Priority	The priority of the task.
Task Type	The type of the task. This is the name of the user task as assigned in the workflow model.
Status	The status of the task.
Reservation	Distinction between reserved and unreserved tasks.
(Not Grouped)	Default value. No grouping is defined.

Table 24.3 Grouping Properties in My Inbox

24.2 Substitution

It is often necessary to be able to set up representatives so that the processing of internal company tasks can be guaranteed even in the event of the absence of responsible persons. In this case, My Inbox offers the option of defining representatives who will have the tasks assigned to their worklist for a certain period of time. When defining substitutions, a distinction is made in My Inbox between two types: planned and unplanned substitutions. The *planned substitutions* are scenarios in which the start and end of the absence to be represented is usually known. If a replacement is entered in this way, the replacing user automatically receives the tasks during this period. After this period, new tasks will be reassigned to the original user.

For *unplanned substitutions*, however, only a date is specified that defines the start of the substitution. If the registered user accepts the substitution, he will be able to see and edit the tasks assigned to him in his role in the internal My Inbox app. Only when this substitution rule is deactivated will no further tasks be delivered to the substitution user.

To define a substitution, proceed as follows in My Inbox. In the header of My Inbox, there is an icon for the profile of the currently logged-in user at the right edge. Clicking this icon opens a menu in which you can navigate to the substitution rules. Select the **Manage My Substitutions** menu item to access the substitution management, as shown in Figure 24.8.

In the substitution rules view, the planned and unplanned substitutions are displayed in separate lists. You can switch between these lists using the icons displayed above the list. To create a planned replacement, click the **Add New Substitute** button in the bottom-right corner of the application, as shown in Figure 24.9.

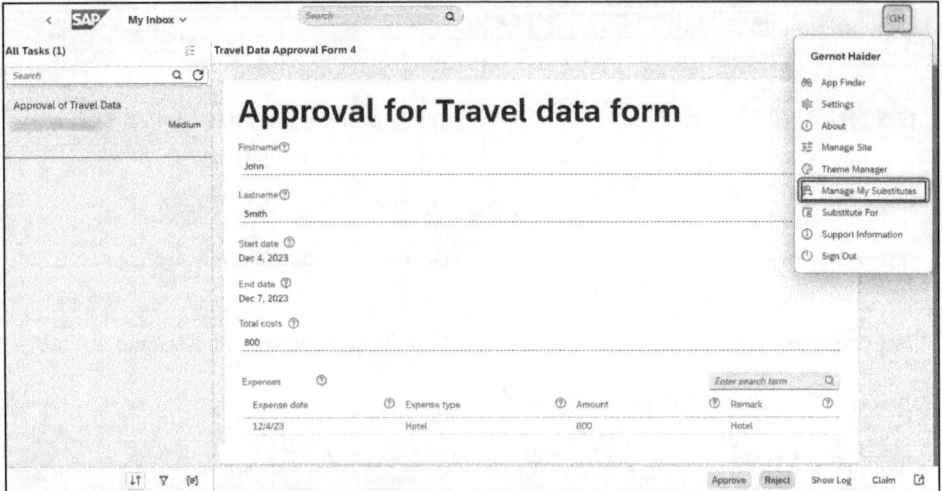

Figure 24.8 Navigate to Substitution Rules

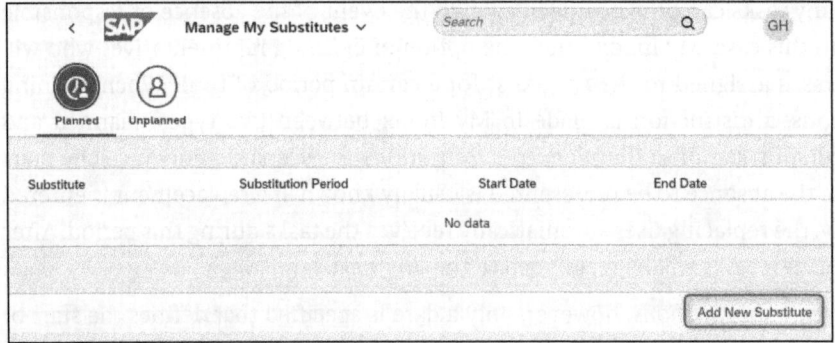

Figure 24.9 Adding New Substitution

As a result, a dialog is displayed in which the user ID of the person representing the responsible person and the period of time for the substitution can be specified. Figure 24.10 shows an example of using the dialog to specify a new substitution.

After clicking the **Save** button, the new substitution rule is entered into the list. If the start date is in the future, the substitution is displayed as inactive. Within the set period, the status changes to active. Figure 24.11 shows the list of planned substitutions after entering a new substitution rule.

Unplanned substitutions are created in the same way, with the difference that in this case only the user ID of the person replacing the responsible user is specified. The substitution rule is active from the moment it is created, and tasks are delivered provided that the defined person agrees to the substitution. The validity of this rule only expires when the corresponding entry is deleted from the list of unplanned substitutions. Figure 24.12 shows a list of unplanned substitutions after an entry was added.

24.2 Substitution

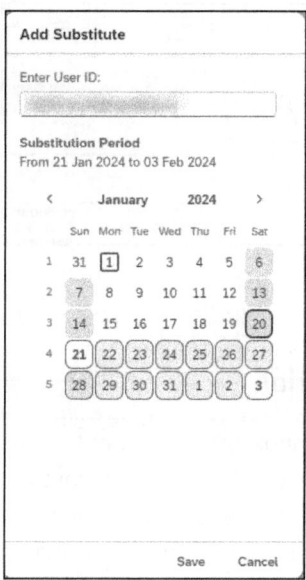

Figure 24.10 Providing Data for New Substitution

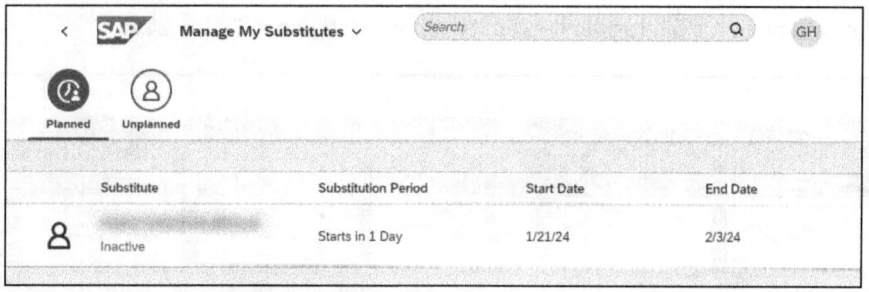

Figure 24.11 List of Planned Substitutions

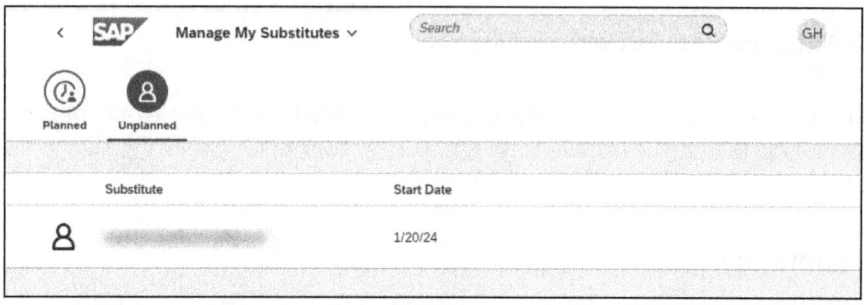

Figure 24.12 List of Unplanned Substitutions

To delete an entry from the lists of planned or unplanned substitutions, select the respective entry and click the **Delete** button in the bottom-right corner of the application, as shown in Figure 24.13. The substitution rule is then deleted from the list.

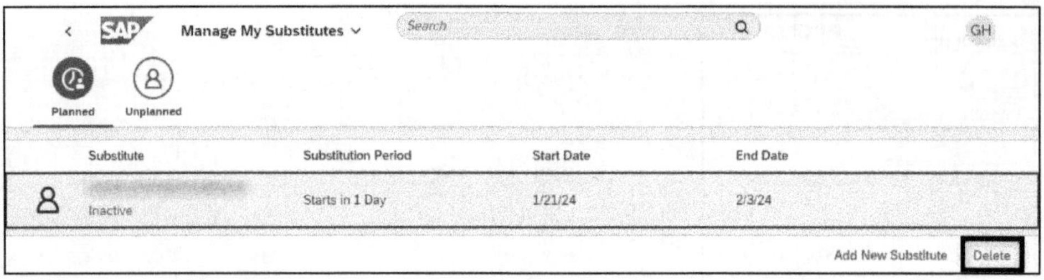

24 My Inbox

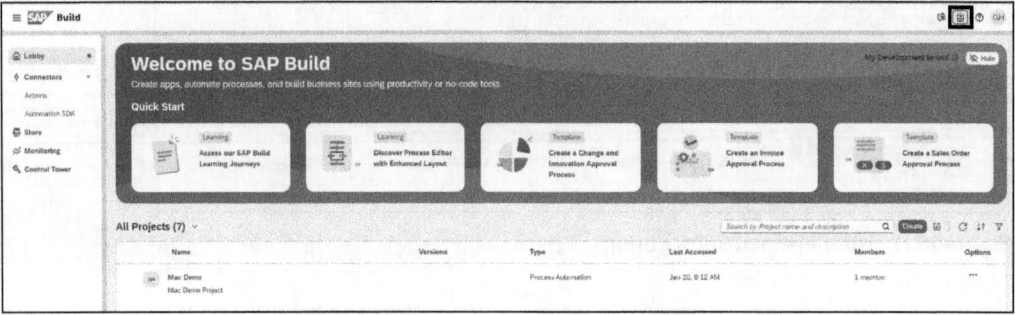

Figure 24.13 Deleting Substitution Rule

24.3 My Inbox and SAP Build Process Automation

The SAP Build Process Automation service includes My Inbox integration out of the box. Process steps that require user interaction and are assigned to a person from the process are automatically added to the task list in My Inbox and can be edited there. The result of this can then be reacted to in the process in order to control further automated processing. During the design phase of a business process, the developer can use the SAP Build lobby to access a version of My Inbox that can be used for testing purposes. To do this, first navigate to the SAP Build lobby. In the header area of the lobby, select the ☒ icon. This will open the testing version of My Inbox, as shown in Figure 24.14.

Figure 24.14 Navigating to Test Version of My Inbox

The end user accesses My Inbox via launchpad integration in SAP Build Work Zone. For more about how this can be installed and set up, refer back to Chapter 10.

24.4 Summary

In this chapter, you learned how the My Inbox application is used and what features it brings. We showed you how tasks can be searched, sorted, filtered, and grouped. We also showed you how substitution rules can be organized and what types of substitutions My Inbox provides. Finally, we discussed how SAP Build Process Automation integrates with My Inbox and how My Inbox can be used during the design phase.

Chapter 25
Monitoring and Administration

Many of the capabilities you read about in the last chapters provide monitoring and configuration functionalities. These are brought together in two separate areas: Monitoring and the control tower. In this chapter, we will introduce the option provided by SAP Build Process Automation to help you monitor the created automations as an administrator. We also will explain the capabilities that support you to configure the features according to your needs.

In the complex landscape of modern systems and technologies, the ability to keep a vigilant eye on operations is paramount. In this chapter, we delve into the area of monitoring—a crucial aspect that ensures the health, performance, and security of digitized processes.

In the second part of this chapter, we will explore the options that SAP Build Process Automation offers for administration in the control tower.

25.1 Monitoring

SAP Build Process Automation provides monitoring for deployed workflows and automation jobs. Events that are raised by the workflows are collected and presented to the user. A user that needs to work with the monitoring environment must be assigned to the `ProcessAutomationAdmin` role collection.

Starting from the landing page of SAP Build Process Automation, monitoring can be reached by selecting the **Monitoring** menu item in the left panel. On the starting page of the monitoring area, tiles can be clicked to enter the corresponding monitoring views. Figure 25.1 shows a screenshot of the landing page for monitoring.

Using this view as the entry for monitoring SAP Build Process Automation runtime artifacts, you can monitor the following objects: instances, automation jobs, acquired events, automations, processes and workflows, triggers, and visibility scenarios, each of which we'll discuss in the following sections.

25   Monitoring and Administration

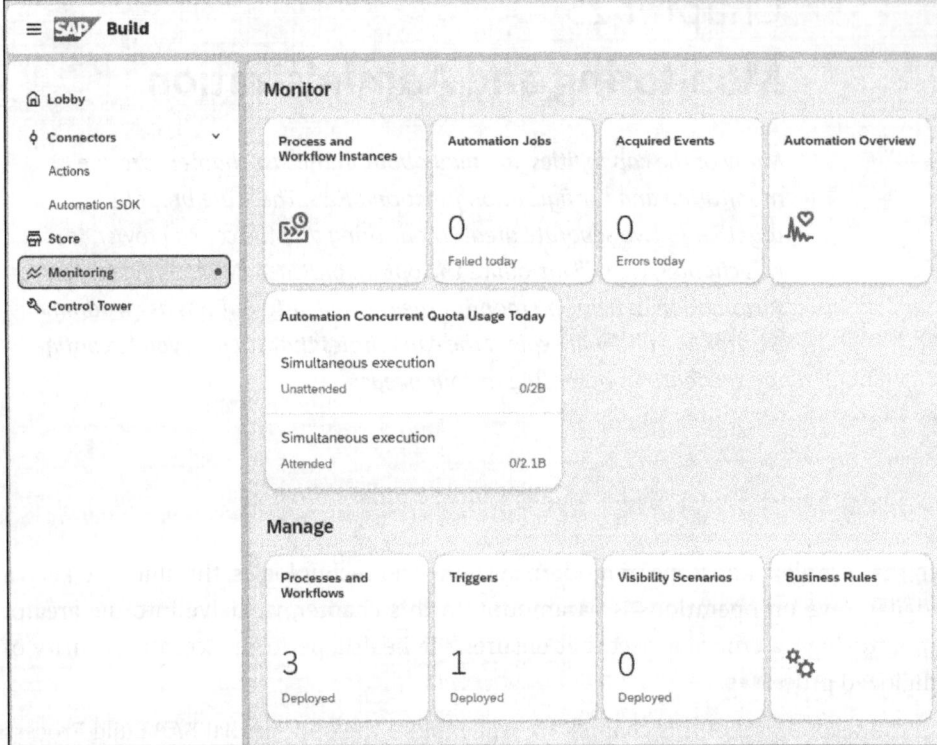

Figure 25.1 Monitoring Landing Page

25.1.1   Process and Workflow Instances

In the process and workflow instances view, you can see a list of all workflow instances that were already processed or are still running. To navigate to this view, click the **Process and Workflow Instances** tile in the **Monitor** section of the **Monitoring** view.

By default, process instances with a completed or canceled status are hidden in the initial overview of the items. Filter options offer the possibility of restricting the list of items displayed according to required criteria. The fields to be used for the filter can be activated and deactivated as required. In this way, a home page can be set up in which the necessary criteria for viewing the required workflow instances are visible at a glance and the requirements can be adjusted. The filters for optimizing the search results can be based on one or more of the parameters shown in Table 25.1.

Filter Parameter	Description
Status	The status of the workflow execution.
Project	The name of the business process project to which the instance belongs.
Definition	The definition ID of the instance.

Table 25.1 Filter Parameters Available on Overview Page

Filter Parameter	Description
Hierarchical Level	Using this parameter, you can decide to filter only the root workflow instances in case of workflows that include subflows.
Instance ID	The instance ID of the workflow.
Started After	Using this parameter, you can limit the search result to instances started on and after the selected date.
Started Before	The search result is limited to instances started on and before the selected state.
Started By	Only workflow instances started by one or more uses are displayed.
Completed Before	The search result is limited to instances that were completed on and before the selected data.
Completed After	Use this parameter if you want to display instances that were completed on and after a selected date.

Table 25.1 Filter Parameters Available on Overview Page (Cont.)

Figure 25.2 shows a screenshot of the overview page for the workflow instances.

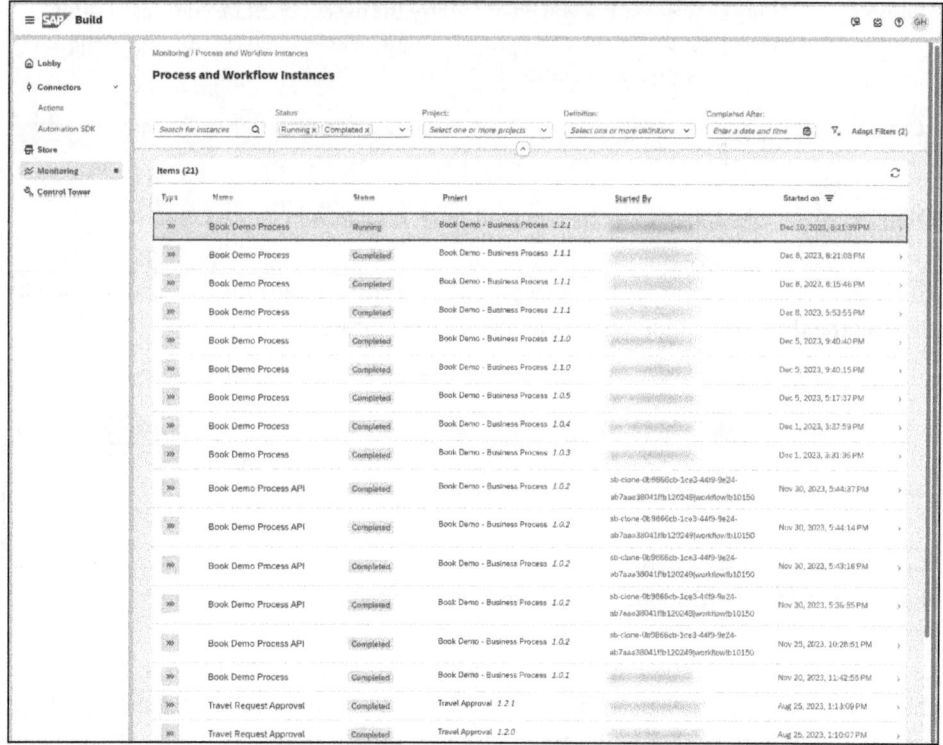

Figure 25.2 Overview of Process and Workflow Instances

25 Monitoring and Administration

By selecting one of the displayed items, you can navigate to the details of the corresponding workflow instance. Depending on the current status of the selected instance, you can change the status according to your needs. To do so, you can select the status to be set in the header section of the details view, as shown in Figure 25.3.

Figure 25.3 Changing Status of Workflow Instance

Based on the current status of the selected instance, the instance statuses listed in Table 25.2 can be selected for change.

Status	Description
Put on Hold	Suspend a running or erroneous workflow instance by putting it on hold.
Resume	The execution of a workflow put on hold can be continued. You can also retry failed workflow steps by choosing this status.
Retry	Choose this status if you want to retry the execution of failed steps or often erroneous workflow. This option is only available when there is an error in executing a workflow instance.
Cancel	By choosing this option, you can terminate the execution of the selected workflow instance.

Table 25.2 Available Options for Changing Status of Workflow Instance

The header section of the details view provides additional options that can be used by clicking the three-dot menu button, as shown in Figure 25.4.

Figure 25.4 Additional Options for Instance Monitoring

Choosing the **Show Subflow Instances** option makes all available subflow instances of the selected workflow instance appear in the item list on the overview page. From there, you can explore the subflow instances by navigating to their detail views. The **Show Tasks** option provides a monitoring page that displays all tasks that belong to the selected workflow instance. You can see how many tasks were created and select a task to see its detailed information. Figure 25.5 shows a screenshot of the task information view.

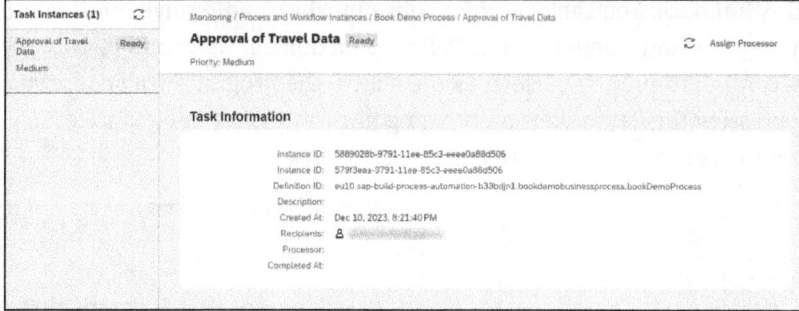

Figure 25.5 Tasks of Workflow Instance

In the **Logs** tab of the details view, you can explore the executed activities and enter details if available. For erroneous instances, the error logs are shown in this tab. If necessary, these logs can be downloaded in JSON format. Underneath the **Logs** tab, the current context of the workflow instance is displayed in the **Context** tab. Figure 25.6 shows a details view that belongs to a running workflow instance.

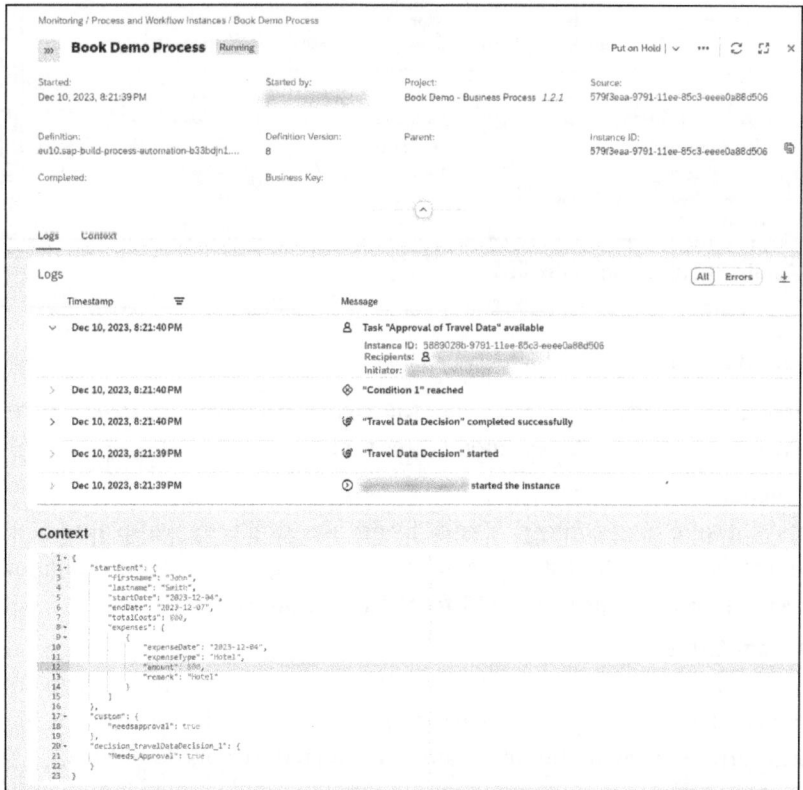

Figure 25.6 Details View of Running Workflow Instance

25 Monitoring and Administration

25.1.2 Automation Jobs

If you use automation jobs in one of your projects, you will find all the information about it in this interface. You can see which jobs ran when and receive information about the processing status and duration. A filter function allows you to restrict the results list according to properties such as the status, the project, or the connected machine. By pressing the **Cancel Jobs** button, ongoing processes can be canceled. Figure 25.7 shows a screenshot of monitoring automation jobs.

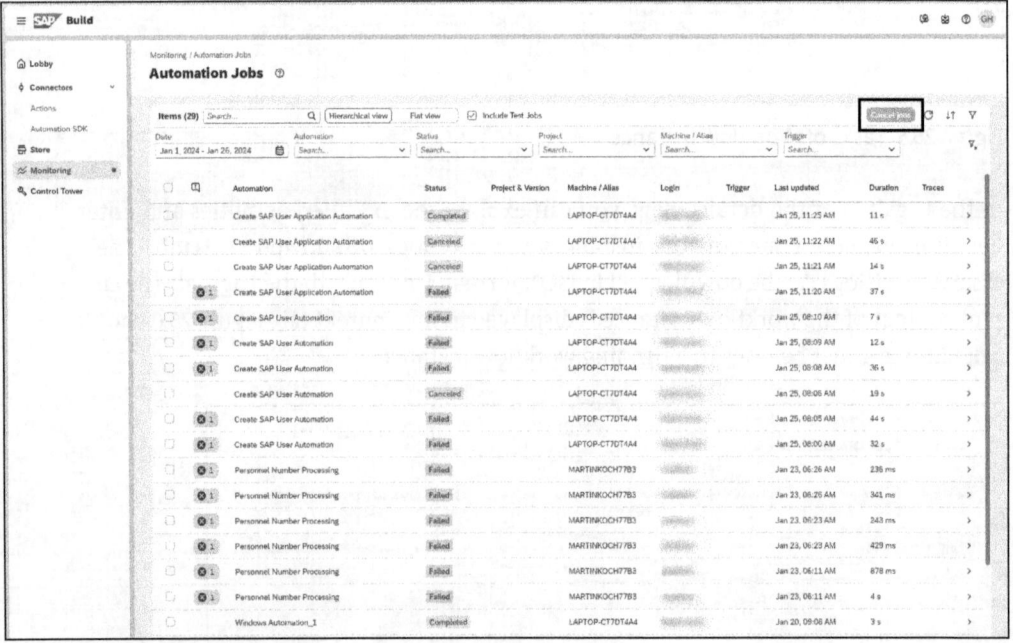

Figure 25.7 Monitoring of Automation Jobs

25.1.3 Acquired Events

In the **Acquired Events** overview, you can view the events that were consumed by process visibility. The view differentiates among three types:

- **Visibility Events**
  The events that were successfully processed by process visibility are displayed here. Using the filter function, this list can be filtered according to the **Process Definition ID**, **Process Instance ID**, **Event Type**, and **Timestamp** attributes.

- **Visibility Event Errors**
  If errors occur when consuming events through process visibility, they will be displayed in this list. In this list, you will see the messages that could not be processed, the resulting error messages, and time stamps for when the errors occurred.

25.1 Monitoring

- **Business Events**
  The third list shows the business events that were consumed by SAP Build Process Automation in the last seven days. This table shows the event types, the sources of the events, when the events were generated, and the contents of the events. This list can also be downloaded in the form of a CSV file.

When you enter this view, no data is initially displayed. Only by defining all the filters will the last 5,000 events be loaded. To display further events, the timestamp must be set accordingly in the filter function. Figure 25.8 shows a screenshot of consumed visibility events.

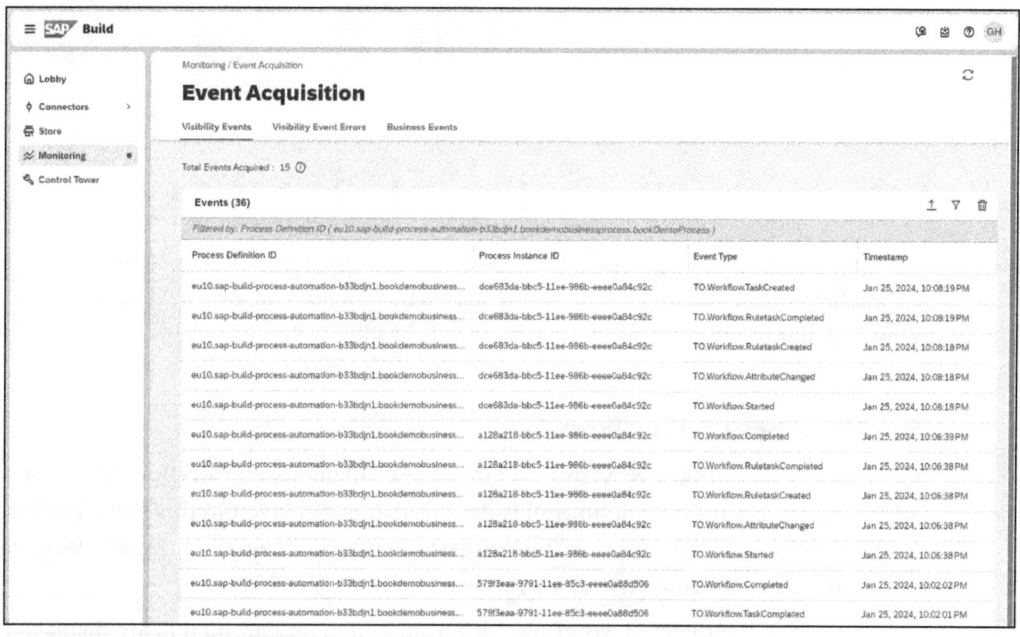

Figure 25.8 Acquired Events

25.1.4 Automation Overview

The **Automation Overview** section provides information about automations. This information is presented graphically and is divided into two groups. The **Status** group displays all running jobs and the number of running agents grouped by status. Past activities for jobs and agents are also shown grouped by status in the **History** section. Figure 25.9 shows an example of historical jobs and agent statistics.

25 Monitoring and Administration

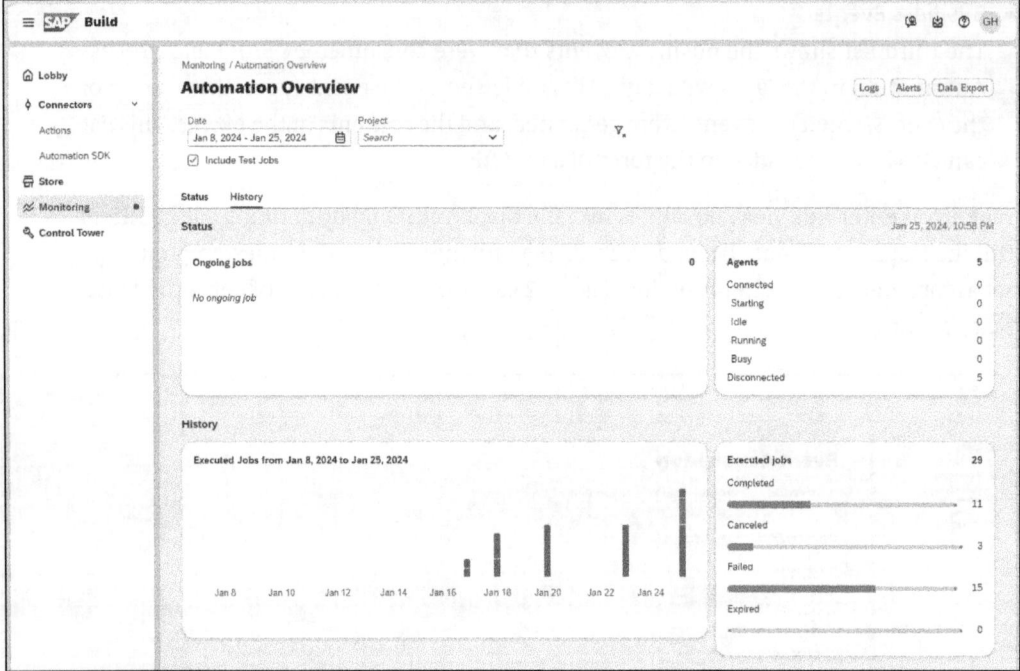

Figure 25.9 Automation Overview

25.1.5 Processes and Workflows

The processes and workflows view provides a list of all deployed workflow definitions. As soon as a new workflow definition is deployed, it is displayed in this view. You can navigate to this view by selecting the **Processes and Workflows** tile from the **Manage** section in the monitoring view, as was shown back in Figure 25.1.

In the details of the selected workflow definition, the main parameters are displayed. By selecting a workflow definition from the list in the left panel, you can choose one of the following options provided by this view:

- **Show Instances**
  By choosing this option, you navigate to the monitoring of the workflow instances, and the filter for the selected workflow definition is automatically set to only show workflow instances that refer to the selected workflow definition.

- **Start New Instance**
  This option enables you to start a new workflow instance of the selected workflow definition. In a displayed popup dialog, you can provide the context for the workflow trigger in JSON format, as shown in Figure 25.10.

- **Download Model**
  Here you can download the data model of the workflow definition in JSON format.

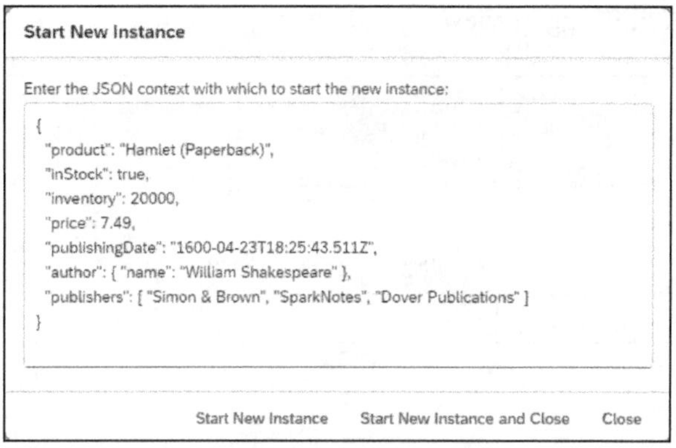

Figure 25.10 Starting New Workflow Instance

25.1.6 Triggers

In the triggers view, you can explore which triggers are deployed and where they are used. To navigate to this view, click the **Triggers** tile in the **Monitoring** area. The view provides a filter option that enables you to filter the list by the trigger type, its status, the name of the project in which the trigger is used, and other matching attributes. In the list of triggers, the columns listed in Table 25.3 can be displayed.

Column	Description
Trigger	The name of the configured trigger and its type.
Executing	The name of the process or the automation that uses this trigger.
Project & Version	The project and its version that contains the deployed automation.
Next Execution	The next planned execution of the deployed automation (only shown for scheduled triggers).
Last Updated	The date of the last update of the automation.
Attributes	Matching attributes associated with the trigger.
Trace Activation	An indicator the traces have been activated for this trigger.
Actions	Depending on the type of your trigger, a dropdown menu is available that provides the following options for managing the trigger: • Enable or Disable the trigger • Run Now (only available for scheduled triggers) • Edit • Add Notifier • Delete • Activate Traces

Table 25.3 Columns of Triggers List

25 Monitoring and Administration

Figure 25.11 shows a screenshot of the **Triggers** view.

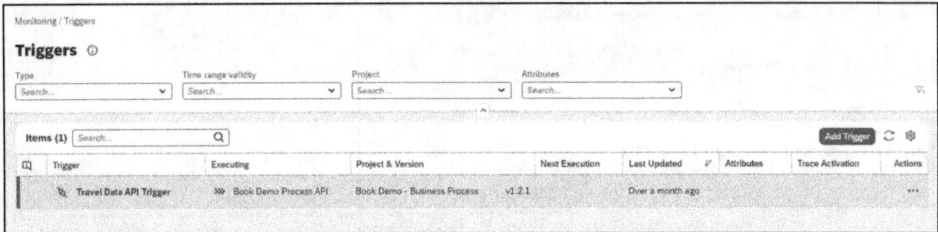

Figure 25.11 Triggers View

25.1.7 Visibility Scenarios

The **Visibility Scenarios** section provides an overview of all deployed scenarios. You can schedule the data determination job or unschedule it, and see a detailed list of when these jobs ran and what data was processed. Here you can also actively initiate data collection. You can use the 🔳 icon to navigate to the dashboard of the respective visibility scenario. Figure 25.12 shows the interface for visibility scenarios using the scenario created in Chapter 22 as an example.

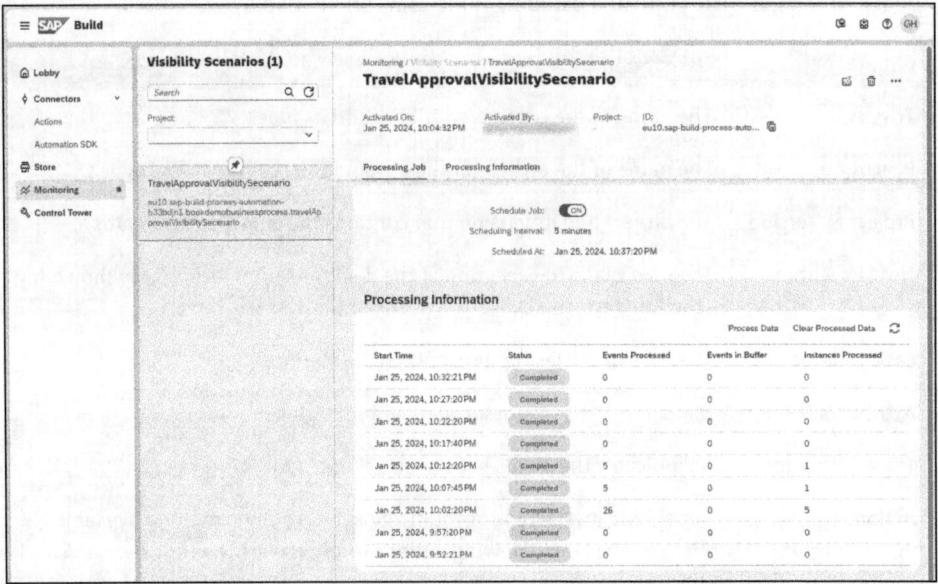

Figure 25.12 Visibility Scenarios

25.2 Administration via the Control Tower

The control tower provides a collection of configuration tasks that can be executed according to your needs. The landing page of the control tower can be reached by clicking the **Control Tower** menu item in the left menu of the lobby. The landing page of the control tower is divided into four sections that provide different tiles to configure specific parts that can be used in your automations:

- **Tenant Configuration**
  In this section, you can configure the tenant details to control the behavior of the tenant.
- **Backend Configuration**
  This section provides configuration options such as configuring the mail server, configuring the connection to SAP Cloud ALM, destinations, and external authentication.
- **Agent Configuration**
  The agent configuration can be used to configure and manage agents to be used for your automations.
- **Others**
  In this section, the control tower provides additional configuration functionality for API keys, alert handlers, and cloud studio variable.

In the following sections, we will dive into the single configuration options and show how the functionality can be used to configure the automation service. Figure 25.13 shows the control tower landing page.

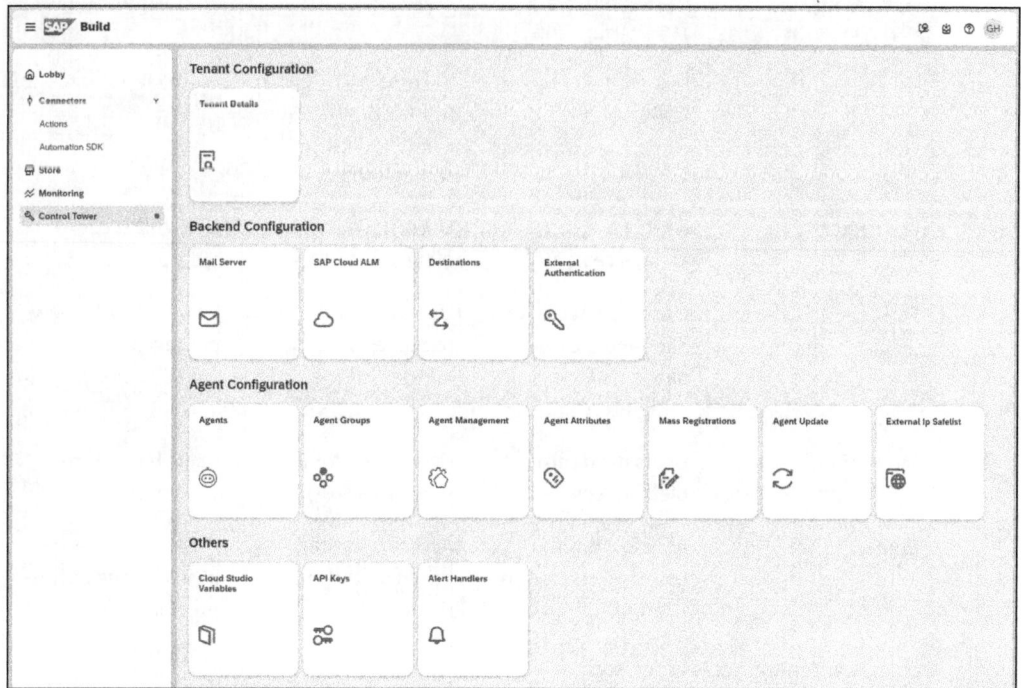

Figure 25.13 Landing Page of Control Tower

25.2.1 Tenant Details

By configuring the tenant details, via the **Tenant Details** tile, you can set the behavior of your installed process automation service by setting the type of the tenant. To do so, you must be assigned the ProcessAutomationAdmin role collection in your subaccount.

Depending on how the tenant is used, according to your business needs, you can set the type to restrict users' access to special functionalities. Therefore, the tenant configuration provides three different types of tenants that can be set using this configuration functionality:

- **Development**
  Use this type for your tenant if it is used to design and customize business scenarios before they are transported to your productive tenant.
- **Test**
  This type can be set if your tenant is used for testing new configurations within assigned projects.
- **Production**
  This type is used for tenants that work with your organization's productive data.

According to the type set for your tenant, the features listed in Table 25.4 can be accessed by the users working on the tenant.

User Access to Features	Development Tenant	Test Tenant	Production Tenant
Lobby	Users can explore all features.	View, export, deploy, promote, and manage members of the shared projects.	Access is limited.
Connectors	Users have complete access.	Users have complete access.	Users have complete access.
Store	Users can add packages and create packages from a template.	Users can only view packages.	Users can only view packages.
Monitoring	Users have complete access.	Users have complete access.	Users have complete access.
Control tower	Access is limited based on the authorization.	Access is limited based on the authorization.	Access is limited based on the authorization.

Table 25.4 Tenant Access for Users

In addition to the type of the tenant, you can set a user-friendly name, a description, and a link that will be shown in the shell bar of the lobby, as shown in Figure 25.14. To set the details of the tenant, click the corresponding tile in the control tower landing page and set the fields according to your needs. Figure 25.15 shows an example configuration.

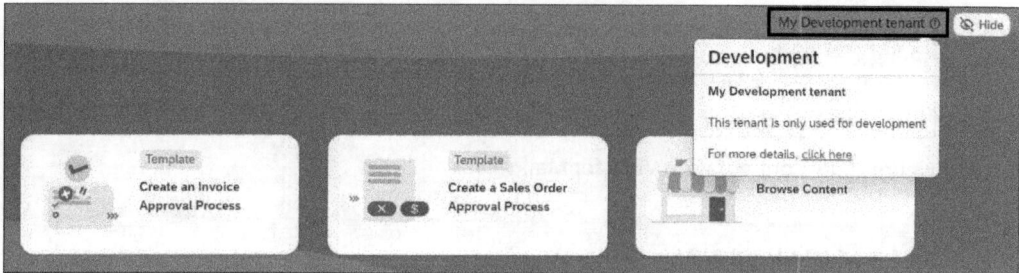

Figure 25.14 Tenant Properties Shown in Bar of Lobby

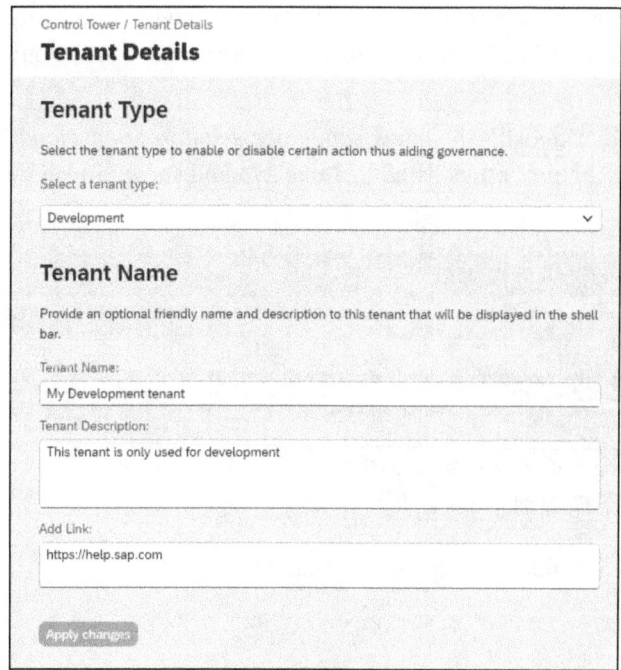

Figure 25.15 Configuration View of Tenant Details

25.2.2 Mail Server

If you want to use email functionality in your business processes, you need to configure an SMTP mail destination that will then be used from the mail step in your processes. The mail server configuration functionality can be used to explore the current settings and to send an email in order to test the configuration. The configuration of the SMTP server itself is done in the SAP BTP configuration. The view in Figure 25.16 provides a forwarding link that enables you to jump directly into this configuration.

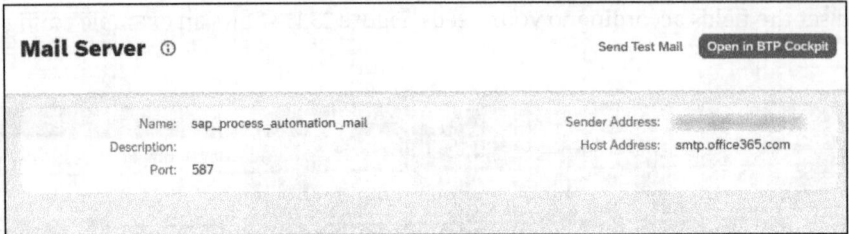

Figure 25.16 Configuration View for Mail Server

25.2.3 SAP Cloud ALM

By integrating SAP Build Process Automation with SAP Cloud ALM, you enable administrators to monitor your process automations in one place. For this integration, SAP Build Process Automation uses a dedicated destination that can be configured to create a connection that is used to transfer data from your process automations to SAP Cloud ALM.

In SAP BTP, create a new destination with the name *sap_process_automation_cloud_alm*. You also need to provide the properties listed in Table 25.5 and shown in Figure 25.17 for the destination.

Property	Value
Type	HTTP
URL	The URL given in your service key instance—for example, "https://<datacenter>.alm.cloud.sap/"
Authentication	OAuth2ClientCredentials
Token Service Url	Your SAP Cloud ALM URL, followed by */oauth/token*
Client ID	The client ID from your service key instance
Client Secret	The client secret from your service key instance

Table 25.5 Parameters to Configure SAP Cloud ALM Destination

Once the destination is configured, you can open the SAP Cloud ALM configuration in the control tower by clicking the corresponding tile, then enter a service name and description. After saving the configuration, click **Register** to activate the monitoring, as shown in Figure 25.18.

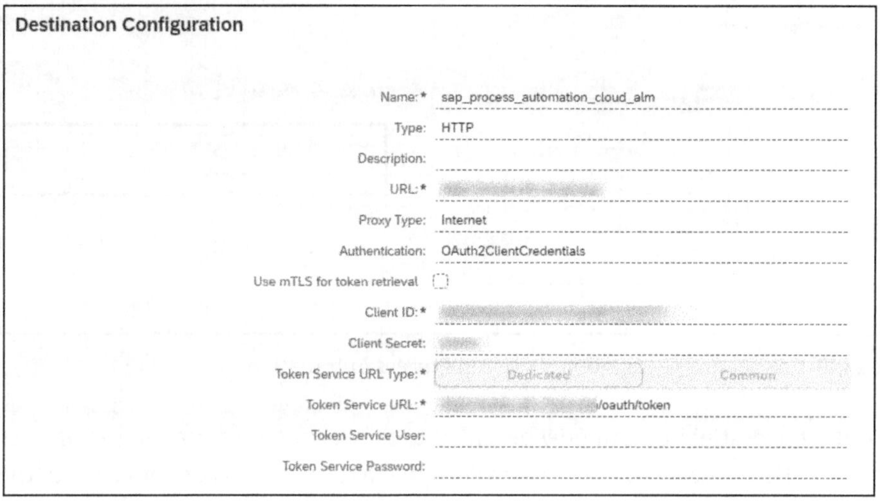

Figure 25.17 Destination for SAP Cloud ALM

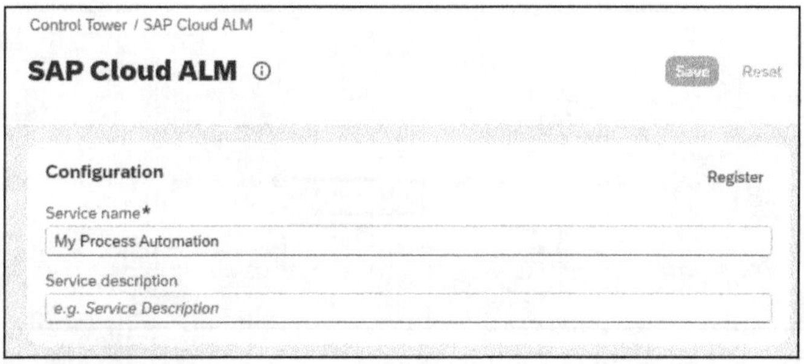

Figure 25.18 Activating Monitoring

25.2.4 Destinations

In SAP BTP, destinations are used to define endpoints and to store sensitive information that is used to connect to foreign systems such as client credentials keys or certificates. Whenever you need to create a connection to another system from your business process, you can use one of these destinations. The destination view of the control tower provides the list of destinations used in your process automation. Therefore, only specific destinations are considered within this view. If you want to use a destination in SAP Build Process Automation, this destination must contain the property sap.processautomation.enabled and the value of this property must be set to true.

If you need to add a destination to SAP Build Process Automation, navigate to the destinations in your SAP BTP subaccount and either add the sap.processautomation.enabled property to an already existing destination or create a new destination that contains this property. Figure 25.19 shows an example for a destination configured to be used in the process automation.

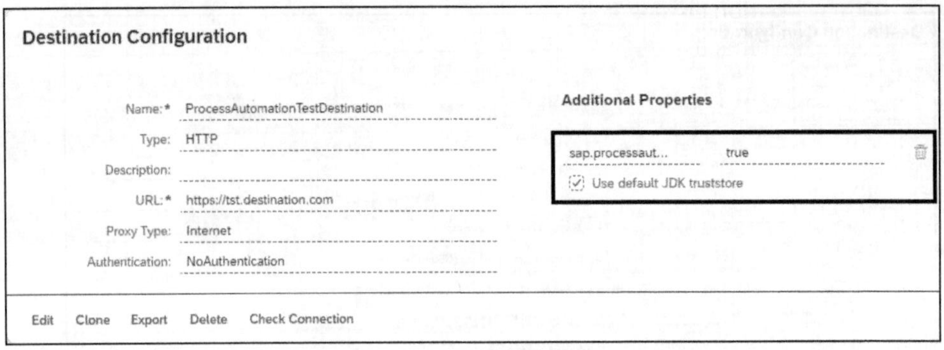

25 Monitoring and Administration

Figure 25.19 Configuring Destination for Use in Process Automation

As a result, the destination can now be added to the destinations used in the process automation. Therefore, select the **Destinations** tile in the backend configuration of the control tower. In the destinations view, click the **New Destination** button, as shown in Figure 25.20.

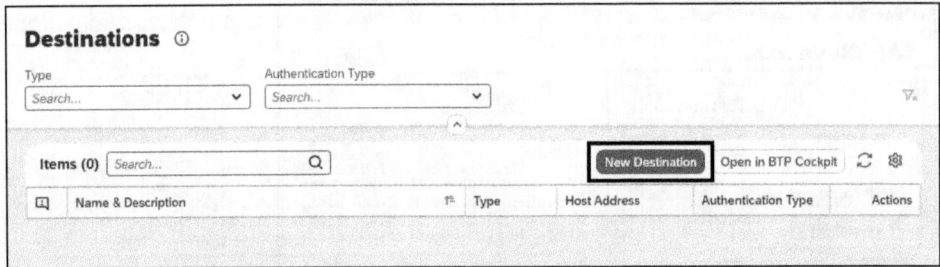

Figure 25.20 Adding New Destination to Process Automation

In the displayed pop-up dialog, all destinations that are enabled for the process automation are shown and can be selected. Figure 25.21 shows a screenshot of this dialogue, including the destination we just configured.

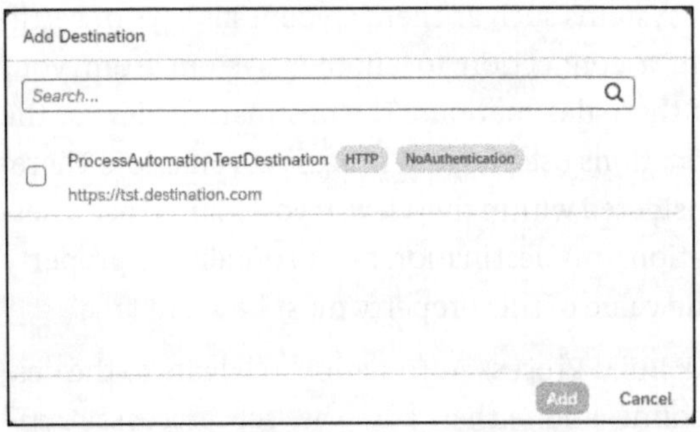

Figure 25.21 List of Destinations to Be Added

Select the destination to be added and click the **Add** button. As a result, the destination is added to the items list and can now be used in the process automation (see Figure 25.22).

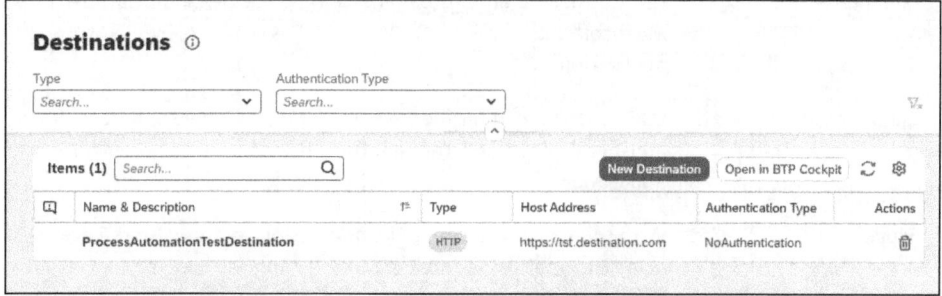

Figure 25.22 Added Destination

25.2.5 External Authentication

External authentication provides the functionality to configure access to Google or Microsoft 365 to run automations. By selecting the tile for **External Authentication**, you can navigate to the view of the configured authentications for the corresponding portal. Figure 25.23 shows a screenshot of this view.

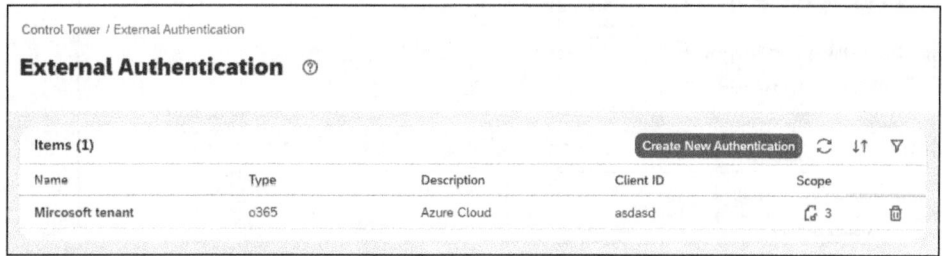

Figure 25.23 List of External Authentications

By selecting the **Create New Authentication** button, you can add a new set of authentication details to access Google or Microsoft 365. In the popup dialog that is displayed, you can enter the details in Table 25.6 to configure the access.

Field	Description
Type of External Portal	Select the type of portal to be used. You can choose **Google** or **Microsoft 365**.
Name	The name of the new authentication configuration. This name must be unique as it is used as an identifier.

Table 25.6 Properties to Configure External Authentication

Field	Description
Description	This field is optional. It can be used to describe the authentication.
Client ID	Use this field to enter the client ID from Google Cloud Platform or Microsoft 365. In Microsoft 365, this ID is obtained by the Azure application.
Client Secret	Use this field to enter the client secret from Google or Microsoft 365. In Microsoft 365, the client secret is obtained by the Azure application.
Scope	With the scope, you define the permissions granted for the agent using external authentication. By selecting the checkboxes you add the corresponding permissions for the user.
Tenant ID	This ID represents the Azure platform and is only needed for external authentications of type Microsoft 365.

Table 25.6 Properties to Configure External Authentication (Cont.)

In the popup dialog, enter the data according to your needs, then click the **Create** button to add a new external authentication to the item list, as shown in Figure 25.24.

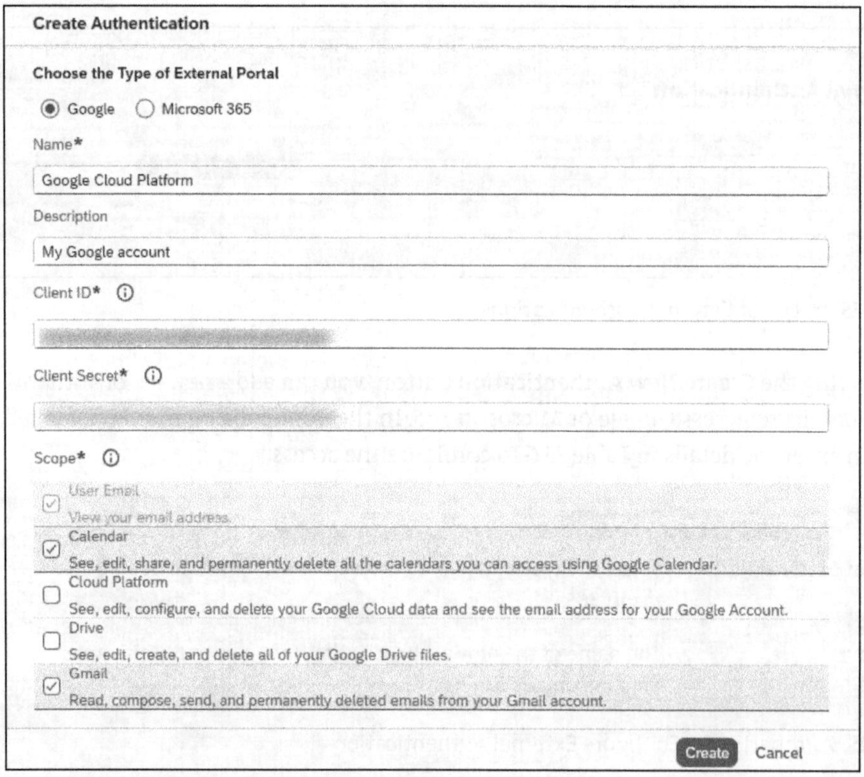

Figure 25.24 Creating New External Authentication

25.2.6 Agent Configuration

The **Agent Configuration** category consists of seven use cases, each shown behind its own tile. The following sections describe most of these: agents (an overview), agent groups, agent attributes, and external IP safelist.

Agents

The agent overview can be opened via the **Agents** tile. Registered agents can be edited here. Figure 25.25 shows a screenshot of this overview.

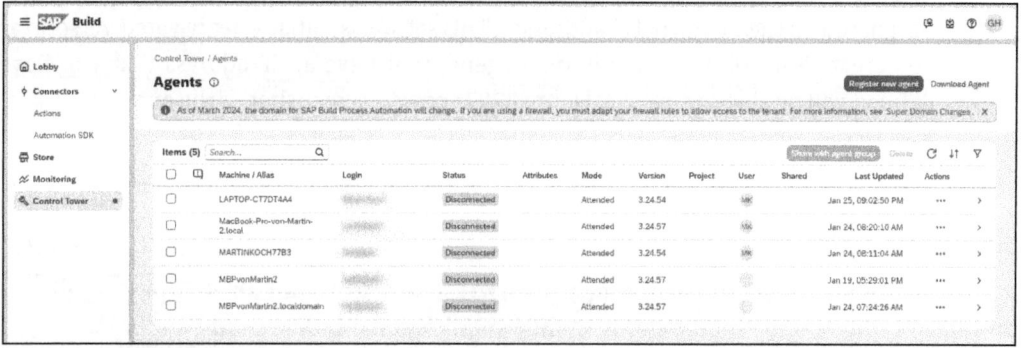

Figure 25.25 Agent Overview

By selecting a list entry, you can navigate to the detailed view, which displays further information about the selected agent. Data on activities and events that have occurred can be viewed here. Figure 25.26 shows an example activity plot for an agent over the last week.

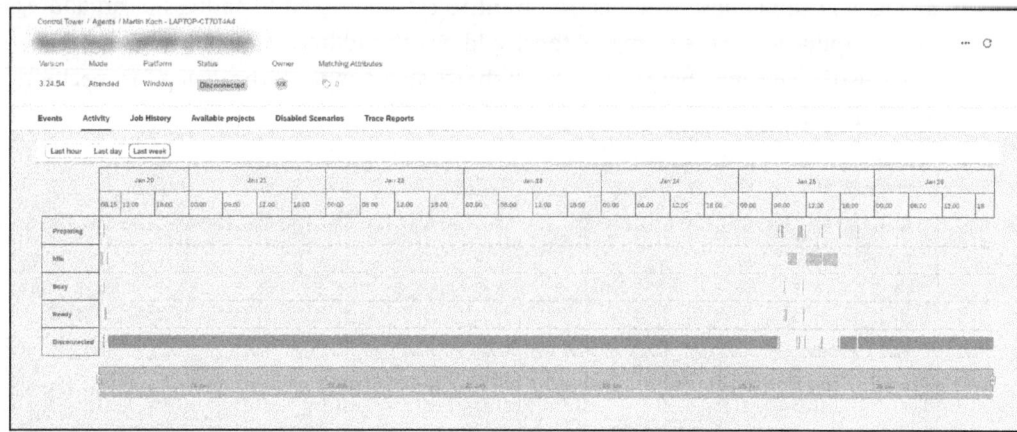

Figure 25.26 Activity of Agent

Agent Groups

If you use a large number of agents, you can group them together to make access easier. In the agent overview, you can add individual machines to groups.

Agent Attributes

Agent attributes help you ensure that only specific agents perform a desired job.

External IP Safelist

You can use the **External IP Safelist** tile to ensure that only agents are allowed to connect based on their external IP addresses. This setting is initially deactivated. As soon as you create a list and activate it, only agents that have an IP address that you have defined will be able to connect. Figure 25.27 shows a screenshot of the external IP safelist.

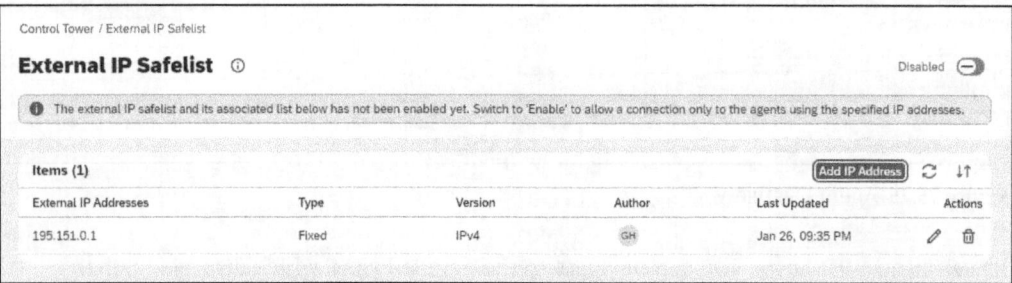

Figure 25.27 External IP Safelist

By clicking the **Add IP Address** button, a new list entry can be created. In the following dialog, you can choose whether you want to set up a fixed IP address or an address range. In addition to the format of the IP address, the address itself or the address range must also be specified here. Figure 25.28 shows an example of the dialog for specifying an IP address.

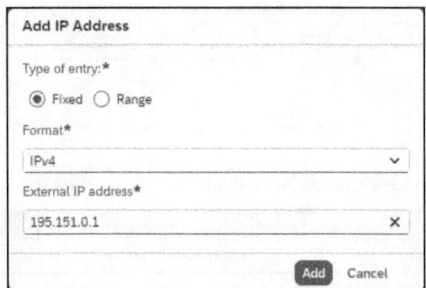

Figure 25.28 Definition of IP Address in External IP Safelist

25.2 Administration via the Control Tower

When specifying IP addresses, the * character can be used as a wildcard. For example, by specifying the IP address *195.151.0.**, all addresses from *195.151.0.0* to *195.151.0.255* can be set up for access. If an agent is not in the safelist based on their IP address, they will receive the following error message when trying to connect: **Your agent cannot connect because it has not been added to the safelist. Please contact your administrator to add your agent to the safelist.**

25.2.7 Cloud Studio Variables

If you use environment variables in one of your projects, these variables are listed in this view. Environment variables can be defined inside the properties of your project, as shown in Figure 25.29.

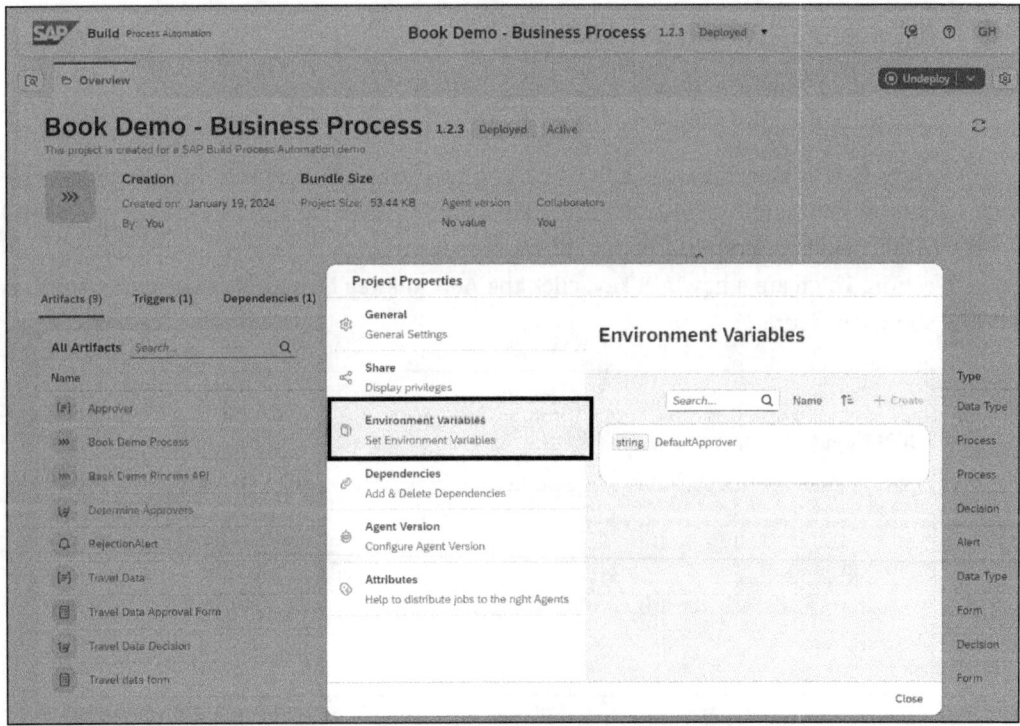

Figure 25.29 Managing Environment Variables for Project

Once an environment variable is defined, the value can be obtained during deployment. After deployment, the new environment variable is displayed in the list of the cloud studio variables, as shown in Figure 25.30. In this list, you get an overview of the variables used in the different versions of your projects and you can explore the value provided for each variable.

785

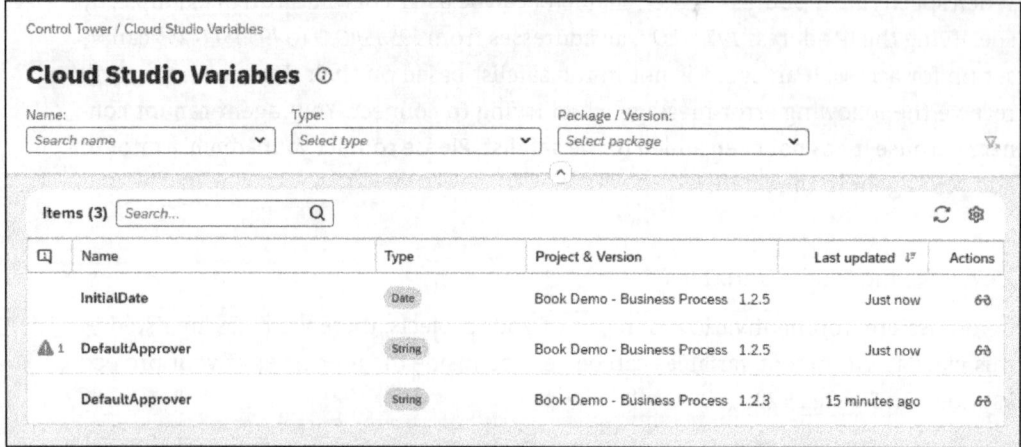

Figure 25.30 List of Environment Variables

25.2.8 API Keys

The API keys view provides a list of tokens that are used during authorization to execute API triggers. Additionally, they are used to gain read access to API triggers, and jobs. These tokens are generated during creation and must be used by the calling application. To create a new API key, click the **Add API Key** button above the items list, as shown in Figure 25.31.

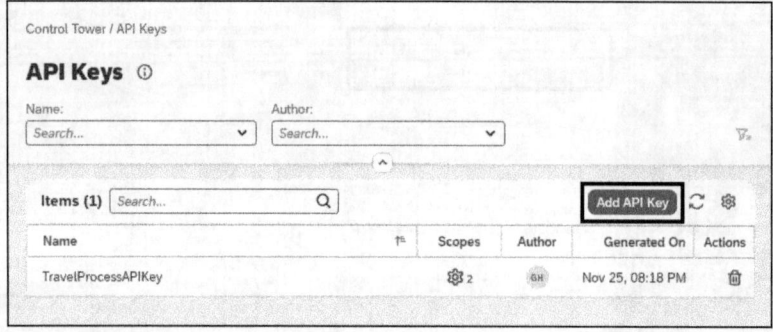

Figure 25.31 Adding New API Key

In the displayed wizard (see Figure 25.32), enter a name for your new API key and an optional description. Click the **Next** button to proceed to the next step.

In the next step, you can set the permissions to be provided by the API key. Figure 25.33 shows a screenshot of this step with the scopes that can be used, and Table 25.7 provides the details.

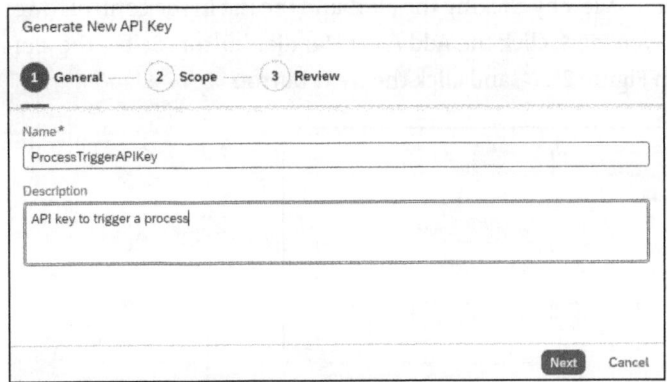

Figure 25.32 Provide Name and Description for New API Key

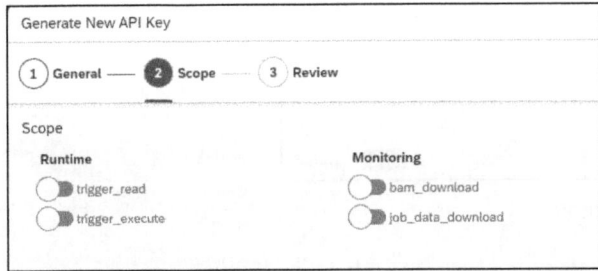

Figure 25.33 Available Scopes of API Keys

Scope	Description
trigger_read	Read access to API triggers and jobs
trigger_execute	Permission to execute API triggers
bam_download	Download business activity monitoring data
job_download_data	Download archived monitoring data for jobs

Table 25.7 Scopes of API Keys

In Chapter 19, Section 19.2.2 you can see how an API key is used to call an API trigger in order to start a business process and provide data from an external application.

25.2.9 Alert Handlers

In SAP Build Process Automation, you can use alerts to define business events. Whenever an alert is created, it can be raised from an existing automation. To react to this alert, you can create alert handlers to send notifications according to the raised alert.

Navigate to the alert handler's view by clicking the alert and the tile in the control tower landing page. In the displayed view, click the **Add Alert Handler** button. Select the alert to be handled, as shown in Figure 25.34, and click the **Next** button.

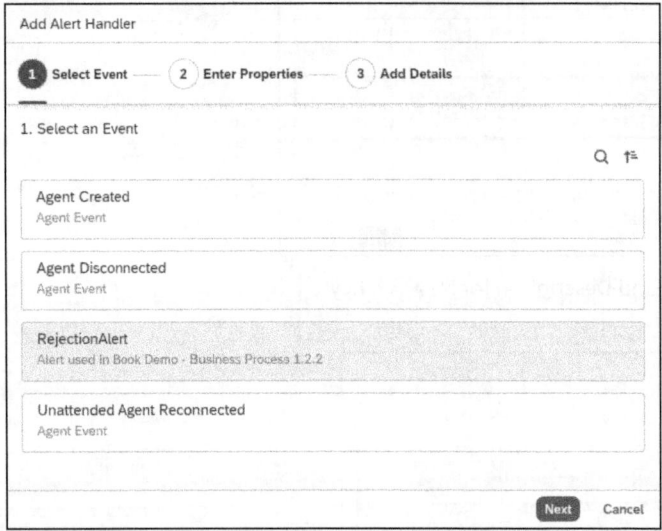

Figure 25.34 Adding New Alert Handler

On the next page of the wizard, enter a name and a description for your new alert handler, then select **Next** (see Figure 25.35).

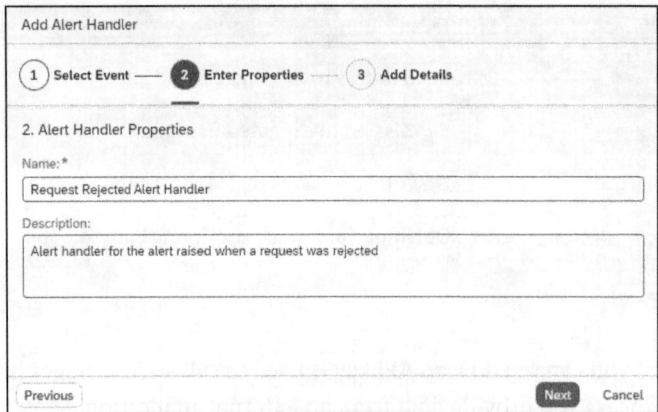

Figure 25.35 Setting Properties for Alert Handler

Finally, define the email details for the notification. In this step, you can use various variables coming from the process automation to define the subject and the content of your notification. These variables can be accessed by using the syntax *${variable name}*. The form also provides a WYSIWYG editor that can be used to create HTML-format emails for your notifications. Figure 25.36 shows an example of the configuration of the email details of an alert handler.

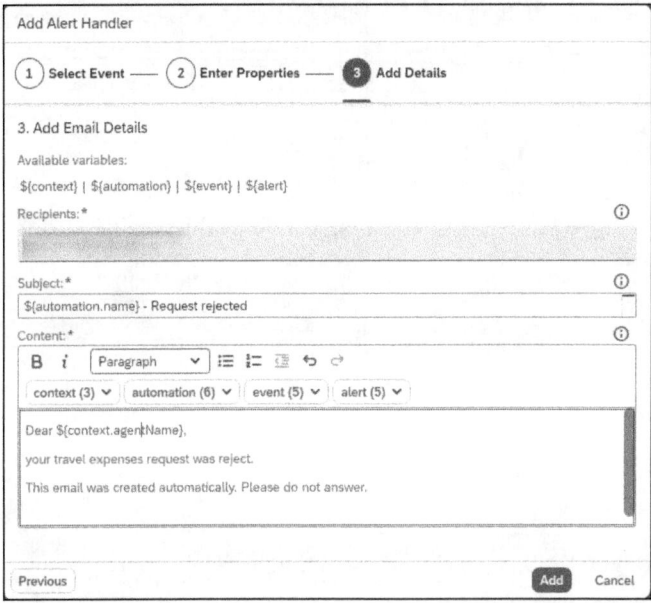

Figure 25.36 Email Details of Alert Handler

After clicking the **Add** button, the alert handler is automatically activated and added to the items list.

25.3 Summary

This chapter provided an overview of the monitoring capabilities provided by SAP Build Process Automation. We showed you the tools that can be used to monitor process instances, jobs, and automations. We also offered insight into the options that are provided to help you during the design phase, like starting a process instance by providing a trigger message. In addition, we showed you how visibility scenarios can be monitored and how data collection can be triggered manually.

In the second part of this chapter, you got to know the tools provided by the control tower, a collection of configuration tasks that can be used to set specific attributes and properties for your environments. You saw how the tenant is specified for development or production to ensure that permissions for users are applied accordingly. We also showed you the options provided for the backend and the agent configuration. Finally, we showed you how API keys can be maintained and how alert handlers are created.

The Authors

Rene Glavanovits is an SAP consultant and developer at CloudDNA GmbH, an SAP partner in Austria. He specializes in the latest SAP technologies, specifically in the development of full-stack applications with SAP Fiori, SAPUI5, OData, core data services (CDS), and SAP Cloud Application Programming Model.

Gernot Haider is the cofounder of sapio GmbH, an Austrian SAP partner that supports a wide variety of customers in the implementation of cloud integration projects and HCM projects with SAP solutions. Previously, he was a middleware architect responsible for the operation of SAP Process Orchestration and Cloud Integration.

Martin Koch is the managing director of CloudDNA GmbH, an SAP partner in Austria. His focus is on SAP Business Technology Platform (SAP BTP), user experience, integration, and security. In addition to his SAP-related work, he is an active champagne importer and dealer in Austria.

Daniel Krancz is a software developer and consultant at CloudDNA GmbH. His focus is on full stack development with SAPUI5/SAP Fiori, OData, SAP Cloud Application Programming Model, and ABAP RESTful application programming model, as well as mobile development. He has been teaching SAP trainings since 2019 and has been named a certified trainer and consultant for his achievements by SAP.

Index

$select parameter 661

A

ABAP 31
ABAP RESTful application programming model 185
Acquired events 770
Action groups 510
Action projects 655
   action editor 666
   API 668
   create 655
   input 659
   OData services 658
   output 662
   publish 673
   release 673
   response message structure 670
   test 664, 668, 672
Actions 623, 655–656
   add 680
   create 707
   deployment 683
   input parameters 681
   navigational 708
   select 679
   trigger workflow 710
   within process 674
Adaptive cards 386
Administration 774
Agents 783
   attributes 784
   configuration 783
   groups 784
Agile development 29
Alerts 158, 510
   handlers 787
Analytical cards 386
Android 292, 303
   build schemes 304
API key 247, 572, 786
   create 572
   sending via request header 581
API Management 245
API specification 656
API triggers 564, 750
   add 564

API triggers (Cont.)
   add properties 566
   create 564
   deploy 569
   new release 569
   process inputs 565, 567
   scope 574
   view 571
Apple Developer Portal 293–294
Application accessibility 290
Application backend 137
Application programming interface (API) 38, 564
Applications 510
Application users 544
Application versioning 285
Approval forms 604
   add 605
   approve/reject 611
   branches 608
   create 606
   due date 608
   input values 609
   process instance 611
   run in parallel 620
Approvals 512
Artifacts 554
Artificial intelligence (AI) 47
Attended mode 716
Attributes 207, 700
   calculated 703
   context 702
   default 701
   entities 706
Authentication 245, 665
   service provider 71
Authorizations 67, 166, 333, 521, 580
Automatic editor 732
Automation 509, 715, 722
   create 724, 732, 736
   create application 724
   data type details 740
   deployment 750
   edit activity 739
   environment variables 747
   Excel 737
   files 734
   jobs 770

Automation (Cont.)
  overview 771
  parameters 724
  project 716, 736
  run 744
  test 745
  UI vs web 725
  workbook 739
Automation SDK 549

B

Backends 168
Bell notifications 457
Boosters 92, 94, 337, 528
  preerquisites check 530
  subaccount selection 530
Branches 613, 619
Breaking changes 239
Build 282
  settings 286
  trigger 283
BuildApps_Administrator role collection 102
BuildApps_Developer role collection 103
Build Work Zone, advanced edition
  target system 357
Business content 321, 403
  add content channel 409
  add roles 410
  assign roles 413
  check content 410
  configure site 411
  create destinations 405
  export 429
  expose content 407
  import 429
  transport 427
Business processes 549
  artifacts 553
  create 549, 551
  description 552
  design console 553
  forms 595
  identifier 552
  Overview tab 553
Buttons 142

C

Capture applications 722
Cards 381, 384
  structure 385
  types 386

Card templates 383
Cells 139
Channel manager 319
Citizen developer 24
Cloud connector 52, 78
  download 80
  system mapping 83
Cloud Foundry 288, 531, 536
  command line interface 288
  organization 58
Cloud Platform Enterprise Agreement
  (CPEA) 56
Cloud studio 734
Cloud studio variables 785
Commercial model
  consumption-based 56
  subscription-based 57
Compliance 488
  alerts 495
  dictionary 492
  history 495
  notifications 496
  reports 497
  unwanted terms 494
Compliance monitor 489
Component cards 386
Component Tap event 190
Condition editor 615
Conditions 613
  define 617
Connectors 549, 656
Containers 138
Content federation 403
Content manager 318, 320, 410, 427
Content packages 373
  actions 377
  create 374
  deploy 377
  install 380
  structure 376
  upload 378
Content providers 409
Content transport 425
  manual 427
Control tower 542, 550, 674, 716, 774
Correlation ID 501

Index

D

Data binding 155, 158, 180, 188, 252, 256
   types 155
Data browser 229
   new record 229
Data entities 128
Data integration 165
   cloud 170
   non-SAP 194
   on-premise 185
Data records 176
Data types 199, 514, 632
   add fields 633, 635
   properties 633
Data variables 175
   ordering 181
Decision editor 627
   diagram 627
   error details 628
Decisions 513, 627
   create new 630
   development tool 631
   element 631
   input binding 642
   output binding 644
   values 629
Decision tables 627, 646
   approvers 646
   binding 651
   condition details 648
   criteria 650
   monitoring 651
   parameters 646
   rules 647
   test 650
Default branch 613
Delay logic block 183
Dependencies 554
Deployment 281
   manual 288
Design-time destinations 405
Desktop agent 716
   activate plugin 719
   add tenant 719
   download 716
   install 718
   register 719
   runtime environments 716
Destinations ... 128, 171, 405, 538, 540, 674, 779
   add 680
   create new 665
   properties 541

Device management 293
Dev spaces 374, 536
DIAG protocol 727
Digitalization 27
Digital transformation 53
Distribution certificate 293–294
Document templates 514
Drag and drop 591
Dropdown fields 199

E

Email notifications 457, 487
Endpoints 288
Entities 172, 174, 204, 258, 705
Entitlements 530
Entity modeling 220
Environment variables 625, 745
   password 747
Error logs 499
   access 500
Event triggers 583
   fields 583
   output 584
Expression editor 742
Extended data entities 224
   configure 225
   create 224
Extensions 243
   build a detail page 254
   build page 246
   build the UI 250
   component tap 264
   create data resource 246
   create entity 246
   data binding 247
   data record 270
   data variable 272
   Delete button logic 274
   input fields 278
   logic blocks 276
   new project 266
   relative path 248
   reponse mapper 250
   Save button logic 268
   schema 248
   test 248
External authentication 781
External IP safelist 784

F

Favicon 284
Feature enablement 483
Feature management 481
Field types 203
Fonts 148
for-each loop 252, 741
Form editor 678
Forms 512, 585
   change element order 590
   create 585, 587, 677
   elements 588
   example 595
   field properties 601
   fields 599
   headline 597
   new step 585
   properties panel 592
   table 602
   table columns 603
   undo/redo 592
Form triggers 555
   add 555
   blank form 556
   deploy 559
   example 557
   form link 561
   monitoring 561
   variables 557
   version 559
   workflow instances 563
Formulas 158, 181
Full stack 40
Functions 231
   create new 231
   input parameters 232
   outcomes 234

G

GET request 661
Global account 70
Global accounts 93
Globally Unique Identifiers (GUIDs) 272
GROW with SAP S/4HANA Cloud 508

H

Horizon theme 145
HTTP codes 662
HTTP protocol 169

Hyperscalers 54
   AWS 55
   Azure 55
   GCP 55

I

Identity Authentication 66–67, 72, 99, 105, 107, 349, 353, 468, 544
   new group 107
   select users 109
   tenant 73
Identity providers 71, 97, 168
Identity Provisioning ... 105, 348, 350, 355, 468
   client secret 355
   secrets 355
   source system 357
   tenant 350
if branch 613
Infrastructure-as-a-service (IaaS) 51
Input fields 141
Input parameters 749
Instances 575
Internet Communication Framework (ICF) 87
iOS 292–293

J

Java Development Kit (JDK) 80
JSON message 582

K

Keystore 303
Keystore app 293

L

Labels 197
Layout tree 120
Legacy systems 27, 35
License costs 539
Lifecycle management 161, 519
List cards 387
Lobby 101, 113, 136, 266, 514, 549
   import project 518
Local data storage 201
Logic blocks 160, 200, 207, 260
Logic canvas 157, 232
Logic flows 237
Log message 742

Low-code 23, 30–31
   restrictions 32
   target group 31
   use cases 31

M

Mail body 682
Mail server configuration 543
manifest.json file 386
Manual capture 727
Mapping 198
Marketplace 188
Microsoft Entra ID 68, 105
Microsoft Excel 734
Microsoft Teams 416
   apps 417
   file upload 419
   integration 417
   SAP Build Work Zone apps 420
   user profile 420
Minimal viable products (MVP) 29
Mobile app 35
Mobile deployment 292
Mobile Development Kit (MDK) 36–37
Mobile services 329
Monitoring 550, 753, 765
MTAR target 283
Multifactor authentication 166
Multitarget archive 283
My Inbox 611, 755
   approve/reject actions 756
   filter 758
   grouping tasks 760
   priorities 759
   search 757
   sort 757
   substitutions 761
   task functionality 755

N

Native entities 221
   add field 222
   configure 222
   field datatypes 224
No-code 23, 30, 32
   development environment 111
   target group 32
   use cases 32
Notifications 455
   activities 457

Notifications (Cont.)
   channels 456
   group 459
   overview 456
   push 460
   types 457

O

OAuth 2.0 580
Object cards 388
OData services 87
OpenID Connect 71, 343
OpenID Connect protocol 105

P

Page parameters 211
Parameters 660, 668
Pay-as-you-go 56
Performance indicators 710
   create 710
   filters 712
   representation 710
PKCS#12 file 293
Planned substitutions 761
Platform-as-a-service (PaaS) 51
Platform users 544
Plugins 284
Postman 579
POST request 569
Principal propagation 84
Process and workflow instances 766
   logs 769
   status 768
   subflows 768
ProcessAutomationAdmin role
   collection 545
ProcessAutomationDeveloper role
   collection 545
ProcessAutomationParticipant role
   collection 545
Processes 676
Processes and workflows 772
Process visibility 687
Pro-code 25, 30
   limitations 31
   target group 31
   use cases 31
Profanity monitor 489
Profile images 487
Project launcher 510
Project list 515

Index

Project members 522
Properties 138
Prototyping 32
Provisioning profile 296

R

React 292
ReactNative 292
Recorder 722, 727, 731
Remote content providers 404
Repeat With property 178–179, 187
REST 43
RISE with SAP 315
Risk-based authentication 73
Roadmap 64
Robotic process automation (RPA) 47
Role collections 101, 342, 413, 442, 532, 545
    assign 325
    attribute mapping 70
Roles 68, 101, 103, 322
Rules 627
Runtime destinations 405

S

SAML 2.0 353
SAML identity provider 68
Sandbox API 244
SAP Ariba 51
SAP Authorization and Trust
    Management 348, 442
SAP BTP, Cloud Foundry environment 54, 94, 507
SAP BTP, Neo environment 53
SAP BTP cockpit 60, 93, 105, 290, 405
SAP Build 25, 39
SAP Build Apps 25, 36, 38–40, 91, 135
    activate authentication 166
    administration 128
    application logic 157
    architecture overview 94
    authentication 130
    build process 152
    bundle settings 300
    Community Edition 111
    configuration 99
    create project 114
    CSS color picker 146
    data center 97
    deployment 281
    develop app 135

SAP Build Apps (Cont.)
    documentation 132
    extensions 243
    help 131
    history 130
    image assets 300
    installation 91
    managing development projects 116
    page variables 154
    plans 92
    preview 150
    project members 116
    role collections 102
    testing 149, 160
    theming 145
    visual cloud functions 42
SAP Build Apps Composer 111
    build service 126
    launch 112, 125
    layout 123
    marketplace 124
    navigation 128
    page layout 120
    pages 124
    properties 120
    themes 126
    UI components 122
    user interface 117
    variables 123
    view 119
SAP Build Apps Preview app 126
SAP Build Process Automation 25, 34, 39, 47, 507, 527
    business processes 549
    collaboration 521
    components 509, 529
    configuration 536
    create decision 629
    extensions 508
    features 48, 508
    installation 527
    license costs 534
    My Inbox 764
    options menu 516
    register desktop agent 719
    security 543
    subscribe 507
SAP Build Work Zone 25, 39, 44, 311
    administration 481
    advanced topics 455
    content federation 403
    extternal integrations 403

Index

SAP Build Work Zone (Cont.)
    functionality 314
    UI integration 373
SAP Build Work Zone, advanced
    edition 38, 44, 46, 327, 426
    add users 341
    administrators 333
    architecture 332
    booster 340
    components 328, 339
    configuration 344
    destination 351
    domain type 351
    end user home page 371
    entitlements 340
    features 46
    functionality 327
    groups 346
    home page 430
    install 337
    jobs 368
    role collections 342
    transformations 359
    trust 344
    trust configuration 343
    user attributes 345
SAP Build Work Zone, standard
    edition 44–45, 315, 426
    apps 321
    browsers 323
    business page 317
    create service instance 323
    error logs 500
    features 45
    functionality 316
    installation 323
    pages 322
    spaces 322
    working environment 318
SAP Business Accelerator Hub 244, 655, 665
    API key 666
SAP Business Application Studio 374
    new project 374
    project details 375
SAP Business Technology Platform
    (SAP BTP) 43, 51, 54, 113, 165
    accounts 57
    authentication 167
    commercial models 55
    content providers 404
    directories 57
    entitlements 59

SAP Business Technology Platform
    (SAP BTP) (Cont.)
    global account 57
    overview 52
    quotas 59
    region 58
    role collections 69
    services 44
    subaccount 57
SAP Cloud ALM 778
SAP Cloud Identity Services 67, 72, 94,
    105, 333, 344, 521
    Identity Authentication 72
SAP Cloud Portal 426
SAP Cloud Transport Management 438
    booster 442
    configuration 448
    content types 450
    destination 439, 442
    import overview 450
    role collections 445
    service key 438, 441
    setup 438
    subaccounts 444
SAP Destination service 43
SAP Discovery Center 61, 91, 524, 531
    missions 62
    services 62
SAP Event Mesh 583
SAP Fiori launchpad 25, 407, 414
SAP GUI 722
SAP HANA 53
SAP HANA enterprise search 462
    enable 464
    results 463
    widget 467
SAP Help Portal 132
SAP ID service 105, 544
SAP Integration Suite 245
SAP Intelligent RPA 545
SAP Launchpad service 44, 312
SAP Learning platform 133
SAP Mobile Cards 330
SAP Mobile Services 97
SAP Mobile Start 474
    app 480
    end user setup 477
    register 479
    setup 475
SAP Road Map Explorer 133, 524
SAP S/4HANA 27, 185, 265, 405
    extensions 265

799

SAP S/4HANA Cloud 51, 170, 265, 583
SAP SuccessFactors 51, 244, 351
SAP SuccessFactors Employee Central 244
SAP Task Center 468, 755
   create destination 472
   integration 468
SAPUI5 .... 31, 284
SAP Work Zone 44, 312
SDK packages 734
Security 65, 105
   access control lists 67
   permissions 67
   roles 67
Service bindings 575
Service collections 171
Service instances 536–537
Set Page Variable logic block 208
Single-page applications 151
Single sign-on (SSO) 66, 71
Site directory 318
Site manager 318
SMTP server 778
Software-as-a-service (SaaS) 51
Status codes 662
Subaccounts 70, 95, 337, 531, 665
   overview 535
   usage analytics 539

T

Table cards 387
Team collaboratio 161
Tenant details 776
Text rules 635
   add to decision 637
   create new 638
   else-then 641
   fixed value 641
   if branch 644
   logical expression 640
   operator 640
   properties 636
   results 638
   test 644
   use in process 642
Timeline cards 388
Tokens 580
Transaction
   /UI2/CDM3_EXP_SCOPE 408
   SICF 407
   SU01 730
Transformations 368

Transport nodes 448
Transport routes 448
Trigger forms 677
Triggers 554–555, 753, 773
   monitor 754

U

UI canvas 137, 157, 207
UI elements 139, 144
UI integration cards 384
   add to work page 395
   configuration 390
   develop 389
   general information 390
   preview 392
   publish 396
   upload 393
UI theme designer 330
Unattended mode 716
Unplanned substitutions 761–762
Unscannable filter 490
Usage analytics 98
Usage strings 300
Use cases 33
   minimum viable product 37
   mobile apps 35
   process automation 33
   software modernization 34
   web sites 38
User accounts 58
User authentication 66, 165–166
   biometric authentication 67
   certificates 67
   multi-factor authentication 66
   social login 67
   two-factor authentication 66
User authorization 66, 68
User interfaces 176
User management 486
Users 730
User tasks 510

V

Versions 520, 750
Virtual fields 226
   formula 226
Visibility dashboard 713
Visibility scenarios 512, 687, 774
   action menu 690
   actions 707

Visibility scenarios (Cont.)
    attributes 700
    configure 690
    configure visibility 693
    correlation 695
    create 687
    events 692
    general 691
    import 690
    performance indicators 710
    phases 696
    processes 691
    project artifacts 689
    state 698
    status 699
    values 688
Visual cloud functions 170, 217
    change types 240
    create 218
    delete/pause deployment 241
    deployment 238
    deployment cards 239
    user interface 219
Visual development 40

W

Web and mobile applications 137
Web app 283
Webhooks 460
Web page cards 387
Widgets 395, 398
    builders 399
    preview 401
Workflows 24
Work pages 381, 395
    edit 381
Workspaces 381, 432
    export 433
    import 433
Workspace templates 436
    import 438
    upload 437

- Walk through the end-to-end development process for full stack applications with CDS, SAPUI5, OData, SAP Fiori elements, and more

- Implement applications based on OData V4 and OData V2

- Extend and deploy applications and configure file upload, form validation, and ETags

Glavanovits, Koch, Krancz, Olzinger

Full Stack Development with SAP

Your guide to end-to-end SAP development is here! From the database to the user interface, you'll learn what it takes to create applications as a full stack developer. See how to use SAPUI5, OData, CDS, TypeScript, and SAP Fiori elements in your applications. Get detailed, step-by-step instructions for working with OData V4 and V2, including developing applications using the ABAP RESTful application programming model and creating projects in Transaction SEGW. Extend and deploy your applications—then you're ready to go!

635 pages, pub. 09/2023
E-Book: $84.99 | **Print:** $89.95 | **Bundle:** $99.99

www.sap-press.com/5733

- Install and configure the cloud connector for SAP
- Set up connections for common use cases, including SAP Business Application Studio, SAP Web IDE, and more
- Run cloud connector securely, monitor errors, configure principal propagation and more

Martin Koch, Siegfried Zeilinger

Cloud Connector for SAP

Establish quick and secure communication between your cloud and on-premise systems with SAP Connectivity service's cloud connector! Set up and configure the cloud connector, from performing sizing to implementing connectivity APIs. Link on-premise SAP products to SAP BTP and its services, including SAP Business Application Studio, SAP Integration Suite's Cloud Integration, and more. With information on creating secure connections, administering the cloud connector, and monitoring, this guide has everything you need!

352 pages, pub. 04/2023
E-Book: $84.99 | **Print:** $89.95 | **Bundle:** $99.99

www.sap-press.com/5683

- Develop, design, deploy, and manage workflows for SAP S/4HANA
- Use the SAP Business Workflow engine and flexible workflow frameworks
- Develop cloud-based workflows with SAP Build Process Automation

Dutta, Ghosh, Goon, Jana, Mukherjee, Rao, Rao, Sane, Veshala, Viswanathan

Workflow for SAP S/4HANA

Whether you're running on-premise SAP S/4HANA or SAP S/4HANA Cloud, get the tools you need to automate your business processes with workflows! Build methods and tasks, define workflows and triggering events, and create notifications with SAP Business Workflow. Walk step by step through standard and custom workflow development using SAP S/4HANA's flexible workflows. Design and deploy cloud-ready workflows with SAP BusinessApplication Studio and SAP Build Process Automation. This is the only workflow guide you need!

686 pages, pub. 11/2023
E-Book: $84.99 | **Print:** $89.95 | **Bundle:** $99.99

www.sap-press.com/5697

- Learn about SAP's new technology platform
- Explore products, services, and tools for data management, application development, integration, analytics, and more
- Walk through customer use cases to see how SAP BTP can bring value to your business

Banda, Chandra, Gooi

SAP Business Technology Platform

An Introduction

What is SAP Business Technology Platform, and what does it offer your organization? Answer these questions and more with this introduction! See how SAP BTP serves as your complete technical foundation and learn about its capabilities for application development, integration, data management, analytics, and more. Identify business use cases and follow practical examples that show how to use SAP BTP's portfolio to its full potential. Envision how SAP BTP enhances your business!

570 pages, pub. 05/2022
E-Book: $74.99 | **Print:** $79.95 | **Bundle:** $89.99

www.sap-press.com/5440

www.sap-press.com

- Build and deploy cloud applications on SAP BTP using Java, Node.js, and ABAP
- Work with SAP Business Application Studio, SAP Cloud Application Programming Model, cloud services, and more
- Manage, monitor, and secure your applications

Acharya, Bajaj, Dhar, Ghosh, Lahiri

Application Development with SAP Business Technology Platform

Every SAP S/4HANA journey begins with a single step—so get all the steps you need for your finance system conversion project! Follow the implementation path through preparation and post-migration testing, with special attention to data migration and functional configuration. From the general ledger to asset accounting and beyond, you'll align your new system with existing finance requirements and go live. Get the nitty-gritty details and pro tips with this go-to-guide and make your brownfield project a success!

574 pages, pub. 12/2022
E-Book: $84.99 | **Print:** $89.95 | **Bundle:** $99.99

www.sap-press.com/5504

- Your all-in-one overview of full stack web development, from design and interactivity to security and operations
- Learn about frontend tools, including HTML, CSS, JavaScript, APIs, and more
- Work with backend technologies, including Node.js, PHP, web services, and databases

Philip Ackermann

Full Stack Web Development

The Comprehensive Guide

Full stack web developers are always in demand—do you have the skillset? Between these pages you'll learn to design websites with CSS, structure them with HTML, and add interactivity with JavaScript. You'll master the different web protocols, formats, and architectures and see how and when to use APIs, PHP, web services, and other tools and languages. With information on testing, deploying, securing, and optimizing web applications, you'll get the full frontend and backend instructions you need!

740 pages, pub. 08/2023
E-Book: $54.99 | **Print:** $59.95 | **Bundle:** $69.99

www.sap-press.com/5704

Interested in reading more?

Please visit our website for all new book and e-book releases from SAP PRESS.

www.sap-press.com